SCIENCE 366

A Chronicle of Science and Technology

SCIENCE 366

A Chronicle of Science and Technology

BIMAN BASU

UNIVERSITIES PRESS (INDIA) PRIVATE LIMITED

Registered Office
3-6-747/1/A & 3-6-754/1, Himayatnagar,
Hyderabad 500 029 (A.P.), India
e-mail: info@universitiespress.com

Distributed by
Orient Blackswan Private Limited

Registered Office
3-6-752, Himayatnagar, Hyderabad 500 029 (A.P.), India
e-mail: centraloffice@orientblackswan.com

Other Offices
Bangalore, Bhopal, Bhubaneshwar, Chennai, Ernakulam, Guwahati,
Hyderabad, Jaipur, Kolkata, Lucknow, Mumbai, New Delhi, Patna

First published 2008
Reprinted 2010

Cover and book design

ISBN 13: 978 81 7371 607 2
ISBN 10: 81 7371 607 2

Set in Georgia 10/12 by
OSDATA, Hyderabad 500 029

Printed at
India Binding House
NOIDA

Published by
Universities Press (India) Private Limited
3-6-747/1/A & 3-6-754/1, Himayatnagar
Hyderabad 500 029 (A.P.), India
e-mail: info@universitiespress.com

To

Dr. Bal Phondke, who has been my friend,
philosopher and guide in my forays into the world of science writing.

CONTENTS

PREFACE

Dates are important in our life. Most of us remember our own birth dates as well as those of our near and dear ones. Many of us also remember our first day in school or the first day in office, the date of our wedding. Some dates bring back pleasant memories while others are not so pleasant. There are also historical dates that we observe every year. But not many of us know the dates when some important scientific discoveries were made or some famous scientists were born. There are also dates that mark important breakthroughs in our understanding of the universe around us – new discoveries and new inventions that have changed our life.

My interest in dates was aroused when I was a student at the Delhi University more than 45 years ago. A chemistry professor of ours always started his class by reading out from a large notebook, which he always carried and which listed the names of famous chemists born on a particular day. I found it quite interesting. It was a nice diversion from the rigid and often boring curriculum!

My interest grew over the years and soon I was collecting important dates concerning scientists and their work from whatever I read – newspapers, magazines, and books. The present volume is a compilation of my collections stretching over a period of more than 30 years. It includes not only dates relating to scientists and their work, but also dates relating to birth and death of inventors, important inventions and discoveries, scientific organisations, and important scientific events. Events up to March 2006 are covered.

The number of scientists and inventors whose contributions have changed our life run into several thousand. So, making a selection for inclusion in this book was quite a difficult task. Of course, there were giants like Galileo, Newton, Archimedes and many others, and inventors like Leonardo da Vinci, Edison, Bell, Faraday and others who could not be left out at any cost. But making a choice from among the rest was not an easy job. Finally I decided to include all the Nobel laureates for their acknowledged eminence in various disciplines of science. Similar criteria were adopted while selecting the Indian entries. All the Indian Fellows of the Royal Society have been included as are those who have significantly contributed to the advancement of science and technology in India. But I admit that the list is not exhaustive and some deserving names may have been left out inadvertently. I hope readers would pardon me for any such omission.

The entries are arranged chronologically so that items can be searched date wise, but an exhaustive index is also given to help in random searching. An entry for the date of birth of a scientist or inventor gives a brief biography of the scientist/inventor, while an entry for the date of founding or inauguration of a scientific institution gives a brief summary of the activities and achievements of the institution. I have tried to include as many Indian scientists and Indian scientific institutions and establishments as possible to give the reader a broad view of the country's S&T infrastructure and its contribution to S&T in general. All the entries are cross-referenced for easy navigation.

I have tried to make the entries as comprehensive as possible. Every person mentioned in the book is identified by his/her nationality while every place name is fully identified by the state and country, to avoid confusion.

Since the book is primarily organised by dates, some important personalities and events could not be included, as the relevant dates of birth or dates of the events could not be ascertained from any source. So the reader may find some important names missing. Similarly, some important scientific institutions had to be left out, as the dates of their founding/inauguration could not be ascertained.

I have gathered all the information given in the book from a wide range of sources including encyclopaedias, biographies, journals, newspapers, web sites, and a variety of other works of reference. I wish I could acknowledge all the references by name, but that being impossible I have appended a brief list of the major works referred to at the end. While every care has been taken to authenticate the dates and verify the facts given in the book, I shall be grateful if readers could bring to my knowledge any factual error they notice.

All my effort would have been in vain but for the keen interest shown by Mr. Madhu Reddy of the Universities Press, Hyderabad, who has been regularly interacting with me and giving me constant encouragement. Ms Jebah S David of the Universities Press did a fantastic job of editing the voluminous manuscript. Nothing escaped her keen eyes and I am grateful to her for bringing to my notice several anomalies and omissions that I could correct in time. Her hard work has certainly contributed to enhancing the worth of the book.

I am grateful to the Nobel Foundation for granting me permission to use photographs of the Nobel laureates from their web site. Many friends have helped in procuring other photographs for the book. I am grateful to them.

In the end I must say, I have immensely enjoyed writing this book, poring through dozens of reference books and browsing through thousands of web pages. I would consider my effort worthwhile if readers find this unique volume of some value.

Biman Basu
New Delhi

January 1

1801
Giuseppe Piazzi, Director of the Palermo Observatory in Italy discovered the first minor planet, which was later named Ceres. Having a diameter of 1,003 kilometres, Ceres is the largest known minor planet and orbits the Sun at an average distance of about 450 million kilometres. It revolves around the Sun in 4.6 terrestrial years. Ceres has now been designated as a dwarf planet. (☞ Jul. 16, Jul. 22)

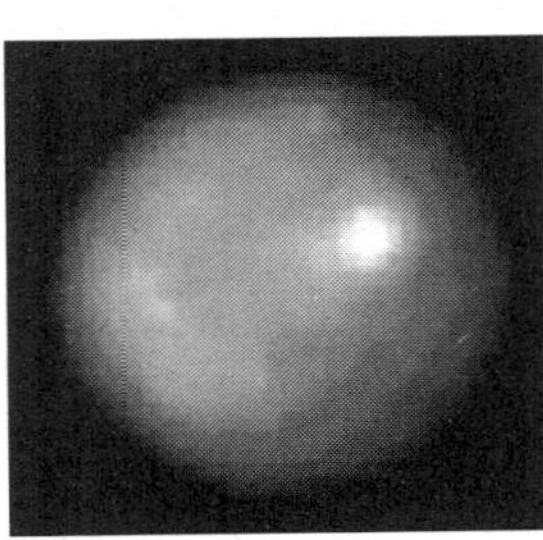
Ceres

1817
Martin Heinrich Klaproth, German chemist who discovered uranium, zirconium and cerium, died in Berlin, Germany. (☞ Dec. 1)

1894
Satyendra Nath Bose, Indian mathematician and theoretical physicist, Fellow of the Royal Society (FRS), was born in Calcutta (now Kolkata), West Bengal, India. He graduated from the University of Calcutta and taught at the Universities of Dacca (1921–45) and Calcutta (1945–56). Bose wrote 'Planck's law and the hypothesis of light quanta' – his most famous paper – which he communicated to the *Philosophical Magazine* in England, but it did not evoke any response. He then sent a duplicate copy to the German physicist, Albert Einstein. Einstein was impressed. He translated it into German and published it in the journal *Zeitschrift fur Physik* in August 1924. Bose collaborated with Einstein in developing a theory regarding the gas-like qualities of electromagnetic radiation, which came to be known as the Bose–Einstein statistics, and gave rise to the concept of particles known as bosons. Einstein took the next logical step of considering a gas, the atoms of which are indistinguishable and obey the same distribution as proposed by Bose in the case of photons. This led to discovery of Bose–Einstein Condensate (BEC), which represents the fifth state of matter, namely coherent matter. Bose's main scientific work related to statistical mechanics, the electromagnetic properties of the ionosphere, the theories of X-ray crystallography and thermoluminescence, and the unified field theory. (☞ Feb. 4)

1894
Heinrich Rudolf Hertz, German physicist who was the first to produce and detect electromagnetic waves, died in Bonn, Germany. (☞ Feb. 22)

1896
German inventor, Wilhelm Röntgen announces his discovery of X-rays. (☞ Jan. 23, Feb. 10, Mar. 27, Nov. 8)

1943
Raghunath Anant Mashelkar, Indian chemical technologist, Fellow of the Royal Society, was born in Mashel, Goa, India.

After obtaining his PhD from Bombay University in 1969, he occupied various academic positions in Europe and USA including that of a Visiting Professor at the Technical University of Denmark and at the University of Delaware in USA. In 1976, he joined the National Chemical Laboratory (NCL), Pune, Maharashtra, as a scientist. He was head of the Chemical Engineering Division at NCL from 1978 to 1989, when he took over as Director of NCL. He was appointed Director-General of the Council of Scientific and Industrial Research (CSIR) in 1995. Mashelkar's main contributions are in the field of polymer science. He has also made path-breaking contributions in the engineering analysis of non-Newtonian fluids, gel science and polymer reaction engineering. Mashelkar is one of only three Indian engineers to have been elected an FRS.

1955
Shanti Swarup Bhatnagar, Indian chemist and science administrator, Fellow of the Royal Society, the first Director of the Council of Scientific and Industrial Research died in New Delhi, India. (☞ Feb. 21)

1966
The National Institute of Oceanography (NIO), a constituent national laboratory of the Council of Scientific and Industrial Research (CSIR), was founded. Located in Dona Paula, Goa, NIO is the largest oceanographic laboratory in the Indian Ocean region. The major R&D programmes of NIO include: activities under the International Geosphere–Biosphere Programme, surveys for undersea polymetallic nodules, oceanographic studies of the Antarctic waters, island development programmes, coastal zone management, resources and parameters mapping of the exclusive economic zone of India, air–sea interaction studies, development of drugs from the sea, marine biotechnology, befouling and corrosion studies, technologies for rural development, development of marine instruments, and development of acoustic and remote sensing techniques for monitoring the oceans. The institute has carried out surveys for polymetallic nodules in an area of about 4 million square kilometres in the Central Indian Ocean, leading to the registration of India as the first 'Pioneer Investor' country in the world under the UN and the allocation of an area of 1,50,0000 square kilometres for developing a mining site. NIO operates two research vessels *Gaveshani* and *Sagar Kanya*, which are well equipped for oceanographic research throughout the year.

1991
The first 220-MW unit of the Narora Atomic Power Station was commissioned at Narora, Uttar Pradesh, India. It is a pressurised heavy water reactor that uses natural uranium as fuel and heavy water as moderator and coolant. (☞ Jul. 1)

1995
Eugene Paul Wigner, Hungarian-born American physicist, Nobel laureate (Physics, 1963), who discovered the fundamental symmetry principles, died in Princeton, New Jersey, USA. (☞ Nov. 17)

January 2

1729

Johann Daniel Titius, Prussian astronomer and physicist, was born in Konitz, Prussia (now Chojnice, Poland). Titius joined the faculty of the University of Wittenberg in 1756 after receiving a degree from the University of Leipzig. He was the first to propose a law relating distances of the planets from the Sun in 1766. It was confirmed by German astronomer Johann E. Bode and later came to be known as the Titius–Bode Law (also called Bode's law). Titius suggested that the mean distances of the planets from the Sun very nearly fit a simple relationship of $A = 4 + (3 \times 2^n)$ giving the series 4, 7, 10, 16, 28*, 52, 100, corresponding to the relative distance of the six planets known then, up to Saturn, and an unassigned value (*) between Mars and Jupiter, which is now known to be occupied by the minor planet belt. The Titius–Bode Law could accurately predict the distance between Uranus, which was discovered later, and the Sun, but it could not accurately predict the distances of Neptune and Pluto. Although best known for his law, Titius also made contributions in physics, concentrating on thermometry, and in biology, classifying plants, animals, and minerals. (☞ Dec. 11)

1822

Rudolf Clausius, German mathematical physicist who formulated the Second Law of Thermodynamics, was born in Köslin, Prussia (now Koszalin, Poland). Clausius graduated from the University of Berlin in 1844, and got his doctorate from the University of Halle in 1848. He then taught in Berlin, Zurich, Wurzburg, and Bonn. Clausius was one of the founders of thermodynamics. His most important paper – on the mechanical theory of heat – published in 1850, first stated the basic ideas of the Second Law of Thermodynamics. In 1865 he introduced the concept of entropy and stated that entropy can never decrease in a physical process and can only remain constant in a reversible process. In all, Clausius wrote eight important papers on the topic. He re-stated the French physicist, Sadi Carnot's principle of the efficiency of heat engines. The Clausius–Clapeyron equation expresses the relation between the pressure and temperature at which two phases of a substance are in equilibrium. With Maxwell, he developed the kinetic theory of gases. In the paper '*Über die Art der Bewegung welche wir Wärme nennen*' (On the kind of motion which we call heat), published in 1857, he provided a full account of the kinetic theory of molecular motions. This was the first systematic treatment of kinetic theory. (☞ Aug. 24)

1839

Louis J. Daguerre, French inventor, took the first photograph of the Moon using his newly invented process for making silver images known as daguerreotypes on a copper plate. (☞ Jul. 10, Nov. 18)

1892

Nil Ratan Dhar, Indian chemist and soil scientist, was born in Jessore, India (now in Bangladesh). After passing the ISc examination in 1909, Dhar joined the Presidency College, Kolkata, getting his BSc Hons. degree in chemistry in 1911. While doing his MSc course in the Presidency College, Dhar experimented with nitrites under Acharya P.C. Ray. He passed his MSc examination in 1913. In 1915, Dhar went to England and was awarded the DSc degree of the London University in 1917. Dhar's main work related to nitrogen fixation and nitrogen loss in soil and the relationship between the nitrogen and phosphate status of soils. His name was considered for the Nobel Prize more than once for his (and his team's) contribution to the study of photochemical fixation of nitrogen. He taught at the University of Allahabad and was the founder of the Sheila Dhar Institute of Soil Science in Allahabad. He was a founder member of the National Institute of Sciences of India (now the Indian National Science Academy). In 1961 he was elected a fellow of the French Academy of Sciences. (☞ Dec. 5)

1900

The first electric trolley bus was operated in New York, USA.

1906

Willis Carrier, American engineer, was awarded a patent for an air conditioning machine. Primarily designed to reduce humidity and temperature, the patent specifically stated that it could be used for ventilating buildings and for other commercial purposes. (☞ Nov. 26)

1920

Isaac Asimov, American biochemist and a highly successful and prolific writer of popular science and science fiction, was born in Petrovichi, Russia. Asimov was brought to the United States at the age of three. He grew up in Brooklyn, New York, and graduated from the Columbia University in 1939. After taking a PhD from the Columbia University in 1948, he joined the faculty of the Boston University and remained associated thereafter. Asimov began contributing stories to science-fiction magazines in 1939 and in 1950 published his first book, *Pebble in the Sky*. His trilogy of novels, *Foundation*, *Foundation and Empire*, and *Second Foundation* (1951–53), which recounts the collapse and rebirth of a vast interstellar empire in the Universe of the future, is his most famous work of science fiction. Among Asimov's books on various topics in science, written with lucidity and humour, are *The Chemicals of Life* (1954), *Inside the Atom* (1956), *The World of Nitrogen* (1958), *Asimov's Guide to Science* (1960), *Life and Energy* (1962), *The Human Brain* (1964), *The Neutrino* (1966), *Science, Numbers and I* (1968), *Our World in Space* (1974), and *Views of the Universe* (1981). (☞ Apr. 6)

1954

Indian Institute of Chemical Technology (IICT), a constituent national laboratory of the Council of Scientific and Industrial Research (CSIR), was inaugurated. Located in Hyderabad, Andhra Pradesh, IICT had its origin as the Central Laboratories for Scientific and Industrial Research (CLSIR), established in 1944 by the then Government of Hyderabad State. After the integration of Hyderabad State with the Indian Union, the laboratory expanded with its growing activities. In 1956, the Central Laboratories came under the aegis of the CSIR, New Delhi and was re-named Regional Research Laboratory, Hyderabad (RRL-H). The RRL-H was re-christened as the Indian Institute of Chemical Technology in 1989. The major areas of research at IICT include natural products chemistry, agrochemicals, drugs and intermediates, speciality and fine Chemicals, gas and energy, and chemical engineering. Process development work, particularly for bulk chemicals is carried out at appropriate pilot plant scales to collect techno-economic and design data. With the help of the excellent design and engineering expertise available, the Institute has been providing engineering designs for commercial plants with standard commercial guarantees. More than 150 technologies developed by the IICT are now in commercial production.

1959
The Russian unmanned Moon-probe *Luna-1* was launched from the Baikanur Cosmodrome in USSR (now Kazakhstan). It flew within 5, 955 kilometres of the Moon two days later, becoming the first spacecraft to break away from Earth's gravity and pass close to the Moon. (☞ Jan. 4)

January 3

1657
William Harvey, English physician and discoverer of the true nature of the circulation of blood and of the function of the heart as a pump, died in London, UK. (☞ Apr. 1)

1823
Robert Whitehead, British engineer who invented the modern torpedo, was born in Bolton, Lancaster, England. The torpedo is a cigar-shaped, self-propelled underwater missile, launched from a submarine or ship and is designed to destroy ships and submarines. Whitehead was interested in the problems connected with trying to remotely control and detonate underwater explosive charges. By 1868 he had successfully demonstrated a practical torpedo design. The British Navy became interested and, after purchasing some of the torpedoes, encouraged Whitehead to set up a manufacturing facility in England. Whitehead's torpedo, which he developed in 1866, was powered by a compressed-air engine driving a single propeller and carried 90 kg of dynamite. (☞ Nov. 14)

1906
William Wilson Morgan, American astronomer, was born in Bethesda, Tennessee, USA. After three years of undergraduate study at the Washington and Lee University, Lexington, Tennessee, Morgan joined the staff of the Yerkes Observatory, the astronomical observatory of the University of Chicago, in Williams Bay, Wisconsin, where he spent his entire career, including three years as the director. He was the first astronomer to demonstrate that the Milky Way galaxy has a spiral structure. While at Yerkes, he earned bachelor's and doctoral degrees at the University of Chicago. He taught at the same university from 1947 until his retirement in 1974. (☞ Jun. 21)

1950
The National Chemical Laboratory (NCL), a constituent national laboratory of the Council of Scientific and Industrial Research (CSIR), was inaugurated. Located in Pune, Maharashtra, NCL is a multidisciplinary research centre that carries out R&D in polymer science, organic chemistry, catalysis, materials chemistry, chemical engineering, biochemical sciences and process development. These activities are directed towards developing a process, material or product that can deliver potential benefits to industry and the people at large. Often market-driven and performed largely through staff, project assistants and sometimes research scholars, the research is mostly funded by the industry and project mode funding from national and multi-national government agencies. NCL maintains a unique collection of experts, instrument facilities and information resources including library and collections of organisms in this part of the world. It is a national resource centre for the chemical and allied sciences.

1957
The world's first battery-powered electric watch was introduced in Lancaster, Philadelphia, USA, by the Hamilton Watch Company.

1983
Technology Policy Statement was announced by the Government of India. It spelt out the basic objectives of the Technology Policy, namely development of indigenous technology and efficient absorption and adaptation of imported technology appropriate to national priorities and resources. It also, among other things, called for measures to attain technological competence and self-reliance, to reduce vulnerability, particularly in strategic and critical areas, making the maximum use of indigenous resources; provide the maximum gainful and satisfying employment to all strata of society, with emphasis on the employment of women and weaker sections of society; use traditional skills and capabilities, making them commercially competitive; ensure the correct mix of mass production technologies and production by the masses; develop technologies which are internationally competitive, particularly those with export potential; reduce demands on energy, particularly energy from non-renewable sources; ensure harmony with the environment, preserve the ecological balance and improve the quality of the habitat; and recycle waste material and make full utilisation of by-products.

January 4

1809
Louis Braille, French educator who developed a system of printing and writing that is extensively used by the blind, was born in Coupvray, near Paris, France. Braille became blind at the age of three after an accident that occurred while he was playing with tools in his father's harness shop. In 1819 he went to Paris to attend the National Institute for Blind Children, and from 1826 he taught there. Braille developed his system of writing when he was only 15 after adapting a scheme developed by Charles Barbier, in which a message coded in dots was embossed on cardboard. Braille's adaptation is much simpler and consists of a six-dot code in various combinations. (☞ Jan. 6)

1885
The first appendectomy – successful surgical removal of an appendix – was performed by American surgeon William W. Grant on a 22-year-old woman named Mary Gartside in Davenport, Iowa, USA.

1940
Brian David Josephson, British physicist, Nobel laureate, was born in Cardiff, England. He was awarded his bachelor's, master's and PhD degrees from the Trinity College, Cambridge, and was elected a fellow of the Trinity College in 1962. While still an undergraduate, Josephson became interested in superconductivity, and began to explore the properties of a junction between two superconductors that later came to be known as a Josephson junction. Josephson extended the earlier work done by the Japanese physicist Leo Esaki and the Norwegian-born American physicist Ivar Giaever

on 'tunnelling', a phenomenon by which electrons can penetrate insulating solids, and discovered the 'Josephson effect', the flow of electric current between two pieces of superconducting material separated by a thin layer of insulating material. Josephson was awarded the 1973 Nobel Prize for Physics, jointly with Esaki and Giaever. Josephson received half the prize money for his theoretical predictions of the properties of a supercurrent through a tunnel barrier, in particular those phenomena which are generally known as the Josephson effects, while Esaki and Giaever shared the other half for their experimental discoveries regarding tunnelling phenomena in semiconductors and superconductors, respectively. (☞ Mar. 12, Apr. 5)

1959
The unmanned Russian Moon-probe *Luna-1* flew within 5, 955 kilometres of the Moon, becoming the first spacecraft to break away from Earth's gravity and pass close to the Moon. (☞ Jan. 2)

1961
Erwin Schrödinger, Austrian theoretical physicist, Nobel laureate (Physics, 1933), who contributed to the wave theory of matter and to other fundamentals of quantum mechanics, died in Vienna, Austria. (☞ Aug. 12)

1981
Swiss scientists announce the first successful cloning of a mammal – three mice.

1996
The first atoms of antimatter were created by a team of European physicists at the CERN research centre in Switzerland. Using the powerful particle accelerator, they created nine atoms of antihydrogen, which survived for 40 nanoseconds.

2004
American robotic rover *Spirit* landed in the Gusev crater on Mars after a seven-month voyage from Earth. Encased in a cocoon of inflated balloons, the rover travelled about 200 metres while bouncing 26 times and rolling after the impact. Soon after landing, the landing site was named 'Columbia Memorial Station,' in honour of the seven astronauts killed in the *Columbia* space shuttle disaster. The six-wheeled rover, the size of a golf cart was designed to crawl over the Martian surface, poring over rocks and soil like robot field geologists and snapping panoramic pictures as it crept along. In the two years since its landing on the Red

Planet, the rover travelled over 5,000 metres and pursued a steady course of solar-driven geologic fieldwork, bringing back some 70,000 images and a new understanding of Mars as a potential habitat. (☞ Jun. 10)

January 5

1855
King Camp Gillette, American inventor of the safety razor and disposable blade, was born in Fond du Lac, Wisconsin, USA. Brought up in Chicago, Gillette became a travelling salesman of hardware after the great Chicago fire of 1871. His penchant

for mechanical tinkering led him to invent 'something that would be used and thrown away,' so that the customer would keep coming back. The result was the safety razor with disposable blades. Gillette's safety razor obviated the need for repeated honing of the permanent, straight edge razor used in those days. (☞ Jul. 9, Nov.15)

1874

Joseph Erlanger, American physiologist, Nobel laureate, was born in San Francisco, USA. Erlanger received his BS degree in chemistry from the University of California, Berkeley, and later went to the Johns Hopkins University, Baltimore, Maryland, to study medicine, where he obtained his MD degree in 1899. After a year of hospital training at the Johns Hopkins Hospital, he was appointed assistant in the Department of Physiology at the Medical School there. Erlanger's research into nerve function was the product of collaboration with his student Herbert Gasser. By 1922, they were able to amplify the electrical responses of a single nerve fibre so that they could analyse the responses using a cathode-ray oscilloscope. In 1932, Erlanger and Gasser found that the fibres of a nerve conduct impulses at different rates, depending on the thickness of the fibre, and that each fibre has a different threshold of excitability; i.e., each requires a stimulus of different intensity to create an impulse. They also discovered that different fibres transmit different kinds of impulses. Erlanger was awarded the 1944 Nobel Prize for Physiology or Medicine, jointly with Gasser, for discovering that fibres within the same nerve cord possess different functions. (☞ Jul. 5, Dec. 5)

1896

The first paper of German physicist and Nobel laureate Wilhelm K. Röentgen on the discovery of X-rays was published in the journal *Wurzburg Physikalisch-Medizinische Gesellschaft*. (☞ Jan. 1, Feb. 10, Mar. 27, Nov. 8)

1940

FM radio was first broadcast. Major Edwin H. Armstrong of Columbia University, inventor of 'frequency modulation,' demonstrated his 'static-less' broadcast system in a relay through five stations in five states in USA. The programme was broadcast using ultra-short radio waves and the technique of frequency modulation, the characteristic of which is freedom from distortion, fading and static interference, which severely affected the quality of conventional amplitude modulated broadcasts.

1969

Soviet unmanned spacecraft *Venera-5* was launched towards Venus from Baikanur Cosmodrome in Kazakhstan. (☞ May 16)

1970

Max Born, German physicist, Nobel laureate (Physics, 1954), whose statistical formulation of the behaviour of subatomic particles led to the replacement of the original quantum theory, died in Göttingen, Germany. (☞ Dec. 11)

1981

Harold C. Urey, American chemist, Nobel laureate (Chemistry, 1934), who was a key figure in the development of the atomic bomb and made fundamental contributions to a widely accepted theory of the origin of the Earth and other planets, died at La Jolla, California, USA. (☞ Apr. 29)

January 6

1655

Jacob Jacques Bernoulli, Swiss mathematician, who introduced the first principles of the calculus of variation, was born in Basel, Switzerland. Bernoulli graduated from the University of Basel with a master's degree in philosophy in 1671 and a licentiate in theology in 1676. While Bernoulli was taking his university degrees, he was also studying mathematics and astronomy. In 1676, after taking his theology degree, Bernoulli moved to Geneva where he worked as a tutor. He then travelled through Europe meeting the leading mathematicians and scientists there. As a result of his travels, Bernoulli began a correspondence with many mathematicians, which he carried on over many years. He returned to Switzerland and taught mechanics at the University in Basel from 1683, giving a series of important lectures on the mechanics of solids and liquids. Bernoulli's first important contributions were a pamphlet published in 1685 on the parallels of logic and algebra and work on probability and geometry in 1687. His work in geometry resulted in a construction which could divide any triangle into four equal parts with two perpendicular lines. By 1689 he had published important work on infinite series and published his law of large numbers in probability theory. He published five treatises on infinite series between 1682 and 1704. Bernoulli also discovered a general method to determine evolutes of a curve as the envelope of its circles of curvature. Bernoulli numbers, a concept that he developed but which was published only in 1713 after his death, play an important role in mathematics and in various topics like analysis, number theory and differential topology. (☞ Aug. 16)

1745

Jacques-Étienne Montgolfier, French pioneer of hot-air ballooning, was born in Annonay, France. His father Pierre Montgolfier had pro-sperous paper factories in the small town of Vidalon, near Annonay. While carrying on their father's paper busi-ness, he and his brother Joseph-Michel maintained their in-terest in scientific experimentation. Jacques-Étienne along with his brother demonstrated the first untethered hot-air balloon flight in 1783. They filled their balloon with heated air by burning straw and wool under the opening at the bottom of the balloon. The balloon rose into the air about 1,000 metres, remained there some 10 minutes, and then settled to the ground about two-and-a-half kilometres from where it rose. (☞ Jun. 4, Aug. 26, Sep. 19, Nov. 21)

1838

Samuel Morse, American painter and inventor, for the first time publicly demonstrated his telegraph in Morristown, New Jersey, USA, but it was not until five years later that the American Congress funded $30,000 to construct an experimental telegraph line from Washington to Baltimore, a distance of 64 kilometres. (☞ Apr. 2, Apr. 27, May 24)

1852
Louis Braille, French educator and inventor of the Braille system of writing for the blind, died in Paris, France. (☞ Jan. 4)

1884
Gregor Mendel, Austrian monk and botanist, who laid the foundation of the science of genetics, died in Brünn, Austria-Hungary (now Brno, Czech Republic). (☞ Jul. 22)

1944
Rolf M. Zinkernagel, Swiss immunologist and pathologist, Nobel laureate, was born in Basel, Switzerland. Zinkernagel was awarded his MD from the University of Basel in 1970 and his PhD from the Australian National University, Canberra, in 1975. He joined the John Curtin School of Medical Research in Canberra in 1973 as a research fellow. His research, in collaboration with Australian immunologist Peter C. Doherty, centred on the white blood cells known as T lymphocytes, which enable the immune system to distinguish virus-infected cells from normal cells and act to destroy invading viruses and virus-infected cells. Zinkernagel and Doherty were awarded the 1996 Nobel Prize for Physiology or Medicine, for their discovery of how the immune system distinguishes virus-infected cells from normal cells. (☞ Oct. 15)

1990
Pavel A. Cherenkov, Soviet physicist, Nobel laureate (Physics, 1958), who discovered that electrons emit light as they pass through a transparent medium at a speed higher than the speed of light in that medium, died in Russia. (☞ Jul. 28)

January 7

1610
Galileo Galilei, Italian astronomer, discovered three moons – Io, Europa, and Callisto – orbiting Jupiter. He discovered a fourth moon – Ganymede – a few weeks later. Galileo was using one of history's first telescopes, which he had designed and constructed in 1609. The four moons were the first bodies discovered to regularly orbit some object in space other than the Earth, and were important evidence that Earth was not the centre of the Universe. Using his telescope Galileo later discovered the rings of Saturn and the phases of Venus. (☞ Jan. 8, Jan. 13, Feb. 15)

1871
Dmitry Ivanovich Mendeleyev, Russian chemist, announced that the gaps in his periodic table represented undiscovered elements. He also predicted the properties of three of those elements. The three new elements – gallium, scandium, and germanium – were subsequently discovered in 1875, 1879, and 1885. (☞ Feb. 2, Feb. 8)

1902
Spark plug with a high-voltage magneto ignition was patented by the German engineer and industrialist Robert Bosch. A glaring deficiency of the early ignition systems was the inability to consistently ignite the air/fuel mixture. The use of the spark plug enabled production of consistent engine power that early auto pioneers were striving for. To this day, the Bosch invention remains one of the most important breakthroughs in automotive history. The spark plug laid the foundation for the worldwide automotive industry. (☞ Mar. 9, Sep. 23)

1925
Gerald M. Durrell, British naturalist and animal lover, was born in Jamshedpur, India. Durrell was a pioneer in wildlife conservation and played a key role in preserving and breeding endangered species by housing them in zoos with the intent of eventually releasing them in the wild. He was also a prolific author and wrote more than 35 amusing and informative books about the animal kingdom and his adventures in pursuit of threatened species. His popular titles include *A Zoo in My Luggage*, *My Family and Other Animals*, and *Birds, Beasts and Relatives*. (☞ Jan. 30)

1935
National Institute of Sciences of India (NISI) was established in Calcutta (now Kolkata). Dr. L.L. Fermor was its first President. The headquarters of the NISI moved to Delhi in May 1946. In February 1970, the name was changed to the Indian National Science Academy (INSA), which is the apex body of Indian scientists representing all branches of science and technology. Its objectives encompass promotion of science in India including its application to national welfare, safeguarding the interests of scientists, and establishing linkages with international bodies.

1941
John E. Walker, British chemist, Nobel laureate, was born in Halifax, Yorkshire, England. Walker received an academic education at Rastrick Grammar School in Halifax, specialising in physical sciences and mathematics. He received his BA degree in chemistry from St. Catherine's College, Oxford, in 1964. In 1965, he began research on peptide antibiotics in the Sir William Dunn School of Pathology, Oxford, and was awarded the DPhil degree in 1969. In 1974 Walker joined the Medical Research Council's Laboratory of Molecular Biology at the Cambridge University where he began a structural study of the adenosine triphosphate (ATP) synthase from bovine heart mitochondria and from eubacteria. ATP synthase helps in the synthesis of ATP, the carrier of chemical energy. These studies resulted eventually in a complete sequence analysis of the complex from several species. In the early 1980s, he determined the sequence of amino acids that make up the synthase's protein units. In the 1990s, working with X-ray crystallographers, Walker was also able to determine the three-dimensional structure of the enzyme. Walker's findings offer insight into the way life-forms produce energy. Walker was awarded the 1997 Nobel Prize for Chemistry, jointly with American chemist, Paul D. Boyer and Jens C. Skou of Denmark, for their explanation of the enzymatic process that creates ATP. (☞ Jul. 31, Oct. 8)

1943
Nikola Tesla, Serbian-born American inventor, who discovered the rotating magnetic field, the basis of most alternating-current machinery, died in New York, USA. (☞ Jul. 9)

1968
The American Moon-probe *Surveyor-7* was launched towards Moon from the Kennedy Space Center, Florida, USA. (☞ Jan. 9)

1984
Alfred Kastler, French physicist, Nobel laureate (Physics, 1966), whose research facilitated the study of atomic structures by means of the radiations that atoms emit under excitation by light and radio waves, died in Bandol, France. (☞ May 3)

1986
India's largest optical telescope with a mirror diameter of 2.34 m at Kavalur Observatory in Tamil Nadu was christened after the Indian astronomer M.K. Vainu Bappu. (☞ Aug. 10, Aug. 19)

1998
Vladimir Prelog, Swiss chemist, Nobel laureate (Chemistry, 1975), whose wide-ranging research on the stereochemistry of alkaloids, antibiotics, enzymes, and other natural compounds contributed to the understanding of stereoisomerism, died in Zurich, Switzerland. (☞ Jul. 23)

January 8

1642
Galileo Galilei, Italian astronomer, who was the first to use a telescope to study the Moon and the planets, died in Arcetri, near Florence, Italy. (☞ Jan. 7, Jan. 13, Feb. 15)

1823
Alfred Russel Wallace, British naturalist, was born in Usk, England. He was the first westerner to describe some of the most interesting natural habitats in the tropics. In 1837 Wallace became an apprentice in his eldest brother's sur-veying business and for approximately 8 of the next 10 years, surveyed and mapped in Bedfordshire and then in Wales. As a surveyor, Wallace spent a great deal of time outdoors. With his naturalist friend Henry Walter Bates, who had introduced him to etymology four years earlier, Wallace travelled to Brazil in 1848 as a self-employed specimen collector. He spent a total of four years travelling, collecting, mapping, drawing, and writing in unexplored regions of the Amazon river basin. He collected butterflies, other insects, and birds; and he searched for clues to solve the mystery of the origin of plant and animal species. However, except for one shipment of specimens sent to his agent in London, most of Wallace's collections were lost on his voyage home when his ship went up in flames and sank. Nevertheless, he managed to save some of his notes before his rescue and return journey. From these he published several scientific articles, two books – *Palm Trees of the Amazon and Their Uses* and *Narrative of Travels on the Amazon and Rio Negro* – and a map depicting the course of the River Negro. Wallace spent eight years in the Malay Archipelago, from 1854 to1862, travelling among the islands, collecting biological specimens for his own research and for sale, and writing scores of scientific articles on mostly zoological subjects. Among these were two extraordinary articles dealing with the origin of new species. Wallace is best known for devising a theory of the origin of species through natural selection made independently of Darwin. He contributed greatly to the scientific foundations of zoogeography, including his proposal, based on his observations in the Malay Archipelago, for the evolutionary distinction between the fauna of Australia and Asia. (☞ Nov. 7)

1851
Jean-Bernard-Leon Foucault, French physicist, gave the first demonstration

of the 'Foucault pendulum'. He used the motion of a heavy iron ball swinging from a wire 67 metres long, to prove that the Earth rotates about its axis. Such a pendulum always swings in the same vertical plane, but as the Earth rotates, this vertical plane slowly changes, at a rate and direction dependent on the geographic latitude of the pendulum. (☞ Feb. 11, Sep. 18)

1889
Herman Hollerith, American inventor, patented his tabulating machine that was the precursor of the electronic computer. His tabulating machine could be used to record statistics by electrically reading and sorting punched cards that had been numerically encoded by perforation position. (☞ Feb. 29, Nov. 17)

1891
Walther Bothe, German physicist, Nobel laureate, was born in Oranienburg, Germany. In 1925 he and fellow German physicist Hans Geiger used two Geiger counters to gather data on the Compton Effect by simultaneously measuring the energies and directions of single photons and electrons emerging from individual collisions. They used the data to show that momentum and energy are conserved at the atomic level. In 1929, applying this coincidence-counting method, Bothe found that cosmic rays are not composed exclusively of gamma rays, as was previously believed. In 1930 he discovered an unusual radiation emitted by beryllium when it is bombarded with alpha particles. The English physicist James Chadwick later identified this radiation as neutrons. During World War II, Bothe was one of the leaders of German research on nuclear energy. He was responsible for the planning and building of Germany's first cyclotron, which was completed in 1943. Bothe was awarded the 1954 Nobel Prize for Physics, jointly with the German physicist Max Born, for his invention of a new method of detecting subatomic particles and for other resulting discoveries. (☞ Feb. 8, Dec. 11)

1935
The spectrophotometer – a device for comparing relative intensities of two light radiations wavelength by wavelength – was patented by Arthur Cobb Hardy. The instrument could detect 2 million shades of colour.

1942
Stephen W. Hawking, English theoretical physicist, was born in Oxford, England. Hawking studied mathematics and physics at the University College, Oxford and Trinity Hall, Cambridge, from where he obtained his PhD. In the early 1960s he contracted amyotrophic lateral sclerosis, an incurable degenerative neuromuscular disease. He continued to work despite the disease's progressively disabling effects, which confined him to a wheelchair and the use of a computer voice synthesiser for speaking. Hawking worked primarily in the field of general relativity and particularly on the physics of black holes. In 1971 he suggested the formation, following the Big Bang, of numerous objects, which he called 'mini black holes', containing as much as 1,000,000,000 tonnes of mass but occupying only the space of a proton. The concept of mini black holes is unique in that their immense mass and gravity require

that the laws of relativity rule them, while their minute size requires that the laws of quantum mechanics apply to them also. In 1974 Hawking proposed that, in accordance with the predictions of quantum theory, black holes emit subatomic particles until they exhaust their energy and finally explode. Hawking is the author of the bestseller *A Brief History of Time*.

1973
Russian Moon-probe *Luna-21* was launched from Baikanur Cosmodrome in USSR (now Kazakhstan). It carried a remotely controlled lunar rover *Lunokhod-2*. The primary objectives of the mission were to collect images of the lunar surface, examine ambient light levels to determine the feasibility of astronomical observations from the Moon, perform laser ranging experiments from Earth, observe solar X-rays, measure local magnetic fields, and study mechanical properties of the lunar surface material. It was powered by solar cells. (☞ Jan. 15)

1997
Melvin Calvin, American biochemist, Nobel laureate (Chemistry, 1961), who discovered the chemical pathways of photosynthesis, died in Berkeley, California, USA. (☞ Apr. 8)

January 9

1839
The invention of the daguerreotype photographic process, developed by the French artist and chemist Louis Jacques Mande Daguerre, was announced before the French Academy of Sciences by the eminent French astronomer and physicist Louis Arago. It was the first photographic process that allowed an image to be chemically fixed as a permanent picture. Joseph Nicéphore Niépce had produced the first long lasting image in 1824 using a process called heliography which required nearly 8 hours of bright sunlight to affix the image. Daguerre discovered the daguerreotype process by accident. In 1835, so the story goes, he put an exposed plate in his chemical cupboard, and some days later found, to his surprise, that the latent image on the plate had developed. Daguerre eventually concluded that this was due to the presence in the cupboard of mercury vapour from a broken thermometer. This important discovery that a latent image could be developed made it possible to reduce the exposure time from some eight hours to thirty minutes. (☞ Jan. 2, Jul. 10, Nov. 18)

An early daguerreotype taken using the process

1922
Har Gobind Khorana, Indian-born American biochemist, Nobel laureate, was born in Raipur, Punjab (now in Pakistan), India. Born into a poor family he attended the Punjab University at Lahore (now in Pakistan) and the University of Liverpool, England, on government scholarships. He obtained his PhD at Liverpool in 1948, and began research on nucleic acids during a fellowship at the University of Cambridge

(1951) under Sir Alexander Todd. Khorana's main contribution has been in deciphering the genetic code using synthetic genes. In the 1960s, while at the University of Wisconsin, USA, Khorana confirmed Marshall Nirenberg's findings that the way the four different types of nucleotides are arranged on the spiral 'staircase' of the DNA molecule determines the chemical composition and function of a new cell. He was the first to synthesise the first wholly artificial gene – found in the bacterium *Escherichia coli*. Khorana was awarded the 1968 Nobel Prize for Physiology or Medicine, jointly with Robert W. Holley and Marshall W. Nirenberg, for research that helped to show how the nucleotides in nucleic acids, which carry the genetic code of the cell, control the cell's synthesis of proteins. (☞ Jan. 28, Apr. 10)

1950

Alec John Jeffreys, British geneticist, Fellow of the Royal Society, was born in Luton, Bedfordshire, England. Jeffreys graduated from the Oxford University and moved to the Leicester University in 1977. He was one of the first to show that mammalian genes are not simple, as in bacteria, but are interrupted by introns. At Leicester, he combined molecular biology and genetics to study variation in human DNA, particularly at highly variable repeat DNA sequences. This culminated in his accidental development of DNA fingerprinting in 1984 and demonstration that it could be used to resolve issues of identification and kinship. DNA fingerprinting uses variations in the genetic code to identify individuals. The technique has been applied in forensics for law enforcement, to resolve paternity and immigration disputes, and can also be applied to non-human species, for example in wildlife population genetics studies. Jeffreys' DNA technique was used in the first regional screen of human DNA to identify the rapist and killer of two girls in Leicestershire, England. The suspect was identified through the screen and convicted of murder after samples taken from him matched semen samples taken from the two dead girls. Jeffreys refined his DNA fingerprinting technique, by developing DNA profiling based on highly variable mini satellites in the human genome. DNA profiling focussed on just a few of these highly variable mini satellites, making the system more sensitive, more reproducible and amenable to computer databasing. With highly automated and sophisticated equipment, the modern-day DNA fingerprinter can process hundreds of samples a day.

1968

The American Moon-probe *Surveyor-7* became the first spacecraft to land undamaged on the lunar surface. The spacecraft landed 29 kilometres north of the crater Tycho. This was the first soft landing of an unmanned vehicle prior to the manned Apollo flights that put men into the Moon. It sent back 21,000 photographs and a great deal of scientific data. (☞ Jan. 7)

1969

Concorde, the world's first supersonic airliner made its first trial flight at Bristol, UK. (☞ Jan. 21, Mar. 2, Jul. 25, Oct. 1, Oct. 24)

1982

A 21-member Indian scientific team led by S.Z. Qasim landed on Antarctica for the first time. The team had started the expedition on

6 December, 1981.They set up a base named 'Dakshin Gangotri' to carry out scientific studies on the icy continent. Since the first expedition in 1981, scientific expeditions to Antarctica have become an annual feature – twenty-four more scientific expeditions have been sent to continue scientific research programmes on atmospheric science, meteorology, biology, oceanography, and earth sciences. In addition to these regular expeditions, India also launched an expedition to the Weddell Sea in 1989 and an expedition for krill exploration in 1995. A second base named 'Maitri' was established in March 1989 after 'Dakshin Gangotri' became buried in ice. 'Maitri' is built on adjustable telescopic legs and has 25 single-occupancy rooms as well as research, supply, power and general purpose buildings. It is located on the East Antarctic mainland south of Capetown and 90 nautical miles from the usual Antarctic coastline. The annual scientific expeditions to Antarctica have helped to prepare a viable ground for front ranking research endeavours in basic and environmental science and have won for India a well-deserved recognition amongst the Antarctic Treaty Nations. India has Consultative Status in the Antarctic Treaty System. ☞ Dec. 6)

1998
Kenichi Fukui, Japanese theoretical chemist, Nobel laureate (Chemistry, 1981), whose theory of frontier orbitals advanced the understanding of the mechanism of chemical reactions, died in Kyoto, Japan. (☞ Oct. 4)

January 10

1778
Carl Linnaeus, Swiss botanist who was the first to frame principles for defining genera and species of organisms and to create a uniform system for naming them, died in Uppsala, Sweden. (☞ May 23)

1863
World's first underground railway system, from Paddington to Farringdon Street, was opened in London. The 6.5-km route had seven stations. The four-carriage trains were drawn by steam locomotives. (☞ Nov. 4)

1916
Sune K. Bergström, Swedish biochemist, Nobel laureate, was born in Stockholm, Sweden. Bergström was educated at the Karolinska Institute, Stockholm, from where he received his doctoral degrees in medicine and biochemistry in 1944. He held research fellowships at the Columbia University, USA, and the University of Basel, Switzerland and then returned to Sweden as a professor of chemistry at the University of Lund. Bergström's research conerned isolation, id-entificaion and analysis of prostaglandins, whi-ch are biochemical com-pounds that influence blood pressure, body temperature, allergic reactions, and other physiological phenomena in mammals. Bergström was the first to demonstrate the existence of more than one such compound and to determine the elemental compositions oftwoofthem. Bergström was awarded the 1982 Nobel Prize for Physiology or Medicine, jointly with fellow Swede, Bengt Ingemar Samuelsson and the English biochemist John Robert Vane, for their isolation, identification, and analysis of numerous prostaglandins. (☞ Mar. 29, May 21)

1936
Robert Woodrow Wilson, American radio astronomer, Nobel laureate, was born in Houston, Texas, USA. Wilson was educated at the Rice University, Houston, and the California Institute of Technology, Pasadena, where he was awarded his doctorate in 1962. While working at the Bell Telephone Laboratories, in collaboration with the German-born American astrophysicist Arno Penzias, he began monitoring radio emissions from a ring of gas encircling the Milky Way Galaxy. The two scientists detected an unusual background radiation at a temperature of 3 K, corresponding to microwave frequencies that seemed to permeate the cosmos uniformly. This radiation appeared to be a remnant of the 'Big Bang', the primordial explosion billions of years ago from which it is believed that the Universe originated. Wilson and Penzias were jointly awarded the 1978 Nobel Prize for Physics with the Russian physicist Pyotr Leonidovich Kapitsa. Penzias and Wilson shared half of the prize money for their discovery of cosmic microwave background radiation, while Kapitsa received the other half for his basic inventions and discoveries in the area of low-temperature physics. (☞ Apr. 26, Jul. 8)

1947
The isolation of the poliovirus was reported by scientists at the Stanford University, USA.

1969
Soviet unmanned spacecraft *Venera-6* was launched towards Venus from Baikanur Cosmodrome in Kazakhstan. (☞ May 17)

1977
Alexander R. Todd, British biochemist, Nobel laureate (Chemistry, 1957), who did research on the structure and synthesis of nucleotides and nucleosides, and elucidated the structure of vitamin B_{12}, died in Cambridge, UK. (☞ Oct. 2)

January 11

1787
William Herschel, German-born British astronomer discovered two moons of Uranus – Titania and Oberon – six years after he had discovered the planet. Titania's diameter is 1,610 km and orbits Uranus at a distance of 436,300 km. Oberon has a diameter of 1,522 km and orbits Uranus at a distance of 582,600 kilometres. (☞ Mar. 13, Aug. 25, Nov. 15)

1882
Theodor Schwann, German physiologist who founded modern histology, died in Cologne, Germany. (☞ Dec. 7)

1922
Insulin was first used to treat a diabetic patient, 14-year-old Leonard Thompson at Toronto General Hospital in Canada. The doctors were Walter A. Campbell and Alma A. Fletcher. The boy lived a normal life with regular administration of insulin. (☞ Feb. 11, Feb.27, Nov. 14)

1924
Roger Charles Louis Guillemin, French-born American physiologist, Nobel laureate, was born in Dijon, France. Educated at the Universities of Dijon, Lyon, and Montreal (Canada), Guillemin taught at the Baylor College of Medicine in Houston, Texas, USA from 1953 to 1970,

except for the years 1960–63, when he was on the faculty of the Collège de France in Paris. He became a US citizen in 1963. In 1970 he became a research professor at the Salk Institute for Biological Studies in La Jolla, California. Guillemin's researches concerned hormones released by the hypothalamus. He and his colleagues discovered, isolated, analysed, or synthesised several hormones including the thyrotropin-releasing hormone (TRH), which regulates thyroid activity; the growth hormone-releasing hormone (GHRH), which causes the pituitary to release gonadotropin; and somatostatin, which regulates the activities of the pituitary gland and the pancreas. Guillemin also discovered an important class of proteins called endorphins that are involved in the perception of pain. Guillemin was awarded the 1977 Nobel Prize for Physiology or Medicine, jointly with Polish-born American endocrinologist Andrew Schally and American medical physicist Rosalyn Yalow, for researches on the hormones produced by the hypothalamus gland. (☞ Jul. 19, Nov. 30)

1939
The element francium (Fr, atomic number 87) was discovered by Marguerite Perey of the Curie Institute, Paris and was named Francium after her native country France. Francium is the heaviest known member of the alkali metals series and exists only in short-lived radioactive forms.

1950
The Institute of Nuclear Physics (now Saha Institute of Nuclear Physics, SINP) was inaugurated by the French chemist Irene Joliot-Curie in Calcutta (now Kolkata). The institute grew out of the Palit Research Laboratory in Physics of the University of Calcutta through the initiative of the Indian physicist Meghnad Saha, who became the Palit Professor of Physics in 1938. After the discovery of nuclear fission by Otto Hahn and Fritz Strassman in 1939, Saha had seen the immense potential of nuclear science for the betterment of India, and by 1940 he had had the syllabus of the post-graduate physics re-organised to include nuclear physics in the physics curriculum. Soon it was deemed necessary to build a small-scale cyclotron to gain first-hand knowledge in this virgin field. Saha decided to have it built locally rather than buying one from abroad. A piece of land was made available within the University College of Science campus for erecting a building to house the machine. Soon after, the Institute of Nuclear Physics was founded. Initially a part of the University of Calcutta, the Institute was accorded autonomy in 1951. The Institute shifted to its new building in Bidhannagar in the eastern outskirts of Kolkata in the late 1980s. It was named after Saha after his death in 1956. The indigenously designed cyclotron became operational in April 1960. The external beam became available in 1966 and was extensively used in the study of nuclei using proton-induced reactions. A 14.8 MeV neutron generator was also successfully installed in the Institute during this period. At present SINP is primarily an institute for basic research and training in physical sciences – condensed matter physics, high-energy physics, microelectronics, and nuclear science – and biophysical sciences. (☞ Oct. 6)

1988
Isidor Isaac Rabi, American physicist, Nobel laureate (Physics, 1944), who developed a method for measuring the

magnetic properties of atoms, atomic nuclei, and molecules, died in New York, USA. (☞ Jul. 29)

1991
Carl David Anderson, American physicist, Nobel laureate (Physics, 1936), who discovered the positron, or positive electron, the first known particle of antimatter, died in San Marino, California, USA. (☞Sep. 3)

January 12

1665
Pierre de Fermat, French lawyer and mathematician, often called the founder of the modern theory of numbers and known for his famous 'last theorem', died in Castres, France. (☞ Aug. 17)

1895
Yellapragada Subbarow, Indian-born American biochemist, was born at Bhimavaram, Andhra Pradesh, India. Not a bright student in school, young Yellapragada studied math, physics, and chemistry before being accepted by the Madras Medical College, where he became interested in medical research. The deaths of two of his brothers from tropical diseases caused him to narrow his field of concentration, and in 1923 he went to the US and entered the Harvard School of Tropical Medicine. Since his Indian medical degree was not accepted in the United States, he earned money by working as a night porter at a Boston hospital. After a stretch as a junior faculty member at Harvard, Subbarow joined Lederle Laboratories in 1940 where he directed

research and did work in the development of synthetic folic acid, tetracycline, and anti-malarial drugs till his death in 1948. (☞ Aug. 9)

1899
Paul Hermann Müller, Swiss chemist, Nobel laureate, was born in Olten, Switzerland. Müller matriculated in 1918 and obtained his doctorate from the Basle University in 1925. He began his career with investigations of dyes and tanning agents as a research chemist at the J.R. Geigy Company in Basel. In 1935 he began his search for an 'ideal' insecticide, one that would show rapid, potent toxicity for the greatest number of insect species but would cause little or no damage to plants and warm-blooded animals, and would also have a high degree of chemical stability, so that its effect would persist for long periods of time. Four years later, Müller tested a substance known as dichlorodiphenyltrichloroethane (DDT) and found that it satisfied these requirements. Müller was awarded the 1948 Nobel Prize for Physiology or Medicine for discovering the potent toxic effects of DDT on insects. (☞ Oct. 12)

1903
Igor Vasilyevich Kurchatov, Soviet nuclear physicist who guided the development of Russia's first atomic bomb, was born in Sim, Russia. He graduated from the Crimean University in Simferopol in 1923 and four years later joined the Physico-Technical Institute of the Academy of Sciences in Leningrad. After initially working on ferroelectricity, he devoted his time on research in nuclear physics. As director of the nuclear physics laboratory at the

Physico-Technical Institute, he supervised the construction of the first Soviet cyclotrons. His team produced the first Soviet atomic bomb in 1949, and in 1953, detonated Russia's first thermonuclear (hydrogen) bomb. (☞ Feb. 7)

1926
Pasteur Institute in Paris announced the discovery of an anti-tetanus serum that could be used for vaccination against tetanus. This anti-tetanus vaccine is routinely used to prevent tetanus in mothers after childbirth. Tetanus, also called lockjaw, is an acute infectious disease caused by toxins produced by a bacterium called *Clostridium tetani*, widely found in soil. Spores of the bacterium enter the body through any wound and produces rigidity and spasms of the voluntary muscles, especially the jaw muscles.

1967
India's Atomic Energy Establishment at Trombay was re-named Bhabha Atomic Research Centre in honour of Homi Jehangir Bhabha. (☞ Jan. 20, Oct. 30)

1997
Charles Brenton Huggins, Canadian-born American surgeon, Nobel laureate (Physiology or Medicine, 1966), who demonstrated the relationship between hormones and certain types of cancer, died in Chicago, Illinois, USA. (☞ Sep. 22)

2005
NASA's *Deep Impact* spacecraft was launched from Cape Canaveral, Florida, USA on a unique mission to Comet Tempel-1. After reaching the comet in July 2005, the spacecraft deployed an impactor that carved out a large crater on the comet. The resulting explosion was equivalent to the explosion of 4 tonnes of TNT, and was bright enough to be seen with the unaided eye back on Earth. *Deep Impact* was at a safe distance of 500 kilometres from the explosion, and got a unique view of the comet's composition, and what lies under its surface. More and more findings are still being recorded. (☞ Jul. 4)

January 13

1610
Italian astronomer Galileo Galilei discovered Ganymede, the fourth moon of Jupiter. Although discovered in the 1600s, it was only in the mid-1800s that the names of the Galilean moons, Io, Europa, Ganymede and Callisto, were officially adopted. Ganymede is the largest moon of Jupiter and is the largest in our solar system with a diameter of 5,262 kilometres. If Ganymede orbited the Sun instead of Jupiter it could be classified as a planet. (☞ Jan. 7, Jan.8, Feb 15)

1858
Oskar Minkowski, German physiologist and pathologist who introduced the concept that diabetes results from suppression of a pancreatic substance, was born in Alekostas, Russia (now in Lithuania). Between 1882 and 1926, Minkowski taught medicine at the Universities of Strasbourg and Cologne in France; Greifswald in Germany; and Breslau in USA. In 1889 he discovered (with the German physiologist, Joseph von Mering) that dogs subjected to removal of the pancreas develop the symptoms of diabetes. This led Minkowski to postulate that the pancreas is the site of

secretion of an 'anti-diabetic' substance, now known to be insulin. (☞ Jul. 18)

1864

Wilhelm Wien, German physicist, Nobel laureate, was born in Gaffken, Prussia (now in Poland). Wien obtained his doctorate at the University of Berlin in 1886 and soon began research on radiation. His study of radiation from a blackbody – a surface that absorbs all radiant energy falling on it – led Wein to formulate a law in 1893, called the Displacement Law, which stated that the intermediate wavelength at which the radiation reaches a maximum value is inversely proportional to the absolute temperature of the body. In 1898, while studying streams of ionised gas, Wien identified a positive particle equal in mass to the hydrogen atom, which was later named the proton. With this work, Wein laid the foundation of mass spectroscopy. Wien was awarded the 1911 Nobel Prize for Physics for his Displacement Law concerning the radiation emitted by a perfectly efficient blackbody. (☞ Aug. 30)

1927

Sydney Brenner, South African-born British biologist, Nobel laureate, was born in Germiston, South Africa. Brenner did his MSc in 1947 at the University of Witwatersrand, South Africa and received his MB from the same university in 1951. After receiving a PhD from the University of Oxford, England in 1954, Brenner began work with the British Medical Research Council (MRC). During 1979–86 he directed the MRC's Laboratory of Molecular Biology and later, during 1986–91, it's Molecular Genetics Unit. In 1996 he founded the Molecular Sciences Institute in California, USA, and in 2000 became a distinguished research professor at the Salk Institute for Biological Sciences in La Jolla, California. Brenner's research mainly concerned cell differentiation and related topics. In the early 1960s he realised that fundamental questions regarding cell differentiation and organ development were hard to tackle in higher animals. Therefore, a genetically amenable and multicellular model organism simpler than mammals was required. He found the common nematode *Caenorhabditis elegans* to be an ideal organism for study. This worm, approximately 1 mm long, has a short generation time and is transparent, which made it possible to follow cell division directly under the microscope. In 1974, Brenner demonstrated that specific gene mutations could be induced in the genome of *C. elegans* by the chemical compound ethyl methane sulphonate (EMS) and that different mutations could be linked to specific genes and to specific effects on organ development. These studies further demonstrated that natural cell death is controlled by a unique set of genes – a process referred to as 'programmed cell death'. Brenner was awarded the 2002 Nobel Prize for Physiology or Medicine, jointly with the American biologist, H. Robert Horvitz and the British biologist, John E. Sulston, for their discoveries concerning genetic regulation of organ development and programmed cell death. (☞ Mar. 27, May 8)

1942

Henry Ford was issued the first US patent for construction of an automobile using plastic. It covered an automobile body construction, an auto body chassis frame made of steel tubes or pipes with a body made of plastics. The first such car manufactured in the US was produced by the Ford Motor Company, Michigan in August 1941. It had fourteen plastic panels mounted on a tubular welded frame and windows and windshield made of acrylic sheets, which reduced the weight of the car by almost 30 per cent. (☞ Jul. 30)

1949
Rakesh Sharma, Indian cosmonaut, was born in Patiala, Punjab. In 1966 he enrolled in the National Defence Academy, Khadakvasla, and after passing out was commissioned into the Indian Air Force. He served in different operational squadrons and flew modern planes. On finishing a course for test pilots, he took part in testing aircraft of different types. In 1982, Major Sharma was selected to become a cosmonaut. He began training at the Gagarin Space Training Centre in Russia and underwent a full course of training for flights aboard the *Soyuz-T* spacecraft and the *Salyut* orbital station. In April 1984 Squadron Leader Sharma became the first citizen of India to go into space when he flew aboard the Russian rocket *Soyuz T-11,* which took off from the Baikanur Cosmodrome. The *Soyuz T-11* docked and transferred the three-member Soviet–Indian international crew, which also included the ship's Commander Y.V. Malyshev and Flight Engineer G.M. Strekalov to the *Salyut-7* orbital station. The crew spent seven days aboard the *Salyut* space station during which they conducted scientific and technical studies, which included 43 experimental sessions. Sharma conducted Earth observation studies concentrating on India. He also did life sciences and materials processing experiments and became the first yoga practitioner in zero-G to combat the debilitating effects of weightlessness. Sharma retired with the rank of Wing Commander in 1984 and joined Hindustan Aeronautics Limited as a test pilot and worked on the indigenous Light Combat Aircraft program. (☞ Apr. 3)

1958
Linus Pauling, American chemist, Nobel laureate, presented a petition signed by 9,000 scientists to Dag Hammarskjöld, Secretary General of the United Nations, asking for a halt in the testing of nuclear bombs. Pauling, together with his wife, was instrumental in collecting signatures from scientists all over the world. (☞ Feb. 28, Aug. 19)

1994
The Hubble Space Telescope started sending sharper pictures after its faulty optical system was repaired. (☞ Apr. 24, Dec. 2)

January 14

1742
Edmond Halley, British astronomer who was the first to calculate the orbit of a comet, later named after him, died at Greenwich, near London, UK. (☞ Nov. 8)

1905
Ernst Abbe, German physicist whose theoretical and technical innovations in optical theory led to great improvements in microscope design, died in Jena, Germany. (☞ Jan. 23)

1929
Fernand-Isidore Widal, French physician and bacteriologist who developed the Widal reaction, a procedure for diagnosing typhoid fever, died in Paris, France. (☞ Mar. 9)

1953
The Central Electrochemical Research Institute (CECRI), a constituent national laboratory of the Council of Scientific and Industrial Research (CSIR), and one of the world's premier R&D laboratories in the field of electrochemical science and technology, was inaugurated. Located in Karaikudi, Tamilnadu, CECRI carries out R&D in the areas of corrosion science and engineering, industrial metal finishing, batteries, electro-metallurgy, electro-pyrometallurgy, electrochemicals, electrochemical instrumentation and pollution control. Research is directed towards development of new processes of products or the novel use of electrochemistry. Upgradation of already-developed technology and basic research is also carried out. Over the years, CECRI has filed 680 patents and developed 190 processes. The institute also conducts short-term refresher courses for the benefit of industry and educational institutions.

2005
Huygens, European Space Agency's probe landed on Titan, Saturn's largest moon. The space probe entered the thick atmosphere of Titan at an altitude of 1,270 kilometres and parachuted down to Titan's surface. The event was the only landing to take place in the outer solar system and the furthest from Earth. The probe found nitrogen to be the main component of Titan's atmosphere, methane being the second most important component. The images sent back by *Huygens* showed what looked like a rock-strewn surface

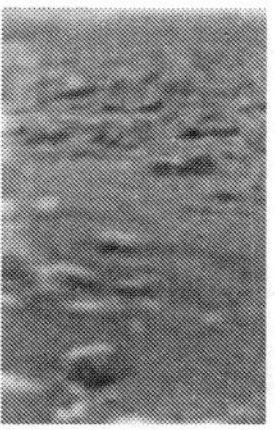

Cassini Huygens space craft (left) and surface of Titan

and ravines that could have been carved by the liquid hydrocarbons thought to cover much of Titan's surface. The ravines, stubby drainage-like channels, appeared to funnel toward what appeared to be a shoreline. The instrument-laden probe was carried to Saturn on board NASA's *Cassini* spacecraft and was released from *Cassini* in late December 2004. (☞ Apr. 26, Jun. 24, Jul. 1, Aug. 18, Oct. 15, Dec. 24, Dec. 30)

January 15

1785
William Prout, English chemist and biochemist noted for his discoveries concerning digestion, metabolic chemistry, and atomic weights, was born in Horton, Gloucestershire, England. Although a physician by training, Prout was also an experimentalist and theorist. Between 1815 and 1827 he published a number of papers that helped establish the barely explored areas of metabolic and physiological chemistry. Prout's classification of food components into water, fats, carbohydrates, and proteins was quickly adopted by his fellow biochemists. In 1815, Prout advanced the idea that the atomic weights of elements are whole-number multiples of the atomic weight of hydrogen, which came to be known as 'Prout's hypothesis' and

proved highly useful for later investigations of atomic weights, atomic theory, and the classification of the elements. (☞Apr. 9)

1861

The safety elevator, commonly known as the 'lift', was patented as a 'hoisting apparatus' by the American inventor, Elisha G. Otis, of New York, USA. His invention used spring-loaded pawls that would release and engage in a mortised track in the walls of the shaft to arrest a fall in case of the lifting rope breaking. (☞ Apr. 8, Aug. 3)

1895

Artturi I. Virtanen, Finnish biochemist and Nobel laureate, was born in Helsinki, Finland. Virtanen was educated at the Classical Lyceum at Viipuri, Finland. After finishing school, he studied chemistry, biology, and physics at the University of Helsinki, where he took his MSc in 1916 and obtained his DSc in 1919. In 1921 he became Director of the Laboratory of Valio, Finnish Cooperative Dairies' Association, and that of the Biochemical Research Institute at Helsinki in 1931. He was appointed professor of biochemistry at the Finland Institute of Technology at Helsinki in 1931 and at the University of Helsinki in 1939. Virtanen's research mainly concerned improving the production and storage of protein-rich green fodder. He studied the fermentation processes that spoiled stores of silage. It was known that the fermentation product, lactic acid increases the acidity of the silage to a point at which destructive fermentation ceases. So, Virtanen thought the fermentation process could be prevented by adding dilute hydrochloric or sulphuric acid to newly stored silage, thereby increasing the acidity of the fodder beyond which fermentation would not occur. In a series of experiments, carried out during 1928–29, he showed that acid treatment did indeed prevent spoilage due to fermentation and that it had no adverse effect on the nutritive value and edibility of the fodder and of products derived from animals which ate the fodder. Virtanen was awarded the 1945 Nobel Prize for Chemistry for his investigations directed toward improving the production and storage of protein-rich green fodder. (☞ Nov. 11)

1907

Lee De Forest was issued a US patent for the three-element vacuum tube triode. It was described as a 'device for amplifying feeble electric currents – such, for example, telephone currents'. It made possible live radio broadcasting and became the key component of all radio, telephone, radar, television, and computer systems before the invention of the transistor in 1947. (☞ Jun. 30, Aug. 26)

1908

Edward Teller, Hungarian-born American nuclear physicist, was born in Budapest, Hungary. Teller participated in the production of the first atomic bomb in 1945 and led the development of the world's first thermonuclear weapon, commonly known as the hydrogen bomb. After attending schools in Budapest, he earned a degree in chemical engineering at the Institute of Technology in Karlsruhe, Germany. He then went to Munich and Leipzig to earn a PhD in physical chemistry in 1930. Teller studied nuclear physics under Niels Bohr in Copenhagen and then taught at the University of Göttingen during 1931–33. In 1935 he immigrated to the US and worked with George Gamow to establish new rules for classifying the ways subatomic particles

escape the nucleus during radioactive decay. When the nuclear physicist J. Robert Oppenheimer set up the secret Los Alamos Scientific Laboratory in New Mexico in 1943 to develop the atom bomb, Teller was among the first men recruited. Teller credited with developing the hydrogen bomb, became known in the United States as 'the father of the H-bomb'. (☞ Sep. 9)

1914

The first meeting of the Indian Science Congress Association was inaugurated in the premises of the Asiatic Society, Calcutta (now Kolkata), India, with Sir Asutosh Mukherjee, the then Vice-Chancellor of the Calcutta University, as President. One hundred and five scientists from different parts of India and abroad attended and 35 papers divided into six sections – Botany, Chemistry, Ethnography, Geology, Physics and Zoology – were presented during the three-day session. The Indian Science Congress Association owes its origin to the foresight and initiative of two British chemists, Prof. J. L. Simonsen and Prof. P.S. MacMahon. It occurred to them that scientific research in India might be stimulated if an annual meeting of research workers somewhat on the lines of the British Association for the Advancement of Science could be arranged. The Association was formed with the following objectives: (i) to advance and promote the cause of science in India; (ii) to hold an annual congress at a suitable place in India; (iii) to publish proceedings, journals, transactions, etc; and (iv) to popularise science.

1915

George Claude of Paris was issued a US patent for the neon sign. Titled a 'System of illuminating by luminescent tubes,' the patent described the technical details of a neon tube. It considered the materials used and size of the electrodes, diameter of the tube, and suitable operating pressure. Tube lengths of 5 to 6 metres were discussed. It was also explained how it was possible to maintain the light output for extensive life. (☞ Sep. 24, Dec. 11)

1948

Henri-Alexandre Deslandres, French physicist and astrophysicist who invented the spectroheliograph, died in Paris, France. (☞ Jul. 24)

1957

Non-conservation of parity was announced by the Columbia University Physics Department in USA. In particle physics, it was long held that charge conjugation and parity by a weak nuclear force, which is responsible for reactions such as the decay of atomic nuclei, are conserved. Charge conjugation is a mathematical operation that changes the sign of the charge and implies that every charged particle has an oppositely charged antimatter counterpart, or antiparticle. Parity conservation means that left and right and up and down is indistinguishable in the sense that an atomic nucleus throws off decay products up as often as down and left as often as right. But experiments conducted at the Columbia University by the Chinese-born American physicist Chien-Shiung Wu based on theoretical work by Chinese-born American physicists Chen Ning Yang and Tsung-Dao Lee showed that parity is not conserved for a weak interaction, such as the decay of cobalt-60. Yang and Lee were awarded the 1957 Nobel Prize for

Physics for their work. (☞ May 29, Sep. 22, Nov. 25)

1962

The Tata Institute of Fundamental Research (TIFR) was inaugurated. The first step in the founding of TIFR was a letter written in 1943 by the Indian physicist Homi Bhabha to the industrialist and philanthropist JRD Tata. In this historic letter, Bhabha expressed his vision of a fundamental research institute in India, and sought JRD's support for its creation. The institute started functioning in June 1945, at the Cosmic Ray Research Unit, which had earlier been set up for Bhabha, on the campus of the Indian Institute of Science in Bangalore, Mysore state (now Karnataka). Within six months it was shifted to Kenilworth, a bungalow on Peddar Road in Bombay that belonged to Homi Bhabha's aunt, and was then shifted to a new location vacated by the Royal Bombay Yacht Club, near the Gateway of India, in 1949. The institute moved to its present location in southern Mumbai in 1962. Meanwhile, the Atomic Energy Commission (AEC) of the Government of India which was set up soon after independence began to carry out cooperative projects with TIFR. In 1953, AEC declared its recognition of TIFR as 'the only laboratory of the Commission for fundamental research in atomic science'. On the technological side, TIFR pioneered activity in the field of computer technology by designing India's first computer (*TIFRAC*). Groups working in other emerging fields such as accelerators, microwave communications, software technology, semiconductor technology and educational research were also set up at the Institute. Some of these later became the nucleus of dedicated organisations.

1969

The first link-up in space of two manned spacecraft, Soviet spacecraft *Soyuz-4* and *Soyuz-5* took place. They docked in orbit and transferred crew. Cosmonaut Vladimir A. Shatalov switched to manual control, rendezvoused and docked with *Soyuz-5* carrying cosmonauts Yevegeny V. Khrunov, Boris V. Volynov, and Aleksey S. Yeliseyev. They remained docked for four and a half hours and completed three orbits of the Earth. During that time, two cosmonauts – Khrunov and Yeliseyev – 'space walked' from *Soyuz-4* to *Soyuz-5*, becoming the first cosmonauts to return to Earth in a different spacecraft from the one in which they went into space.

1973

Russian Moon-probe *Luna-21*, carrying the second remotely controlled lunar rover *Lunokhod-2*, landed in Le Monnier crater at the eastern edge of the Mare Serenitatis on Moon. The rover spent four months taking photographs and conducting experiments while travelling 37 kilometres. It sent back 86 panoramic images and over 80,000 TV pictures of the lunar surface. It also conducted mechanical tests of the surface, laser ranging measurements, and other experiments. (☞ Jan. 8, Jan. 16)

2006

After a seven-year journey, NASA's *Stardust* spacecraft dropped a capsule containing a pinch of comet dust into the Earth's atmosphere and it landed gently in the Utah desert in USA. Comet particles were captured using a material called aerogel – a special type of foamed glass, made so lightweight that it is barely visible and almost floats in air. The material harvested from Comet Wild-2 is thought to be a pristine sample of the gas and dust that formed the solar system. The dust sample is to be studied by some 180 scientists across the world. (☞ Feb. 7)

January 16

1767
Anders Gustav Ekeberg, Swedish chemist who in 1802 discovered the element tantalum, was born in Stockholm, Sweden. Ekeberg graduated from the University of Uppsala in 1788 and travelled in Germany. He returned to Uppsala in 1794 and began teaching chemistry. Though he was partly deaf from a childhood infection and had been blinded in one eye by an exploding flask, he carried on admirably. Perhaps his greatest contribution to chemistry was the discovery of the talent of his student Jöns Jacob Berzelius. (☞ Feb. 11)

1806
Nicolas Leblanc, French surgeon and chemist who in 1790 developed the process for making soda ash (sodium carbonate) from common salt, died in Saint-Denis, France. (☞ Dec.6)

1909
Ernest Henry Shackleton, British explorer, discovered the magnetic South Pole. (☞ Jan.5, Feb 15)

1932
Dian Fossey, American zoologist who became the world's leading authority on the mountain gorillas of Central Africa, was born in San Francisco, California, USA. Fossey graduated from the San Jose State College in 1954 and worked as an occupational therapist at a children's hospital in Louisville, Kentucky, for several years. In 1963 she took a trip to eastern Africa, where she met the anthropologist Louis Leakey and had her first glimpse of mountain gorillas. Persuaded by Leakey, she established the Karisoke Research Centre in 1967 in Rwanda's Virunga Mountains, which was one of the last bastions of the endangered mountain gorilla. Living a solitary life for many years, she observed the gorillas' habits and gradually gained their acceptance. Fossey left Africa in 1970 to complete work for a doctorate at the University of Cambridge in England. In 1974 she received her degree in zoology with the completion of her dissertation, *The Behavior of the Mountain Gorilla.* She returned to Rwanda with student volunteers who made broader kinds of research possible. Fossey generated international media coverage in 1978 in her battle against poachers. In 1980 Fossey returned to the United States to accept a visiting associate professorship at the Cornell University, Ithaca, New York. Back in Rwanda, Fossey resumed her campaign against poachers, taking increasingly drastic measures to protect the Virunga gorillas. In 1985, Fossey's mutilated body, hacked by a machete, was found near the Karisoke Research Centre. It was suspected that poachers, whose devastating attacks on the gorillas she had tried to stop, were responsible for her murder, although this was never proved. (☞ Dec. 26)

1953
The element fermium (Fm, atomic number 100) was isolated and identified at the University of California, Berkeley, USA. Only about 200 atoms of the element were isolated by ion-exchange chromatography from the debris of the November 1952 test of the hydrogen bomb conducted at Eniwetok Atoll in the Pacific Ocean. Samples of debris were collected by drone aircraft flying through the clouds. Fermium was the eighth transuranium element of the actinide series to be discovered, and was named in honour of Enrico Fermi.

1967
Robert J. Van de Graaff, American physicist and inventor of the Van de Graaff generator, died in Boston, USA. (☞ Dec. 20)

1973
Radio-controlled Russian Moon-rover *Lunokhod-2* began exploration of the Moon. (☞ Jan. 8, Jan. 15)

1980
The US–Swiss firm Biogen produced interferon, a natural virus-fighting substance, through genetic engineering. It was produced by a team led by the Swiss biochemist Charles Weissmann who had cloned human interferon genes and spliced them for mass culture into bacteria as interferon producers. Recombinant interferons opened up new possibilities in the treatment of fatal diseases like leukaemia and hepatitis B.

2003
American space shuttle *Columbia*, carrying astronauts Rick Husband, William McCool, David Brown, Indian-born Kalpana Chawla, Laurel Clark, Michael Anderson, and Ilan Ramon, was launched on its 28th flight from the Kennedy Space Center, Florida, USA. The 16-day mission was dedicated to research investigating human physiology, fire suppression, and other areas of research relevant to people across the globe. Fifteen investigations focussed on the effects of space flight factors such as weightlessness on the functioning of basic body systems. Their subjects included a wide range of living organisms, including viruses, bacteria, fungi, rats and even the astronauts themselves. (☞ Feb. 1, Apr. 12)

January 17

1706
Benjamin Franklin, American scientist, inventor, printer, publisher, and statesman, was born in Boston, USA. At the early age of 12, Benjamin was apprenticed to his brother James, a printer. His mastery of the printer's trade, of which he was proud to the end of his life, was achieved between 1718 and 1723. In the same period he read tirelessly and taught himself to write effectively. In 1749 he explained updrafts of air as due to local heating of the atmosphere by the Sun. In the field of science, Franklin is better known for proving the electrical nature of lightning in 1752 with his famous kite experiment. He also invented the lightning conductor and coined many terms still used in discussing electricity, such as positive, negative, battery, conductor, and so on. He invented a stove, still being manufactured, to give more warmth than open fireplaces. He was also the first to design and use bifocal eyeglasses. (☞ Apr. 17, Jun. 15)

1997
Clyde W. Tombaugh, American astronomer who discovered the dwarf planet Pluto in 1930, died in Las Cruces, USA. (☞ Feb. 4, Feb. 18)

January 18

1773
British explorer, James Cook's *Resolution* became the first ship to cross the Antarctic

Circle. Between July 1772 and July 1775 Cook made what ranks as one of the greatest sailing ship voyages, with his *Resolution*, and a consort ship, the *Adventure*. He found no trace of the southern continent, the so-called Terra Australis, though he sailed beyond latitude 70°S in the Antarctic. But he successfully completed the first west–east circumnavigation in high latitudes, charted Tonga and Easter Island during the winters, and discovered New Caledonia in the Pacific and the South Sandwich Islands and South Georgia Island in the Atlantic. He showed that a real Terra Australis existed only in the landmasses of Australia, New Zealand, and whatever land might remain frozen beyond the ice rim of Antarctica. Throughout the whole voyage, not one of his crew died of scurvy, a scourge faced by the early maritime explorers. Back in England, he was promoted to captain and elected a fellow of the Royal Society. He was awarded one of its highest honours, the gold Copley Medal, for a paper that he prepared on his work against scurvy. (☞ Feb. 14, Oct. 27)

1825

Edward Frankland, English chemist who was one of the first investigators in the field of structural chemistry, was born in Churchtown, Lancashire, England. Frankland learned to perform chemical experiments while apprenticed to a druggist. He got his PhD from the University of Marburg in 1849, and became the first professor of chemistry at the Owens College, Manchester. In 1863 he succeeded Michael Faraday as professor of chemistry at the Royal Institution of Great Britain, London. Since 1850 he was involved in studying small organic molecules containing such metals as zinc. These studies led him to the idea that an atom of an element can combine only with a certain fixed number of atoms of other elements. In 1852, he established a theory of valency, which became the basis of modern structural chemistry. (☞ Aug. 9)

1908

Jacob Bronowski, Polish-born British mathematician, who wrote extensively on the humanistic aspects of science, was born in Lodz, Poland. Bronowski's family immigrated to Germany and then to England, when he was still a child. He studied mathematics at the University of Cambridge, and not only achieved high honours in mathematics but was also awarded critical acclaim for his poetry and prose. After receiving his PhD from Cambridge in 1933, he taught mathematics at the University College of Hull. While on a scientific mission to Japan after World War II, Bronowski saw first-hand the ruins of Nagasaki. From that time on, he concentrated on the ethical as well as the technological aspects of science, and he shifted his attention from mathematics to the life sciences, the study of human nature, and the evolution of culture. Among his most popular books are *The Common Sense of Science* (1951), *Science and Human Values* (1956) and *The Identity of Man* (1965). His last major project was the BBC television series *'The Ascent of Man'* (1973). (☞ Aug.22)

1995

Adolf Butenandt, German biochemist, Nobel laureate (Chemistry, 1939), who isolated and identified the sex hormones androsterone and progesterone, died in Bremerhaven-Lehe, Germany. (☞ Mar. 24)

January 19

1736

James Watt, Scottish instrument maker and inventor, was born in Greenock, Scotland. James wanted to be a mathematical-instrument maker. In 1757, he opened a shop within the Glasgow University and made mathematical instruments such as, quadrants, compasses and scales. In 1764, while repairing a Newcomen steam engine, which worked by atmospheric pressure and needed the cylinder to be cooled after every stroke, Watt was appalled by its waste of steam. In May 1765, after wrestling with the problem of improving it, he suddenly came upon a solution – the separate condenser, his first and greatest invention. In 1796, Watt took out the famous patent for 'A new invented method of lessening the consumption of steam and fuel in fire engines'. In 1782, he patented the double-acting steam engine, in which the piston pushed as well as pulled. Six years later his application of the centrifugal governor for automatic control of the speed of the engine, and in 1790 his invention of a pressure gauge, virtually completed the Watt engine. (☞ Aug. 25)

1747

Johann E. Bode, German astronomer, was born at Hamburg, Germany. Bode confirmed the empirical law formulated by Johann D. Titius in 1766. This law gave a mathematical expression to calculate the mean distances between the Sun and its planets, which later came to be known as Bode's law, or the Titius–Bode rule. Titius suggested that the mean distances of the planets from the Sun very nearly fit a simple relationship of $A = 4 + (3 \times 2^n)$ giving the series 4, 7, 10, 16, 28*, 52, 100, corresponding to the relative distance of the then six known planets, up to Saturn, and an unassigned value (*) between Mars and Jupiter, which is now known to be occupied by the minor planet belt. This law was an important factor in the discovery of the minor planets, most of which are located between Mars and Jupiter. Bode founded in 1774 the well-known *Astronomisches Jahrbuch* (Astronomic Yearbook) 51 issues of which were published. He became director of the Berlin Observatory in 1786. Among his other publications was *Uranographia*, a collection of 20 star maps accompanied by a catalogue of 17,240 stars and nebulae. (☞ Nov. 23)

1813

Henry Bessemer, English inventor and engineer, was born at Charlton, Hertfordshire, England. He developed the first process for manufacturing steel inexpensively, leading to the development of the Bessemer converter. The Bessemer converter is a cylindrical steel pot approximately 6 metres high, lined with a siliceous refractory. Air is blown in through openings near the bottom, creating oxides of silicon and manganese, which become part of the slag, and of carbon, which are carried out in the stream of air. An ingot of steel can be produced, within a few minutes, ready for the forge or rolling mill. Other inventions of Bessemer included movable stamps for dating deeds and other government documents, the improvement of a typesetting machine, the manufacture of 'gold' powder from brass for use in paints, and sugarcane-crushing machinery of advanced design. Bessemer was elected a

fellow of the Royal Society and was knighted in 1879. (☞ Mar. 15, Oct. 17)

1894
Solid air was first produced by British chemist and physicist James Dewar at the Friday meeting of the Royal Institution, London. He had previously exhibited liquid air obtained at the temperature of −192°C. (☞ Mar. 27, Sep. 20)

1959
First high-purity uranium ingot was produced at Bhabha Atomic Research Centre, Trombay.

1996
The first successful liver transplant in India was performed at the Government Stanley Medical College, Chennai on a patient named Selvaraj. Earlier, two attempts at Apollo Hospital, Chennai and All India Institute of Medical Sciences, New Delhi had failed as the patients died within a few days.

2006
New Horizons, the first ever space probe to Pluto, was launched by the American space agency NASA from Cape Canaveral, Florida, USA. Scheduled to reach Pluto in 2015 using Jupiter gravity assist, the space probe will cross the entire span of the solar system and conduct fly-by studies of Pluto and its only moon Charon. The seven science instruments on the piano-sized probe will shed light on the two bodies' surface properties, geology, interior makeup and atmospheres.

January 20

1930
Edwin E. Aldrin, American astronaut, was born in Montclair, New Jersey, USA. A graduate of the US Military Academy, Aldrin became an air force pilot and in 1966 joined James A. Lovell, on the four-day *Gemini-12* flight. Aldrin's 5 $^1/_2$-hour walk in space proved that man could function effectively in the vacuum of space. He was a member of the three-man *Apollo-11* crew with Neil Armstrong and Michael Collins and was the second man, after Armstrong, to set foot on the lunar soil, after their lunar module *Eagle* landed on the Moon. Aldrin retired from the National Aeronautics and Space Administration in 1971 to become commander of the Aerospace Research Pilots' School at Edwards Air Force Base, California. In March 1972 he retired from the Air Force to enter private business. He wrote an autobiographical book, *Return to Earth*, which was published in 1973. (☞ Jul. 21, Oct. 31)

1931
David M. Lee, American physicist, Nobel laureate, was born in Rye, New York, USA. After graduating from Harvard University, Cambridge, Massachusetts in 1952, Lee was awarded a PhD in physics from Yale University in 1959. He joined the faculty of the Cornell University in 1959, becoming a full professor there in 1968. At Cornell, Lee along with fellow American Robert C. Richardson built a special cooling apparatus for their research in low-temperature. While studying liquid helium, which they had cooled to within a few thousandths of a degree above absolute zero (−273°C), a student named Douglas D. Osheroff noticed an odd phenomenon – an almost complete lack of friction to flow – termed superfluidity. The discovery of

superfluidity in helium-3 enabled scientists to directly study the strange quan-tum mechanical effects in macroscopic systems that previously could only be studied indirectly in molecules, atoms, and subatomic particles. Lee was awarded the 1996 Nobel Prize for Physics, jointly with Richardson and Osheroff, for the discovery of superfluidity in the isotope helium-3. (☞ Jun. 26, Aug. 1)

1957
The Trombay centre of India's Atomic Energy Establishment was inaugurated by Prime Minister Jawaharlal Nehru. In 1967 Prime Minister Indira Gandhi re-named it as Bhabha Atomic Research Centre (BARC) in memory of its founder Homi Bhabha who died in an air crash in 1966. By that time the Trombay establishment had become a unique nuclear research centre where technology for building nuclear reactors, fuel fabrication, reprocessing of irradiated fuel and for the widespread application of radioisotopes in a number of areas like medicine, agriculture and industry had been successfully developed. The major objective of the broad spectrum of activities at BARC is primarily to provide R&D support needed to sustain India's nuclear power programme. BARC also initiates and sustains research and development activities in all newly emerging areas of nuclear and allied sciences and technology to keep India abreast with the advanced nations in these frontier areas. (☞ Jan. 12, Oct. 30)

1957
Apsara, India's first nuclear reactor was inaugurated at the Atomic Energy Establishment (re-named Bhabha Atomic Research Centre), Trombay, Maharashtra, India. (☞ Aug. 4)

January 21

1892
John Couch Adams, British mathematician and astronomer, who first correctly estimated the position of the planet Neptune, died in Cambridge, England. (☞ March 11, June 5, Sep. 23)

1912
Konrad E. Bloch, German-born American biochemist, Nobel laureate, was born in Niesse, Germany. Bloch attended elementary school in the same city and in 1930 went to Munich to study chemistry at the Technische Hochschule. In 1936 he immigrated to USA where he joined the Department of Biochemistry, College of Physicians and Surgeons, Columbia University in New York, and received his PhD in 1938. In 1946 Bloch moved to the University of Chicago in Chicago, Illinois as assistant professor of biochemistry and later became professor in 1950. In 1954 he was appointed Higgins Professor of Biochemistry in the Department of Chemistry, Harvard University in Cambridge, Massachusetts, and in 1968 he became Chairman of the Department. Bloch's research mainly concerned the synthesis of cholesterol and fatty acids in the body. He identified the chemical process by which the body turns acetic acid into cholesterol. He discovered that high levels of cholesterol in the bloodstream cause fatty deposits on the inner walls of arteries, which may lead to constricted blood flow

and increase the chances of blood clotting and heart attack. He was awarded the 1964 Nobel Prize for Physiology or Medicine, jointly with the German biochemist Feodor Lynen, for discoveries concerning the natural synthesis of cholesterol and fatty acids. (☞ Apr. 6, Oct. 15)

1926

Camillo Golgi, Italian physician and cytologist, Nobel laureate (Physiology or Medicine, 1906), who devised the silver nitrate method of staining nerve tissue for studies, died in Pavia, Italy. (☞ Jul. 7)

1950

National Physical Laboratory (NPL), a constituent national laboratory of the Council of Scientific and Industrial Research (CSIR), was inaugurated. Located in New Delhi, the primary responsibility of NPL is to continuously establish, maintain and improve by research, the national standards of measurements and to realise the units based on the International System (SI units). NPL also has the statutory obligation to establish, maintain and update the calibration facilities for different parameters. Over the years the laboratory has built up considerable expertise enabling it to shift from 'artefact' standards to quantum standards. Currently, the 'second' is realised through a bank of caesium atomic clocks, 'metre' is based on a few recommended stabilised frequency laser radiations and 'farad' is realised through a calculable capacitor linked with the length standard, and having 'ohm' and 'henry' traceable to it. Josephson voltage standard at 1mV level has also been established and work on a primary standard of resistance based on the Quantum Hall effect is in progress. Some new standards established are interferometric manometer – primary standard of pressure and a self-calibrating electrically compensated absolute radiometer-standard of optical radiation in the visible and UV region. NPL broadcasts standard time and frequency signals (STFS) under the call sign ATA round-the-clock on HF band. The laboratory also broadcasts coded STFS via INSAT on a radio networking channel. Other major areas of research at NPL include materials science, materials characterisation, radio and atmospheric science, and superconductivity and cryogenics. Technologies developed at NPL include piezoelectric accelerometer, carbon composite half rings for orthopaedic applications, re-usable heat pad/pack, basic Sodar (sonic detection and ranging) operating in monostatic/Doppler mode, long-afterglow phosphor, and a blood glucose digital analyser.

1954

The first nuclear-powered submarine *USS Nautilus* was launched on Thames River in Groton, Connecticut, USA. The $65 million submarine was built by the Electric Boat Company, a division of General Dynamics Corporation. Because of the nuclear power plant, the *Nautilus* could stay submerged for months at a time, unlike diesel-fuelled submarines, whose engines required vast amounts of oxygen. With the power generated by nuclear fuel about the size of a baseball, *Nautilus* could travel more than 160,000 kilometres under water. On one 265-hour underwater voyage, it averaged 19.1 knots – much faster than the top speed of conventional submarines. In 1958, the *Nautilus* sailed from Honolulu in the Pacific to Portland, England, passing under the Arctic ice cap and was under the polar ice cap for 95 hours. (☞ Mar. 3, Aug. 3)

1959
Jnan Chandra Ghosh, eminent Indian chemist and the first Director of the Indian Institute of Technology, Kharagpur, died in New Delhi, India. (☞ Aug. 18, Sep. 4)

1970
The first commercial flight of the wide-bodied jumbo jet Boeing-747 started between New York and London.

1976
Regular supersonic passenger flights of the *Concorde* started. The first two *Concorde*s with commercial passengers took flight simultaneously – from London's Heathrow Airport to Bahrain in the Persian Gulf, and from Orly Airport outside Paris to Rio de Janeiro, Brazil via Senegal in West Africa. At their cruising speeds, the *Concorde*s flew at almost 2,200 kph, well over the speed of sound, cutting air travel time by more than half. As the only supersonic passenger aircraft *Concorde* remained a brilliant technological achievement till it was taken off service in 2003, but its impact on internati-onal air travel has been limited. Airlines feared that the high cost of buying and operating the aircraft

would not be recovered and there was widespread opposition from environmental groups on the grounds of the *Concorde*'s noise on takeoff and its fuel consumption. Only British Airways and Air France have operated the aircraft. (☞ Jan. 9, Mar. 2, Jul. 25, Oct. 1, Oct. 24)

January 22

1561
Francis Bacon, English philosopher and scientist, was born in London, England. From 1573 to 1575 Bacon was educated at the Trinity College, Cambridge, but his weak constitution caused him to suffer ill health there. While at Cambridge he developed a distaste for what he termed 'unfruitful' Aristotelian philosophy. Bacon had a more rational mind; he characterised the new century after Renaissance as an age distinguished by 'the opening of the world by navigation and commerce and the further discovery of knowledge'. Bacon was among the first to view science as a powerful tool to understand and conquer nature. In his *Novum Organum*, published in 1620, he recommended induction and experimentation as the bases of the scientific method. Bacon performed his first scientific experiment in 1626, just before his death. He stuffed a dead chicken with snow to see if it would decay less rapidly. The chill he caught during this experiment is thought to have led to his death. Bacon's last work, *The New Atlantis*, published in 1627 after his death, is a utopian tale that predicts robots, telephones, tape recorders, and the electric motor, while emphasising the importance of experimentation over deduction. (☞ Apr. 9)

1775
André-Marie Ampère, French mathematician and physicist, was born in Marseille, France. He founded the science of electrodynamics. By the time he was 12 years old, Ampère had mastered all mathematics

known at his time. He became a professor of physics and chemistry at Bourg-en-Bresse in 1801 and a professor of mathematics at the École Polytechnique in Paris in 1809. When he learned of Danish physicist Hans Christian Ørsted's discovery in 1820 that a magnetic needle is deflected when the current in a nearby wire varies, Ampère prepared within a week the first of several papers expounding the theory of this new phenomenon, establishing a relationship between electricity and magnetism. He formulated a law of electromagnetism, known as Ampère's Law, which describes mathematically the magnetic force between two electric currents. Ampère also built the first galvanometer using a free-moving magnetic needle to measure the flow of electricity. (☞ Jun. 10)

1908

Lev Davidovich Landau, Soviet physicist, Nobel laureate, was born in Baku, Azerbaijan. Landau graduated from the Leningrad (now St. Petersburg) State University in 1927 and got a travelling fellowship to go abroad. After brief stays in Göttingen and Leipzig, he went to Copenhagen to work in Niels Bohr's Institute for Theoretical Physics. In 1937 the Soviet physicist Pyotr L. Kapitsa, who worked on low-temperatures, persuaded Landau to move to Moscow. There, at the S I Vavilov Institute of Physical Problems, which had been created by the USSR Academy of Sciences, Landau's close interest in experimental physics led to his explanation of the peculiar behaviour of helium known as superfluidity when the gas is cooled below 2.18K (−270.97°C). Kapitsa had found that liquid helium is superfluid; that is, it has less resistance against moving through a tube than any other known liquid, which even makes it spontaneously flow out of a container. Landau was awarded the 1962 Nobel Prize for Physics for his work explaining superfluidity of helium. (☞ Apr. 1)

1936

Alan J. Heeger, American chemist, Nobel laureate, was born in Sioux City, Iowa, USA. Heeger grew up in Omaha, Nebraska. He received his BSc from the University of Nebraska where he majored in physics and mathematics, and his PhD from the University of California at Berkeley in 1961. He joined the University of Pennsylvania as assistant professor in 1962 and became professor in 1967. Heeger moved to the University of California at Santa Barbara in 1982 where he is at present professor of physics and also serves as Director of the Institute for Polymers and Organic Solids. Heeger's research work mainly concerns the science and technology of semiconductive and metallic polymers. It was at the University of Pennsylvania that Heeger began his work on plastics. Together with the New Zealand-born American chemistry professor, Alan G. MacDiarmid, who had some expertise in synthesis, and Japanese scientist, Hideki Shirakawa, who had conducted his own work on polyacetylene, Heeger did a series of experiments during the mid-1970s that led to the historic discovery of metallic conductivity in an organic polymer. The discovery opened up the possibilities of a wide variety of commercial applications of these unique polymers, such as antistatic substances for photographic film, and for 'smart' windows that can exclude sunlight. In addition, semi-conductive polymers have recently been developed into light-

emitting diodes, solar cells and as displays in mobile telephones and mini-format television screens. Heeger was awarded the 2000 Nobel Prize for Chemistry, jointly with MacDiarmid and Shirakawa, for the discovery and development of conductive polymers. (☞ Apr. 14, Aug. 20)

1939
The uranium atom was split for the first time at the Columbia University in New York City, USA using the cyclotron. It heralded the beginning of the Manhattan Project that led to the construction of the first atom bomb. (☞ Jul. 16, Aug. 6, Aug. 9)

1965
India's first plutonium plant was inaugurated at the Trombay centre of India's Atomic Energy Establishment (now BARC). Plutonium was extracted by re-processing spent uranium fuel rods from the experimental reactors *Apsara* and *CIRUS*.

2003
The spacecraft *Pioneer-10* sent its last, faint signal from a distance of 12,000 million kilometres. The spacecraft was launched in March 1972. It was the first to be sent to the outer solar system and the first to investigate the planet Jupiter, after which it followed an escape trajectory from the solar system. (☞Mar. 2, Jun. 13, Dec. 3)

January 23

1796
Karl Karlovich Klaus, Russian chemist who discovered ruthenium, was born in Dorpat, Russia (now Tartu, Estonia). Klaus was educated at Dorpat, where he became a pharmacist. Later he taught chemistry and pharmacy at the Universities of Dorpat and Kazan. Klaus's researches mainly concerned the platinum metals osmium, palladium, iridium, and rhodium. It was in the course of investigating the waste residues of the platinum refinery in St. Petersburg that Klaus discovered ruthenium, in 1844. (☞ Mar. 24)

1814
Alexander Cunningham, British army officer and archaeologist who excavated many sites in India, including Sarnath and Sanchi, and served as the first director of the Indian Archaeological Survey, was born in London, England. Cunningham joined the Bengal Engineers at the age of 19 and spent 28 years in the British service in India, retiring as major general in 1861. In 1837 he excavated at Sarnath, outside Varanasi, site of one of the most sacred Buddhist shrines, and carefully prepared drawings of the sculptures discovered there. In 1850 he excavated Sanchi, site of some of the oldest surviving buildings in India. He also studied the temple architecture of Kashmir and Buddhist sites in Ladakh. In 1854 he published *The Bhilsa Topes*, the first serious attempt to trace Buddhist history through its architectural remains. (☞ Nov. 28)

1840
Ernst Abbe, German physicist whose theoretical and technical innovations in optical theory led to great improvements in microscope design, was born in Eisenach, Germany. In 1863 Abbe joined the University of Jena, where he taught mathematics and physics. In 1866 he became the research director of the Zeiss optical works and two years later invented the apochromatic lens system for microscopes,

which eliminates colour distortion of light. He discovered the optical formula now called the 'Abbe sine condition' that a lens must satisfy to form a sharp image, free from the blurring or distortion caused by coma and spherical aberration. In 1891 Abbe set up and endowed the Carl Zeiss Foundation for research in science and social improvement. (☞ Jan. 14)

1857

Andrija Mohorovicic, Croatian meteorologist and geophysicist who discovered the boundary between the Earth's crust and mantle, was born in Volosko, Croatia. Mohorovicic studied under the physicist and philosopher Ernst Mach at the University of Prague and graduated in 1875 in mathematics and physics. He taught in a secondary school for seven years, before receiving an appointment to the Royal Nautical School at Bakar, near Rijeka. He taught meteorology and oceanography, and established a meteorological station there in 1887. In 1891 he became a professor at the Main Technical School at Zagreb, where he continued his studies. He became director of the meteorological observatory in Zagreb in 1892, and received his PhD from the University of Zagreb in 1897. By 1908 Mohovicic had acquired a new and improved seismograph for the Zagreb observatory and on October 8, 1909 recorded a destructive earthquake in the Kulpa Valley. From the readings recorded there and those of other stations, Mohorovicic observed that certain seismic waves arrived at detecting stations sooner than anticipated. He deduced that the earthquake was centred in an outer layer of Earth – the crust – and that the fast waves had travelled through an inner layer – the mantle. He postulated that between them lay a region that was later named the Mohorovicic discontinuity or 'Moho' in short. (☞ Dec. 18)

1872

Paul Langévin, French physicist and inventor of the sonar, was born in Paris, France. Contemporary to Marie Curie, Albert Einstein and Hendrik Lorentz, Langévin is also known for his work on the molecular structure of gases, and for his theory of magnetism. He suggested that the alignment of molecular moments in a paramagnetic substance would be random in the absence of an externally applied magnetic field, but would be non-random in its presence. In later years, Langévin became increasingly involved with the study of Einstein's work on space and time. He was a firm supporter of the theory of the equivalence of energy and mass. Einstein later wrote that Langévin had all the tools for the development of the special theory of relativity at his disposal before Einstein proposed it himself and that if he had not proposed the theory, Langévin would have done so. Langévin's interest in the study of piezoelectricity and piezoceramics led him to invent the underwater sonar that uses echo of ultrasound waves for submarine detection. (☞ Dec. 19)

1876

Otto Paul Hermann Diels, German organic chemist, Nobel laureate, was born in Hamburg, Germany. Diels studied chemistry at the University of Berlin under Emil Fischer and became professor of chemistry at the University of Kiel in 1916. In 1906 Diels discovered a highly reactive substance, carbon suboxide (the acid anhydride of malonic acid), and determined its properties and chemical composition. He also devised an easily controlled method of removing some of the hydrogen atoms from certain organic molecules by the use of metallic selenium. Diel's most important work concerned the diene synthesis, in which organic compounds with two carbon–carbon double bonds were used for synthesising many cyclic organic substances. Diels developed this method

in collaboration with his student, Kurt Alder. Known as the Diels–Alder reaction, this method proved especially important in the production of synthetic rubber and plastics. Diels was awarded the 1950 Nobel Prize for Chemistry, jointly with Alder, for their discovery and development of the diene synthesis. (☞ Mar. 7, Jul. 10)

1896
Wilhelm Conrad Röntgen, German physicist, made the first public lecture-demonstration of his X-ray device, in Würzburg, Germany. (☞ Jan. 1, Feb. 10, Mar. 27, Nov. 8)

1907
Hideki Yukawa, Japanese physicist, Nobel laureate, was born in Tokyo, Japan. Yukawa was brought up in Kyoto and graduated from the local university in 1929 and became a lecturer there. His investigations in theoretical physics, particularly in the theory of elementary particles started then. In 1933 he moved to the Osaka University, where he earned his PhD in 1938. From 1939 to 1950 he was professor of theoretical physics at the Kyoto University and in 1953 he became director of the Research Institute for Fundamental Physics in Kyoto. While a lecturer at Osaka University, in 1935, he published a paper entitled 'On the interaction of elementary particles' in which he proposed a new field theory of nuclear forces and predicted the existence of the meson. Encouraged by the discovery of one type of meson in cosmic rays made by American physicists in 1937, he devoted himself to the development of the meson theory, on the basis of his original idea. Since 1947 he worked mainly on the general theory of elementary particles in connection with the concept of the 'non-local' field. Yukawa was awarded the 1949 Nobel Prize for Physics for research on the theory of elementary particles. (☞ Sep.8)

1918
Gertrude Belle Elion, American pharmacologist, Nobel laureate, was born in New York, USA. Elion received her bachelor's degree in chemistry from the Hunter College in 1937, and a master's degree from the New York University in 1941. During World War II, she was hired as an analytical chemist; her job included the measurement of the acidity of pickles and the colour of mayonnaise. In 1944, she joined the Burroughs Wellcome Laboratories as a biochemist to work with the American pharmacologist George Herbert Hitchings. She spent the next 39 years there, becoming head of the Department of Experimental Therapy in 1967. By 1949, Elion had synthesised a purine that inhibited growth in mouse leukaemia. The purine was used at the Sloan Kettering Institute in New York, to treat four patients with chronic granulocytic leukaemia, two of whom went into remission. She then started making compounds of it and eventually ended up

with the first effective drug that induced remission in childhood leukaemia. She also collaborated with Hitchings to develop drugs to treat bacterial and viral infections and prevent rejection of transplanted organs. Elion was awarded the 1988 Nobel Prize for Physiology or Medicine, jointly with Hitchings and the British pharmacologist James W. Black, for their discoveries of important principles for drug treatment. (☞ Feb. 21, Apr. 18, Jun. 14)

1929

John C. Polanyi, Canadian chemist and educator, Nobel laureate, was born in an expatriate Hungarian family in Berlin, Germany. Polanyi was educated in England and was awarded his PhD and DSc. from the Manchester University. He began teaching at the University of Toronto, Canada in 1956, and became a professor in 1974. Polyanyi is credited with the development of a technique called 'infrared chemiluminescence', which is based on the observation that molecules, when excited, emit infrared light. By means of spectroscopic analysis of the changes in emitted light that take place during a chemical reaction, Polanyi was able to trace the exchange of chemical bonds. The technique is useful in tracing the exchange of chemical bonds during a chemical reaction and thus helps in the study of chemical-reaction dynamics. Polyanyi was awarded the 1986 Nobel Prize for Chemistry, jointly with the American chemist Dudley R. Herschbach and Taiwanese-born American chemist Yuan T. Lee, for development of methods for analysing basic chemical reactions. (☞ Jun. 18, Nov. 29)

1960

The deepest dive by a manned vessel was achieved when the specially constructed bathyscaphe, the *Trieste*, carrying Swiss marine engineer Jacques Piccard and Lieutenant Don Walsh of the US Navy descended 10,912 m. in the Pacific Ocean into Challenger Deep the deepest point known to exist on Earth, near the island of Guam. During the five-hour descent, Piccard and Walsh saw a 30-cm-long fish, which proved that life could exist at such depths of the ocean. (☞ Jan. 28, Mar. 24)

January 24

1838

Samuel Morse, American inventor, transmitted the first telegraphic message using the Morse code at New York University. The message was: 'Calling the Universe'. (☞ Apr. 2, Apr. 27, May 24)

1950

Percy Spencer, a self-taught engineer with the Raytheon Corporation in Waltham, Massachusetts, USA, was granted the first patent for a microwave oven. Spencer got the idea during a radar-related research project around 1946 when he noticed something very unusual. He was testing a new vacuum tube called a magnetron for use in radar, when he discovered that the chocolate bar in his pocket had melted. This intrigued Dr. Spencer, so he tried another experiment. This time he placed some popcorn kernels near the tube and watched the popcorn sputter, crack and pop all over his lab. In 1947, Raytheon demonstrated the world's first microwave oven and called it a 'Radarange'. These primitive units where gigantic and enormously expensive, standing $5^1/_2$ feet (about 1.7 metres) tall, weighing over

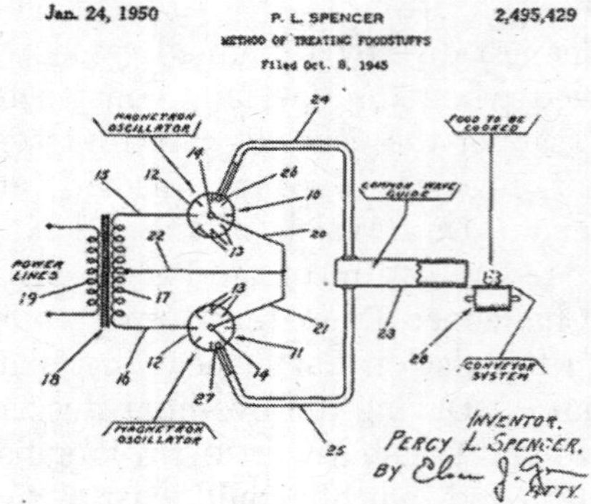

Microwave patent

750 pounds (340 kilogrammes), and costing about $5,000 each. The magnetron tube had to be water-cooled; so plumbing installations were also required.

1966

Homi Jehangir Bhabha, Indian physicist, Fellow of the Royal Society, the founder and prime architect of the Indian atomic energy programme, was killed in an air crash over Mont Blanc in the Alps. (☞ Oct. 30)

1986

The American unmanned spacecraft *Voyager-2* became the first space probe to fly-by Uranus at a distance of 81,593 kilometres from the planet, taking dozens of photographs, and finding new moons and rings. (☞ Jul. 9, Aug. 20, Aug. 25)

2002

INSAT-3C, the second of India's third generation geostationary satellites in the INSAT series was launched by the European Space Agency's Ariane rocket from Kourou in French Guyana. Designed and built by the Indian Space Research Organisation, *INSAT-3C* is an exclusively communications satellite and carries 24 C-band transponders, six extended C-band transponders, two S-band transponders, and a mobile satellite service transponder to augment the present INSAT capacity for communication and broadcasting, besides providing continuity of the services of *INSAT-2C* when it reaches the end of its mission life. *INSAT-3C* is co-located with *INSAT-1D* at 74°east longitude.

January 25

1627

Robert Boyle, Anglo-Irish chemist and natural philosopher, was born in Lismore, Ireland. In 1635, Boyle was sent to the Eton College, after which he spent the years from 1639 to 1644 with a tutor, mostly in Switzerland. While at the University of Oxford during 1656–1668, Boyle carried out experiments using an air pump and demonstrated the physical characteristics of air and the necessary role played by air in combustion, respiration, and the transmission of sound. In 1662, he propounded the relationship, now known as Boyle's Law, that at a constant temperature, the volume of a gas is inversely proportional to the pressure. Boyle was opposed to the Aristotelian theory of the four elements (earth, air, fire, and water) and also the three principles (salt, sulphur, and mercury) proposed by Paracelsus. Instead, he developed the concept of a corpuscular view of matter that was a forerunner of the modern theory of chemical elements. He was a founding member of the Royal Society of London. (☞ Dec. 30)

1798

Benjamin Thompson Rumford, American-born British physicist, presented a paper to the Royal Society, London, titled 'Enquiry concerning the source of heat which is excited by friction', in which he presented the idea that heat represents a form of motion, as opposed to the prevailing idea of it being a fluid. While drilling out cannons in the Munich Munitions Works in Germany, he noticed that as long as the friction of boring continued, the canon became and stayed hot. Since more heat was being released than could have been originally contained in the metal, these observations were an outright contradiction to the caloric theory. Rumford was therefore led to conclude that it was the mechanical process of boring which was producing the heat. (☞ Mar. 26. Aug. 21)

1917

Ilya Prigogine, Russian-born Belgian physical chemist, Nobel laureate, was born in Moscow, Russia. Prigogine moved to Belgium as a child. He was awarded his PhD in 1941 from the Free University in Brussels, where he became a professor in 1947. In 1962 he became director of the International Institute of Physics and Chemistry at Solvay, Belgium. He also served as director of the Center for Statistical Mechanics and Thermodynamics at the University of Texas in Austin, USA from 1967 until his death. The Center was later re-named the Ilya Prigogine Center for Statistical Mechanics and Complex Systems in his honour. Prigogine's work primarily dealt with the application of the second law of thermodynamics, especially the irreversible movement toward a state of disorder, or higher entropy, to complex systems, including living organisms. The second law states that physical systems tend to move spontaneously and irreversibly toward a state of disorder, a property known as entropy, but it does not explain how complex systems such as living things could have arisen spontaneously from less-ordered states and maintained themselves, apparently in defiance of the second law. Prigogine argued that as long as systems receive energy and matter from an external source, non-linear systems could go through periods of instability and then self-organisation, resulting in more complex systems whose characteristics cannot be predicted except as statistical probabilities. Prigogine's work was influential in a wide variety of fields, from physical chemistry to biology, and was fundamental to the new disciplines of chaos theory and complexity theory. Prigogine was awarded the 1977 Nobel Prize for Chemistry for contributions to non-equilibrium thermodynamics. (☞ May 28)

1923

Arvid Carlsson, Swedish pharmacologist, Nobel laureate, was born in Uppsala, Sweden. Carlsson began his medical studies at the University of Lund in 1941 but his education was interrupted by several years of service in the Swedish Armed Forces during World War II. In 1951, he received both his ML degree (the equivalent of the American MD) and his MD (the equivalent of the American PhD) from the University of Lund. He then became a professor at the same university. In 1959 he became a professor at the University of Gothenburg in Sweden. In the 1950s, Carlsson demonstrated that dopamine, a neurotransmitter in the brain, was not just a precursor for norepinephrine, as had been previously believed. He developed a method for measuring the amount of dopamine in brain tissues and found that dopamine levels in the basal ganglia, a brain area important for movement, were particularly high. His research led to the realisation that Parkinson's disease is caused by a lack of dopamine in certain parts of the brain and that an efficient remedy such as L-dopa for this disease could be developed. This led other doctors to try L-dopa on patients suffering from Parkinson's disease and found it to alleviate some of the symptoms in the early stages of Parkinson's. Carlsson made a number of subsequent discoveries, which have further clarified the role of dopamine in the brain. He demonstrated the mode of action of drugs used for the treatment of schizophrenia. Carlsson was awarded the 2000 Nobel Prize for Physiology or Medicine, jointly with the American neurobiologist Paul Greengard and Austrian-born American neurobiologist Eric Kandel, for their discoveries concerning signal transduction in the nervous system. (☞ Nov. 7, Dec. 11)

1949

Paul M. Nurse, British biologist, Nobel laureate, was born in London, England. Nurse did his BSc from the University of Birmingham in 1970. In 1972 he joined the British microbiologist J. Murdoch Mitchison at the Institute of Cell, Animal and Population Biology, University of Edinburgh, to work on the problem of identifying the rate controlling steps during the cell cycle. He received his PhD from the University of East Anglia, Norwich, Norfolk, in 1973. Nurse was a professor at the University of Oxford from 1987 to 1993. In 1996 he became the Director-General of the Imperial Cancer Research Fund (ICRF; now Cancer Research UK). Since 2003 he has been President of the Rockefeller University in New York. In the middle of the 1970s, Nurse discovered the gene cdc2 in the yeast, *Schizosaccharomyces pombe*, which he showed to have a key function in the control of cell division. Later he found that cdc2 was identical to the gene ('start') that the American biologist Leland H. Hartwell had earlier identified in baker's yeast. In 1987, Nurse identified, cloned and characterised with the help of genetic and molecular methods, one of the key regulators of the cell cycle, called CDK (cyclin dependent kinase). On the basis of these findings, half a dozen different CDK molecules have been found in humans. Nurse was awarded the 2001 Nobel Prize for Physiology or Medicine, jointly with Hartwell and the British biologist Richard Timothy Hunt, for their discoveries of key regulators of the cell cycle. (☞ Feb. 19, Oct. 30)

1983

The Infrared Astronomical Satellite (*IRAS*), a joint scientific project of the US, the UK and the Netherlands, was launched from California's Vandenberg Air Force Base in USA on a Delta rocket. The satellite orbited Earth from north to south at an altitude of 900 kilometres and operated for 10 months until its coolant was depleted. The mission of *IRAS* was to map the entire sky at infrared wavelengths. It was equipped with a special infrared telescope to scan the sky. *IRAS* increased the number of catalogued astronomical sources by about 70%, detecting about 350,000 infrared sources. Among the many new discoveries were the dust discs encircling several stars, including Beta Pictoris and Vega, which may be an early stage in the

formation of new solar systems. The most famous discovery of *IRAS* was that of a new type of galaxy, called a starburst galaxy, in which new stars form more rapidly than in other types of galaxies. *IRAS* also revealed for the first time the core of our galaxy, the Milky Way.

2002

India test-fired a short-range variant of the Agni ballistic missile called *Agni-1*, which can strike targets up to a distance of 700 kilometres. The 15-metre-tall 12-ton surface-to-surface missile is powered by a single solid-fuel motor. It was fired from a mobile launcher from the coastal testing range at Chandipur-on-sea in Orissa on the east coast of India. (☞ Apr. 11, May 22)

2004

American robotic rover *Opportunity* landed on Mars. Encased in a cocoon of inflated balloons, the rover travelled about 200 metres, bouncing 26 times and rolling after the impact. It came to rest inside a small crater. *Opportunity* landed close to a thin outcrop of rocks and its scientific instruments found evidence in those rocks that once a body of salty water flowed gently over the area. Preliminary interpretations point to a past environment that could have been hospitable to life and that could also have preserved fossil evidence of it. (☞ Jul. 7)

January 26

1823

Edward Jenner, English surgeon and discoverer of vaccination for smallpox, died in Berkeley, Gloucestershire, UK. (☞ May 17)

1886

Karl Benz, German mechanical engineer, patented the first auto powered by an internal combustion engine. The engine had one cylinder with a 91.4 mm bore, 150 mm stroke and a 984 cc displacement. Its output at 400 rpm was 0.7 kW (0.9 hp), producing a top speed of 16 km/h. (☞ Apr. 4, Nov. 25)

1891

Nikolaus August Otto, German engineer who developed the four-stroke internal combustion engine – the first practical alternative to the steam engine as a power source – died in Cologne, Germany. (☞ June 10)

1911

Polykarp Kusch, German-born American physicist, Nobel laureate, was born in Blankenburg, Germany. Kusch came to the United States in 1912 and became a citizen in 1922. In 1937, he worked with the physicist Isidor I. Rabi at the Columbia University on the effects of magnetic fields on beams of atoms. In 1947, through precise atomic beam studies, Kusch demonstrated that the magnetic properties of the electron were not in agreement with existing theories. Although the discrepancy he found was minute, this anomaly was of great significance to theories of the interactions of electrons and electromagnetic radiation, now known as quantum electrodynamics. Kusch was awarded the 1955 Nobel Prize for Physics, jointly with American physicist Willis E. Lamb, for his accurate determination that the magnetic moment of the electron is greater than its theoretical value. (☞ Mar. 20)

January 27

1880

Thomas Alva Edison, American inventor, was awarded a patent for 'an electric lamp for giving light by incandescence', the first electric incandescent lamp. His first lamps had a filament made from carbonised bamboo fibre. Edison's invention of the light bulb had a major impact on the electronics and computer industries. (☞ Feb. 11, Feb. 19, Apr. 15, Aug. 8, Aug. 9, Aug. 27, Oct. 18, Oct. 21, Dec. 6)

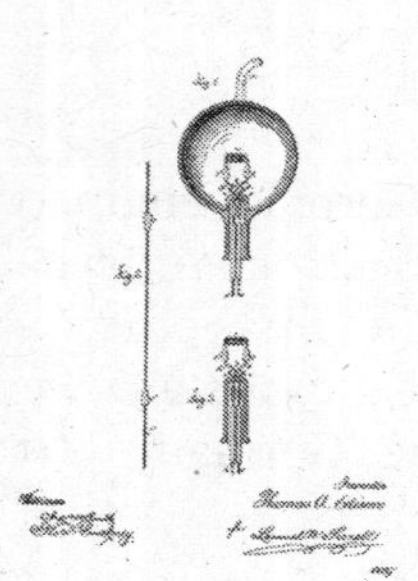

Edison's lamp patent

1903

John Carew Eccles, Australian research physiologist, Nobel laureate, was born in Melbourne, Australia. Eccles graduated from the University of Melbourne in 1925, and was then awarded his PhD from the Oxford University in 1929. He returned to Australia in 1937. While working at the Australian National University, Canberra during 1951–66, he demonstrated that one nerve cell communicates with a neighbouring cell by releasing chemicals into the synapse. This finding settled a long-standing controversy over whether nerve cells communicate with each other by chemical or by electric means. His work had a profound influence on the medical treatment of nervous diseases and research on kidney, heart, and brain function. Eccles was awarded the 1963 Nobel Prize for Physiology or Medicine, jointly with the English physiologists Alan Hodgkin and Andrew Huxley, for their discovery of the chemical means by which impulses are communicated or repressed by nerve cells. (☞ Feb. 5, May 2, Nov. 22)

1926

John Logie Baird, Scottish inventor, gave the first public demonstration of a true television system to members of the Royal Institution in London. Baird's invention, a pictorial transmission machine he called a 'televisor', used mechanical rotating discs to scan moving images into electronic impulses. This information was then transmitted by cable to a screen where it showed up as a low-resolution pattern of light and dark. Baird's first television program showed the heads of two ventriloquist dummies, which he operated in front of the camera apparatus out of view of the audience. (☞ Jun. 14, Aug. 13)

1936

Samuel Chao Chung Ting, American physicist, Nobel laureate, was born in Ann Arbor, Michigan, USA. Ting grew up in Mainland China and Taiwan and immigrated to the United States at the age of 20. He studied at the University of Michigan, Ann Arbor, where he was awarded his doctorate in 1962. Ting taught briefly at the Columbia University and was briefly at a nuclear facility at Hamburg, West Germany, before joining the faculty of the Massachusetts Institute of Technology, Cambridge, in 1967. In 1974, in experiments conducted at the Brookhaven National Laboratory in New York, Ting discovered a new subatomic particle that he called the J-particle (now usually called the J/psi

particle). The discovery of this particle, the first of a new class of very massive, long-lived mesons, led to a significant expansion and refinement of the quark model. Ting was awarded the 1976 Nobel Prize for Physics, jointly with the American physicist Burton Richter, for his discovery of the J/psi particle. (☞ Mar. 22)

1950
The discovery of a new broad-spectrum antibiotic, terramyacin (oxytetracycline) was announced in the journal *Science*. Made by Charles Pfizer & Co., it was isolated from soil, and found effective against pneumonia, dysentery, and other infections. Acting on theories that bacteria-fighting organisms could be found in soil, Pfizer solicited soil samples worldwide, received 135,000 soil samples and conducted more than 20 million tests before the scientists found a substance that was effective against a wide range of bacteria.

1967
Three American astronauts, Edward H. White, Virgil I. Grissom and Roger B. Chaffee were killed in a fire during a flight simulation test of *Apollo-1* capsule. White was the first American astronaut to walk in space during the flight of *Gemini-4* in June 1965.

1984
The first 170-MW unit of the Madras Atomic Power Station at Kalpakkam, near Madras (now Chennai), Tamil Nadu, India, began commercial operation. It is a pressurised heavy water reactor that uses natural uranium as fuel and heavy water as moderator and coolant. (☞ Mar. 21)

1998
The first direct observation of an extrasolar planet, made by the Hubble Space Telescope, was announced by Al Schultz of the Space Science Institute in Baltimore, Maryland, USA. The giant planet, ten times the size of Jupiter, was observed orbiting the star Proxima Centauri, the closest star to the Sun. (☞ Oct. 6)

January 28

1884
Auguste Piccard, Swiss-born Belgian physicist notable for his exploration of both the upper stratosphere and the depths of the sea, was born in Basel, Switzerland. Piccard studied at the Swiss Federal Institute of Technology, in Zürich, from where he was awarded his doctorate in physics. Piccard became interested in balloon ascents as a means of making experiments.

Having studied cosmic rays, he conceived of an experiment for observing them at 16,000 metres or higher in the upper atmosphere. To reach that height, Piccard built a balloon in 1930, with an airtight cabin, and equipped with pressurised air. In May 1931, Piccard and a friend reached an altitude of 15,781 m, where the atmospheric pressure is about $^{1}/_{10}$th that at sea level. In the late 1940s, Piccard designed the bathyscaphe for deep diving under the sea. In August 1953, when Piccard was 69, his bathyscaphe named *Trieste* went down to 3,150 m, the deepest at that time. (☞ Mar. 24)

1903
Kathleen Lonsdale, British crystallographer who developed several X-ray techniques for the study of crystal structure, was born in Newbridge, Ireland. Kathleen attended

school at Ilford, and then joined Bedford College in London where she completed her BSc in physics in 1922. From 1922 to 1927 and from 1937 to 1942, she was research assistant to Sir William Henry Bragg. In 1929, in Leeds, using the technique of X-ray diffraction she was able to prove that all the ring C–C bonds in benzene were of the same length and that all the internal C–C–C bond angles were 120°. She later developed an X-ray technique with which she obtained an accurate measurement of the distance between carbon atoms in diamond. In 1946 she moved back to London, where she became professor of chemistry and head of the Department of Crystallography at the University College. In 1945, when after 285 years, the Royal Society decided to allow women to become members, Kathleen Lonsdale was the first woman to be elected as a fellow of the society. In 1968, she became the first woman President of the British Association for the Advancement of Science. (☞ Apr. 1)

1922

Robert William Holley, American biochemist, Nobel laureate, was born in Urbana, Illinois, USA. Holley obtained his PhD in organic chemistry from the Cornell University, Ithaca, in 1947. By 1960, Holley and others had shown that small molecules of ribonucleic acids, called transfer RNAs, were involved in the assembly of amino acids into proteins. Holley and his collaborators developed techniques to separate the different transfer RNAs from the mixture in the cell. By 1965 he had determined the composition of the transfer RNA that incorporates the amino acid alanine into protein molecules. It has since been shown that all transfer RNAs have similar structures. Holley was awarded the 1968 Nobel Prize for Physiology or Medicine, jointly with the American biochemist Marshall Warren Nirenberg and Indian-born American biochemist Har Gobind Khorana for researches that helped to decipher the genetic code chemically and explain how the genetic information stored in the DNA of a cell controls the synthesis of proteins in cells. (☞ Jan. 9, Feb. 11, Apr. 10)

1925

Raja Ramanna, Indian nuclear physicist and the leading architect of India's nuclear programme, was born in Tumkur, Karnataka. Ramanna was educated at the Bishop Cotton Boys' School in Bangalore and graduated in science from the Madras Christian College. He completed his PhD in physics at King's College London, and on his return home in 1949 joined the Tata Institute of Fundamental Research in Mumbai at the invitation of Homi Bhaba. Ramanna rose to become head of the Bhabha Atomic Research Centre (BARC) in Mumbai (formerly Bombay), a post he held for two periods (1972–78 and 1981–83). He was also chairman of India's Atomic Energy Commission, and secretary of the Department of Atomic Energy from 1983 to 1987. He was instrumental in conducting India's first nuclear test at Pokhran in the western desert of Rajasthan in 1974. Ramanna had also served as director-general of the Defence Research and Development Organisation and as scientific adviser to

India's defence minister. Ramanna was a gifted musician, and could play the piano as dextrously as he could speak about atomic energy. His research interests included nuclear physics, European music and Buddhist philosophy. He remained director emeritus of the Bangalore-based National Institute of Advanced Studies until his death. (☞ Sep. 25)

1986
American space shuttle *Challenger* exploded 73 seconds after take-off, killing the seven-member crew of Michael J. Smith, Francis R. Scobee, Judith A. Resnik, Ellison S. Onizuka, Ronald E. McNair, Gregory B. Jarvis, and Sharon Christa McAuliffe. Space shuttle flights were suspended until 1988. An independent US commission blamed the disaster on unusually cold temperatures that morning and the failure of the O-rings, a set of gaskets in the rocket boosters. (☞ Apr. 4)

1996
The Indian remote sensing satellite *IRS-1C* was launched from Baikanur Cosmodrome in Kazakhstan. It was placed in a polar, Sun-synchronous orbit at an altitude of 817 kilometres. The satellite, fourth in the *IRS* series, carried three sensors, namely the panchromatic camera (PAN), the linear imaging and self-scanning sensor (LISS-III) and the wide field sensor (WiFS).

January 29

1838
Edward Williams Morley, American chemist who is best known for his collaboration with the German-born American physicist, Albert Abraham Michelson in what came to be known as the Michelson–Morley experiment, was born in Newark, USA. Morley's research centred on precise determination of the density and atomic weight of various gases, especially oxygen. He assisted Michelson in the latter's pursuit of measurements of the greatest possible accuracy to detect a difference in the speed of light through an omnipresent 'aether.' Yet the hypothetical aether could not be detected and the velocity of light was found to be the same irrespective of the velocity of the source. This result was a major step leading toward Albert Einstein's special theory of relativity. (☞ Feb. 24, Dec. 19)

1839
John Herschel, English astronomer, discovered that the chemical sodium thiosulphate could dissolve light-sensitive silver salts and stop them from darkening any further on exposure to light. It could thus act as a 'fixing agent' for photographic images. (☞ Mar. 7, Mar. 14, May 11)

1886
Karl Friedrich Benz, German mechanical engineer, was awarded the first patent for a petrol-driven vehicle. The vehicle was a three-wheeler. Benz built his first four-wheeled car in 1891. By 1900, Benz & Company, the company started by Benz in Mannheim, became the world's largest

Benz three-wheeler

manufacturer of automobiles. (☞ Apr. 4, Nov. 25)

1897
The Indian physicist, Jagadis Chandra Bose delivered at the Royal Institution, London, England, the famed Evening Discourse on 'The electromagnetic radiation and the polarisation of the electric ray', in which he described the optical behaviour of microwaves (millimetre-waves), such as reflection, refraction, total reflection and polarisation diffraction. (☞ Nov. 30)

Bose at Royal Institution, London

1926
Abdus Salam, Pakistani nuclear physicist, Nobel laureate, was born in Jhang Maghiana, Punjab (now in Pakistan). Salam attended the Government College at Lahore, and in 1952 he was awarded his PhD in theoretical physics from the University of Cambridge. In 1957, he became professor of theoretical physics at the Imperial College of Science and Technology, London, where he carried out his Nobel Prize-winning research. His hypothetical equations, which demonstrated an underlying relationship between the electromagnetic force and the weak nuclear force, postulated that the weak force must be transmitted by hitherto-undiscovered particles known as weak vector bosons. In 1964 he helped found the International Centre for Theoretical Physics at Trieste, Italy, in order to provide support for physicists from Third World countries. Salam was awarded the 1979 Nobel Prize for Physics, jointly with the American physicists Steven Weinberg and Sheldon Lee Glashow for their work in formulating the electroweak theory, which explains the unity of the weak nuclear force and electromagnetism. (☞ May 3, Nov. 21, Dec. 5)

1934
Fritz Haber, German physical chemist, Nobel laureate (Chemistry, 1918), who developed a method of synthesising ammonia directly from nitrogen and hydrogen, died in Basel, Switzerland. (☞ Dec. 9)

1947
Linda B. Buck, American biologist, Nobel laureate, was born in Seattle, Washington, USA. Buck received her BS degrees in psychology and microbiology from the University of Washington in 1975 and her PhD in immunology from the University of Texas, Dallas, Texas, in 1980. In 1984 she joined the Howard Hughes Medical Institute, Columbia University, New York as an associate, becoming assistant professor of neurobiology at the Harvard Medical School, Boston in 1991. Since 2001 she has been a full investigator at Howard Hughes Medical Institute. Buck did her post-doctoral work at the Columbia University under fellow American biologist, Richard Axel. Her primary research interest has been on how pheremones and odours are

detected in the nose and interpreted in the brain. In 1991, Buck and Axel cloned for the first time, olfactory receptors, showing that they belong to the family of 'G-protein-coupled receptors'. By analysing rat DNA, they estimated that there were approximately one thousand different genes for olfactory receptors in the mammalian genome. This research opened the door to the genetic and molecular analysis of the mechanisms of olfaction. Buck was awarded the 2004 Nobel Prize for Physiology or Medicine, jointly with Axel, for their discoveries of odorant receptors and the organisation of the olfactory system. (☞ Jul. 2)

January 30

1894

Charles B. King of Detroit, Michigan, USA was awarded a patent for the pneumatic hammer. King exhibited his hammer at the World's Columbian Exhibition in Chicago, Illinois in 1893. A pneumatic hammer contains an air-operated piston that delivers successive blows to a chisel or forming tool at the end of the hammer. The tool also has a separate mechanism to control the airflow to the piston, thus allowing the operator to control the speed and force of the blows.

1899

Max Theiler, American microbiologist, Nobel laureate, was born in Pretoria, South Africa. Theiler attended local schools except for one year in Basle, Switzerland and then went on to the Rhodes University College, Grahamstown and the University of Capetown Medical School. He had his medical training at St. Thomas's Hospital, London, and the London School of Hygiene and Tropical Medicine. After graduation in 1922 he joined the Department of Tropical Medicine at the Harvard Medical School, Boston, USA, where he carried out important studies on amoebic dysentery and began work on yellow fever. In 1930, Theiler joined the Rockefeller Institute (now Rockefeller University) in New York City, where he continued his research on infectious diseases, including yellow fever. In the late 1930s, Theiler developed the first attenuated, or weakened, strain of the yellow fever virus, which he used to develop a vaccine widely used for human immunisation against yellow fever. Theiler was awarded the 1951 Nobel Prize for Physiology or Medicine for his research on yellow fever. (☞ Aug. 11)

1928

Johannes Andreas Grib Fibiger, Danish pathologist, Nobel laureate (Physiology or Medicine, 1926), who achieved the first controlled induction of cancer in laboratory animals, died in Copenhagen, Denmark. (☞ Apr. 23)

1948

Orville Wright, American inventor and aviation pioneer who with his brother, Wilbur, achieved the first powered, sustained, and controlled airplane flight, died in Dayton, Ohio, USA. (☞ Apr. 16, Aug. 19, Dec. 17)

1949

Peter Courtland Agre, American biological chemist, Nobel laureate, was born in Northfield, Minnesota, USA. Agre went to the Theodore Roosevelt High School in Minneapolis and in 1970 earned his bachelor's degree in chemistry from the Augsburg College in that city. He received his MD from the Johns Hopkins University School of Medicine, Baltimore, Maryland, in 1974. In 1981, after post-graduate medical training and then a fellowship at the University of North Carolina at Chapel Hill, North Carolina, Agre returned to John Hopkins where in 1993 he became a professor in the department of biological chemistry, a position he still holds. Agre's main work related to the long-sought 'channels' that regulate and facilitate water molecule transport through cell membranes, a process essential to all living organisms. That the body's cells must contain specific channels for transporting water was suspected as early as the middle of the nineteenth century. However, it was not until 1988 that Agre succeeded in isolating the long-sought-after water channel. This decisive discovery, dubbed 'water pore' or aquaporin, opened the door to a whole series of biochemical, physiological and genetic studies of water channels in bacteria, plants and mammals. Today, researchers can follow in detail a water molecule on its way through the cell membrane and understand why only water and not other small molecules or ions, can pass. Working on this basic knowledge, scientists are searching for drugs that can specifically target water channel defects.

Agre was awarded the 2003 Nobel Prize for Chemistry, jointly with American biophysicist Roderick MacKinnon, for discoveries concerning channels in cell membranes. Agre received half the prize money for the discovery of water channels, while MacKinnon received the other half for structural and mechanistic studies of ion channels. (☞ Feb. 19)

1959
The first high-purity nuclear-grade uranium ingot was produced at the Uranium Metal Plant at the Trombay centre of India's Atomic Energy Establishment (later re-named Bhabha Atomic Research Centre). (☞ Jan. 20, Jul. 10)

1991
John Bardeen, American physicist, Nobel laureate (Physics, 1956, 1972), who was the co-inventor of the transistor and co-author of the BCS theory of superconductivity, died in Boston, Massachusetts, USA. (☞ May 23)

1995
Gerald Malcolm Durrell, British naturalist and prolific writer, known internationally for his pioneering role in preserving and breeding endangered species in zoos, died in St. Helier, Jersey, UK. (☞ Jan. 7)

January 31

1868
Theodore William Richards, American chemist, Nobel laureate, was born in Germantown, Pennsylvania, USA. Richards graduated from the Haverford College, Pennsylvania, in 1885 and took advanced degrees at the Harvard University, Cambridge, Massachusetts, where he became instructor in chemistry in 1891 and full professor in 1901. In his doctoral work Richards made an accurate measurement of the ratio of the atomic weight of oxygen to that of hydrogen. He spent most of his

career making accurate determination of atomic weights. His figures were more accurate than those that were accepted as standard. In 1913, his team showed that lead present in uranium had a lower atomic weight than normal specimens of lead, thus supporting the idea that it was formed by radioactive decay and also confirming the existence of isotopes. Richards was awarded the 1914 Nobel Prize for Chemistry for his research on the atomic weights of about 60 elements that indicated the existence of isotopes. (☞Apr. 2)

1881

Irving Langmuir, American physical chemist, Nobel laureate, was born in Brooklyn, New York, USA. Langmuir studied metallurgical engineering at the Columbia University, and then went to Germany to work under Walther H. Nernst, a pioneer physical chemist, at the University of Göttingen, where he took his PhD in 1906. After returning to USA, he conducted research for the General Electric Company where he investigated electrical discharges in gases, electron emission, and the high-temperature surface chemistry of tungsten.

He also worked on the phenomenon of adsorption and the application of this to catalysis. In 1913, Langmuir found that the life of the tungsten vacuum bulbs then in use could be extended considerably if they were filled with a mixture of nitrogen and argon. He also developed a vacuum pump, high-vacuum tubes used in radio broadcasting, and an atomic hydrogen welding torch capable of temperatures up to 3,000°C. He made first use of the term plasma. Langmuir was awarded the 1932 Nobel Prize for Chemistry for his studies of molecular films on solid and liquid surfaces. (☞ Aug. 16)

1929

Rudolf Ludwig Mössbauer, German physicist, Nobel laureate, was born in Munich, Germany. Mössbauer passed his intermediate degree examinations from the Technical University in Munich in 1952. From 1955 to 1957 he worked on his thesis for the doctorate and carried out a series of investigations at the Institute for Physics of the Max Planck Institute for Medical Research in Heidelberg. From the year 1953 onwards his main work was directed towards the study of absorption of gamma rays in matter, in particular the study of nuclear resonance absorption. This led to the discovery, in 1957, of recoil-less nuclear resonance absorption and its theoretical interpretation. The Mössbauer effect refers to the resonant and recoil-less emission and absorption of gamma rays by atoms bound in a solid.

Mössbauer was appointed Professor of Physics at the California Institute of Technology in 1961. Mössbauer was awarded the 1961 Nobel Prize for Physics, jointly with the American physicist Robert Hofstadter, for his researches concerning the resonance absorption of gamma rays and his discovery of the Mössbauer effect. (☞ Feb. 5)

1958

The United States entered the space age by launching the first successful orbiting satellite, *Explorer-I*, in orbit around Earth. The orbit had a period of 114.9 minutes. The satellite measured cosmic radiation, and led to the discovery of the Van Allen radiation belt.

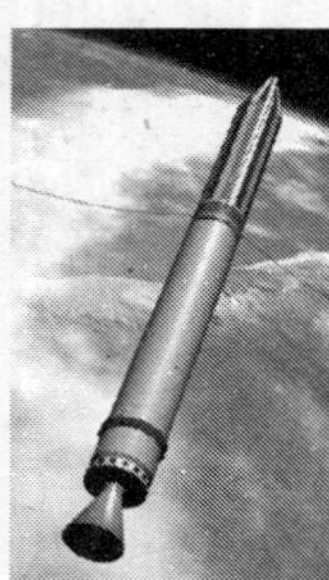

Explorer –1

1966

Soviet Moon-probe *Luna-9* was launched from Baikanur Cosmodrome in USSR (now Kazakhstan). (☞ Feb. 3)

1971

American manned lunar mission *Apollo-14*, carrying astronauts Alan Shephard, Edgar Mitchell, and Stuart Roosa was launched towards Moon by a Saturn-V rocket from the Kennedy Space Center, Florida, USA. (☞ Feb. 5, Feb. 9)

February 1

1905

Emilio Gino Segrè, Italian-born American physicist, Nobel laureate, was born in Tivoli, Italy. In 1922, Segrè began studies in engineering at the University of Rome, but later switched over to physics. He studied under Enrico Fermi and was awarded his doctorate in physics in 1928. Two years later he and Fermi performed neutron bombardment experiments, in which many elements heavier than uranium were created. In 1935, they discovered slow neutrons, which turned out to be vital for the operation of nuclear reactors. In 1937, while working at the University of Palermo in Italy, he discovered technetium, the first man-made element. In 1938, he moved to the United States where he discovered the element astatine in 1940. In 1955, while working as a professor of physics at Berkeley, Segrè along with the American physicist Owen Chamberlain discovered antiprotons and thus set the stage for the discovery of many additional antiparticles. Segrè was awarded the 1959 Nobel Prize for Physics, jointly with Chamberlain, for their discovery of the antiproton. (☞ Apr. 22, Jul. 10)

1923

The world's first leaded gasoline went on sale in Ohio, USA. The fuel was called Ethyl after its new additive, tetraethyl lead. It developed to stop engine knock, a common problem in petrol engines. Since the 1970s, toxic lead has been gradually replaced with lead-free gasoline because of environmental concerns.

1944

Three Rockefeller University scientists, Oswald Avery, Colin MacLeod and Maclyn McCarty, reported in the *Journal of Experimental Medicine*, the first substantial research evidence proving that DNA is the chemical basis of heredity. The identification of DNA settled a contentious scientific debate – at that time, some scientists argued that DNA was too simple to make up genes of humans and other organisms. The Rockefeller discovery paved the way for the research of Francis Crick, Rosalind Franklin, James D. Watson, Maurice Wilkins and others, who determined the double helix structure of DNA and showed how it could account for the encoding of genetic information and how that information could be copied in every cell division.

1958

Clinton Joseph Davisson, American physicist, Nobel laureate (Physics, 1937), who discovered the wave character of electrons, died in Charlottesville, USA. (☞ Oct. 22)

1976

Werner Karl Heisenberg, German physicist, Nobel laureate (Physics, 1932), who discovered a way to formulate quantum mechanics in terms of matrices, died in Munich, Germany. (☞ Dec. 5)

1976

George Hoyt Whipple, American pathologist, Nobel laureate (Physiology or

Medicine, 1934), who discovered that raw liver could be used to reverse the effects of anaemia, died in Rochester, New York, USA. (☞ Aug. 28)

2003
American space shuttle *Columbia* broke up just minutes before it was scheduled to land at the Kennedy Space Center, Florida, USA, killing all seven astronauts on board. The astronauts aboard *Columbia* were Rick Husband, William McCool, David Brown, Indian-born Kalpana Chawla, Laurel Clark, Michael Anderson, and Ilan Ramon of Israel. The shuttle was travelling about 20,000 kilometres per hour when it broke up. Later investigations revealed that damage to heat-shielding tiles on the shuttle's left wing by falling thermal insulation at the time of launch led to excess heat build-up and ultimate disintegration of *Columbia*. All shuttle flights were subsequently grounded. (☞ Jan. 16, Apr. 12, Jul. 1)

February 2

1907
Dmitry Ivanovich Mendeleyev, Russian chemist, who developed the periodic classification of the elements, died in St. Petersburg, Russia. (☞ Jan. 7, Feb. 8)

1947
Edwin H. Land, American inventor, first demonstrated his invention of instant photography, which came to be known as Polaroid photography, at a meeting of the Optical Society of America. He demonstrated a camera, known as the Polaroid Land Camera, which produced a finished print in 60 seconds. (☞ Feb.10, Mar. 1, May 7)

1970
Bertrand Russell, English logician and philosopher, Nobel laureate (Literature, 1950), best known for his work in mathematical logic and for his social and political campaigns, including his advocacy of both pacifism and nuclear disarmament, died in Merioneth, Wales, UK. (☞ May 18)

1980
William Howard Stein, American biochemist, Nobel laureate (Chemistry, 1972), who discovered the functioning of pancreatic enzyme ribonuclease, died in New York, USA. (☞ Jun. 25)

February 3

1468
Johann Gutenberg, German inventor, printer, who invented moveable type, died in Mainz, Germany. Nothing is known about his birth and early life, but it is believed that he was born in the late 14th century. Gutenberg is best known for his invention of movable metal type and the printing press. Around 1430 he is believed to have established himself in Strassburg (now Strasbourg, France), where he began trying out new ways of printing. He combined parts of presses used to make wine and paper to build a printing press. He used his metalworking skills to cast metal numbers and letters. He then arranged these numbers and letters to form words and sentences. Next he placed them in rows on the printing press. Gutenberg could move the numbers and letters around as needed, which allowed him to print many books faster than ever before. His was a new method of printing never used before. He also invented a printing ink – a mixture of oil, copper, and lead. Gutenberg's best-known printed

work is his 1,200-page Bible, variously known as the *Gutenberg Bible*, or *Mazarin Bible*, which was printed around 1455. The three-volume work, in Latin text, is also known as the *42-line Bible* because it is printed in two 42-line columns on each page. It was the first book printed with movable type. Gutenberg's printing press and movable type changed communication forever. By the end of the 1400s, books had become available to more people and new ideas in science, religion, and art spread quickly across Europe.

1900
Tiruvenkata Rajendra Seshadri, Indian chemist, Fellow of the Royal Society, was born at Kulitalai, Tamil Nadu, India. Seshadri completed his school education at Tiruchi, and moved to Madras (now Chennai) to join the Presidency College. After receiving BSc Hons. degree in chemistry, Seshadri worked with the Ramakrishna Mission for one year before joining the chemistry department of the Presidency College as a university research scholar. He went to England to work for his PhD at the University of Manchester under the British chemist Robert Robinson. Seshadri specialised in the chemistry of natural products, especially flavones, flavanones, anthocyanins and quinines, and held senior teaching positions in several Indian universities. After returning to India in 1930, Seshadri became a research fellow at the University of Madras and after some months joined the Agricultural College and Research Institute at Coimbatore, Tamil Nadu. In 1949 Seshadri joined the University of Delhi to head the Chemistry Department. He turned it into one of the most active centres in the country for research on the chemistry of natural products. Seshadri inspired a large number of young students, setting the pattern of work by his own example. The laboratories were kept open from morning till late in the night, and he was always available for guidance. (☞ Sep. 27)

1925
Oliver Heaviside, English physicist, who predicted the existence of the ionosphere, died in Torquay, Devon, UK. (☞ May 18)

1925
The first electric train service in India was started between Bombay (now Mumbai) and Kurla.

1966
Unmanned Russian Moon-probe *Luna-9* landed safely on the Moon in the region of Oceanus Procellarum. It was the first ever soft landing on another celestial body, and by demonstrating that the lunar surface was firm enough for spacecraft landings, opened the way for manned trips to the Moon. (☞ Jan. 31)

1975
William David Coolidge, American engineer and physical chemist whose improvement of tungsten filaments was essential for the development of the modern incandescent lamp bulb and the X-ray tube, died in Northampton, Massachusetts, USA. (☞ Jul. 4, Oct. 23, Dec. 30)

1984
American space shuttle *Challenger* was launched for the 4th time. During this mission, astronaut Bruce McCandless

performed the first untethered space walk using the manned manoeuvring unit (MMU). The MMU is a one-man propulsion backpack that snaps onto the back of the space suit's portable life-support system, allowing the astronaut to work outside without a tether and to move up to a distance of 100 metres from the spacecraft. The MMU can support an astronaut outside the spacecraft for as much as six hours at a time. (☞ Jan. 28, Apr. 4)

February 4

1906
Clyde W. Tombaugh, American astronomer who discovered Pluto, was born in Streator, Illinois, USA. After finishing high school, Tombaugh built his own telescope according to specifications published in a 1925 issue of *Popular Astronomy*. Using this instrument, he made observations of Jupiter and Mars and sent sketches of these planets to Lowell Observatory, when he was awarded a job offer. Tombaugh's assignment at the Observatory was to locate the ninth planet, a search instigated in 1905 by astronomer Percival Lowell. Ultimately Tombaugh did find Pluto on a set of photographic plates. Pluto was the only planet (now designated as a dwarf planet) discovered in the twentieth century, after a systematic search instigated by the predictions of other astronomers. Tombaugh also discovered several clusters of stars and galaxies, studied the apparent distribution of extragalactic nebulae, and made observations of the surfaces of several planets and the Moon. (☞ Jan. 17, Feb. 18, Mar. 13)

1928
Hendrik Antoon Lorentz, Dutch physicist, Nobel laureate (Physics, 1902), whose refinement of the electromagnetic theory gave rise to Albert Einstein's special theory of relativity, died in Haarlem, Netherlands. (☞ Jul. 18)

1941
Roy Plunkett, American chemist, was awarded a US patent for tetrafluoroethylene polymers, which is commonly known as Teflon, used in non-stick coatings of kitchenware.

1974
Satyendra Nath Bose, Indian mathematician and physicist, Fellow of the Royal Society, known for his collaboration with Albert Einstein in developing a theory regarding the gas-like qualities of electromagnetic radiation, died in Kolkata, India. (☞ Jan. 1)

February 5

1840
John Boyd Dunlop, Scottish inventor who was a pioneer of the pneumatic tyre, was born in Dreghorn, North Ayrshire, Scotland. Dunlop was educated at the Edinburgh University and became a veterinary surgeon when he was only 19. He worked as a vet in Edinburgh for nearly ten years before moving to Belfast. He built up a large practise in Ireland but found that the rough roads and the iron, wood or solid rubber wheels made travel uncomfortable. In 1887, when his 9-year-old son complained of the rough

ride he experienced on his tricycle over the cobbled streets of Belfast, Dunlop came up with a design based on an inflated rubber tube and patented it the following year. (☞ Oct. 23, Dec. 7)

1914

Alan Lloyd Hodgkin, British physiologist, Nobel laureate, was born in Banbury, Oxfordshire, England. Hodgkin was educated at the Trinity College, Cambridge.

After conducting radar research for the British Air Ministry during the war years, he joined the faculty at Cambridge, where he worked with fellow English physiologist Andrew Fielding Huxley on measuring the electrical and chemical behaviour of individual nerve fibres. By inserting microelectrodes into the nerve fibres of the giant Atlantic squid *Loligo forbesi*, they were able to show that, contrary to the accepted theory, which postulated a breakdown of the nerve membrane during impulse conduction, the electrical potential of a fibre during conduction of an impulse exceeded the potential of the fibre at rest. It was known that the activity of a nerve fibre depends on the fact that a large concentration of potassium ions is maintained inside the fibre, while a large concentration of sodium ions is found in the surrounding solution. Their experimental results showed that the nerve membrane allows only potassium to enter the fibre during the resting phase but allows sodium to penetrate when the fibre is excited. Hodgkin was awarded the 1963 Nobel Prize for Physiology or Medicine, jointly with Andrew Huxley and Australian physiologist John Eccles, for the discovery of the chemical processes involved in nerve conduction. (☞ Jan. 27, Nov. 22, Dec. 20)

1915

Robert Hofstadter, American scientist, Nobel laureate, was born in New York, USA. Hofstadter was educated at the Princeton University, where

he earned a PhD in 1938. He joined the faculty at Princeton in 1946, where his researches mainly dealt with the study of infrared rays, photoconductivity, and crystal and scintillation counters. While teaching at the Stanford University from 1950 to 1985, Hofstader used a linear electron accelerator to study the constituents of atomic nuclei and discovered that protons and neutrons have a definite size and form; they were not structure-less particles. He was able to determine the precise size of the proton and neutron and provide the first reasonably consistent picture of the structure of the atomic nucleus. He also made substantial contributions to gamma ray spectroscopy, leading to the use of radioactive tracers to locate tumours and other disorders. Hofstadter was awarded the 1961 Nobel Prize for Physics, jointly with the German physicist Rudolf Ludwig Mössbauer, for his investigations of protons and neutrons, which revealed the hitherto unknown structure of these particles. (☞ Jan. 31, Nov. 17)

1967

American spacecraft *Lunar Orbiter-3* was launched on a mission to photograph smooth areas of the lunar surface in order to select safe landing sites for the Apollo missions. The spacecraft was also equipped to collect data on radiation intensity and micrometeoroid impact on the Moon. The spacecraft photographed the Moon from February 15 to 23. In all five *Lunar Orbiter* missions were sent and all were successful, photographing 99% of the Moon's surface with a resolution of 60 m or better.

1971
Apollo-14 astronauts Alan Shepard and Edward Mitchell landed on the Moon on board lunar module *Antares*. After landing near Fra Mauro, the two astronauts spent a total of 9.2 hours on the Moon outside the lunar module and carried out scientific studies. (☞ Jan. 31)

1974
The American space probe *Mariner-10* flew by Venus at a distance of 5,800 kilometres and sent back the first close-up photographs of the planet's cloud structure. (☞ Mar. 16, Mar. 29, Sep. 21, Nov. 3)

February 6

1802
Charles Wheatstone, English physicist, Fellow of the Royal Society, who popularised the Wheatstone bridge, a device that accurately measured electrical resistance and became widely used in laboratories, was born in Barnwood, England. Wheatstone did not have any formal scientific education. He worked as a musician and instrument-maker into his early thirties and then pursued a career as an academic and scientist for the rest of his life. His early experiments in acoustics, optics, and electricity won him recognition in the scientific community while his work with the telegraph made his name a household word. Wheatstone did not actually invent the "Wheatstone Bridge". His contemporary, Samuel Hunter Christie, came up with the idea of the bridge circuit, but Wheatstone set the precedent for using it in the way in which it has been most commonly used for measuring electrical resistance. Over time, the device became associated with him and took on his name. He did, however, invent the musical instrument called concertina, the stereoscope, and an early form of the telegraph. In 1834 Wheatstone was appointed professor of experimental physics at King's College, London. In 1836 he became a Fellow of the Royal Society. (☞ Oct. 19)

1804
The world's first steam railway locomotive, built by the English mechanical engineer, Richard Trevithick, made its maiden run on a cast railway line 15 kilometres long, carrying 25 tonnes at the speed of 6 km/hr. With it, Trevithick demonstrated that a smooth wheel could move on a smooth track. Until then it was considered impossible to achieve sufficient traction without using gear-like wheels on a slotted track (☞ Apr. 13, Apr. 22)

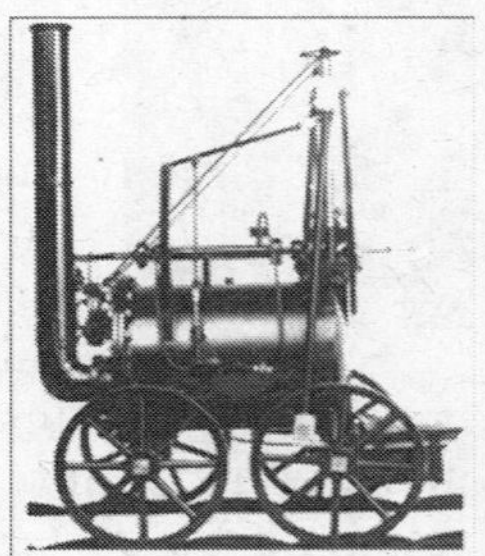

1804
Joseph Priestley, English clergyman and physical scientist, who discovered oxygen, died in Northumberland, Pennsylvania, USA. (☞ Mar. 8, Mar. 13)

1886
The element germanium (Ge, atomic number 32) was discovered by the German chemist, Clement Winkler while analysing the sulphide mineral argyrodite. He found that all the known elements it contained made up only 93 per cent of its weight. Tracking down the remaining seven per cent, he found the new element, which he called germanium, for Germany. The new element turned out

to be the eka-silicon predicted by Dmitry I. Mendeleyev in 1871.

1892
William P. Murphy, American physician, Nobel laureate, was born in Stoughton, Wisconsin, USA. Murphy was awarded his MD from the Harvard University, Cambridge, Massachusetts, and then joined the staff of Peter Bent Brigham Hospital, Boston in 1923, where he began his collaboration with fellow physician George R. Minot. In the early 1920s, the American pathologist George H. Whipple had demonstrated that liver in the diet sharply raised red blood cell counts in anaemic patients. Murphy and Minot carried the work further and found that feeding raw liver to their pernicious anaemic patients gave amazing results. Their discovery laid the groundwork for the development of vitamin B_{12} therapy for anaemia. Murphy was awarded the 1934 Nobel Prize for Physiology or Medicine, jointly with Minot and Whipple, for their discoveries concerning liver therapy in cases of anaemia. (☞ Aug. 28, Oct. 9, Dec. 2)

1946
Upendranath Brahmachari, Indian physician, who discovered a treatment for *Kala Azar*, died in Kolkata, India. (☞ Jun. 7)

1991
Salvador Luria, Italian-born American biologist, Nobel laureate (Physiology or Medicine, 1969), whose research provided experimental evidence that phage-resistant bacteria were the result of spontaneous mutations rather than a direct response to changes in the environment, died in Lexington, Massachusetts, USA. (☞ Aug. 13)

2002
Max Ferdinand Perutz, Austrian-born British biochemist, Nobel laureate (Chemistry, 1962), who did pioneering work on globular proteins, died in Cambridge, England. (☞ May 19)

February 7

1865
Gregor Mendel, Austrian botanist and plant breeder, who first discovered the laws of genetics, read his first scientific paper to the Brünn Society for the Study of Natural Sciences in Moravia. (☞ Jan. 6, Jul. 22)

1905
Ulf von Euler, Swedish physiologist, Nobel laureate, was born in Stockholm, Sweden. After his graduation from the Karolinska Institute in Stockholm, Euler served on the faculty of the institute from 1930 to 1971. Euler's outstanding achievement was his identification of noradrenaline, the key neurotransmitter in the sympathetic nervous system. He also found that noradrenaline is stored within the nerve fibres themselves. Von Euler had earlier, in 1935, discovered the substance he named prostaglandins, which play active roles in stimulating human muscle contraction and in the regulation of the cardiovascular and nervous systems. Von Euler was awarded the 1970 Nobel Prize for Physiology or Medicine, jointly with British bio-physicist Bernard Katz and American biochemist

Julius Axelrod, for their independent study of the mechanics of nerve impulses. (☞ Mar. 9, Mar. 26, May 30)

1906

Oleg Konstaninovitch Antonov, Soviet aviation engineer and aircraft designer, was born near Moscow, Russia. Antonov designed his first glider by the time he was 18. After graduation from the Kalinin Polytechnical Institute in Leningrad in 1930, Antonov was directed to Moscow to set up the Central Design Bureau for Gliders. In Tushino, near Moscow, a new glider-manufacturing factory was under construction, and in 1933, when construction of the factory was over, Antonov was appointed its Chief Designer. During World War II he worked on military aircraft. In 1946, he was appointed head of the aircraft design bureau, which was later moved to Kiev. Antonov established his own design bureau and in 1947 produced the rugged *AN-2* agricultural biplane that turned out to be one of the most versatile and enduring general-purpose aircraft in the world. In 1952, he moved his bureau to Kiev in the Ukraine and began work on a series of multi-engine turboprop transport aircraft. His four-engine *AN-12* earned him and his design team the coveted Lenin Prize. Antonov excelled at producing durable transport aircraft for the harsh weather and primitive operating conditions in far-flung Soviet Union. His huge *AN-22* was the largest aircraft in the world when it was introduced at the 1965 Paris Air Show, and earned him another Lenin Prize. The highly successful *AN-124*, the world's largest production aircraft, first flew in 1982. During his lifetime, Antonov designed 52 glider and 22 aircraft types including the world's largest and the heaviest-carrying ones and became the founder of an original school of design. He was the founder of Antonov ASTC, a world-famous aircraft company in Ukraine. (☞ Mar. 4, Dec. 21)

1932

The 'neutron' was described in an article in the journal *Nature* by its discoverer, James Chadwick, who coined the name for this neutral particle found in the nucleus of atoms. (☞ July 24, Oct. 20)

1960

Igor Vasilyevich Kurchatov, Soviet nuclear physicist who guided the development of Russia's first atomic bomb, died in Moscow, Russia. (☞ Jan. 12)

1984

The Manned Manoeuvring Unit (MMU) was used for the first time during the 4th mission of the space shuttle *Challenger*. Astronaut Bruce McCandless floated free in space without a tether outside the space shuttle in an MMU. The MMU is propelled by small nitrogen thrusters controlled by the astronaut's hands. Because no umbilical cord attaches the astronaut to the spacecraft, it gives much greater mobility than was available to early spacewalkers. (☞ Feb. 3)

1999

American spacecraft *Stardust* was launched by NASA on a mission to collect samples of cometary dust. Comets were formed at the same time as the solar system and are made up of primitive condensates and grains incorporated into

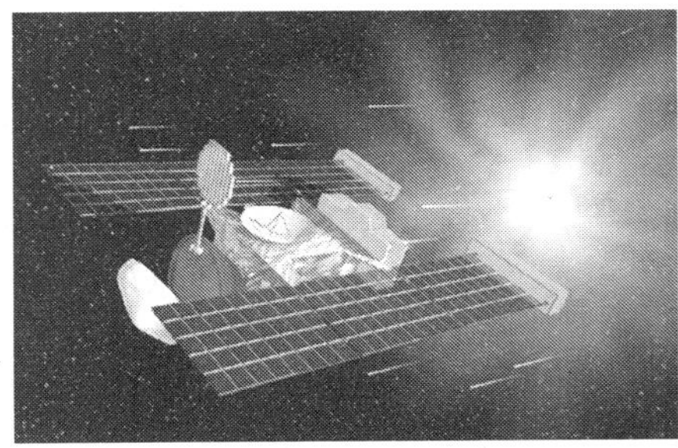

them at this time. The primary objective of the *Stardust* mission was to fly-by Comet Wild-2. It was to collect samples of dust and volatiles in the coma of the comet and return these samples to Earth for detailed study. The mission also collected samples of interstellar dust grains. The sample material capsule from *Stardust* returned to Earth in January, 2006. The samples from the coma are expected to provide insights into the composition and dynamics of the early solar system. The craft has been put in hibernation mode waiting to be sent to another comet or asteroid (☞ Jan. 15)

February 8

1700
Daniel Bernoulli, Swiss mathematician and nephew of Jacob Jacques Bernoulli, was born in Groningen, Netherlands. He was the most distinguished of the second generation of the Bernoulli family of Swiss mathematicians. Bernoulli studied not only mathematics but also physics, mechanics, astronomy, philosophy, logic, biology, and medicine at the Universities of Heidelberg, Strasbourg, and Basel. He received an MD degree in 1721 from the Basel University in Switzerland. Failing to get any position at the Basel University, Bernoulli went to Venice, Italy to study practical medicine. His medical work on the flow of blood and blood pressure gave him an interest in fluid flow. While in Venice, Bernoulli designed an hourglass for use at sea so that the trickle of sand was constant even when the ship was rolling in heavy seas. He won a prize from the Paris Academy for this work. Bernoulli also attained fame through his work *Mathematical Exercises* and on the strength of this he was invited to take up the chair of mathematics at St. Petersburg. The most important work which Bernoulli did while at St. Petersburg was his work on hydrodynamics. Bernoulli's theorem, which he derived in 1738, is named after him. The theorem states, in effect, that the total mechanical energy of a flowing fluid, comprising the energy associated with fluid pressure, the gravitational potential energy of elevation, and the kinetic energy of fluid motion, remains constant. Bernoulli's theorem is the principle of energy conservation for ideal fluids in steady, or streamline, flow. Bernoulli's theorem is the basis for many engineering applications, such as aircraft-wing design. The air flowing over the upper curved surface of an aircraft wing moves faster than the air beneath the wing, so that the pressure underneath is greater than that on the top of the wing, causing lift. This must be considered while designing the aircraft. (☞ Mar. 17)

1828
Jules Verne, French author whose writings laid much of the foundation of modern science fiction, was born in Nantes, France. Verne studied law in Paris but afterward chose to follow his interest in literature. In 1863 he published a story in a magazine titled 'Five Weeks in a Balloon'. The great success of the story encouraged him to produce other similar stories and novels,

with increasingly deft depictions of fantastic but nonetheless carefully conceived imaginary scientific wonders. The works included *A Journey to the Centre of the Earth*, *From the Earth to the Moon*, *Twenty Thousand Leagues Under the Sea*, and *The Mysterious Island*, in which he foresaw a number of scientific devices and developments, including the submarine, the Aqua-Lung, television, and space travel. Verne's novels were enormously popular throughout the world and many of them were made into highly successful movies. (☞ Mar. 24)

1834

Dmitry Ivanovich Mendeleyev, Russian chemist who developed the periodic classification of the elements, was born in Tobolsk, Russia. In 1855 Mendeleyev qualified as a teacher, winning a gold medal for his academic achievements. He was posted to the Crimea, where he continued his chemical studies at Odessa; he returned to St. Petersburg in 1856 to obtain an advanced degree in chemistry. In 1864 he became professor of chemistry at the Technical Institute at St. Petersburg. Since he could not find a textbook that met his needs, he set about writing his own and produced the classic textbook *The Principles of Chemistry*. It was while writing the book that the idea of devising a system of classifying the elements came to Mendeleyev's mind. Ultimately, he formulated the periodic law, according to which, when all known elements are arranged in order of increasing atomic weight, the resulting table shows a periodicity of properties. Mendeleyev's table of elements had gaps, but he predicted that they would be filled by elements not yet discovered; three were discovered within 20 years, and they possessed the properties he had predicted. (☞ Jan. 7, Feb. 2, Mar. 6)

1841

William Henry Fox Talbot, English inventor, produced the first photographic negative, which made possible multiple copies of positive prints on paper. (☞ Feb. 11)

1906

Chester Floyd Carlson, American physicist who invented the dry copying process known as xerography, was born in Seattle, USA. Carlson earned a college degree from the California Institute of Technology, Pasadena, in 1930 and, after a short time spent with the Bell Telephone Company, obtained a position with the patent department of PR Mallory Company, a New York electronics firm. Frustrated with the slow mimeograph machine and the cost of photography for getting copies of patent drawings and specifications, Carlson began in 1934 to think of a quick, convenient way to copy line drawings and text. Rather than photographic or chemical copying processes, he turned to electrostatics for a solution to the problem. Four years later he succeeded in making the first xerographic copy, and obtained the first of many patents for the xerographic process in 1940. In 1947 a small firm in Rochester, New York, which later became the Xerox Corporation, obtained the commercial rights to xerography. In 1958, Xerox introduced its first office copier. (☞ Sep. 19, Oct. 22)

1957

John von Neumann, Hungarian-American mathematician who made important contributions in quantum physics, logic, meteorology, and computer science, died in Washington, DC, USA. (☞ Dec. 3)

1957
Walther Bothe, German physicist, Nobel laureate (Physics, 1954), who developed the coincidence method of detecting the emission of electrons by X-rays, died in Heidelberg, Germany. (☞ Jan. 8)

1974
'Skylab-4', the last of the *Skylab* missions, ended. The mission carried Gerald P. Carr, William R. Pogue, and Edward G. Gibson and lasted 84 days and 1 hour, completing 1,214 orbits of the Earth. The mission included observation of the Comet Kohoutek among numerous experiments and four extra-vehicular activities totalling 22 hours and 13 minutes. (☞ May 14, May 25, Jun. 22, Jun. 28, Jul. 11, Sep. 25, Nov. 16)

1975
Robert Robinson, British chemist, Nobel laureate (Chemistry, 1947), who did research on the structure and synthesis of many different organic compounds, notably the naturally occurring, nitrogen-containing compounds called alkaloids, died in Buckinghamshire, near London, UK. (☞ Sep. 13)

1979
Dennis Gabor, Hungarian-born British electrical engineer, Nobel laureate (Physics, 1971), who invented holography, died in London, UK. (☞ Jun. 5)

February 9

1811
Robert Fulton, American inventor, engineer, and artist, was granted a patent for the practical steamboat. Foulton's first steamboat – a 45-metre-long boat – had a single-cylinder condensing steam engine that drove two 5-metre-diameter side paddlewheels to propel the boat. The 240-km trial run from New York to Albany required 32 hours, at an average of almost 7.6 km/hr. (☞ Aug. 14, Nov. 14)

1910
Jacques Lucien Monod, French biochemist, Nobel laureate, was born in Paris, France. It was through his father, who read the works of Darwin, that Monod developed his interest in biology very early in life. In 1928, he began his higher education, and registered at the Faculty of Paris for a degree in natural sciences. Monod obtained his degree in 1931 and his doctorate in natural sciences in 1941. In 1945 he started studies at the Pasteur Institute, together with French biologists André Lwoff and François Jacob on the gene function. Monod and Jacob did much to elucidate how genes regulate cell metabolism by directing the biosynthesis of enzymes. They proposed the existence of messenger RNA. They then suggested that certain genes, which they called 'operons' or operator genes, regulate the activity of other, protein-synthesising genes by affecting the formation of messenger RNA. In 1953, Monod became director of the Department of Cellular Biochemistry at the Pasteur Institute and also a professor at the University of Paris. In 1971, he was appointed director of the Pasteur Institute. Monod was awarded the 1965 Nobel Prize for Physiology or Medicine, jointly with Jacob and Lwoff for their discoveries concerning genetic control of enzyme and virus synthesis. (☞ May 8, May 31, Jun. 17)

1928
The first transmission of TV images across the Atlantic Ocean was achieved by the Scottish engineer and inventor John Logie Baird. The transmission from Purley in England was received at Hartsdale, New York, USA. (☞ Jan. 27, Jun. 14, Aug. 13)

1971
Apollo-14 astronauts Alan Shephard, Edgar Mitchell, and Stuart Roosa splashed down safely in the Pacific Ocean after their journey to the Moon. (☞ Jan. 31, Feb. 5)

1986
Comet Halley reached its perihelion – that is, its closest approach to the Sun – when it returned 75 years after its previous

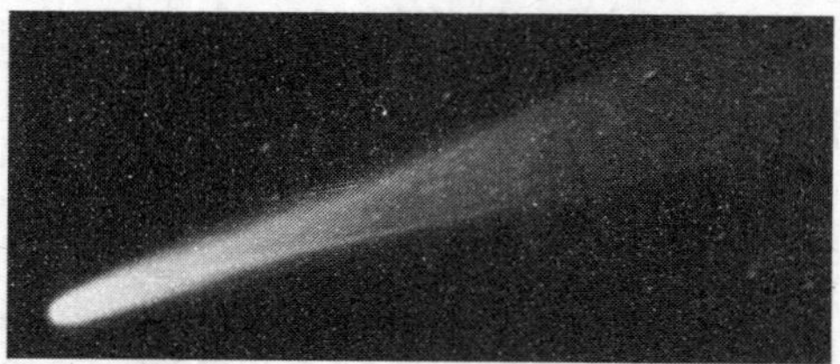

appearance in 1910. At perihelion, the comet was 88 million kilometres from the Sun, between the orbits of Mercury and Venus. (☞ Mar. 14 , Mar. 30, Nov. 8, Dec. 25)

1994
Howard Martin Temin, American virologist, Nobel laureate (Physiology or Medicine, 1975), who discovered the enzyme reverse transcriptase, died in Madison, Wisconsin, USA. (☞ Dec. 10)

1996
A single atom of element 112, the heaviest element known till date, was created by an international research team led by Peter Armbruster and Sigurd Hofmann at the Institute for Heavy Ion Research at Darmstadt, Germany, by bombarding lead-208 atoms with zinc-70 ions accelerated to high energies using a heavy ion accelerator. The new atom with a nucleus consisting of 112 protons and 165 neutrons, giving it an atomic mass of 277, was named ununbium, symbol Uub, according to an internationally adopted system for naming new elements. Being highly reactive, it lasted a fraction of a thousandth of a second before decaying into atoms of smaller nucleus.

February 10

1840
Per Teodor Cleve, Swedish chemist and geologist who discovered the elements holmium and thulium, was born in Stockholm, Sweden. After graduating from the Stockholm Gymnasium in 1858, Cleve matriculated at the University of Uppsala in 1858, and received his PhD in 1863. After a brief employment with the university in Uppsala and travels in Europe and North America, he became assistant professor of chemistry at the University of Uppsala in 1868. In addition, he taught at the Technological Institute in Stockholm from 1870 to 1874. In 1879 Cleve showed that the newly discovered element scandium was the element previously predicted by Dmitry Ivanovich Mendeleyev, who called it eka-boron. In that same year working with a sample of oxide called erbia that had all traces of scandia and ytterbia removed, he found two new rare earth elements, which he named holmium, after Stockholm, and thulium, after the old name for Scandinavia. (☞ Jun. 18)

1885
Edward J. Claghorn of New York was issued the first patent for automobile seat belts. The device was described in the patent as ‘designed to be applied to the

person and provided with hooks and other attachments for securing the person to a fixed object'.

1897
John Franklin Enders, American virologist and microbiologist, Nobel laureate, was born in West Hartford, Connecticut, USA. Enders was a student of English literature at Harvard University before he turned to bacterial studies. In 1929 he joined the Harvard faculty as an assistant in the department of bacteriology and immunology, and got a PhD, in 1930. From 1945 to 1949, Enders served the US Army as a civilian consultant on infectious diseases. During this period, Enders, with his co-workers Thomas H. Weller and Frederick C. Robbins, began research into new methods of producing the polio virus in large quantities for preparation of a vaccine. Until that time the only effective method of growing the virus had been in the nerve tissue of living monkeys, and the vaccine thus produced was dangerous to humans. The Enders, Weller, and Robbins team succeeded in culturing the virus in test tubes using cultures of non-nerve tissue from human embryos, which led to the development of the Salk vaccine for polio in 1954. Enders was awarded the 1954 Nobel Prize for Physiology or Medicine, jointly with Weller and Robbins, for their discovery of the ability of poliomyelitis viruses to grow in cultures of various types of tissue. (☞ Jun. 15, Aug. 25, Sep. 8)

1902
Walter Houser Brattain, American scientist, Nobel laureate, was born in Amoy, China. Brattain spent his childhood and youth in Washington, USA and was awarded a BS degree from the Whitman College in 1924. He then got his MA degree from the University of Oregon in 1926 and the PhD degree from the University of Minnesota in 1929. In 1929 he became a research physicist for the Bell Telephone Laboratories, Murray Hill, New Jersey. His chief field of research was solid-state physics, involving the study of the surface properties of solids, particularly the atomic structure of a material at the surface. Inspired by the ideas of William B. Shockley at the Bell Labs, Brattain and John Bardeen built the first transistor in 1947. Soon, the transistor replaced the bulkier vacuum tube in electronic devices and was the forerunner of microminiature electronic parts. Brattain was awarded the 1956 Nobel Prize for Physics, jointly with Shockley and Bardeen for their researches on semiconductors and their discovery of the transistor effect. (☞ Feb. 13, May 23, Oct. 13)

1912
Joseph Lister, British surgeon and medical scientist, founder of antiseptic surgery and a pioneer in preventive medicine, died in Walmer, Kent, UK. (☞ Mar. 16, Apr. 5, Jun. 17)

1923
Wilhelm Conrad Röntgen, German physicist, who was the first recipient of the Nobel Prize for Physics (1901) for his discovery of X-rays, died in Munich, Germany. (☞ Jan. 1, Jan. 23, Mar. 27, Nov. 8)

1948
Edwin Herbert Land, American inventor and physicist, patented the Polaroid

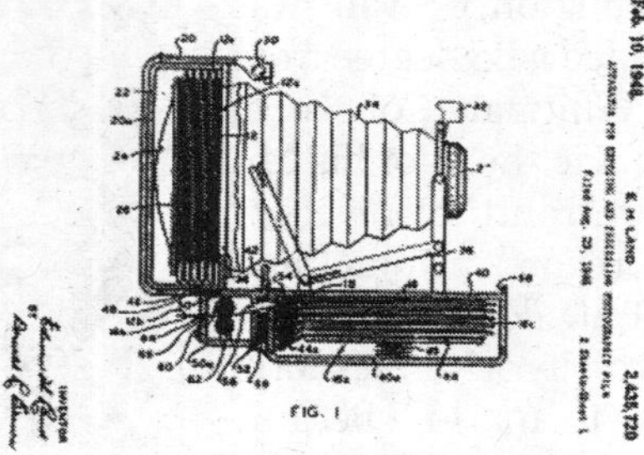

Polaroid camera patent

camera that produced a finished print in 60 seconds. Land's camera used a film pack that came with its own developing chemicals contained in flexible capsules that ruptured as the film passed between rollers after exposure, triggering the development process. (☞ Feb. 2, Mar. 1, May 7)

February 11

1650
René Descartes, French mathematician, scientist, and 'the father of modern philosophy', died in La Haye, Touraine, France. (☞ Mar. 31)

1800
William Henry Fox Talbot, English mathematician, physicist, chemist who invented the negative–positive photographic process, was born in Dorset, England. Talbot was educated at Harrow and at the Trinity College, Cambridge, and published many articles in the fields of mathematics, astronomy, and physics. In 1835 he published his first article describing a paper 'negative' from which multiple prints could be made. He modified and improved this process in 1840 and called it the 'calotype.' Talbot patented the process in 1841. In 1842 he was awarded a medal from the British Royal Society for his experiments with the calotype. In 1844 he published *The Pencil of Nature*, the first photographically illustrated book. (☞ Feb. 8, Sep. 17)

1813
Anders Gustav Ekeberg, Swedish chemist who in 1802 discovered the element tantalum, died in Uppsala, Sweden. (☞ Jan. 16)

1847
Thomas Alva Edison, American inventor who, singly or jointly, held a world record of 1,093 patents, was born in Milan, Ohio, USA. Edison began selling newspapers along the railroad at an early age and learned how to operate a telegraph. In his time the only source of electricity for telegraphy was primitive batteries giving a low-voltage current. By the time he died, in 1931, he had played a critical role in introducing the modern age of electricity, with central generating stations sending out electric power over wide distances. In addition, he created the world's first industrial research laboratory. In 1868, he made his first invention, an electric vote-recording machine. In 1876 he moved his laboratory to Menlo Park, in New Jersey where in 1879 he invented the first commercially practical incandescent electric light bulb. Among his other notable inventions are the phonograph, the carbon-button transmitter for the telephone earpiece and microphone, a revolutionary generator of unprecedented efficiency, which powered the first

commercial electric light and power system. By the late 1880s he had developed the key elements of motion-picture apparatus, which revolutionised entertainment. (☞ Jan. 27, Feb. 19, Apr. 15, Aug. 8, Aug. 9, Aug. 27, Oct. 18, Oct. 21, Dec. 6).

1868

Jean-Bernard-Léon Foucault, French physicist who introduced and helped develop a technique of measuring the absolute velocity of light with extreme accuracy and proved the rotation of Earth using a pendulum, died in Paris, France. (☞ Jan. 8, Sep. 18)

1898

Leo Szilard, Hungarian-born American physicist who, with Enrico Fermi, designed the first nuclear reactor, was born in Budapest, Hungary. In 1919, Szilard started studying electrical engineering and physics in Budapest before moving to Berlin to study under Albert Einstein, Max Planck and Max von Laue. In 1923 he began carrying out X-ray diffraction experiments at the Kaiser-Wilhelm Institute for Chemistry, Berlin. In 1922, Szilard was awarded his PhD from the University of Berlin and joined the staff of the Institute of Theoretical Physics as Laue's assistant. Szilard was appointed a lecturer in physics in 1929. He also worked with Albert Einstein and they eventually developed a home refrigerator without moving parts. In 1933, he left Berlin and went to London, where he conceived the neutron chain reaction that forms the basis of the fission bomb and all nuclear reactors. He conducted the first fission experiments in 1938 while at the Columbia University in USA. From 1942 until the end of World War II, he conducted nuclear research at the University of Chicago, where he helped Enrico Fermi construct the first nuclear reactor that sustained nuclear chain reaction. (☞ May 30, Dec. 2)

1922

Canadian surgeon Frederick Banting and his assistant Charles Best announced the use of insulin – the hormone responsible for controlling blood sugar levels –to treat diabetes in a dog in their first paper published on the subject. The previous year, Best had made an extract from pancreatic tissue and injected the pinkish solution into a diabetic dog. After an hour, the blood glucose had dropped from 0.2 to 0.1%. Banting and Best continued their research and improved the purity of insulin. The discovery was one of the most revolutionary moments in medicine. (☞ Feb. 27, Nov. 14)

1924

Jacques Loeb, German-born American biologist, noted chiefly for his experimental work on artificial parthenogenesis (reproduction without fertilisation), died in Hamilton, Bermuda. (☞ Apr. 7)

1973

Johannes Hans Jensen, German physicist, Nobel laureate (Physics, 1963), who proposed a shell model of the nucleus, died in Heidelberg, Germany. (☞ Jun. 25)

1993

Robert William Holley, American biochemist, Nobel laureate (Physiology or Medicine, 1968), whose research helped explain how the genetic code works, died in Urbana, Illinois, USA. (☞ Jan. 28)

February 12

1809

Charles Robert Darwin, English naturalist renowned for his theory of evolution, was born in Shrewsbury, England. When he was 16, he was sent to study medicine at the University of Edinburgh, where he was repelled by surgery performed without anaesthetics. Later he joined the Cambridge University to study divinity. In August 1831, Darwin was invited to sail as the unpaid naturalist on *HMS Beagle*, which was to survey the east and west coasts of South America and continue to the Pacific islands to establish a chain of chronometric stations. During this voyage Darwin made detailed study of the animals and plants of the places he visited. When he returned to England in 1836 he was promptly made a fellow of the Geological Society. Around this time, Darwin began a remarkable series of notebooks in which he initiated a set of questions and answers about 'the species problem'. He proceeded to collect facts about species through letters and discussions with breeders, gardeners, naturalists, and zookeepers, as well as through extensive reading. Finally, his famous work on the *Origin of Species by Means of Natural Selection* was published in 1859. The first edition sold out immediately, and by 1872 the work had run through six editions. The theory was accepted quickly in most scientific circles. Darwin's other famous work *The Descent of Man* came out in 1871. (☞ Apr. 19, Jul. 1, Aug.1, Oct. 1, Nov. 24, Dec. 27)

1884

Lewis Edson Waterman, American inventor, was granted a patent for the fountain pen. Waterman's pen had a steel nib that was partly slit in the middle. By pressing on the nib while writing, the thin gap between the halves of the nib widened to allow the ink to flow. Unlike earlier pens, which had to be dipped frequently into an inkpot, the fountain pen had a reliable supply of ink within the pen itself.

1918

Julian Seymour Schwinger, American physicist, Nobel laureate, was born in New York, USA. Schwinger had an intense awareness of physics at an early age and its study became an all-engrossing activity for him. He attended school in New York City and had his first publication as a professional physicist at the age of sixteen. After receiving his PhD from the Columbia University in 1939, at the age of 21, Schwinger worked under physicist J. Robert Oppenheimer at the University of California, Berkeley. In 1945 Schwinger joined the faculty of the Harvard University, Cambridge, Massachusetts, where in 1947 he became one of the youngest professors in that university's history. It was at Harvard that he developed equations that served to unite electromagnetic theory with quantum mechanics to form the new field of quantum electrodynamics, which reconciled quantum mechanics with Albert Einstein's special theory of relativity. Schwinger was awarded the 1965 Nobel Prize for Physics, jointly with fellow American Richard P. Feynman and Japanese physicist Shin'ichiro Tomonaga, for his work in formulating quantum electrodynamics. (☞ Mar. 31, May 11, Jul. 16)

1941
The first injection of penicillin was administered to a human test subject, a policeman named Albert Alexander, by the German-born British biochemist Ernst Boris Chain and Australian pathologist Howard Walter Florey at the Radcliffe Infirmary in England. (☞ Feb. 13, Jun. 19, Sep. 24)

1961
Soviet unmanned spacecraft *Venera-1*, the first spacecraft to be sent to Venus, was launched. (☞ May 19)

2001
The American unmanned *Near Earth Asteroid Rendezvous* (*NEAR*) spacecraft (later re-named *NEAR Shoemaker*) landed on the minor planet Eros, becoming the first spacecraft to orbit, land on and send data from the surface of a minor planet. Eros is approximately 33×13×13 kilometres in size and resembles a 'fat banana'. It is the second largest near-Earth minor planet and spins on its axis once every 5 hours, 16 minutes. Eros was 316 million kilometres from Earth when the spacecraft landed on it. The spacecraft obtained 69 high-resolution images before touchdown. The *NEAR Shoemaker* mission was the first in NASA's Discovery Program of low-cost planetary missions. (☞ Feb. 17)

February 13

1910
William B. Shockley, English-born American engineer and teacher, Nobel laureate, was born in London, England. Shockley studied physics at the California Institute of Technology from where he took his BSc degree in 1932. He obtained his PhD from the Massachusetts Institute of Technology, Cambridge, Massachusetts, in 1936. He joined the technical staff of the Bell Telephone Laboratories in Murray Hill, New Jersey the same year and began experiments with semiconductors. During World War II, he served as director of research for the Anti-submarine Warfare Operations Research Group of the US Navy. After the war, he returned to Bell Telephone as director of its research program on solid-state physics. Working with John Bardeen and Walter H. Brattain, he invented the point-contact transistor in 1947 and a more effective device, the junction transistor, in 1948. Shockley was awarded the 1956 Nobel Prize for Physics, jointly with Bardeen and Brattain for their researches on semiconductors and their discovery of the transistor effect. (☞ Feb. 10, May 23, Aug. 12)

1929
Scottish biologist Alexander Fleming published the results of his investigations on the antagonistic effect of his mould by-product penicillin on bacteria, noting that his discovery might have a therapeutic value if it could be produced in quantity. In 1941, the German-born British biochemist Ernst Boris Chain together with Australian pathologist Howard Walter Florey isolated and purified penicillin and performed the first clinical trials of the antibiotic. (☞ Feb. 12, Jun. 19, Aug. 6, Sep. 24)

1946
The world's first programmable general-purpose electronic digital computer, the

ENIAC

Electronic Numerical Integrator and Calculator, or ENIAC, was demonstrated at the Moore School of Electrical Engineering at the University of Pennsylvania, USA, by John W. Mauchly and J. Presper Eckert. Built out of some 17,468 electronic vacuum tubes, 70,000 resistors, 10,000 capacitors, and 6,000 switches ENIAC was in its time the largest single electronic apparatus in the world and occupied a room 16 metres long and 10 metres wide. It weighed 30 tons and consumed 100 kW of power. The ENIAC demonstrated that high-speed digital computing was possible using the vacuum tube technology then available.

February 14

1779

James Cook, English seaman, first of the really scientific navigators, who discovered New Zealand and Australia, died in Kealakekua Bay, Hawaii. (☞ Oct. 27)

1869

Charles Thomson Rees Wilson, Scottish physicist, Nobel laureate, was born in Glencorse, Scotland. Wilson initially intended to become a physician and took up biology. But later he became interested in the physical sciences, especially physics and chemistry and obtained a degree in science from Cambridge in 1892. Wilson began studying clouds in 1895. He tried to duplicate the effects of certain clouds and devised a way of suddenly expanding moist air in a closed chamber, which came to be known as the Wilson cloud chamber. The expansion cooled the air so that it became supersaturated and moisture condensed on dust particles or even, electrically charged atomic particles. When he passed ionising radiations such as alpha, beta or X-rays through such supersaturated moisture, the formation of droplets was so dense that photographs showed their tracks as white lines. Early in 1911, he was the first person to see and photograph the tracks of individual alpha and beta particles and electrons. Perfected by 1912, his chamber proved indispensable in the study of nuclear physics. Wilson was awarded the 1927 Nobel Prize for Physics, jointly with American physicist Arthur H. Compton, for his method of making the paths of electrically charged particles visible by condensation of vapour. (☞ Sep. 10, Nov. 15)

1917

Herbert A. Hauptman, American mathematician and crystallographer, Nobel laureate, was born in New York, USA. Hauptman graduated from the City College of New York in 1937 and went on to further study mathematics at the Columbia University, completing his MA in 1939. He got his PhD from the University of Maryland in 1955. After World War II, Hauptman joined the Naval Research Laboratory in Washington, DC where he met his college mate, crystallographer Jerome Karle, and began collaborating on the study of crystal structures. Together they developed mathematical methods for

deducing the molecular structure of chemical compounds from the patterns formed when X-rays are diffracted by their crystals. Hauptman became a professor of biophysics at the State University of New York at Buffalo in 1970. Hauptman was awarded the 1985 Nobel Prize for Chemistry, jointly with Karle, for their outstanding achievements in the development of direct methods for the determination of crystal structures. (☞ Jun. 18)

1929

Devendra Lal, Indian physicist, Fellow of the Royal Society, was born in Benaras, Uttar Pradesh, India. Lal did his BSc (1947), MSc (1949) from the Benaras Hindu University and his PhD from the Bombay University in 1960. He then went to USA to do postdoctoral research at the University of California, San Diego, California. His professional experience covers positions from research fellow to senior professor with the Tata Institute of Fundamental Research, Mumbai; Director of the Physics Research Laboratory, Ahmedabad, India; and Professor of Nuclear Physics in the Geological Research Division of the Scripps Institution of Oceanography, University of California, San Diego, USA. The primary thrust of Lal's work has been in the fields of cosmic ray physics, nuclear physics, earth sciences and solar system physics. His early work was in the fields of composition and energy spectrum of primary cosmic radiation, and elementary particle physics. Lal made extensive studies of meteorites, of lunar rocks brought by the Apollo and Luna Missions and of material from the ocean bed. When cosmic rays, which come from space, strike the surface of the Earth, the ocean bed, the Moon or a meteorite, they produce some nuclear changes and leave fine tracks – 'signatures' – in the rocks they traverse. Lal evolved techniques to understand past events in the solar system from these cosmic ray signatures. One of his most significant findings is that the intensity of cosmic rays during the last few million year has been the same as it is today.

1950

Karl Guthe Jansky, American electrical engineer, who discovered cosmic radio emissions in 1932, died in Red Bank, New Jersey, USA. (☞ Oct. 22)

1961

Lawrencium (Lr, atomic number 103), was first produced in at the University of California, Berkeley, California, USA. The 14th member of the actinide series in Group IIIb of the periodic table, lawrencium does not occur in nature; it was produced artificially by Albert Ghiorso, T. Sikkeland, A.E. Larsh, and R.M. Latimer by bombarding isotopes of californium (atomic number: 98) with boron ions (atomic number: 5) accelerated in a heavy ion linear accelerator. The element was named after the scientist Ernest O. Lawrence, the inventor of the cyclotron.

1975

Julian Huxley, English biologist, philosopher, educator, and author who greatly influenced the modern development of embryology, and studies of behaviour and evolution, died in London, UK. (☞ Jun. 22)

1980
The *Solar Maximum Mission* satellite, or *Solar Max* was launched by the American space agency NASA. The primary objective of the mission was to provide coordinated observations of solar activity, in particular solar flares, during a period of maximum solar activity. The payload was made up of seven instruments, specifically selected to study the short - wavelength and coronal manifestations of flares. After a malfunction in the satellite in January 1981 cut short the original mission, it was recovered by the crew of the space shuttle *Challenger* in April 1984 and repaired in orbit. After it was repaired, *Solar Max* went on collecting data until it burnt up in the Earth's atmosphere in December 1989. (☞ Dec. 2)

Solar Max

2003
Dolly, the world's most famous cloned sheep, was put to sleep. Dolly was less than seven years old but had been suffering from a progressive lung disease that normally afflicts sheep twice that age. (☞ Jul. 5)

February 15

1564
Galileo Galilei, Italian natural philosopher, astronomer, and mathematician who made fundamental contributions to the sciences of motion, and astronomy, was born in Pisa, Italy. Galileo was educated at a monastery near Florence and went on to study medicine at the University of Pisa and mathematics with a private tutor. Galileo's experimental work in dynamics paved the way for Kepler's and Newton's laws of motion, and he is often credited with being one of the first scientists to fully exploit the experimental method and to insist on a mathematical description of the laws of nature. His study of balls rolling down inclined planes convinced him that falling objects are accelerated independent of their mass, and that objects retain their velocity unless a force acts on them. Galileo also described that a pendulum's swings always take the same amount of time, independent of the amplitude, a discovery that later made precise clocks possible. Galileo was one of the first persons to use a telescope to observe the sky. He discovered the four largest moons of Jupiter and also observed that the planet Venus exhibited a full set of 'phases' like the Moon. Both of these discoveries lent support to the Sun-centric model of the solar system developed by Copernicus. For his support of the Sun-centred solar system propounded by the Polish astronomer Nicolaus Copernicus, Galileo had to face the Inquisition and was put under life-long house arrest (1634–1642) in his own villa in Arcetri near Florence, where he died. In 1992, that is, 359 years after the Galileo trial, Pope John Paul II issued an apology, lifting the edict of the Inquisition against Galileo. (☞ Jan. 7, Jan. 8, Jun. 22, Oct. 31)

1861
Charles Édouard Guillaume, French physicist, Nobel laureate, was born in Fleurier, Switzerland. Guillaume had his early education in Neuchâtel before going to the Zurich Polytechnic where he obtained his doctor's degree. He

spent a short time as an officer in the artillery before entering the International Bureau of Weights and Measures at Sèvres, near Paris, as an assistant, in 1883. He became Associate Director in 1902 and from 1915 until his retirement in 1936 he was Director of the Bureau. At the Bureau, Guillaume was concerned with developing the international standards for the metre, kilogram, and litre. A chance observation by Guillaume on the coefficient of expansion of nickel–iron alloys led to a systematic investigation of a whole series of ferronickel alloys and the discovery of invar, a nickel–steel alloy with a very low coefficient of expansion. The alloy is widely used in instruments of precision, such as thermostats and pendulums of astronomic clocks. He also produced elinvar, with an elasticity that remains nearly constant over a wide range of temperatures and was ideal for making watch hairspring. Guillaume was awarded the 1920 Nobel Prize for Physics in recognition of the service he had rendered to precision measurements in physics by his discovery of anomalies in nickel–steel alloys. (☞ Jun. 13)

1873

Hans von Euler-Chelpin, German-born Swedish biochemist, Nobel laureate, was born in Augsburg, Germany. Euler-Chelpin attended school in Munich, Wurzburg and Ulm, and then studied art from 1891 until 1893 at the Munich Academy of Painting. His desire to study problems of colour and especially the colour

of the spectrum led Euler-Chelpin to begin, in 1893, the study of science. He studied chemistry and physics at the University of Berlin from where he took his doctorate in 1895. In 1906 he was appointed professor of general and organic chemistry at the Royal University, Stockholm, and in 1929 was appointed Director of the newly set up Vitamin Institute and Institute of Biochemistry. After doing his doctorate from the University of Berlin in 1895, Euler-Chelpin worked with the German physical chemist Walther Nernst and in 1897 became Assistant to the Swedish physical chemist Svante Arrhenius at the Royal Institute of Technology in Stockholm. Euler-Chelpin's research mainly concerned the working of enzymes. He worked out the structure of the yeast coenzyme and showed that the molecule is made up from a nucleotide similar to that found in nucleic acid. He also helped determine the chemical structure of several vitamins, which form portions of coenzymes. Euler-Cheplin was awarded the 1929 Nobel Prize for Chemistry, jointly with the English biochemist Arthur Harden for their investigations on the fermentation of sugar and fermentative enzymes. (☞ Oct. 12, Nov. 7)

1874

Ernest Henry Shackleton, Irish explorer, was born in Kilkea, County Kildare, Ireland. Shackleton joined the Merchant Navy after attending school at the Dulwich College. After 10 years he gave that up and joined a British expedition led by Captain Robert Falcon Scott to try and be the first to reach the South Pole. In the first expedition in December 1902, Scott, Shackleton and Wilson reached within 640 kilometres of the South Pole, the furthest south yet achieved by anybody. On the return journey, Shackleton was invalided and sent home early. Still, his experience on this expedition spurred him on, and inspired him to have a go at reaching the South Pole himself. In March 1907 Shackleton outlined his trip, which he organised himself with the minimum of official support. The magnetic South Pole was first reached during Shackleton's British Antarctic Expedition (1907–1909). At that time the magnetic pole lay within the Antarctic continent at latitude 71.6°S and longitude 152°E. Today it lies far out to sea at latitude 65°S and longitude 139°E. (☞ Jan. 5, Jan. 16)

1951
The first cancer radiation therapy was performed using a beam from an atomic reactor at the US Atomic Energy Commission's unit in the Brookhaven National Laboratory at Upton, New York, USA.

1959
Owen Willans Richardson, English physicist, Nobel laureate (Physics, 1928), who proved that electrons are emitted from hot metal and not from the surrounding air, the basic principle used in vacuum tubes, died in Alton, Hampshire, UK. (☞ Apr. 26)

1988
Richard P. Feynman, American theoretical physicist, Nobel laureate (Physics, 1965), who re-made quantum electrodynamics and thus altered the way science explains the nature of waves and particles, died in Los Angeles, California, USA. (☞ May 11)

1999
Henry Way Kendall, American nuclear physicist, Nobel laureate (Physics, 1990), who contributed to the development of the quark model in particle physics, died in Wakulla, Florida, USA. (☞ Dec. 9)

February 16

1948
Gerard Kuiper, Dutch-born American astronomer, discovered Miranda – a moon of Uranus. Miranda has a diameter of almost 471 kilometres and orbits Uranus at an average distance of about 260,000 kilometres. Observations made by the American spacecraft *Voyager-2* showed that Miranda is geologically the most active moon of Uranus. (☞ Dec. 7)

1956
Meghnad Saha, Indian astrophysicist, Fellow of the Royal Society, noted for his development in 1920 of the thermal ionisation equation, died in New Delhi, India. (☞ Oct. 6)

1960
The American nuclear submarine *USS Triton* set off to circumnavigate the globe under water. Among the largest submarines and propelled by two nuclear reactors, *Triton* covered a distance of 78,858 kilometres, or 42,552 nautical miles in 84 days.

1970
Peyton Rous, American pathologist, Nobel laureate (Physiology or Medicine, 1966), who discovered cancer-inducing viruses, died in New York, USA. (☞ Oct. 5)

1997
Chien-Shiung Wu, Chinese-born American physicist, who provided the first experimental proof that the principle of parity conservation does not hold in weak subatomic interactions, died in Manhattan, New York, USA. (☞ May 29)

2001
William Howell Masters, American gynaecologist who is known for his studies of human sexuality, died in Tucson, Arizona, USA. (☞ Dec. 27)

February 17

1781
René Théophile Hyacinthe Laënnec, French physician who invented the stethoscope and is generally considered the father of chest medicine, was born in Quimper, France. While working at the Necker Hospital in Paris, Laënnec used a foot-long wooden cylinder, placed on the chests of his patients, to hear the various sounds made by the lungs and heart. This was the earliest form of the stethoscope. Using this method for three years, he studied patients' chest sounds and correlated them with the diseases found in autopsy. He described his methods and findings in his classic book *De l'auscultation médiate* (1819), which form the basis of the modern diagnostic technique. (☞ Aug. 13)

1888
Otto Stern, German-born American physicist, Nobel laureate, was born in Sohrau, Germany (now Zory, Poland). Stern began to study physical chemistry in 1906, receiving his PhD degree from the University of Breslau in 1912. In 1914 he went to the University of Frankfurt am Main as professor of theoretical physics. His earliest work was in the field of theoretical physics, especially that of statistical thermodynamics and quantum theory.

After 1919, his attention was directed more towards experimental physics. His development and application of the molecular beam method proved to be a powerful tool for investigating the properties of molecules, atoms and atomic nuclei. He collaborated with German physicist Walther Gerlach to work on the deflection of atoms by the action of magnetic fields on their magnetic moment, and then went on to measure the magnetic moments of subatomic particles, including the proton. In 1933 he moved to the United States, being appointed as Research Professor of Physics at the Carnegie Institute of Technology, Pittsburgh. Stern was awarded the 1943 Nobel Prize for Physics for his contribution to the development of the molecular ray method and his discovery of the magnetic moment of the proton. (☞ Aug. 17)

1932
The neutron was discovered by James Chadwick, English physicist, at the Cavendish Laboratory in Cambridge, England. In 1920, the British physicist Ernest Rutherford predicted the existence of a neutral particle in the atomic nucleus, with a mass almost that of the proton. He had also predicted that because it had no charge, there would be no electrical repulsion of the neutron as it passed through matter. This would make it more penetrating than the proton, but also more difficult to detect. The discovery the neutron came after Chadwick repeated the experiments of the French scientist couple Irène and Frédéric Joliot-Curie, who had found that when the penetrating radiation emitted from beryllium bombarded with alpha particles, hit a block of paraffin wax, it caused the wax to emit protons. After detailed study of the experiment, Chadwick had conclusive proof of the existence of the neutron in the alpha particles. (☞ Oct. 20)

1938
John Logie Baird, Scottish engineer and inventor, gave the first public demonstration of a colour television. The 'high definition' colour pictures were transmitted from the Crystal Palace South Tower via radio link to the Dominion Theatre in London where three thousand people saw them on a large screen. (☞ Jan. 27, Feb. 9, Jun. 14, Aug. 13)

1951
The Central Drug Research Institute (CDRI), a constituent national laboratory of the Council of Scientific and Industrial Research (CSIR), India was inaugurated. Located in Lucknow, Uttar Pradesh, CDRI is a multidisciplinary research laboratory that conducts research in various areas of biomedical sciences. It has developed more than a dozen new drugs, apart from diagnostics for early detection of diseases. Many cheaper and more convenient technologies for known drugs have also been developed at pilot plant level, which have been transferred to the drug industry where they have been successfully employed. New drugs developed by CDRI include Centchroman, a novel non-steroidal agent unrelated to any conventionally used contraceptive; Centbucridine, a local anaesthetic safer and more effective than the currently available drug lignocaine; Arteether, a semi-synthetic derivative of artemisinine, an anti-malarial drug effective against falciparum malaria; Isaptent, a cervical dilator for carrying out medical termination of pregnancy or abortion as well as for performing minor gynaecological surgeries; antidepressant drugs Centpropazine, and Picroliv; and Gugulipid, a cholesterol-lowering drug taking the lead from the ancient Indian system of medicine.

1965
American space probe *Ranger-8* was launched towards the Moon. (☞ Feb. 20)

1988
A new minor planet was discovered, using the 45cm Schmidt telescope at the Kavalur Observatory of the Indian Institute of Astrophysics in Tamil Nadu. It has been named 'Ramanujan' after the Indian mathematical genius Srinivasa Ramanujan. This was the first such discovery made in India in the 20th century.

1996
The American unmanned *Near Earth Asteroid Rendezvous* (*NEAR*) spacecraft was launched from the Cape Canaveral Air Station, Florida, USA, on a mission to answer important questions about the origin and composition of minor planets, comets, and our solar system. It was later re-named *NEAR Shoemaker* in honour of Eugene M. Shoemaker, the legendary geologist who influenced decades of research on the role of minor planets and comets in shaping the planets. The spacecraft was equipped with an X-ray/gamma ray spectrometer, a near-infrared imaging spectrograph, a multi-spectral camera fitted with a CCD imaging detector, a laser altimeter, and a magnetometer. In June 1997 the spacecraft sent back spectacular images of the minor planet Mathilde. In January 1998, *NEAR* flew-by Earth as part of its trajectory toward the minor planet Eros to determine its mass, structure, geology, composition, gravity, and magnetic field. The spacecraft went into orbit around Eros in February 2000. (☞ Feb. 12)

February 18

1745
Alessandro Volta, Italian physicist whose invention of the electric battery provided the first source of continuous, reliable current, was born in Como, Italy. Volta

became professor of physics at the Royal School of Como in 1774 and discovered and isolated methane gas in 1778. In 1775 Volta's interest in electricity led him to invent the electrophorus, a device used to generate static electricity. In 1780 Volta's friend Luigi Galvani discovered that contact of two different metals with the muscle of a frog resulted in the generation of an electric current. Volta began experimenting in 1794 with metals alone and found that animal tissue was not needed to produce a current. By 1799 he had developed the so-called 'voltaic pile,' a forerunner of the electric battery, which produced a steady stream of electricity. The voltaic pile consisted of an alternating column of zinc and silver disks separated by porous cardboard soaked in brine. In 1801 in Paris, he gave a demonstration of his battery's generation of electric current before Napoleon Bonaparte, Emperor of France. The volt, a unit of the electromotive force that drives current, was named in his honour in 1881. (☞ Mar. 5)

1871

Harry Brearley, English metallurgist who invented stainless steel, was born in Sheffield, England. At the age of twelve, Brearley started work as a cellar lad at the same crucible steel workshop as his father. Shortly afterwards, he was moved to the firm's laboratories, where he worked as general assistant. By the age of twenty, he was apprenticed as a laboratory assistant. In 1907 he was made in charge of the Brown-Firth Research Laboratory in Sheffield. Noticing that the internal diameter of steel rifle barrels was quickly eroded by discharge gases and high temperatures, he started investigating new steels which could resist corrosion. As a result of his investigations, using first the crucible process, and then more successfully an electric furnace, he developed, in 1912, an alloy of steel with chromium and nickel, which was much more rust resistant than the steel then in common use. Thus it was termed 'stainless steel'. The added metals in stainless steel produce a surface film of metal oxides, which resists rusting. Stainless steel, with its sleek, shiny surface and tremendous strength, is a marvel of technology. It has revolutionised most modern industries, including food, medicine, and transportation. It has also revolutionised the cutlery industry. Earlier, table cutlery had to be either silver or nickel-plated. Cutting knives of carbon steel had to thoroughly washed and dried after use, and even then rust stains would have to be rubbed off. The non-corrosive and rust-resistant properties of stainless steel have made it essential in the preparation, delivery and storage of food. (☞ Aug. 12, Sep. 5)

1894

Aleksandr Ivanovich Oparin, Russian biochemist noted for his studies on the origin of life from chemical matter, was born in Uglich near Moscow, Russia. Oparin was greatly influenced by the work of the English naturalist Charles Darwin. He majored in plant physiology at Moscow State University. By drawing on the insights of chemistry, Oparin extended the Darwinian theory of evolution backward in time to explain how simple

organic and inorganic materials might have combined into complex organic compounds. According to his hypothesis, the earliest organisms were heterotrophic; i.e., they obtained their nutrition ready-made from compounds that had already been formed by simple chemical reactions. He first introduced his concept of the origin of life at a meeting of the Russian Botanical Society in 1922. He organised the first international meeting on the origin of life in Moscow in 1957 at which representatives from 16 countries participated. A second conference was held in 1963 and a third in Pont-à-Mousson, France in 1970. Oparin put down his theory in the book *The Origin of Life on Earth.* He was awarded the UNESCO Kalinga Prize for science popularisation, jointly with the English chemist George Porter, in 1976. (☞ Apr. 21)

1913
Frederick Soddy, English chemist, introduced the term 'isotope' from Greek words meaning 'same place' to refer to different elements produced in different radioactive transformations. He suggested that the elements produced in such transformations were capable of occupying the same place on the periodic table. Such elements, he said, would differ in atomic weight while being indistinguishable and inseparable chemically. (☞ Sep. 2, Sep. 22)

1930
Clyde Tombaugh, American astronomer, discovered the ninth planet of the solar system, Pluto (now designated as a dwarf planet). The discovery culminated three decades of work at the Lowell Observatory in Flagstaff, Arizona, USA. The search for a ninth planet was started before Tombaugh was born, by the American astronomer Percival Lowell, who sought to explain deviations in the positions of Uranus and Neptune by some unknown planet-like object's gravity. Tombaugh did a systematic search by taking photographs of small sections of the night sky and then sifting through the millions of star images by day until he found one dim dot that appeared to 'move' when two photographs were compared. The new object was named Pluto after the Roman god of the underworld. (☞ Jan. 17, Feb. 4, Mar. 13)

1967
J. Robert Oppenheimer, American theoretical physicist and science administrator, who was the director of the Los Alamos laboratory during the development of the atomic bomb, died in Princeton, New Jersey, USA. (☞ Apr. 22)

1971
India's first satellite communication link was established between Arvi, Maharashtra and Goonhilly in UK.

1971
Plutonium fuel for the experimental reactor *PURNIMA* was fabricated at the Bhabha Atomic Research Centre in Trombay, Maharashtra, India.

1977
The American space shuttle *Enterprise* took off on its 'maiden flight', atop a Boeing-747 jumbo jet from the Edwards Air Force Base in California, USA. *Enterprise* was the first space shuttle orbiter that was built as a test vehicle and not equipped for space flight. It never flew in space. After rolling out of the assembly facility at Rockwell's Air Force Plant in Palmdale, California in September 1976, *Enterprise* was put through a series of approach and landing tests at Edwards Air Force Base during February–October 1977. In five free flights the astronaut crew separated the shuttle from the carrier aircraft and manoeuvred a landing at the Edwards Air Force Base.

The free flights were intended to verify the air-worthiness of the space shuttle design, test on-board systems, and manual and automatic space shuttle landing methods under conditions simulating a return from space. The tests were intended to validate aspects of the shuttle program before the actual shuttle flights began four years later. The *Enterprise* in now on display at the National Air and Space Museum's Steven F. Udvar-Hazy Center located to the east of Dulles, Washington, DC (☞ Apr. 12)

February 19

1473

Nicolaus Copernicus, Polish astronomer who proposed the Sun-centred model of the solar system, was born in Torun, Poland. Copernicus studied mathematics, philosophy, and astronomy at the University of Krakow. This experience stimulated the young Copernicus to further study liberal arts at Bologna, medicine at Padua, and law at the University of Ferrara in Italy from where he did his doctorate in canon law in 1503. Shortly afterward he returned to Poland and eventually settled permanently as a canon at the cathedral in Frauenberg where, in addition to his regular church duties, he turned his attention to astronomy – a subject in which he had long been interested. By May 1514, Copernicus had written and discreetly circulated in manuscript form the first outline of his classic work *De revolutionibus orbium coelestium* (On the Revolutions of the Heavenly Spheres), in which he challenged the Earth-centred universe that had been

dogmatically accepted since the time of Aristotle. Copernicus proposed that a rotating Earth revolving with the other planets about a stationary central Sun could explain, in a simpler way than the prevalent theory, the observed phenomena of the daily rotation of the heavens, the annual movement of the Sun through the ecliptic, and the periodic retrograde motion of the planets. Although it is assumed that Copernicus delayed publishing his book for fear of reprisal from the Church, the more probable reason could be that the work required both regular astronomical observations and intricate mathematical proofs. His administrative duties certainly interfered with both the research and the writing. The first printed copy of *De revolutionibus* is said to have reached the astronomer on his deathbed, in 1543. (☞ May 24)

1859

Svante Arrhenius, Swedish physical chemist, Nobel laureate, was born in Vik, Sweden. Arrhenius attended the Cathedral School at Uppsala and went on to the university there, where he studied physics, mathematics, and chemistry. He did his doctorate from the Academy of Sciences, Stockholm where he worked on electrolysis and published his first paper in 1883. In May 1884, at Uppsala, he defended his doctoral thesis containing the basics of his dissociation theory, which stated that electrolytes – substances that dissolve in water to yield a solution that conducts electricity – are separated, or dissociated, into electrically charged particles, or ions, even when there is no current flowing through the solution. Although ridiculed at first, his work was finally recognised and he was given a

lectureship in physical chemistry at Uppsala. In 1895 he became professor of physics at the Royal Institute of Technology in Stockholm. In 1911 he was elected as a foreign member of the Royal Society of London. Arrhenius was awarded the 1903 Nobel Prize for Chemistry in recognition of the extraordinary services he had rendered to the advancement of chemistry by his electrolytic theory of dissociation. (☞ Oct. 2)

1878

Thomas A. Edison, American inventor, received a patent for his phonograph that could reproduce recorded sound. The device used a cylinder covered with tin foil and a free-floating needle that marked a track on the revolving foil to record sound. The recorded sound could be played back by running the needle over the tracks. His first recording was made by reciting 'Mary had a little lamb' into a large horn, which transmitted vibrations to the needle that scribed a recording on the tin foil covered cylinder rotated by hand. The recorded sound came out of the horn when the cylinder was rotated again. (☞ Jan. 27, Feb. 11, Apr. 15, Aug. 8, Aug. 9, Aug. 27, Oct. 18, Oct. 21, Dec. 6)

1941

David J. Gross, American physicist and string theorist, Nobel laureate, was born in Washington, DC, USA. Gross received his bachelor's degree and master's degree from the Hebrew University of Jerusalem, Israel, in 1962, and his PhD in physics from the University of California, Berkeley, USA in 1966. After completing a fellowship at the Harvard University, Cambridge, Massachusetts, Gross accepted a position at the Princeton University in 1969, becoming professor of physics in 1972. In 1997 he joined the faculty of the University of California at Santa Barbara, where he is now a professor and the director of the Kavli Institute for Theoretical Physics. While working at Princeton University in 1973, Gross, with his first graduate student, Frank Wilczek, set out to work on the problem of how quarks are bound within the nucleus. Physicists had known that quarks, in pairs and triplets, were bound inside the atomic nucleus by the action of the strong nuclear force, but no quark had ever been released from the nucleus and observed on its own. With a series of calculations, Gross and Wilczek developed a novel explanation that seemed to run counter to logic. They theorised that the binding force between quarks in the atomic nucleus actually decreases as the quarks draw together and increases as the particles move farther apart. The concept can be likened to a rubber band, which exerts increasing amounts of energy, as the ends are pulled farther apart. At greater distances, the strong force is so powerful that no quark can break free, and that is why they have never been observed in isolation. The two termed this phenomenon 'asymptotic freedom', the discovery of which was important for the development of quantum chromodynamics. Gross was awarded the 2004 Nobel Prize for Physics, jointly with Wilczek and fellow American physicist H. David Politzer for the discovery of asymptotic freedom in the theory of the strong interaction. (☞ May 15, Aug. 31)

1943

Richard Timothy Hunt, British biochemist, Nobel laureate, was born in Neston, near Liverpool, England. Hunt attended the Clare College, Cambridge to read natural sciences and received his BA from the University of Cambridge in 1964. After receiving his PhD in 1968 from Cambridge,

Hunt went to New York, USA, and joined the Albert Einstein College of Medicine in Bronx to carry out studies on the inhibition of protein synthesis in the cell by double-stranded RNA. Since 1991 he has been the Principal Scientist at the Imperial Cancer Research Fund, London. Hunt discovered the first cyclin molecule in the early 1980s. Cyclins are proteins formed and degraded during each cell cycle. They were called cyclins because the levels of these proteins vary periodically during the cell cycle. The discovery of cyclin, which was made using sea urchins, *Arbacia*, as a model system, was the result of Hunt's finding that this protein was degraded periodically in the cell cycle. Periodic protein degradation is an important general control mechanism of the cell cycle. Hunt later discovered cyclins in other species and also found that the cyclins were conserved during evolution. Today around ten different cyclins have been found in humans. Hunt was awarded the 2001 Nobel Prize for Physiology or Medicine, jointly with the American biologist Leland H. Hartwell and British biologist Paul M. Nurse, for their discoveries of key regulators of the cell cycle. (☞ Jan. 25, Oct. 30)

1956

Roderick MacKinnon, American biophysicist, Nobel laureate, was born in Burlington, Massachusetts, USA. After high school, MacKinnon entered the Brandeis University, Waltham, Massachusetts, and received a BA in biochemistry in 1978. He received his MD from the Tufts University, Medford, Massachusetts in 1982. In 1989, MacKinnon joined the faculty at the Harvard Medical School, Boston, Massachusetts, as an assistant professor, where he continued his research and rose to the rank of full professor. In 1996, he left Harvard to join the Rockefeller University, New York, as professor and head of the Laboratory of Molecular Neurobiology and Biophysics at the Howard Hughes Medical Institute. At Rockefeller, MacKinnon used X-ray crystallography to define the structure of the ion channel vital for transportation of ions across cell membranes. In 1998, he published the first three-dimensional structure of the potassium ion channel. The discovery had immediate implications for pharmacological treatments of diseases like epilepsy and cardiac arrhythmia. MacKinnon was awarded the 2003 Nobel Prize for Chemistry, jointly with fellow American biological chemist Peter Courtland Agre, for discoveries concerning channels in cell membranes. MacKinnon received half the prize money for structural and mechanistic studies of ion channels, while Agre received the other half for the discovery of water channels. (☞ Jan. 30)

1984

The Centre for Advanced Technology (CAT) was founded. A unit of the Department of Atomic Energy, Government of India, CAT is located in the outskirts of Indore in Madhya Pradesh, and is primarily engaged in R&D in non-nuclear front line research areas of lasers, particle accelerators and related technologies. The first batch of scientists from BARC, Mumbai, moved to CAT in June 1986. Since then the centre has rapidly grown into a premier institute for work related to R&D in lasers, accelerators and their applications. In the early 1990s, CAT took up construction of the Synchrotron Radiation Sources (SRS) *Indus-1* and *Indus-2*. *Indus-1*, a 450MeV SRS, was commissioned in April 1999. The much more powerful *Indus-2*, a 2.5 GeV SRS, is expected to have a critical wavelength of about 4 angstroms. Under the laser programme, the major achievements of CAT have been in the design and development of a variety of laser systems for use in basic research, industrial applications, medicine and surgery.

1988

André F. Cournand, French-American physician and physiologist, Nobel laureate (Physiology or Medicine, 1956), who developed techniques for the catheterisation of the heart, died in Great Barrington, Massachusetts, USA. (☞ Sep. 24)

February 20

1844

Ludwig Eduard Boltzmann, Austrian physicist, was born in Vienna, Austria. Boltzmann commenced his undergraduate work at Linz and continued his study at the University of Vienna, where he earned his doctorate in 1866 under the tutelage of Josef Stefan, working on the kinetic theory of gases. After receiving his doctorate from the University of Vienna in 1866, he held professorships in mathematics and physics at Vienna, Graz, Munich, and Leipzig. Boltzmann is known as the founder of statistical mechanics – the branch of physics that makes theoretical predictions about the behaviour of macroscopic systems on the basis of statistical laws governing its component particles. In the 1870s, Boltzmann in a series of papers explained in statistical terms the second law of thermodynamics, i.e., that entropy must in general increase for an isolated system, and showed how such 'irreversible' behaviour could follow from the time-symmetric laws of mechanics. During these investigations he worked out the general law for the distribution of energy among the various parts of a system at a specific temperature. He also derived the theorem of equipartition of energy known as the Maxwell–Boltzmann law, which states that the average amount of energy used for each different direction of motion of an atom is the same. He also worked out a kinetic theory of gases, and the Stefan–Boltzmann law concerning a relationship between the temperature of a body and the radiation it emits. The fundamental constant, *k*, occurring in nearly every statistical formulation of both classical and quantum physics is named after Boltzmann. Having the dimensions of energy per degree of temperature, the Boltzmann constant has a value of 1.380662×10^{-23} joule per kelvin (K), or 1.380662×10^{-16} erg per kelvin. (☞ Sep. 5)

1873

India's first tramcar, pulled by six horses on steel rails, ran in Calcutta (now Kolkata) from Sealdah to Armenian Ghat Street, covering a distance of 2.5 kilometres.

1901

René Dubos, French-born American microbiologist, environmentalist, and author, was born in Saint-Brice, France. Dubos graduated from the Institut National Agronomique in Paris in 1921. Three years later he emigrated to the United States and continued his studies at the Rutgers University from where he obtained his PhD in 1927. He then joined the Rockefeller Institute for Medical Research in New York City, where he spent most of his career, becoming a professor in 1957 and professor emeritus in 1971. Dubos pioneered research in isolating antibacterial substances from certain soil micro-organisms and expanded his original studies of soil bacteria to include investigations of bacterial enzymes and toxins, infectious diseases, and the relationship between microbes and other life on Earth. In 1939 he isolated tyrothricin, the first commercially produced antibiotic. Dubos was a prolific writer of both scientific and popular books. In 1969 he won the Pulitzer Prize for *So Human an Animal*. Dubos' other works include *Bacterial and Mycotic Infections in Man* (1948), *Pasteur and Modern Medicine* (1960), and *Man, Medicine, and Environment* (1968). (☞ below)

1907

Henri Moissan, French chemist, Nobel laureate (Chemistry, 1906), who isolated the highly reactive gaseous element, fluorine and invented the arc furnace, died in Paris, France. (☞ Sep. 28)

1937

Robert Huber, German biochemist, Nobel laureate, was born in Munich, Germany. Huber received a diploma in chemistry in 1960 and was awarded his doctorate from the Munich Technical University. In 1972 he joined the staff of the Max Planck Institute for Biochemistry at Martinsried, Germany, where he conducted his award-winning research with the German biochemists Johann Deisenhofer and Hartmut Michel. Their research mainly concerned the use of X-ray diffraction to determine the atomic structure of complex molecules such as proteins. Huber and his team used this technique to determine the three-dimensional structure of a protein complex called the 'photosynthetic reaction centre' that is essential to photosynthesis in certain bacteria. By 1985, the three scientists had succeeded in describing the complete atomic structure of the protein, which has significantly increased our understanding of the mechanisms of photosynthesis in plants, too. Huber was awarded the 1988 Nobel Prize for Chemistry, jointly with Deisenhofer and Michel, for the determination of the three-dimensional structure of a photosynthetic reaction centre. (☞ Jul. 18, Sep. 30)

1962

The first American orbital space mission was launched from the Kennedy Space Center, Florida, with *Friendship-7*, the Mercury spacecraft carrying astronaut John Glenn as the one-man crew. The spacecraft spent 4 hours 30 minutes in orbit, completing three orbits around the Earth, at a maximum altitude of about 260 kilometres.

Friendship-7

1965

American space probe *Ranger-8* crashed on the Moon in the Mare Tranquillitatis. It sent back 7,137 photographs of the lunar surface before impact. (☞ Feb. 17)

1972
Maria Goeppert-Mayer, German-born American physicist, Nobel laureate (Physics, 1963), who proposed the shell model of the nucleus, died in San Diego, California, USA. (☞ Jun. 28)

1982
René Dubos, French-born American microbiologist, environmentalist and author, whose pioneering research in isolating antibacterial substances from certain soil micro-organisms led to the discovery of major antibiotics, died in New York, USA. (☞ above)

1986
The first component of the Russian space station *Mir* – a 20-tonne core module with six docking ports for transport craft along with add-on modules – was launched. The complete station was built up by piecing together additional mo-dules over the next ten years. *Mir*'s first crew, Leonid Kizim and Vladimir Solovyou, made space history when they left *Mir* to travel in a *Soyuz* spacecraft to another orbiting station *Salyut-7* for two months. They later returned to *Mir*. It was the first ever space-station-to-space-station flight. The second component of the space station – the *Kvant-1* module – was joined in March 1987. It contained astrophysics equipment, more living space and gyroscopes for *Mir*'s stability. It eventually became a base for two extendable truss structures and additional solar panels. The rest of the space station was built up gradually till 1996. More than 100 cosmonauts from 12 countries occupied *Mir* almost continuously. One Russian cosmonaut, Valeri Polyakov, stayed aboard for 438 consecutive days. Another, Sergei Avdeyev, spent a total of more than two years in space over three flights. *Mir* remained 15 years in orbit – 10 years longer than originally planned. During its stay in orbit, *Mir* circled the Earth about 88,000 times, covering a distance of 3.6 billion kilometres. But on-board problems and near-fatal accidents made the Russian space authorities abandon the station and guide it to a safe re-entry in March 2001. Although most of the space station burnt up on re-entry, 25 tonnes of wreckage finally streaked across the sky and hit the intended target area in the South Pacific, 5,800 kilometres off the eastern coast of Australia. (☞ Mar. 23)

Mir

February 21

1894
Shanti Swarup Bhatnagar, Indian chemist and science administrator, Fellow of the Royal Society, was born in Bhera, Punjab, India. Educated at the Punjab University from where he received his BSc and MSc degrees, Bhatnagar obtained his DSc from the University of London, England, in 1921. After serving as professor of chemistry at the newly opened Benaras Hindu University for three years, Bhatnagar was appointed Director of the University Chemical Laboratories at Punjab University in 1924, a position he held till 1940. Bhatnagar's research spanned colloids, magneto-chemistry, industrial, and practical chemistry. Early in his career he realised that the

application of magnetism to chemistry had great possibilities. He used magnetic susceptibility measurements to study the properties of organic compounds, solutions, films and colloids. Bhatnagar was appointed the first Director of the Board of Scientific and Industrial Research, which was to later become the Council of Scientific and Industrial Research. By the time he died in 1955, Bhatnagar had created one of the finest science and technology edifices of independent India – a chain of twelve discipline-based laboratories ranging from physics and chemistry to leather and electrochemicals. (☞ Jan. 1)

1895

Carl Peter Henrik Dam, Danish biochemist, Nobel laureate, was born in Copenhagen, Denmark. Dam graduated in chemistry from the Polytechnic Institute, Copenhagen in 1920, and was appointed instructor in chemistry at the School of Agriculture and Veterinary Medicine the same year. In 1925 Dam studied microchemistry in Graz (Austria) and then became Associate Professor at the Institute of Biochemistry, Copenhagen University in 1929. During 1929–1934, Dam and his associates at the Copenhagen University demonstrated a deficiency disease in chicks characterised by a tendency to bleed and an increased blood-clotting time. He ascribed the disease to lack of an anti-haemorrhagic vitamin, which he later showed to be fat-soluble and present in green leaves. He named it vitamin K for '*Koagulations Vitamin*'. In 1940, he went to the United States, lecturing and continuing his research, mainly at the University of Rochester, New York. After his return to the Polytechnic Institute, Copenhagen in 1946, Dam's main research subjects were vitamin K, vitamin E, fats, cholesterol, and, in later years, nutritional studies in relation to gall-stone formation. Dam was awarded the 1943 Nobel Prize for Physiology or Medicine, jointly with the American biochemist Edward Adelbert Doisy, for his discovery of vitamin K. (☞ Apr. 17, Nov. 13)

1926

Heike Kamerlingh Onnes, Dutch physicist, Nobel laureate (Physics, 1913), who discovered superconductivity, died in Leiden, Netherlands. (☞ Jul. 10, Sep. 21)

1927

Ashesh Prosad Mitra, Indian radio physicist, Fellow of the Royal Society, was born in Calcutta (now Kolkata), West Bengal, India. After graduating from Presidency College, Mitra completed his D.Phil. in 1955 from the University of Calcutta. From 1952–1968, he was Visiting Assistant Professor, Engineering Research, at the Pennsylvania State University, USA. Mitra's major work has been in the field of Earth's near-space environment, through ground-based and space techniques. His pioneering work on cosmic radio noise for studying the upper atmosphere led to a series of discoveries in ionosphere, solar physics and cosmic rays. Mitra also introduced tropospheric radio research in India in the 1970s, and has contributed to major improvements in radio communication capabilities in the country. He initiated new experimental techniques for ionospheric and tropospheric monitoring, developed an ionospheric prediction system that has supported broadcasting and point-

to-point communication systems over the past three decades. Mitra was the driving force behind the Indian programme of the International Geophysical Year 1957–58, the International Quiet Sun Year 1964–1965, and the International Geosphere–Biosphere Program in the 1990s. He initiated research in ionospheric physics at the National Physical Laboratory (NPL) in New Delhi. Mitra was the Director, NPL (1982–86) and Director-General of the Council of Scientific and Industrial Research (1986–91).

1938
George Ellery Hale, American astronomer known for his development of important astronomical instruments, including the 200-inch (508-cm) Hale telescope of the observatory on Mt. Palomar, died in Pasadena, California, USA. (☞ Jun. 29)

1941
Frederick Grant Banting, Canadian physician, Nobel laureate (Physiology or Medicine, 1923), who, with the American physiologist Charles Herbert Best, extracted the hormone insulin from pancreas to treat diabetes, died in Newfoundland, Canada. (☞ Nov. 14)

1968
Howard Walter Florey, Australian pathologist, Nobel Laureate (Physiology or Medicine, 1945), who isolated and purified penicillin for clinical use, died in Oxford, UK. (☞ Sep. 24)

1999
Gertrude Belle Elion, American pharmacologist, Nobel laureate (Physiology or Medicine, 1988), who synthesised and co-developed two of the first successful drugs for the treatment of leukaemia, died in Chapel Hill, North Carolina, USA. (☞ Jan. 23)

February 22

1828
Friederich Wöhler, German biochemist, for the first time synthesised the organic chemical, urea from ammonium cyanate, an inorganic compound. This was a landmark event, for it was the first time a material previously only associated with the body function of a living organism was made from inorganic chemicals of non-living origin. (☞ Jul. 31, Sep. 23)

1857
Heinrich Rudolf Hertz, German physicist who was the first to broadcast and receive radio waves, was born in Hamburg, Germany. Hertz studied under the German physicists Gustav Robert Kirchhoff and Hermann von Helmholtz in Berlin, and became professor at Bonn in 1889. Hertz clarified and expanded the electromagnetic theory of light that had been put forth by the British physicist James Clerk Maxwell in 1884. He also proved that electricity could be transmitted in electromagnetic waves, which travel at the speed of light and which possess many other properties of light. Hertz generated electric waves by means of the oscillatory discharge of a condenser through a loop provided with a spark gap, and then detected them with a similar type of circuit. His experiments with these electromagnetic waves led to the development of the wireless telegraph and the radio. The unit of frequency, Hertz, which is one cycle per second, is named after him. (☞ Jan. 1)

1879

Johannes Nicolaus Brønsted, Danish physical chemist, Fellow of the Royal Society, known for the widely applicable acid–base concept, was born in Varde, Denmark. Brønsted received a degree in chemical engineering in 1899 from the Polytechnic Institute in Copenhagen and began studies in pure science at the University of Copenhagen.

He took his master's degree in chemistry in 1902. In 1905 he started to work at his 'affinity' measurements on binary mixtures of sulphuric acid and water, which formed his doctoral thesis. He received his PhD in 1908 from the University of Copenhagen and was immediately appointed professor of inorganic and physical chemistry. In 1923, within several months of each other, Brønsted and the English chemist Thomas Martin Lowry published essentially the same theory about how acids and bases behave. Since they came to their conclusions independently of each other, both names have been used for the theory name. According to the Brønsted–Lowry theory, an acid is defined as a compound tending to give up a proton (or hydrogen ion), while a base is one tending to combine with a proton. Brønsted was also an authority on the catalytic properties and strengths of acids and bases and did important work with electrolyte solutions. (☞ Dec. 17)

1914

Renato Dulbecco, Italian virologist, Nobel laureate, was born in Catanzaro, Italy. Dulbecco graduated from high school at 1930 and obtained his MD from the University of Turin in 1936 where he remained for several years as a member of its faculty. He went to the United States in 1947 and studied viruses, first with Italian-born American virologist Salvador Luria at the Indiana University, then at the California Institute of Technology. He moved from Caltech to the Salk Institute in 1962 and in 1972 to the Imperial Cancer Research Fund Laboratories in London to work on human cancer. While at the Salk Institute, Dulbecco worked with Marguerite Vogt, a professor in the Molecular and Cell Biology Laboratory, and pioneered the growth of animal viruses in culture. They investigated how certain viruses gain control of the cells they infect and found that virus replication either led to a destruction of cells concomitant with a release of newly produced virus particles, or a transformation of cells by incorporation of the viral DNA. Dulbecco suggested that human cancers could be caused by similar reproduction of foreign DNA fragments. Dulbecco was awarded the 1975 Nobel Prize for Physiology or Medicine, jointly with American virologists Howard M. Temin and David Baltimore, for their discoveries concerning the interaction between tumour viruses and the genetic material of the cell. (☞ Mar. 7, Dec. 10)

1936

John Michael Bishop, American virologist, Nobel laureate, was born in York, Pennsylvania, USA. Bishop graduated from the Gettysburg College in 1957 and from the Harvard Medical School in 1962. Between 1964 and 1968, he worked as a researcher at the

National Institutes of Health in Bethesda, Maryland, before joining the faculty of the University of California at San Francisco in 1968. In 1970 Bishop teamed up with fellow virologist Harold Varmus, and set out to test the theory that healthy body cells contain dormant viral oncogenes that, when triggered, cause cancer. Working with the Rous sarcoma virus, known to cause cancer in chickens, Bishop and Varmus found that a gene similar to the cancer-causing gene within the virus was also present in healthy cells. Subsequent research showed that even without viral involvement, certain chemical carcinogens could convert these genes into a form that allows uncontrolled cellular growth and cause cancer. Bishop was awarded the 1989 Nobel Prize for Physiology or Medicine, jointly with Varmus, for their discovery of the cellular origin of retroviral oncogenes. (☞ Dec. 18)

1946
Selman Abraham Waksman, Ukrainian-born American biochemist, Nobel laureate (Physiology or Medicine, 1952), announced his discovery of the antibiotic streptomycin. (☞ Jul. 22, Aug. 16)

2004
The capsule from the American spacecraft *Genesis*, carrying samples of solar wind particles, returned to Earth. The original plan to receive the capsule in air did not work, as the parachutes failed to deploy and the capsule crashed in the deserts of Utah, New Mexico. (☞ Aug. 8)

February 23

1855
Carl Friedrich Gauss, German mathematician, astronomer and physicist who developed the theory of numbers and who applied mathematics to electricity, magnetism, astronomy and geodesy, died in Göttingen, Germany. (☞ Apr. 30)

1886
Charles Martin Hall, American chemist, successfully extracted aluminium from its ore by an electrolytic process. He dissolved the alumina ore in a bath of molten cryolite, a mineral containing fluorine, sodium and aluminium, and passed electric current through the solution. He patented the process, which was the first to become an inexpensive, commercial process for extracting aluminium. (☞ Dec. 6, Dec. 27)

1904
Mahendra Lal Sircar, Indian doctor and philanthropist, who founded the Indian Association for the Cultivation of Science, died in Kolkata, West Bengal, India. (☞ Nov. 2)

1924
Allan MacLeod Cormack, South African-born American physicist, Nobel laureate, was born in Johannesburg, South Africa. After completing his bachelor's and master's degrees at Cape Town, Cormack went to St. John's College, Cambridge, UK, as a research student. In 1950 he returned to Cape Town as a lecturer and remained there till 1956. Then, after a year's research fellowship at the Harvard University, Cambridge, Massachusetts, he was appointed assistant professor of physics at the Tufts University in Massachusetts, USA, in 1958. It was during a part-time position as physicist for the radiology department of the Groote Schuur Hospital in Cape Town that Cormack's

interest was first aroused in the problem of X-ray imaging of soft tissues or layers of tissue of differing densities. Cormack's master's thesis was in X-ray crystallography, and his interest in X-ray technology led him to develop the theoretical foundations that formed the essence of 'computerised axial tomography'. He published his results in two papers in the *Journal of Applied Physics* in 1963 and 1964. The papers generated little interest until 1972 when the first CT scan machine was built under the leadership of the English electrical engineer Godfrey Hounsfield in England. Cormack was awarded the 1979 Nobel Prize for Physiology or Medicine, jointly with Hounsfield, for the development of computer-assisted tomography. Cormack was unique among the Nobel laureates because he never earned a doctorate degree in medicine or any other field of science. (☞ May 7, Aug. 28)

1944
Leo Hendrik Baekeland, American industrial chemist who helped found the modern plastics industry through his invention of Bakelite, died in New York, USA. (☞ Nov. 14)

1973
Dickinson Woodruff Richards, American physiologist, Nobel laureate (Physiology or Medicine, 1956), who developed a technique for catheterisation of the heart, died in Lakeville, Connecticut, USA. (☞ Oct. 30)

February 24

1582
The Gregorian calendar was introduced by Pope Gregory XIII, replacing the old Julian calendar. In the Gregorian calendar, the tropical year is approximated as 365 $^{97}/_{400}$ days = 365.2425 days. Thus it takes approximately 3,300 years for the tropical year to shift one day with respect to the Gregorian calendar. The approximation 365 $^{97}/_{400}$ is achieved by having 97 leap years every 400 years. In shifting from the Julian to the Gregorian system, 10 days were dropped from October 1582 so that October 15 followed immediately after October 4, and from then on the reformed calendar has been used.

1663
Thomas Newcomen, English engineer and inventor of the atmospheric steam engine was born in Dartmouth, Devon, England. While working as an ironmonger in his native town, Newcomen corres-ponded with the English physicist Robert Hooke about the applicability of steam-power for the purpose of driving machinery. He then built an engine with a piston working in a vertical cylinder and a massive overhead rocking beam connec-ted to the mine pumps. In 1712 the first practical steam engine in the world was set to work at a South Staffordshire colliery in England. It had a piston connected to one end of a large crossbeam; the other end was connected to a very heavy pump piston. On each stroke, water chilled and condensed the steam in the cylinder, dropping the piston, thus moving the crossbeam and operating the pump. It worked at twelve strokes a minute, raising ten gallons (about 45 litres) of water from a depth of 156 feet (about 48 metres). Although it was slow and inefficient, Newcomen's engine was built in almost

Newcomen's engine

every mining area of Britain. Their use quickly spread to other countries in Europe and Scandinavia. The engines were also employed for water supply purposes. (☞ Aug. 5)

1810
Henry Cavendish, English physicist and chemist, who conducted experiments in diverse fields and discovered the composition of air, died in London, UK. (☞ Oct. 10)

1815
Robert Fulton, American inventor, engineer, and artist, who invented the first practical steamboat, died in New York, USA. (☞ Feb. 9, Aug. 14, Nov. 14)

1871
Charles Darwin's *Descent of Man* was published in London. In his Introduction, Darwin wrote, 'by this work light would be thrown on the origin of man and his history; and this implies that man must be included with other organic beings in any general conclusion respecting his manner of appearance on this earth'. (☞ Feb. 12, Apr. 19)

1923
Edward Williams Morley, American chemist who is best known for his collaboration with the German-born American physicist Albert Abraham Michelson in what has come to be known as the Michelson–Morley experiment, died in West Hartford, Connecticut, USA. (☞ Jan. 29)

1938
First commercial use of nylon began, with DuPont in USA starting commercial production of nylon toothbrush bristles. (☞ Sep. 20)

1968
The discovery of a 'pulsar' was announced by Cambridge graduate student Jocelyn Bell (now Burnell) and British astrophysicist Antony Hewish in a paper published in the journal *Nature*. Bell and Hewish were using a special radio telescope to look for radio scintillation – fluctuations in the signals from distant radio sources produced by turbulence in interplanetary space. In November 1967, Bell noticed a strange regular 'beep' coming from the constellation of Vulpecula every 1.3 seconds. Puzzled by what sort of object could produce such a regular pattern, Bella and Hewish initially dubbed it 'LGM' – standing for 'little green men,' because the only source that they could imagine that could be so regular was some sort of extraterrestrial technical civilisation. After a few weeks, however, three more rapidly pulsating sources were detected, all with different periods. They were name 'pulsars'. These were later identified as rapidly spinning neutron stars – remnants of supernova explosions – that radiate beams of radio waves like a cosmic lighthouse. (☞ May 11, Nov. 28)

1969
The American unmanned spacecraft *Mariner-6* was launched from the Kennedy Space Center, Florida, USA, towards the planet Mars. Weighing 413 kilograms, *Mariner-6* was sent to fly-by

Mariner-6

Mars in order to collect data about the surface and atmosphere of the Red Planet. The spacecraft was designed to fly over the equator and southern hemisphere of Mars and was capable of continuous telemetry transmission. (☞ Jul. 31)

1987
Supernova 1987A in Tarantula Nebula in the Large Magellanic Cloud, the galaxy closest to us outside our Milky Way, was discovered by Canadian astronomer Ian Shelton, at Las Campanas Observatory, Chile. The brightest of the supernovae in the 20th century, it was the first supernova visible to the naked eye since 1604.

February 25

1950
George Richards Minot, American physician, Nobel laureate (Physiology or Medicine, 1934), who introduced the raw liver diet in the treatment of anaemia, died in Brookline, Massachusetts, USA. (☞ Dec. 2)

1971
Theodor H.E. Svedberg, Swedish chemist, Nobel laureate (Chemistry, 1926), who invented the ultracentrifuge for the study of colloids, died in Örebro, Sweden. (☞ Aug. 30)

1988
The first test launch of India's indigenous *Prithvi* missile was carried out at Sriharikota, Andhra Pradesh. The 8.5-metre-tall *Prithvi* missile is powered by two liquid propellant rocket engines designed and developed by the Defence Research and Development Organisation (DRDO). The missile has a maximum range of 250 kilometres with a 500 kg warhead, and a reduced range of 150 kilometres with an enhanced 1000 kg warhead. It can carry conventional and nuclear warheads. Its uniqueness lies in the fact that a range of warheads can be fitted on to the missile, and can be changed under battlefield conditions. There are three separate versions of the *Prithvi* for the army, navy and air force. *Prithvi*'s navy and air force versions have an extended range of 250 kilometres. The development of the 150-km-range army version of *Prithvi* was completed successfully after 16 flight trials. It has already been inducted into the Indian army. *Prithvi* is part of the prestigious Integrated Guided Missile Development Programme under the auspices of DRDO.

1999
Glenn T. Seaborg, American nuclear chemist, Nobel laureate (Chemistry, 1951), who produced nine transuranic elements, died in Lafayette, California, USA. (☞ Apr. 19)

February 26

1903
Giulio Natta, Italian chemist, Nobel laureate, was born in Imperia, near Genoa, Italy. Natta obtained his doctorate in chemical engineering from the Milan Polytechnic in 1924 and then held chairs in chemistry at Pavia, Rome and Turin universities, before returning to the Milan

Polytechnic in 1938 as professor and research director of industrial chemistry. Here, he developed new uses for the commercial synthesis of methanol, formaldehyde, succinic acid and butyraldehyde. His intensive interest in and studies of polymers in conjunction with the Italian firm Montecatini led in 1953 to his use of special catalysts for the polymerisation of propylene and production of polypropylenes of highly regular molecular structure. The contribution of Natta to the development of high polymers had a profound impact on the manufacture of films, fibres, synthetic rubber. Natta was awarded the 1963 Nobel Prize for Chemistry, jointly with the German chemist Karl Ziegler, for their discoveries in the field of the chemistry and technology of high polymers. (☞ May 2, Nov. 26)

1931
Otto Wallach, German chemist, Nobel laureate (Chemistry, 1910), whose work laid the scientific basis for the modern perfume industry, died in Göttingen, Germany. (☞ Mar 27)

1935
Radar (*RA*dio *D*etection *A*nd *R*anging) was first demonstrated by the Scottish physicist Robert Watson-Watt at Daventry, England. The first successful trial took place using a British Broadcasting Service short wave radio transmitter to track a Royal Air Force bomber. By 1939 England had established a chain of radar stations along its south and east coasts to track enemy aircraft. In the same year two British scientists, physicist Henry Boot and biophysicist John T. Randall, invented an electron tube called the resonant-cavity magnetron, which could be used to generate high-frequency radio pulses with large amounts of power, thus permitting the development of microwave radar that was the most important advance made in the technology of radar during World War II.

1946
Ahmed Zewail, Egyptian chemist, Nobel laureate, was born in Damanhur, Egypt. Zewail received his BS degree in 1967 and MS degree in 1969 from the University of Alexandria. He then went to USA and joined the University of Pennsylvania, where he earned a PhD in 1974. In 1976 he joined the faculty at the California Institute of Technology at Berkeley and in 1990 was selected as the institute's first Linus Pauling professor of chemical physics. Zewail also served as a visiting professor at a number of institutions, including the Texas A&M University, the University of Iowa, and the American University at Cairo. Zewail's research mainly concerned the study of fast chemical reactions. At the end of the 1980s he performed a series of experiments that were to lead to the birth of a research area called 'femtochemistry'. It involved using a high-speed camera to image molecules in the actual course of chemical reactions and trying to capture pictures of them in the transition state. The camera was based on new laser technology with light flashes of some tens of femtoseconds (one femtosecond is one quadrillionth of a second or one thousandth of a nanosecond). The technique enabled chemists to study substances, termed as intermediates, formed along the way from the original one to the final product during a chemical reaction. Zewail was awarded the 1999 Nobel Prize for Chemistry

for his studies of the transition states of chemical reactions using femtosecond spectroscopy.

February 27

1879
Saccharin, the artificial sweetener, was discovered by the American chemists Ira Remsen and Constantin Fahlberg at Johns Hopkins University in Baltimore, Maryland, USA, while investigating the oxidation of o-toluenesulphonamide.

1899
Charles Herbert Best, American physiologist who, with Canadian physician Frederick Banting, was the first to obtain insulin hormone from pancreatic extract in a form that controlled diabetes in dogs, was born in West Pembroke, Maine, USA. In May 1921, while still an undergraduate, Best became a laboratory assistant to Banting at the University of Toronto, where they carried on their studies on pancreatic extracts. Best also discovered the vitamin, choline and the enzyme, histaminase. He was the first to introduce anticoagulants in the treatment of thrombosis (blood clots). Because Best did not receive his medical degree until 1925, he did not share the 1923 Nobel Prize for Physiology or Medicine, which was jointly awarded to Banting and the Scottish physiologist John James R. Macleod for their researches on the use of insulin in the treatment of diabetes. (☞ Mar. 31, Apr. 15, Sep. 6, Nov. 14)

1900
Aspirin was patented by German chemist Felix Hoffmann. Aspirin or acetylsalicylic acid is a derivative of salicylic acid that is a mild, non-narcotic analgesic useful in the relief of headache and muscle and joint aches. The drug works by inhibiting the production of prostaglandins, body chemicals that are necessary for blood clotting and which also sensitise nerve endings to pain. Hoffmann, who worked for a German company called Bayer, gave acetylsalicylic acid to his father who was suffering from arthritic pain, with good results. Hoffmann then convinced Bayer to market the new wonder drug. The name Aspirin is derived from the 'A' in acetyl chloride, the 'spir' in *Spiraea ulmaria* (the plant salicylic acid is derived from) and the 'in', a then familiar name ending for medicines.

1926
David Hunter Hubel, Canadian-born American neurobiologist, Nobel laureate, was born in Ontario, Canada. Hubel studied at the McGill University in Montreal. In 1958 he moved to the Wilmer Institute, Johns Hopkins Hospital, where he began collaboration with Swedish neurobiologist Torsten Wiesel. A year later the entire laboratory moved to the Harvard Medical School in Boston, where Hubel became a professor in 1965. At Harvard, Hubel and Wiesel began experiments implanting electrodes into the brain of anaesthetised cats, correlating the anatomical structure of the visual cortex of the brain with the physiological responses to different types of visual stimulation. Their most famous studies were in the area of visual perception, with particular emphasis

on the nerve impulses mediating between the retina and the brain. They found that various nerve cells were responsible for different types of visual stimuli. Hubel was awarded the 1981 Nobel Prize for Physiology or Medicine, jointly with Wiesel, for mapping the path of nerve impulses from the eye to various centres of the brain. (☞ Jun. 3)

1932
A report on the discovery of the neutron by English physicist James Chadwick was published in the journal *Nature*. (☞ Jul. 24, Oct. 20)

1936
Ivan Petrovich Pavlov, Russian physiologist, Nobel laureate (Physiology or Medicine, 1904), known chiefly for his development of the concept of the conditioned reflex, died in Leningrad (now St. Petersburg), USSR. (☞ Sep. 14)

1942
British electronic engineer J.S. Hey discovered radio waves from the Sun while working on radars with the British Army during World War II.

1989
Konrad Lorenz, Austrian zoologist, Nobel laureate (Physiology or Medicine, 1973), founder of modern ethology, died in Altenburg, Austria. (☞ Nov. 7)

1997
Ian Wilmut, Scottish researcher at the Roslin Institute, announced the birth of a cloned adult sheep. The resulting lamb named Dolly died six years later. (☞ Jul. 5)

1998
George Herbert Hitchings, American pharmacologist, Nobel laureate (Physiology or Medicine, 1988), who developed several drugs essential in the treatment of several major diseases, died in Chapel Hill, North Carolina, USA. (☞ Apr. 18)

February 28

1849
Joseph Freiherr von Mering, German physician, pharmacologist, and experimental pathologist, whose work led to the discovery that insulin is synthesised in the pancreas, was born in Cologne, Germany. While at the University of Strassburg, in 1889, Mering and fellow physiologist Oskar Minkowski were trying to disrupt fat digestion in dogs by removing the pancreas, the only organ known to secrete a fat-degrading enzyme. But what they found surprised them; rather than effecting fat digestion, the animals exhibited the symptoms of diabetes, characterised by a rise in blood sugar level. This led Mering and Minkowski to propose that the pancreas secreted some anti-diabetic substance that enabled the body to use blood sugar properly, though they did not isolate it. The term insulin was coined by the British physiologist Edward Albert Sharpley-Schafer in 1915, some years before its eventual isolation by the Canadian physiologists Frederick Banting and Charles S. Best in 1921. Now it is known that insulin, the hormone synthesised in the pancreas, is important for the body to make proper use of sugar. As director of the medical clinic in Halle from 1900 to 1908, Mering also worked with the German chemist Emil Fischer to develop the barbiturates known as barbital, veronal, and proponal. (☞ Jan. 5, Jan. 13)

1896
Philip Showalter Hench, American physician, Nobel laureate, was born in Pittsburgh, Pennsylvania, USA. After attending local schools, Hench entered the

Lafayette College, Easton, Pennsylvania, where he graduated in 1916. In 1920 he received his doctorate in medicine from the University of Pittsburgh. After a year as an intern at the Saint Francis Hospital, Pittsburgh, he became a Fellow of the Mayo Foundation, the graduate school of the University of Minnesota's Department of Medicine. His association with the Mayo Clinic, Rochester, Minnesota, began in 1923 when he became first an assistant, then, three years later, head of its Department of Rheumatic Diseases. At the Mayo Clinic he specialised in arthritic disease. In the course of his work he observed the favourable effects of jaundice on arthritic patients which caused a remission of pain. Other bodily changes, for example pregnancy, produced the same effect. These and other observations led him gradually to the conclusion that the pain-alleviating substance was a steroid. In the period 1930–1938, the American chemist Edward Calvin Kendall had isolated several steroids from the adrenal gland cortex. After several years of collaboration with Kendall, it was decided in 1948–1949 to try the effect of one of these substances, later named cortisone, on arthritic patients. The result turned out to be highly successful. Hench was awarded the 1950 Nobel Prize for Physiology or Medicine, jointly with Kendall and the Swiss chemist Tadeus Reichstein, for their discoveries relating to the hormones of the adrenal cortex, their structure and biological effects. (☞ Mar. 8, Mar. 30, Jul. 20)

1901

Linus Pauling, American chemist, Nobel laureate, was born in Portland, Oregon, USA. After receiving his PhD from the California Institute of Technology, Pasadena, Pauling worked in the laboratories of such noted scientists as Arnold Sommerfeld in Munich, Niels Bohr in Copenhagen, Erwin Schrödinger in Zürich, and Sir William Henry Bragg in London as a postdoctoral fellow. He returned to the California Institute of Technology as assistant professor of chemistry in 1927, becoming full professor in 1931. Pauling was among the first to apply the principles of quantum mechanics to the structure of molecules and effectively utilise X-ray diffraction, electron diffraction, and magnetic effects for the calculation of inter-atomic distances and the angles between chemical bonds. His work also dealt with the many aspects of molecular structure, ranging from simple molecules to proteins. Pauling was awarded the 1954 Nobel Prize for Chemistry, for his research into the nature of the chemical bond and its application to the elucidation of the structure of complex substances. Pauling was also awarded the 1962 Nobel Prize for Peace in recognition of his efforts on behalf of the international control of nuclear weapons and his campaigns against nuclear testing. (☞ Jan. 13, Aug. 19)

1915

Peter B. Medawar, Brazilian-born British zoologist, Nobel laureate, was born in Rio de Janeiro, Brazil. Medawar moved to England as a young boy and in 1935 took a degree in zoology from the Magdalen College, Oxford. Thereafter, Medawar worked for a time at Sir Howard Florey's School of

Pathology at Oxford. It was here that he became interested in research in fields of biology that were related to medicine. During World War II at the Burns Unit of the Glasgow Royal Infirmary in Scotland, he carried out research on tissue transplants, particularly skin grafting. This work led him to recognise that graft ejection is an immunological response. After the war, Medawar continued his transplant research and learned of the work done by Australian immunologist Frank Macfarlane Burnet, who first advanced the theory of acquired immunological tolerance. According to Burnet's hypothesis, during early embryological development and soon after birth, vertebrates develop the ability to distinguish between 'self' and 'non-self.' Medawar lent support to Burnet's theory when he found that fraternal cattle twins accept skin grafts from each other. Medawar was jointly awarded the 1960 Nobel Prize for Physiology or Medicine with Burnet for discovery of acquired immunological tolerance. (☞ Sep. 3, Oct. 2)

1928
Chandrasekhara Venkata Raman, Indian physicist, Nobel laureate, announced his discovery of the effect named after him. While studying the phenomenon of scattering of monochromatic light by different transparent media at the Indian Association for the Cultivation of Science in Calcutta (now Kolkata), Raman made the unexpected and highly surprising discovery that the scattered light showed not only the wavelengths of the original monochromatic light but also wavelengths longer and shorter than the primary light. The effect came to be known as the 'Raman effect'. The Raman effect opened up new routes to the study of the structure of matter. (☞ Nov. 7, Nov. 21)

1930
Leon N. Cooper, American physicist, Nobel laureate, was born in New York, USA. After receiving his PhD degree from the Columbia University, Cooper spent a year at the Institute for Advanced Study and taught at the University of Illinois and the Ohio State University before joining the faculty of the Brown University in Rhode Island in 1958. He carried out his studies on the theory of superconductivity while he was still in his 20s. His main contribution to the BCS theory (named after its creators, Bardeeb, Cooper and Schrieffer) was the discovery in 1956 that electrons, which under normal conditions repel each other, are attracted to each other in superconductors, a phenomenon termed as 'Cooper pairs.' Cooper was awarded the 1972 Nobel Prize for Physics, jointly with fellow Americans John Bardeen and John Robert Schrieffer, for his role in developing the BCS theory of superconductivity. (☞ May 23, May 31)

1936
Charles-Jules-Henri Nicolle, French bacteriologist, Nobel laureate (Physiology or Medicine, 1928), who discovered that typhus is transmitted by the body louse, died in Tunis, Tunisia. (☞ Sep. 21)

1939
Daniel C. Tsui, Chinese-born American physicist, Nobel laureate, was born in Honan province, China. Tsui graduated from the Augustana College, Rock Island, Illinois, USA, in 1961, and obtained a PhD in physics from the University of Chicago in 1967. He then joined the Bell

Laboratories at Murray Hill, New Jersey, where he and the German-born American physicist Horst L. Störmer discovered in 1982 that the electrons in a powerful magnetic field at very low temperatures could form a quantum fluid with particles having fractional electric charges. Tsui and Störmer's research at Bell Laboratories was based on the 'Hall effect', which is the development of a crosswise electric field in a current-carrying ribbon whose surface lies perpendicular to a strong magnetic field. This transverse electric field results from the force that the magnetic field exerts on the moving particles of the electric current. This effect discovered by Tsui and Störmer came to be known as the 'fractional quantum Hall effect.' Tsui was awarded the 1998 Nobel Prize for Physics, jointly with Störmer and American physicist Robert B. Laughlin for their discovery of a new form of quantum fluid with fractionally charged excitations. (☞ Apr. 6, Nov. 1)

1948

Steven Chu, American physicist, Nobel laureate, was born in St. Louis, USA. Chu was awarded his doctorate in physics in 1976 from the University of California, Berkeley, where he was a postdoctoral fellow from 1976 to 1978. He joined the staff at Bell Laboratories in 1978 and became the head of the quantum electronics research department at AT&T Bell Laboratories in 1983. He joined the faculty of the Stanford University in 1987. In 1985 Chu and his co-workers used an array of intersecting laser beams to slow down atoms. As a result, the atoms cooled down to temperatures as low as –273.15°C. Chu and his colleagues also developed an atomic trap using lasers and magnetic coils that enabled them to capture and study the chilled atoms. By 1985, he had cooled sodium atoms to a temperature of about 240 millionth of a degree above absolute zero. Chu was awarded the 1997 Nobel Prize for Physics, jointly with fellow American physicist William D. Phillips and French physicist Claude Cohen-Tannoudji for his pioneering research in cooling and trapping atoms using laser light. (☞ Apr. 1, Nov. 5)

2005

American millionaire businessman and pilot Steve Fossett took off from the Salina Municipal Airport in Kansas, USA, on Virgin Atlantic *GlobalFlyer* – a single-engine jet plane – on a mission to complete the first solo non-stop flight around the world without re-fuelling. *GlobalFlyer* was powered by a single turbofan jet engine, fed by thirteen separate fuel tanks on the aircraft. The plane's unique structure made of lightweight material permitted the take-off weight to comprise 82% of fuel. The stiffest carbon fibres were used in the construction of the craft's long wings, and the skin was a sandwich of graphite/epoxy and Aramid honeycomb. (☞ Mar. 3)

Global flyer

February 29

1860

Herman Hollerith, American inventor of the tabulating machine that was an important precursor of the modern electronic computer, was born in Buffalo, USA. For the first time in 1890, Hollerith used punched cards for computation when he was working for the US Census Bureau. He developed a device that could automatically read census information, which had been punched onto the card. The machine, which read the cards, used a pin going through a hole in the card to make an electrical connection with mercury placed beneath. The resulting electrical current activated a mechanical counter. It saved the US 5 million dollars for the 1890 census by completing the analysis of the data in a fraction of the time it would have taken without it and with a smaller amount of manpower than would have been necessary otherwise. As a result of his invention, reading errors were consequently greatly reduced, workflow was increased and, more important, stacks of punched cards could be used as an accessible memory store of almost unlimited capacity. Furthermore, different problems could be stored on different batches of cards and worked on as needed. Hollerith's tabulator became so successful that he started his own firm, the Tabulating Machine Company, to market the device; this company eventually became the International Business Machines (IBM). (☞ Jan. 8, Nov. 17)

1892

Rudolf Diesel, French engineer, obtained a German patent for the first high-pressure compression–ignition engine – the diesel engine. A diesel engine does not use spark plugs to ignite the fuel mixture. The heat generated by compression of air ignites the fuel, which is injected directly into the cylinder. The most outstanding feature of the diesel engine is its efficiency. By compressing air, rather than using an air–fuel mixture, the diesel engine is not limited by the pre-ignition problems that plague spark-ignition engines. Thus diesel engines can be operated at compression ratios high enough to produce efficiencies greater than those attainable with spark-ignition systems. Diesel demonstrated his engine at the Exhibition Fair in Paris, France in 1898. This engine stood as an example of Diesel's vision because it was fuelled by peanut oil – the 'original' biodiesel. (☞ Mar. 18, Sep. 29)

1932

Conjeevaram Srirangachari Seshadri, Indian mathematician, Fellow of the Royal Society, was born in Kancheepuram, Tamil Nadu. Seshadri did his BA (Hons) from the Loyola College, Madras (now Chennai) in 1953, and obtained his PhD from the Bombay University in 1958. He joined the Tata Institute of Fundamental Research, Mumbai in 1953 as a lecturer and rose to be senior professor of mathematics in 1975. In 1975 he joined the Institute of Mathematical Studies in Chennai, and from 1986–89 he was dean, School of Mathematics at SPIC Science Foundation in Chennai where he organised a strong group of young and active mathematicians. A large part of Seshadri's work belongs to the theory of vector bundles on curves. He proved that an algebraic vector bundle on the affine 2-space is trivial. His famous papers, with M. S. Narasimhan, ushered in the modern theory of holomorphic vector bundles on compact Riemann surfaces. He then extended the whole theory to the algebraic context in any characteristic.

March 1

1896

Antoine-Henri Becquerel, French physicist, accidentally discovered radioactivity. Becquerel was performing experiments with uranium minerals that he found could fog photographic film covered in black paper after exposure to the Sun. He thought sunlight was necessary to stimulate the mineral to emit the penetrating radiation. But when he tried developing a photographic plate he had left in a desk drawer for some days because of cloudy weather, he found it had fogged, meaning that some spontaneous and penetrating rays like X-rays was coming out of the uranium mineral independent of any input of energy. This was the first recorded evidence of radioactivity. (☞ Aug. 25, Dec. 15)

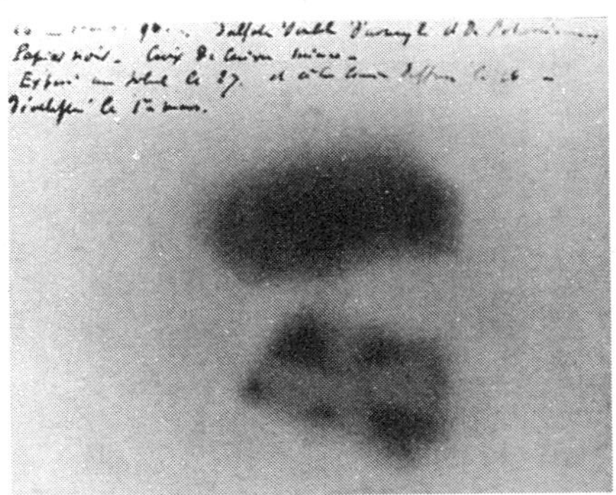

Photographic plate fogged by radioactive emissions

1910

Archer J.P. Martin, British biochemist, Nobel laureate, was born in London, UK. Martin attended the Bedford School in London from 1921 to 1929 and graduated from the Cambridge University in 1932. He obtained his PhD from the University of Cambridge in 1936 and worked as a research chemist for the Wool Industries Research Association in Leeds from 1938 to 1946.

From 1946 to 1948 he was Head of the Biochemistry Division of the Research Department of Boots Pure Drug Company at Nottingham and in 1948 he joined the staff of the Medical Research Council, first at the Lister Institute and later at the National Institute for Medical Research. He was appointed Head of the Division of Physical Chemistry at the Institute in 1952 and was a chemical consultant from 1956 to 1959 when he took over as research director of Abbotsbury Laboratories Ltd. He later worked in the Netherlands, Switzerland and USA, as well as in the UK. He was made a Fellow of the Royal Society in 1950. It was at Leeds, while working with fellow British biochemist Richard L.M. Synge, that Martin developed the technique of partition chromatography. Martin and Synge used the new technique for separating and determining the mono-amino monocarboxylic acids present in wool. Partition chromatography, in which separation is based mainly on differences between the solubility of the components in the mobile and stationary phases, is a quick and economical analytical technique for separating different components of a mixture, permitting their quick identification and analysis. As a new tool, it led to extensive advances in chemical, medical, and biological research. Martin was awarded the 1952 Nobel Prize for Chemistry, jointly with Synge, for their invention of partition chromatography. (☞ Jul. 28, Oct. 28)

1911
Jacobus Henricus van't Hoff, Dutch physical chemist and first winner of the Nobel Prize for Chemistry in 1901 for work on rates of reaction, chemical equilibrium and osmotic pressure, died in Berlin, Germany. (☞ Aug. 30)

1966
Soviet unmanned spacecraft *Venera-3* crash-landed on Venus, becoming the first spacecraft to land on another planet. (☞ Nov. 16)

1982
Soviet unmanned spacecraft *Venera-13* landed on Venus and sent back the first colour pictures from the surface and data on the soil. It transmitted signals from the surface for 127 minutes. (☞ Oct. 30)

1991
Edwin Herbert Land, American inventor and physicist, who developed the Polaroid technique of instant photography, died in Cambridge, Massachusetts, USA. (☞ May 7)

1995
Georges J.F. Köhler, German immunologist, Nobel laureate (Physiology or Medicine, 1984), who developed the technique for producing unlimited amounts of extremely pure proteins known as monoclonal antibodies, died in Freiburg, Germany. (☞ Apr. 17)

1999
The hot air-cum-gas balloon *Breitling Orbiter-3*, carrying the British pilot Brian Jones and the Swiss psychiatrist Bertrand Piccard, took off from the Swiss Alpine village of Chateau d'Oex on the first non-stop round-the-world balloon flight. The *Breitling Orbiter-3* stood 55 metres tall when fully inflated. It combined the advantages of hot air and helium technologies to create a balloon capable of remaining aloft for extended periods of time. The envelope was constructed of a nylon fabric welded to a helium-tight membrane, covered with an outer protective skin coated with aluminum on both sides to provide improved thermal control. The shape and special features of the envelope insured maximum temperature stability in order to maintain a constant altitude, conserve helium and reduce propane consumption. (☞ Mar. 21)

March 2

1840
Heinrich Wilhelm Olbers, German astronomer and physician who discovered the minor planets (asteroids), Pallas and Vesta as well as five comets, died in Bremen, Germany. (☞ Mar. 29, Oct. 11)

1902
Edward Uhler Condon, American physicist after whom the Franck–Condon Principle was named, was born in Alamogordo, New Mexico, USA. As an undergraduate, Condon had worked as a reporter for *The Oakland Inquirer*, thinking he might pursue

a career in journalism. But the intellectual challenge of physics, after a brief flirtation with chemistry, caught his fancy. Condon made important contributions while still a student. The basis for his papers on the separability of electronic and vibrational motions in molecules (the Franck–Condon Principle) was in his thesis. The Frank–Condon Principle states that the molecule does not change geometry during an electronic transition. During World War II, Condon made valuable contributions to the development of both atomic energy and radar. In 1943 he helped J. Robert Oppenheimer recruit the group that made the first atomic bombs at Los Alamos, New Mexico, USA. Condon was director of the US National Bureau of Standards (1945–51) and president of both the American Physical Society (1946) and the American Association for the Advancement of Science (1953). (☞ Mar. 26)

1958

The British Commonwealth Trans-Antarctic expedition, led by Dr. Vivian Fuchs, completed the first surface crossing of the South Pole.

1969

French-built supersonic airliner *Concorde* made its maiden flight from Toulouse, France. (☞ Jan. 9, Jan. 21, Jul. 25, Oct. 1, Oct. 24)

1972

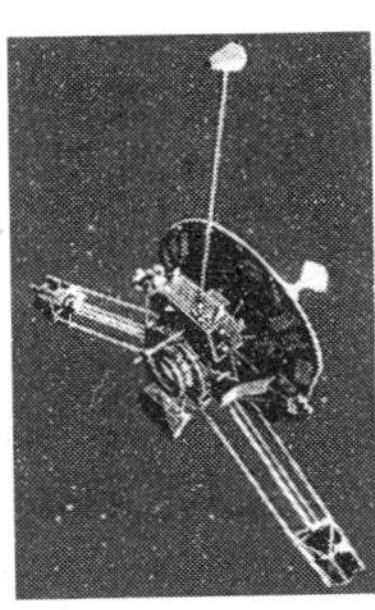
Pioneer-10

American spacecraft *Pioneer-10* was launched towards Jupiter. It passed close by Jupiter before leaving the solar system. It was the first spacecraft to travel through the Minor planet belt, and the first spacecraft to make direct observations and obtain close-up images of Jupiter. *Pioneer-10* also made valuable scientific investigations in the outer regions of our solar system until the end of its science mission on March 1997. The spacecraft was tracked till its power source became too feeble and the last contact was made in January 2003, when the spacecraft was more than 12 billion kilometres from the Earth. (☞ Jan. 22, Jun. 13, Dec. 3)

1995

Physicists at the US Department of Energy's Fermi National Accelerator Laboratory announced the discovery of the subatomic particle called the 'top' quark in proton–antiproton collisions at the Fermilab Tevatron Collider. The top quark is the last undiscovered quark of the six predicted by current scientific theory. The search began in 1977 when physicists found the fifth, 'bottom', quark at Fermilab. It took nearly two decades because the top quark was much more massive than was originally imagined, and so it required a more powerful accelerator to create it. The top quark is 40 times heavier than the next heaviest quark and about 35,000 times heavier than the up and down quarks that make up most of the matter we see around us. Since a top quark appears only once in several billion collisions, it was necessary to perform trillions of collisions.

2004

European Space Agency's *Rosetta* spacecraft was launched by an Ariane-5 rocket from Kourou, French Guiana on a mission to Comet 46P/Wirtanen. The 2,900-kg spacecraft takes its name from the Rosetta stone, which provided the key to deciphering Egyptian hieroglyphs. To gain enough orbital velocity to reach its target, *Rosetta* will make three fly-bys of Earth and one of Mars to get gravity

Rosetta

assists. Upon reaching the comet in 2014, the spacecraft will go into orbit around it and make observations as the comet becomes more active, the closer it moves towards the Sun. The spacecraft will drop a lander, named *Philae*, which will attempt to make the first ever controlled landing on a comet. On its 10 year journey *Rosetta* will hopefully fly-by at least one asteroid.

March 3

1703
Robert Hooke, English physicist who discovered the law of elasticity, known as Hooke's Law, and coined the term 'cell', died in London, UK. (☞ Jul. 18)

1847
Alexander Graham Bell, Scottish-born American audiologist and inventor of the telephone, was born in Edinburgh, Scotland. Bell was mainly educated at home. However, he did spend two years in the Edinburgh Royal High School. In 1864 Bell began work as a teacher at the Elgin's Western House Academy. Four

years later he moved to London where he worked as an assistant to his father, who was a leading authority in elocution and speech correction. Bell gave lectures on visible speech, a method of teaching speech to the deaf that had been developed by his father. In 1871 he was invited to the United States where he opened a school for teachers of the deaf in Boston and in 1873 became professor of vocal physiology at the city's university. After experimenting with various acoustical devices Bell produced the first working telephone, which he patented in March 1876. In 1883 Bell invented the graphophone, the first practical system of recording sound. The laboratory also experimented with flat disc records, electroplating records, and impressing permanent magnetic fields on records (an early type of tape recorder). (☞ Mar. 7, Mar. 10, Aug. 2)

1918
Arthur Kornberg, American biochemist and physician, Nobel laureate, was born in Brooklyn, New York, USA.

Kornberg was awarded his medical degree from the University of Rochester. His career ranged from the US Navy, to the National Institutes of Health in Bethesda, to the Department of Microbiology at the Washington University School of Medicine in St. Louis, and finally to the Department of Biochemistry at the Stanford University School of Medicine. From his early studies of the mechanisms of the enzymatic synthesis of coenzymes and inorganic pyrophosphate, he extended his interest to the biosynthesis of nucleic acids, particularly DNA. After elucidating key steps in the pathways of pyrimidine and purine nucleotide synthesis, he found the enzyme that assembles the building blocks into DNA, named DNA polymerase. This ubiquitous class of enzymes makes genetically precise DNA and is essential for the replication, repair and rearrangement of DNA. Kornberg's discovery of DNA polymerase paved the way to subsequent discovery, in his and other laboratories, of many other enzymes needed for DNA and RNA metabolism. These enzymes formed the basis for the invention of recombinant DNA, which helped ignite the biotechnology revolution. Kornberg was awarded the 1959 Nobel Prize for Physiology or Medicine, jointly with Spanish-born American biochemist Severo Ochoa, for their discovery of the mechanisms in the biological synthesis of ribonucleic acid and deoxyribonucleic acid. (☞ Sep. 24, Dec. 14)

1969
American manned lunar mission *Apollo-9*, carrying astronauts James A. McDivitt, David R. Scott, and Russell R. Schweikart, was launched by Saturn-V rocket from the Kennedy Space Center, Florida, USA. The *Apollo-9* mission was the first mission with the entire Saturn-V/Apollo configuration and it was the heaviest load ever launched into space. Although *Apollo-9* did not go to the Moon, it was the first manned flight in the Earth orbit of all lunar hardware, including the lunar module. Astronaut Schweickart performed 37 minutes of 'space walk.' During the 152 orbits of the spacecraft, human reactions to space and weightlessness were tested. (☞ Mar. 13)

1980
The American nuclear-powered submarine *USS Nautilus* was decommissioned, after a career spanning 25 years during which it covered almost 800,000 kilometres. (☞ Jan. 21, Aug. 3)

1993
Albert Bruce Sabin, Polish-born American physician and microbiologist who developed the first oral polio vaccine, died in Washington, DC, USA. (☞ Aug. 26)

1999
Gerhard Herzberg, German-born Canadian physicist, Nobel laureate (Chemistry, 1971), known for his contribution in determining the electronic structure and geometry of molecules, died in Ottawa, Canada. (☞ Dec. 25)

2005
American millionaire businessman and pilot Steve Fossett landed the single-engine jet plane Virgin Atlantic *GlobalFlyer* at the Salina Municipal Airport, Kansas, USA, after flying non-stop for 67 hours and 2 minutes, covering a distance of 37,000 kilometres and creating a world record in solo non-stop flight around the world without refuelling. (☞ Feb. 28)

March 4

1904
George Gamow, Russian-born American nuclear physicist, cosmologist and writer, was born in Odessa, Russia (now in Ukraine). Gamow became interested in astronomy at an early age, examining the starry sky through a little telescope, his father's present on his thirteenth birthday. He then decided to become a scientist and began his study of mathematics, physics, and astronomy. Gamow was awarded his PhD degree from the University of Leningrad. Later, he went to

the Institute of Theoretical Physics in Copenhagen, working on a stipend from the Royal Danish Academy. There he proposed a hypothesis that atomic nuclei can be treated as little droplets of so-called 'nuclear fluid'. These views led ultimately to the present theory of nuclear fission and fusion. Gamow spent a year working with Lord Rutherford at Cambridge, a second year in Copenhagen, and later became a professor at the University of Leningrad (1931–33). In 1934 joined the George Washington University, Washington, DC as professor. There he developed a theory explaining the internal structure of red giant stars. With Ralph Alpher, he developed a theory of the origin of chemical elements in stars by the process of successive neutron capture. Gamow was one of the foremost advocates of the Big Bang theory, which describes the origin of the universe as a colossal explosion that took place billions of years ago. In 1954, he expanded his interests into biochemistry and his work on deoxyribonucleic acid (DNA) made a basic contribution to modern genetic theory. (☞ Aug. 19)

1952
Charles Scott Sherrington, English physiologist, Nobel laureate (Physiology or Medicine, 1932), whose work on the nervous system laid the foundations for an understanding of the integrated nervous function in higher animals, died in Eastbourne, Sussex, UK. (☞ Nov. 27)

1984
Oleg Konstaninovitch Antonov, Soviet aviation engineer and aircraft designer who designed 52 glider and 22 aircraft types, died in Kiev, Ukrain. (☞ Feb. 7)

1995
Second 220-megawatt unit of the Kakrapar Atomic Power Project in Gujarat, India, was connected to the grid. The unit is a pressurised heavy water reactor that uses natural uranium as fuel and heavy water as moderator and coolant.(☞ May 6, Sep. 1)

March 5

1616
The Catholic Church decreed that the Copernican theory, which proclaimed that the Earth moved around the Sun, was 'false and erroneous'.

1827
Alessandro Giuseppe Antonio Anastasio Volta, Italian physicist, whose invention of the electric battery provided the first source of continuous current, died in Como, Italy. (☞ Feb. 18)

1827
Pierre-Simon, Marquis de Laplace, French mathematician, astronomer, and physicist, who is best known for his investigations into the stability of the solar system, died in Paris, France. (☞ Mar. 23)

1979
The American unmanned spacecraft *Voyager-1* made its closest approach to Jupiter. Images sent back by *Voyager-1* and its twin, *Voyager-2*, showed the complicated, swirling turbulence of Jupiter's atmosphere in exquisite detail. They showed a giant storm three times the size of Earth raging in Jupiter's upper atmosphere, surrounded by rippling currents that rotated around it. The spacecraft returned spectacular photos of Jupiter's moons Amalthea, Io, Europa, Ganymede and Callisto that opened up a completely new world for planetary scientists. One of the most interesting find was on Io, where images showed a bizarre yellow, orange and brown world with at least eight active volcanoes spewing material into

space, making it one of the most geologically active planetary bodies in the solar system. The spacecraft also discovered two new moons, Thebe and Metis. (☞ Sep. 5, Nov. 5, Nov. 12)

1979
The first magnetar, a strange and rare kind of super-dense, highly magnetic neutron star, was detected. Gamma ray detectors on nine spacecraft across the solar system recorded an intense spike of gamma radiation. It was just $^{2}/_{10}$th of a second long – with as much energy as the Sun releases in 1,000 years. This radiation formerly referred to as soft gamma repeaters (SGRs) puzzled scientists for two decades until later analysis of the signals showed the object to be a spinning neutron star with an intense magnetic field of about 800 trillion Gauss, a field strength similar to that for so-called 'magnetars' predicted by previous theoretical work done by Duncan and Thompson in 1992. By comparison, Earth's magnetic field is a mere 0.6 Gauss at the poles. The strongest field achieved in laboratories on Earth is 1 million Gauss – and that too only in a small volume. Normal radio pulsars reach about 1 trillion to 5 trillion Gauss. The term 'magnetar' is a blend of *magne*tic and *star*; it was coined by American astrophysicists Robert Duncan of the University of Texas, Austin and Christopher Thompson of the University of North Carolina, USA, in a paper in the *Astrophysical Journal* in 1992. In February 2005 astronomers announced the detection of the most powerful burst of gamma rays from an exploding magnetar named SGR 1806-20, which unleashed about 10,000 trillion trillion trillion (10^{36}) watts of energy. It was 100 times more powerful than any other similar eruption witnessed. Even at a distance of 50,000 light-years, the flare disrupted the Earth's ionosphere. If such a blast took place within 10 light-years of the Earth, it would destroy much of the ozone layer, and the resulting increased radiation would have caused mass extinction.

1982
Soviet unmanned spacecraft *Venera-14* landed on Venus. After entering the Venusian atmosphere, a parachute was deployed. At an altitude of about 50 kilometres the parachute was released and simple air braking was used the rest of the way to the surface. The lander survived for 57 minutes in an environment with a temperature of 465°C and a pressure of 94 Earth atmospheres. After landing, *Venera-14* took photographs of the surface and obtained a sample of the soil using a mechanical drilling arm and deposited it in a hermetically sealed chamber. Composition of the sample was determined by the X-ray fluorescence spectrometer, which showed it to be similar to oceanic basalts. (☞ Nov. 4)

March 6

1787
Joseph von Fraunhofer, German physicist who first studied the dark lines of the Sun's spectrum, now known as Fraunhofer lines, was born in Straubing, Bavaria (Germany). Fraunhofer did an apprenticeship as a lens- and mirror-maker, after which he was employed in a Munich company making scientific instruments. He was active in the Bavarian Academy of Sciences, where his skills in applied optics attracted attention. In 1823 he was appointed director of the Physics Museum in Munich. Fraunhofer discovered the dark lines in the Sun's spectrum while

studying the properties of glass and their suitability for prisms. The dark lines are caused by selective absorption of those wavelengths by gaseous elements in the Sun. Over 25,000 have now been found in the solar spectrum. (☞ Jun. 7)

1869
Dmitry I. Mendeleyev, Russian chemist, published his first version of the periodic table of the elements. At first, Mendeleyev's ordering of the elements was based on atomic weight – the weight of a single atom of the element. This method of ordering broke down in a few cases, such as tellurium and iodine. The atomic weight of tellurium is actually higher than that of iodine, suggesting that tellurium belongs to group 17 while iodine belongs to group 16. Judging by their chemical properties, however, tellurium belongs in group 16 and iodine in 17. This anomaly was resolved in 1914, when it was discovered that atomic numbers provide a better basis than atomic weight for ordering the elements. Tellurium has an atomic number of 52 and iodine 53. Mendeleyev's version of the periodic table was the first to accommodate all the known elements and also successfully predict the existence of several elements that had not yet been discovered. These undiscovered elements included gallium, scandium, and germanium. (☞ Feb. 2, Feb. 8)

1937
Valentina Tereshkova, Soviet cosmonaut, the first woman to travel into space, was born in Maslennikovo, Russia. Soon after starting work in a textile mill at the age of 18, Valentina joined an amateur parachuting club. Six years later, she applied to become a cosmonaut. Valentina, three other women parachutists, and a female pilot were selected to train as cosmonauts in 1962. Valentina was ultimately chosen to fly in space because of her expert skill in parachuting and went into space on board *Vostok-6* spacecraft, which completed 48 orbits in 71 hours. In space at the same time was Valery F. Bykovsky, another Soviet cosmonaut who had been launched two days earlier in *Vostok-5*. (☞ Jun. 16, Jun. 19)

1986
Soviet spacecraft *Vega-1* passed Comet Halley at a distance of 8,890 kilometres and sent back photographs and data. *Vega-1* was the first spacecraft to reach Comet Halley. (☞ Jun. 11, Dec.15)

2005
Hans Albrecht Bethe, German-born American theoretical physicist, Nobel laureate (Physics, 1967), who discovered the mechanism of energy production in stars, died in Ithaca, New York, USA. (☞ Jul. 2)

March 7

1765
Joseph Nicéphore Niepce, French chemist and inventor who was the first to make a permanent photographic image, was born in Chalon-sur-Saône, France. In 1826 he produced the first known photograph, which he called a heliograph, using bitumen on a pewter plate. From 1829 he worked with Louis Daguerre, who perfected the process after the death of Niepce. A nephew, Claude Felix Abel Niepce de Saint-Victor, also a chemist, was the first

to use albumen in photography and also produced photographic engravings on steel. (☞ Jul. 5)

1792

John Herschel, English astronomer, was born in Slough, Buckinghamshire, England. Herschel graduated from the Cambridge University in 1913. In 1820 he was among the founders of the Royal Astronomical Society. In later years, Herschel made important contributions to chemistry, the physics of light and particularly to mathe-matics. Herschel's first major task in astronomy was the re-observation of the double stars catalogued by his father, the famous astronomer William Herschel. He served as secretary of the Royal Society during 1824–27. In 1833 Herschel and his family set sail for the Cape of Good Hope in South Africa with a large reflecting telescope to survey the southern skies not visible in England. Before returning to England in 1838 Herschel had recorded the locations of 68,948 stars and had amassed long catalogues of nebulae and double stars. He had also described many details of the Great Nebula in the constellation Orion, as well as the Magellanic Clouds – actually two galaxies visible only in the southern hemisphere – and had observed Comet Halley and the moons of Saturn, seven of which he also named. Herschel was also a highly accomplished chemist. In 1839 he discovered the solvent power of sodium thiosulphate for light-sensitive salts of silver leading to its use as a fixing agent in photography. (☞ Jan. 29, Mar. 14, May 11)

1857

Julius Wagner-Jauregg, Austrian psychiatrist and neurologist, Nobel laureate, was born in Wels, Austria. Wagner-Jauregg was a pioneer in fever therapy. While a member of the psychiatric staff (1883–89) at the University of Vienna, he noted that persons suffering from certain nervous disorders showed a marked improvement after contracting infections that caused fever. In 1887 he suggested that such infections be deliberately induced as a treatment of insanity, especially recommending malaria because it could be controlled with quinine. He introduced the treatment of *dementia paralytica*, caused by syphilitic meningoencephalitis, by inoculation with the organisms causing malaria, attributing the success of the procedure to the induced malarial fever. The method brought a previously incurable fatal disease under medical control. Although malaria treatment of the disease was later supplanted by administration of antibiotics, his work led to the development of fever therapy and shock therapy for a number of mental disorders. Wagner-Jauregg was awarded the 1927 Nobel Prize for Physiology or Medicine for his discovery of the therapeutic value of malaria inoculation in the treatment of *dementia paralytica*. (☞ Sep. 27)

1876

Patent for the telephone was issued to the Scottish-born American audiologist and inventor Alexander Graham Bell. (☞ Mar. 3, Aug. 2)

1938

David Baltimore, American virologist, Nobel laureate, was born in New York, USA. Baltimore was awarded his bachelor's degree from the Swarthmore College in

1960 and his PhD from the Rockefeller University in 1964. In 1968, he returned to MIT, where he had held a postdoctoral position, as an associate professor and became a professor in 1972. At MIT, during investigations on the relationship between DNA and RNA in a cell's internal functions – specifically, infection of healthy cells by cancer-causing RNA viruses – Baltimore identified the enzyme reverse transcriptase. The existence of reverse transcriptase had been hypothesised some years earlier, but the theory was considered far-fetched until June 1970, when Baltimore and Caltech alumnus Howard Temin published papers about their independent and simultaneous identification of the enzyme. The work of Baltimore and Temin led to the discovery of retroviruses – viruses with an RNA genome. Diseases such as AIDS and feline leukaemia are caused by retroviruses. The discovery led to the isolation of the AIDS virus and to the development not only of many modern biological techniques but also the concept of an oncogene – a gene that when mutated can produce uncontrolled cell proliferation. Batimore was awarded the 1975 Nobel Prize for Physiology or Medicine, jointly with Temin and the Italian-born American virologist Renato Dulbecco, for their discoveries concerning the interaction between tumour viruses and the genetic material of the cell. (☞ Feb. 22, Dec. 10)

1954

Fritz Wolfgang London, German-born American physicist, was born in Breslau, Germany (now Wroclaw, Poland). London received his education at several universities including those of Bonn, Frankfurt, Göttingen, and Munich in Germany, and Paris in France. He was a Rockefeller research fellow at Zürich, Switzerland, and Rome, Italy, and a lecturer at the University of Berlin, Germany. From 1933 to 1936 he was a research fellow at the University of Oxford, England and then went to the University of Paris as master and director of research. London immigrated to the United States in 1939 to become professor of theoretical chemistry at the Duke University, Durham, North Carolina. While working at the University of Zurich in 1927, London and the German physicist Walter Heitler developed a wave equation for the hydrogen molecule with which it was possible to calculate approximate values of the molecule's ionisation potential, heat of dissociation, and other constants. These predicted values were reasonably consistent with empirical values obtained by spectroscopic and chemical means. This theory of the chemical binding of homopolar molecules is considered one of the most important advances in modern chemistry. The approach was later called the valence-bond theory. (☞ Mar. 30)

1954

Otto Paul Hermann Diels, German organic chemist, Nobel laureate (Chemistry, 1950), who with fellow German chemist Kurt Alder developed a method of preparing cyclic organic compounds, died in Kiel, Germany. (☞ Jan. 23, Jul. 10)

1979

Scientists discovered a ring around Jupiter while examining photographs taken by the *Voyager-1* spacecraft. The ring is now known to be composed of three major components. The main ring is about 7,000 kilometres wide and has an abrupt outer boundary 129,130 kilometres from the centre of the planet. The main ring encompasses the orbits of two small moons, Adrastea and Metis, which may act as the source for the dust that make up most of the ring. (☞ Sep. 5)

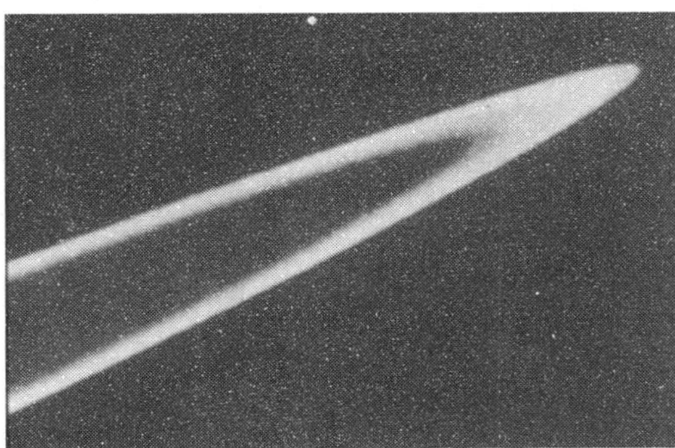

Jupiter ring

1996

The first surface photos of the planet Pluto (now designated as a dwarf planet), taken by the Hubble Space Telescope, were released. Pluto is the only solar-system dwarf planet that has never been visited by a spacecraft. (☞ Apr. 25)

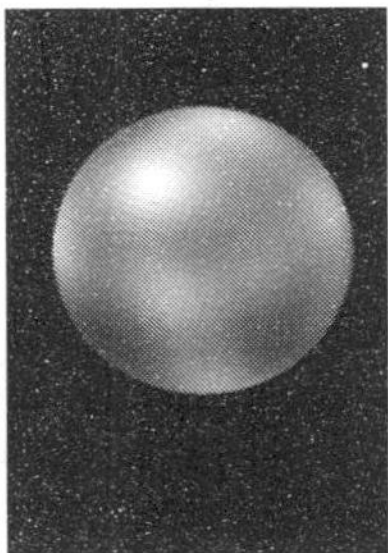

Pluto

1997

Edward Mills Purcell, American physicist, Nobel laureate (Physics, 1952), who independently developed the technique of nuclear magnetic resonance (NMR), died in Cambridge, Massachusetts, USA. (☞ Aug. 30)

March 8

1618

German astronomer Johannes Kepler formulated his Third Law of Planetary Motion. The Third Law can be found in his main philosophical work, *Harmonice Mundi* (Harmonies of the World), published in 1619, which states that the squares of the periods of revolution of any two planets are proportional to the cubes of their mean distances from the Sun. The famous English physicist Isaac Newton used Kepler's third law to derive the universal law of gravitation. (☞ Nov. 15, Dec. 27)

1775

Joseph Priestley, English chemist, after discovering oxygen, carried out an experiment in his home laboratory with mice to study whether it was necessary to support life. He heated mercuric oxide in a closed glass chamber with concentrated Sun's rays using a convex lens to produce oxygen. He wrote, 'I procured a mouse, and put it into a glass vessel, containing two ounce-measures of the air from mercurius calcinatus. Had it been common air, a full-grown mouse, as this was, would have lived in it for about a quarter of an hour. In this air, however, my mouse lived a full hour'. (☞ Feb. 6, Mar. 13, Aug. 1)

1839

James Mason Crafts, American chemist, was born in Boston, Massachusetts, USA. Crafts graduated from the Lawrence Scientific School of Harvard in 1858, after which he spent seven years abroad, studying chemistry at the Freiberg Mining School, the University of Heidelberg in Germany, and at the Ecolé des Mines in Paris, France. On his return to the United States, he was professor of chemistry at the Cornell University from 1868 to 1870, and from 1870 till 1880 a member of the faculty in the Massachusetts Institute of Technology in Boston. In 1872 he was elected a member of the National Academy of Sciences. In 1874, Crafts went to Paris, France where, for 17 years, he carried out most of his chemical investigations, largely in collaboration with the French organic chemist Charles Friedel. Their work includes researches on silicon and its combinations with compound radicals, and investigations on the vapour densities of halogen compounds, principally iodine. (☞ Mar. 12, Jun. 20)

1879

Otto Hahn, German chemist, Nobel laureate, was born in Frankfurt am Main, Germany. Hahn studied chemistry at Marburg and Munich, taking his doctorate examination in 1901 at Marburg. While working under Professor Ernest Rutherford at the Physical Institute of McGill University, Montreal, Canada, during 1905–06, he discovered radioactinium and conducted investigations on α-rays of radiothorium and radioactinium. On his return to Europe, Hahn moved to Berlin where he worked with Lise Meitner on β-rays, their absorbability, magnetic spectra, etc., and the use of the radioactive recoil to obtain new radioactive transformation products. His most spectacular discovery came at the end of 1938 when, while working jointly with Fritz Strassmann, Hahn discovered the fission of uranium and thorium in medium heavy atomic nuclei. Hahn was awarded the 1944 Nobel Prize for Chemistry for his discovery of the fission of heavy nuclei. Element 105 carries the name hahnium in recognition of his work. (☞ Jul. 28)

1886

Edward Calvin Kendall, American chemist, Nobel laureate, was born at South Norwalk, Connecticut, USA. He was educated at the Columbia University, where he obtained his bachelor's degree of science in 1908 and his master's, specialising in chemistry, in 1909. From 1910 until 1911 he was a research chemist at Parke, Davis and Co., where he did research on the thyroid gland. Kendall's name will always be associated with the isolation of thyroxine, the active principle of the thyroid gland, but his greatest achievement was his work on the hormones of the cortex of the adrenal glands. Kendall was awarded the 1950 Nobel Prize for Physiology or Medicine, jointly with fellow American chemist Philip S. Hench and Swiss chemist Tadeus Reichstein, for their discoveries relating to the hormones of the adrenal cortex, their structure and biological effects. (☞ Feb. 28, May 4, Jul. 20)

1953

The heart–lung machine was first used on a human patient during heart surgery. The machine, developed by the American physician John Heysham Gibbon, had a pump to circulate the blood around the body and was able to add oxygen to the blood and remove carbon dioxide from it. It was used during open-heart bypass surgery on a patient named Cecelia Bavolek, with the machine totally supporting her heart and lung functions during the operation.

1973

Kohoutek's comet was discovered by the Czeck astronomer Lubos Kohoutek from Hamburg, Germany, when it was 700 million kilometres away from the Sun. It attained perihelion in December 1973. Initially, it was predicted as being the comet of the century – a comet so bright that it can be noticed by a casual observer – but it failed to meet expectations. (☞ Dec. 26)

1979

Volcanoes on Jupiter's moon Io were discovered by the American spacecraft

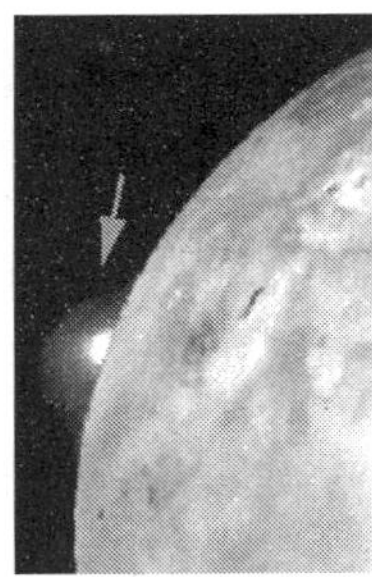

Voyager-1. This moon of Jupiter is literally bursting with volcanoes. Dozens of active vents pepper the landscape with volcanic rings stretching hundreds of kilometres across. The volcanoes themselves are the hottest spots in the solar system with temperatures exceeding 1,527°C; that is, about $^1/_3$rd the temperature of the surface of the Sun. The plumes, which rise as much as 500 kilometres into space, are so large they can be seen by the Hubble Space Telescope. (☞ Sep. 5)

2004
Sivaramakrishna Chandrasekhar, Indian liquid-crystal physicist, Fellow of the Royal Society, died in Bangalore, Karnataka, India. (☞ Aug. 6)

March 9

1851
Hans Christian Ørsted, Danish physicist and chemist who discovered that electric current in a wire can deflect a magnetised compass needle, died in Copenhagen, Denmark. (☞ Aug. 14)

1862
Fernand-Isidore Widal, French physician and bacteriologist, was born in Dellys, Algeria. Widal studied medicine in Paris; became an intern in 1884 and received his doctorate of medicine in 1889. He was appointed professor of internal pathology at the University of Paris in 1911; in 1918 he became professor for clinical medicine, remaining in office until 1929. He was a member of the Académie de Médecine from 1906, and of the Académie des Sciences from 1919. Widal is best known for the test he developed for diagnosing typhoid fever. In 1896 he developed a procedure for diagnosing the disease based on the fact that antibodies in the blood of an infected individual cause the bacteria to bind together into clumps (the Widal reaction). He was also the first to recognise that both nephritis (inflammation of the kidney) and cardiac oedema (accumulation of excessive fluid in tissues as a result of heart disease) cause retention of sodium chloride in the body, and to recommend salt deprivation in the treatment of both diseases. (☞ Jan. 14)

1923
Walter Kohn, Austrian-born American physicist, Nobel laureate, was born in Vienna, Austria. Kohn did not take an interest in mathematics or science until high school, when the outbreak of World War II interrupted his studies, forcing him to escape from Austria. When Kohn first arrived in England in 1939, he wanted to be a farmer since jobs for intellectuals were rare, but later became interested in physics and mathematics. Kohn was finally sent to Canada where, as a refugee, he worked as a lumberjack and used his 20 cents a day to support a habit of buying chemistry and mathematics textbooks.

He studied mathematics for two years at the University of Toronto. His hard work won him a scholarship to Harvard. In 1960, Kohn went to the University of California, San Diego where he founded the physics department and began research on the electronic structure of alloys. He developed the 'density functional theory', which is an extremely successful mathematical approach for describing ground state properties of metals, semiconductors, and insulators. Kohn was awarded the 1998 Nobel Prize in Chemistry, jointly with the British chemist John A. Pople, for his development of the density functional theory. (☞ Oct. 31)

1923
Johannes Diederik van der Waals, Dutch physicist, Nobel laureate (Physics, 1910), whose work made the study of temperatures near absolute zero possible, died in Amsterdam, Netherlands. (☞ Nov. 23)

1934
Yuri Alekseyevich Gagarin, Soviet cosmonaut who was the first human to travel into space, was born at Klushino, Russia. After graduating from the Soviet Air Force cadet training school at Chkalov in 1957, Gagarin volunteered to become a cosmonaut and joined a group of test pilots for training. On April 12, 1961, he was launched into space on board *Vostok-1* spacecraft. He orbited the Earth once in 1 hour 29 minutes at a maximum altitude of 301 kilometres, before returning to Earth. In 1968, Gagarin was killed in the crash of a two-seat jet aircraft while on what was described as a routine training flight. (☞ Mar. 27, Apr. 12)

1942
Robert Bosch, German engineer and industrialist who invented the spark plug and magneto for automobiles, died in Stuttgart, Germany. (☞ Jan. 7, Sep. 23)

1948
Brazilian physicist César Mansueto Giulio Lattes and American physicist Eugene Gardner at the University of California, Berkeley, USA, confirmed the existence of heavy and light mesons formed during the bombardment of carbon nuclei with alpha particles using the 184-inch cyclotron at the university's Radiation Laboratory. The existence of mesons was proposed by the Japanese theoretical physicist Hideki Yukawa in 1935 to explain nuclear forces. He suggested the existence of a new particle, with a mass about 200 times larger than that of the electron. The unknown particle was supposed to be emitted and absorbed by protons and neutrons. The exchange of those particles between the constituents of the atomic nucleus would produce a short-range attraction between them that would explain the stability of the nucleus. This particle was given the name 'meson' (from the Greek *mesos* meaning intermediate) because its mass was intermediate between those of the electron and of the proton. (☞ Jan. 23)

1961
Russian spacecraft *Sputnik-9* was launched carrying a black dog named Chernushka, which was safely brought back to Earth after the one-orbit mission. Also onboard the spacecraft was a dummy cosmonaut, mice and a guinea pig.

1974
Earl W. Sutherland, Jr., American pharmacologist and physiologist, Nobel laureate (Physiology or Medicine, 1971), who isolated cyclic adenosine monophosphate (cyclic AMP) and demonstrated its involvement in numerous metabolic processes that occur in animals, died in Miami, Florida, USA. (☞ Nov. 19)

1981
Max Delbrück, German-born American biologist, Nobel laureate (Physiology or Medicine, 1969), who pioneered the study of molecular genetics through his work on bacteriophages (viruses that infect bacteria), died in Pasadena, California, USA. (☞ Sep. 4)

1983
Ulf von Euler, Swedish physiologist, Nobel laureate (Physiology or Medicine, 1970), who identified the key neurotransmitter noradrenaline in the sympathetic nervous system, died in Stockholm, Sweden. (☞ Feb. 7)

1986
The Soviet spacecraft *Vega-2* travelled through Comet Halley's coma, passing within 8,000 kilometres of the nucleus, and sent back photographs. (☞ Jun. 15, Dec. 21)

March 10

1628
Marcello Malpighi, Italian physician and biologist who founded the science of microscopic anatomy, was born in Crevalcore, near Bologna, Italy. Malpighi entered the University of Bologna in 1646 and received doctorates in both medicine and philosophy in 1653. He was appointed as a teacher in the same university, whereupon he immediately dedicated himself to further studies in anatomy and medicine. Drawing on the work of William Harvey, he studied the circulatory and respiratory systems of all living things and arranged them in a hierarchical system, beginning with plants, above which were insects, then fishes, mammals, then man. In 1661 he identified and described the pulmonary and capillary network connecting small arteries with small veins; it was one of the major discoveries in the history of science. He was one of the first scientists to study such anatomical structures as the lungs, kidneys, spleen, brain, tongue, and skin at the microscopic level. (☞ Nov. 30)

1876
First human voice was transmitted over the telephone. The Scottish-born American audiologist and inventor Alexander Graham Bell spoke the words, 'Mr. Watson, come here, I want you.' (☞ Mar. 3, Aug. 2)

1891
Almon Brown Strowger, American inventor, was issued a US patent for his electromechanical switch to automate a telephone exchange. Strowger did not invent the idea of automatic switching but he was the first to put it to effective use. Using his automatic switch, the user could connect to any number by turning a rotary dial on the phone. His electromechanical selector used electromagnets and pawls to move a wiper (with contacts on the end) vertically and around a bank of many other contacts, able to make a connection with any one of them. Strowger formed his company, Strowger Automatic Telephone Exchange, in October 1891. (☞ Nov. 3)

1923
Val Logsdon Fitch, American particle physicist, Nobel laureate, was born in Merriman, Nebraska, USA. In the mid-1940s, as a member of the US Army, Fitch was sent to Los Alamos, to work on the Manhattan Project. After getting a PhD in physics from the Columbia University

in 1954, he then joined the faculty of the Princeton University, New Jersey, USA. In experiments conducted at the Brookhaven National Laboratory in 1964, Fitch and fellow American particle physicist James Watson Cronin proved that certain subatomic reactions do not adhere to fundamental symmetry principles. Specifically, they proved, by examining the decay of neutral K-mesons, that a reaction run in reverse does not merely re-trace the path of the original reaction, which showed that the reactions of subatomic particles are not indifferent to time. This experiment implied that reversing the direction of time would not precisely reverse the course of certain reactions of subatomic particles. Fitch was awarded the 1980 Nobel Prize for Physics, jointly with Cronin, for the discovery of violations of fundamental symmetry principles in the decay of neutral K-mesons. (☞ Sep. 29)

1932

Udipi Ramachandra Rao, Indian space scientist, was born in Admar, Karnataka, India. Rao started his career as a cosmic ray scientist, under the Indian physicist Vikram Sarabhai. He continued his research at the Massachusetts Institute of Technology, USA. His experiments on a number of American *Pioneer* and *Explorer* spacecrafts led to a complete understanding of the solar cosmic ray phenomena and the electromagnetic state of the interplanetary space. Rao and his group have also carried out extensive work in X-ray and gamma ray high-energy astronomy using rocket and satellite-borne payloads. After taking charge as Chairman, Space Commission and Secretary, Department of Space, India in 1984, Rao accelerated the development of rocket technology, leading to the successful launch of the ASLV rocket in 1992. He was also responsible for the development of the operational PSLV launch vehicle, which successfully launched an 850-kg satellite into a polar orbit in 1996. Rao initiated the development of the geostationary launch vehicle GSLV and the development of cryogenic technology in 1991.

1966

Frits Zernike, Dutch scientist, Nobel laureate (Physics, 1953), who invented the phase-contrast microscope, an instrument that permits the study of internal cell structure without the necessity of staining, died in Groningen, Netherlands. (☞ Jul. 16)

1977

The rings of planet Uranus were discovered from Earth by stellar occultation experiments made when Uranus passed in front of a star. The star was observed to blink out briefly five times before the planet passed in front of it, and again five times afterward, indicating that five narrow rings encircled the planet. This data suggested that Uranus was surrounded by at least five rings. Subsequent Earth-based observations indicated that there were actually nine major rings. Working outward from the planet, they are named 6, 5, 4, Alpha, Beta, Eta, Gamma, Delta, and Epsilon. The epsilon ring is about 100 kilometres wide. All the other rings are much narrower. Two additional rings were found by space probe *Voyager-2*, bringing the total to 11. Direct observation of the rings from Earth is almost impossible, because the rings are

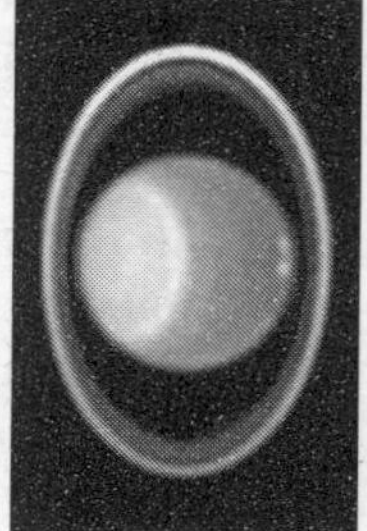

extremely thin and are lost in the planet's glare. (☞ Jan. 24)

2006
The NASA spacecraft *Mars Reconnaissance Orbiter* arrived at Mars. Six instruments on board the craft are to carry out extreme close-up photography of the Martian surface, analyse minerals, look for subsurface water, and trace how much dust and water are distributed in the atmosphere. The images are expected to show Martian features as small as a kitchen table from the spacecraft's low orbital altitude. These studies would help determine if there are deposits of minerals that form in water over long periods of time, detect any shorelines of ancient seas and lakes, and analyse deposits placed in layers over time by flowing water. (☞ Aug. 12)

March 11

1811
Urbain Le Verrier, French astronomer, who predicted the existence of the planet Neptune, was born in Saint-Lô, France. Leverrier was educated at the Ecole Polytechnique, Paris and began to teach astronomy there in 1837. French physicist François Arago, then the director of the Paris Observatory, was impressed by Leverrier's ability, and Leverrier was to spend most of his time at the Paris Observatory. Le Verrier made the prediction only on the basis of mathematical calculations. Although the English astronomer John C. Adams was independently working on the same problem, Le Verrier was more fortunate in that his prediction was confirmed by the German astronomer Johann Gottfried Galle, in 1846. (☞ Sep. 23)

1920
Nicolaas Bloembergen, Dutch-born American physicist, Nobel laureate, was born in Dordrecht, The Netherlands. Bloembergen graduated from the University of Utrecht in 1943. In 1946 he entered Harvard UniversityCambridge, Massachusetts in USA, where he did fundamental research on nuclear magnetic re-onance. After receiving his PhD from the University of Leiden in the Netherlands in 1948, he returned to Harvard, where he became a professor of applied physics in 1951. He became an American citizen in 1958. Bloembergen's early research on nuclear magnetic resonance led him to an interest in masers. He designed a three-stage crystal maser that was dramatically more powerful than earlier gaseous masers and that has now become the most widely used microwave amplifier. Bloembergen then developed laser spectroscopy, which allows high-precision observations of atomic structure. His laser spectroscopic investigations led him in turn to formulate non-linear optics, a new theoretical approach to the analysis of how electromagnetic radiation interacts with matter. Bloembergen was awarded the 1981 Nobel Prize for Physics, jointly with American physicist Arthur Leonard Schawlow and Swedish physicist Kai Manne Börje Siegbahn for their revolutionary spectroscopic studies of the interaction of electromagnetic radiation with matter. (☞ Apr. 20, May 5)

1955
Alexander Fleming, Scottish bacteriologist, Nobel laureate (Physiology or Medicine, 1945), who discovered the antibiotic penicillin, died in London, UK. (☞ Aug. 6)

1960
Space probe *Pioneer-5* was launched from Cape Canaveral, Florida, USA, in one of the first in-depth attempts to study the solar system. It was designed to send back scientific data on interplanetary space while in a planet-like orbit about the Sun. After its launch *Pioneer-5* entered an orbit around the Sun between Earth and Venus. The spacecraft functioned for a record 106 days and provided a wealth of new data on interplanetary space including measurements of magnetic fields, cosmic radiation, electrical fields and micrometeorites.

March 12

1824
Gustav Robert Kirchhoff, German physicist who, with the chemist Robert Bunsen, firmly established the theory of spectrum analysis, was born in Königsberg, Prussia (now Kaliningrad, Russia). Kirchhoff graduated from the University of Koenigsberg in 1847, and spent a few years lecturing in Berlin and Breslau. In 1845, Kirchhoff developed a pair of laws governing current and resistance in electrical networks. The first law states that at any given instant, the sum of the voltages around any closed path, or loop, in the network is zero. The second law states that at any junction of paths in a network, the sum of the currents arriving at any instant is equal to the sum of the currents flowing away. In 1852 he took on a position at the University of Heidelberg where he remained till 1875. Kirchhoff's most celebrated contributions to physics were in the field of spectroscopy. In collaboration with Robert Bunsen, Kirchhoff founded the science of spectroscopy. Kirchhoff and Bunsen began by effectively inventing the spectroscope, a prism-based device that separated light into its spectrum, with which they began studying the spectral 'signature' of various chemical elements in gaseous form. The key observation made by Kirchhoff and Bunsen was that the spectral lines emitted by a gas occurred at the same wavelength as the absorption lines observed when incandescent light (provided by Bunsen's now famous gas burner) shone through the same gas heated at the same temperature. Kirchhoff took the gigantic leap of then identifying the dark spectral lines observed in the solar spectrum, known as Fraunhofer lines, with the emission lines observed by various heated chemical substances. In this manner, Kirchhoff demonstrated the existence, in the Sun, of many chemical elements isolated on Earth. (☞ Mar. 31, May 10, Oct. 17)

1832
Charles Friedel, French organic chemist and mineralogist, was born in Strasbourg, France. Friedel studied in Strasbourg and qualified in both chemistry and mineralogy. In 1856 he was made curator of the collection of minerals at the Ecolé des Mines. In 1871 he became an instructor at the Ecole Normale and from 1876 was professor of mineralogy at the Sorbonne, Paris. In 1877 along with the American chemist James Mason Crafts, he discovered the chemical process known as the Friedel–Crafts reaction, which uses aluminium chloride

as a catalyst to facilitate the addition of an alkyl halide (halogenoalkane) to an aromatic compound. The reaction is now employed in the industrial preparation of triphenylamine dyes. From 1879 to 1887, Friedel worked on the attempted synthesis of minerals, including diamonds, using heat and pressure. He established the similarity in properties and structure between carbon and silicon. (☞ Mar. 8, Apr. 20)

1838

William Henry Perkin, English chemist and inventor, was born in London, England.

Perkin was enthused about chemistry at a young age by attending Faraday's public lectures. At the age of 18, while attempting to synthesise quinine from a coal tar chemical, Perkin mixed aniline and sodium dichromate and unexpectedly found a deep purple-coloured dye, which he extracted with alcohol. He became the first chemist to create a synthetic dye – the aniline dye, better known as mauveine. Perkin's dye provided a valuable alternative for dyeing textiles, which in his time were coloured from natural sources such as mollusks, guano, and Madder root. After he patented the dye, his father invested in his efforts to manufacture the dye. It went on sale in 1857, and became popular in France. In the late 1860s, Perkin synthesised two new dyes, 'Britannia Violet' and 'Perkin's Green'. (☞ Jul. 14)

1910

Dorothy Mary Crowfoot Hodgkin, English chemist, Nobel laureate, was born in Cairo, Egypt. Hodgkin became interested in chemistry and in crystals at about the age of 10. She went to Oxford and the Somerville College from 1928–32, where she began her crystallographic studies. She and a co-worker took the first X-ray diffraction photograph of the protein pepsin while she was studying for her doctorate at the University of Cambridge in the mid-1930s; later she made a survey of the sterols. She returned to Oxford in 1935. From 1942 to 1949 Hodgkin worked on the structural analysis of penicillin. The structure was determined in 1944 but only published in 1949. She and her colleagues made the first X-ray photograph of vitamin B_{12}, one of the most complex non-protein compounds, in 1948, and eventually they completely determined its three-dimensional molecular structure. She completed a similar three-dimensional analysis of insulin in 1969. Hodgkin was awarded the 1964 Nobel Prize for Chemistry for her determinations by X-ray techniques of the structures of important biochemical substances. (☞ Jul. 29)

1925

Leo Esaki, Japanese physicist, Nobel laureate, was born in Osaka, Japan. Esaki was awarded his PhD in 1959 from the University of Tokyo. In 1956 he became chief physicist of the Sony Corporation, where he carried out research in the field of quantum mechanics and concentrated on the phenomenon of tunnelling – a phenomenon in which the wave-like character of matter enables electrons to pass through barriers that the laws of classical mechanics say are impenetrable.

He devised ways to modify the behaviour of solid-state semiconductors by adding impurities, or 'doping' them. This work led to his invention of the double diode, which became known as the Esaki diode, an important component of solid-state physics with practical applications in high-speed circuits found in computers and communications networks. Esaki was awarded the 1973 Nobel Prize for Physics, jointly with Norwegian-born American physicist Ivar Giaever and British physicist Brian David Josephson, for his experimental discoveries regarding tunnelling phenomena in semiconductors. (☞ Jan.4, Apr. 5)

1942
William Henry Bragg, English physicist, Nobel laureate (Physics, 1915), who pioneered X-ray crystallography, died in London, UK. (☞ Jul. 2)

1991
Ragnar Arthur Granit, Finnish-born Swedish physiologist, Nobel laureate (Physiology or Medicine, 1967), who discovered the primary physiological and chemical visual processes in the eye, died in Stockholm, Sweden. (☞ Oct. 30)

March 13

1733
Joseph Priestley, English clergyman, and physical scientist who is best remembered for his discovery of the element oxygen, was born near Leeds, Yorkshire, England. Joseph was a brilliant student and with the help of local teachers, he became proficient in physics, philosophy, algebra, mathematics and several different languages. In 1767, Priestley made his first major discovery – that graphite can conduct electricity. That same year he published *The History of Electricity*. Priestley also isolated and described the properties of carbon dioxide, nitrous oxide, and oxygen; invented soda pop, identified the gases involved in plant respiration, and observed photosynthesis for the first time. He also isolated and described for the first time the properties of ammonia, sulphur dioxide, hydrogen sulphide, and carbon monoxide. (☞ Feb. 6, Mar. 8, Aug. 1)

1781
Uranus was discovered by William Herschel, German-born English astronomer. At first he thought that the faint object he saw in the constellation of Gemini was a comet, but he later discovered that the computed path of this object was not the elongated orbit of a comet, but more like the orbit of a planet. He named it *Georgium Sidum*, meaning George's Star, in honour of the then King George III. The name Uranus was suggested by the German astronomer, Johann Bode, who proposed that the planet be named after the father of Saturn, in line with classical tradition, although it was called the 'Georgian Planet' and indeed 'Herschel' until 1850. Uranus was the first planet to be discovered beyond Saturn and with the aid of a telescope. (☞ Aug. 25, Nov. 15)

1840
James H. Young and Adrian Delcambre of Lille, France were issued a patent for the first typesetting machine. Using a keyboard, a single operator could compose about 6,000 characters per hour on the machine, which was named the Pianotype.

1855

Percival Lowell, American astronomer who predicted the existence of the planet Pluto and initiated the search that ended in its discovery, was born in Boston, Massachusetts, USA. Lowell graduated from the Harvard University, Cambridge, Massachusetts, in 1876 with distinction in mathematics. In 1894, he founded an observatory that would change the course of astronomy. Atop Mars Hill, at a height of 2,000 metres in Flagstaff, Arizona, Lowell launched research on a number of projects from the observatory. Flagstaff's sky conditions were perfect for night observation. The ongoing work at Lowell's observatory led to discoveries such as the planet Pluto and the first evidence of the expanding universe. Lowell was an amateur astronomer passionately committed to finding proof of intelligent life on Mars and spent 15 years intensively studying Mars. By looking at the surface of Mars through a 24-inch telescope, Lowell drew in intricate detail, several dark patches and networks of several hundred fine, straight lines, which he thought were canals, constructed by intelligent beings who once flourished on Mars. Lowell believed that the bright areas were deserts, the dark areas were patches of vegetation and that water from the melting polar cap were brought down the canals toward the equatorial region to revive vegetation. He published his views in three books: *Mars* (1895), *Mars and its Canals* (1906), and *Mars as the Abode of Life* (1908), which were responsible for arousing public curiosity about intelligent life on Mars. (☞ Nov. 12)

1899

John Hasbrouck Van Vleck, American physicist and mathematician, Nobel laureate, was born in Middletown, Connecticut, USA. Van Vleck was educated at the University of Wisconsin, Madison, and at the Harvard University, Cambridge, Massachusetts, where he was awarded a PhD in 1922. He joined the faculty of the University of Minnesota, Minneapolis, in 1924 and later taught at Wisconsin from 1928 to 1934. He then went to Harvard, where he eventually served as chairman of the physics department. Van Vleck developed during the early 1930s the first fully articulated quantum mechanical theory of magnetism. He was a chief architect of the 'ligand field theory' of molecular bonding, which is one of several theories that describe the electronic structure of coordination or complex compounds, notably transition metal complexes, which consist of a central metal atom surrounded by a group of electron-rich atoms or molecules called ligands. The ligand field theory deals with the origins and consequences of metal–ligand interactions as a means of elucidating the magnetic, optical, and chemical properties of these compounds. His publications include *Quantum Principles and Line Spectra* (1926) and *The Theory of Electric and Magnetic Susceptibilities* (1932). Van Bleck was awarded the 1977 Nobel Prize for Physics, jointly with fellow American physicist Philip Warren Anderson and English physicist Nevill Francis Mott, for their fundamental theoretical investigations of the electronic structure of magnetic and disordered systems. (☞ Oct. 27)

1930
The discovery of a ninth planet was announced by American astronomer Clyde W. Tombaugh at the Lowell Observatory, Flagstaff, USA. It was found to be only one-tenth as large as Earth and 6,500 million kilometres away. Tombaugh was looking for the 'planet X' that the American astronomer Percival Lowell had tried to find for more than seven years till his death. Tombaugh began his search in December 1929, as an amateur astronomer at the Observatory. He took photographs of the sky every one or two weeks and looked for anything that shifted against the background of stars. This method was very successful and finally led to the discovery of the elusive planet X, which was later named Pluto after the Greek and Roman god of the underworld. (☞ Jan. 17, Feb. 4, Aug.24)

1969
The command module of *Apollo-9* carrying astronauts James McDivitt, David Scott, and Russell Schweickart safely splashed down in the Pacific Ocean after spending ten days in the Earth orbit testing out the command module and the lunar module that would be used later for Moon landings. (☞ Mar. 3)

March 14

1835
Giovanni Virginio Schiaparelli, Italian astronomer who is remembered for his observations of Mars, was born in Savigliano, Italy. Schiaparelli entered the Turin University in 1850, and graduated in 1854. Two years later he went to Berlin to study astronomy under the German astronomer Johann Franz Encke. Between 1862 and 1900 he was director of the Brera Observatory, Milan. During 1877, while observing Mars at its closest approach to Earth he saw on the surface of the planet the markings that he called *canali* (channels), which were later misinterpreted as 'canals'. Schiaparelli made extensive studies, both observational and theoretical, of comets, determining from the shapes of their tails that there was a repulsive force from the sun. He showed that meteor swarms travel through space in cometary orbits and was the first to explain that the regular meteor showers seen from Earth were the result of the gradual disintegration of comets and proved it for the Perseids meteor shower. He also suggested that Mercury and Venus rotate on their axes. He discovered the minor planet Hesperia and was a major observer of double stars. (☞ Jul. 4)

1839
The word 'photography' was first used by the English astronomer John Herschel in a lecture he gave to the Royal Society of London. The word is derived from the Greek words *photos* for 'light' and *graphos* for 'writing'. (☞ Mar. 7, May 11)

1854
Paul Ehrlich, German medical scientist, Nobel laureate, was born near Breslau, Germany (now Wroclaw, Poland). Ehrlich was educated at the Gymnasium at Breslau and subsequently at the Universities of Breslau, Strassburg, Freiburg-im-Breisgau and Leipzig. In 1878 he obtained his doctorate of medicine

and joined the Berlin Medical Clinic where he worked on staining of animal tissue with aniline dyes discovered by the English chemist William Henry Perkin. In 1882 Ehrlich published his method of staining the tubercle bacillus, which later led to the Gram method of staining bacteria so much used by modern bacteriologists. In 1897 Ehrlich was appointed Public Health Officer at Frankfurt-am-Main and when, in 1899, the Royal Institute of Experimental Therapy was established at Frankfurt, Ehrlich became its Director. In Frankfurt he continued to look for chemical agents to use against disease. In 1906 Georg-Speyerhaus, a research institute for chemotherapy was established next-door to Ehrlich's Institute, also under Ehrlich's directorship. The research programs were guided in part by Ehrlich's theory that the germicidal capability of a molecule depended on its structure, particularly its side-chains, which could bind to the disease-causing organism. The most successful products of this quest were Salvarsan, and Neosalvarsan, the most effective drugs for treating syphilis until the advent of antibiotics in the 1940s. Ehrlich was awarded the 1908 Nobel Prize for Physiology or Medicine, jointly with Russian zoologist and microbiologist Ilya Ilyich Mechnikov in recognition of their work on the theory of immunity. (☞ May 16, Aug. 20)

1879

Albert Einstein, German-born American physicist, Nobel laureate, was born in Ulm, Württemberg, Germany. In 1896 he entered the Swiss Federal Polytechnic School in Zurich to be trained as a teacher in physics and mathematics. In 1901, the year he gained his diploma, he acquired Swiss citizenship and, as he was unable to find a teaching post, he accepted a position as technical assistant in the Patent Office at Berne, Switzerland. It was while working here that Einstein completed an astonishing range of theoretical physics papers, written in his spare time without the benefit of close contact with scientific literature or colleagues. In the first of the three papers, all written in 1905, Einstein examined the phenomenon discovered by the German physicist Max Planck, according to which electromagnetic energy seemed to be emitted from radiating objects in discrete quantities, or quanta, contradicting the classical electromagnetic theory. Einstein used Planck's quantum hypothesis to describe the electromagnetic radiation of light and in the process laid the foundation of the photon theory of light, which is at the root of the photoelectric process. Einstein's second paper of 1905 proposed what is today called the 'special theory of relativity.' He based his new theory on a re-interpretation of the classical principle of relativity, namely that the laws of physics had to have the same form in any frame of reference. As a second fundamental hypothesis, Einstein assumed that the speed of light remained constant in all frames of reference, as required by Maxwell's theory. Later in 1905, Einstein showed how mass and energy was equivalent, a concept that forms the basis of nuclear power and energy of the stars. In 1916, he derived a field equation that, along with the rule that a freely falling body moves along a curved space–time, or geodesic, forms the comprehensive treatment of gravitation known as the 'general theory of relativity.' His theories of relativity and gravitation were a profound advance over the old Newtonian physics and revolutionised scientific and philosophic inquiry. Einstein

was awarded the 1921 Nobel Prize for Physics for his services to theoretical physics, and especially for his discovery of the law of the photoelectric effect. (☞ Apr. 18)

1915
George Eastman, American inventor, who invented the photographic roll film that revolutionised amateur photography, died in Rochester, New York, USA. (☞ Jul. 12, Sep. 4)

1986
European Space Agency's *Giotto* spacecraft approached within 596 kilometres of the nucleus of Comet Halley and sent back the first close-up images of the nucleus. The spacecraft carried 10 instruments including a multi-colour camera, and returned data to Earth until shortly before the closest approach, when the downlink was temporarily lost. *Giotto* was severely damaged by high-speed dust encountered during the fly-by and its operation was suspended temporarily. *Giotto* was re-activated in April 1990 and re-targeted for a successful fly-by of comet Grigg-Skjellerup in 1992. The spacecraft was named after Giotto di Bondone, Italian painter and architect, who is considered as one of the artists of the Italian Renaissance. (☞ Jul. 2)

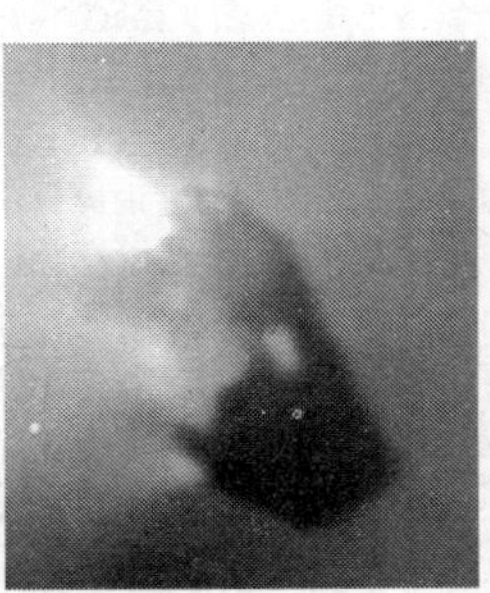

Nucleus of Comet Halley

1995
William Alfred Fowler, American nuclear astrophysicist, Nobel laureate (Physics, 1983) who worked out how elements are formed in stars, died in Pasadena, California, USA. (☞ Aug. 9)

March 15

1813
John Snow, English physician, who discovered that cholera is transmitted through contaminated water, was born at York, England. Snow was educated at a private school in his native city until the age of fourteen, when he was apprenticed to a surgeon living at Newcastle-on-Tyne. After serving for a short time as a colliery surgeon and unqualified assistant during the cholera epidemic of 1831–32, he joined the Hunterian School of Medicine, London in 1836. He began to attend medical practice at the Westminster Hospital in the following year, and in 1838 he became a licentiate of the Society of Apothecaries, having been admitted as a member of the Royal College of Surgeons of England. In 1844 he earned an MD from the University of London and in 1850 he was admitted as a licentiate of the Royal College of Physicians. Snow is best known for his theory that cholera is communicated by means of a contaminated water supply. He published his essay on the mode of communication of cholera in 1849. After meticulously studying the sources of water in the affected areas, he was able to demonstrate that water contaminated by sewage was the main causative factor of cholera epidemics. Cholera was earlier believed to be caused by 'miasma', or bad air. Snow also made great improvements in the method of administering ether as an anaesthetic. (☞ Jun. 16)

1854
Emil Adolf von Behring, German bacteriologist, winner of the first Nobel

Prize for Physiology or Medicine in 1901, was born in Hansdorf, West Prussia (now in Poland). Behring was awarded his medical degree in 1878 from the Army Medical College in Berlin. After serving 10 years with the Army Medical Corps, he became an assistant (1889) at the Institute for Hygiene, Berlin, where while working with the Japanese bacteriologist Kitasato Shibasaburo, he showed that it was possible to provide an animal with passive immunity against tetanus by injecting it with the blood serum of another animal infected with the disease. The infected animal's blood serum contained what they called 'antitoxin' against the infectious agent. Behring applied this antitoxin technique to achieve immunity against diphtheria. Administration of diphtheria antitoxin, developed with the help of German medical scientist Paul Ehrlich and first successfully marketed in 1892, became a routine part of the treatment of the disease. Behring was awarded the 1901 Nobel Prize for Physiology or Medicine for his work on serum therapy, especially its application against diphtheria, by which he opened a new road in the domain of medical science and thereby placed in the hands of the physician a victorious weapon against illness and deaths. (☞ Mar. 31)

1892

Jesse Wilford Reno, American inventor, patented his moving stairs, or inclined elevator as he called it, that came to be known as the escalator. Reno submitted his first patent application for a 'new and useful endless conveyor or elevator' in 1891. It became effective 15 months later. The machine was built and installed at Coney Island, Brooklyn, as an amusement ride in September 1895.

1898

Henry Bessemer, English inventor and engineer who developed the first process for manufacturing steel inexpensively, died in London, UK. (☞ Jan. 19, Oct. 17)

1920

Edward Donnall Thomas, American physician, Nobel laureate, was born in Mart, Texas, USA. Thomas studied at the University of Texas, and the Harvard Medical School from where he was awarded his MD in 1946. He served at a few hospitals and a research centre before becoming a professor of medicine at Columbia University's College of Physicians and Surgeons (1955–63), and the University of Washington School of Medicine (from 1963) in Seattle. In 1956, he successfully transplanted bone marrow to a leukaemia patient from his identical twin. He also developed drugs to suppress the immune system that led to rejection of foreign tissue transplants. In 1969 these refinements enabled him to perform the first successful bone marrow transplant in a leukaemia patient from a relative who was not an identical twin. Before his work, leukaemia had invariably been a fatal disease. Thomas was awarded the 1990 Nobel Prize for Physiology or Medicine, jointly with American surgeon Joseph E. Murray, for their discoveries concerning organ and cell transplantation in the treatment of human disease. (☞ Apr. 1)

1930

Zhores Ivanovich Alferov, Russian physicist, Nobel laureate, was born in Vitebsk, Belorussia, USSR (now in Belarus). Alferov graduated from the Department

of Electronics of the Ulyanov-Lenin Electrotechnical Institute in Leningrad. He earned doctorates in physics and mathematics from the Ioffe Physico-Technical Institute in St Petersburg in 1970. In 1973 Alferov took over the chair of optoelectronics at the St Petersburg State Electrotechnical University (former Ulyanov-Lenin Electrotechnical Institute) and in 1988 he was appointed as Dean of the Faculty of Physics and Technology at the St Petersburg Technical University. Alferov was elected a member of the USSR Academy of Sciences in 1979. From 1989 onward, he has been Vice-President of the USSR (Russian) Academy of Sciences and President of its St Petersburg Scientific Center. Alferov's main contribution has been in the area of semiconductor heterostructures. His outstanding contributions to physics and the technology of semiconductor heterostructures, especially investigations of injection properties, development of lasers, solar cells, LEDs, and epitaxy processes have led to the creation of modern heterostructure physics and electronics. Alferov was awarded the 2000 Nobel Prize for Physics, jointly with American physicists Herbert Kroemer and Jack S. Kilby, for basic work on information and communication technology. Alferov and Kroemer shared half of the prize money for developing semiconductor heterostructures used in high-speed- and optoelectronics, while Kilby received the other half for his part in the invention of the integrated circuit. (☞ Aug. 25, Nov. 8)

1955
A new subatomic particle called a negative proton (antiproton) was discovered by the American physicist Owen Chamberlain and Italian-born American physicist Emilio Segrè at the University of California, Berkeley, USA. (☞ Feb. 1, Jul. 10)

1962
Arthur Holly Compton, American physicist, Nobel laureate (Physics, 1927), who discovered the Compton Effect, died in Berkeley, California, USA. (☞ Sep. 10)

2004
John A. Pople, British mathematician and theoretical chemist, Nobel laureate (Chemistry, 1998), who developed computational methodology to study the quantum mechanics of molecules, their properties and reactions, died in Chicago, Illinois, USA. (☞ Oct. 31)

March 16

1867
Joseph Lister, British surgeon and medical scientist, published an article in *The Lancet* on his discovery of antiseptic surgery, the first of a series of such articles that he wrote. (☞ Feb. 10, Apr. 5, Jun. 17)

1879
Georg Simon Ohm, German physicist who discovered the law, named after him, was born in Erlangen, Bavaria (now in Germany). In early childhood, Ohm was taught by his father who imparted to Ohm a high standard of knowledge in mathematics, physics, chemistry and philosophy. As a high school teacher, Ohm started his research with the electrochemical cell, recently invented by the

Italian Count Alessandro Volta. Using equipment of his own creation, in 1827, he was able to show from his experiments that there was a simple relationship between resistance, current and voltage. He determined that the current that flows through a wire is proportional to its cross-sectional area and inversely proportional to its length. In a paper published in 1927, Ohm stated the famous Ohm's Law, which gave a mathematical relationship between the potential difference (voltage), current, and resistance at a given temperature. Thus, if V is the voltage in volts between two ends of a wire and I the current flowing through it, in amperes, then the quotient $^{V}/_{I}$, which is the resistance, R, remains constant. Unfortunately, when Ohm published his finding, his colleagues dismissed his ideas. In 1849, just five years before his death, Ohm's life-long dream was realised when he was given a professorship of Experimental Physics at the University of Munich. The unit of electrical resistance, the ohm, is named after him. (☞ Jul. 6)

1918

Frederick Reines, American physicist, Nobel laureate, was born in Paterson, New Jersey, USA. Reines earned his ME and MS degrees from the Stevens Institute of Technology in Hoboken, New Jersey, and his PhD from the New York University in 1944. He was a member and then Group Leader in the theoretical division of the Los Alamos Scientific Laboratory from 1944 to 1959. Reines started experiments to detect the elusive particles called neutrinos along with his colleague Clyde L. Cowan, Jr. in the early 1950s. The Austrian-born American physicist Wolfgang Pauli had postulated the existence of neutrinos about twenty years before as a means of preserving fundamental conservation laws in radioactive decays of nuclei. The Italian-born American physicist Enrico Fermi had suggested the name *neutrino* (the little neutral one), to distinguish it from the neutron, and formulated a successful theory of weak nuclear processes with the neutrino as a central participant. However, the neutral character of the particle and its lack of any mass made it interact very feebly with matter, making it almost impossible to detect. Reines and Cowan used a novel detection device to detect the elusive neutrinos emitted from nuclear reactors, comprising a 400-litre tank of water solution of cadmium chloride fitted with a set of scintillation detectors. Reines subsequently built other neutrino detectors underground and helped pioneer the field of neutrino astronomy. He was awarded the 1995 Nobel Prize for Physics, jointly with American physicist Martin Lewis Perl. Perl received half the prize money for the discovery of the tau lepton, while Reines received the other half for the detection of the neutrino. (☞ Jun. 24, Aug. 26)

1926

The first liquid-fuel rocket flight was launched by Robert Hutching Goddard in a field in Auburn, Massachusetts, USA. His rocket used gasoline fuel and oxygen. The rocket burned about 20 seconds before reaching sufficient thrust for taking off. During that time it melted part of the nozzle. It took off to a

Goddard with his rocket

height of 12.5 metres, levelled off and within 2.5 seconds hit the ground about 62 metres away, reaching a maximum speed of 100 kph. (☞ Jul. 14, Oct. 5)

1935
John James R. Macleod, Scottish physiologist, Nobel laureate (Physiology or Medicine, 1923) who did research on diabetes and named the hormone insulin, died in Aberdeen, Scotland, UK. (☞ Sep. 6)

1966
The first manned docking of two spacecraft in orbit was accomplished. The American spacecraft *Gemini-VIII* with command pilot astronaut Neil A. Armstrong and pilot astronaut David R. Scott, docked with an unmanned *Gemini Agena* target vehicle. The final manoeuvre for docking was done manually by Armstrong. (☞ Jul. 21)

Gemini Agena

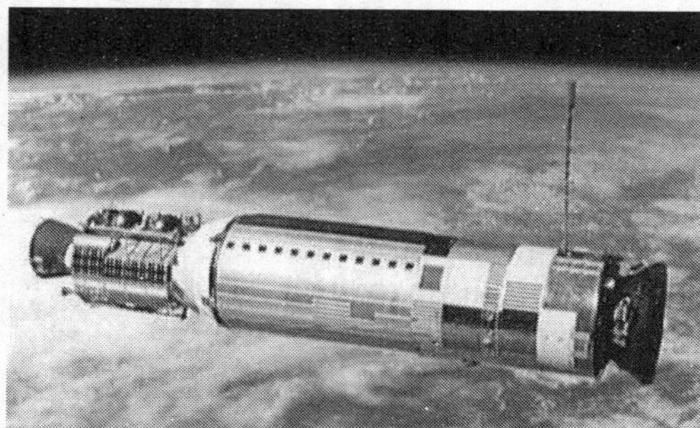

Gemini-VIII

1975
The American unmanned spacecraft *Mariner-10* made the third and last close approach to Mercury. During three fly-bys at intervals of six months, *Mariner-10* could cover only half of the surface of Mercury, as the planet's slow rotation left the other half always in the dark when *Mariner-10* returned. (☞ Feb. 5, Mar. 29, Sep. 21, Nov. 3)

1998
Derek Harold Richard Barton, British chemist, Nobel laureate (Chemistry, 1969), whose work helped establish conformational analysis – the study of the 3-D geometric structure of complex molecules, died in Texas, USA. (☞ Sep. 8)

2000
The 220-MW first unit of the Kaiga Atomic Power Station at Kaiga in Karnataka, India, was commissioned. It is a pressurised heavy water reactor that uses natural uranium as fuel and heavy water as moderator and coolant. (☞ Nov. 16)

March 17

1782
Daniel Bernoulli, Swiss mathematician, who developed the Bernoulli's theorem, relating pressure, velocity, and elevation in a moving fluid, died in Basel, Switzerland. (☞ Feb. 8)

1846
Friedrich Wilhelm Bessel, German astronomer, who established exact positions for thousands of individual stars and was the first to determine the distance of a star using the parallax method, died in Königsberg, Prussia (now Kaliningrad, Russia). (☞ Jul. 22)

1853
Christian Doppler, Austrian physicist who first described how the observed frequency of sound waves is affected by the relative motion of the source and the detector, the phenomenon known as the 'Doppler effect,' died in Venice, Italy. (☞ Nov. 29)

1881
Walter Rudolf Hess, Swiss physiologist, Nobel laureate, was born in Frauenfeld, Switzerland. Hess completed a course at the Gymnasium in 1900. He took his

Doctor of Medicine degree in 1906 from the University of Zurich. Although his aim was always to be a physiologist, due to various reasons he became a practising ophthalmologist. He began a lucrative private practice in Repperswil, Switzerland as an ophthalmologist in 1908, but gave it up in 1912 to devote time to the study of physiology at the University of Bonn. He was appointed professor of physiology and later director of the Physiological Institute at the University of Zürich. He became interested in the study of the autonomic nervous system – those nerves originating at the base of the brain and extending throughout the spinal cord that control and regulate the internal organs and functions such as digestion and excretion without any conscious recognition or effort by the organism. Hess perfected a method of implanting electrodes in the brains of rats and was thus able to localise centres of the brain associated with certain instincts. Using fine electrodes to stimulate or destroy specific areas of the brain in cats and dogs, he found that the seat of autonomous function lies at the base of the brain, in the medulla oblongata and that part of the interbrain known as the hypothalamus. Hess was awarded the 1949 Nobel Prize for Physiology or Medicine, jointly with the Portuguese neurologist António Egas Moniz, for his discovery of the functional organisation of the interbrain as a coordinator of the activities of the internal organs. (☞ Aug. 12, Nov. 29)

1950
Scientists at the University of California at Berkeley, USA announced the creation of a new radioactive element, element 98, named 'californium' (symbol Cf). Californium is a synthetic chemical element of the actinide series in Group IIIb of the periodic table. The scientists – Stanley G. Thompson, Kenneth Street, Jr., Albert Ghiorso, and Glenn T. Seaborg – produced the isotope californium-245 by bombarding curium-242 (atomic number 96) with helium ions in the 60-inch cyclotron at Berkeley. Since then, longer-lived isotopes have been created, including californium-251 with a half-life of 800 years. Californium-252, with a half-life of 2.65 years, has industrial and medical applications as a very intense point source of neutrons. (☞ Apr. 19)

1956
Irène Joliot-Curie, French physical chemist, Nobel laureate (Chemistry, 1935), who with her husband Frédéric Joliot-Curie, prepared radioactive isotopes artificially, died in Paris, France. (☞ Sep. 12, Mar. 19)

1983
Haldan Keffer Hartline, American physiologist, Nobel laureate (Physiology or Medicine, 1967), whose work led to an understanding of the chemical mechanism of vision, died in Fallston, Maryland, USA. (☞ Dec. 22)

1988
IRS-1A, the first satellite in India's IRS series was launched by a Semiorka rocket from Tyouratam in Khazakistan. Designed and built by the Indian Space Research Organisation, the satellite carried two linear imaging and self scanner (LISS) cameras for Earth imaging and was placed in a Sun-synchronous polar orbit

at an inclination of 99 degrees and an altitude of 904 kilometres. It takes 103 minutes to complete one orbit around the Earth, making 14 orbits per day. The satellite returns periodically over the same geographical area, giving a complete coverage of the globe in 22 days.

March 18

1858
Rudolf Diesel, German thermal engineer who invented the high-compression internal-combustion engine that bears his name, was born in Paris, France. Diesel was educated at the Munich Polytechnic. After graduation he was employed as a refrigerator engineer, but he was more interested in engine

design. After studying the four-stroke internal combustion engines developed by German engineer Nikolaus August Otto, Diesel conceived of an engine that would not need a spark plug for ignition, but would be pressure-ignited. In his engine, air would be compressed so much that the temperature would far exceed the ignition temperature of the fuel. In 1894, Diesel filed for a patent of his new invention, dubbed the diesel engine. He operated his first successful engine in 1897. By 1898, Diesel was a millionaire. His engines were used to power pipelines, electric and water plants, automobiles and trucks, and marine craft, and soon after were used in mines, oil fields, factories, and ocean-going liners. (☞ Feb. 29, Sep. 29)

1964
Norbert Wiener, American mathematician, who established the science of cybernetics, died in Stockholm, Sweden. (☞ Nov. 26)

1965
Russian cosmonaut Aleksey Leonov became the first human to perform a space walk. The spacecraft *Voskhod-2* was launched into space carrying Leonov and Pavel Belyayev aboard. On the second orbit, Leonov left the spacecraft through the air lock while still tethered to the vessel. He was the first man to climb out of a spacecraft in space. While outside, he took motion pictures and practised moving outside the spacecraft for 20 minutes. *Voskhod-2* made 17 orbits at about 177 kilometres above Earth. (☞ May 30, Jul. 17)

1987
The discovery of 'high-temperature' superconductivity was announced at a meeting of the American Physical Society (APS) in New York, USA. The phenomenon of superconductivity, discovered in 1911, was at first known to occur at only 4 degrees

above absolute zero (–269°C), when all electrical resistance in a metal sample disappeared. In April 1986, two researchers at IBM in Switzerland, K. Alex Muller and George Bednorz, detected superconductivity in a ceramic material at a temperature of 35 degrees above absolute zero (–238°C). When published in September 1986, the news stirred the wider scientific community into action. By January 1987, M. K. Wu at the University of Houston had discovered materials that showed superconductivity at 100 degrees above absolute zero (–173°C), which was announced at the APS meeting. The discovery of high-temperature superconductors opened up enormous possibilities in the use of superconductors as a means of saving costs, since inexpensive liquid nitrogen (b.p.: –196°C) could be used for cooling instead of the prohibitively expensive liquid helium (b.p.: –269°C) used earlier. (☞ Apr. 20, May 16)

March 19

721 BCE
The first ever recorded solar eclipse was seen from Babylon.

1876
John Hubert Marshall, English archaeologist, director-general of the Archaeological Survey of India (1902–31), was born in Chester, Cheshire, England. In the 1920s, Marshall was responsible for the large-scale excavations that revealed Harappa and Mohenjo-daro, the two largest cities of the previously unknown Indus Valley Civilisation, a civilisation which he firmly believed was comparable in every way with the ancient civilisations of Egypt and Mesopotamia. Marshall re-organised the Archaeological Survey of India and greatly expanded its scope of activity. Initially, his chief task was to save and conserve the standing Indian temples, sculptures, paintings, and other ancient remains, many of which had been long neglected and were in a sad state of decay. His energetic efforts resulted in the preservation of ancient buildings all over British India. He was also responsible for the excavation of ancient sites at Taxila, Vaisali, Nalanda, Rajagriha and Sarnath. (☞ Aug. 17)

1883
Walter Norman Haworth, British chemist, Nobel laureate, was born in Chorley, Lancashire, England. Haworth attended the local school until the age of fourteen when he joined his father's linoleum factory to learn linoleum design and manufacture. His interest in chemistry was aroused through the use of dye stuffs in his work. He graduated from the University of Manchester in 1906, was awarded a doctorate from Göttingen in 1910, and returned to Manchester to get his DSc degree in 1911. After holding teaching positions in Scotland and Durham, Haworth was appointed Professor and Director of the Department of Chemistry in the University of Birmingham in 1925 and he remained in this position until his retirement in 1948. Haworth's early researches involved investigations on the constitution of terpenes but later he turned his attention to carbohydrates, extending the German chemist Emil Fischer's method of reacting sugars with methanol to an elegant preparation of methylated derivatives which were, in turn, used to characterise the constitution of sugars. By 1928, Haworth had evolved and confirmed, among others, the structures of several sugars and the glucoside

ring structure of normal sugars. He was the first to prepare vitamin C, or ascorbic acid, artificially, which considerably reduced the cost of its commercial production. Haworth was awarded the 1937 Nobel Prize for Chemistry, jointly with the Swiss chemist Paul Karrer, for his investigations on carbohydrates and vitamin C. (☞ Mar. 19, Apr. 21)

1900

Frédéric Joliot-Curie, French physical chemist, Nobel laureate, was born in Paris, France. After graduating from the Ecole de Physique et Chimie in Paris, Frédéric became assistant of Marie Curie in the Radium Institute, Paris. He obtained his Doctor of Science degree in 1930, having prepared a thesis on the electrochemistry of radioelements. He became a lecturer in the Paris Faculty of Science in 1935, and was appointed professor of the Collège de France in 1937. Jointly with his wife Irene Joliot-Curie he carried out considerable research on the structure of the atom, but their greatest discovery was artificial radioactivity. By bombarding boron, aluminium, and magnesium with alpha particles, they produced the isotopes nitrogen-13, phosphorus-30, silicon-27 and aluminium-28. Frédéric was awarded the 1935 Nobel Prize for Chemistry, jointly with his wife Irène, in recognition of their synthesis of new radioactive elements. (☞ Aug. 14, Sep. 12)

1943

Mario J. Molina, Mexican-born American chemist, Nobel laureate, was born in Mexico City, Mexico. Molina was awarded a degree in chemical engineering in 1965 from the Universidad Nacional Autonoma de Mexico and a postgraduate degree in 1967 from the University of Freiburg, Germany. He received a PhD in physical chemistry in 1972 from the University of California, Berkeley, where he worked for a year before joining the American chemist F. Sherwood Rowland at the University of California, Irvine. The pair conducted experiments on pollutants in the atmosphere, discovering that man-made substances such as chlorofluorocarbons (CFCs) used in spray cans and refrigerators rise into the stratosphere, where ultraviolet radiation breaks them into their component elements of chlorine, fluorine, and carbon. Their researches showed that CFCs could harm the fragile ozone layer that protects the world from the dangerous ultraviolet radiation of the Sun. The discoveries of Molina, Rowland, and the German chemist Paul Crutzen, that some industrially manufactured gases deplete the ozone layer, led to the signing of the Montreal Protocol in 1987 to limit the widespread use of CFCs. Molina joined the Massachusetts Institute of Technology, Cambridge, Massachusetts as Professor of Earth, Atmospheric, and Planetary Sciences in 1989. Molina was awarded the 1995 Nobel Prize for Chemistry, jointly with Rowland and Crutzen, for their work in atmospheric chemistry, particularly concerning the formation and decomposition of ozone. (☞ Jun. 28, Dec. 3)

1950

Walter Norman Haworth, British chemist, Nobel laureate (Chemistry, 1937), who was the first to synthesise vitamin C, died in Birmingham, UK. (☞ Mar. 19)

1987
Louis-Victor de Broglie, French physicist, Nobel laureate (Physics, 1929), best known for his research on quantum theory and for his discovery of the wave nature of electrons, died in Paris, France. (☞ Aug. 15)

March 20

1727
Isaac Newton, English physicist and mathematician, who formulated the law of universal gravitation and discovered the composition of white light, laying the foundation for modern physical optics, died in London, UK. (☞ Dec. 25)

1916
Einstein's 'general theory of relativity' was published as an academic paper in the journal *Annalen der Physik*, titled 'Die Grundlagen der allgemeinen Relativitästheorie'. At the time of publication, this theory was thought to, though not proven to, account for the slow rotation of the elliptical path of the planet Mercury, which Newtonian gravitational theory failed to do. Einstein's general theory of relativity was publicly verified during the total solar eclipse of 1919. Photographs taken during the eclipse by the Royal Society of London clearly showed the displacement of stars caused by the bending of light by Sun's gravitation, as predicted by Einstein's theory. (☞ Mar. 14, Apr. 18)

1934
The first test of a practical radar apparatus was made in Kiel Harbour, Germany, by Rudolf Kuhnold, Chief of the German Navy Signals Research Department. His 700-watt transmitter worked on a frequency of 600 megahertz. Using a receiver and disk reflectors, it could receive echoes from signals bounced off the battleship *Hesse*, anchored more than half-a-kilometre away. Although it was invented in Germany, during World War II, it was the US and Great Britain that took Kuhnold's pioneering work and developed it into a reliable detection system.

1944
Erwin Neher, German biophysicist, Nobel laureate, was born in Landsberg, Germany. Neher earned a degree in physics from the Technical University of Munich and then attended the University of Wisconsin at Madison, USA, where he obtained an MSc degree in 1967. From 1968 to 1972, Neher did graduate and postdoctoral work at the Max Planck Institute for Psychiatry in Munich. He first developed the idea of the 'patch-clamp' technique – a laboratory method that can detect the very small electrical currents produced by the passage of ions through the cell membrane – in his doctoral thesis and earned his PhD from the Technical University of Munich in 1970. In 1972 Neher went to the Max Planck Institute for Biophysical Chemistry, Göttingen, and two years later began his collaboration with Bert Sakmann. They used a thin glass pipette, one-thousandth of a millimetre in diameter, fitted with an electrode which when in contact with a cell membrane, could detect the flow of individual ions through the ion channels in the cell membrane. The technique was used to study a broad range of cell functions. Neher and Sakmann presented their patch-clamp findings at a scientific gathering in 1976. Neher was awarded the 1991 Nobel Prize for Physiology or Medicine, jointly with Sakmann, for their discoveries concerning the function of single ion channels in cells. (☞ Jun. 12)

1962
Biresh Chandra Guha, Indian biochemist and nutritionist, who made valuable contribution to the study of biosynthesis of vitamin C, died in Lucknow, Uttar Pradesh, India. (☞ Jun. 8)

1993
Polykarp Kusch, German-American physicist, Nobel laureate (Physics, 1955), who accurately determined the magnetic moment of the electron, died in Dallas, Texas, USA. (☞ Jan. 26)

March 21

1925
Wolfgang Pauli, Austrian physicist, published his 'exclusion principle'. In an article in the journal *Zeitschrift für Physik*, Pauli introduced the idea that two nearby electrons cannot be in exactly the same state at the same time. (☞ Apr. 25, Dec. 15)

1932
Walter Gilbert, American molecular biologist, Nobel laureate, was born in Boston, Massachusetts, USA. Gilbert studied chemistry and physics at the Harvard University, Cambridge, Massachusetts. After graduating, he went to the Cambridge University, England, where he pursued his PhD in physics. There he made acquaintance with another young American, James D. Watson, who was working on the structure of DNA. By 1957, Gilbert was back at Harvard. In 1960, he started spending his free time in biology laboratories, where he collaborated with talented scientists from various disciplines in the search for the 'messenger' that relayed information from DNA to the areas in the cell where proteins are manufactured. In 1961, Gilbert published his first paper on messenger RNA in the journal *Nature*, and gained immediate respect for his work in experimental biology. In 1964, he published the last of his papers in theoretical physics and officially became a tenured biophysicist at Harvard. Gilbert then got involved in solving the biggest mystery of the moment in molecular biology: why different cells produced different proteins, even though they all had exactly the same information in their DNA. This line of questioning led to his laboratory's discovery of the lac repressor in the bacterium *Escherichia coli.* His group developed a method for determining the sequence of nucleotide links in the chain-like molecules of nucleic acids (DNA and RNA). Gilbert was awarded the 1980 Nobel Prize for Chemistry, jointly with the American biochemist Paul Berg and English biochemist Frederick Sanger, for their contributions concerning the determination of base sequences in nucleic acids. (☞ Jun. 30, Aug. 13)

1965
American space probe *Ranger-9* was launched towards the Moon. (☞ Mar. 24)

1986
The 170-MW second unit of the Madras Atomic Power Station at Kalpakkam, near Madras (now Chennai), Tamil Nadu, India, began commercial operation. It is a pressurised heavy water reactor that uses natural uranium as fuel and heavy water as moderator and coolant. (☞ Jan. 27)

1988

Patrick Christopher Steptoe, British obstetrician and gynaecologist, who together with fellow British medical scientist Robert Edwards perfected the technique of *in vitro* fertilisation of the human egg to enable infertile couples to have babies, died in Canterbury, Kent, UK. (☞ Jun. 9, Jul. 25, Sep. 27)

1996

The Indian remote sensing satellite, *IRS-P3*, was launched from Sriharikota, Andhra Pradesh, India using a polar satellite launch vehicle PSLV-D3. The satellite was put in a polar, Sun-synchronous orbit at an altitude of 817 kilometres. It carried an X-ray astronomy and two remote sensing payloads, namely the wide field sensor (WiFS) and the modular optoelectronics scanner (MOS), which was primarily meant for ocean related studies.

1999

A hot air-cum-gas balloon carrying human passengers achieved the first non-stop round-the-world balloon flight. After 20 days of flight, the experimental *Breitling Orbiter-3* balloon, flown by the British pilot Brian Jones and the Swiss psychiatrist Bertrand Piccard, touched down in the Egyptian desert. The *Breitling Orbiter-3* had travelled 40,814 kilometres on a non-stop global journey that took 19 days, 21 hours and 47 minutes to complete, reaching an altitude of 11,737 metres. The two balloonists had set off from the Swiss Alps on March 1, and then drifted southwest to North Africa. Once over Mauritania in North Africa, the wind speed and direction changed and the balloon started to head across Mali, Algeria, Libya and Egypt before crossing into Asia. After crossing India, Bangladesh and Myanmar, the *Breitling Orbiter-3* crew had a crucial advantage over rival balloonists: They got permission to pass over Southern China, a move that saved them vital time. After crossing China, picking up a fast jet stream over the Pacific, the Swiss–English duo made good time until grinding to a near halt over Central America due to uncooperative winds. The *Orbiter* finally picked up a jet stream 24 hours later and continued eastward toward North Africa. Technically, they would have broken the record when they crossed over northern Mauritania. But rather than land in the remote wastelands of Mauritania, Jones and Piccard opted to head for a more accessible landing zone in Egypt. (☞ Mar. 1)

March 22

1394

Ulugh Beg, Mongol scientist, mathematician, and the greatest astronomer of his time, was born in Soltaniyeh, Timurid, Persia (now Iran). In 1417, to push forward the study of astronomy, Ulugh Beg began building a madrasah as a centre for higher education and appointed the best scientists he could find to positions there as lecturers. In addition to the madrasah, Ulugh Beg built an observatory at Samarkand, the construction of which began in 1428. The observatory, which was circular in shape, had three levels. It was over 50 metres in diameter and 35 metres high. In his observations, Ulugh Beg discovered a number of errors in the stellar catalogue computed by the 2nd-century Alexandrian astronomer Ptolemy, whose figures were still being used.

Beg led scientific meetings where problems in astronomy were freely discussed. His *Catalogue of the Stars* was the first comprehensive stellar catalogue since that of Ptolemy. (☞ Oct. 27)

1868

Robert Andrews Millikan, American physicist, Nobel laureate, was born in Morrison, Illinois, USA. Millikan led a rural existence in his childhood, attending the Maquoketa High School in Iowa. During his undergraduate course at the Oberlin College in Ohio, his favourite subjects were Greek and mathematics; but after his graduation in 1891 he took, for two years, a teaching post in elementary physics. It was during this period that he developed his interest in the subject in which he was later to excel. In 1893, after obtaining his master's degree in physics, he was appointed Fellow in Physics at the Columbia University. He received his PhD in 1895 for research on the polarisation of light emitted by incandescent surfaces. Millikan then spent a year in Germany, at the Universities of Berlin and Göttingen, and returned to the USA at the invitation of physicist Albert Abraham Michelson, to become an assistant at the newly established Ryerson Laboratory at the University of Chicago. Millikan was an eminent teacher, and became professor at the University of Chicago in 1910, a post that he retained till 1921. During his early years at Chicago he spent much time preparing textbooks and simplifying the teaching of physics. Millikan made numerous momentous discoveries, chiefly in the fields of electricity, optics, and molecular physics. His earliest major success was the accurate determination of the charge carried by an electron, using the elegant 'falling-drop method'; he also proved that this quantity was a constant for all electrons, thus demonstrating the atomic structure of electricity. Next, he experimentally verified Einstein's all-important photoelectric equation, and made the first direct photoelectric determination of Planck's constant *h*. He coined the term 'cosmic rays' during his study of the radiation from outer space. Millikan was awarded the 1923 Nobel Prize for Physics for his work on the elementary charge of electricity and on the photoelectric effect. (☞ Dec. 19)

1895

The first motion picture was shown on a screen by French inventors Auguste and Louis Lumière in Paris. An invited audience at 44 Rue de Rennes in Paris viewed the film *La Sortie des ouvriers de l'usine Lumière*. The film showed workers leaving the Lumières' own factory in Lyon. The workers streamed out, most on foot, some with their bicycles, followed by those with cars. The Lumières soon began opening cinemas in London, Brussels, Berlin and New York. (☞ Apr. 10, Jun. 6, Oct. 5, Oct. 19, Dec. 28)

1931

Burton Richter, American physicist, Nobel laureate, was born in Brooklyn, New York, USA. Richter studied at the Massachusetts Institute of Technology, Cambridge, USA, from where he received his doctorate in 1956. He then did postdoctoral work at the Stanford University, Stanford, California, to become a professor in 1967, and director of the Stanford Linear Accelerator Centre in 1984. A pioneer in colliding beam technology,

Richter's research centered on experimental particle physics with high energy electrons and electron–positron colliding beams. With financial support from the Atomic Energy Commission, in 1973 he completed construction of the Stanford Positron–Electron Asymmetric Ring, a colliding-beam accelerator with which he discovered a new subatomic particle that he called the J-particle (now called the J/psi particle), the first of a new class of very massive, long-lived mesons. Richter was awarded the 1976 Nobel Prize for Physics, jointly with fellow American physicist Samuel Chao Chung Ting, for their pioneering work in the discovery of a heavy elementary particle of a new kind. (☞ Jan. 27)

1960

The first laser was patented by American physicists Arthur L. Schawlow and Charles Hard Townes, who had done the research in the Bell Telephone Laboratories, Murray Hill, New Jersey, USA. The invention of the laser, which stands for Light Amplification by Stimulated Emission of Radiation, can be dated to 1958 with the publication of the scientific paper, 'Infrared and optical masers', by Schawlow, then a Bell Labs researcher, and Townes, a consultant to Bell Labs. That paper, published in *Physical Review*, the journal of the American Physical Society, launched a new scientific field and opened the door to a multibillion-dollar industry. (☞ May 5)

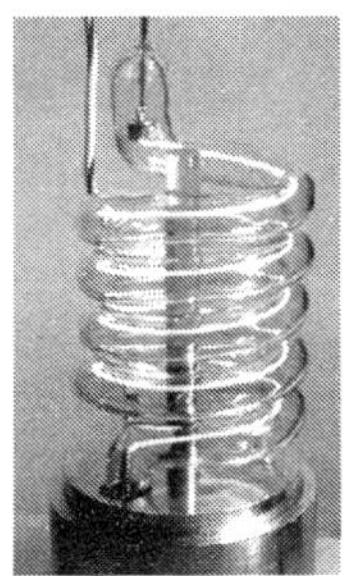

First laser

1982

The American space shuttle *Columbia*, carrying a two-man crew of Jack R. Lousma, and C. Gordon Fullerton, was launched on its third flight. The mission included testing of a remote manipulator system and measurements of thermal response of the orbiter in various attitudes to Sun. (☞ Feb. 1, Mar. 30, Apr. 12)

1990

Three Indian scientists, R. S. Nadgauda, V. A. Parasharami and A. F. Mascarenhas, announced the successful induced flowering of bamboo in a paper published in the journal *Nature*. The scientists, working in the National Chemical Laboratory (NCL), Pune, Maharashtra, were able to induce flowering *in vitro* in two bamboo varieties, viz., *Bambusa arundinaceae* and *Dendrocalamus brandisii*, within a few months. Normally bamboo flowers only once during its lifetime, after a lapse of 12 to 120 years, which makes breeding from seeds extremely time-consuming. Therefore the NCL success came as a major breakthrough. It opened up a possibility of hybridisation and improvement of bamboos, which is difficult by conventional methods. (☞ Jan. 3)

1996

American space shuttle *Atlantis*, carrying astronauts Kevin P. Chilton, Richard A. Searfoss, Shannon W. Lucid, Linda M. Godwin, Michael R. Clifford and Ronald M. Sega was launched on its 16th flight from Kennedy Space Center, Florida, USA. During the mission *Atlantis* docked with the Russian space station *Mir*, for the third time. Also during this mission American astronaut, Shannon W. Lucid, was transferred to *Mir* for an extended stay. The linkup was intended as preparation for the planned internationally-operated space station. (☞ Mar. 24, Mar. 31)

1997

Comet Hale-Bopp made its closest approach to Earth, passing within

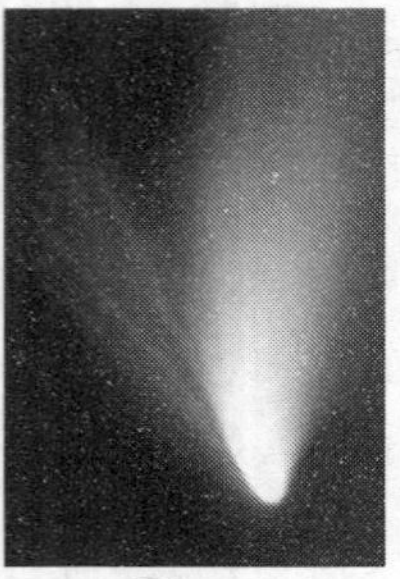

197 million kilometres of our planet. Its icy nucleus is estimated to be 40-km wide, making it at least ten times larger than that of Comet Hyakutake and twice the size of Comet Halley.

2000

INSAT-3B, the first satellite of India's third generation INSAT series was launched by the Ariane-505 vehicle of the European launch provider Arianespace from Kourou, French Guyana. Designed and built by the Indian Space Research Organisation, *INSAT-3B* carries 12 extended C-band transponders, three Ku-band transponders, and S-band mobile satellite service payloads. The 2,070-kg satellite is primarily intended for business, developmental and mobile communication. It is located at 83° East longitude in the geostationary orbit.

March 23

1749

Pierre-Simon, Marquis de Laplace, French mathematician, astronomer, and physicist who is best known for his investigations into the stability of the solar system, was born in Beaumount-en-Auge, Normandy, France. At the age of 18 Laplace went to Paris, where he was appointed as a professor in the military school, and admitted as a member of the Academy of Science. He had by this time mastered the whole range of mathematical science, as then known, and had besides, solved several problems, which had for many years defied the attempts of astronomers. Among Laplace's achievements was an improvement in the mathematical description of the solar system, which suggested that the system could be self-sustaining – without the need Newton saw for occasional divine intervention to keep it stable. Laplace investigated the variations of the Moon's motions, especially as affected by the eccentricity of the Earth's orbit; the inequalities in the motions of Jupiter and Saturn; the motion of the moons of Jupiter; the aberration in the movements of comets; and the theory of the tides. He was elected to the French Academy in 1816. (☞ Mar. 5)

1840

Englishman J.W. Draper took the first successful photograph of the Moon. He made a daguerreotype, a precursor of the modern photograph. In a daguerreotype, which is a direct-positive process, the image is created on a sheet of copper plated with a thin coat of silver without the use of a negative. Exposure times for the earliest daguerreotypes ranged from three to fifteen minutes.

1881

Hermann Staudinger, German chemist, Nobel laureate, was born in Worms, Germany. Staudinger attended school in Worms, and after matriculation in 1899 continued his studies first at the University of Halle and later at Darmstadt and Munich. He graduated from Halle in 1903. In November 1907 he was appointed

associate professor at the Polytechnic in Karlsruhe, moved to Zurich, Switzerland, in 1912, and finally to the University of Freeburg, Germany, in 1926. Staudinger's first discovery was that of the highly reactive organic compounds known as ketenes. His work on polymers began with the research he conducted for the German chemical firm BASF on the synthesis of isoprene, the monomer of which natural rubber is composed. While working on the synthesis of isoprene in 1910, he recognised that the current theories for the structure of natural rubber were incorrect. In May 1922, Staudinger introduced the concept of macromolecules, a term which he coined, and polymerisation, in a report which appeared in *Helvetica Chimica Acta*. His pioneering work provided the theoretical basis for the polymer chemistry and greatly contributed to the development of modern plastics and laid the foundation for the great expansion of the plastics industry later in the 20th century. His researches on polymers eventually contributed to the development of molecular biology. Staudinger was awarded the 1953 Nobel Prize for Chemistry for his discoveries in the field of macromolecular chemistry. (☞ Sep. 8)

1907

Daniel Bovet, Swiss-French-Italian pharmacologist, Nobel laureate, was born in Neuchâtel, Switzerland. After completing his secondary education at Geneva, Bovet graduated from the University of Geneva in 1927. From 1929 until 1947 he worked at the Pasteur Institute in Paris. Here he worked first as assistant and later as Chief of the Laboratory of Therapeutic Chemistry. In 1944, Bovet discovered pyrilamine (mepyramine), the first antihistamine, which was effective against allergic reactions. In 1947, a search for a synthetic substitute for curare (a muscle relaxant) led to his discovery of gallamine and other muscle relaxants. In 1947 he joined the Istituto Superiore di Sanità in Rome, Italy to organise a Laboratory of Therapeutic Chemistry. He then became Chief of the Laboratory of Therapeutic Chemistry of the Institute. Bovet was awarded the 1957 Nobel Prize for Physiology or Medicine for his discoveries relating to synthetic compounds that inhibit the action of certain body substances, and especially their action on the vascular system and skeletal muscles. (☞ Apr. 8)

1912

Wernher von Braun, German-born American rocket engineer who played a prominent role in all aspects of rocketry and the early stages of space exploration, was born in Wirsitz, Germany. Braun studied engineering at Berlin's Charlottenburg Institute of Technology and after reading *The Rocket into Interplanetary Space* by Hermann Oberth he became interested in rocket technology and helped form the German Society for Space Travel.

In 1934 Braun successfully built two rockets for the Ordnance Department of the German Army that rose vertically for more than 2.4 kilometres. In 1937 Braun became the technical director of the establishment and began to develop a long-range ballistic missile and a supersonic anti-aircraft missile. During World War II, Braun began working on a new secret weapon, the *V2* rocket. This 15-m-long, liquid-fuelled rocket carried a one-ton warhead, was capable of supersonic speed and could fly at an altitude of over 80 kilometres. As a result it could not be effectively stopped once launched. The *V2* rocket was first used in September 1944. Over 5,000 *V2*s were fired on Britain. In 1952 Braun moved to USA and became technical director of the US Army's Ballistic Missile Agency and was chiefly responsible for the manufacture and successful launching of Redstone, Jupiter-C, Juno and Pershing missiles. After the Soviet Union successfully launched *Sputnik* in October 1957, Braun concentrated on the development of space rockets and in January 1958 helped launch the first American satellite *Explorer-I*. In 1960 Braun became director of the Marshall Space Flight Center where he developed the Saturn rocket that helped the United States to land on the Moon in 1969. (☞ Jun.16)

1929

Govind Swarup, Indian radio astronomer, Fellow of the Royal Society, was born in Thakurdwara, Uttar Pradesh, India.

After getting his master's degree from the Allahabad University, Swarup did his doctorate from the Stanford University, USA. Swarup's main contributions have been in the field of solar and extragalactic radio astronomy and cosmology. He guided and supervised the construction of the Ooty Radio Telescope and was the driving force behind the setting up of India's largest radio telescope – the Giant Metre-wave Radio Telescope – near Pune in Maharashtra.

1946

Gilbert Newton Lewis, American chemist whose theory of the electron pair fostered understanding of the covalent bond and extended the concept of acids and bases, died in Berkeley, California, USA. (☞ Oct. 23)

1950

The UN World Meteorological Organisation (WMO) was established. WMO is an intergovernmental organisation with a membership of 187 member states and territories. It originated from the International Meteorological Organisation (IMO), which was founded in 1873. WMO is the specialised agency of the United Nations for meteorology (weather and climate), operational hydrology and related geophysical sciences. Since its establishment, WMO has played a unique and powerful role in contributing to the welfare of humanity. Under the WMO leadership and within the framework of WMO programmes, the National Meteorological and Hydrological Services have contributed substantially to the protection of life and property against natural disasters, to safeguarding the environment and to enhancing the economic and social well being of all sectors of society in areas such as food security, water resources and transport. March 23 is thus observed as World Meteorology Day.

1959

The maser was patented by American physicist Charles H. Townes. 'Maser' is an acronym for 'microwave amplification by the stimulated emission of radiation'. The invention is an apparatus for amplifying and producing electromagnetic energy directly from excited molecules or atoms. The concept grew out of research in microwave spectroscopy following World War II. The maser ultimately led to the development of the laser. (☞ Jul. 28)

1965
America's first two-person space flight began as *Gemini-3* blasted off from Cape Kennedy with astronauts Virgil I. Grissom and John W. Young aboard. The spacecraft landed safely after completing three orbits of the Earth with altitude ranging from 122 to 175 kilometres.

1983
Barney C. Clark, the first recipient of a permanent artificial heart, *Jarvik-7*, died at the University of Utah's Medical Center after 112 days with the device. Doctors at the University of Utah Medical Center said that the death resulted from a massive circulatory collapse 'resulting from a multitude of causes'. (☞ Dec. 2)

1989
Electrochemists Stanley Pons and Martin Fleischmann at the University of Utah, USA, claimed to measure a production of heat that could only be explained by a nuclear process – termed 'cold fusion'. They used just a pair of palladium electrodes connected to a battery and immersed in a jar of heavy water. They claimed they had sustained a controlled nuclear fusion reaction in a test-tube that generated up to 100 per cent more energy than they put in. They stated that heavy hydrogen absorbed by the palladium electrodes underwent fusion. Although there were sporadic sightings of excess heat, which Fleischmann said could not be accounted for by chemistry alone, efforts to replicate the experiment by other scientists did not succeed. Crucially, there were no telltale signs of nuclear processes, notably subatomic particles called neutrons.

2001
After fifteen triumphant years in orbit, the Russian space station *Mir* ended its record-breaking mission with a flawless re-entry into Earth's atmosphere. (☞ Feb. 20)

March 24

1776
John Harrison, English clock designer, who developed and built the world's first successful maritime clock, died in London, UK. (☞ Mar. 31)

1864
Karl Karlovich Klaus, Russian chemist who discovered ruthenium died in Dorpat, Russia (now Tartu, Estonia). (☞ Jan. 23)

1882
Robert Koch, German physician, announced before the Physiological Society of Berlin that he had isolated and grown the tubercle bacillus, which he believed to be the cause of all forms of tuberculosis. Three weeks later, he published an article entitled 'The etiology of tuberculosis'. He had observed the bacillus in association with all cases of the disease and grown the organism outside the body of the host. He could also reproduce the disease in a susceptible host inoculated with a pure culture of the isolated organism. This day is observed every year as 'World TB Day'. (☞ May 27, Dec. 11)

1884
Peter Joseph Wilhelm Debye, Dutch physical chemist, Nobel laureate, was born in Maastricht, Netherlands. Debye received his early education at the elementary and secondary schools in his hometown. He continued his studies at the Aachen Institute of Technology from where he received a degree in electrical technology in 1905. In 1911, Debye became professor of theoretical physics at the

Zurich University in Switzerland, where he remained for two years. He returned to The Netherlands in 1912 when he was appointed professor of theoretical physics at the Utrecht University, and in 1914 he moved to the University of Göttingen, to take charge of the Theoretical Department of the Physical Institute. Debye's first major scientific contribution was the application of the concept of dipole moment to the charge distribution in asymmetric molecules in 1912. He developed equations relating dipole moments to temperature, dielectric constant, etc. In consequence, molecular dipole moments are measured in 'debyes,' a unit named in his honour. Debye established a theory of specific heat with some improvements on that proposed by Einstein. He also performed important work in the analysis of crystalline powders using X-ray diffraction techniques and determined the dimensions of gaseous molecules and interatomic distances using X-rays. Debye was awarded the 1936 Nobel Prize for Chemistry for his contributions to our knowledge of molecular structure through his investigations on dipole moments and on the diffraction of X-rays and electrons in gases. (☞ Nov. 2)

1903

Adolf Butenandt, German biochemist, Nobel laureate, was born in Bremerhaven-Lehe, Germany. Butenandt went to school at Bremerhaven and studied chemistry at the Universities of Marburg and Göttingen. In 1927 he graduated from the University

of Göttingen. From 1927 until 1930 he was scientific assistant at the Institute of Chemistry, Göttingen. From 1936 until 1960 he was professor in the University of Berlin and Director of the Max Planck Institute for Biochemistry, Berlin-Dahlem, which later moved to Tübingen and then to Munich. Between 1945 and 1956 he taught at the Universities of Tübingen and Munich. From 1956 until 1960 he was Director of the Institute of Physiological Chemistry in the University of Munich. Butenandt's research mainly concerned sex hormones. In 1929 he isolated oestrone (a hormone that influences development of the female reproductive tract) in pure, crystalline form. In 1931 he isolated the male sex hormone androsterone in pure, crystalline form and in 1934 he isolated the hormone progesterone from the corpus luteum. He obtained testosterone from androsterone in 1939. In addition to these researches, Butenandt carried out investigations of the interrelationships of the sex hormones and on the possible carcinogenic properties of some of them. His work on the sex hormones was largely responsible for the production of cortisone on a large scale. Butenandt was awarded the 1939 Nobel Prize for Chemistry, jointly with the Swiss chemist Leopold Ružička. Butenandt received half the prize money for his work on sex hormones, while Ružička received the other half for his work on polymethylenes and higher terpenes. (☞ Jan. 18, Sep. 3)

1905

Jules Verne, French author whose writings laid the foundation of modern science fiction, died in Amiens, France. (☞ Feb. 8)

1917

John Cowdery Kendrew, British biochemist, Nobel laureate, was born in Oxford, Oxfordshire, England. Kendrew was educated at the Dragon School, Oxford and Clifton College, Bristol. He went to the Trinity College, Cambridge in 1936 as a major scholar and graduated in chemistry in 1939. He spent the first few months of

the war doing research on reaction kinetics in the Department of Physical Chemistry at Cambridge. During World War II, he was engaged in operational research at the Royal Air Force headquarters. In the war years his interests became more biological, and he decided to work on the structure of proteins. He returned to Cambridge in 1946 and, in the Cavendish Laboratory, began collaboration with Austrian-born British bio-chemist Max Ferdinand Perutz. Kendrew and Perutz's work mostly centred on the X-ray analysis of the muscle protein myoglobin, which stores oxygen and gives it to the muscle cells when needed. This project culminated in the production of a three-dimensional model of myoglobin at 6Å resolution in 1957, and an almost complete structure in 1960. Kendrew and Perutz were jointly awarded the 1962 Nobel Prize for Chemistry for their studies of the structures of globular proteins. (☞ May 19, Aug. 23)

1941

Joseph H. Taylor, Jr., American radio astronomer and physicist, Nobel laureate, was born in Philadelphia, USA. Taylor graduated from the Haverford College, Philadelphia in 1963 and earned a PhD in astronomy from the Harvard University, Cambridge, Massachusetts, in 1968. He taught at the University of Massachusetts, Amherst, from 1969 to 1981 and then joined the faculty at the Princeton University. In 1974, while at Amherst, using the large radio telescope at Arecibo, Puerto Rico, Taylor along with his graduate student Russell A. Hulse discovered a pulsar (a rapidly spinning neutron star) emitting radio pulses at intervals that varied in a regular pattern, decreasing and increasing over an eight-hour period. They concluded from these signals that the pulsar must be alternately moving toward and away from the Earth – i.e., that it must be orbiting around a companion star, which the two men deduced, was also a neutron star. Their discovery of this first binary pulsar, PSR 1913+16, provided an unprecedented test of Albert Einstein's theory of gravitation, which, according to the general theory of relativity, predicts that objects accelerated in a strong gravitational field, will emit radiation in the form of gravitational waves. Taylor and Hulse concluded that the energy drain due to emission of gravitational waves by the binary pulsar should reduce the orbital distance between the two stars, which could in turn be measured by a slight, gradual reduction in the timing of the pulsar's distinctive radio emissions, which they indeed found. Taylor and Hulse were jointly awarded the 1993 Nobel Prize for Physics for the discovery of a new type of pulsar, a discovery that has opened up new possibilities in the study of gravitation. (☞ Nov. 28)

1962

Auguste Piccard, Swiss-born Belgian physicist notable for his exploration of both the upper stratosphere and the depths of the sea in ships of his own design, died in Lausanne, Switzerland. (☞ Jan. 28)

1965

American space probe *Ranger-9* crashed on the Moon inside the Alphonsus crater. It sent back 5,814 photographs of the lunar surface before impact. (☞ Mar. 21)

1994
Comet 'Shoemaker-Levy 9' was discovered photographically by the husband and wife scientific team of Carolyn S. and Eugene M. Shoemaker and David H. Levy using the 0.46-m Schmidt telescope at Palomar Observatory in California, USA. Calculation of the orbit of the comet showed that it had come as close as 120,000 kilometres of Jupiter in July 1992 and had broken up into as many as 17 separate fragments 'strung out like pearls on a string'. The fragments were again predicted to return near Jupiter – to within 45,000 kilometres of the planet – in July 1994. Since this distance was less than the radius of Jupiter, it was predicted that at least parts of the comet could very well hit Jupiter, which it did, providing the first direct observation of the collision of two solar system objects. (☞ Jul. 17)

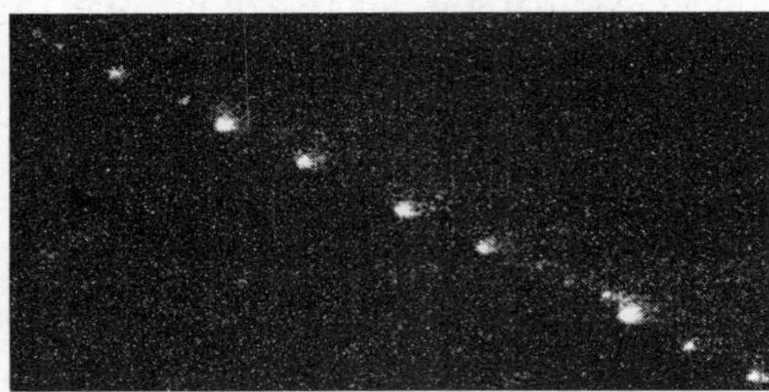

Fragments of Comet 'Shoemaker-Levy 9'

1996
The American space shuttle *Atlantis* docked with the Russian space station *Mir*, for the first time transferring an American

astronaut for an extended stay on *Mir*. The linkup was the third in a series of dockings intended as preparation for a planned internationally operated space station. (☞Mar. 22, Mar. 31)

2002
César Milstein, Argentine immunologist, Nobel laureate (Physiology or Medicine, 1984), who discovered the principle for production of monoclonal antibodies, died in Cambridge, England. (☞ Oct. 8)

March 25

1655
Christiaan Huygens, Dutch mathematician, physicist and astronomer, discovered Titan, Saturn's largest moon and determined its period of revolution. However, the moon was not named until almost two centuries later when the English astronomer John Herschel assigned names to the seven moons of Saturn that were known at that time. Saturn's largest moon was named Titan, because the word means 'one that is great in size, importance, or achievement'. (☞ Apr. 14, Jul. 8)

1903
The discovery of radium by the French physicist couple Pierre and Marie Curie was announced in *The Times*. The paper reported that the Curies had communicated to the Academy of Sciences that the recently discovered element radium 'possesses the extraordinary property of continuously emitting heat, without combustion, without chemical change of any kind, and without any change to its molecular structure, which remains spectroscopically identical after many months of continuous emission of heat ... such that the pure radium salt

would melt more than its own weight of ice every hour ... A small tube containing radium, if kept in contact with the skin for some hours ... produces an open sore, by destroying the epidermis and the true skin beneath ... and cause the death of living things whose nerve centres do not lie deep enough to be shielded from their influence'. (☞ Apr. 19, May 15, July 4, Nov. 7)

1914

Norman Ernest Borlaug, American agricultural scientist, Nobel laureate, was born in Cresco, Iowa, USA. After completing his primary and secondary education in Cresco, Borlaug, he enrolled in the University of Minnesota where he studied forestry. After receiving his BSc degree in 1937, he worked for the US Forestry Service at stations in Massachusetts and Idaho. Returning to the University of Minnesota to study plant pathology, he received his master's degree in 1939 and the doctorate in 1942. From 1944 to 1960 he served as research scientist at the Rockefeller Foundation's Cooperative Mexican Agricultural Program in Mexico. At a research station at Campo Atizapan he developed 'dwarf' strains of wheat that dramatically increased crop yields. Wheat production in Mexico multiplied threefold. This dwarf variety of wheat exported to the east in the mid-1960s was responsible for a 60 per cent increase in harvests in Pakistan and India. He also created a wheat–rye hybrid known as triticale. Later, Borlaug also collaborated with scientists from other parts of the world, especially from India and Pakistan, to adapt the new wheat varieties to new lands and in gaining acceptance for their production. Borlaug was awarded the 1970 Nobel Prize for Peace for giving us a well-founded hope, an alternative of peace and of life – the Green Revolution.

1989

Cray X MP-14, the first supercomputer acquired by India, was inaugurated in New Delhi. The supercomputer was installed in the headquarters of the Indian Meteorological Department and was used for medium-range weather forecasting by the National Centre for Medium Range Weather Forecasting, specialising in numerical weather prediction models. The Cray X MP-14 was upgraded to X/MP-216 in 1992. The Cray X/MP-216 computer system had a guaranteed life span till 1996 but continued to work till 2000.

March 26

1753

Benjamin Thompson Rumford, American-born British physicist, and a founder of the Royal Institution of Great Britain, London, was born in Woburn, Massachusetts, USA. Rumford moved to London in 1776 and began investigating gunpowder. In 1784, he became an *aide-de-camp* to the Elector of Bavaria. He spent eleven years in Bavaria, re-organising the Bavarian army and establishing workhouses for the poor. In 1791 he was named a Count of the Holy Roman Empire. Rumford returned to England in 1795 and introduced improved methods of heating and cooking. He developed a more accurate theory of heat. In 1798 he presented a paper to the Royal Society in which he overturned the theory that heat is a

liquid form of matter and established the beginnings of the modern theory that heat is a form of motion. Rumford was an active inventor and developed improvements for chimneys and fireplaces. He invented the double boiler, a kitchen range, and a drip coffeepot. (☞ Aug. 21)

1911

Bernard Katz, German-born British physiologist, Nobel laureate, was born in Leipzig, Germany. Katz had his schooling at the Albert Gymnasium in Leipzig and then studied medicine at the University of Leipzig, obtaining his MD in 1934. He received a PhD from the London University in 1938. In 1939, he joined the laboratory of the Australian physiologist John Crew Eccles at the Sydney Hospital, Australia, as a Carnegie Research Fellow. His research uncovered fundamental properties of synapses, the junctions across which nerve cells signal to each other and to other types of cells. Katz discovered in the early 1950s that the release of acetylcholine, a signalling molecule with which synapses linking 'motor nerves' to muscles stimulate contraction, occurs continuously and spontaneously, although at low levels when neurons are at rest. Further, he found that acetylcholine is released in discrete packets, later called vesicles. In the late 1960s, Katz determined that the amount of acetylcholine in a vesicle was related to the electrical potential at the terminal of an axon (the long extension of a neuron that transmits the impulse). Katz's work had immediate influence on the study of organophosphates and organochlorines, the basis of new post-war study for nerve agents and pesticides, as he found that nerve agents easily disrupted the complex enzyme cycle. Katz was awarded the 1970 Nobel Prize for Physiology or Medicine, jointly with the American biochemist Julius Axelrod and Swedish physiologist Ulf von Euler, for their discoveries concerning the humoral transmitters in the nerve terminals and the mechanism for their storage, release and inactivation. (☞ Feb. 7, May 30)

1916

Christian Boehmer Anfinsen, American biochemist, Nobel laureate, was born in Monessen, Pennsylvania, USA. After receiving his BS degree in chemistry in 1937, Anfinsen pursued graduate study at the University of Pennsylvania, where he worked toward a MS degree in organic chemistry in 1939 while serving as an assistant instructor. In 1939, the American Scandinavian Foundation awarded him a fellowship to develop new methods for analysing the chemical structure of complex proteins, namely enzymes, at the Carlsberg Laboratory in Copenhagen, Denmark. After the outbreak of World War II, Anfinsen returned to USA and received a doctorate in biochemistry from the Harvard University, Cambridge, Massachusetts, in 1943. In 1950 he joined the staff of the National Institutes of Health (NIH), Bethesda, Maryland, USA. Upon his arrival at the NIH, Anfinsen wanted to expand his study of the amino acid structure of proteins. In an important paper in the *Journal of Biological Chemistry* in 1954, Anfinsen showed that the sequence of amino acids in a peptide chain determines the folding pattern. The mysteriously complex process of protein folding could be entirely explained simply by

the physical and chemical interactions among the amino acid side groups. In 1954, Anfinsen returned to the Carlsberg Laboratory in Copenhagen under a Rockefeller Foundation postdoctoral fellowship. There, working with the Danish biochemist Kaj Linderstrøm-Lang, he conducted in-depth physical analyses of the structure of ribonuclease (the enzyme that hydrolyses RNA), and published a preliminary study of his findings on the topic in *Biochimica et Biophysica Acta* in 1955. He found that ribonuclease is made up of a single peptide chain folded into a sphere bound together by four disulphide bonds. Anfinsen was awarded the 1972 Nobel Prize for Chemistry, jointly with fellow American biochemists Stanford Moore and William H. Stein, for his work on ribonuclease, especially concerning the connection between the amino acid sequence and the biologically active conformation. (☞ May 14, Jun. 25, Sep. 4)

1938

Anthony J. Leggett, British-born American physicist, Nobel laureate, was born in Camberwell, South London, England. Leggett attended the local Catholic elementary school, and later studied at the College of the Sacred Heart in Wimbledon. He received a BA degree in *literae humaniores* from the Balliol College, Oxford University in 1959 and another BA degree in physics from the Merton College, Oxford University in 1961. He received his PhD in theoretical physics from the Magdalen College, Oxford University in 1964. From 1967 to 1983 Leggett taught physics at the University of Sussex. Leggett's research interests lay mainly within the fields of

theoretical condensed matter physics and the foundations of quantum mechanics. In the early 1970s, while in Sussex, Leggett made decisive contributions to the understanding of the complex superfluidity of helium-3, shortly after its discovery. Leggett cleverly combined the Landau theory of Fermi liquids, the BCS theory of metallic superconductivity and the theory of nuclear magnetic resonance to explain and predict the main properties of the new superfluid phases. Leggett was awarded the 2003 Nobel Prize for Physics, jointly with the Russian-born American physicist Alexei A. Abrikosov and Russian physicist Vitaly L. Ginzburg, for pioneering contributions to the theory of superconductors and superfluids. (☞ Jun. 25, Oct. 4)

1951

Carl E. Wieman, American physicist, Nobel laureate, was born in Corvallis, Oregon, USA. Wieman did his BS from the Massachusetts Institute of Technology, Cambridge, Massachusetts, in 1973 and received his PhD from the Stanford University, California in 1977. From 1977–1979 he worked as an assistant research scientist in the department of physics, University of Michigan where he later became assistant professor of physics. In 1984 he joined the University of Colorado in Boulder, Colorado as associate professor of physics where he became professor of physics in 1987. Wieman carried out the first experiments on laser cooling and trapping of atoms in 1984 and later, with the American physicist Eric A. Cornell, succeeded in achieving a new state of matter known as the Bose–Einstein condensate, first predicted by the Indian physicist

Satyendra Nath Bose and the German-born American physicist Albert Einstein in 1924. Cornell and Wieman produced a pure condensate of about 2,000 rubidium atoms at a temperature of 20 degrees nano kelvin, i.e., 20 billionth of a degree above absolute zero, using the technique of laser cooling. Wieman was awarded the 2001 Nobel Prize for Physics, jointly with Cornell and the German physicist Wolfgang Ketterle, for achieving Bose–Einstein condensation in dilute gases of alkali atoms, and for early fundamental studies of the properties of the condensates. (☞ Oct. 21, Dec. 9)

1953
American virologist Jonas Edward Salk announced a new vaccine to immunise people against polio in a paper in the *Journal of the American Medical Association*. Earlier, in July 1952, Salk had tried the refined vaccine made of killed poliovirus on children who had already had polio and recovered. After the vaccination, their antibodies increased. He then tried it on volunteers who had not had polio, including himself, his wife, and their children. The volunteers all produced antibodies, and none got sick. (☞ June 23, Oct. 28)

1959
The Central Institute of Medicinal and Aromatic Plants (CIMAP), a constituent national laboratory of the Council of Scientific and Industrial Research (CSIR), was set up at the CSIR Headquarters in New Delhi. Now located at Lucknow, Uttar Pradesh, the CIMAP is primarily engaged in research aimed at genetic improvement, cultivation, production and chemical processing of economically important medicinal, nutraceutical, essential oil, dye, and gum yielding plants. It also undertakes characterisation of biodiversity and conservation of genetic resources of medicinal and aromatic plants and other plants of interest. Development of agro-technologies and chemical and processing technologies for economically important medicinal and aromatic plants, both indigenous and exotic, is an important goal of the institute. The main achievements of CIMAP include domestication and development of improved cultivars and agronomic practices for indigenous medicinal and aromatic plants, introduction of exotic medicinal and aromatic plants of commercial importance, such as the introduction of Bulgarian rose in the Kashmir valley and geranium and pyrethrum in the north Indian plains.

1972
Soviet unmanned spacecraft *Venera-8* was launched towards Venus from the Baikanur cosmodrome in Kazakhstan. *Venera-8* was an atmospheric probe and carried temperature, pressure, and light sensors as well as radio transmitters. (☞ Jul. 22)

1974
Edward Uhler Condon, American physicist after whom the Franck–Condon principle was named and who applied quantum mechanics to an understanding of atomic, molecular and nuclear structure, died in Boulder, Colorado, USA. (☞ Mar. 2)

1994
The US space agency NASA released a picture showing the first moon discovered in orbit around a minor planet. The potato-

Ida and moon

shaped minor planet Ida and its newly-discovered moon, named Dactyl, was imaged by NASA's *Galileo* spacecraft, about 14 minutes before its closest approach to the minor planet in August 1993. Ida appears to be about 60 kilometres long and 23 kilometres wide. The tiny moon is about 1.5 kilometres across. (☞ Aug. 28, Oct. 18)

March 27

1845

Wilhelm Conrad Röntgen, German physicist, recipient of the first Nobel Prize for Physics, in 1901, was born in Lennep, Prussia (now Remscheid, Germany). Röntgen entered the University of Utrecht in The Netherlands in 1865 to study physics. Not having attained the credentials required for a regular student, and hearing that he could enter the Polytechnic at Zurich, Switzerland, by passing its examination, he passed this and began studies there as a student of mechanical engineering. In 1869 he received his PhD from the University of Zurich. Röntgen's research work covered many different fields including the specific heats of gases; thermal conductivity of crystals; the electrical and other characteristics of quartz; the influence of pressure on the refractive indices of various fluids; and the modification of the planes of polarised light by electromagnetic influences. Röntgen's name, however, is chiefly associated with his discovery of the invisible rays that he called X-rays. In 1895, at the University of Würzburg, Germany, Röntgen's attention was drawn to a glowing fluorescent screen on a nearby table. Röntgen immediately realised that some invisible rays originating from the partially evacuated glass Crookes tube he was using to study cathode rays caused the fluorescence. Several scientists had already carried out previous work in this field, and by the work of these scientists the properties of cathode rays had become well known. Unlike cathode rays, however, these mysterious rays penetrated the opaque black paper wrapped around the tube. Röntgen's work on cathode rays thus led him to the discovery of the new and different kind of ray, which he called X-ray. Röntgen's discovery of X-rays revolutionised the field of physics and medical diagnosis. Röntgen was awarded the 1901 Nobel Prize for Physics in recognition of the extraordinary services he had rendered by the discovery of the remarkable rays subsequently named after him. (☞ Jan. 1, Jan. 23, Feb. 10, Nov. 8)

1847

Otto Wallach, German chemist, Nobel laureate, was born in Königsberg, Germany. During his early school years at the humanistic 'Gymnasium' at Potsdam, Wallach had a profound liking for history and art – in those days subjects like chemistry were hardly taught at secondary-school level. In 1867 he went to Göttingen to study chemistry and obtained his doctor's degree in 1869. His thesis dealt with the position isomers in the toluene series. From 1872 onwards Wallach resided at Bonn. It was here that the German chemist Friedrich August Kekulé drew his attention to the existence of an old forgotten cupboard full of bottles containing

essential oils, and invited him to make a study of the contents. Wallach thus entered a field of study in which he was to be the eminent pioneer for more than a decade, and which was to be his main life's work, crowned with the highest possible distinction. In his first publication in 1884, he raised the question of the diversity of the various members of the $C_{10}H_{16}$ group, which in the then prevailing practice came under a multitude of names ranging from terpene to camphene, citrene, carvene, cinene, etc. Utilising common reagents such as hydrogen chloride and hydrogen bromide, he succeeded in characterising the differences between the structures of these compounds. A year later he could establish that many of them were indeed identical. In 1909 he published the results of his extensive studies in his book *Terpene und Campher*, a volume of 600 pages. Wallach was awarded the 1910 Nobel Prize for Chemistry for analysing fragrant essential oils and identifying the compounds known as terpenes. (☞ Feb. 26)

1923
James Dewar, British chemist and physicist, who was the first to liquefy several gases and who invented the vacuum flask, died in London, England. (☞ Sep. 20)

1942
John E. Sulston, British biologist, Nobel laureate, was born in Cambridge, England. Sulston studied natural sciences at the Pembroke College in Cambridge and received his BA in 1963 and PhD in 1966 from the University of Cambridge. After three years of postdoctoral work in USA, he joined the Medical Research Council in England in 1969 to work with the South African-born British biologist Sydney Brenner. Sulston extended Brenner's work with *C. elegans* and developed techniques to study all cell divisions in the nematode, from the fertilised egg to the adult organism. In a publication in 1976, Sulston described the cell lineage for a part of the developing nervous system. He showed that the cell lineage is invariant, i.e., every nematode underwent exactly the same programme of cell division and differentiation. Thus he showed that specific cells undergo programmed cell death as an integral part of the normal differentiation process, and identified the first mutation of a gene participating in the cell death process. From 1992 to 2000 Sulston was director of the Sanger Institute in Cambridge. Sulston was awarded the 2002 Nobel Prize for Physiology or Medicine, jointly with the American biologist H. Robert Horvitz and the South African-born British biologist Sydney Brenner, for their discoveries concerning genetic regulation of organ development and programmed cell death. (Jan. 13, May 8)

1967
Jaroslav Heyrovsky, Czech chemist, Nobel laureate (Chemistry, 1959), who discovered and developed polarography, died in Prague, Czechoslovakia (now Czech Republic). (Dec. 20)

1968
Yuri Alekseyevich Gagarin, Soviet cosmonaut and the first human to travel into space, died near Moscow, Russia, in a plane accident. (☞ Mar. 9, Apr. 12)

1969
The American unmanned spacecraft *Mariner-7* was launched from the Kennedy Space Center, Florida, USA, towards the planet Mars. Weighing 413

kilograms, *Mariner-7* was sent to fly-by Mars in order to collect data about the surface and atmosphere of the Red Planet. The spacecraft was designed to fly over the equator and southern hemisphere of Mars and was capable of continuous telemetry transmission. (☞ Jul. 31, Aug. 4)

March 28

1802

German amateur astronomer Heinrich Wilhelm Olbers discovered the minor planet Pallas. Having a diameter of about 480 kilometres, Pallas is the second largest asteroid and orbits the Sun at a distance of 417 million kilometres with a period of 1,684 days. (☞ Oct. 11)

1892

Corneille Heymans, Belgian physiologist, Nobel laureate, was born in Ghent, Belgium. Heymans had his medical education at the University of Ghent, where he obtained his doctor's degree in 1920. After his graduation he worked at several colleges before becoming a lecturer in pharmacodynamics at the University of Ghent in 1922. In 1930 he succeeded his father, J. F. Heymans, as Professor of Pharma-cology, and was also appointed head of the department of pharmacology, pharmacodynamics, and toxicology. At the same time he became Director of the J. F. Heymans Institute of Pharmacology and Therapeutics, founded by his father. The scientific investigations carried out by Heymans at the Institute were mainly directed towards the physiology and pharmacology of respiration, blood circulation, metabolism, and numerous pharmacological problems. These studies led, in particular, to the discovery of different types of receptors. Experimenting with anaesthetised dogs, Heymans demonstrated the existence of a set of sensory organs, known as 'pressoreceptors,' in the wall of the carotid sinus – a slight enlargement of the carotid artery in the neck, at the point where it divides into the external and internal carotids. He showed that these receptors monitor blood pressure and help to regulate heart rate and respiration. He also found near the pressoreceptors, and at the base of the aorta, a set of 'chemoreceptors' that monitor the oxygen content of the blood and help to regulate breathing through the medulla, the respiratory centre at the base of the brain. Heymans was awarded the 1938 Nobel Prize for Physiology or Medicine for the discovery of the role played by the sinus and aortic mechanisms in the regulation of respiration. (☞ Jul. 18)

1930

Jerome Isaac Friedman, American physicist, Nobel laureate, was born in Chicago, Illinois, USA. Friedman was educated at the University of Chicago, from where he received his PhD in 1956. After conducting research at Chicago and at the Stanford University, where he met Canadian

physicist Richard E. Taylor and American nuclear physicist Henry W. Kendall, he began teaching at the Massachusetts Institute of Technology in 1960. He became a full professor there in 1967 and head of the physics department in 1983. At the Stanford Linear Accelerator Center of Stanford University, Friedman, Kendall and Taylor conducted a series of experiments from 1967 to 1973, in which they used the particle accelerator to direct a beam of high-energy electrons at target protons and neutrons. Their results showed that protons and neutrons are not fundamental particles of matter but are composed of smaller particles known as quarks, the existence of which had been hypothesised by the American physicists Murray Gell-Mann and George Zweig in 1964. This evidence allowed scientists to develop the 'standard model' theory of matter, which states that all matter is made up of combinations of six quarks and six leptons that interact with five types of force particles. Friedman was awarded the 1990 Nobel Prize for Physics, jointly with Taylor and Kendall, for their pioneering investigations concerning deep inelastic scattering of electrons on protons and bound neutrons, which have been of essential importance for the development of the quark model in particle physics. (☞ Sep. 15, Nov. 2, Dec. 9)

1979

A nuclear accident occurred at the Three Mile Island nuclear power plant outside Harrisburg, Pennsylvania, USA. Resulting from human and mechanical errors, a cooling system malfunctioned and caused a partial meltdown of the reactor's core. Some radioactivity leaked into the atmosphere. Although it is the worst American nuclear accident, disaster was avoided.

1982

William Francis Giauque, Canadian-born American physical chemist, Nobel laureate (Chemistry, 1949), who proposed a new method of achieving extremely low temperatures using a process called adiabatic demagnetisation, died in Berkeley, California, USA. (☞ May 12)

1998

India's most powerful supercomputer *Param-10000* was unveiled by the Centre for Development of Advanced Computing (C-DAC) in Pune, Maharashtra. Based on an open-frame architecture, and with a processing power of 6.4 gigaflop, it was at that time considered to be the most powerful supercomputer in Asia, outside Japan. *Param-10000* is a parallel processing computer, which is scalable up to teraflop range. It is capable of applications in as diverse areas as long-range weather forecasting, drug design, molecular modelling, and remote sensing. (☞ Apr. 1)

March 29

1807

Vesta-4, the only minor planet visible to the naked eye, was first observed by the German amateur astronomer Heinrich Wilhelm Olbers from Bremen, Germany. Vesta is a main belt minor planet with a diameter of 525 kilometres and a rotation period of 5.34 hours. Pictures taken by the Hubble Space Telescope in 1995 show Vesta's complex surface, with a geology similar to that of terrestrial worlds – such as Earth or Mars – showing an exposed mantle, ancient lava flows and impact basins. It is no bigger than the state of Uttar Pradesh and it once had a molten interior. This contradicts conventional ideas that minor planets are essentially cold, rocky fragments left behind from the early days of planetary formation. (☞ Oct. 11)

1904

A patent for a solid-state detector of electromagnetic waves, the first in the world, was granted to the Indian biophysicist Jagadis Chandra Bose. Patented as a 'detector for electrical disturbances', the device used a crystal of lead sulphide (galena) as detector that did away with mechanical means for effecting the recovery of the coherer which was used as detectors at that time. Galena is known to be the most efficient wideband semiconductor diode. Bose's detector could be used both for short wavelength radio waves and for white and ultraviolet light. In his application, Bose explained, "... with a glass lens, the instrument will detect and record lights not only some way beyond the violet, but also in the regions of infrared in the invisible regions of electric radiation. We may thus style this apparatus a 'tejometer' ... or universal radiometer." (*tej*: Sanskrit for radiation.)

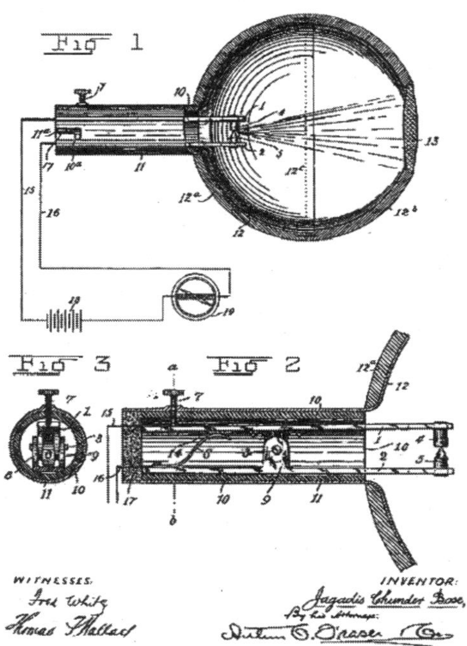

1927

John Robert Vane, English biochemist, Nobel laureate, was born in Tardebigg, Worcestershire, England. Vane graduated from the University of Birmingham and earned a doctorate at the University of Oxford in 1953. He spent two years as a faculty of the Yale University, Connecticut, USA, before returning to England to join the Institute of Basic Medical Sciences of the University of London in 1955, where he stayed for 18 years, progressing from senior lecturer to reader to professor of experimental pharmacology. He became research director of the Wellcome Research Laboratories, Beckenham, Kent, in 1973. Vane's research interests mainly concerned blood constituents called platelets and substances known as prostaglandins – biochemical compounds that influence blood pressure, body temperature, allergic reactions, and other physiological phenomena in mammals. Vane and his group developed the 'cascade superfusion' bioassay technique for measurement of the release and fate of hormones that act on blood vessels in the circulation of blood. In the mid-1960s, the research focussed on

prostaglandins, leading in 1971 to the forging of the link between aspirin and the prostaglandins. Vane discovered that the action of aspirin was due to the inhibition of the formation of prostaglandins associated with pain, fever, and inflammation. Vane was awarded the 1982 Nobel Prize for Physiology or Medicine, jointly with the Swedish biochemists Sune K. Bergström and Bengt Ingemar Samuelsson for their discoveries concerning prostaglandins and related biologically active substances. (☞ Jan. 10, May 21)

1974
The American spacecraft *Mariner-10* took the first close-up pictures of Mercury. On its way to Mercury, *Mariner-10* made its first fly-by of Venus and discovered evidence of rotating clouds on the planet. The mission required more course corrections than any previous mission and was the first spacecraft to use the gravitational pull of one planet to help it reach another planet. This craft was also the first to use the solar wind as a means of locomotion; when the probe's thruster fuel ran low, scientists used the solar panels as sails to make course corrections. In three fly-bys past Mercury, it mapped about half the planet's surface and sent back 674 pictures. It found a thin atmosphere and a magnetic field. (☞ Feb. 5, Mar. 16, Sep. 21, Nov. 3)

March 30

239 BCE
The perihelion passage of Comet Halley was recorded for the first time by Chinese astronomers in the *Shih Chi* and *Wen Hsien Thung Khao* chronicles. At that time, of course, it was not known as Comet Halley. It was only after its periodic nature was recognised by the English astronomer Edmond Halley in 1705, and it actually returned in 1759 as predicted, that it was named posthumously after Halley. The comet goes round the Sun in a highly elliptical orbit and returns to the vicinity of Earth every 75 years or so; it was last seen in 1986 and will be next seen in 2061. (☞ Feb. 9, Nov. 8, Dec. 25)

1791
The unit of 'metre' was finally defined by the French National Assembly to be 1/10,000,000th of the distance between the Earth's North Pole and the equator. The proposal was made by the French Academy of Sciences. (☞ Aug. 1)

1842
Crawford W. Long, American physician first used ether as an anaesthetic during a minor operation in Jefferson, Georgia. His patient was a student named James Venable. Long placed an ether-soaked towel over the face of his student, making him unconscious, and removed a tumour from his neck. Long's work came four years before the American dental surgeon William Morton made a public demonstration of ether as an anaesthetic. Long's discovery of surgical anaesthesia went unrecognised for years, till his work was reported in 1849 in the *Southern Medical Journal*. It is now widely considered to represent the discovery of surgical anaesthesia. (☞ Jun. 16, Oct.16, Nov. 1, Nov. 4)

1891
Arthur William Sidney Herrington, American engineer and manufacturer who developed a series of military vehicles, the best known of which was the jeep, was born in Coddenham, East Suffolk, England. Herrington immigrated to the United States with his family at the age of five and studied at the Stevens Preparatory

School and the Stevens Institute of Technology, Hoboken, New Jersey. He was first employed by the Harley-Davidson Motor Company of Milwaukee. During World War I, appalled by the problems of conventional drive vehicles in off-the-road manoeuvring, he experimented with and designed a series of trucks with four- and six-wheel drives for rough terrain, the smallest of which was the quarter-ton 'Jeep' with a four-wheel drive that became the prototype for various models built in the 1930s and 1940s. The Jeep has turned out to be one of the most versatile multi-purpose vehicles ever designed. It has served as a machine gun firing mount, a reconnaissance vehicle, a pickup truck, a front line limousine, an ammunition bearer, a wire-layer, and a taxi. (☞ Sep. 6)

1949
Friedrich Bergius, German chemist, Nobel laureate (Chemistry, 1931), who invented the process of conversion of coal dust and hydrogen directly into gasoline and lubricating oils without isolating intermediate products, died in Buenos Aires, Argentina. (☞ Oct. 11)

1954
Fritz Wolfgang London, German-born American physicist who, with the German physicist Walter Heitler, devised the first quantum mechanical treatment of the hydrogen molecule, died in Durham, North Carolina, USA. (☞ Mar. 7)

1965
Philip Showalter Hench, American physician, Nobel laureate (Physiology or Medicine, 1950), who discovered the cortisone treatment for arthritis, died in Ocho Rios, Jamaica. (☞ Feb. 28)

1982
The American space shuttle *Columbia*, carrying astronauts Jack R. Lousma, and C. Gordon Fullerton, landed at Edwards Air Force Base, California, USA, after its third mission. (☞ Feb. 1, Mar. 22, Apr. 12)

March 31

1596
René Descartes, French mathematician, scientist, was born in Touraine, France. At the age of eight, Descartes was enrolled in the Jesuit school of La Flèche in Anjou, where he remained for eight years. Besides the usual classical studies, he received instruction in mathematics and in scholastic philosophy. Upon graduation from school, he studied law at the University of Poitiers, graduating in 1616, but he never practiced law. Between 1623 and 1628 Descartes made a pilgrimage to Italy and parts of France. In 1628, having sold his properties in France, he moved to the Netherlands, where he spent most of the rest of his life. Descartes was one of the most important and influential thinkers in human history and is sometimes called the founder of modern philosophy. In addition to his accomplishments as a philosopher, Descartes was an outstanding mathematician, who invented analytic geometry and attempted to devise the simple universal laws that governed all physical change. He is famous for having made an important connection between

geometry and algebra, which allowed for the solving of geometrical problems by way of algebraic equations. In establishing the ground for science, Descartes sought to overthrow a system of natural philosophy that had been established for centuries – a qualitative, Aristotelian physics. Because he lived at a time when traditional ideas were being questioned, he sought to devise a method for reaching the truth. This concern and his method of systematic doubt had an enormous impact on the subsequent development of philosophy. Descartes introduced the now famous Latin phrase *cogito ergo sum*, or in English 'I think, therefore I am'. (☞ Feb. 11, Jun. 8)

1693

John Harrison, English clock designer, who developed and built the world's first successful maritime clock, was born in Foulby, Yorkshire, England. A carpenter by initial trade, Harrison built and repaired clocks in his spare time. He was a man of many skills and used these to improve on the way clocks were built. He developed the gridiron pendulum, consisting of alternating brass and steel rods assembled so that the different expansion and contraction rates cancelled each other. In 1728 Harrison packed up full-scale models of his inventions and drawings for a proposed maritime clock to compete for the 'Longitude Prize' and headed for London seeking financial assistance. It took Harrison seven years to build Harrison Number One or H1. He presented it to members of the Royal Society to be submitted to the Board of Longitude. The Board was so sceptical of any such design, after 14 years of failures, that they demanded a sea trial. Harrison boarded a small ship to Lisbon and back, and on their return the captain and navigator both vouched for the design. But the terms of the prize needed a successful transatlantic crossing. With more funds from the Board, and 13 years of hard work, Harrison ultimately produced the version named H4, which was ultimately sent on a transatlantic trial. H4 was an instrument of beauty, being of the shape of a large pocket watch. Although the first trial proved the accuracy of H4, a second trial was ordered which, too, proved it to be astonishingly accu-rate. But Harrison had to build another version, H5, and wait till 1773 to receive the full prize money of 20,000 pounds. (☞ Mar. 24)

The H_4

1811

Robert Wilhelm Bunsen, German chemist, who is best known for the laboratory gas burner named after him, was born in Göttingen, Westphalia, Germany. Bunsen was educated at the University of Göttingen, where he received his doctorate in 1830, at the age of 19. He served on the faculties of several universities and was at Heidelberg from 1852 to 1889. His first important contribution to chemistry came with his investigation of certain organic compounds of arsenic, in the process of which he discovered that ferric oxide could be used as an antidote to arsenic poisoning. From his studies of the gaseous products of blast furnaces he evolved a method of gas analysis. While at Heidelberg, with the German physicist Gustav Kirchhoff, Bunsen first observed that, when put in a gas flame, each element emits a light of

characteristic wavelength, which is basis of the common lab 'flame test'. Their studies opened the field of spectrum analysis, which became of great importance in the study of the Sun and stars and also led Bunsen almost immediately to his discovery of two new elements: caesium and rubidium. Interestingly, Bunsen did not invent the essential piece of laboratory equipment that has immortalised his name; he only improved the gas burner to make the flame colourless to aid his endeavours in spectroscopy. (☞ Mar. 12, May 10, Aug. 16)

1890

William Lawrence Bragg, Australian-born British physicist, Nobel laureate, was born in Adelaide, Australia. Bragg graduated in mathematics from the Adelaide University in 1908. He moved to England with his father William Henry Bragg in 1909 and entered the Trinity College, Cambridge, as an Allen Scholar, taking first-class honours in the Natural Science Tripos (a course in which most streams of science are taught in the first year with increasing specialisations in the following years) in 1912. In 1914 he was appointed as fellow and lecturer in natural sciences at the Trinity College. From 1912 to 1914 Bragg worked with his father, William Henry Bragg. Bragg and his father played a seminal role in X-ray crystallography, which studied the phenomenon of diffraction of X-rays by a crystal. William Henry's original interest was in what diffraction showed about the nature of X-rays, while William Lawrence was more concerned with what X-rays revealed about the crystalline state. Simple inorganic crystals like sodium chloride were the subjects of their original studies in X-ray crystallography. The surprising discovery was that in the solid state these ionic compounds did not exist as paired positive and negative ions. Sodium chloride, for instance, did not exist as NaCl units; but rather, Na^+ and Cl^- alternated in a regular fashion in the crystal lattice. The results of their work were published in an abridged form in *X-rays and Crystal Structure* in 1915. William Lawrence Bragg was awarded the 1915 Nobel Prize for Physics, at the young age of 25, jointly with his father William Henry Bragg, for their services in the analysis of crystal structure by means of X-rays. (☞ Mar. 12, Jul. 2)

1906

Shin'ichiro Tomonaga, Japanese physicist, Nobel laureate, was born in Kyoto, Japan. Tomonaga obtained his bachelor's degree in physics from the Kyoto Imperial University in 1929. He was engaged in graduate work for three years at the same university and was then appointed a research associate at the Institute of Physical and Chemical Research, Tokyo, where he started to work in a newly developed frontier of theoretical physics, quantum electrodynamics. Tomonaga stayed in Leipzig, Germany, from 1937 to 1939, to study nuclear physics and the quantum field theory in collaboration with the theoretical group of the German physicist Werner Heisenberg. In 1940, Tomonaga directed his attention to the meson theory and developed the intermediate coupling theory – the idea that interactions between two particles take place through the exchange of a third (virtual) particle – in order to clarify the structure of the meson cloud around the nucleon. He joined the faculty of the Tokyo Bunrika University as professor of physics in

1941, when he began his investigations of the problems of quantum electrodynamics. Tomonaga's theoretical work made quantum electrodynamics (the theory of the interactions of charged subatomic particles with the electromagnetic field) consistent with the theory of special relativity. He used this concept to develop a quantum field theory (1941–43) that was consistent with the theory of special relativity. Tomonaga was awarded the 1965 Nobel Prize for Physics, jointly with American physicists Richard P. Feynman and Julian Seymour Schwinger, for their fundamental work in quantum electrodynamics, with deep-ploughing consequences for the physics of elementary particles. (☞ Feb. 12, May 11, Jul. 8)

1917
Emil Adolf von Behring, German bacteriologist, Nobel laureate (Physiology or Medicine, 1901), who is considered the founder of the science of immunology, died in Marburg, Germany. (☞ Mar. 15)

1934
Carlo Rubbia, Italian physicist, Nobel laureate, was born in Gorizia, Italy. Rubbia was educated at the Normal School of Pisa and got a doctorate from the University of Pisa in 1957. He taught in Pisa for two years before moving to the Columbia University, New York, USA, as a research fellow. He joined the faculty of the University of Rome in Italy in 1960 and was appointed senior physicist at the European Centre for Nuclear Research (CERN; now the European Organization for Nuclear Research), Geneva, Switzerland, in 1962. In 1970 he was appointed professor of physics at the Harvard University in Cambridge, Massachusetts, USA, and thereafter divided his time between Harvard and CERN. In 1973 a research group under Rubbia's direction provided one of the experimental clues that led to the formulation of the electroweak theory. The electroweak theory embodied the idea that the so-called weak force involved in the radioactive decay of atomic nuclei can be transmitted by any of three particles called intermediate vector bosons. Furthermore, it indicated that these particles (W^+, W^-, and Z^0) should have masses nearly 100 times that of the proton. In 1976, Rubia suggested adapting CERN's Super Proton Synchrotron (SPS) to collide protons and antiprotons in the same ring and the world's first antiproton factory was built. It started running in 1981. In January 1983, came the announcement, that W particles had been created. A couple of months later the even more elusive Z particles were also observed. Rubia was awarded the 1984 Nobel Prize for Physics, jointly with the Dutch engineer Simon van der Meer, for their decisive contributions to the large project, which led to the discovery of the communicators of weak interaction, the W and Z field particles. (☞ Nov. 24)

1945
Hans Fischer, German biochemist, Nobel laureate (Chemistry, 1930), who determined the constitution of haemin and chlorophyll and synthesised haemin, died in Munich, Germany. (☞ Jul. 27)

1966
Russian unmanned Moon-probe *Luna-10*, the first spacecraft to orbit the Moon, was launched from the Baikanur cosmodrome in USSR (now Kazakhstan). The 540-kg spacecraft entered lunar orbit three days later and completed its first orbit three hours later. Scientific instruments on board *Luna-10* included a gamma-ray spectrometer, a magnetometer, a meteorite

detector, instruments for solar-plasma studies, and devices for measuring infrared emissions from the Moon. (☞ Apr. 4)

1978
Charles Herbert Best, American physiologist who, with the Canadian physician Frederick Banting, was the first to obtain a pancreatic extract of insulin in a form that controlled diabetes in dogs, died in Toronto, Ontario, Canada. (☞ Feb. 27, Nov. 14)

1981
Ananda Mohan Chakrabarty, Indian-born American microbiologist, employee of General Electric, USA, was issued a US patent on a genetically engineered *Pseudomonas* bacterium – a bug that has a voracious appetite for oil. This landmark decision set a precedent for patenting of life and life forms, including seeds, plants, mice, sheep, cows and even parts of human beings.

1996
American space shuttle *Atlantis*, carrying astronauts Kevin P. Chilton, Richard A. Searfoss, Shannon W. Lucid, Linda M. Godwin, Michael R. Clifford and Ronald M. Sega landed at the Edwards Air Force Base, California, USA, after completing its 16th flight during which it docked with the Russian space station *Mir* and transferred an American astronaut to *Mir*. (☞ Mar. 22, Mar. 24)

2001
Clifford G. Shull, American physicist, Nobel laureate (Physics, 1994), who developed the neutron scattering technique, died in Medford, Massachusetts, USA. (☞ Sep. 23)

April

April 1

1578
William Harvey, English physician and discoverer of the true nature of the circulation of the blood and of the function of the heart as a pump, was born in Folkestone, Kent, England. Harvey received his bachelor's degree in 1597 from the Cambridge University. He continued his studies at the University of Padua, Italy, the foremost medical school of the time, where he studied under the Italian scientist and surgeon, Hieronymus Fabricius, who had observed the one-way valves in veins, but had not figured out exactly what their role was. Harvey was fascinated by the way blood flowed through the human body. Functional knowledge of the heart and the circulation had remained almost at a standstill ever since the time of the Greco-Roman physician Galen – 1,400 years earlier. Most people of the day believed that food was converted into blood by the liver, and then was consumed as fuel by the body. Harvey returned to England in 1602, and in 1618 was appointed as a physician to the court of King James I. By studying animals given to him by his regal employer, Harvey eventually developed an

accurate theory of how the heart and circulatory system operated. He published his theories in 1628 in his famous book *On the Motion of the Heart and Blood in Animals,* which made him famous throughout Europe. Harvey was intrigued by almost everything about the body, and at some point turned his attention to reproduction. He speculated that reproduction in humans and other mammals must be through the union of an egg and sperm. Although it was 200 years before a mammalian egg was finally observed, Harvey's theory was so compelling and so well thought out that the world assumed he was right long before the discovery was finally made. (☞ Jan. 3)

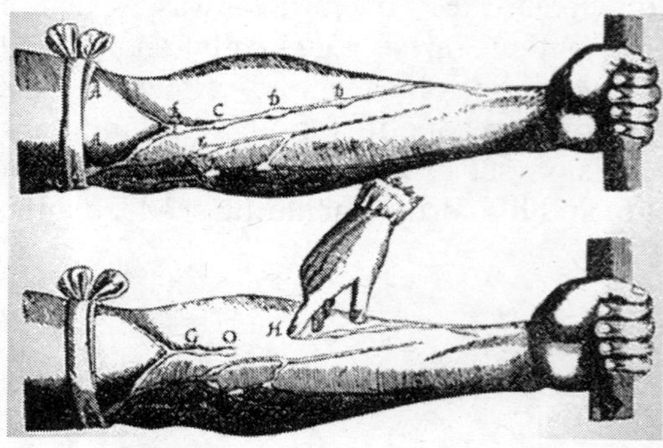

Harvey's drawings of arteries and veins, which he showed to be part of the human circulatory system.

1865
Richard Zsigmondy, Austrian-born German chemist, Nobel laureate, was born in Vienna, Austria. Zsigmondy's interest in chemistry and physics developed at an early age, and he carried out many experiments in his own small laboratory at home. After receiving his doctorate from the University of Munich, Germany, in 1889, Zsigmondy did research at Berlin and then joined the faculty of the University of Graz, Austria. His work on lustre colours for glass and porcelain induced him to study more closely the chemistry of colloids, and led

to an appointment with the glass works Schott und Genossen in Jena, Germany, where he remained until 1900. He left in order to exclusively pursue scientific research. During this period he discovered how to prepare a water suspension of gold. He theorised that much could be learned about the colloidal state of matter from studying the manner in which the particles scatter light. To facilitate such study, he and Heinrich Siedentopf developed in 1903, the ultra-microscope, which he used to investigate various aspects of colloids, including Brownian motion. His work proved particularly helpful in biochemistry and bacteriology. In 1907 Zsigmondy was appointed professor and director of the Institute of Inorganic Chemistry at the University of Göttingen, Germany, where he remained until his retirement in February 1929. Zsigmondy was awarded the 1925 Nobel Prize for Chemistry for his demonstration of the heterogeneous nature of colloid solutions and for the methods he used, which have since become fundamental in modern colloid chemistry. (☞ Sep. 23)

1905

Indian Agricultural Research Institute (IARI) was founded. Originally known as the Agricultural Research Institute and located in Pusa, Bihar, the institute was re-named as the Imperial Agricultural Research Institute in 1918. After the great Bihar Earthquake in January 1934, the institute was shifted to New Delhi in 1935. After Independence in 1947 the name was again changed to the Indian Agricultural Research Institute. IARI is India's premier national institute for agricultural research, education and extension. It has served the cause of science and society with distinction through first-rate research, generation of appropriate technologies and development of human resources. The Green Revolution in India began in the fields of IARI. During the 1950s, the advancement of scientific disciplines constituted the core programme and provided the base for its fast expansion in the 1960s and 1970s in all its three interactive areas, namely, research, education and extension. The basic areas of research at IARI include utilisation of global plant genetic resources, including conservation of agriculturally important microbial, cyanobacterial and insect resources, production of efficient, productive and stable genotypes of crops, especially hybrids, and improvment in bioenergenetics. Besides basic research, applied and commodity research gained great importance resulting in the development of several popular high yielding varieties of almost all major crops and their associated management technologies, which in turn brought about an unprecedented increase in the national food and agricultural production. Emphasis is given to post-harvest technology, agro-processing, product development, value addition and utilisation research on agricultural commodities, by-products, agricultural wastes and renewable energy resources.

1919

Joseph Edward Murray, American surgeon, Nobel laureate, was born in Milford, Massachusetts, USA. Murray attended a small liberal arts college, College of the Holy Cross, and concentrated on Latin, Greek, Philosophy and English. He then spent four years at the Harvard Medical School. His interest in the biology of tissue and organ transplantation arose from his military experience at the Valley Forge General Hospital in Pennsylvania,

which was a major plastic surgical centre. During his three years there, Murray spent all his available spare time in the plastic surgical wards, which were jammed with hundreds of battle casualties. Some were so extensively burned that donor sites for skin autografts were not available. He frequently observed how skin taken from one soldier to replace the badly burned skin of another soldier would gradually be rejected by the burned soldier's body. He was intent on discovering why this occurred. By observing a colleague successfully trade sections of skin between a pair of identical human twins, Murray became encouraged to study a similar procedure with human organs. His efforts resulted in the first successful transplantation of a kidney from one dog to another, and the first human kidney transplant in 1954 between two men who were identical twins. He continued to develop the process, creating new drugs that made it easier for non-relatives to be donors. Following the development of anti-rejection drugs, which allowed an organ recipient's body to accept another person's donated organ, Murray showed that it was possible to transplant organs between non-identical relatives and from deceased persons to the living. Murray was awarded the 1990 Nobel Prize for Physiology or Medicine, jointly with the American physician Edward Donnall Thomas for their discoveries concerning organ and cell transplantation in the treatment of human disease. (☞ Mar. 15)

1933

Claude Cohen-Tannoudji, French physicist, Nobel laureate, was born in Constantine, Algeria. Cohen-Tannoudji was educated at the École Normale Supérieure, Paris, receiving his doctorate in 1962. After graduating he continued to work as a research scientist in the department of physics at Ècole Normale. He also taught at the University of Paris from 1964 to 1973, and was appointed as a Professor in the Collège de France in 1973. The main work of Cohen-Tannoudji and his colleagues at the École Normale involved the slowing down of atoms to study them more closely using lasers. They expanded on the work of the American physicists Steven Chu and William D. Phillips, and devised new methods for cooling and trapping atoms with laser. Early efforts at cooling had found that laser traps also tended to agitate atoms causing them to move out of the beam. Cohen-Tannoudji proposed that laser traps operate by a process of what has since been called 'Sisyphus cooling'.

The laser beams, he argued, produce a series of standing waves of light polarised in different directions. As the atoms pass through the various fields their energy levels and thereby temperature is lowered. In 1995 Cohen-Tannoudji's group was able to cool helium atoms to within eighteen-millionths of a degree above absolute zero (–273.15°C), slowing them down to a speed of about two centimetres per second. Their work and that of Chu and Phillips, furthered scientists' understanding of how light and matter interact. Cohen-Tannoudji was awarded the 1997 Nobel Prize for Physics, jointly with Chu and Phillips, for development of methods to cool and trap atoms with laser light. (☞ Feb. 28, Nov. 5)

1948

The famous letter giving the details of the Big Bang theory of the Universe, written by

American physicists Ralph Alpher, Hans Bethe and George Gamow was published in the journal *The Physical Review*. The Big Bang theory had previously been around as a competing theory with the Steady State theory for a while. The paper titled 'The origin of chemical elements' gave a mathematical analysis of atomic events during a 'hot Big Bang' creation of the universe, and explained the relative abundances of the light elements (particularly helium) in the universe. The paper was written by Alpher and Gamow. Bethe was persuaded to lend his name as a co-author in order to create the whimsical author list of Alpher, Bethe, Gamow, which is a play on the Greek letters: Alpha, Beta, Gamma. Their new theory became popularly known as the 'alpha-beta-gamma' theory. Bethe did actually make later contributions to discussions of the theory. (☞ Mar. 4, Jul. 2)

1960

Tiros-I, the first weather observation satellite, was launched from Cape Kennedy, USA. The satellite carried two television cameras – one low-resolution and one high-resolution. The images were stored in two onboard videotape recorders while the satellite was out of range of the ground station network. The video systems relayed thousands of pictures containing cloud-cover views of the Earth.

Early photographs provided information about the structure of large-scale cloud regimes. *Tiros-I* was operational for only 78 days, but proved that satellites could be useful tools for surveying global weather conditions from space.

1968

Lev Davidovich Landau, Soviet physicist who worked in the field of low-temperature physics and atomic physics and explained the peculiar superfluid behaviour of liquid helium at very low temperature, died in Moscow, Russia. (☞ Jan. 22)

1971

Kathleen Lonsdale (née Yardley), British crystallographer who developed several X-ray techniques for the study of crystal structure, died in London, UK. (☞ Jan. 28)

1981

The 200-MW second unit of the Rajasthan Atomic Power Project at Rawatbhata, near Kota, Rajasthan, was commissioned. It is a pressurised heavy water reactor that uses natural uranium as fuel and heavy water as moderator and coolant. (☞ Jun. 1, Dec. 16, Dec. 23)

2003

India's most advanced third generation supercomputer *Param Padma* was launched by the Centre for Development of Advanced Computing (C-DAC). The *Param Padma* has 1 teraflop of power, which means it can perform 1 trillion operations per second. Such power has previously only been available to countries such as the USA and Japan. The *Param Padma* is powered by C-DAC's flexible and scalable high performance computing and communication (HPCC) software environment. The storage system of *Param Padma* has been designed to provide a primary storage of 5 terabytes scalable to

22 terabytes. The *Param Padma* system is also accessible by users from remote locations and can be used for research in a variety of fields including defence, space and meteorology. (☞ Mar. 28)

April 2

1845
French physicists Hippolyte-Louis Fizeau and Jean-Bernard-Léon Foucault took the first photograph of the Sun. (☞ Sep. 18, Sep. 23)

1872
Samuel Finley Breese Morse, American painter and inventor, who developed the Morse code for telegraphy, died in New York, USA. (☞ Apr. 27)

1889
Charles Martin Hall, American chemist, received a patent for an inexpensive electrolytic process to extract aluminium from its ore, which brought the metal into wide commercial use. Although aluminium is the most abundant metal in the Earth's crust, it is not found naturally in its pure form, and thus must be separated from its ore. In the electrolytic process, the purified oxide called alumina, obtained from bauxite ore, is dissolved in molten cryolite and electrolysed with direct current. Under the influence of the current, the oxygen of the alumina is deposited on the carbon anode and is released as carbon dioxide, while free molten aluminium – which is heavier than the electrolyte – is deposited on the carbon lining at the bottom of the cell. Because of the strong affinity of aluminium for oxygen, chemical extraction of aluminium is extremely difficult. (☞ Dec. 6)

1928
Theodore William Richards, American chemist, Nobel laureate (Chemistry, 1914), whose research on the atomic weights of approximately 60 elements indicated the existence of isotopes, died in Cambridge, Massachusetts, USA. (☞ Jan. 31)

1953
The Central Building Research Institute (CBRI), a constituent national laboratory of the Council of Scientific and Industrial Research (CSIR), was inaugurated. Located in Roorkee, Uttaranchal, CBRI is engaged in R&D on all aspects of building and housing. It assists the building industry in solving problems of planning, designing, foundations, materials and construction including disaster mitigation in all kinds of buildings with a view to achieving economy, comfort, functional efficiency, speed, productivity in construction, environment preservation and energy conservation. Among the notable technologies developed at CBRI are innovative bricks and blocks for building construction, designs for earthquake-proof buildings, water- and fire-proofing technologies for thatched roofs, erosion-resistant mud plaster techniques for improving mud dwelling units, low-cost sanitation systems for housing units; partially pre-fabricated roofing components, ferro-cement boards for replacement of timbers in doors and window shutters, and energy-efficient lime and brick kilns. As well as creating functional housing units at cheaper prices using, by and large, locally available resources, the technologies selected were capable of generating considerable employment thus augmenting the earning opportunities of the local population.

1978
Velcro, the loop-and-hook fastener invented by the Swiss engineer and inventor

George de Mestral, was released for public use. Velcro uses two nylon tapes, one with stiff 'hooks' like burrs, which clings to the second tape with soft 'loops' like the fabric of Mestral's clothes. Because it is easy to use, maintenance free, and totally safe, Velcro fasteners have been used for just about every conceivable application where a temporary bond is required, particularly in clothing where it replaces buttons or zippers. It is also commonly used as a fastener in children's shoes since it is easier than tying shoelaces. (☞ Sep. 13)

1995
Hannes Alfvén, Swedish astrophysicist, Nobel laureate (Physics, 1970), one of the founders of the field of plasma physics, died in Djursholm, Stockholm, Sweden. (☞ May 30)

April 3

1984
The Soviet–Indian international crew comprising Indian cosmonaut–researcher Rakesh Sharma along with Russian commander Y. V. Malyshev and flight engineer G. M. Strekalov was launched into orbit on board the spacecraft *Soyuz T-11*. They docked with the *Salyut-7* orbital station to conduct scientific and technical studies and experiments.

Sharma conducted an Earth observation program concentrating on India. He also did life sciences and materials processing experiments and became the first yoga practitioner in zero-G. (☞ Jan. 13, Apr. 11, Apr. 19)

1999
INSAT-2E, the fifth satellite of India's *INSAT-2* series, was launched from Kourou in French Guiana, by European Space Agency's Ariane rocket into a geosynchronous transfer orbit with a perigee of 200 kilometres and an apogee of 35,980 kilometres. Designed and built by ISRO, *INSAT-2E* is a multi-purpose satellite for telecommunication, television broadcasting and meteorological services. The satellite is located at 83° east longitude in geostationary orbit. The satellite carries seventeen transponders – 12 operating in the normal C-band frequency and 5 in the lower extend C-band. Seven of the normal C-band transponders have wide beam coverage and the rest have zonal coverage.

April 4

1821
Linus Yale, American inventor and designer of the compact cylinder pin-tumbler lock, was born in Salisbury, New York, USA. As a youngster, Yale wanted to become a portrait painter, but he became interested in lock making after his father, who was a successful inventor, and was also interested in new lock designs, set up a lock factory in Newport, New York. Yale's best-known lock design is the cylinder pin-tumbler lock. The basic concept was used in ancient Egypt around 4,000 years ago, in locks that were very large and made out of wood. The design has a main barrel, which

is drilled so it has usually five or six parallel cylindrical slots that are set close together. A metal pin, or 'tumbler', fits closely to the walls of each of the slots. A second metal pin, or 'driver', sits above the tumbler and is pushed down on the tumbler by a very small coil spring, which is permanently compressed as it sits inside the lock's enclosed case. Yale's lock used a smaller, flat key with serrated edges like the ones we still use today. When inserted into the lock, the key pushes the bottom pins into the right position, which allows the user to turn the key and unlock the lock. Yale patented his cylinder pin-tumbler lock in 1861. Four years later he patented an improved cylinder lock, which is one of the most secure key-operated locks ever invented and which forms the basis of the Yale lock still in use. (☞ Dec. 24)

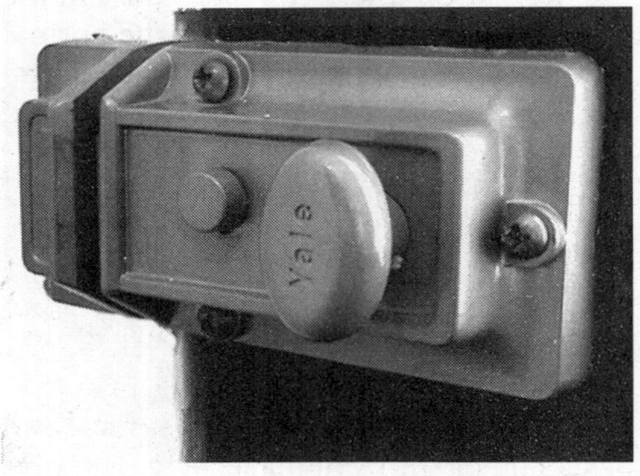

1823
William Siemens (also Carl Wilhelm Siemens), German-born English engineer and inventor, pioneer in undersea cables, was born in Lenthe, near Hanover, Germany. After being educated in the polytechnic school of Magdeburg and the University of Göttingen, he visited England at the age of nineteen, in the hope of popularising a process in electroplating invented by himself and his brother Werner. In 1859, Siemens became a naturalised British subject. Siemens' achievements are mainly in two distinct fields, the applications of heat and the applications of electricity. In 1856 he introduced the regenerative furnace, which used waste gases to preheat fuel gases. This became part of the open-hearth furnace, which he used to purify steel. In the field of electricity, Siemens' name is closely associated with the growth of land and submarine telegraphs, the invention and development of the dynamo, and the application of electricity to lighting and locomotion. He was head of the English branch of the Siemens firm, which made telegraphic and other electrical apparatus and handled electrical engineering projects. In 1869, he helped design and build the London–Calcutta (now Kolkata) telegraph line, which was a milestone in the history of communications engineering. In 1883, he was one of the first to apply electric power to railways. (☞ Nov. 19)

1887
William Cumming Rose, American biochemist who researched the role of amino acids in nutrition determining which were essential, and calculated the minimum daily requirement for each of them, was born in Greenville, South Carolina, USA. Rose attended an assortment of local schools until the age of fourteen, when the inadequacy of the education system made his father remove him from school and tutor him at home. He was well prepared by the time

he entered college. He had already been introduced to Latin, Greek, and Hebrew and had acquired an interest in chemistry. Rose entered graduate school at the age of 19 when he enrolled in the Sheffield Scientific School at Yale. Four years later he received his PhD under L. B. Mendel for his study that was part of a series on the origin of creatine (a nitrogenous organic acid that is found in the muscle tissue of vertebrates mainly in the form of phosphocreatine and supplies energy for muscle contraction), and its dehydration product creatinine (found in increased levels in advanced kidney disease). In the 1930s Rose undertook experiments that introduced the idea of an essential amino acid into nutrition in both human and rodent diets. Nutritionists had known for a long time that rats fed on a diet in which the only protein was zein, which is found in corn, would inevitably die. Rose worked with the constituent amino acids rather than proteins, but found that the rats died regardless of the combination of amino acids he tried. However, if the milk protein casein was added to their diet, the ailing rats recovered. Rose concluded that casein must contain an unknown amino acid that was essential to life. In a long series of experiments extracting and testing various fragments of casein, Rose discovered threonine, the essential amino acid that saved the rodents when added to other amino acids in their diet. Over the years, he manipulated the rodent diet to establish the primary importance of ten amino acids: lysine, tryptophan, histidine, phenylalanine, leucine, isoleucine, methionine, valine, and arginine in addition to the newly discovered threonine. In the 1940s he undertook a ten-year research project on the human diet, including the nutritive properties of amino acids. This work led to the determination of the minimum daily requirements of amino acids for humans. Rose was elected to the National Academy of Sciences in 1936 and received the National Medal of Science in 1966. He was also awarded the Willard Gibbs Medal from the American Chemical Society in 1952. (☞ Sep. 25)

1910
William Crookes, British chemist and physicist noted for his discovery of the element thallium and for his cathode-ray studies, died in London, UK. (☞ Jun. 17)

1929
Karl Benz, German automobile engineer who built the world's first practical automobile to be powered by an internal-combustion engine, died in Ladenburg, Germany. (☞ Jan. 29, Nov. 25)

1932
Wilhelm Ostwald, German chemist, Nobel laureate (Chemistry, 1909), who almost single-handedly organised physical chemistry into a nearly independent branch of chemistry and did pioneering work on catalysts, died near Leipzig, Germany. (☞ Sep. 2)

1932
Vitamin C was isolated by American biochemist Charles Glen King of the University of Pittsburgh. Through painstaking research spread over five years, extracting components from lemon juice – requiring untold thousands of lemons – King and his colleagues isolated a crystalline substance, identified, and later synthesised vitamin C. Their discovery led to the prevention of scurvy, a long time source of human suffering.

1966
Russian unmanned Moon-probe *Luna-10* went into orbit around the Moon, becoming the Moon's first man-made moon. (☞ Mar. 31)

1983
The American space shuttle *Challenger*, carrying astronauts Paul J. Weitz, Karol J. Bobko, Donald H. Peterson, and F. Story Musgrave was launched into orbit on its maiden flight from the Kennedy Space Center, Florida, USA. The mission saw the first space walk of the space shuttle program, as well as the deployment of the first satellite in the tracking and data relay system. Named after the British Naval research vessel *HMS Challenger* that sailed the Atlantic and Pacific oceans during the 1870s, *Challenger* flew nine successful space shuttle missions till it was lost in an accident soon after launch in 1986. (☞ Jan. 28, Apr. 9)

April 5

1827
Joseph Lister, British surgeon and medical scientist who was the founder of antiseptic medicine, was born in Upton, Essex, England. After getting his bachelor of medicine degree from the University College, London, Lister entered the Royal College of Surgeons. After six years he got a professorship of surgery at Glasgow. As professor of surgery at the Glasgow University, he was very aware that many people survived the trauma of an operation but died afterwards of what was known as 'ward fever.' At the time the usual explanation for wound infection was that chemicals in the air damaged the exposed tissues. Facilities for washing hands or the patient's wounds did not exist and it was even considered unnecessary for the surgeon to wash his hands before he saw a patient. Lister became aware of a paper published by the French bacteriologist Louis Pasteur, which showed that rotting and fermentation could occur only if micro-organisms were present. He believed that it was microbes carried in the air that caused diseases to be spread in wards. People who had been operated on were especially vulnerable as their bodies were weak and their skin had been cut open so that germs could get into the body with more ease. Lister decided that the wound itself needed to be thoroughly cleaned. He also tested the results of spraying instruments, the surgical incisions, and dressings with a solution of carbolic acid (phenol) that was known to kill micro-organisms and found that carbolic acid solution swabbed on wounds markedly reduced the incidence of gangrene and subsequently published a series of articles on the 'Antiseptic principle of the practice of surgery' describing this procedure in the medical journal *The Lancet*. Lister also made surgeons wear clean gloves and wash their hands before and after operations with 5% carbolic acid solutions. (☞ Feb. 10, Mar. 16, Jun. 17)

1899
Alfred Blalock, American surgeon, was born in Culloden, Georgia, USA. Blalock graduated from the University of Georgia in 1918, and in the same year joined the Johns Hopkins University School of Medicine, Baltimore, Maryland. He earned his MD from Johns Hopkins in 1922. From 1923 to 1925 he was hospital physician and assistant resident surgeon at Hopkins. Unable to obtain a residency at Hopkins,

he went to Nashville to become the first resident in surgery in the new Vanderbilt University Hospital. While at Vanderbilt he did pioneering work on the nature and treatment of haemorrhagic and traumatic shock. He demonstrated that surgical shock resulted primarily from the loss of blood, and he encouraged the use of plasma or whole blood transfusions as treatment following the onset of shock. But Blalock's most significant achievement was the development of a surgical technique, together with Vivien T. Thomas and Helen B. Taussig, to treat infants born with the condition known as the 'blue baby' syndrome. The technique involved the joining, of the subclavian artery to the pulmonary artery in the affected child, allowing the blood another chance to become oxygenated. The procedure is now known as a 'Blalock–Taussig shunt'. Infants with blue baby syndrome are born with a congenital heart defect that causes a bluish colouration of the skin as a result of the mixing of deoxygenated blood with oxygenated blood in the heart chamber. The colour is most noticeable around the lips and at the tips of the fingers and toes. After many animal experiments, Blalock, Thomas, and Taussig decided to try it out on a human baby, a 15-month-old girl named Eileen Saxon. The operation was a success, although Saxon later died. In the following months and years, Blalock performed the procedure on thousands of children, often accompanied by Thomas. The operation not only directly saved thousands of lives, it also marked the start of the modern era of cardiac surgery. Blalock's contributions to surgery were recognised nationally and internationally. He was awarded membership in the National Academy of Sciences and the American Philosophical Society. In his last years Blalock devoted much of his time to the establishment of a centre for child medicine and child surgery in Baltimore. (☞ Sep. 15)

1929

Ivar Giaever, Norwegian-born American physicist, Nobel laureate, was born in Bergen, Norway. Giaever graduated in 1952 with a degree in mechanical engineering and immigrated to Canada in 1954. After a short period as an architect's aide he joined the Canadian General Electric's Advanced Engineering Program. In 1956 he immigrated to USA and joined the General Electric Research and Development Center in 1958. He concurrently started to study physics at the Rensselaer Polytechnical Institute, in Troy, New York, where he obtained a PhD degree in 1964. From 1958 to 1969, Giaever worked in the fields of thin films, tunnelling and superconductivity. He pursued the possible applications of the work of the Japanese solid-state physicist, Leo Esaki, on 'tunnelling,' to superconductor technology, eventually 'marrying', as he put it, the two concepts to produce superconductor devices that flouted previously accepted limitations and allowed electrons to pass like waves of radiation through 'holes' in solid-state devices. In 1960, using a sandwich consisting of an insulated piece of superconducting metal and a normal one, Giaever achieved new tunnelling effects that led to greater understanding of superconductivity. Giaever was awarded the 1973 Nobel Prize for Physics, jointly with Esaki and the British physicist Brian Josephson; Giaever

and Esaki received the prize for their experimental discoveries regarding tunnelling phenomena in semiconductors and superconductors, respectively, while Josephson received it for his theoretical predictions of the properties of a supercurrent through a tunnel barrier, in particular those phenomena which are generally known as the Josephson effects. (☞ Jan. 4, Mar. 12)

1967
Hermann Joseph Müller, American geneticist, Nobel laureate (Physiology or Medicine, 1946), best remembered for his demonstration that mutations and hereditary changes can be caused by X-rays, died in Indianapolis, Indiana, USA. (☞ Dec. 21)

1973
American spacecraft *Pioneer-11* was launched from the Kennedy Space Center, Florida, USA, on a mission to Jupiter and Saturn. During its fly-by of Jupiter in December 1974, *Pioneer-11* obtained dramatic images of the Great Red Spot, made the first observation of the immense polar regions, and determined the mass of Jupiter's moon, Callisto. Accelerated by Jupiter's gravity and looping high above the ecliptic plane and across the solar system, *Pioneer-11* raced toward Saturn, flying to within 21,000 kilometres of Saturn in September 1979 and took the first close-up pictures of the planet. Instruments on board the spacecraft located two previously undiscovered small moons and an additional ring, charted Saturn's magnetosphere and magnetic field, and found its planet-size moon, Titan, to be too cold for life. Hurtling underneath the ring plane, *Pioneer-11* sent back amazing pictures of Saturn's rings. (☞ Sep. 1, Dec. 2)

1991
The American space shuttle *Atlantis*, carrying astronauts Steven R. Nagel, Kenneth D. Cameron, Jerry L. Ross, Jay Apt, and Linda M. Godwin, was launched on its 8th flight from the Kennedy Space Center, Florida, USA. It carried the *Compton Gamma Ray Observatory* (*CGRO*) for observing the universe in gamma rays. (☞ Apr. 8, Apr. 11)

April 6

1911
Feodor Lynen, German biochemist, Nobel laureate, was born in Munich, Germany. Lynen completed his primary and secondary schooling in Munich, and in 1930 matriculated from the chemistry department of the Munich University. After several years as a lecturer in the same department (1942–47), he became a professor. Lynen's main research interest was the metabolism of cholesterol and fatty acids. His work aimed at elucidating the chemical details of metabolic processes in living cells, and of the mechanisms of metabolic regulation. In 1951, while at Munich, he succeeded in isolating the coenzyme acetyl CoA from yeast. This research led to his demonstration that the acetylation, of coenzyme A, i.e., the introduction of an acetyl group into the molecule, is the key first step in a chain of reactions that result in the formation of cholesterol and fatty acids. In 1954 Lynen became director of the Max Planck Institute for Cell Chemistry in Munich. Lynen

was awarded the 1964 Nobel Prize for Physiology or Medicine, jointly with the German-born American biochemist Konrad Bloch, for their discoveries concerning the mechanism and regulation of the cholesterol and fatty acid metabolism. (☞ Jan. 21, Aug. 6)

1920

Edmond H. Fischer, American biochemist, Nobel laureate, was born in Shanghai, China. Fischer earned a PhD in chemistry from the University of Geneva, Switzerland, in 1947 and conducted research there until 1953. That year he emigrated to the United States, where he joined the American biochemist Edwin G. Krebs on the faculty of the University of Washington, Seattle. He became a full professor there in 1961. Soon after joining the University of Washington, Fischer and Krebs started work on reversible phosphorylation – the attachment or detachment of phosphate groups to cell proteins. The two scientists were the first to purify and characterise one of the enzymes, called phosphorylase, involved in the process of phosphorylation. They also discovered the enzymes that catalyse the attachment and detachment of phosphate groups, known as protein kinases and phosphatase, respectively. In the decades following these initial discoveries, scientists were able to identify many other enzymes that regulate specific processes in cells, leading to explanations of the mechanisms controlling basic activities in all living cells. Fischer and Krebs were jointly awarded the 1992 Nobel Prize for Physiology or Medicine for their discoveries concerning reversible protein phosphorylation as a biological regulatory mechanism. (☞ Jun. 6)

1928

James Dewey Watson, American geneticist and biophysicist, Nobel laureate, was born in Chicago, Illinois, USA. Watson received his BS in 1947 from the University of Chicago and a PhD in 1950 from the Indiana University, both in zoology. In 1952, Watson joined the Cavendish laboratories in Cambridge, England, at a time when a race was on to determine the structure of DNA – the key molecule of heredity. The X-ray crystallography experiments of British crystallographer Rosalind Franklin and New Zealand-born British biophysicist Maurice Wilkins provided much information about DNA – in particular that DNA was a molecule in which two 'strands' formed a tightly linked pair. He soon met the British biophysicist Francis Crick and discovered their common interest in solving the DNA structure. They thought it should be possible to correctly guess its structure, given both the experimental evidence already available plus careful examination of the possible stereochemical configurations of polynucleotide chains. Watson and Crick made the intuitive leap: in 1953, they proposed that the structure of DNA was a winding helix in which pairs of bases (adenine paired with thymine, and guanine paired with cytosine) held the two strands together. The Watson–Crick model of the DNA double helix provided enormous impetus for research in the emerging fields of molecular genetics and biochemistry. In 1955 Watson joined the faculty of the Harvard University in Cambridge, Massachusetts, and became professor in 1961. In 1976 he resigned from Harvard, to become full-time director of the Cold Spring Harbor Laboratory in New York. In 1988 he was appointed Associate

Director for Human Genome Research of the National Institutes of Health (NIH). The next year he was appointed Director of the National Center for Human Genome Research at the NIH. Watson was awarded the 1962 Nobel Prize for Physiology or Medicine, jointly with Crick and Wilkins, for their discoveries concerning the molecular structure of nucleic acids and its significance for information transfer in living material. (☞ Apr. 25, Jun. 8, Dec. 15)

1938

Roy J. Plunkett, researcher at the Du Pont Company's laboratory and his technician Jack Rebok accidentally discovered the chemical compound polytetrafluoroethylene (PTFE) that was later marketed as 'Teflon.' Working with cylinders filled with different types of the refrigerant freon, Plunkett found that although the gas had disappeared from some of the cylinders, their weight was intact. When Plunkett cut open the empty cylinders he found a wax-like white substance coating the inside of the cylinders. On analysing the substance, it was found to be a polymer of tetrafluoroethylene. Plunkett found it had unusual properties – it was a wonderful solid lubricant in powdered form, was chemically inert and had a very high melting point. He realised it was formed by an unexpected polymerisation. Teflon is widely used as a coating for making frictionless and non-stick surfaces, as in non-stick kitchenware.

1949

Horst Ludwig Störmer, German-born American physicist, Nobel laureate, was born in Frankfurt-am-Main, Germany. Störmer studied physics at the University of Stuttgart in Germany and received his PhD from the Max-Planck Institute of Solid-State Physics at Stuttgart. He met the Chinese-born American physicist Daniel Tsui from Bell Labs in Grenoble, France. He joined the Bell Laboratories at Murray Hill, New Jersey, USA, in 1982 and became director of its physics laboratory in 1992. At Bell Labs, Störmer and Tsui carried on research on the Hall effect, which denotes the voltage that develops between the edges of a thin current-carrying ribbon placed flat between the poles of a strong magnet. This effect had long been familiar to science, but in 1980 the German physicist Klaus von Klitzing discovered that at extremely low temperatures, even if the strength of an applied magnetic field is increased continuously, the corresponding change in the voltage of the deflected current occurs in discrete jumps or steps, thereby exhibiting quantum properties. Störmer and Tsui extended Klitzing's work, observing the Hall effect in semiconductors at temperatures close to absolute zero (–273.16°C) and under extremely powerful magnetic fields. In 1982 they observed that under these conditions, the Hall effect varies not only stepwise but also in fractional increments, implying that the charge carriers in the current carry exact fractions of an electron's charge, which came to be known as the 'quantum Hall effect.' In 1983 the American physicist Robert B. Laughlin provided an explanation of this puzzling phenomenon. Störmer was awarded the 1998 Nobel Prize in Physics, jointly with Tsui and Laughlin, for their discovery of a new form of quantum fluid with fractionally charged excitations. (☞ Feb. 28, Nov. 1)

1961
Jules Bordet, Belgian immunologist, Nobel laureate (Physiology or Medicine, 1919), who discovered bacteriolytic and haemolytic substances in blood, died in Brussels, Belgium. (☞ Jun. 13)

1965
Early Bird, the world's first commercial communications satellite, was launched by the American space agency NASA. Placed in the geosynchronous orbit, 36,000 kilometres above the equator, the satellite provided line-of-sight communications between Europe and North America. As a communications repeater, *Early Bird* handled communications of all type, including telephone, television, telegraph, and facsimile transmissions. With a capability of 240 telephone circuits or one TV channel, *Early Bird* successfully demonstrated the concept of geosynchronous satellites for commercial communications. Although designed for an operational lifetime of only 18 months, the satellite was in continuous, full-time service for nearly four years. (☞ Jun. 28)

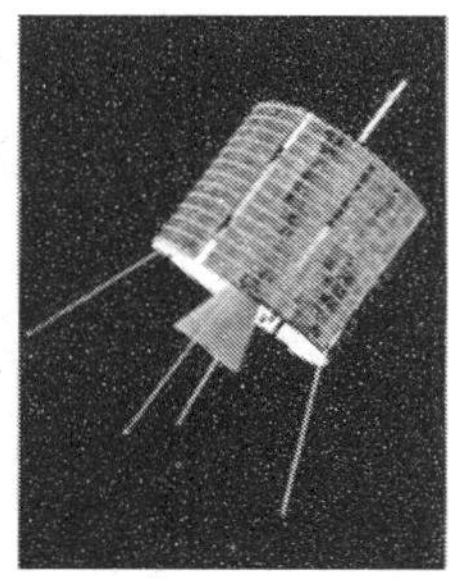

1984
Space shuttle *Challenger* was launched on its 5th mission, with American astronauts Robert L. Crippen, Francis R. Scobee, George D. Nelson, James D. van Hoften, and Terry J. Hart. The primary objective was retrieving the malfunctioning *Solar Max* satellite from orbit, repairing it and putting it back in orbit, which was successfully achieved. This was the first on-orbit repair job carried out by a shuttle crew. After it was repaired, *Solar Max* went on collecting data until it burnt up in the Earth's atmosphere in December 1989. (☞ Apr. 4)

1992
Isaac Asimov, American author and biochemist and a highly successful and prolific writer of popular science and science fiction, died in New York, USA. (☞ Jan. 2)

April 7

1823
Jacques-Alexandre-César Charles, French mathematician, physicist, and inventor, who invented the hydrogen balloon and formulated the gas law known after him, died in Paris, France. (☞ Nov. 12)

1859
Jacques Loeb, German-born American biologist, noted chiefly for his experimental work on artificial parthenogenesis (reproduction without fertilisation), was born in Mayen, near Koblenz, Prussia (now in Germany). Loeb received his MD from the University of Strasbourg in 1884. In 1891, he emigrated to the United States and taught at the University of Chicago, and the University of California. From 1910 he was a member of the Rockefeller Institute (now Rockefeller University). Before he emigrated to the USA, Loeb's work mainly concerned the movements of simple animals in response to light, gravity, etc., a phenomenon known as 'tropism.' But he was best known for his experimental work on reproduction without fertilisation in some lower animals. Loeb's work showed that the larvae of invertebrates called echinoderms could be chemically stimulated to develop even without fertilisation of the egg. He also

propounded the mechanistic philosophy that all ethics were the outgrowth of man's inherited tropisms. (☞ Feb. 11)

1923
The first operation for a brain tumour under local anaesthetic was performed at the Beth Israel Hospital in New York City, USA, by K. Winfield Ney. The skull of the patient, Henry A. Brown, under only a local anaesthetic (cocaine), was opened by Dr. Winfield Ney and assisting surgeons. A large brain tumour was successfully removed. Brown was fully conscious throughout the operation, was able to answer the doctors' questions, and cooperated cheerfully.

1947
Henry Ford, American industrialist who introduced assembly line manufacturing for mass production, died in Dearborn, Michigan, USA. (☞ Jul. 30)

1948
The World Health Organization (WHO), the United Nations specialised agency for health, was established. WHO's objective, as set out in its Constitution, is the attainment by all peoples of the highest possible level of health. Health is defined in WHO's Constitution as a state of complete physical, mental and social well-being and not merely the absence of disease or infirmity. The day is designated as World Health Day and celebrated every year by the 191 member countries of the WHO to emphasise significant issues in public health of worldwide concern. World Health Day offers a unique opportunity to raise global awareness of a specific health theme.

2001
American space agency NASA launched the *2001 Mars Odyssey* spacecraft from Cape Canaveral. *Odyssey* travelled 460 million kilometres before entering orbit around the red planet in October 2001. Its primary mission was to look for water in the form of ice under the Martian surface and to create a thermal map of the planet. From an altitude of about 400 kilometres, the spacecraft searched for traces of hydrogen, which could point out the existence of water. Another goal of the *Odyssey* mission was to analyse the radiation environment to determine its potential effects on human health. (☞ Oct. 24)

2001
Gopalasamudram Narayana Iyer Ramachandran, Indian biophysicist, Fellow of the Royal Society, who determined the structure of collagen and developed the 'Ramachandran map' for determination of protein conformation, died in Chennai, Tamil Nadu, India. (☞ Oct. 8)

April 8

1732
David Rittenhouse, American astronomer, instrument maker and inventor who was an early observer of the transit of Venus, was born in Germantown, Philadelphia USA. Rittenhouse showed mathematical and mechanical ability at an early age soon mastering Newton's *Principia* in an English translation. At 19 he started an instrument shop on his father's farm where he showed great skill in clock repairing. He built an orrery – a model showing relative motions of the Moon and the planets – which could show the solar and lunar eclipses and other phenomena for a period of 5,000 years, either forward or backward. In 1769 Rittenhouse was asked by the American Philosophical Society to observe the transit of Venus that was to take place on the 3rd of June that year. For observing the

transit, Rittenhouse constructed a high-precision pendulum clock, an astronomical quadrant, an equal altitude instrument, and an astronomical transit. He was the first to observe, during the transit, that Venus has an atmosphere. He was professor of astronomy in the University of the State of Pennsylvania, from 1779 to 1782. One of Rittenhouse's major contributions to astronomy was the introduction in 1786 of the use of cross-hairs made of spider silk in the focus of a transit instrument. (☞ Jun. 3, Jun. 8, Jun. 26, Nov. 24)

1818

August Wilhelm von Hofmann, German chemist, who studied the extraction and exploitation of coal-tar derivatives, and whose research on aniline helped lay the basis of the aniline-dye industry, was born in Giessen, Hesse-Darmstadt, Germany. In 1858, while he was a professor at the Royal College of Chemistry in London, Hofmann obtained the dye known as fuchsine or magenta by the reaction of carbon tetrachloride with aniline. Later he isolated from it a compound, which he called rosaniline and used as a starting point for other aniline dyes, including aniline blue. With alkyl iodides he obtained a series of violet dyes, which he patented in 1863. These became known as 'Hofmann's violets' and were a considerable commercial and financial success. In the field of organic chemistry, Hofmann is best known for his studies of the organic derivatives of ammonia

and phosphine and for his subsequent discovery of the Hofmann degradation reaction. He also developed the Hofmann method of finding the vapour densities, and from these the molecular weights, of liquids. Hofmann devised much of the terminology of paraffins (alkanes) and their derivatives that was accepted at the 1892 Geneva Conference on nomenclature. He founded the German Chemical Society. (☞ May 2)

1861

Elisha Graves Otis, American inventor of the automatic safety brake for elevators, died in Yonkers, New York, USA. (☞ Jan. 15, Aug. 3)

1911

Melvin Calvin, American biochemist, Nobel laureate, was born in St. Paul, Minnesota, USA. Calvin earned his undergraduate degree from the Michigan College of Mining and Technology in 1931 and his PhD in chemistry from the University of Minnesota in 1935. Following post-doctoral studies in England, Calvin was invited to join the faculty of the University of California at Berkeley. At Berkeley, Calvin organised a team of researchers to study photosynthesis – the process by which green plants convert sunlight energy into chemical energy. Using the radioactive carbon-14 isotope, which had been discovered in 1940 at the 60-inch cyclotron at Berkeley, as a tracer, Calvin and his team mapped the complete route that carbon travels through a plant during photosynthesis, starting from its absorption as atmospheric carbon dioxide to its conversion into carbohydrates and other organic compounds. In doing so, the Calvin group showed that sunlight acts on the chlorophyll in a plant rather than on carbon dioxide, as was previously believed, to stimulate the synthesis of organic compounds. They also demonstrated

the 'dark reactions' of photosynthesis that occur through the night turning carbon dioxide into sugar. The whole sequence of reactions came to be known as the 'Calvin cycle'. In his final years of active research, Calvin studied the use of oil-producing plants as renewable sources of energy. Calvin was awarded the 1961 Nobel Prize for Chemistry for his research on the carbon dioxide assimilation in plants. (☞ Jan. 8)

1936
Robert Bárány, Austrian otologist, Nobel laureate (Physiology or Medicine, 1914), who investigated physiology and pathology of the vestibular (balancing) apparatus of the inner ear, died in Uppsala, Sweden. (☞ Apr. 22)

1968
The British De Havilland *Comet*, the world's first jet-propelled airliner, took off from Hatfield, England, on its last flight. The aircraft was taken off service after a series of crashes due to structural failure. (☞ May 2, Jul. 27)

1981
Gopal Chandra Bhattacharya, Indian naturalist, researcher, and science populariser died in Kolkata, India. (☞ Aug. 1)

1984
Pyotr Leonidovich Kapitsa, Russian physicist, Nobel laureate (Physics, 1978), who made pioneering discoveries in low-temperature physics and discovered the superfluidity of helium II, died in Moscow, Russia. (☞ Jul. 8)

1991
The *Compton Gamma Ray Observatory* (*CGRO*) was launched on board the space shuttle *Atlantis* during its 8th flight. Weighing over 17 tonnes, the *Compton Observatory* was the heaviest astrophysical payload ever flown at the time of its launch. This orbiting observatory saw the sky in gamma-ray photons – light invisible to the human eye. These photons are blocked by the Earth's atmosphere from reaching the Earth's surface. Results from *CGRO* have shown the entire universe to be a violent and rapidly changing place – when viewed in gamma rays. Astronomers using *CGRO* data continue to make monumental discoveries, including discovery of a whole new class of star-like objects, and of objects so strange that astronomers cannot yet

figure out what they are. The Observatory was named in honour of Arthur Holly Compton, who won the 1927 Nobel Prize for Physics for work on scattering of high-energy photons by electrons. Due to one of its gyroscope failing, it was de-orbited and entered the Earth's atmosphere on 4th June, 2000. (☞ Apr. 5, Apr. 11, Oct. 3)

1992
Daniel Bovet, Swiss-French-Italian pharmacologist, Nobel laureate (Physiology or Medicine, 1957), who is credited with the discovery of several chemotherapeutic agents, namely, sulpha drugs, anti-histamines, and muscle relaxants, died in Rome, Italy. (☞ Mar. 23)

April 9

1626
Francis Bacon, English philosopher, remembered for his influence in promoting the scientific method, died in London, UK. (☞ Jan. 22)

1770
Thomas Johann Seebeck, German physicist who discovered that an electric current flows between different conductive materials that are kept at different temperatures, was born in Tallinn, Estonia. Seebeck studied medicine in the Universities of Berlin and Göttingen and received his medical degree in 1802, but since he preferred research in physics to the practice of medicine he went into education and research as a career. After graduation he joined the faculty of the University of Jena where he carried out research on the theory of colour and the effect of coloured light. Seebeck then engaged in research on the solar spectrum. He returned to the University of Berlin around 1818 and worked independently on the magnetisation of iron and steel when electrical currents were passed through conductors. While trying to find a relation between electricity and heat in 1821, Seebeck joined two wires of dissimilar metals (copper and bismuth) to form a loop and discovered that if he heated one junction to a high temperature, and the other junction remained at a cooler temperature, a magnetic field appeared around the circuit. Although he did not realise that an electrical current was also generated, the phenomenon has come to be known as the 'Seebeck effect.' It forms the basis of the thermocouple and is considered the most accurate method of measurement of temperature. (☞ Dec. 10)

1800
Humphry Davy, English chemist, discovered the physiological effects of nitrous oxide (N_2O), which became known as laughing gas. It is said that he 'breathed 16 quarts of the gas in seven minutes' and became 'completely intoxicated' with it. Nitrous oxide was discovered by the English chemist Joseph Priestley in 1772. (☞ Mar. 13, May 29, Dec. 17)

1850
William Prout, English chemist and biochemist noted for his discoveries concerning digestion, metabolic chemistry, and atomic weights, died in London, England. (☞ Jan. 15)

1903
Gregory Pincus, American endocrinologist whose work on the antifertility properties of steroids led to the development of the first effective oral contraceptive, was born in Woodbine, New Jersey, USA. Pincus received his bachelor's degree in agriculture from the Cornell University, Ithaca, New York, in 1924 and then went to the Harvard University in Cambridge, Massachusetts, where he was an instructor in zoology while also working toward his master's and doctorate degrees. From 1927 to 1930 he moved from the Harvard to the Cambridge University in England to the Kaiser Wilhelm Institute in Berlin, Germany, conducting research. He became an instructor in general physiology at Harvard in 1930 and was promoted a year later to assistant professor. In the late 1940s, Pincus focussed his attention on the role of hormones in reproduction. His experiments involved the use of progesterone, a female steroidal hormone secretion that is triggered by ovulation.

Progesterone was also known to safeguard against overlapping pregnancies by preventing further fertilisation and thus served as a natural birth-control agent. In 1956, Pincus, his research colleague Min-Cheuh Chang, and a Harvard gynaecologist John Rock began experiments on hundreds of women in Brookline, Massachusetts, Puerto Rico, and Haiti using progesterone in pill form as a birth-control agent, and reported success. In 1957, the Food and Drug Administration of USA authorised the marketing of steroids for miscarriage and certain menstrual disorders, and in 1960, the Federal agency licensed the first synthetic progesterone–oestrogen contraceptive pill. Pincus is also credited with the development of estrone, a hormone used in the treatment of breast cancer and pregnancy complications. (☞ May 9, Aug. 22)

1922
Patrick Manson, Scottish parasitologist, who was a pioneer in the study of transmission of diseases caused by parasites and founded the London School of Tropical Diseases, died in London, England. (☞ Oct. 3)

1981
The journal *Nature* published the longest scientific name in history. With 16,569 nucleotides, the systematic name for human mitochondrial DNA is 207,000 letters long.

1983
The American space shuttle *Challenger*, carrying astronauts Paul J. Weitz, Karol J. Bobko, Donald H. Peterson, and F. Story Musgrave landed at the Edwards Air Force Base, California, USA, after completing its maiden flight. (☞ Apr. 4)

April 10

1849
Walter Hunt, of New York, USA, a mechanic was issued a patent for the safety pin. Hunt's pin was made from one piece of wire, which was coiled into a spring at one end and a separate clasp and point at the other end, allowing the point of the wire to be forced by the spring into the clasp. Hunt invented the safety pin in order to pay a debt of $15; he eventually sold the rights to his patent for $400.

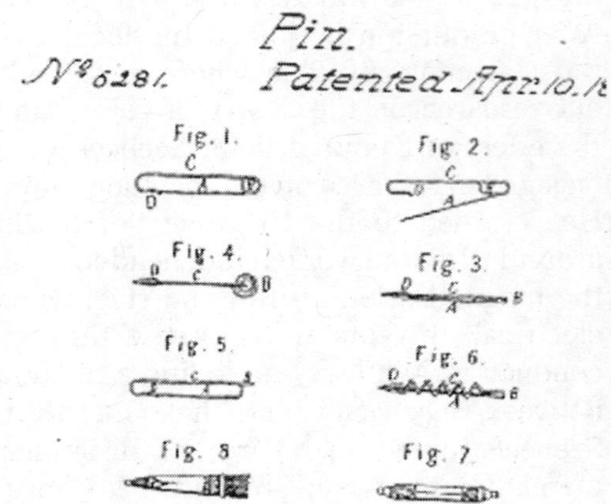

1863
Paul-Louis-Toussaint Héroult, French metallurgist who invented the electric arc furnace, was born in Thury-Harcourt, France. Héroult spent his childhood in England but studied in Paris at the School of Mines. In 1886, he invented an electrolytic process for producing aluminium starting from bauxite. This process was announced just three months after the American chemist Charles Martin Hall had received

the patent for an identical process. Héroult's process differed from Hall's in that the former used one large, central graphite electrode in the graphite cell holding the molten material. This process, finally called the Hall–Héroult process, was one of the first commercial applications of the electrical furnace, and led to a steep fall in the price of aluminium, which cost as much as silver before the electrolytic process was developed. Héroult's second most important invention was the electric arc furnace for steel in 1900. The Heroult furnace gradually replaced the giant smelters for the production of a variety of steels. (☞ May 9)

1887

Bernardo Alberto Houssay, Argentine physiologist, Nobel laureate, was born in Buenos Aires, Argentina. Houssay was a child prodigy, entering college at the age of 9 and becoming a hospital intern at 13. Working with dogs that had been rendered diabetic by the excision of their pancreas, Houssay found that the removal of the pituitary gland that lies at the base of the brain, or even a part of it, greatly relieved the symptoms of the disease and made the animal unusually sensitive to insulin. He demonstrated that injection of pituitary extracts into normal animals induces diabetes by increasing the amount

of sugar in the blood, thus indicating that the secretions of the pituitary gland oppose the action of insulin. Houssay founded Argentina's leading physiological research centre, the Institute of Biology and Experimental Medicine, in Buenos Aires, in 1944 and became its director in 1946. Houssay was awarded the 1947 Nobel Prize for Physiology or Medicine, jointly with the Czech-born American biochemist couple Carl and Gerty Cori, for his discovery of the part played by the hormone of the anterior pituitary lobe in the metabolism of sugar. (☞ Aug. 15, Sep. 21, Dec. 5)

1917

Robert Burns Woodward, American chemist, Nobel laureate, was born in Boston, Massachusetts, USA. Woodward received his BS degree in 1936 and completed his doctorate in 1937. Soon afterwards he joined the department of chemistry at the Harvard University, Cambridge, Massachusetts, where he remained until his death in 1979. Woodward's basic strength lay in his skill of structure analysis and syntheses of complex molecules. He demonstrated that the understanding of chemical reaction mechanisms made possible the planning and successful execution of an extended sequence of reactions to build up complex compounds in the laboratory. He was one of the first to determine the structure of such organic chemical compounds as the alkaloid strychnine, and the antibiotics penicillin, terramycin, and aureomycin. Woodward is best known for his chemical synthesis of several organic substances, beginning with his synthesis of quinine the well-known antimalarial, in 1944, followed by the steroids, cholesterol and cortisone, and vitamin B_{12}. He also synthesised lanosterol, the famous poison strychnine, and the antihypertensive drug reserpine. Woodward's synthesis of reserpine in 1956 has become a model of elegant technique and has been used

commercially for reserpine production. Woodward was awarded the 1965 Nobel Prize for Chemistry for his outstanding achievements in the art of organic synthesis. (☞ May 3, Jul. 8)

1927

Marshall Warren Nirenberg, American biochemist, Nobel laureate, was born in New York, USA. Nirenberg was educated at the University of Florida at Gainesville from where received his BSc degree in 1948 and his MSc degree in zoology in 1952. He continued studies in biochemistry at the University of Michigan, Ann Arbor, and received his PhD degree in 1957. From 1957 to 1959 he obtained postdoctoral training at the National Institutes of Health in Bethesda, Maryland as a fellow of the American Cancer society. In 1959 he began to study the steps that relate deoxyribonucleic acid (DNA), ribonucleic acid (RNA) and protein. These investigations led to the demonstration that messenger RNA is required for protein synthesis and that synthetic messenger RNA preparations can be used to decipher various aspects of the genetic code. Nirenberg demonstrated that, with the exception of 'nonsense codons', each possible triplet (called a codon) of four different kinds of nitrogen-containing bases found in DNA and, in RNA for some viruses, ultimately cause the incorporation of a specific amino acid into a cell protein. Thus Nirenberg played a major role in deciphering the genetic code. In 1962 he became head of the section of biochemical genetics at the National Institutes of Health. Nirenberg was awarded the 1968 Nobel Prize for Physiology or Medicine, jointly with the American biochemist Robert William Holley and Indian-born American biochemist Har Gobind Khorana, for their interpretation of the genetic code and its function in protein synthesis. (☞ Jan. 9, Jan. 28)

1949

Birbal Sahni, Indian palaeobotanist, Fellow of the Royal Society, best known for his study of fossil plants of India, died in Lucknow, Uttar Pradesh, India. (☞ Nov. 14)

1954

Central Salt and Marine Chemicals Research Institute (CSMCRI), a constituent national laboratory of the Council of Scientific and Industrial Research (CSIR), was founded. Located in Bhavnagar, Gujarat, CSMCRI is a multidisciplinary research institute with a blend of scientists and technologists having specialisation in chemistry, biology and engineering sciences. The institute carries out R&D on salt, inorganic chemicals, marine algae, non-traditional agriculture on wasteland, separation technologies based on membrane, renewable energy resources, and environmental monitoring. The institute focusses on specific seaweeds that are a source of commercially important polysaccharides such as carrageenan and bacteriological agar. *Eucheuma* (source of carrageenan) cultivation technology developed by the institute has already been commercialised. Saline wasteland development and saline water irrigation is yet another thrust area of CSMCRI. The institute has produced vegetable salt from *Salicornia brachiata* (a halophyte) – the invention attracted worldwide interest. As a pioneer in Indian solar thermal technologies, CSMCRI has embarked on an ambitious plan to develop a low cost solar concentrator. CSMCRI has installed the world's first animal powered RO desalination plant that can treat water having up to 5000 ppm total dissolved solids to produce 600–700 litres per hour of drinking water. The unit can be run both with oxen and camel power. The institute has developed 34 processes covering various products of marine origin including interpolymer membranes for water desalination.

1954
Auguste Lumière, French inventor, who, with his brother Louis, invented and pioneered the manufacturing of photographic equipment, died in Lyon, France. (☞ Mar. 22, Oct. 5, Oct. 19)

1982
India's first multipurpose geostationary satellite *INSAT-1A* was launched by NASA's Thor Delta rocket from Cape Canaveral, USA. The satellite was designed by the Indian Space Research Organisation and built by Ford Aerospace, USA. The satellite carried 12 C-band transponders and 3 S-band transponders, and was meant to provide telecommunication, networking and meteorological services, but it had to be abandoned in orbit because its solar panel failed to open.

1992
Peter Dennis Mitchell, British chemist, Nobel laureate (Chemistry, 1978), who clarified how adenosine diphosphate (ADP) is converted into the energy-carrying compound adenosine triphosphate (ATP) in the mitochondria of living cells, died in Bodmin, Cornwall, UK. (☞ Sep. 29)

2003
INSAT-3A, the third of India's third generation INSAT series of satellites, was launched by the Ariane-5 launch vehicle of the European launch provider Arianespace from Kourou, French Guyana. Designed and built by the Indian Space Research Organisation, the 2,950-kg multipurpose satellite carries 12 C-band transponders, six upper extended C-band transponders, six Ku-band transponders, and a satellite aided search and rescue (SAS&R) transponder. The meteorological payloads include a very high resolution radiometer (VHRR) with a 2-km resolution in the visible spectral band and an 8-km resolution in infrared and water vapour bands, a charge-coupled device (CCD) camera operating in visible, near infrared and short wave infrared bands with a 1-km resolution, and a data relay transponder. *INSAT-3A* is positioned at 93.5° east longitude in a geostationary orbit.

April 11

1755
James Parkinson, English physician and amateur palaeontologist, who was the first to describe the neuromuscular disease now known by his name, was born in London, England. Parkinson was educated in Hoxton and practised there. Early on he assisted and then took over the practice of his father in general medicine. His hobbies included geology and palaeontology. He wrote a little known medical monograph titled *Observations on the Nature and Cure of Gout* in 1805. He was the first to recognise a burst appendix as a cause of death, and wrote the first scientific article on appendicitis in 1812. But it was his *Essay on the Shaking Palsy*, published in 1817, which gained him immortality in the annals of medicine. In it he

described the symptoms of the movement disorder that came to be called Parkinson's disease, as 'a generalised slowness of movement, a tremor or slight shaking on one side of the body when at rest, some stiffness of the limbs, and problems of gait or balance'. Parkinson was a Fellow of the Royal College of Surgeons, and one of the original members of the Geological Society of London. (☞ Dec. 21)

1798

Macedonio Melloni, Italian physicist, who was the first to show the similarity between infrared radiation and light, was born in Parma, Italy. From 1824 to 1831 Melloni was a professor at Parma. In 1839 he went to Naples and was appointed director of the Vesuvius observatory, a post that he held until 1848. Melloni's reputation as a physicist rests especially on his discoveries in radiant heat, made with the aid of a combination of thermopile and galvanometer. He developed the thermopile to make quantitative measurements of the intensity of radiation. In 1846, from an observation point high on Mount Vesuvius, he measured the slight heating effect of moonlight. He also showed that rock salt, being transparent to infrared, made suitable lenses and prisms to demonstrate the similar properties of reflection, refraction, polarisation and interference of infrared radiation with visible light. (☞ Aug. 11)

1817

Charles Messier, French astronomer, who prepared the first catalogue of deep sky objects, died in Paris, France. (☞ Jun. 26)

1875

Samuel Heinrich Schwabe, amateur German astronomer who discovered the 10-year sunspot activity cycle, died in Dessau, Germany. (☞ Oct. 25)

1890

'Elephant Man' Joseph Carey Merrick died in London, England. (☞ Aug. 5)

1908

Masaru Ibuka, Japanese industrialist who patented the transistor and launched the modern electronics industry, was born in Nikko City, Tochigi Prefecture, Japan. Ibuka graduated from the Waseda University's School of Science and Engineering in 1933, and joined Photo-Chemical Laboratory, a company which processed movie films. In 1945, he left the company and founded a radio repair shop in Tokyo. In 1946, he co-founded Sony Corporation (known as Tokyo Telecommunications Engineering Corporation till 1958) with Akio Morita. Early research into magnetic powders and tape base material led to the development of magnetic recording tape in 1949. A year later, the company developed and marketed the first tape recorder in Japan. Several important developments followed, including Japan's first transistor radio in 1955 and the world's first transistor television in 1960, to name a few. He was instrumental in securing the licensing of transistor technology to Sony from Bell Labs in the 1950s, thus making Sony one of the first companies to apply transistor technology to non-military uses. In 1967, Ibuka managed the R&D project team that was responsible for the development of the Trinitron colour television system. Ibuka is acclaimed as the innovator who led Sony to create new and unique products that

broke from the Japanese corporate tradition of merely copying technology and ideas from the west. In doing so he not only succeeded in creating Sony Corporation, but changed the entire face of Japanese industry as well. (☞ Dec. 19)

1952

First surgical success in treating Parkinson's disease was achieved by Dr. Irving Spencer Cooper, an Assistant Professor of Surgery at the New York University, New York, USA. While operating on a patient named Raymond Walker to correct an aneurysm (a sac-like widening of an artery) in the brain, Cooper accidentally ruptured the aneurysm and was forced to clip the vessel called the anterior choroidal artery. When the patient came out of anaesthesia the Parkinsonian tremor had disappeared and his motor and sensory functions restored. Since 1968, surgery has been replaced by the use of the drug L-Dopa as the preferred treatment of Parkinson's disease.

1970

American manned lunar mission *Apollo-13*, with a three-man crew of James Lovell, Fred Haise, and John Swigert, was launched by a Saturn-V rocket from the Kennedy Space Center, Florida, USA. Two days later and 320,000 kilometres from the Earth, disaster struck. A liquid oxygen tank exploded, disabling the normal supply of oxygen, electricity, light, and water. The lunar landing was aborted. After circling the Moon, the crippled spacecraft began a long, cold journey back to Earth, carefully guided by the mission control, ultimately ending with the crew of the crippled craft safely splashing down in the Pacific Ocean. (☞ Apr. 17)

1984

Astronauts on board the space shuttle *Challenger* completed the first in-space satellite repair. Astronauts George Nelson and James Van Hoften retrieved the ailing *Solar Max* astronomy satellite. It had originally been launched in 1980, but had been malfunctioning. The rescued satellite was brought into the shuttle cargo bay to carry out the repair. Its attitude control mechanism and the main electronics system of the coronagraph instrument were serviced, and *Solar Max* was released back into orbit, ready to resume normal operation. (☞ Feb. 14, Apr. 4, Apr. 6)

1984

The Soviet–Indian international crew comprising Indian cosmonaut-researcher Rakesh Sharma along with Russian commander YV Malyshev and flight engineer GM Strekalov returned to Earth after a 7-day stay on board the Russian space station *Salyut-7*. (☞ Apr. 3, Apr. 19)

1986

Comet Halley made its closest approach to Earth during the 1985–86 return, when it came within 63 million kilometres of our planet on its way out of the solar system. But unlike its previous returns, when it appeared much brighter, the famous comet was barely visible to the naked eye. Its perihelion (the closest point to the Sun) occurred earlier in the year, in February 1986, when it was 88 million kilometres from the Sun, between the orbits of Mercury and Venus. The comet is scheduled to return in 2061. (☞ Feb. 9, Mar. 6)

1991

The American space shuttle *Atlantis* carrying astronauts Steven R. Nagel, Kenneth D. Cameron, Jerry L. Ross, Jay Apt, and Linda M. Godwin safely landed at the Edwards Air Force Base in California, USA after completing its 8th flight. During the flight the *Compton Gamma Ray*

Observatory (CGRO) was deployed in orbit for observing the universe in gamma rays. (☞ Apr. 5)

1999

India test-fired the ballistic missile, *Agni-II*, an advanced version of *Agni*, from the missile test range at Chandipur-on-sea in Orissa on the east coast. It is a two-stage intermediate-range ballistic missile (IRBM) with a mobile launch capability and can cover a range of over 2,000 kilometres. The 20-metre-tall, 16-tonne, *Agni-II* has a solid propulsion system and is capable of delivering a 1,000-kg payload. The *Agni-II* has a much faster deployment time than the *Agni-I*, due to its two-stage solid fuel propulsion system compared to the *Agni-I*'s first-stage solid fuel and second-stage liquid fuel system. When inducted into the Indian army, *Agni-II* will always be in a ready-to-fire mode and can be launched within 15 minutes as compared to almost half a day of preparation required for the *Agni-I*. Another major feature in terms of operational configuration is that *Agni-II* is designed to be launched from a mobile launcher; one that can move on road or a standard broad-gauge rail system. A rail- and road-based missile system reduces vulnerability and allows for greater operational flexibility. The indigenously developed missile is capable of carrying both conventional and nuclear weapons. (☞ Jan. 25, May 22)

2006

European Space Agency's *Venus Express* went into orbit around Venus, which will bring it to within 400 km of the planet's north pole. The spacecraft will orbit Venus for about 500 Earth days to study its atmosphere, which has undergone runaway greenhouse warming due to abundant carbon dioxide, with temperatures reaching 460°C on the surface. The mission is expected to shed light on the mechanisms of climate change on Earth. (☞ Nov. 9)

April 12

1884

Otto Fritz Meyerhof, German biochemist, Nobel laureate, was born in Hanover, Germany. Meyerhof went to the Wilhelms Gymnasium in Berlin, leaving school at the age of 14. In 1909 he graduated in medicine with a thesis on a psychiatric subject and devoted himself for a time to psychology and philosophy. After working for a short time on physical chemistry at Heidelberg, Meyerhof spent some time in the laboratory of the Heidelberg Clinic and at the Zoological Station at Naples, Italy. From 1912 he worked at the University of Kiel, becoming professor in 1918. He headed a department specially created for him at the Kaiser Wilhelm Institute for Biology in Berlin from 1924 to 1929, and then moved to Heidelberg. In 1929 Meyerhof was asked to take charge of the newly founded Kaiser Wilhelm Institute for Medical Research at Heidelberg. In 1938 he left Germany and went to France as Director of Research at the Institut de Biologie Physico-chimique at Paris. In 1940 he immigrated to America and joined the University of Pennsylvania as Research Professor of Physiological Chemistry. Meyerhof's most important contribution was his demonstration that, in anaerobic conditions, the amounts of glycogen metabolised and of lactic acid produced in a contracting

muscle are proportional to the tension in the muscle. He also demonstrated that 20–25% of the lactic acid is oxidised during the muscle's recovery period and that energy produced by this oxidation is used to convert the remainder of the lactic acid back to glycogen. Meyerhof introduced the term glycolysis to describe the anaerobic degradation of glycogen to lactic acid, and showed the cyclic nature of energy transformations in living cells. Meyerhof was awarded the 1922 Nobel Prize for Physiology or Medicine, jointly with the British biochemist Archibald Vivian Hill. Meyerhof received half the prize money for his discovery of the fixed relationship between the consumption of oxygen and the metabolism of lactic acid in the muscle, while Hill received the other half for his discovery relating to the production of heat in the muscle. (☞ Sep. 26, Oct. 6)

1937
Frank Whittle, English aviation engineer, tested the first prototype jet engine. (☞ May 15, Jun. 1)

1961
The Russian air force pilot Yuri Gagarin became the first man to orbit the Earth. He went into space in a capsule called *Vostok-1*, which was put into orbit by a rocket from the Baikonur launch pad in Kazakhstan. The payload included life-support equipment and radio and television to relay information on the condition of the pilot. Gagarin made a single orbit of Earth in 108 minutes, travelling at 27,400 kph, and returned safely. Gagarin ejected after re-entry and descended with his own parachute, as planned. (☞ Mar. 9, Mar. 27)

1971
Igor Yevgenyevich Tamm, Soviet physicist, Nobel laureate (Physics, 1958), who developed the theoretical interpretation of the Cherenkov effect, died in Moscow, Russia. (☞ Jul. 8)

1981
American space shuttle *Columbia* was launched on its maiden flight from Cape Canaveral, Florida, USA. It carried astronauts John W. Young and Robert L. Crippen. The main objective of the mission was to test the major systems. The orbiter sustained tile damage on launch; 16 tiles were lost and 148 damaged. (☞ Feb. 1, Apr. 14)

1997
George Wald, American biochemist, Nobel laureate (Physiology or Medicine, 1967), who discovered that vitamin A is a vital ingredient of the pigments in the retina and elucidated the chemistry of vision, died in Cambridge, Massachusetts, USA. (☞ Nov. 18)

April 13

1771
Richard Trevithick, English mechanical engineer and inventor who successfully harnessed high-pressure steam and constructed the world's first steam railway locomotive, was born in Illogan, Cornwall, England. Trevithick was educated at Camborne School but he was more interested in sports than academic learning. He went to work with his father at the Wheal Treasury Mine and soon revealed an aptitude for engineering. While at a nearby mine, he developed a successful high-pressure steam engine that was soon in great demand for raising

the ore and refuse from mines. In February 1804, Trevithick produced the world's first steam engine to run successfully on rails. The locomotive, with its single vertical cylinder, 2.5-metre flywheel and long piston rod, managed to haul ten tonnes of iron, seventy passengers and five wagons from the ironworks at Penydarren to the Merthyr-Cardiff Canal. During the 15-km journey the locomotive reached speeds of nearly 8 km/h. In 1805 Trevithick adapted his high-pressure engine for driving an iron-rolling mill and propelling a barge with the aid of paddle wheels. In 1812, he invented the Cornish boiler in which the hot flue gasses could also be used to heat the water, thereby improving efficiency. (☞ Feb. 6, Apr. 22)

1869
George Westinghouse, American engineer and inventor, was granted a patent for compressed air brake for trains. In this system an auxiliary source is used to compress air, which is passed through the couplings to each carriage. While the system is active the brakes are prevented from working but if a coupling breaks, or the driver applies the brakes, the air is lost or cut off and the brakes are instantly applied in each carriage.

1905
Bruno Rossi, Italian pioneer in the study of cosmic radiation, was born in Venice, Italy. Rossi studied at the University of Padua and received his doctorate in physics at the University of Bologna in 1927. He started his academic career at the University of Florence and held the chair in physics in Padua from 1932 to 1938. He then spent short periods in Copenhagen, Denmark; Manchester, England; and Chicago, USA; and joined the faculty of the Cornell University in USA in 1940. From 1943 to 1946 he was on the staff of the Los Alamos Laboratory in New Mexico, where the atomic bomb was developed. Rossi's experimental investigations of cosmic rays and their interactions with matter in the 1930s laid the foundation for high-energy particle physics. He was one of the first to use rockets to study cosmic rays above the Earth's atmosphere. He was a pioneer in the study of high-energy astrophysics, and was largely responsible for starting X-ray astronomy, as well as the study of interplanetary plasma. (☞ Nov. 21)

1912
Cosmic rays were discovered by Austrian physicist Victor Francis Hess. He ascended in a balloon carrying an electroscope during an almost total eclipse of the Sun and found that, as the height increased, the degree of ionisation decreased at first and then rapidly increased. At a height of 5,000 metres it was many times greater than at the Earth's surface, and it was not diminished by the eclipsed Sun. He, therefore, concluded that the unknown rays of high penetration which entered the Earth's atmosphere from space could not be emitted by the Sun. He gave this phenomenon the name 'cosmic radiation'. It was later called 'cosmic rays' by the American physicist Robert Andrews Millikan. The term 'cosmic rays' refers to elementary particles, nuclei, and electromagnetic radiation of extra-terrestrial origin, which include exotic, short-lived particles such as muons, pi-mesons or lambda baryons. These are believed to originate in supernovae explosions. The cosmic rays are accelerated as the shock waves from these exploding stars travel through the surrounding interstellar gas. Hess was awarded the 1936 Nobel Prize for Physics, jointly with the American physicist Carl David Anderson, for his discovery of cosmic rays. (☞ Jun. 24, Sep. 3)

1941
Michael Stuart Brown, American molecular geneticist, Nobel laureate, was born in New York, USA. Brown graduated in 1962 from the College of Arts and Sciences of the University of Pennsylvania, with chemistry as his major subject. In 1966 he received his MD degree from the University of Pennsylvania School of Medicine. In 1968 Brown joined the National Institutes of Health as clinical associate in gastroenterology and hereditary disease. He then joined the Laboratory of Biochemistry, where he learned the techniques of enzymology and the fundamental principles of metabolic regulation. In 1971 he joined the division of gastroenterology in the Department of internal medicine at the University of Texas Southwestern Medical School in Dallas. Soon after his arrival in Dallas, Brown succeeded in solubilising and partially purifying a substance called '3-hydroxy-3-methylglutaryl coenzyme A reductase,' a previously enigmatic enzyme that catalyses the rate-controlling enzyme in cholesterol biosynthesis. He and fellow American molecular geneticist Joseph L. Goldstein developed the hypothesis that abnormalities in the regulation of this enzyme were the cause of familial hypercholesterolemia, a genetic disease in which excess cholesterol accumulates in blood and tissues. They compared cells of normal persons with those of persons who inherited a tendency to have abnormally high blood cholesterol levels and circulatory ailments. They traced a genetic defect that led to their being deficient in cell receptors for fatty molecules called LDL (low-density lipoproteins), which normally help remove cholesterol from the bloodstream. Brown was awarded the 1985 Nobel Prize for Physiology or Medicine, jointly with Goldstein, for their discoveries concerning the regulation of cholesterol metabolism. (☞ Apr. 18)

April 14

1629
Christiaan Huygens, Dutch mathematician, astronomer, and physicist, who founded the wave theory of light, was born in The Hague, Netherlands. Tutored at home by private teachers until he was 16 years old, Huygens learned geometry and how to make mechanical models. He studied law and mathematics at the University of Leiden from 1645 until 1647. For the next two years he continued to study law and mathematics at the College of Orange at Breda. Huygens' earliest publications in 1651 and 1654 dealt with mathematical problems, but he soon turned his attention to lens grinding and telescope construction. Around 1654 he devised a new and better way of grinding and polishing lenses. Using one of his own telescopes, Huygens detected, in 1655, the largest moon of Saturn and was the first to clearly see its rings and explain their appearance over time. He also studied and drew the first maps of Mars. Work in astronomy required accurate timekeeping and this prompted Huygens to tackle this problem. In 1656, while on a visit to Paris, France, he patented the first pendulum clock, which greatly increased the accuracy of time measurement. He also invented an improved type of two-lens eyepiece, now named after him. In 1666 Huygens became

one of the founding members of the French Academy of Sciences. Apart from occasional visits to the Netherlands, he lived from 1666 to 1681 in Paris. In 1678 Huygens's *Traité de la lumiere* appeared, in which he argued in favour of a wave theory of light. Huygens stated that an expanding sphere of light behaves as if each point on the wave front were a new source of radiation of the same frequency and phase. (☞ Mar. 25, Jul. 8)

1927

Alan G. MacDiarmid, New Zealand-born American chemist, Nobel laureate, was born in Masterton, New Zealand. After obtaining his higher education at the University of New Zealand, MacDiarmid earned PhDs in chemistry from the University of Wisconsin at Madison, USA in 1953, and the University of Cambridge, England in 1955. He then began teaching at the University of Pennsylvania in USA where he became Blanchard Professor of Chemistry in 1988. In 1973, MacDiarmid began research on an unusual polymeric material made of sulphur and nitrogen $(SN)_x$ with metallic conductivity. His interest in organic conducting polymers began in 1975 when he was introduced to a new form of polyacetylene by the Japanese chemist Hideki Shirakawa at the Tokyo Institute of Technology. Later, with the American chemist Alan J. Heeger and Shirakawa, he demonstrated that certain plastics could be chemically altered to be almost as conductive as metals. The discovery led scientists to uncover other conductive polymers, which contributed to the growing field of molecular electronics. It opened up the possibilities of a wide variety of commercial applications of these unique polymers, such as antistatic substances for photographic film, and for 'smart' windows that can exclude sunlight. In addition, semi-conductive polymers have recently been developed into light-emitting diodes, solar cells and as displays in mobile telephones and mini-format television screens. MacDiarmid was awarded the 2000 Nobel Prize for Chemistry, jointly with Heeger and Shirakawa, for the discovery and development of conductive polymers. (☞ Jan. 22, Aug. 20)

1964

Rachel Louise Carson, American biologist well known for her writings on environmental pollution and the natural history of the sea and author of the bestseller *Silent Spring*, died at Silver Spring, Maryland, USA. (☞ May 27)

1981

The first test flight of America's first operational space shuttle, *Columbia*, carrying astronauts John W. Young and Robert L. Crippen ended successfully as the orbiter landed at the Edwards Air Force Base in California, USA. (☞ Apr. 12)

April 15

1452

Leonardo Da Vinci, Italian artist, sculptor, engineer and scientist, was born in Vinci, Italy. Leonardo grew up in Florence where he became an apprentice in the workshop of the great Florentine sculptor Andrea del Verrocchio. His training included grinding and mixing paint pigments, the mixing of colours, studying their uses and preparing panel surfaces. He studied geometry and optics and also learnt the technical skills of foundry and metalwork. Throughout his career, Leonardo explored

the art and science of flight. Extensive research led him to design several types of flying machines, including what we know today as a parachute, glider and a kind of helicopter. Leonardo also showed great interest in anatomy, his first anatomical sketches dating back to 1487. He looked at the human body as a machine. His drawings of the human body record superb detail in not only the appearance, but also the function. As early as 1510, Leonardo showed that there was no direct connection between maternal and foetal placental blood vessels. The sketches in Leonardo's notebooks show a spirit of scientific inquiry and a mechanical inventiveness that is regarded as being centuries ahead of their time. Leonardo's scientific and technical observations are found in his handwritten manuscripts, of which over 4,000 pages survive. The manuscripts are difficult to read: not only did Leonardo write in mirror-image script from right to left, but he used peculiar spellings and abbreviations, nor are his notes arranged in any logical order. While portions of Leonardo's technical treatises on painting were published as early as 1651, the scope and calibre of much of his scientific work remained unknown until the 19th century. Yet his geological and palaeontological observations and theories foreshadow many later breakthroughs. (☞ May 2)

1874

Johannes Stark, German physicist, Nobel laureate, was born in Schickenhof, Bavaria, Germany. Stark was educated at the Gymnasium in Bayreuth and in Regensburg. He later studied physics, mathematics, chemistry and crystallography at the Munich University from where he graduated in 1897. He worked as an assistant at the Physics Institute at the Munich University from 1897 until 1900 and then became a lecturer of physics at the University of Göttingen. After brief stints as professor of physics at the Technische Hochschule in Hannover, Technische Hochschule in Aachen, and at the University of Greifswald, Stark moved to the Physics Institute of the University of Würzburg, in 1920. Stark's scientific works covered three large fields: electric currents in gases, spectroscopic analysis, and chemical valency. His spectroscopic work dealt with the connection between the alteration in the structure and in the spectrum of chemical atoms. In 1902 he predicted, correctly, that high-velocity rays of positive ions called canal rays would demonstrate the Doppler effect, and in 1913 he showed that a strong electric field can alter the wavelength of light emitted by atoms, a phenomenon that came to be known as the 'Stark effect.' In 1913, Stark modified the photo-equivalence law proposed by the German-born American physicist Albert Einstein in 1906. Now called the 'Stark–Einstein law,' it states that each molecule involved in a photochemical reaction absorbs only one quantum of the radiation that causes the reaction. Stark was awarded the 1919 Nobel Prize for Physics for his discovery of the Doppler effect in canal rays and the splitting of spectral lines in electric fields. (☞ Jun. 21)

1894

Thomas Alva Edison, American inventor, gave the first public demonstration of his kinetoscope – a moving picture machine, which was the forerunner of the modern movie. Unlike the modern movie, which is projected on a screen, the kinetoscope had peepholes that allowed one person at a time to view the moving pictures. An electric motor moved a perforated filmstrip to an eyepiece, where a slotted disc exposed the images to the viewer at the rate of approximately 40 frames per second. (☞ Jan. 27, Feb. 11, Feb. 19, Aug. 8, Aug. 9, Aug. 27, Oct. 18, Oct. 21, Dec. 6)

1896

Nikolay Nikolaevich Semenov, Russian physical chemist, Nobel laureate, was born in Saratov, Russia. Semenov graduated from the Petrograd University in 1917 and in 1920 he took charge of the electron phenomena laboratory of the Leningrad Physico-Technical Institute. In 1931, he became Director of the Institute of Chemical Physics of the USSR Academy of Sciences. Semenov's outstanding work on the mechanism of chemical transformation includes an exhaustive analysis of the application of the chain theory to varied reactions and, more especially, to combustion processes. He proposed a theory of degenerate branching, which led to a better understanding of the phenomena associated with the induction periods of oxidation processes. He was the first to show that chain reactions are the norm in chemical transformations of matter. He also carried out investigations on electron phenomena, dielectric breakdown and the propagation of explosive waves. Semenov was awarded the 1956 Nobel Prize for Chemistry, jointly with the British chemist Cyril Hinshelwood, for their researches into the mechanism of chemical reactions. (☞ Jun. 19, Sep. 25)

1907

Nikolaas Tinbergen, Dutch-born British zoologist and ethologist, Nobel laureate, was born in The Hague, Netherlands. From early childhood Tinbergen was fond of camping and bird watching and had two aquaria in his backyard where he could study nest building and other fascinating behaviours of Sticklebacks. After receiving a PhD degree from the University of Leiden in 1932, he taught there until 1949. Tinbergen gained prominence in the late 1930s as one of the founders of ethology, the branch of biology that studies animal behaviour. He emphasised the importance of both instinctive and learned behaviour for survival and used animal behaviour as a basis for speculations about the nature of human violence and aggression. He is especially well known for his long-term observations of sea gulls, which led to important generalisations on courtship and mating behaviour. In 1949 Tinbergen joined the faculty of the University of Oxford in England, where he organised a research department for the study of animal behaviour. He became a British citizen in 1955. Tinbergen was awarded the 1973 Nobel Prize for Physiology or Medicine, jointly with the German zoologist Konrad Lorenz and the Austrian zoologist Karl von Frisch, for their discoveries concerning organisation and elicitation of individual and social behaviour patterns. (☞ Nov. 7, Nov. 20, Dec. 21)

1922

Insulin was discovered by Canadian physician Frederick Grant Banting and Canadian physiologist Charles Best in the laboratory of the Scottish physiologist John James Rickard Macleod at the University of Toronto, Canada. The three scientists extracted the hormone from the pancreas of dogs. (☞ Feb. 27, Sep. 6, Nov. 14)

April 16

1838

Ernest Solvay, Belgian industrial chemist who invented the Solvay process for producing washing soda, or sodium carbonate, was born in Rebecqu-Rognon, near Brussels, Belgium. From a very young age Solvay exhibited a passion for physics, chemistry and natural history, though an illness prevented him from going on to university. He worked in his uncle's chemical factory from the age of 21. In 1861, he developed a process using ammonia, obtained as a by-product in the coke industry, and carbon dioxide for the manufacture of anhydrous sodium carbonate from common salt. The process came to be known as the 'Solvay process.' He established his first factory at Couillet, Belgium, in 1863 and further perfected the process until 1872, when he patented it. Sodium carbonate is an important industrial chemical used in the manufacture of glass and ceramics and in the pulping of wood to make paper. It is also used in petroleum refining, as a water softener, as a cleaner and degreaser in washing compounds, and in the manufacture of other sodium-containing compounds, such as sodium hydroxide. Solvay used the wealth earned from his patents for philanthropic purposes, including the establishment of the Institute for Sociology at the University of Brussels, as well as international institutes for research in chemistry, physics, and sociology. (☞ May 26)

1853

The first railway train in India ran between Bori Bunder and Thane (Maharashtra), a distance of 34 kilometres.

1867

Wilbur Wright, American inventor, who with his brother Orville achieved the first powered flight, was born in Millville, Indiana, USA. Wilbur was a brilliant student, and would have graduated from high school had his family not moved during his senior year. A skating accident, his mother's illness and subsequent death kept him from attending college. As he grew up, Wilbur joined his brother to set up a printing firm and a bicycle shop – ventures that showcased their mechanical aptitude, business sense, and originality. Wilbur began to voraciously read everything he could about aviation, from material provided by the Smithsonian Institute to newspapers articles. As all independent thinkers and inventors do, he imagined something completely novel to solve the problem that had plagued other would-be flyers – by emphasising on control of the aircraft instead of increased power. By late 1903, the Wright brothers had constructed a powered airplane with 12-metre long wings, weighing about 340 kilograms with the pilot. They designed and built their own light-weight petrol engine for the airplane. Then in December, near Kitty Hawk, North Carolina, they made the world's first flight in a powered, heavier-than-air machine. With Orville at the controls, the plane flew 37 metres in 12 seconds. The brothers made three more flights that day.

The longest, by Wilbur, was 260 metres in 59 seconds. (☞ May 30, Aug. 19, Dec. 17)

1958
Rosalind Franklin, British scientist whose X-ray diffraction photographs of proteins were crucial for the determination of the molecular structure of deoxyribonucleic acid (DNA) by the American geneticist James D. Watson and British biophysicist Francis Crick, died in London, England. (☞ Apr. 6, Jun. 8, Jul. 25)

1972
American manned lunar mission *Apollo-16*, carrying astronauts John W. Young, Thomas K. Mattingly II, and Charles M. Duke, Jr. was launched by a Saturn-V rocket from the Kennedy Space Center, Florida, USA. (☞ Apr. 21, Apr. 27)

April 17

1790
Benjamin Franklin, American printer and publisher, author, inventor and scientist, who discovered the electrical nature of lightning and invented the lightning rod, died in Philadelphia, USA. (☞ Jan. 17)

1942
Jean-Baptiste Perrin, French physicist, Nobel laureate (Physics, 1926), who, with his studies of the Brownian motion of minute particles suspended in liquids, confirmed the atomic nature of matter, died in New York, USA. (☞ Sep. 30)

1946
Georges J.F. Köhler, German immunologist, Nobel laureate, was born in Munich, Germany. Köhler studied biology at the University of Freiburg, obtaining a diploma in 1971 and a PhD in 1974. In 1974, he joined the laboratory of Argentinean-born British immunologist César Milstein at the University of Cambridge in England where he invented a method of forcing immune system cells to make pure antibodies against a chosen antigen.

He injected a mouse with a powerful antigen and extracted antibody-producing white blood cells from the spleen. He then fused the extracted white blood cells with a kind of cancerous cell called a myeloma, producing what is called a 'hybridoma'. A hybridoma is a cell line that has both the antibody-producing capacity of B-cells of the immune system and the indefinite replicating capacity of cancer cells. Thus each hybridoma can produce infinite copies of its own particular antibody, making it possible to obtain specific antibodies, called monoclonal antibodies, in any amount after screening the hybridoma in culture. These antibodies and the subsequent developments revolutionised the diagnosis and treatment of many immunological diseases, including some forms of cancer and AIDS. Köhler worked at the Basel Institute of Immunology in Switzerland from 1977 to 1984, and in 1985 he was appointed a director of the Max Planck Institute of Immunobiology in Freiburg. Köhler was awarded the 1984 Nobel Prize for Physiology or Medicine, jointly with Milstein and the Danish immunologist Niels K. Jerne, for theories concerning the specificity in development and control of the immune system and the discovery of the principle for production of monoclonal antibodies. (☞ Mar. 1, Oct. 8, Dec. 23)

1967
The American Moon-probe *Surveyor-3* was successfully launched from the Kennedy Space Center, Florida, USA. It was the second American spacecraft to make a

soft landing on the Moon, where it studied the lunar surface and sent more than 6,300 pictures back to Earth. A total of seven *Surveyor*s were sent to the Moon. *Surveyor-3* landed in the southeast part of the region called Oceanus Procellarum, the same site where *Apollo-12* landed three years later. (☞ Apr. 20, Nov. 19)

1970
American manned lunar mission *Apollo-13*, crippled by an onboard explosion, landed safely with a splashdown in the Pacific Ocean, four days after the spacecraft aborted its mission. About 56 hours after launch, when the spacecraft was four-fifths of the way to the Moon, a tank containing liquid oxygen exploded, cutting off vital life support for the three-man crew of James A. Lovell, John L. Swigert, Jr., and Fred W. Haise, Jr. With the oxygen stores depleted, the command module was unusable, and the crew had to be transferred to the lunar module. The crippled craft was guided by ground controllers on a trajectory that made it loop around the Moon and head back to Earth. (☞ Apr. 11)

1976
Carl Peter Henrik Dam, Danish biochemist, Nobel laureate (Physiology or Medicine, 1943), who discovered the antihaemorrhagic vitamin K, died in Copenhagen, Denmark. (☞ Feb. 21)

1983
India's satellite launch vehicle SLV-3 was launched on its second developmental flight from Sriharikota in Andhra Pradesh. It placed the *RS D-2* satellite in near-Earth orbit.

1994
Roger Wolcott Sperry, American neurobiologist, Nobel laureate (Physiology or Medicine, 1981), who discovered the functional specialisation in the two hemispheres of the human brain, died in Pasadena, California, USA. (☞ Aug. 20)

April 18

1873
Justus von Liebig, German chemist who made many important contributions to the early systematisation of organic chemistry and biochemistry, died in Munich, Germany. (☞ May 12)

1905
George Herbert Hitchings, American pharmacologist, Nobel laureate, was born in Hoquiam, Washington, USA. Hitchings received his bachelor's and master's degrees from the University of Washington and earned a PhD in biochemistry at the Harvard University, Cambridge, Massachusetts, in 1933. He taught at Harvard until 1939, and in 1942 joined the Burroughs Wellcome Laboratories, where he did most of his research until his retirement in 1975. At Burroughs Wellcome, Hitchings worked with fellow researcher Gertrude B. Elion to design a variety of new drugs that achieved their effects by interfering with the replication or other vital functions of specific disease-causing agents. The drugs they developed include those for treating leukaemia, rheumatoid arthritis, gout, and AIDS. Other important drugs that were developed by Hitchings and Elion include pyrimethamine, an antimalarial agent; trimethoprim, a treatment for urinary and respiratory tract infections; and acyclovir,

the first effective treatment for viral herpes. Hitchings was awarded the 1988 Nobel Prize for Physiology or Medicine, jointly with Elion and the British pharmacologist James W. Black, for their discoveries of important principles for drug treatment. (☞ Jan. 23, Jun. 14)

1940

Joseph L. Goldstein, American molecular geneticist, Nobel laureate, was born in Sumter, South Carolina, USA. Goldstein received the BS degree in chemistry from Lee University in Lexington, Virginia, in 1962. He then attended the Southwestern Medical School of the University of Texas Health Science Center in Dallas. After receiving the MD degree in 1966, Goldstein moved to Boston where he was an intern and resident in medicine at the Massachusetts General Hospital. It was here that Goldstein first met and developed a friendship with the American molecular geneticist Michael Stuart Brown. After completion of his medical training, Goldstein spent two years at the National Institutes of Health, where he also served as a clinical associate at the National Heart Institute. At the NHI, Goldstein served as physician to patients with disorders of lipid metabolism. His curiosity about hypercholesterolemia – a disorder in which the body's tissues are incapable of removing cholesterol from the bloodstream – was aroused when he took care of patients with the striking clinical syndrome of familial hypercholesterolemia. As a special NIH Fellow in medical genetics at the University of Washington in Seattle, Goldstein initiated and completed a population genetic study to determine the frequency of the various hereditary lipid disorders in an unselected population of heart attack survivors. He and his colleagues discovered a predisposing genetic factor that caused excess build up of fat in the blood of 20% of all heart attack survivors. In 1972, Goldstein returned to the University of Texas Health Science Center at Dallas and joined up with Brown to study the processes affecting the accumulation of cholesterol in the bloodstream. They found that the genetic absence of low-density lipoproteins (LDL) was a major cause of familial hypercholesterolemia. These findings led to the successful use of drugs and the manipulation of diet in lowering blood cholesterol levels. Goldstein was awarded the 1985 Nobel Prize for Physiology or Medicine, jointly with Brown, for their discoveries concerning the regulation of cholesterol metabolism. (☞ Apr. 13)

1955

Albert Einstein, German-born American physicist, Nobel laureate (Physics, 1921), who developed the special and general theories of relativity, died in Princeton, New Jersey, USA. (☞ Mar. 14)

2001

India's geostationary satellite launch vehicle (GSLV-D1) was launched on its first flight from the Satish Dhawan Space Centre at Sriharikota, Andhra Pradesh. It placed a 1,540-kilogram *GSAT-1* satellite in the geosynchronous transfer orbit ranging in height from 181 kilometres to 32,051 kilometres. GSLV-D1 used a cryogenic engine supplied by Russia in the third stage. The GSLV project was initiated in 1990 with a goal to achieve self-reliance in launching *INSAT*-class satellites.

2002
Thor Heyerdahl, Norwegian ethnologist and adventurer who organised and led the famous *Kon-Tiki* and *Ra* transoceanic scientific expeditions, died in Colla Micheri, Italy. (☞ Oct. 6)

April 19

1882
Charles Robert Darwin, English naturalist renowned for his theory of evolution and his theory of its operation, known as Darwinism, died in Kent, UK. (☞ Feb. 12)

1906
Pierre Curie, French physical chemist, Nobel laureate (Physics, 1903), who, jointly with his wife Marie Curie, discovered radium and polonium, died in Paris, France. (☞ May 15, Nov. 7)

1912
Glenn T. Seaborg, American nuclear chemist, Nobel laureate, was born in Ishpeming, Michigan, USA. After graduating from David Starr Jordan High School in Los Angeles in 1929, Seaborg was educated at the University of California at Los Angeles, and the University of California at Berkeley, from where he received his PhD in chemistry in 1937. In 1939, Seaborg was appointed an instructor in chemistry at the University of California, Berkeley, where he became assistant professor in 1941, and professor of chemistry in 1945. In 1946, he also took responsibility for directing nuclear chemical research at the Lawrence Radiation Laboratory at Berkeley. Seaborg is perhaps best known for his role in the discovery of plutonium in February 1941. During 1940–58, Seaborg and his co-workers, using the 60-inch cyclotron at Berkeley, produced nine of the transuranic elements (plutonium to nobelium) by bombarding uranium and other elements with nuclei in a cyclotron. He coined the term 'actinide' for the elements in this series. Seaborg was chairman of the US Atomic Energy Commission 1962–71. Element 106, seaborgium, was named in his honour. Seaborg was awarded the 1951 Nobel Prize for Chemistry, jointly with his early collaborator, American nuclear physicist Edwin McMillan, for their discoveries in the chemistry of transuranium elements. (☞ Feb. 25, Nov. 16)

1943
The hallucinogenic effect of lysergic acid diethylamide (LSD) was first experienced by Swiss chemist Albert Hofmann who had deliberately ingested 250 micrograms of the drug he had synthesised at the Sandoz Laboratories in Basel, Switzerland. Hofmann experienced restlessness, dizziness and 'extreme activity of imagination'. LSD is known to be the most common hallucinogen and is one of the most potent mood-changing chemicals. The user may feel several different emotions at once or swing rapidly from one emotion to another. If taken in large enough doses, the drug produces delusions and visual hallucinations. Hofmann prepared the drug from lysergic acid, which is found in ergot, a fungus that grows on rye and other grains. Although LSD was originally developed as a drug for relieving respiratory ailments, it was never used for its intended purpose.

1967
American lunar probe *Surveyor-3* landed in Oceanus Procellarum on the Moon,

carried out soil physics studies, and sent back 6,315 photographs. (☞ Apr. 17)

1971

Salyut-1, the first space station in history, was launched into orbit by the erstwhile Soviet Union. The *Salyut-1* station was 16 metres long with two major sections, each with different diameters. It weighed about 19 tonnes and had a single docking port in the front end to receive the *Soyuz* crew delivery spacecraft. But the crew on board *Soyuz-10* – the first spacecraft sent to *Salyut-1* – was unable to enter the station because of a docking mechanism problem. The *Soyuz-11* crew, cosmonauts Dobrovolsky, Volkov, and Patsayev stayed on board for 24 days – from June 6 to June 29 – and conducted

experiments to prove that humans could stay in orbit for relatively long periods, but died during return to the Earth because the air escaped from their *Soyuz* spacecraft. The Soviet Union subsequently placed six more *Salyut* space stations into orbit, which made it possible for the much larger *Mir* space station to be a success and paved the way for the upcoming *International Space Station*. While the name of all seven stations bears the name Salyut, each station was radically different. *Salyut-2, -3,* and *-5* were also known as the Almaz project. These were primarily military missions that incorporated a large optical telescope into the design. The telescope was used for reconnaissance of military installations. *Salyut-6* and *-7* were the longest of the *Salyut* flights. Both stations were occupied for over four years. *Salyut-6* was occupied from December 10, 1977 to May 22, 1981. Cosmonauts from erstwhile Communist countries such as Poland, East Germany, Hungary, Vietnam, Cuba, Mongolia, and Rumania also visited *Salyut-6* as guest-cosmonauts during weeklong missions. *Salyut-7* was occupied from May 13, 1982 to June 25, 1986. It hosted five long-duration crews that were visited for short periods by *Soyuz* ships carrying cosmonauts from France and India as well as the second Soviet woman cosmonaut, Svetlyana Savitskaya. These two stations prepared the Russians for running and maintaining the *Mir* space station and for their future role as essential partners in the *International Space Station*. (☞ Apr. 3, Apr. 11)

1975

Aryabhata, India's first satellite was launched by a Russian rocket from Kapustin Yar, Russia. Named after the 5th century Indian mathematician Aryabhata, the satellite was placed in a low-Earth orbit, with an apogee of 619 kilometres and a perigee of 563 kilometres, and completed one revolution around Earth in 96.3 minutes. *Aryabhata* was designed and built

by the Indian Space Research Organisation (ISRO) to conduct experiments in X-ray astronomy, aeronomics, and solar physics. The spacecraft was a 26-sided polygon 1.4 m in diameter, with all faces (except the top and bottom) covered with solar cells. All signals from the spacecraft were lost after 5 days of operation.

April 20

1786
John Goodricke, English astronomer, Fellow of the Royal Society, who was the first to notice that some variable stars were periodic, died in York, Yorkshire, UK. (☞ Sep. 17)

1862
The first test of pasteurisation was completed by the French microbiologist Louis Pasteur and French physiologist Claude Bernard. Jars, containing dog's blood and urine, sealed for 48 days, were opened at a meeting of the French Academy of Sciences in Paris. Even though the jars were maintained at an elevated temperature of 30°C, neither liquid showed any observable decay or fermentation. This demonstrated that if food were heated sufficiently to kill germs without significantly altering their chemical composition and then kept sealed, they might be preserved without refrigeration. The process came to be known as 'pasteurisation'. Originally used by brewers, pasteurisation is widely used for preserving milk and other products. (☞ Dec. 27)

1899
Charles Friedel, French organic chemist and mineralogist who, with the American chemist James Mason Crafts, discovered in 1877 the chemical process known as the Friedel–Crafts reaction, died in Montauban, France. (☞ Mar. 12)

1902
The French scientist couple Marie and Pierre Curie isolated one gram of radium (element 88), the first sample of the radioactive element. They had refined it from eight tonnes of pitchblende, an ore of uranium. (☞ May 15, Nov. 7)

1918
Kai Manne Börje Siegbahn, Swedish physicist, Nobel laureate, was born in Lund, Sweden. Siegbahn was the son of Karl Manne Georg Siegbahn, who received the 1924 Nobel Prize for Physics for his discoveries relating to X-ray spectroscopy. Siegbahn received his PhD from the University of Stockholm in 1944, and was professor of physics at the Royal Institute of Technology, Stockholm until 1954. He then moved to the Uppsala University, where he devised a technique of electron spectroscopy for chemical analysis, which works on a fundamental phenomenon, the photoelectric effect – the emission of electrons that occurs when electromagnetic radiation strikes a material. Siegbahn also worked on the related technique of ultraviolet photoelectron spectroscopy. Siegbahn was awarded the 1981 Nobel Prize for Physics, jointly with the Dutch-born American physicist Nicolaas Bloembergen and the American physicist Arthur Leonard Schawlow, for his contribution to the development of high-resolution electron spectroscopy. (☞ Mar. 11, May 5)

1918
Karl Ferdinand Braun, German physicist, Nobel laureate (Physics, 1909), who

developed the cathode-ray oscilloscope, died in New York, USA. (☞ Jun. 6)

1927

Karl Alex Müller, Swiss physicist, Nobel laureate, was born in Basel, Switzerland. Müller received his PhD in physics from the Swiss Federal Institute of Technology in 1958. He spent five years as a project manager with the Battelle Institute in Geneva before joining the IBM Zürich laboratory in 1963, where he worked in solid-state physics. In 1973, Müller became manager of the laboratory's physics department and in 1982 he was appointed an IBM fellow. Soon afterwards he met and recruited the German physicist Johannes Georg Bednorz, and together they began a systematic search for superconductivity in non-metallic substances, especially ceramic metal oxides. In 1986 they reported superconductivity in a layered, ceramic material at a then record-high temperature of 35 Kelvin (about –238°C). Their discovery started new research worldwide into related materials that yielded dozens of new superconductors, eventually reaching a transition temperature of 135 Kelvin. Before this discovery, superconductivity at temperatures more than 22 Kelvin was believed to be unattainable. Müller and Bednorz were jointly awarded the 1987 Nobel Prize for Physics for their important breakthrough in the discovery of superconductivity in ceramic materials. (☞ May 16)

1940

The electron microscope was first publicly demonstrated by the Radio Corporation of America (RCA) at Camden, New Jersey, USA. Earlier, two postgraduate students working in the physics department of the University of Toronto, Canada, with their physics professor, between 1937 and 1939, developed the first ever transmission electron microscope. Then, in 1940, one of the students, James Hillier, joined RCA, where he modified the design of the University of Toronto microscope and developed it into the prototype for the RCA production model. The commercial electron microscope for RCA was supervised under the watchful eye of a Russian-born American electronic engineer and vice president of the Corporation, Vladamir Kosma Zworykin. It was almost 10 metres high, weighed half a tonne and was able to produce a magnification of 100,000 times. (☞ Jul. 20, Dec. 25)

1967

The American Moon-probe *Surveyor-3* made a soft-landing on the Moon in the southeast part of the region called Oceanus Procellarum. It sent back more than 6,300 picture of the lunar surface back to Earth. (☞ Apr. 17)

April 21

1882

Percy Williams Bridgman, American physicist, Nobel laureate, was born in Cambridge, Massachusetts, USA. Bridgman received his early education in public schools in the nearby city of Newton until 1900, when he entered the Harvard University in Cambridge from where he

graduated in 1904, and did his MA in 1905. After doing his PhD at Harvard in 1908, he joined the Faculty of the University. Bridgman's researches mainly concerned the effects of high pressures on materials and their thermodynamic behaviour. He carried out extensive investigations on the properties of matter at pressures up to 100,000 atmospheres including a study of the compressibility, electric and thermal conductivity, tensile strength and viscosity of more than 100 different compounds. He developed a method of packing which eliminated leak, and later introduced various methods of external support to pressure vessels for working at higher pressures. Bridgman was awarded the 1946 Nobel Prize for Physics for the invention of an apparatus to produce extremely high pressures, and for the discoveries he made therewith in the field of high-pressure physics. (☞ Aug. 20)

1889

Paul Karrer, Swiss chemist, Nobel laureate, was born in Moscow, Russia. Karrer received his early education at Wildegg and at the grammar school in Lenzburg, Switzerland, where he matriculated in 1908. He studied chemistry at the University of Zürich and after getting his PhD in 1911, he spent a further year as assistant in the Chemical Institute. In 1912 Karrer joined the German medical scientist Paul Ehrlich at the Georg Speyer Haus, Frankfurt-am-Main, Germany, as a chemist. He left Frankfurt six years later to become a reader at University of Zürich. In 1919 he became professor of chemistry and director of the Chemical Institute. Karrer's early researches involved the preparation and investigation into the properties of complex metal compounds, but his most important work concerned plant pigments, particularly the yellow carotenoids. He determined the chemical structure of the carotenoids and showed that some of these substances are transformed in the animal body into vitamin A. His work led to the establishment of the correct constitutional formula for β-carotene, the chief precursor of vitamin A. Karrer also confirmed the constitution of vitamin C proposed by the Hungarian biochemist Albert Szent-Györgyi; showed lactoflavin to be part of the complex originally described as vitamin B_2; and studied vitamin E. Karrer was awarded the 1937 Nobel Prize for Chemistry, jointly with the British chemist Walter Norman Haworth, for his investigations on carotenoids, flavins and vitamins A and B_2. (☞ Mar. 19, Jun. 18)

1965

Edward Victor Appleton, British physicist, Nobel laureate (Physics, 1947), who discovered the layer of the ionosphere named after him, died in Edinburgh, Scotland. (☞ Sep. 6)

1972

The *Apollo-16* lunar module *Orion*, carrying astronauts John W. Young and Charles M. Duke, Jr., landed at the western edge of the Descartes Mountains on Moon. This was the first landing in the central lunar highlands, and the crew successfully explored and sampled a kind of terrain never before visited on the lunar surface. Thomas K. Mattingly remained in orbit in the command module *Casper*. During their 71-hour stay on the Moon, Young and Duke covered a distance of approximately

27 kilometres on the lunar surface and collected 95 kg of lunar rocks for study. (☞ Apr. 16, Apr. 27)

1980
Aleksandr Oparin, Russian biochemist noted for his studies on the origin of life from chemical matter, died. (☞ Feb. 18)

1994
Discovery of planets outside of our solar system was announced by the Polish astronomer Alexander Wolszczan, professor of astronomy at the University of Pennsylvania in Philadelphia, USA. He discovered two planets orbiting a star (a pulsar) in the constellation Virgo while making observations from the Arecibo observatory in Puerto Rico in 1990. The planets were found to have masses 3.4 and 2.8 time the mass of the Earth. Till date, more than 60 planets have been discovered orbiting stars outside our solar system.

April 22

1833
Richard Trevithick, English mechanical engineer and inventor who successfully harnessed high-pressure steam and constructed the world's first steam railway locomotive, died in Dartford, Kent, UK. (☞ Feb. 6, Apr. 13)

1834
Gaston Planté, French physicist who produced the first electric storage battery, or accumulator, was born in Orthez, France. Planté began his academic career in Paris as a lecture assistant in physics at the Conservatory of Arts and Crafts in 1854. In 1860, he became professor of physics at the Polytechnic Association for the Development of Popular Instruction. Planté began experiments on a battery for storing electrical energy in 1859. His first model contained two sheets of lead, separated by rubber strips, rolled into a spiral, and immersed in a solution containing about 10 per cent sulphuric acid, forming a 'cell'. A year later he presented a battery to the French Academy of Sciences, consisting of nine cells, housed in a protective box with the terminals connected in parallel. His battery could deliver remarkably large currents. Later Planté's battery was improved by coating the positive lead plates with a paste of lead sulphide. Variants of it are still used in automobiles. (☞ May 21)

1876
Robert Bárány, Austrian physiologist, Nobel laureate, was born in Vienna, Austria. Bárány passed his matriculation in Vienna and graduated from the Vienna School of Medicine in 1900. He trained in internal medicine, neurology, and psychiatry in Germany and finally joined the General Hospital of Vienna Ear Clinic in 1903 as a surgeon. He devised new tests for detecting diseases of the inner ear that leads to loss of balance, and for examining activities of the cerebellum and their relation to disturbances of equilibrium. He described surgical procedures for otosclerosis (hardening of the inner ear that leads to deafness) and worked as a war surgeon during World War I, when he was taken prisoner by the Russians. After release, he lived the rest of his life in Sweden working as chief of otological services at the University of Uppsala. Bárány was awarded the 1914 Nobel Prize for Physiology or Medicine for his work on the physiology and pathology of the vestibular apparatus. (☞ Apr. 8)

1904
J. Rober Oppenheimer, American theoretical physicist and science administrator, was born in New York, USA. After graduating from the Ethical Culture School of New York, Oppenheimer enrolled at the Harvard University, Cambridge, Massachusetts in 1922, where he took an intense program that ranged from mathematics and sciences to philosophy and Eastern religions and French and English literature. Not deeming himself good as an experimental researcher, Oppenheimer resolved to turn to theoretical physics. In 1926 he studied with the German physicist Max Born at the University of Göttingen in Germany, from where he received his doctoral degree in March 1927. In 1929, he returned to USA to teach at the California Institute of Technology, Pasadena, and the University of California at Berkeley and for many years divided his time between Pasadena and Berkeley. Oppenheimer's early research was devoted in particular to energy processes of subatomic particles, including electrons, positrons, and cosmic rays. In 1927, working closely with Born, he solved a knotty problem involving calculations pertaining to subatomic particles. During the early 1930s Oppenheimer and his students first predicted the existence of the positron – an elementary particle having the same mass as the electron but with a positive charge. During World War II, Oppenheimer was made the director of the Los Alamos Laboratory in New Mexico, which developed the first atomic bomb under the Manhattan Project. After the War, he chaired the US Atomic Energy Commission. In 1953, Oppenheimer was accused of being a communist agent because he opposed the building of the hydrogen bomb. Although he was not found guilty of treason, his security access was taken away and his contract as adviser to the Atomic Energy Commission was terminated. From 1947 to 1966, Oppenheimer was director of the Institute for Advanced Study, Princeton, New Jersey. (☞ Feb. 18)

1909
Rita Levi-Montalcini, Italian neurologist, Nobel laureate, was born in Turin, Italy. Levi-Montalcini graduated from medical school in Turin with a degree in medicine and surgery, and enrolled in a three-year specialisation course in neurology and psychiatry, during which she studied the effects of peripheral tissues on nerve cell growth. After a short period spent in Brussels, Belgium, as a guest of a neurological institute, she returned to Turin in 1940 to set up a small research unit at home, but had to go into hiding in Florence during the German occupation of Italy. She went to USA in 1947 to work with the German zoologist Viktor Hamburger, who was studying the growth of nerve tissue in chick embryos at the Washington University in St. Louis, Missouri. In course of their research, Levi-Montalcini and Hamburger discovered the presence of a substance in mouse tumour that stimulated nerve growth when implanted into chick embryos. They named the substance nerve-growth factor (NGF). Meanwhile, the American biochemist Stanley Cohen, who had joined Levi-Montalcini in the Washington University, isolated the nerve-growth factor from the tumour, the first of many cell-growth factors to be found in the

bodies of animals. Levi-Montalcini and Cohen were jointly awarded the 1986 Nobel Prize for Physiology or Medicine for their discoveries of growth factors. (☞ Nov. 17)

1919
Donald James Cram, American chemist, Nobel laureate, was born in Chester, Vermont, USA. Cram was educated at the Rollins College, Florida, and at the University of Nebraska, Lincoln. After receiving his doctorate in organic chemistry from the Harvard University, Cambridge, Massachusetts in 1947, he joined the faculty of the University of California at Los Angeles, where he became a full professor in 1956. Cram's research mainly centred on molecules called crown ethers – two-dimensional cyclic polyethers that can surround specific metal ions to form cyclic complexes – that were earlier synthesised by the American chemist Charles J. Pedersen. Cram extended Pedersen's work and synthesised molecules that were three-dimensional, creating an array of differently shaped molecules that could interact selectively with other molecules because of their complementary three-dimensional structures. His work represented a major step toward the synthesis of laboratory-made mimics of enzymes and other natural molecules whose special chemical behaviour depended on their characteristic three-dimensional structure. Cram was awarded the 1987 Nobel Prize for Chemistry, jointly with Pedersen and the French chemist Jean-Marie Lehn, for their development and use of molecules with structure-specific interactions of high selectivity. (☞ Jun. 17, Sep. 30, Oct. 3)

1950
The Central Fuel Research Institute (CFRI), a constituent national laboratory of the Council of Scientific and Industrial Research (CSIR), was inaugurated. Located at Dhanbad, Bihar, CFRI carries out both fundamental and applied research on India's fuel resources, especially coal and lignite, to assess their quality and potential uses in the most efficient manner. CFRI has contributed vast knowledge in the basic studies on fuel science and has conducted pilot plant demonstrations in the area of coal gasification, fluid bed combustion, oil agglomeration, conversion to liquid fuels by hydrogenation and by the Fischer–Tropsch route, formation of coke, solvent refined coal, coal oil and coal–water slurry. CFRI's contribution has been mostly relating to the quality assessment of coal resources; generation of the basic process know-how for the operating coking coal washeries; prescribing coal blends for coke making in the steel industry; designing non-recovery coke ovens for foundry and domestic coke; processing of coal tar; and utilisation of fly ash for building bricks and also as a soil conditioner in agriculture.

1989
Emilio Segrè, Italian-born American physicist, Nobel laureate (Physics, 1959), who discovered antiprotons – an antiparticle having the same mass as a proton but opposite electrical charge – died in Lafayette, California, USA. (☞ Feb. 1)

April 23

1858
Max Planck, German theoretical physicist, Nobel laureate, was born in Kiel, Germany. Planck studied at the Universities of Munich and Berlin, and received his doctorate

of philosophy from the Munich University in 1879. He held a teaching post without pay in Munich from 1880 to 1885, and was then appointed associate professor of theoretical physics at Kiel. In 1889, he succeeded physicist Gustav Kirchhoff as professor at the Berlin University, where he remained until his retirement in 1926. Planck's earliest work was on thermodynamics, an interest he acquired from his studies under Kirchhoff. At the same time the problems of radiation processes also engaged his attention and he showed that these were to be considered electromagnetic in nature. Planck experimentally found that wavelength distribution of the energy emitted by a black body as a function of temperature were at variance with the predictions of classical physics. To explain the anomaly he derived a relationship based on the revolutionary idea that the energy emitted by a resonator was not continuous, but could only take on discrete values or quanta. The energy for a resonator of frequency ν is $h\nu$ where h is a universal constant, now called Planck's constant. Planck's work on the quantum theory, as it came to be known, was published in the *Annalen der Physik*. Planck was awarded the 1918 Nobel Prize for Physics in recognition of the services he rendered to the advancement of physics by his discovery of energy quanta. (☞ Oct. 4)

1867

Johannes Andreas Grib Fibiger, Danish pathologist, Nobel laureate, was born in Silkeborg, Denmark. Fibiger received his bachelor's degree in 1883 and qualified as a doctor in 1890. After a period of working in hospitals and studying under the bacteriologists Robert Koch and Emil von Behring he was, from 1891 to 1894, assistant to Professor C.J. Salomonsen at the department of bacteriology of the Copenhagen University. He completed his doctorate thesis on 'Research into the bacteriology of diphtheria' while serving as an Army reserve doctor at the Hospital for Infectious Diseases in Copenhagen. He received his doctorate from the University of Copenhagen in 1895, and was subsequently appointed prosector at the University's Institute of Pathological Anatomy. He became professor of pathological anatomy at the University of Copenhagen in 1900. In 1907, while dissecting some rats, he discovered that inflammation of stomach tissue caused by the larvae of a worm caused cancerous tumours in the stomach of the rats. The worms had infected cockroaches eaten by the rats. By 1913 he was able to induce gastric tumours consistently in mice and rats by feeding them cockroaches infected with the worm. Fibiger was awarded the 1926 Nobel Prize for Physiology or Medicine for achieving the first controlled induction of cancer in laboratory animals, which was of profound importance to cancer research. (☞ Jan. 30)

1960

Max von Laue, German physicist, Nobel laureate (Physics, 1914), who discovered diffraction of X-rays in crystals, died in Berlin, Germany. (☞ Oct. 9)

April 24

1744

Anders Celsius, Swedish astronomer, physicist and mathematician, who developed the centigrade temperature scale, died in Uppsala, Sweden. (☞ Nov. 27)

1948
The Central Leather Research Institute (CLRI), a constituent national laboratory of the Council of Scientific and Industrial Research (CSIR), and the world's largest leather research institute, was founded. Located in Chennai, Tamil Nadu, CLRI carries out R&D on modernisation of tanneries; development of environment-friendly chemicals and technologies including enzymatic options; computer-aided designs for footwear and garments; split leathers and quality upgradation of lower ends; molecular biology of collagen and metal–collagen interactions; wastewater management; tannery and slaughterhouse by-products including carcass utilisation, and hazard and risk analysis. State-of-art facilities at CLRI support innovation in leather processing, creative designing of leather products, viz., leather garment, leather goods and footwear and development of novel environmental technologies for the leather sector. CLRI has developed a number of technologies in the areas of leather processing, wastewater treatment, and utilisation of tannery and slaughterhouse by-products.

1964
Gerhard Domagk, German bacteriologist and pathologist, Nobel laureate (Physiology or Medicine, 1939), who discovered the antibacterial effects of the red dye prontosil, died in Burgberg, near Königsfeld, Germany. (☞ Oct. 30)

1967
Vladimir Komarov, Soviet cosmonaut, was killed during landing, becoming the first person to die in a space mission.

1970
The People's Republic of China's first satellite, *Mao-1* was launched by a 'Long March' rocket from Jiuguan into a highly elliptical orbit ranging from 431 kilometres to 2,163 kilometres in altitude, with an orbital period of 111.5 minutes. The satellite transmitted a popular Communist Chinese song, 'The East is Red'.

1987
An American biotech company Advanced Genetic Sciences Inc. conducted the first field trial of free-living genetically engineered bacteria – a frost inhibitor – on a strawberry field in USA.

1990
American space shuttle *Discovery*, carrying astronauts Loren J. Shriver, Charles F. Bolden, Steven A. Hawley, Bruce McCandless II, and Kathryn D. Sullivan, was launched on its 10th flight from the Kennedy Space Center, Florida, USA. The primary payload of the mission was the Hubble Space Telescope, which was deployed in orbit. (☞ Apr. 25, Apr. 29, Aug. 30)

April 25

1874
Guglielmo Marconi, Italian physicist and inventor, Nobel laureate, was born in Bologna, Italy. Marconi received his education first in Bologna and later in Florence. He then went to the technical school in Leghorn, where he studied physics. In 1894 Marconi began experimenting on radio communication using comparatively crude apparatuses: an induction coil for increasing voltages, with a spark discharger controlled by a Morse key at the sending end

and a simple coherer (a device designed to detect radio waves) at the receiver. After preliminary experiments over a short distance, he found that he could increase the range of signalling by using a vertical aerial with a metal plate or cylinder at the top of a pole connected to a similar plate on the ground. By this method he could increase the range of signalling to about 2.4 km, enough to convince him of the potentialities of this new system of communication. He made another important discovery that the simultaneous use of two earths, one for sending and one for receiving signals, greatly increased the range of transmission. He also discovered that increasing the height of the transmitting antenna also increased the range of communication. In 1896, Marconi took out a patent for his system of wireless telegraphy in England. In December 1901, Marconi was able to send the first wireless signal across the Atlantic. Marconi was awarded the 1909 Nobel Prize for Physics, jointly with the German physicist Karl Ferdinand Braun, in recognition of their contributions to the development of wireless telegraphy. (☞ Jun. 2, Jul. 20, Dec. 12)

1900

Wolfgang Pauli, Austrian-born American physicist, Nobel laureate, was born in Vienna, Austria. Pauli received his early education in Vienna before joining the University of Munich from where he obtained his doctor's degree in 1921. He then spent a year at the University of Göttingen in Germany as assistant to the German physicist Max Born, and a further year with the Danish physicist Niels Bohr at Copenhagen, Denmark. During 1923–1928 he was a lecturer at the University of Hamburg in Germany before his appointment as professor of theoretical physics at the Federal Institute of Technology in Zurich, Switzerland. In 1940 he was appointed visiting professor of theoretical physics at the Institute for Advanced Study, Princeton, New Jersey, USA. When he was still a student, Pauli wrote a 200-page encyclopaedia article on the theory of relativity. In 1925, he propounded his 'exclusion principle', which states that in an atom no two electrons can occupy the same quantum state simultaneously, and immediately made clear the reason for the structure of the periodic table of the elements. Pauli was the first to recognise the existence of the neutrino, an uncharged and mass-less particle, to explain the missing energy and momentum observed in radioactive beta-decay. Pauli was awarded the 1945 Nobel Prize for Physics for the discovery of the exclusion principle, also called the Pauli principle. (☞ Dec. 15)

1953

The British journal *Nature* published the first paper on the structure of DNA – a

one-page article by the American geneticist James D. Watson and British biophysicist Francis Crick. The structure explained how DNA passed hereditary information from cell to cell, and from generation to generation. This work earned Watson and Crick the 1962 Nobel Prize for Physiology or Medicine. (☞ Apr. 6, Jun. 8)

1961
Robert Noyce, the American electrical engineer was issued the first patent for an integrated circuit, which was described as a 'semiconductor device-and-lead structure', while Jack Kilby's application was still being analysed. Today, both men are acknowledged as having independently conceived the idea. (☞ Nov. 8, Dec. 12)

1983
The American space probe *Pioneer-10* crossed the orbit of Pluto, the outermost planet, to continue its voyage into the universe beyond our solar system. (☞ Jan. 22, Mar. 2, Jun. 13, Dec. 3)

1990
The Hubble Space Telescope was deployed from the American space shuttle *Discovery* into an orbit 613 kilometres above Earth. It is the first major orbiting observatory, carrying a 240-cm primary mirror with several instrument packages and cameras able to record various parts of the electromagnetic spectrum. The telescope is named after the American astronomer, Edwin Powell Hubble whose work gave the first proofs of the expansion of the universe. Overall, the telescope measures 13.2 m long and 4.2 m in diameter – nearly the size of a large school bus – but it can fit inside a space shuttle cargo bay. Despite an investment of US$1.5 billion, the telescope was plagued with problems from the start. The primary mirror was flawed, blurring images and

substantially reducing the telescope's ability to see distant stars or objects. In December 1993, astronauts of the space shuttle *Endeavour* captured the Hubble telescope, corrected the flaw in its primary mirror, and added new instruments. During the second servicing mission by space shuttle *Discovery* in February 1997 some of Hubble's instruments were exchanged to improve its performance. Two more servicing missions were sent in December 1999 and March 2002 during which a new computer and data recorder and an advanced camera for surveys were installed. The fourth servicing mission scheduled for mid-2006 was cancelled by NASA, making the future of the most sophisticated optical observatory in orbit uncertain. During its decade-and-a-half years in orbit, the Hubble telescope has seen farther and sharper than any optical, ultraviolet, or infrared telescope before it. In total, by observing more than 14,000 astronomical targets, Hubble has contributed significantly to essentially all topics of current astronomical research, covering objects from our own solar system to the most distant galaxies. (☞ Apr. 24)

April 26

1879
Owen Willans Richardson, English physicist, Nobel laureate, was born

in Dewsbury, Yorkshire, England. Richardson graduated from the Trinity College, Cambridge, in 1900 and began to investigate the emission of electricity from hot bodies at the Cavendish Laboratory as a student of physicist J.J. Thomson. In 1906 he was appointed professor of physics at the Princeton University in USA, where he remained until the end of 1913, working on thermionic emission, photoelectric action, and the gyromagnetic effect. In 1913 he was elected a Fellow of the Royal Society. While at Princeton, in 1911, he proved that electrons are emitted from hot metal and not from the surrounding air, as earlier thought. That same year he proposed a mathematical equation that relates the rate of electron emission to the absolute temperature of the metal. This equation, called 'Richardson's law' or the Richardson–Dushmann equation became an important aid in electron-tube research and technology. In 1914 Richardson became professor of physics and, ten years later, director of research at the King's College of the University of London, a post he held till his retirement in 1944. Richardson was awarded the 1928 Nobel Prize for Physics for his work on the thermionic phenomenon and especially for the discovery of the law named after him. (☞ Feb. 15)

1900
Charles Francis Richter, American seismologist, was born in Hamilton, Ohio, USA. Richter was educated at the University of Southern California, Stanford. He obtained his PhD from the California Institute of Technology, Pasadena, California, in 1928 and worked for the seismological laboratory of Carnegie Institution of Washington, Pasadena, California, from 1927 to 1936 before being appointed to the staff of the California Institute of Technology, where he became professor of seismology in 1952. Richter developed his scale to measure the intensity of earthquakes in 1935. The scale assigns numerical ratings to the energy released by earthquakes using a seismograph to record actual earth motion during an earthquake. Richter's scale is an absolute one, based on the amplitude of the waves produced by the earthquake. The scale also takes into account the instrument's distance from the epicentre. The scale is logarithmic so, for example, a quake of magnitude 7 would be ten times stronger than a quake of magnitude 6. A one-unit increase would mean about 30 times more energy being released. The strongest earthquake so far recorded had a Richter-scale value of 8.6. (☞ Sep. 30)

1920
Srinivasa Ramanujan, Indian mathematician, Fellow of the Royal Society, whose contributions to the theory of numbers include pioneering discoveries of the properties of the partition function, died in Madras (now Chennai), Tamil Nadu, India. (☞ Dec. 22)

1932
Michael Smith, British-born Canadian biochemist, Nobel laureate, was born in Blackpool, England. Smith did his doctorate from the University of Manchester in England, in 1956. He became a Canadian citizen in 1964. After holding a number of positions in Canada and the United States, he joined the faculty of the University

of British Columbia in 1966, becoming director of the university's biotechnology laboratory in 1987. Smith's early work with the Indian-born American biochemist Har Gobind Khorana in Vancouver, British Columbia, Canada, in 1956 involved development of a general, efficient procedure for the chemical synthesis of nucleoside-5' triphosphates based on the synthesis of ATP by Khorana in 1954. This study led to more extensive investigations of the reactions of carbodiimides with acids, including phosphoric acid esters and to a general procedure for the preparation of nucleoside-3',5' cyclic phosphates, a class of compounds of great biological significance. Smith devoted several years in the early 1970s to working out the details of the technique of oligonucleotide-based, 'site-directed mutagenesis', which provided researchers with a new way to study protein function. The technique aimed at modifying protein synthesis by inducing selective mutations in the nucleotide sequence of its gene, and then comparing its structure and function with those of the natural protein. Before the advent of Smith's method, mutation was a difficult and time-consuming task because biochemical researchers used imprecise and haphazard methods. Smith was awarded the 1993 Nobel Prize for Chemistry, jointly with the American biochemist Kary B. Mullis, for his fundamental contributions to the establishment of oligonucleotide-based, site-directed mutagenesis and its development for protein studies. (☞ Oct. 4, Dec. 28)

1933
Arno Penzias, German-born American astrophysicist, was born in Munich, Germany. Penzias was educated at the City College of New York in New York, and the Columbia University, where he received his doctorate in 1962. In 1961 Penzias joined the Bell Telephone Laboratories in Holmdel, New Jersey, where, in collaboration with the American radio astronomer Robert Woodrow Wilson, he began monitoring radio emissions from a ring of gas encircling the Milky Way Galaxy. The two scientists detected an unusual background radiation at a temperature of 3 K, corresponding to microwave frequencies that seemed to permeate the cosmos uniformly. This radiation appeared to be a remnant of the 'big bang', the primordial explosion billions of years ago from which the universe originated. Penzias and Wilson were jointly awarded the 1978 Nobel Prize for Physics with the Russian physicist Pyotr Leonidovich Kapitsa. Penzias and Wilson shared half of the prize money for their discovery of cosmic microwave background radiation, while Kapitsa received the other half for his basic inventions and discoveries in the area of low-temperature physics. (☞ Jan. 10, Jul. 8)

1940
Carl Bosch, German industrial chemist, Nobel laureate (Chemistry, 1931), who developed the high-pressure Haber–Bosch process for synthesis of ammonia, died in Heidelberg, Germany. (☞ Aug. 27)

1986
A huge explosion ripped apart the Chernobyl nuclear plant in Pripyat, Ukraine, which sent a cloud of radioactive dust over Europe. It was the result of an experiment gone wrong, causing the fourth reactor to explode and melt down. The graphite-moderated

water-cooled reactor exploded due to build-up of inflammable water gas formed by the reaction of water on hot graphite inside the reactor chamber. Thirty-one people, mostly firemen, were killed immediately after the explosion, and several thousand more – those involved in the clean up – have since died from radiation-related illnesses. It was the worst nuclear accident in history.

1993
The American space shuttle *Columbia* was launched from the Kennedy Space Center, Florida, USA, on its 14th flight on the second German sponsored D-2 Spacelab Mission. The crew consisted of astronauts Steven Nagel, Terence Henricks, Charles Precourt, Bernard Harris, Jerry Ross, Hans Schlegel, and Ulrich Walter. Some 88 experiments covering materials and life sciences, technology applications, Earth observations, astronomy and atmospheric physics were carried out. About 240 tadpoles and 240 fish larvae were flown to test how their bodies adjusted to weightlessness in space; most of the specimens died in orbit. During the 10-day mission, Harris set up the first intravenous line in space, injecting Schlegel with saline as part of a study to replace body fluids lost during adaptation to weightlessness. (☞ Feb. 1, Apr. 12, May 6)

1998
The *Cassini–Huygens* spacecraft made the first fly-by of Venus at a distance of 234 kilometres, en route to Saturn. (☞ Jan. 14, Jun. 24, Jul. 1, Aug. 18, Oct. 15, Dec. 24, Dec. 30)

April 27

1791
Samuel Finley Breese Morse, American painter and inventor, was born in Charlestown, Massachusetts, USA. Morse graduated from the Yale University, New Haven, Connecticut, in 1810 and became a clerk for a Boston book publisher. But painting continued to be his main interest, and in 1811 his parents helped him to go to England in order to study that art. In 1832, while returning by ship after studying art in Europe, Morse conceived the idea of an electric telegraph after hearing a conversation about the newly discovered electromagnet. He probably made his first working model by 1835. Within two years, he turned his full attention to the new invention. A colleague at the university showed him a detailed description of an alternative model, and a friend offered to provide materials and labour to build models in his family's ironworks. These two became partners in Morse's telegraph rights. By 1838 he had developed the system of dots and dashes that became known throughout the world as the Morse code. The first telegraph line in the United States, from Baltimore to Washington, was completed and Morse sent the first message, 'What hath God wrought!' in 1844, ushering in a new age of long-distance communication. (☞ Jan. 24, Apr. 2, May 24)

1880
Francis D. Clarke and M.G. Foster were issued the first US patent for an electric hearing aid. Titled a 'device for aiding the deaf to hear', the device employed bone conduction by which sound is passed through vibration of the skull bone. It was not until 1901 that the first commercially available and practical hearing aid using a carbon microphone was produced. It was invented by Miller Reese Hutchinson, an engineer who worked with the American inventor Thomas Alva Edison.

1896

Wallace Hume Carothers, American chemist who developed nylon, was born in Burlington, Iowa, USA. Carothers was educated in the public schools of Des Moines, Iowa. He first studied accounting and secretarial courses then entered the Tarkio College in Missouri as a science student while simultaneously holding assistantships in English and commercial studies. After receiving a BS from Tarkio, Carothers obtained his master's and doctor's degrees from the University of Illinois, Urbana-Champaign. He held teaching positions briefly at the University of South Dakota, University of Illinois, and Harvard University before joining E.I. du Pont de Nemours & Company in 1928 as head of fundamental research in organic chemistry. At Du Pont, Carothers first worked on the polymerisation of acetylene and its derivatives; this led to the development by other scientists of neoprene. His most outstanding work involved the theory of linear polymerisation, which he tested by synthesising a large number of polymers structurally similar to cellulose and silk. This work culminated in the production of the polyamide fibre nylon, the first synthetic polymer fibre to be spun from a melt, by condensation of adipic acid and hexamethylenediamine. Nylon went into production in 1939, and laid the foundation of the synthetic-fibre industry. With the onset of World War II, nylon was extensively used for war purposes – for example, to make parachute canopies. After the war, nylon found a wide variety of applications including apparel, carpeting, home furnishings, and industrial products. (☞ Feb. 24, Apr. 29, Sep. 20)

1970

The discovery of element 105 was announced at the American Physical Society meeting in Washington, D.C. The element was synthesised by a team led by the American nuclear physicist Albert Ghiorso at the Lawrence Radiation Laboratory at the University of California, Berkeley, California, by bombarding a target californium-249 atom with a beam of 84 MeV nitrogen nuclei in a particle accelerator. The artificially produced element had an atomic mass of 260, and a half-life of less than 40 seconds. Although the Berkeley team suggested the name 'hahnium' for the new element after the German physicist Otto Hahn, the International Union of Pure and Applied Chemistry (IUPAC) approved the name 'dubnium' (symbol Db) after the city of Dubna in Russia, where the element is claimed to have been produced first.

1972

The command module of American spacecraft *Apollo-16*, carrying astronauts John W. Young, Thomas K. Mattingly II, and Charles M. Duke, Jr., safely splashed down in the Pacific Ocean. (☞ Apr. 16, Apr. 21)

2005

The world's largest passenger plane, the *Airbus A380*, completed its maiden flight. The giant plane, which can carry up to 840 passengers on commercial flights, took off from the airport at Blagnac in France and touched down at the same airport after a flight of nearly four hours. The

double-decked, 308-tonne plane carried on board a crew of six and 22 tonnes of test instruments during the flight.

April 28

1900

Jan Hendrik Oort, Dutch physicist and astronomer, was born in Franeker, Netherlands. Oort graduated from the University of Groningen, and was appointed astronomer to the Leiden Observatory in 1924; he became director in 1945, a position he held until 1970. In 1927 he was able to confirm through his own direct observations of star velocities that the Milky Way rotates in its own plane around the centre of the galaxy. After World War II, he led the Dutch group that used the 21-cm radio waves to map hydrogen gas in the Milky Way galaxy and discovered its large-scale spiral structure, the galactic centre, and gas cloud motions. In 1950 Oort proposed that comets originate from a vast cloud of small bodies that orbit the Sun at a distance of about one light-year from which comets are dislodged occasionally by the approach of other stars. The comets then move towards the Sun, developing long tails when they become visible from Earth. The existence of this region, which was named the Oort cloud, is now accepted by most astronomers. (☞ Nov. 5)

1941

Karl Barry Sharpless, American chemist, Nobel laureate, was born in Philadelphia, Pennsylvania, USA. Sharpless attended a Quaker school in Philadelphia city and received his BA from the Dartmouth College, Hanover, New Hampshire in 1963. He received his PhD from the Stanford University, Stanford, California. After postdoctoral years at the Stanford University and Harvard University, he became assistant professor of chemistry at the Massachusetts Institute of Technology, Cambridge, Massachusetts in 1970, and in 1975 was named professor of chemistry. In 1977 he became a professor of chemistry at the Stanford University and in 1980 returned to MIT as professor of chemistry. In 1990 he became W.M. Keck Professor of Chemistry at the Scripps Research Institute in La Jolla, California. Sharpless' research focussed on chiral catalysts for oxidations, a broad family of chemical reactions. 'Chirality', or 'handedness', describes how organic molecules often appear in two forms, one the mirror image of the other, much as a person's two hands mirror each other. In nature, one form often dominates, so human cells will naturally bind to one form of a molecule over the other, while the other form is ignored or may even be harmful. In 1980, working at MIT, Sharpless carried out key experiments that led to a practical method based on catalytic asymmetrical oxidation for producing epoxide compounds, used in the synthesis of heart medicines such as beta blockers and other products. Sharpless was awarded the 2001 Nobel Prize for Chemistry, jointly with the American chemist William S. Knowles and the Japanese biochemist Ryoji Noyori. Sharpless received half the prize money for his work on chirally catalysed oxidation reactions, while Knowles and Noyori shared the other half. (☞ Jun. 1, Sep. 3)

1947
Norwegian ethnologist and adventurer Thor Heyerdahl and five companions left Peru on a balsa raft, the *Kon Tiki*, for Polynesia, which they reached 101 days later. Heyerdahl wanted to demonstrate that the ancient Polynesians could in the same way have originated from South America. (☞ Apr. 18, Aug. 7, Oct. 6)

1991
The American space shuttle *Discovery* was launched on its 12th flight, with a payload related to research on the Strategic Defence Initiative, commonly known as 'Star Wars'. The crew comprised astronauts Michael Coats, Blaine Hammond, Guion Bluford, Gregory Harbaugh, Richard Hieb, Donald McMonagle, and Charles Veach. The payload included an infrared telescope for scanning the upper parts of the Earth's atmosphere, mainly aimed at understanding what future space-based sensors would see when they searched for missile warheads against the background of space and Earth. (☞ Aug. 30)

1999
Arthur Leonard Schawlow, American physicist, Nobel laureate (Physics, 1981), co-inventor of the laser, died in Palo Alto, California, USA. (☞ May 5)

2001
Dennis Tito, a billionaire businessman from California, USA, became the first paying passenger to go to outer space. Aged 60, Tito set off from the Baikanur cosmodrome in Kazakhstan for an eight-day holiday aboard the *International Space Station*. He paid \$20 million for his trip of a lifetime.

April 29

1893
Harold Clayton Urey, American chemist, Nobel laureate, was born in Walkerton, Indiana, USA. Urey received his bachelor's degree from the University of Montana in 1917 and spent two years as an instructor. He received his PhD in chemistry from the University of California, Berkeley in 1923 and went to Copenhagen, Denmark, as a fellow at Niel Bohr's Institute of Theoretical Physics where he did basic research on the theory of atomic structure. In 1924 he returned to USA and taught at the Johns Hopkins University, Baltimore, and at the Columbia University, New York till 1945, when he joined the Institute for Nuclear Studies at the University of Chicago. In the 1920s, Urey achieved separation of deuterium, the heavy isotope of hydrogen, from liquid hydrogen by distillation. In 1931 he and his associates announced their discovery of heavy water, composed of an atom of oxygen and two atoms of deuterium. He also examined the chemical properties and separation of radioactive isotopes of carbon, oxygen, nitrogen, and sulphur. During World War II, as director of research for the Substitute Alloys Materials Laboratories at the Columbia University, Urey led a group responsible for separating uranium isotopes for the Manhattan Project, which developed the atomic bomb. In 1953, Urey and one of his students Stanley L. Miller simulated the effect of lightning in the prebiotic atmosphere of Earth with an electrical discharge in a mixture of hydrogen, methane, ammonia, and water.

This produced a rich mixture of aldehydes and carboxylic and amino acids, as found in proteins, adenine and other nucleic acid bases. Urey was awarded the 1934 Nobel Prize for Chemistry for his discovery of heavy hydrogen. (☞ Jan. 5, May 15)

1937
Wallace Hume Carothers, American chemist who developed nylon, the first synthetic fibre to be spun from a melt, committed suicide by taking cyanide in a hotel room in Philadelphia, USA. (☞ Apr. 27)

1990
American space shuttle *Discovery*, carrying astronauts Loren J. Shriver, Charles F. Bolden, Steven A. Hawley, Bruce McCandless II, and Kathryn D. Sullivan landed at the Edwards Air Force Base, California, USA, after completing its 10th flight during which the Hubble space telescope was deployed in orbit. (☞ Apr. 24, Apr. 25, Aug. 30)

April 30

1777
Carl Friedrich Gauss, German mathematician, physicist and astronomer, was born in Brunswick, Germany. Gauss was extremely precocious, and at an early age he overturned the theories and methods of 18th-century mathematics. In school he distinguished himself in mathematics and languages and when he was fourteen the Duke of Brunswick decided to subsidise his further studies. From 1795 to 1798 he studied mathematics at the University of Göttingen. In 1799 he was awarded a doctorate in absentia by the University at Helmstedt, on a dissertation that was a proof of the fundamental theorem of algebra, which states that every algebraic equation with complex coefficients has complex solutions. Gauss developed the theory of numbers and applied mathematics to electricity and magnetism and astronomy and geodesy. When the minor planet Ceres (now designated as a dwarf planet) was discovered in 1801, Gauss was the first to calculate its orbit. He also published a major work, telling how to determine a planet's orbit. In 1807 he was appointed professor of mathematics and director of observatory at Göttingen. In the 1830s he turned to research on electricity and magnetism and devised an electric telegraph in 1833. (☞ Feb. 23)

1897
The English physicist Joseph John Thomson, English physicist announced the existence of the electron at the Royal Institution Friday Evening Discourse in London, England. It was the first elementary particle to be discovered. Thomson told his audience that he had found a particle of matter a thousand times smaller than the atom. He called it a corpuscle, meaning 'small body,' but later adopted the name 'electron.' His discovery was the result of an attempt to solve a long-standing controversy regarding the nature of cathode rays, which occur when an electric current is driven through a vessel from which most of the air or other gas has been pumped out. Thomson proved that the cathode rays were made up of the same particles irrespective of the kind of gas in the discharge tube and suggested that the electron was a universal component of matter. (☞ Aug. 30, Dec. 18)

1955

Synthesis of the element 101 was announced by a team of scientists at the University of California, Berkeley, USA. The team, comprising Albert Ghiorso, Bernard G. Harvey, Gregory R. Choppin, Stanley G. Thompson, and Glenn T. Seaborg, produced it by bombarding atoms of einsteinium-253 (atomic number 99) with helium ions in a particle accelerator. The new element was named mendelevium after the Russian chemist Dmitry Ivanovich Mendeleyev, who developed the periodic table of elements. In about a dozen repetitions of the experiment, the team of scientists produced 17 atoms of mendelevium-256, which were identified by the ion-exchange adsorption–elution method Other isotopes of mendelevium, all radioactive, have been discovered, the most stable of which is mendelevium-258, with a half-life of two months.

May 1

1753
Carolus Linnaeus, Swedish botanist and explorer, published the first edition of his *Species Plantarum* in which he gave systematic names to plants that are still in use today. He was the first to frame principles for defining genera and species of organisms and to create a uniform system for naming them. Thus, he is often called the father of taxonomy. He extended the familiar scheme of dual Latin names to identify animals in 1758. His system for naming, ranking, and classifying organisms is still in wide use today (though with many changes). His ideas on classification have influenced generations of biologists during and after his own lifetime. The *Species Plantarum* was taken by international consent in 1905 as the starting point for modern botanical nomenclature. (☞ Jan. 10, May 23)

1852
Santiago Ramón y Cajal, Spanish histologist, Nobel laureate, was born in the village Petilla, Aragon, Spain. As a boy Cajal was apprenticed first to a barber and then to a cobbler. He himself wished to be an artist, but his father persuaded him to study medicine. Cajal graduated from the University of Saragossa medical school in 1873 and obtained his MD from Madrid in 1883. He also taught at the universities of Valencia, Barcelona, and Madrid. In 1889 he discovered the mechanisms governing the structure and connective processes of nerve cells in the grey matter of the brain and established the neuron, or nerve cell, as the basic unit of nervous structure. This finding was instrumental in the recognition of the neuron's fundamental role in nervous function. During the next two years he demonstrated the basic changes that the neuron undergoes during the functioning of the nervous system. He devised a way to stain nerve tissues for microscopic studies. He was also the first to isolate the nerve cells, called Cajal's cells that are located near the surface of the brain. Cajal was awarded the 1906 Nobel Prize for Physiology or Medicine, jointly with the Italian physician and cytologist Camillo Golgi, in recognition of their work on the structure of the nervous system. (☞ Jul. 7, Oct. 17)

1888
Nikola Tesla, Serbian-born American inventor was granted a patent for an alternating current (AC) induction motor. Tesla had worked out that two AC sources, out of phase, could cause a rotating magnetic field that could make an AC motor possible. (☞ Jul. 9)

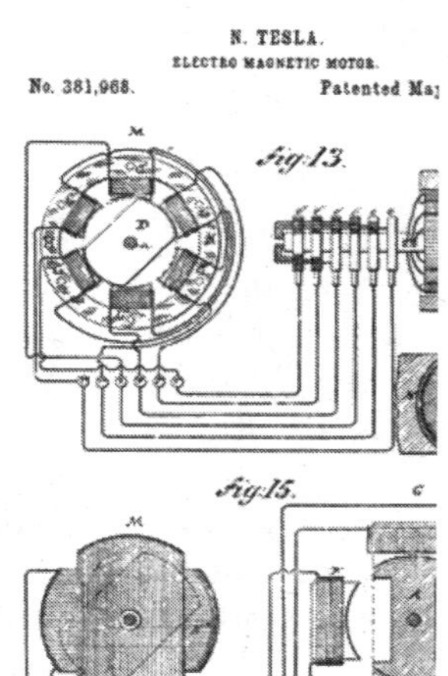

1895
The world's first electric passenger train on mainline started service on a 6-kilometre stretch of track in Baltimore City, Maryland, USA.

1937
Manmohan Sharma, Indian chemical engineer, Fellow of the Royal Society, was born in Jodhpur, Rajasthan, India. Sharma obtained his bachelor's degree in chemical engineering in 1958, his MSc (Tech) degree in 1960 from the Bombay University, and earned his PhD from the Cambridge University, England, in 1964. He was lecturer in chemical engineering at the Bombay University from 1959 to 1961, and professor and head of the chemical engineering section in the department of chemical technology in the same university after his return from Cambridge, from 1964 to 1989. Sharma's research brought out some novel features of gas–liquid, liquid–liquid, and liquid–solid reactions. His greatest contribution was developing the University Department of Chemical Technology as a role model in successfully linking academic programmes to applications in industry.

1949
Nereid, the outermost and the third largest of Neptune's known moons, was discovered by the Dutch-born American astronomer Gerard Kuiper. The irregular-shaped moon with an average diameter of 340 kilometres, Nereid orbits Neptune at an average distance of 5,513,400 kilometres. Nereid's orbit is the most highly eccentric of any planet or moon in the solar system; its distance from Neptune varies from 1,353,600 to 9,623,700 kilometres. Nereid's odd orbit indicates that it may be a captured minor planet or Kuiper Belt object. The name, Nereid refers to the mythological sea nymphs who dwell in the Mediterranean Sea. (☞ Dec. 7, Dec. 23)

1958
The discovery of the powerful Van Allen radiation belts that surround Earth was reported in the *Washington Evening Star*. The discovery was announced by the American physicist James A. Van Allen at a joint symposium of the National Academy of Sciences and the American Physical Society in Washington DC, USA. He used data from the first American satellite *Explorer-I*, which carried a radiation counter, to reveal the existence of bands of electrically charged particles surrounding our globe. Qualitatively, there are two belts around Earth, the inner radiation belt and

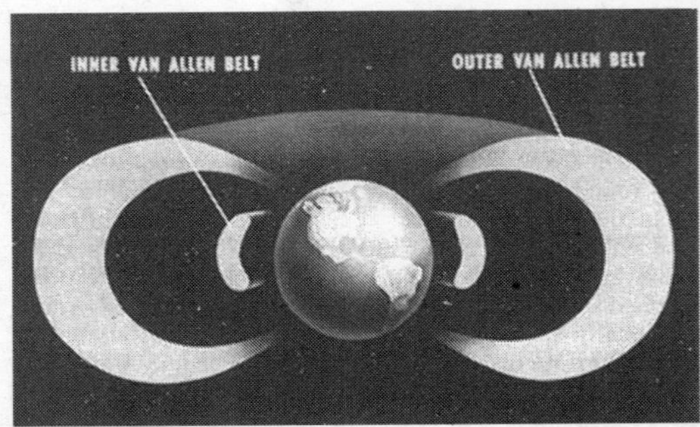

the outer radiation belt. The particles are distributed such that the inner belt consists mostly of protons while the outer belt consists mostly of electrons. The discovery launched the new era of exploration of the magnetospheres of the Earth and other planets. (☞ Jan. 31, Sep. 7)

May 2

1519
Leonardo Da Vinci, Italian artist, sculptor, engineer and scientist, died in Cloux, France. (☞ Apr. 15)

1892
August Wilhelm von Hofmann, German chemist whose research on aniline, helped to lay the basis of the aniline-dye industry, died in Berlin, Germany. (☞ Apr. 8)

1952
The British De Havilland *Comet*, the world's first jet-propelled airliner, made its first commercial flight. (☞ Jul. 27)

1979
Giulio Natta, Italian chemist, Nobel laureate (Chemistry, 1963), who contributed to the development of high polymers useful in the manufacture of films, plastics, fibres, and synthetic rubber, died in Bergamo, Italy. (☞ Feb. 26)

1997
John Carew Eccles, Australian neurophysiologist, Nobel laureate (Physiology or Medicine, 1963), who discovered many aspects of the vertebrate nervous system, notably how nerve impulses are transmitted between neurons, or nerve cells, died in Contra, Switzerland. (☞ Jan. 27)

May 3

1892
George Paget Thomson, English physicist, Nobel laureate, was born in Cambridge, England. George was the son of J.J. Thomson and, like his father, was a highly talented and brilliant man. After receiving a first class honours degree in mathematics and physics, he went on to do research and teaching. During the mid-1920s, Thomson carried out a series of experiments using an apparatus called an electron diffraction camera to test French physicist Louis de Broglie's hypothesis that electrons possess duality, acting both as particles and as waves. The experiment involved bombarding very thin metal foils of aluminium, gold, platinum, and celluloid foils with a narrow beam of electrons. The beam was scattered into a series of rings. Applying mathematical formulas to measurements of the rings, together with knowledge of the crystal lattice, Thomson showed in 1927 that all the readings were in complete agreement with de Broglie's theory. The process of electron diffraction, which his experiments established, has been widely used in the investigation of the surfaces of solids. Thomson was awarded the 1937 Nobel Prize for Physics, jointly with the American physicist Clinton J. Davisson, for their experimental discovery of the diffraction of electrons by crystals. (☞ Sep. 10, Oct. 22)

1902
Alfred Kastler, French physicist, Nobel laureate, was born in Guebwiller, Germany (now in France). While at school, Kastler was interested in mathematics and science. After graduating from the prestigious Ecole Normale Superieure with a teaching certificate, Kastler held a series of teaching jobs. He then received an appointment in 1931 as a research assistant at the University of Bordeaux with permission to begin his graduate studies. There he prepared his

thesis on the step-wise excitation of mercury atoms. In 1937 he became interested in the luminescence of sodium atoms in the upper atmosphere. For many years, he worked in the field of optical spectroscopy, particularly on atomic fluorescence and Raman spectroscopy. He discovered and developed the so-called 'double resonance' method, using both a light beam and a radio-frequency electromagnetic wave to identify the sublevels associated with the main energy levels in atoms. His method of stimulating atoms in a particular substance so that they attain higher energy states was called 'optical pumping'. Since the light energy used to stimulate the atoms was re-emitted, optical pumping marked an important step toward the development of the maser and the laser. Kastler was awarded the 1966 Nobel Prize for Physics for the discovery and development of optical methods for studying Hertzian resonance in atoms. (☞ Jan. 7)

1933
Steven Weinberg, American nuclear physicist, Nobel laureate, was born in New York, USA. Weinberg's early interest in science was encouraged by his family and by his teachers at the Bronx High School of Science. He went to Cornell University, studied at the Niels Bohr Institute in Denmark, and got his PhD from Princeton, USA. Thereafter he embarked on a career of research and teaching that took him to some of the best centres for physics research in the USA: Columbia, Berkeley, MIT, and Harvard. Weinberg's scientific interests were always broad, but his most noted work has been in unified field theory. Four fundamental forces were believed to drive the laws of physics: gravity, electromagnetism, the strong force (which holds an atom's nucleus together), and the weak force (which breaks an atom apart, as in radioactivity). Around 1967, Weinberg theorised that the electromagnetic and the weak forces are the same at extremely high energy levels. This electroweak theory was confirmed by particle accelerator experiments in 1973. This was one giant step closer to physicists' long-dreamed of goal of finding a single elegant equation to explain all the matter and forces in nature. Weinberg was awarded the 1979 Nobel Prize for Physics, jointly with his former schoolmate Sheldon Lee Glashow and the Pakistani nuclear physicist Abdus Salam for their contributions to the theory of the unified weak and electromagnetic interaction between elementary particles, including among other things, the prediction of the weak neutral current. (☞ Jan. 29, Dec. 5)

1944
The first synthetic quinine was produced by the American chemists Robert Burns Woodward and William von Eggers Doering at the Converse Memorial Laboratory, Harvard University, Cambridge, Massachusetts, USA. Quinine, an antimalarial drug, is an organic chemical with the formula $C_{20}H_{24}N_2O_2$. (☞ Apr. 10)

May 4

1825
Thomas Henry Huxley, English biologist, educator and advocate of Darwinism, was born in Ealing, Middlesex, England. After

receiving his preliminary education, Huxley studied medicine. In 1846 he took the diploma of MRCS in order to qualify himself for the medical service of the Royal Navy, which he entered as assistant surgeon. His next appointment was as assistant surgeon in *HMS Rattlesnake*, and he spent the greater part of his time from 1847–51 on the eastern and southern coasts of Australia. The results of his studies in natural history, for which the cruise afforded facilities, appeared in a work entitled *Oceanic Hydrozoans*. Upon his return to England in 1854, Huxley resigned from the navy and took a professorship in the Royal School of Mines, where he turned to the study of fossils, especially of fishes and reptiles. Huxley's writings span a remarkable range, reflecting his broad interests, intellectual passions, and social commitment. Among his most notable scientific works is his *Evidence as to Man's Place in Nature*, published in 1863, which went beyond the *Origin of Species*, by explicitly extending the idea of evolution to the human species. Huxley is best remembered today for his prominent role in defending evolution against attacks from scientists, theists, and philosophers. Huxley coined the word agnostic to describe his own beliefs. (☞ Jun. 29)

1967

American spacecraft *Lunar Orbiter-4* was launched on a photographic mission to the Moon. Between May 11 and 26 *Lunar Orbiter-4* photographed the entire nearside and 95% of the far side of the Moon. In all five *Lunar Orbiter* missions were sent and all were successful, photographing 99% of the Moon's surface with a resolution of 60 m or better.

1972

Edward Calvin Kendall, American chemist, Nobel laureate (Physiology or Medicine, 1950), who unravelled the structure and studied the biological effects of adrenal cortex hormones, died in Princeton, New Jersey, USA. (☞ Mar. 8)

1976

The first laser geodynamics satellite (*Lageos-1*) was launched from the Vandenberg Air Force Base, California, USA, and placed in a very stable circular orbit at an altitude of 5,900 kilometres. A joint program between the American space agency NASA and the Italian Space Agency ASI, *Lageos-1* is a passive satellite that carries no electronic equipment or moving parts. The spherical satellite takes on a golf-ball appearance due to the 'retrocollectors', three-dimensional prisms that reflect light back to the source regardless of the angle of the light received by the satellite. It is primarily used as an aid in accurate measurement of the distances between points on the Earth's surface through precise laser ranging, which help in precision Earth-dynamics measurements such as crustal motions, regional strains, fault motions, polar motion and Earth-rotation variations, solid Earth tides, and other parameters associated with earthquake assessment and alleviation. (☞ Oct. 22)

1989

The *Magellan* spacecraft, named after the sixteenth-century Portuguese explorer whose expedition first circumnavigated the

Earth, was launched on a mission to Venus from the American space shuttle *Atlantis* during its 4th flight. *Magellan* was the first planetary spacecraft to be launched by a space shuttle. After release from *Atlantis*, a solid-fuel motor called the inertial upper stage was fired, sending *Magellan* on a 15-month cruise looping around the Sun 1½ times before it arrived at Venus in August 1990. A solid-fuel motor on *Magellan* was then fired, placing the spacecraft in orbit around Venus. (☞ May 8, Aug. 10, Sep. 15, Oct. 3)

1994

SROSS-C2, the fourth satellite of India's Stretched Rohini Satellite Series, was launched by ISRO's augmented satellite launch vehicle (ASLV) from the Satish Dhawan Space Centre at Sriharikota, Andhra Pradesh. The satellite carried a gamma-ray burst experiment that monitored celestial gamma-ray bursts, and an experiment that measured temperature, density, and characteristics of electrons in the ionosphere.

2003

The first cloned equine, a mule foal was born at the University of Idaho, USA. Researchers cloned the mule, named Idaho Gem, using a cell from a mule foetus and an egg from a horse. Mules are almost always sterile because donkeys have 62 chromosomes, horses have 64 and mules end up with 63. Equines such as horses, donkeys and mules are difficult to reproduce in the lab using *in vitro* fertilisation.

May 5

1881

Louis Pasteur, French chemist and microbiologist, tested inoculations against anthrax upon an ox, several cows and 25 sheep. His experiment proved successful, and was a milestone in the treatment of the disease. (☞ Sep. 28, Dec. 27)

1921

Arthur Leonard Schawlow, American physicist, Nobel laureate, was born in Mount Vernon, New York, USA. As a boy, Schawlow was interested in scientific things – electrical, mechanical, or astronomical – and read nearly everything that the local library could provide on these subjects. He studied radio engineering at the University of Toronto in Canada and continued his studies with a postdoctoral fellowship at the Columbia University, New York, where he worked with the American physicist Charles H. Townes, an established leader in the field of microwave spectroscopy. The two scientists were trying to amplify when Schawlow came up with the idea of a long chamber with a mirror at each end, a design that was key to further laser development. Schawlow and Townes received a patent for the laser in 1960, but they never profited from it because Schawlow was working for Bell Labs and Townes was a Bell Labs consultant at the time. Schawlow was awarded the 1981 Nobel Prize for Physics, jointly with the American physicist Nicolaas Bloembergen and the Swedish physicist Kai Manne Börje Siegbahn.

Schawlow and Bloembergen shared the prize for their contribution to the development of laser spectroscopy, while Siegbahn received it for his contribution to the development of high-resolution electron spectroscopy. (☞ Mar. 11, Apr. 20, Apr. 28)

1961

The first American manned space mission carrying astronaut Alan B. Shephard Jr. on board the spacecraft he named *Freedom-7*, was launched by a Redstone rocket. It was a sub-orbital flight lasting only 15 minutes and carried the spacecraft only to a height of 160 kilometres.

1963

The world's first human liver transplant was performed in USA by Dr. Thomas E. Starzl at the University of Colorado Medical School; however, lack of effective immunosuppressives limited the success. The patient, a 48-year-old man, survived for 22 days. Four years later, the availability of more effective immunosuppressive enabled Dr. Starzl to perform the first successful liver transplant. Since that time, many developments have led to the wide acceptance of this operation as a therapy for patients suffering from cirrhosis and advanced liver disease.

1986

Two Russian cosmonauts Leonid Kizim and Vladimir Solovyou shifted from the *Mir* orbital station to another orbital station *Salyut-7* on board *Soyuz T-15*. It was the first ever space-station-to-space-station transfer. (☞ Feb. 20, Apr. 19)

1988

NHK (Nippon Hōsō Kyōkai – Japan Broadcasting Corporation) of Japan conducted the first live TV transmission from the summit of Mount Everest, the highest point on Earth.

2005

Cartosat-1, India's 11th remote sensing satellite, along with a smaller satellite *Hamsat*, was launched by the polar satellite launch vehicle (PSLV-C6) of the Indian Space Research Organisation from the Satish Dhawan Space Centre, Sriharikota, Andhra Pradesh. The two satellites were placed in a north–south polar orbit ranging in height from 621 to 632 kilometres. Primarily designed to help in map-making, *Cartosat-1* carries cameras for stereo imaging that will help in generation of three-dimensional maps, which will reveal the height of dams, buildings, hills and slopes. *Cartosat-1* is the only satellite of its kind in the world to provide maps with a resolution of 2.5 metres for civilian use. The *Hamsat* is designed to provide satellite-based amateur radio services to the national as well as the international community of amateur radio operators known as ham operators.

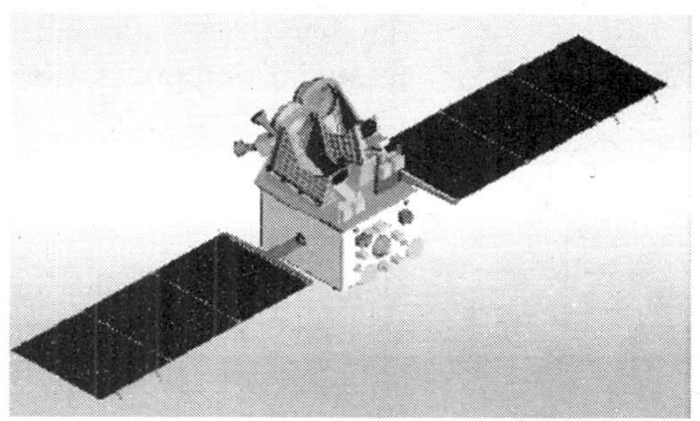

May 6

1856

Sigmund Freud, Austrian neurologist, founder of psychoanalysis, was born in Freiberg, Moravia, (now Príbor, Czech Republic). Freud entered the University of Vienna medical school, but had no intention of practising medicine. He hoped to go into neurophysiological research, but pure research was hard to manage in those

days without adequate financial support. So he decided to enter into private practice with a specialty in neurology. In 1882 he joined the General Hospital in Vienna as a clinical assistant to train with the psychiatrist Theodor Meynert. In 1885 he was appointed lecturer in neuropathology, having concluded important research on the brain's medulla. In 1886, he opened a private practice in Vienna, specialising in nervous and brain disorders. He tried hypnotism with his hysteric and neurotic patients, but gradually discarded the practice. He found he could get patients to talk just by putting them in a relaxing position and encouraging them to say whatever came into their heads. He would then analyse what they had remembered or expressed and determine what traumatic events in their past had caused their current suffering. In 1900, Freud published *The Interpretation of Dreams*, and introduced the wider public to the notion of the unconscious mind. In 1901, he published *The Psychopathology of Everyday Life*, in which he theorised that forgetfulness or slips of the tongue (now called 'Freudian slips') were not accidental at all, but the 'dynamic unconscious' revealing something meaningful. In 1933 he published the *New Introductory Lectures on Psychoanalysis*. By the publication of these, and numerous other works, he revolutionised the field of psychotherapy. One of his most controversial theories was that the sexual drive was the most powerful shaper of a person's psychology, and that sexuality was present even in infants. He shocked society when he published these ideas in 1905. His most well-known theory is that of the 'Oedipus complex' which states that in boys, there is a sexual attraction towards the mother and a sense of jealousy to the point of hatred of the father, and vice versa, for girls. Freud not only influenced the professional practice of psychology and psychiatry, but he changed the way people (in Western cultures) view themselves and think about their lives. (☞ Sep. 23)

1861

American physician John Gorrie was issued a US patent for his invention of an 'ice-making machine'. This was the first US patent issued for a mechanical refrigerator. Gorrie gave up medical practice to engage in time-consuming experimentation with ice making. He designed and built the air-cooling apparatus for treating yellow-fever patients. His basic principle – that of compressing a gas, cooling it by sending it through radiating coils, and then expanding it to lower the temperature further – is the one most often used in refrigerators today. Gorrie is considered the father of air conditioning and refrigeration.

1871

Victor Grignard, French chemist, Nobel laureate, was born in Cherbourg, France. In 1898, while a student at Lyon, Grignard began his prize-winning work with a study of the alkylzinc compounds. While trying to repeat some experiments on the preparation of a tertiary alcohol from a mixture of methyl heptyl ketone, magnesium, and methyl iodide, Grignard hit upon the idea of treating the iodide with the magnesium first and carried out the reaction in ether. This first of the Grignard reagents was a complete success, which Grignard described in his doctoral dissertation in 1901. The Grignard reagents are extremely

important in organic syntheses. They are very versatile and permit the synthesis of a large number of different classes of compounds, particularly secondary and tertiary alcohols, hydrocarbons, and carboxylic acids. In 1910, Grignard obtained a professorship at the University of Nancy, and in 1919 moved to the University of Lyon as professor. Grignard was awarded the 1912 Nobel Prize for Chemistry, jointly with the French chemist Paul Sabatier, for the discovery of the so-called Grignard reagent, which has greatly advanced the progress of organic chemistry. (☞ Nov. 5, Dec. 13)

1929

Paul C. Lauterbur, American chemist, Nobel laureate, was born in Sidney, Ohio, USA. Lauterbur received his BS degree in chemistry from the Case Institute of Technology, Cleveland, Ohio, and his PhD in chemistry from the University of Pittsburgh, Pennsylvania. In 1969 he joined the radiology department of the New York University at Stony Brook as professor of chemistry, and then moved to the College of Medicine, University of Illinois, Chicago, Illinois, in 1985. Since 1985 he has also been professor and director, Biomedical Magnetic Resonance Laboratory at the University of Illinois, College of Medicine at Urbana-Champaign, Illinois. Lauterbur's research mainly concerned magnetic resonance imaging. In the beginning of the 1970s he discovered the possibility of creating a two-dimensional picture by introducing gradients in the magnetic field in magnetic resonance imaging. By analysis of the characteristics of the emitted radio waves, he could determine their origin. This made it possible to build up two-dimensional pictures of structures that could not be visualised with other methods. His findings provided the basis for the development of magnetic resonance into a useful imaging method, which later led to the applications of magnetic resonance in medical imaging. Lauterbur was awarded the 2003 Nobel Prize for Physiology or Medicine, jointly with the British physicist Peter Mansfield, for their discoveries concerning magnetic resonance imaging. (☞ Oct. 9)

1936

The German passenger airship (also called the zeppelin), the *Hindenburg*, took off from the Friedrichshafen airfield in Stuttgart, Germany to begin the first round trip to the United States. There were 50 passengers and 55 crewmen on board, including three American observers. The ship carried 1,125 kilograms of mail and about 150 kilograms of freight. The *Hindenburg* was 245 metres long, was powered by four 1,100-horsepower diesel engines, giving it a maximum speed of 135 km/h. In 1936 this airship carried a total of 1,002 passengers on 10 scheduled round trips between Germany and USA. (☞ see below)

1937

The German passenger airship (also called the zeppelin), the *Hindenburg*, went down in a fiery blaze while trying to land at Lakehurst, New Jersey, USA, after a flight from Frankfurt, Germany. The cause was and is still unknown. It carried 36 passengers and a crew of 61. Thirty-six lost their lives: 13 passengers, 22 crewmen, and one civilian member of the ground crew. Described as a luxurious flying hotel, faster than any ship, the *Hindenburg* was the largest airship ever built. It measured 245 metres in

length and 41 metres in diameter and used almost 200 million litres of highly inflammable hydrogen gas for lift-off. It pioneered the first transatlantic air service, carried hundreds of passengers and travelled thousands of kilometres before it was destroyed in the fire. No flight of a passenger-carrying airship ever took place after the *Hindenburg* accident. (☞ May 6)

1993
The 220-MW first unit of the Kakrapar Atomic Power Station at Kakrapar, near Surat, Gujarat, India, was commissioned. It is a pressurised heavy water reactor that uses natural uranium as fuel and heavy water as moderator and coolant. (☞ Mar. 5, Sep. 1)

1993
The American space shuttle *Columbia*, carrying astronauts Steven Nagel, Terence Henricks, Charles Precourt, Bernard Harris, Jerry Ross, Hans Schlegel, and Ulrich Walter, landed at the Edwards Air Force Base, California, USA after completing its 14th flight during which it formed part of the second German sponsored D-2 Spacelab Mission. (☞ Feb. 1, Apr. 12, Apr. 26)

1909
Edwin Herbert Land, American inventor and physicist, who developed the Polaroid technique of instant photography, was born in Bridgeport, Connecticut, USA. While a student at the Harvard University, Cambridge, Massachusetts in 1926, Land became interested in polarised light – light oriented in a plane with respect to the source – and developed a new kind of polariser, which oriented light by aligning and embedding submicroscopic crystals of iodoquinine sulphate in a plastic sheet. He called it a Polaroid. Later, he applied the polarising principle to light filters, sunglasses, optical devices, and motion picture processes. Land founded the Polaroid Corporation, in Cambridge, Massachusetts, in 1937 and four years later developed the technique of 3-D motion picture using an optical polariser. In 1947 he demonstrated a camera (known as the Polaroid Land Camera) that produced a finished print in 60 seconds. Land's camera used a film pack that came with its own developing chemicals contained in flexible capsules that ruptured as the film passed between rollers after exposure, triggering the development process. The Land photographic process soon found numerous commercial, military, and scientific applications. Many innovations were made in the following years, including the development of a colour process. In the early 1970s, Land attempted to explain the previously known phenomenon of colour constancy with his 'retinex theory' of colour vision. According to the retinex theory, we decide

the colour of a surface by comparing its ability to reflect different colours against that of adjacent surfaces. That is why colour constancy breaks down when such comparisons cannot be made. Land received more than 500 patents for his innovations in light and plastics. (☞ Feb. 2, Feb. 10, Mar. 1)

1939

Sidney Altman, Canadian-American molecular biologist, Nobel laureate, was born in Montreal, Quebec, Canada. Altman did his BS in physics from Massachusetts Institute of Technology, Cambridge, Massachusetts, in 1960, and obtained his PhD in biophysics from the University of Colorado in 1967. He joined the Harvard University, Cambridge, Massachusetts in 1967 to study a DNA endonuclease involved in the replication and recombination of T4 DNA. Two years later he joined the group of the British biophysicist Francis Crick at the Medical Research Council Laboratory of Molecular Biology in Cambridge, England, where he started the work that led to the discovery of RNase P and the enzymatic properties of the RNA subunit of that enzyme. In 1971 Altman joined the Yale University, New Haven, Connecticut, USA, as an assistant professor. His work over the next two decades led to the discovery of catalytic RNA. Altman's revolutionary discovery was that RNA, traditionally thought to be simply a passive carrier of genetic codes between different parts of the living cell, could also take on active enzymatic functions. He became a full professor at Yale in 1980, was the department chairman from 1983 to 1985, and served as dean of the undergraduate Yale College from 1985 to 1989. Altman was awarded the 1989 Nobel Prize for Chemistry, jointly with the American biochemist and molecular biologist Thomas Robert Cech, for their discovery of the catalytic properties of RNA. (☞ Dec. 8)

1963

The communication satellite *Telstar-2* was launched from Cape Canaveral, Florida, USA to relay TV and telephone signals between the United States and Europe. It was placed in an elliptical orbit ranging in height from 972 km to 10,802 km, and was used to transmit the first transatlantic TV programme in colour. Primarily designed as a communications satellite, *Telstar-2* carried an experiment designed to measure the energetic proton and electron distribution in the Van Allen belts. *Telstar-2* replaced *Telstar-1*, which had ceased operating in 1962 due to transistor damage caused by radiation from a high-altitude nuclear test. Telstar II was built with shielding against such radiation. (☞ Jul. 10)

1992

The American space shuttle *Endeavour* blasted off on its maiden voyage from the Kennedy Space Centre, Florida, USA, with a crew of seven, comprising Daniel C. Brandenstein, Kevin P. Chilton, Pierre J. Thuot, Kathryn C. Thornton, Richard J. Hieb, Thomas D. Akers, and Bruce E. Melnick. *Endeavour*, named after the first ship commanded by James Cook, the 18th century British explorer, navigator and astronomer, was a replacement for *Challenger*, which was lost in January 1986 during takeoff. The *Endeavour* crew captured and corrected the orbit of the *Intelsat-4* satellite, which was stranded in an unusable orbit after launch. The satellite was fitted with a new perigee kick motor and released into orbit. The new motor put the spacecraft into a geosynchronous orbit for operational use. (☞ May 16)

1998
Allan MacLeod Cormack, South African-born American physicist, Nobel laureate (Physiology or Medicine, 1979), who formulated the mathematical algorithms that made possible the development of computerised tomography (CT) scanning, died in Winchester, Massachusetts, USA. (☞ Feb. 23)

May 8

1794
Antoine-Laurent Lavoisier, French scientist, usually regarded as the founder of modern chemistry, who discovered the law of conservation of mass and devised the modern method of naming compounds, died in the guillotine in Paris, France. (☞ Aug. 26)

1902
André Michel Lwoff, French biologist, Nobel laureate, was born in Ainay-le-Château, France. Lwoff was educated at the University of Paris and spent most of his research career at the Pasteur Institute in Paris. His first investigations were on parasitic ciliates, their developmental cycle, and morphogenesis. Later, he worked on the problems involved in the nutrition of protozoans. He obtained his MD in 1927 and his PhD in 1932. Lwoff was appointed head of the department at the Pasteur Institute in 1938. He was professor at the Sorbonne, Paris, from

1959 to 1968 and head of the Cancer Research Institute in Villejuif, near Paris, from 1968 to 1972. In the 1920s, Lwoff demonstrated the coenzyme nature of vitamins, and in the late 1940s, worked out the mechanism of lysogeny in bacteria, in which the DNA of a virus becomes attached to the chromosome (DNA) of a bacterium, behaving almost like a bacterial gene, replicating and multiplying as part of the host's DNA. But certain agents such as ultraviolet radiation could turn the 'latent' viral DNA, called the prophage, into a vegetative form that multiplies, destroys its host, and is then released to infect other bacteria. Lwoff was awarded the 1965 Nobel Prize for Physiology or Medicine, jointly with the French biologist François Jacob and biochemist Jacques Monod, for their discoveries concerning genetic control of enzyme and virus synthesis. (☞ Feb. 9, Jun. 17, Sep. 30)

1947
H. Robert Horvitz, American biologist, Nobel laureate, was born in Chicago, Illinois, USA. Horvitz did his MA in biology in 1972 and received his PhD in biology in 1974 from the Harvard University, Cambridge, Massachusetts. He joined the Massachusetts Institute of Technology in the same city as assistant professor of biology in 1978 and became a full professor in 1986. In 1988 he joined the Howard Hughes Medical Institute of MIT as investigator. Horvitz's main research concerned the genetics and cell lineage of the common nematode *Caenorhabditis elegans*. In a series of elegant experiments that started during the 1970s, Horvitz used *C. elegans* to investigate whether there was a genetic program controlling cell death. He identified the first two bona fide 'death genes', ced-3 and ced-4 and showed that functional ced-3 and ced-4 genes were

a pre-requisite for cell death to be executed. Later, Horvitz showed that another gene, ced-9, protects against cell death by interacting with ced-4 and ced-3. He also identified a number of genes that direct how the dead cell is eliminated. Horvitz showed that the human genome contains a ced-3-like gene. Horvitz was awarded the 2002 Nobel Prize for Physiology or Medicine, jointly with the South African-born British biologist Sydney Brenner and British biologist John E. Sulston, for their discoveries concerning genetic regulation of organ development and programmed cell death. (☞ Jan. 13, Mar. 27)

1980

The 33rd assembly of the World Health Organisation in Geneva, Switzerland, officially declared smallpox as eradicated. A young man in Merka Town, Somalia contracted the last naturally occurring case of smallpox in the world in October 1977. He survived, and no new cases were reported in Somalia or elsewhere. But, in 1978 two more cases popped up in Birmingham, England, from smallpox virus escaped from a research lab. One of the patients died. The director of the laboratory committed suicide. These were the last victims of smallpox. In 1979, a global commission certified that smallpox had been eradicated.

1989

The American space shuttle *Atlantis* carrying astronauts David M. Walker, Ronald J. Grabe, Norman E. Thagard, Mary L. Cleave, and Mark C. Lee, landed at the Edwards Air Force Base, California, USA, after completing its 4th flight during which it deployed the spacecraft *Magellan* on a mission to Jupiter. (☞ May 4, Oct. 3)

2003

India's geostationary satellite launch vehicle (GSLV-D2) was launched on its second developmental flight from the Satish Dhawan Space Centre at Sriharikota, Andhra Pradesh. It placed a 1,800-kilogram *GSAT-2* satellite in the geosynchronous transfer orbit, which was later moved to the geosynchronous orbit. GSLV-D2 used a cryogenic engine supplied by Russia in the third stage. The GSLV project was initiated in 1990 with a goal to achieve self-reliance in launching INSAT-class satellites.

May 9

1850

Joseph Louis Gay-Lussac, French chemist best known for his work on gases, died in Paris, France. (☞ Dec. 6)

1914

Paul-Louis-Toussaint Héroult, French metallurgist who invented the electric-arc furnace, died near Antibes, France. (☞ Apr. 10)

1927

Manfred Eigen, German biochemist and physicist, Nobel laureate, was born in Bochum, Germany. Eigen studied at the Göttingen University from where he obtained his PhD in 1951 and later worked at the university's Institute of Physical Chemistry as an assistant lecturer in

chemistry. In 1953 he joined the Max Planck Institute for Biophysical Chemistry, Göttingen, where he became director in 1964 and later chairman. Eigen began his work on the problem of fast ionic reactions in solution in 1951. During the following years he developed a series of measuring techniques involving times down to the order of a nanosecond (10^{-9} s), bridging the gap between classical reaction kinetics and spectroscopy. In 1954, Eigen introduced the relaxation techniques for the study of extremely fast chemical reactions. His general method was to take a solution in equilibrium at a given temperature and pressure, and apply a short disturbance to the solution, thereby destroying the equilibrium very briefly and immediately leading to a new equilibrium. Eigen then studied exactly what happened in this very short time by means of absorption spectroscopy. Using this technique he was able to investigate a large variety of fast reactions including the rate of hydrogen ion formation through dissociation in water, diffusion-controlled protolytic reactions, and the kinetics of keto-enol tautomerism. Eigen was awarded the 1967 Nobel Prize for Chemistry, jointly with the British chemists Ronald Norrish and George Porter, for their studies of extremely fast chemical reactions, caused by disturbing the equilibrium by means of very short pulses of energy. (☞ Nov. 9, Dec. 6)

1931
Albert Abraham Michelson, German-born American physicist, Nobel laureate (Physics, 1907), who established the speed of light as a fundamental constant, died in Pasadena, California, USA. (☞ Dec. 19)

1960
Enovid-10, the world's first contraceptive (birth-control) pill, was approved by the Food and Drug Administration of USA. The pill was a combination of synthetic progesterone and oestrogen hormones that suppressed the release of eggs from a woman's ovaries. (☞ Apr. 9)

May 10

1829
Thomas Young, English physician and physicist who reinforced the wave theory of light with his study of interference of light, died in London, UK. (☞ Jun. 13)

1852
Edward Frankland, English chemist announced the theory of valence in a communication to *Philosophical Transactions of the Royal Society* in which he described the fixed combining powers of elements as 'atomicity.' This term was re-named valence. The theory states that any atom can combine with a certain, limited number of other atoms, which still remains fundamental to the understanding of chemical structure. Frankland's work on valence, along with an international agreement on atomic weight values signed shortly thereafter united various theoretical schools of chemists into the field of structural chemistry. (☞ Jan. 18)

1860
The discovery of the element caesium (Cs, atomic number 55) was announced by the German chemist Robert Wilhelm Bunsen and physicist Gustav Roberts Kirchoff in Heidelberg, Germany. The element derives its name from its characteristic blue spectral lines. (☞ Mar. 12, Mar. 31)

1901

John Desmond Bernal, Irish physicist and X-ray crystallographer who studied the atomic structures of solid compounds, was born in Nenagh, County Tipperary, Ireland. Bernal studied physics at the University of Cambridge and embarked on a career in crystallography. In 1937 he became professor of physics, and later professor of crystallography, at the Birkbeck College, London. Bernal is best remembered for his pioneering work in X-ray crystallography. His research included work on the tobacco mosaic virus, sex hormones, the structure and property of water and the chemical processes involved in the origin of matter. He also performed research into molecular biology, the origin of life and the structure and composition of the Earth's crust. In a 1947 lecture, 'The physical basis of life', he suggested that clays may have played a role in concentrating organic molecules and pondered on the nature of life in a cosmic context. In a 1952 lecture to the British Interplanetary Society he further helped bring the possibility of extraterrestrial life into the province of mainstream biology by arguing that 'the biology of the future would not be confined to our own planet, but would take on the character of cosmobiology'. Following the discovery of organic substances in the Orgeuil meteorite which fell in France, Bernal suggested that if contamination by substances on Earth could be ruled out, the organic substances may have been formed inorganically in early space or as the result of living things on the parent body of which the meteorite was originally part of. As long ago as 1929, in his remarkably perceptive book *The World, the Flesh and the Devil*, Bernal discussed the possibility of colonies in space. He was particularly fascinated by the role of science in society and published books and pamphlets on the subject, which include *The Social Functions of Science* and *Science in History*. (☞ Sep. 15)

1910

Stanislao Cannizzaro, Italian chemist, teacher, and legislator who recognised the distinction between atomic and molecular weights, and discovered the Cannizzaro reaction, died in Rome, Italy. (☞ Jul 13)

1956

The Central Mining Research Institute (CMRI), a constituent national laboratory of the Council of Scientific and Industrial Research (CSIR), was established. Located in Dhanbad, Jharkhand, the primary activities of CMRI is R&D in the field of mining with the objective of making mining safe, productive, environment-friendly and economical as well as to reduce health hazards to miners and environmental pollution due to mining and associated industrial activities. The mission of the institute is to provide scientific and technological inputs to the mineral sector for optimisation of mining technologies and for better safety, economy, and environmental management. The institute is also involved in development of technologies for the exploitation of coal deposits in different geo-mining conditions, numerical modelling and application of computer and information technology in mines and tunnels, and environmental management, reclamation and development of innovative eco-friendly mining practices. Apart from mining and allied fields, the institute has diversified and extended its activities in the areas of development of underground space technology and technology for the construction of caverns, hydel projects, dams and tunnels.

1997
IBM's *Deep Blue* Supercomputer played a fascinating match with the reigning World Chess Champion, Garry Kasparov, and beat him 3.5 to 2.5. Out of six games, *Deep Blue* won two and Kasparov one; three games were drawn.

May 11

1871
John Herschel, English astronomer, who named seven moons of Saturn and made an extensive study of the southern sky, died in Collingwood, Kent, England. (☞ Mar. 7)

1918
Richard Phillips Feynman, American theoretical physicist, Nobel laureate, was born in New York, USA. After completing his schooling at the Far Rockaway High School in New York, Feynman obtained his BSc in 1939 from the Massachusetts Institute of Technology, Cambridge, Massachusetts, where his undergraduate thesis proposed an original and enduring approach to calculating forces in molecules. He received his doctorate from the Princeton University, New Jersey, in 1942. Feynman worked on the atomic bomb project at the Princeton University during 1941–42, and then at Los Alamos during 1943–45. At Los Alamos he became the youngest group leader in the theoretical division of the Manhattan Project.

With the head of that division, German-born American theoretical physicist Hans Bethe, he devised the formula for predicting the energy yield of a nuclear explosive. After World War II, in 1945, Feynman was appointed a professor of theoretical physics at the Cornell University, Ithaca, New York. In 1950 he accepted a position as professor of theoretical physics at the California Institute of Technology, Pasadena, California. Feynman's main contribution was to quantum mechanics, following on from his doctoral thesis. In 1952 he developed the rules that all quantum field theories must obey. In the process, he discovered how to renormalise the theory of quantum electrodynamics – a generalisation of quantum mechanics to include special relativity. He also invented an elegant way of representing quantum interactions, now called Feynman diagrams – graphic analogues of the mathematical expressions needed to describe the behaviour of systems of interacting particles. Feynman wrote a three-volume textbook, *The Feynman Lectures on Physics*, which has become a classic for students of physics. He also published a number of popular books on physics, including *QED: The Strange Theory of Light and Matter*. Feynman was awarded the 1965 Nobel Prize for Physics, jointly with fellow American physicist Julian Seymour Schwinger and the Japanese physicist Shin'ichiro Tomonaga, for their fundamental work in quantum electrodynamics, with deep-ploughing consequences for the physics of elementary particles. (☞ Feb. 12, Feb. 15, Mar. 31)

1924
Antony Hewish, British astrophysicist, Nobel laureate, was born in Fowey, Cornwall, England. Hewish was educated at the King's College, Taunton, and went to the

University of Cambridge in 1942. He was engaged in war service from 1943–46 at the Royal Aircraft Establishment, Farnborough, and also at the Telecommunications Research Establishment, Malvern, where he worked on airborne radar counter-measure devices. In 1946 he joined the radio astronomy group at the Cambridge University led by the British radio astronomer Martin Ryle and obtained his PhD in 1952. In 1961 he joined the Churchill College in Cambridge as director of studies in physics. He was lecturer in the University during 1961–69, reader during 1969–71 and professor of radio astronomy from 1971 until his retirement in 1989. In 1977 Hewish took up leadership of the Cambridge radio astronomy group and was head of the Mullard Radio Astronomy Observatory at Cambridge from 1982–88. In 1967, while using his newly designed radio telescope, Hewish and his graduate student Jocelyn Bell observed an unusual signal corresponding to sharp bursts of radio energy at regular intervals of approximately one second. The source of these unusual bursts was traced to a class of rapidly rotating objects called neutron stars, which came to be known as pulsating stars, or 'pulsars'. Pulsars are small and extremely dense remnants formed after supernova explosions in the last life stages of massive stars. They emit highly directional, rotating beams of radio waves that appear as pulsating signals on Earth. Hewish was awarded the 1974 Nobel Prize for Physics, jointly with Ryle, for their pioneering research in radio astrophysics. Ryle received half of the prize money for his observations and inventions, in particular of the aperture synthesis technique, and Hewish received the other half for his decisive role in the discovery of pulsars. (☞ Feb. 24, Sep. 27)

1941

Robert Koffler Jarvik, American inventor of the first permanently implantable artificial heart, was born in Michigan, USA. Jarvik studied at the University of Utah, from where he graduated with his MD in 1976. He demonstrated his mechanical aptitude early, having invented such useful devices as a surgical stapler and other medical tools when he was just a teenager. It was when his father became ill with heart disease and had to have an open-heart surgery that Jarvik started studying heart transplants. In some cases, he learnt, the condition of the patient was so critical that the wait for a donor heart was fatal. Jarvik began to think about possible designs for artificial hearts that could help people with grave heart disorders to survive till a suitable donor heart could be found. In 1979 he was granted a patent for the first implantable artificial heart, which he called *Jarvik-7*. Made of Dacron polyester, plastic, and aluminium, *Jarvik-7*, which was basically a pump, had an internal power system that regulated it through a system of compressed air hoses that entered the heart through the chest. The air hoses were connected to the chambers. The heart's power system drove the pumps, which pumped blood through the patients' body. The first recipient of Jarvik's artificial heart was a 62-year-old retired dentist named Barney Clark, who

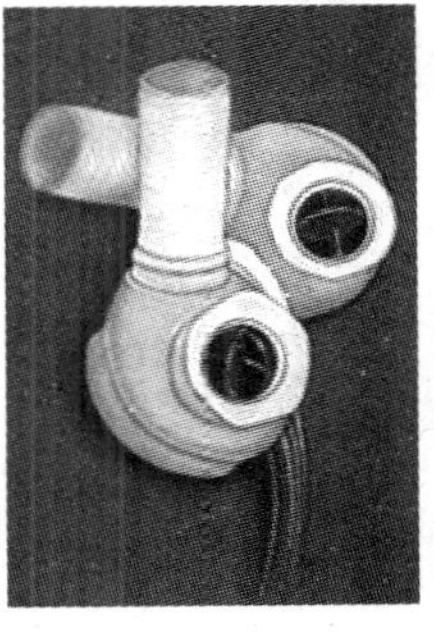

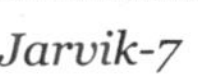

Jarvik-7

lived for 112 days with the new heart. (☞ Nov. 13, Nov. 25, Dec. 2)

1963
Herbert Spencer Gasser, American physiologist, Nobel laureate (Physiology or Medicine, 1944), who developed methods to study nerve impulses using the oscillograph, died in New York, USA. (☞ Jul. 5)

1981
Odd Hassel, Norwegian physical chemist, Nobel laureate (Chemistry, 1969), who discovered the existence of two forms of cyclohexane and established conformational analysis, died in Oslo, Norway. (☞ May 17)

1998
Three nuclear tests were carried out by India at Pokhran in the deserts of Rajasthan. These included a 45-kiloton thermonuclear device (hydrogen bomb), a 15-kiloton fission device, and a 0.2-kiloton device. (☞ May 13, May 18)

1998
The first test flight of India's indigenously designed all-composite trainer aircraft the *Hansa-3* was conducted successfully

at Bangalore, Karnataka. Designed by the National Aerospace Laboratory of the Council of Scientific and Industrial Research, the two-seater trainer aircraft has been certified for day and night flying.

May 12

1803
Justus von Liebig, German chemist, Fellow of the Royal Society, was born in Darmstadt, Germany. His teachers described Liebig as a 'hopelessly useless' student, and his academic performance was poor. But he learned much through assisting his father in the family chemical business, and developed a passion for conducting his own experiments. Liebig studied at the University of Bonn, and received his doctorate from the University of Erlangen, in 1822. After he finished his university studies in Germany, Liebig went to Paris, France, to work with the French chemist Joseph Louis Gay-Lussac, who was in the forefront of chemical research at that time. In 1824 at the age of 21, Liebig was appointed professor at the University of Giessen, where he immediately set about creating a model chemistry laboratory as we know it today. He also invented various kinds of chemical apparatus, including the 'Liebig condenser' used in laboratory distillation. After 25 years at Giessen he moved on to the University of Munich as a professor. Liebig succeeded in institutionalising the independent teaching of chemistry, which in German universities had been taught till then as an adjunct to pharmacy for apothecaries and phy-sicians. One of his most famous accomplishments was the invention of a nitrogen-based fertiliser. (☞ Apr. 18)

1895
William Francis Giauque, Canadian-born American physical chemist, Nobel laureate, was born in Niagara Falls, Ontario, Canada.

Giauque attended public grammar schools in Michigan, USA, but later studied in the Niagara Falls Collegiate Institute and took up employment in a chemical plant before

joining the University of California, Berkeley, from where he received his BS degree in 1920 and his PhD in 1922. He was appointed instructor of chemistry in 1922 in the same department in 1922 and became professor of chemistry in 1934. Giauque carried out exhaustive and meticulous thermodynamic studies, which utilised the third law of thermodynamics while also developing a large body of evidence for its validity. In 1927, while investigating the effect of magnetic fields on the entropies of paramagnetic substances, Giauque proposed a new method of achieving extremely low temperatures using a process called adiabatic demagnetisation. By 1933 he had a working apparatus that obtained a temperature within one-tenth of a degree of absolute zero (−273.15° C). In the course of his low-temperature studies of oxygen, Giauque, together with the American chemist Herrick L. Johnston, discovered the oxygen isotopes of mass 17 and 18. Giauque was awarded the 1949 Nobel Prize for Chemistry for his contributions in the field of chemical thermodynamics, particularly concerning the behaviour of substances at extremely low temperatures. (☞ Mar. 28)

1910

Dorothy Crowfoot Hodgkin, English chemist, Nobel laureate, was born in Cairo, Egypt. Hodgkin was interested in chemistry and in crystals from a young age and began her crystallographic studies while at the Somerville College, Oxford, UK, during 1928–32. She and a co-worker took the first X-ray diffraction photograph of the protein pepsin, when she was studying

for her doctorate at the University of Cambridge in the mid-1930s. Sometime later she studied sterols and other biologically important molecules, including insulin, using the same technique. From 1942 onwards Hodgkin worked on the structural analysis of penicillin, which she completed in 1946. In 1948 she took the first X-ray diffraction photograph of vitamin B_{12}, one of the most complex non-protein compounds, and by 1956 determined the complete three-dimensional atomic arrangement in the vitamin B_{12} molecule. Apart from these structure determinations and the scientific insight they provided, Hodgkin also developed methods that made such structure determinations possible. Hodgkin was awarded the 1964 Nobel Prize for Chemistry for her determinations of the structures of important biochemical substances by X-ray techniques. (☞ Jul. 29)

1931

The frozen body of Alfred Wegener, German scientist, meteorologist, geophysicist, who formulated the first complete statement of the continental drift hypothesis, was found by a search party in Greenland, where he had been on his fourth expedition since 1906 to study the ice cap and its climate. He was last seen alive by his colleagues on his 50th birthday in 1930. (☞ Nov. 1)

May 13

1857
Ronald Ross, British bacteriologist, Nobel laureate, was born in Almora, United Provinces (now Uttaranchal), India. Ross graduated from the St. Bartholomew's Hospital in London in 1875 and entered the Indian Medical Service in 1881. He began the study of malaria in 1892 and in 1894 decided to investigate the hypothesis of the French physician Alphonse Laveran and British parasitologist Patrick Manson that mosquitoes are the carrier of the malarial parasite. After initial failure, Ross discovered the presence of the malarial parasite within the *Anopheles* mosquito in 1897. Using birds that were sick with malaria, he was soon able to ascertain the entire life cycle of the malarial parasite, including its presence in the mosquito's salivary glands and its transmission from infected birds to healthy ones by the bite of a mosquito, thus establishing the hypothesis of Laveran and Manson. In 1899 he returned to England and joined the Liverpool School of Tropical Medicine, but was immediately sent to West Africa to continue his investigations, and there he found the species of mosquitoes, which transmit the deadly African fever. In 1901 Ross was elected a Fellow of the Royal College of Surgeons of England and a Fellow of the Royal Society, of which he became vice-president in 1911. During

his active career, Ross carried out surveys and initiated schemes in many places, including West Africa, the Suez Canal zone, Greece, Mauritius, and Cyprus. He also initiated organisations for the prevention of malaria within the plantations of India and Ceylon (now Sri Lanka). He made many contributions to the epidemiology of malaria and to methods of its survey and assessment, but perhaps his greatest was the development of mathematical models for the study of its epidemiology. Ross was awarded the 1902 Nobel Prize for Physiology or Medicine for showing how malaria enters the organism thereby laying the foundation for successful research on this disease and on methods of combating it. (☞ Aug. 20, Sep. 16)

1913
The *Grand*, the world's first four-engine airplane, took off on its maiden flight in St. Petersburg, Russia. Designed by Igor Sikorsky, the Russian pioneer in aircraft design, it was the largest aircraft

built till then. Until that historic flight, aircraft experts throughout the world had repeatedly stated that a large craft like the *Grand* was too big and much too heavy to rise from the ground. (☞ May 25, Oct. 26)

1940
Russian inventor Igor Sikorsky's helicopter *VS-300* made its first free flight. (☞ May 25, Sep. 14, Oct. 26)

1998
Two tests of sub-kiloton nuclear devices were carried out by India at Pokhran in the deserts of Rajasthan. (☞ May 11, May 18)

May 14

1686

Gabriel Fahrenheit, German physicist, Fellow of the Royal Society, who invented the Fahrenheit scale mercury thermometer, was born in Gdansk, Poland. Fahrenheit settled in Amsterdam in 1701 and devoted himself to the study of physics and the manufacture of precision meteorological instruments. He travelled around Europe, meeting scientists and other instrument makers before returning to Amsterdam to begin trading in 1717. He published his method for making thermometers in the *Philosophical Transactions* in 1724 and was admitted to the Royal Society the same year. Fahrenheit built the first mercury-in-glass thermometer in 1714, which was more accurate than the alcohol-in-glass thermometers used earlier, because mercury possesses a more constant rate of expansion than alcohol and could be used over a wider range of temperatures. He originally took as the zero of his scale the temperature of an equal ice–salt mixture, 30° for the freezing point of water and 90° for normal body temperature. Later, he adjusted the freezing point of water to 32° and the boiling point to 212°, the interval between the two being divided into 180 parts. Using his thermometer, Fahrenheit was able to determine the boiling points of liquids and show that liquids other than water also had fixed boiling points that varied with atmospheric pressure. He also researched the phenomenon of supercooling of water – the cooling of water below freezing without solidification – and discovered that the boiling point of liquids varied with atmospheric pressure. (☞ Sep. 16)

1796

The first vaccination against smallpox was administered to an 8-year-old boy named James Phipps by the English physician Edward Jenner. Jenner carried out the inoculation with pus from the sores of a dairymaid who had a mild case of cowpox. He subsequently tested the boy's resistance to smallpox, which came out to be good. By 1798, he had inoculated several children with cowpox, and subsequently with small pox. Delighted with his success, Jenner published his observations in a little book entitled *An Inquiry into Cause and Effects of the Variolae Vaccinae*. (☞ Jan. 26, May 17, Jul. 1)

1872

Mikhail Semyonovich Tsvet, Russian botanist, was born in Asti, Italy. His mother was Italian, and his father a Russian official. His mother died soon after his birth and he grew up in Geneva, Switzerland. Tsvet received his BS degree from the department of physics and mathematics at the University of Geneva in 1893, but later decided to pursue botany. He received his PhD degree in 1896 for his work on cell physiology. Tsvet was noted for his research on plant pigments, especially chlorophyll, of which he discovered several new forms, and the carotenoids, a term he first coined. He invented chromatography in 1901 during his research on plant pigments. He used liquid-adsorption column chromatography with calcium carbonate as adsorbent and petrol ether/ethanol mixtures as eluent to separate chlorophylls and carotenoids extracted from leaves. The principles underlying Tsvet's chromatography could be applied in many different ways, giving rise to the different varieties of chromatography. In 1902 Tsvet became a laboratory assistant at the University of Warsaw in Poland, and in 1908 he began teaching botany and microbiology at the Warsaw Technical University. In 1917 he became professor and director of the botanical garden at the University of Yuryev (later Tartu) in Estonia, which was moved to Voronezh, a large city in the south of Central Russia, in 1918 when German troops occupied Yuryev. (☞ Jun. 26)

1973

The first American space station *Skylab* was launched from the Kennedy Space Center, Florida, USA. Almost immediately, technical problems developed due to vibrations during lift-off. A critical meteoroid shield ripped off taking one of the craft's two solar panels with it. Because of the loss of the meteoroid shield, the temperatures of the space station rose to 52°C. The launch of the first crew to *Skylab* (Skylab-2) was postponed while NASA engineers, in an intensive 10-day period, developed procedures and trained the crew to make the workshop habitable. (☞ Feb. 8, May 25, Jun. 22, Jun. 28, Jul. 11, Sep. 25, Nov. 16)

1995

Christian Boehmer Anfinsen, American biochemist, Nobel laureate (Chemistry, 1972), who discovered the connection between the amino acid sequence and the biologically active conformation of the enzyme ribonuclease, died in Randallstown, Maryland, USA. (☞ Mar. 26)

May 15

1836

Francis Baily, British astro-nomer, first observed 'Baily's beads' during a total solar eclipse. His vivid description aroused new interest in the study of eclipses. Baily's beads are the bright points of light that appear around the edge of the Moon's disc during a total solar eclipse. The beads are created by sunlight passing through the lunar valleys as the Moon covers the disc of the Sun. Just before totality sets in, the last bead is the brightest, resembling a diamond on a brilliant ring.

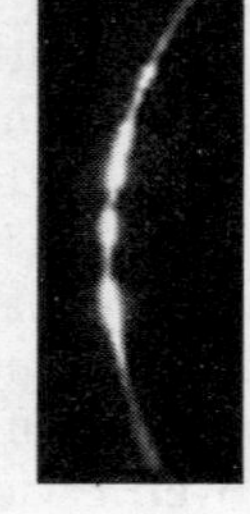

1859

Pierre Curie, French physical chemist, Nobel laureate, was born in Paris, France. Curie received his early education at home before entering the Faculty of Sciences at the Sorbonne in Paris. In 1878 he was appointed as a laboratory assistant at the Sorbonne where he carried out important studies on crystals and (with his brother Jacques) discovered piezoelectricity. Curie studied ferromagnetism,

paramagnetism, and diamagnetism for his doctoral thesis, and discovered the effect of temperature on paramagnetism, now known as Curie's law, which states that magnetic susceptibility is inversely proportional to absolute temperature. After getting his PhD degree in 1895, he was appointed professor of physics at the University of Paris. Curie's studies of radioactive substances were made together with his wife, Marie whom he married in 1895. They announced the discovery of the radioactive elements polonium and radium by fractionation of pitchblende in 1898, and later elucidated the properties of radium and its transformation products. Their work formed the basis for much of the subsequent research in nuclear physics and chemistry. The Curies were awarded the 1903 Nobel Prize for Physics, jointly with the French physicist Antoine-Henri Becquerel, in recognition of the extraordinary services they have rendered by their joint researches on the radiation phenomena discovered by Becquerel. (☞ Apr. 19, Nov. 7, Dec. 15)

1869

Celluloid, the first synthetic plastic material, developed by the American inventor John Wesley Hyatt from a homogeneous colloidal dispersion of cellulose nitrate and camphor, was patented in USA. A tough material, with great tensile strength, resistance to water, oils, and dilute acids, and capable of being produced at a low cost in a variety of colours, celluloid found a great variety of applications in combs, films, toys, and many other mass-produced consumer goods; but due its high flammability celluloid has been replaced by newer synthetic polymers for many of these uses.

1941

The first flight of a jet-propelled aircraft, the *Gloster-Whittle E.28/39*, took place. The plane, piloted by Flight Lieutenant P E G. Sayer, took off from RAF Cranwell in England and landed after a 17-minute flight. Its jet engines were designed by Frank Whittle, 'the father of the jet engine'. (☞ Jun. 1)

1951

Frank Wilczek, American physicist, Nobel laureate, was born in New York City, New York, USA. Wilczek was educated in public schools in New York, and received his bachelor's degree in mathematics from the University of Chicago in 1970. He received his master's degree in mathematics from the Princeton University in 1972 and a PhD in physics from the same university in 1974. In 1973, as a graduate student working with the American physicist David J. Gross at the Princeton University, Wilczek set out to work on the problem of how quarks are bound within the nucleus. Physicists had known that quarks, in pairs and triplets, were bound inside the atomic nucleus by the action of the strong nuclear force, but no quark had ever been released from the nucleus and observed on its own. With a series of calculations, Gross and Wilczek developed a novel explanation that seemed to run counter to logic. They theorised that the binding force between quarks in the atomic nucleus actually decreases as the quarks draw together and increases as the particles move farther apart. The concept can be likened to a rubber band, which exerts increasing amounts of energy, as the ends are pulled farther apart. At greater distances, the strong force is so powerful that no quark can break free, and that is why they have never been observed in isolation. The two termed this phenomenon 'asymptotic freedom' the

discovery of which was important for the development of quantum chromodynamics. Wilczek was awarded the 2004 Nobel Prize for Physics, jointly with Gross and fellow American physicist Hughs David Politzer for the discovery of asymptotic freedom in the theory of the strong interaction. (☞ Feb. 19, Aug. 31)

1953
Stanley L. Miller, an University of Chicago graduate student published his paper on the synthesis of amino acids under conditions that simulated primitive Earth's atmosphere in *Science*. Miller had passed electric current through a chamber containing a mixture of methane, ammonia, hydrogen and water (which was believed at the time to be the atmospheric composition of early Earth). Instead of producing a random mixture of organic molecules, the experiment yielded organic compounds including amino acids, the building blocks of life. Miller's experiment changed the approach of scientific investigation into the origin of life. (☞ Apr. 29)

1958
The Soviet unmanned spacecraft *Sputnik-3* was launched from the Baikanur cosmodrome in Kazakhstan. *Sputnik-3* weighed more than 1,327 kg and carried the first space laboratory, a wide array of instruments that could transmit information about the environment outside the satellite. Solar cells were used for the first time by *Sputnik-3* to power its instruments and transmitters. It was designed to perform experiments on the Earth's magnetic field, radiation belt, and ionosphere and spent two years in orbit measuring solar radiation, cosmic rays, and magnetic fields. The data was used in conjunction with the International Geophysical Year activities.

May 16

1845
Ilya Ilich Mechnikov, Russian zoologist and microbiologist, Nobel Laureate, was born in a village near Kharkoff in Russia (now Kharkiv, Ukraine). Mechnikov went to school at Kharkoff and from an early age was interested in natural history, on which he used to give lectures to his small brothers and other children. He graduated from the University of Kharkoff in 1864 and received his doctorate from the University of St. Petersburg in 1868. In 1870 he was appointed as a titular professor of zoology and comparative anatomy at the University of Odessa. In 1882, he set up a private laboratory at Messina, Italy, to work on comparative embryology, and it was here that he discovered the phenomenon of phagocytosis with which his name will always be associated. While working with the larvae of starfishes, he observed that certain cells unconnected with digestion surrounded and engulfed carmine dye particles and splinters that he had

introduced into the bodies of the larvae. Knowing that when inflammation occurs in animals that have a blood vascular system, leucocytes escape from their blood vessels, it occurred to him that these leucocytes might take up and digest bacteria that get into the body. He called these cells phagocytes (from Greek words meaning 'devouring cells') and named the process 'phagocytosis'. In 1886 Mechnikov was appointed director of the Bacteriological Institute in Odessa, which he left in 1888 to join the Pasteur Institute in Paris, France. Mechnikov helped in many important discoveries about the immune response. Perhaps his most notable achievement was his recognition that the phagocyte is the first line of defense against acute infection in most animals, including humans. This work formed the basis of Mechnikov's cellular theory of immunity, which he published in 1892. Mechnikov was awarded the 1908 Nobel Prize for Physiology or Medicine, jointly with the German medical scientist Paul Ehrlich, in recognition of their work on immunity. (☞ Mar. 14, Jul. 16)

1888

Emile Berliner, German immigrant working in Washington, DC., gave the first demonstration of flat disc recording and reproduction before the Franklin Institute in Philadelphia, USA. The earlier techniques of recording and reproducing sound by Thomas Edison and Alexander Graham Bell all used a cylindrical recording surface – tin foil or wax cylinder – for recording sound that could not be multiplied or mass-produced. In Berliner's flat disc record the sounds were 'read' by a needle, which transmitted the pattern of vibrations to a diaphragm that reproduced the original sounds. It could be used over and over again, and from it many copies of the original recording could be made. Berliner patented the gramophone and founded 'The Gramophone Company' to mass manufacture his sound discs (records) and the gramophone that played them. (☞ Sep. 26)

1947

Frederick Gowland Hopkins, British biochemist, Nobel laureate (Physiology or Medicine, 1929), who co-discovered growth-stimulating vitamins, died in Cambridge, England. (☞ Jun. 20)

1950

J. Georg Bednorz, German physicist, Nobel laureate, was born in Neuenkirchen, Germany. Bednorz graduated from the University of Munster in 1976 and joined the Swiss Federal Institute of Technology at Zurich, Switzerland, where he worked under the Swiss physicist Karl Alex Müller on the crystal growth of perovskite-type solid solutions and investigated them with respect to structural, dielectric and ferroelectric properties, and obtained his PhD in 1982. The same year he joined the IBM Zürich Research Laboratory, and along with Müller, began systematically testing newly developed ceramic materials known as oxides in the hope that such substances could act as superconductors. In 1986 they startled the world by reporting superconductivity in a layered, ceramic material at a then record-high temperature of 33 Kelvin (that is, 33 degrees above absolute zero, or roughly −243 degrees Celsius), much higher than temperatures

at which superconductivity was known to exist till then. Their discovery started a flurry of research worldwide into related materials that yielded dozens of new superconductors, eventually reaching a transition temperature as high as 135 Kelvin (that is, −141 degrees Celsius). Bednorz was awarded the 1987 Nobel Prize for Physics, jointly with Müller, for their important breakthrough in the discovery of superconductivity in ceramic materials. (☞ Apr. 20)

1960
The world's first laser, a ruby laser, was demonstrated by the American physicist Theodore Maiman. He used a ruby cylinder with the ends polished flat, parallel and silvered. Energy was fed in using a flash lamp. He got an output of monochromatic coherent beam of radiation. The laser has since been applied in a very wide range of uses, including eye surgery, dentistry, range-finding, manufacturing, business, entertainment, even for measuring the distance between the Earth and the Moon. (☞ Jul. 11)

1969
Soviet unmanned spacecraft *Venera-5* arrived at Venus and plunged into its thick atmosphere. The spacecraft returned data down to within 26 kilometres of surface and was then lost – crushed by the atmospheric pressure on Venus. (☞ Jan. 5)

1992
The American space shuttle *Endeavour* carrying astronauts Daniel C. Brandenstein, Kevin P. Chilton, Pierre J. Thuot, Kathryn C. Thornton, Richard J. Hieb, Thomas D. Akers, and Bruce E. Melnick landed at the Edwards Air Force Base, California, USA, after its maiden flight. (☞ May 7)

May 17

1730
Dark and light cloud bands of Jupiter were first observed by the Italian astronomer Niccolò Zucchi.

1749
Edward Jenner, English physician and inventor of vaccination, was born in the small village of Berkeley in Gloucestershire, England. From an early age Jenner was a keen observer of nature and after nine years as a surgeon's apprentice he went to the St George's Hospital, London to study anatomy and surgery. After completing his studies, he returned to Berkeley to set up a medical practice. Jenner worked in a rural community and most of his patients were farmers or worked on farms with cattle. In the 18th century smallpox was a very common disease and was a major cause of death. In 1788 an epidemic of smallpox hit Gloucestershire and during this outbreak Jenner observed that those of his patients who worked with cattle and had come in contact with the much milder disease called cowpox never came down with smallpox. The thought occurred to him that the germs from cowpox might be protecting those who contracted it against the more dangerous smallpox. He believed that if he could inject someone with the pus taken from a cowpox pustule it would make the body able to defend itself against the dangerous smallpox germs, which he would inject later. He tried out his theory with an 8-year-old boy named James Phipps and came out successful.

Jenner's discovery in 1796 that inoculation with cowpox gave immunity to smallpox was an immense medical breakthrough and has saved countless lives. In 1798 the results were finally published and Jenner coined the word vaccine from the Latin *vaccinia* for cowpox, and called the process vaccination. (☞ Jan. 26, May 14, Jul. 1)

1865
The first International Telegraph Convention was signed in Paris by the 20 founding members, and the International Telegraph Union (ITU, which was renamed the International Telecommunication Union in 1934) was established to decide on common rules to standardise equipment to facilitate international interconnection, adopt uniform operating instructions and enable subsequent amendments to the initial agreement. The day is observed every year as World Telecommunication Day.

1897
Odd Hassel, Norwegian physical chemist, Nobel laureate, was born in Kristiania (now Oslo), Norway. Hassel studied at the University of Oslo and received his doctorate from the University of Berlin, Germany, in 1924. He joined the faculty of the University of Oslo in 1925, and from 1934 to 1964 was a professor of physical chemistry and director of the physical chemistry department. Hassel's research during the first years of his teaching at the University of Oslo dealt with inorganic chemistry, but from 1930 onwards he concentrated on problems connected with molecular structure, particularly the structure of cyclohexane, its derivatives and other substances containing six-membered rings related to that of cyclohexane. Hassel introduced new methods such as the measurements of electric dipole moments and electron diffraction by vapours for the determination of the three-dimensional geometric structure of molecules, and by 1943 he was able to show the transition between two forms of cyclohexane, thus establishing conformational analysis (the study of the three-dimensional geometric structure of molecules). During the early 1950s Hassel opened a new field of structure investigation, namely that of charge-transfer compounds such as compounds formed by organic electron-donor molecules like ethers and amines and electron acceptors as halogen molecules or organic halides. After some years work he was able to set up rules for the geometry of this kind of addition compounds. Hassel was awarded the 1969 Nobel Prize for Chemistry, jointly with the British chemist Derek H R. Barton, for their contributions to the development of the concept of conformation and its application in chemistry. (☞ May 11, Sep. 8)

1969
Soviet spacecraft *Venera-6* arrived at Venus and plunged into its thick atmosphere. The spacecraft returned data down to within 11 kilometres of surface and was then lost – crushed by the atmospheric pressure on Venus. (☞ Jan. 10)

May 18

1850
Oliver Heaviside, English physicist, who predicted the existence of the ionosphere, was born in London, England. Heaviside suffered from scarlet fever during his

youth, which had a lasting impact on him, leaving him partly deaf. Although he was good at his studies, he left school at 16 and began learning Morse code and electromagnetism. He became a telegraph operator at the Great Northern Telegraph Company in Denmark. In 1872, while working as a chief operator in Newcastle upon Tyne in England, he started an analysis of electricity. He left this position in 1874 and carried on research to develop his 'transmission line theory'. In 1891 the Royal Society of England recognised Heaviside's contributions to the mathematical description of the electromagnetic phenomena by electing him a Fellow of the Royal Society. Heaviside proposed the existence of the reflecting layer of the ionosphere, which bears his name, in 1902 and proposed that radio signals are transmitted around the Earth's curvature due to this layer. The existence of the ionosphere was confirmed in 1923. (☞ Feb. 3)

1872

Bertrand Russell, English logician and philosopher, Nobel laureate, was born in Trelleck, Monmouthshire, England. Russell studied at the Trinity College, Cambridge, and obtained a first-class honours degree in mathematics and philosophy. From his work on the logical foundations of mathematics, Russell derived an enormous confidence in the possibility of resolving philosophical problems by offering careful analyses of the logical structure (rather than the grammatical form) of what we say. He sought to bring philosophy into closer alliance with science. As an empiricist, Russell assumed that all human knowledge must begin with sensory experience. Sense data provide the primitive content of our experience, and for Russell these sense data are not merely mental events, but rather the physical effects caused in us by external objects. Although each occurs immediately within the private space of an individual perceiver, he argued, classes of similar sense data in various perceivers constitute a public space from which even unperceived sensibilia may be said to occur. Thus, the contents of sensory experience are both public and objective. From this beginning, according to Russell, all else follows by logical analysis. Russell is also known for his social and political campaigns, including his advocacy of both pacifism and nuclear disarmament and as a prolific writer. Between 1910 and 1913, he published, with the British mathematician, logician and philosopher Alfred North Whitehead, the landmark work *Principia Mathematica*. His other major works include *The Bomb and Civilization*, *New Hopes for a Changing World*, *Why I Am Not a Christian*, *A Free Man's Worship*, *History of Western Philosophy*, and his *Autobiography*. Russell was awarded the Nobel Prize for literature in 1950. (☞ Feb. 2)

1901

Vincent du Vigneaud, American biochemist, Nobel laureate, was born in Chicago, Illinois, USA. Du Vigneaud studied at the University of Illinois from where he got his BSc in 1923 and MSc in 1924. During 1924–1925 he served on the staff of the Graduate School of Medicine of the Pennsylvania University. In 1927 he submitted a thesis to the School of Medicine of the Rochester University, which earned him the PhD. Du Vigneaud's researches were mainly centred on biochemically important sulphur-containing compounds. He worked with the hormone insulin and two hormones of the posterior pituitary gland. These hormones are built up from amino acids in the same way as proteins, but with a far lower molecular weight. Du Vigneaud and his staff helped identify the chemical structure of insulin in the late 1930s, and in the early 1940s they established the structure of the sulphur-bearing vitamin biotin. Du Vigneaud isolated and synthesised two pituitary hormones: vasopressin, which acts on the muscles of the blood vessels to cause elevation of blood pressure and oxytocin, the principal agent causing contraction of the uterus and secretion of milk. He also studied intermediary metabolism, amino acids and peptides, transmethylation and metabolism of one-carbon compounds, and trans-sulphuration. Du Vigneaud was awarded the 1955 Nobel Prize for Chemistry for his work on biochemically important sulphur compounds, especially for the first synthesis of a polypeptide hormone. (☞ Dec. 11)

1922
Alphonse Laveran, French physician, pathologist, and parasitologist, Nobel laureate (Physiology or Medicine, 1907), who discovered that a parasite causes human malaria, died in Paris, France. (☞ Jun. 18)

1955
A patent for the first atomic pile was finally issued – nearly 11 years after it had been filed. Work on the initial patent application had begun six months before the world's first nuclear reactor was completed. Filed with the US Patent Office in December 1944, the patent application listed the Italian-born American physicist Enrico Fermi and the Hungarian-born American physicist Leo Szilard as co-inventors and described the method by which a self-sustaining nuclear chain reaction had been achieved. Fermi and his team of scientists at the University of Chicago's Metallurgical Laboratory achieved the world's first controlled, self-sustaining nuclear chain reaction in December 1942. By the time the patent was issued, Fermi had been dead for six months. (☞ Feb. 11, Sep. 29, Dec. 2)

1966
Panchanan Maheshwari, Indian botanist, Fellow of the Royal Society, who did pioneering work on the embryology of seed-bearing plants, died in New Delhi, India. (☞ Nov. 9)

1969
American manned lunar mission *Apollo-10*, carrying astronauts Thomas P. Stafford, John W. Young, and Eugene A. Cernan, was launched by the Saturn-V rocket from the Kennedy Space Center, Florida, USA, towards the Moon. This mission was meant to be a dress rehearsal for the lunar landing, without actually landing on the Moon. The mission was the second to orbit the Moon and the first to travel to the Moon with the entire Apollo spacecraft configuration. During the mission, Stafford and Cernan descended inside the lunar module *Snoopy* to within 14 kilometres of the lunar surface. (☞ May 26)

1974
India carried out its first underground nuclear test by exploding a plutonium device at Pokhran in the deserts of Rajasthan. (☞ May 11)

May 19

1910
The Earth passed through the tail of Comet Halley. It was the most intimate contact between the Earth and any comet in recorded history. A few years earlier, astronomers had found traces of the poisonous gas cyanogen in a comet. Although astronomers explained that the gas molecules within the tail were so tenuous that absolutely no ill effects would be noticed, people were afraid that if the Earth passed through a comet's tail everyone would die. Some opportunists even sold 'comet pills' to the panicked portion of the public to counter the effects of the cyanogen gas. But after the Earth had passed through Halley's tail, everyone was still alive – with or without taking the pills! (☞ Feb. 9)

1914
Max Ferdinand Perutz, Austrian-born British biochemist, Nobel laureate, was born in Vienna, Austria. After attending school at the Theresianum, Perutz entered the Vienna University in 1932, where he was attracted to organic chemistry. In 1936, he became a research student at the Cavendish Laboratory in Cambridge, England, to do his PhD under the British molecular physicist John Desmond Bernal. At Cambridge Perutz took the first X-ray diffraction pictures of haemoglobin crystals. In 1947, along with the British biochemist John C. Kendrew, Perutz founded the Medical Research Council Unit for Molecular Biology at Cambridge. There the two scientists continued their investigation of haemoproteins. While Kendrew concentrated on determination of the molecular structure of myoglobin – the oxygen-carrying haeme protein found in muscles, Perutz concentrated on the haemoglobin molecule itself. By 1959 Perutz had shown that the haemoglobin molecule is composed of four separate polypeptide chains that form a tetrameric structure, with four haeme groups near the molecule's surface. Perutz subsequently showed that in oxygenated haemoglobin the four chains are rearranged, a discovery that led to the full determination of the molecular mechanism of oxygen transport and release by haemoglobin. Perutz was also interested in studying glaciers, making measurements, which were the first to show different rates of flow in different parts of the same glacier. Perutz and Kendrew were jointly awarded the 1962 Nobel Prize for Chemistry for their studies of the structures of globular proteins. (☞ Feb. 6, Mar. 24)

1961
Soviet unmanned spacecraft *Venera-1*, the first spacecraft to be sent to Venus, passed within 100,000 kilometres of Venus and went into orbit around the Sun. *Venera-1* provided the first verification that solar wind was uniformly present in deep space. (☞ Feb 12)

1971
Soviet unmanned spacecraft *Mars-2* carrying a descent module with a tiny rover was launched towards Mars from the Baikanur cosmodrome in Kazakhstan. It arrived at Mars in November and released the descent module to carry out studies from the surface. (☞ Nov. 27)

May 20

1860
Eduard Buchner, German biochemist, Nobel laureate, was born in Munich, Germany. Buchner matriculated from the Grammar School in Munich and after a short period of study at the Munich Polytechnic, he started work in a preserve and canning factory, with which he later moved to Mombach on Mainz. He took his doctor's degree at the University of Munich in 1888. The following year he was appointed as an assistant lecturer in the laboratory of the German chemist Adolf von Baeyer, and in 1891 he was appointed as a lecturer at the University. With a financial grant from von Baeyer, Buchner established a small laboratory for studying the chemistry of fermentation, especially on the rupture of yeast cells. In 1896, while continuing his studies on fermentation at the University of Tübingen, Buchner discovered that alcoholic fermentation of sugars is caused by the action of different enzymes contained in yeast and not by the yeast cells themselves. In 1903, he showed that an enzyme, zymase, can be extracted from yeast cells and can be used to ferment sugar without adding yeast. Buchner was awarded the 1907 Nobel Prize for Chemistry for his biochemical researches and his discovery of cell-free fermentation. (☞ Aug. 13)

1875
The International Bureau of Weights and Measures (Bureau International des Poids et Mesures, or BIPM) was established in Sèvres, France, by the International Metric Convention. The Bureau is the depository for the international prototypes of the metre, the kilogram and secondary standards. Its task is to ensure worldwide uniformity of measurements and their traceability to the International System of Units (SI). It does this with the authority of the International Metric Convention, a diplomatic treaty between 51 nations. It operates through a series of consultative committees, whose members are the national metrology laboratories of the Member States of the Convention, and through its own laboratory work. Every year the day is observed as World Metrology Day.

1910
Comet Halley passed by the Earth at a distance of 21 million kilometres. It returned again in 1986. (☞ Feb. 9)

1918
Edward B. Lewis, American developmental geneticist, Nobel laureate, was born in Wilkes-Barre, Pennsylvania, USA. Lewis studied biostatistics at the University of Minnesota, and received his BA degree in 1939. He did his PhD from the California Institute of Technology (Caltech) in 1942. After serving as a captain in the United States Army Air Force as a meteorologist and oceanographer in the Pacific Theatre during World War II, he joined the Caltech faculty in 1946 as an instructor. In 1956 he was appointed as professor of biology. Lewis' research was mainly based on studies of the fruit fly, *Drosophila melanogaster*. By crossbreeding thousands of flies he was able to show that genes are generally arranged on the chromosome in the same order as their corresponding body segments, following what is known as the 'colinearity principle.' Lewis' work on the fruit

fly helped explain mechanisms of general biological development and the causes of congenital deformities in humans and other higher organisms. Lewis was a Rockefeller Foundation Fellow at the Cambridge University, England during 1947–48, and Guest Professor at the Institute of Genetics, University of Copenhagen, Denmark during 1975–76. He was elected to the National Academy of Sciences, USA, in 1968. Lewis was awarded the 1995 Nobel Prize for Physiology or Medicine, jointly with the German developmental geneticist Christiane Nüsslein-Volhard and fellow American developmental biologist Eric F. Wieschaus, for their discoveries concerning the genetic control of early embryonic development. (☞ Jun. 8, Jul. 21, Oct. 20)

1947
Philipp Lenard, German physicist, Nobel laureate (Physics, 1905), who discovered the properties of cathode rays, died in Messelhausen, Germany. (☞ Jun. 7)

1956
The first hydrogen bomb to be dropped from an airplane – a B-52 bomber of the US Air Force – exploded over Namu Atoll at the northwest edge of the Bikini Atoll in the Pacific Ocean. The 3.8 megaton bomb produced a fireball six-and-a-half kilometres in diameter. Designated as 'Cherokee', it was the first test of a deliverable hydrogen bomb. (☞ Nov. 1)

1978
American space probe *Pioneer-Venus 1* was launched by NASA. The probe carried a radar altimeter for making the first detailed mapping of the surface of Venus. It went into orbit around Venus in December 1978; its highly elliptical orbit brought it to within 160 kilometres of the planet's surface allowing it to do radar mapping, cloud studies and take magnetometer readings. (☞ Dec. 4)

1990
The Hubble Space Telescope sent its first photograph from space, an image of a double star 1,260 light years away. (☞ Apr. 25)

1992
SROSS-C, the third satellite of India's Stretched Rohini Satellite Series, was launched by ISRO's augmented satellite launch vehicle (ASLV) from the Satish Dhawan Space Centre at Sriharikota, Andhra Pradesh. The first two satellites of the series were lost due to launch failure. *SROSS-C* carried a gamma-ray burst experiment that monitored celestial gamma-ray bursts, and an experiment that measured temperature, density, and characteristics of electrons in the ionosphere.

2002
Stephen Jay Gould, American palaeontologist, evolutionary biologist, and science writer, died in New York, USA. (☞ Sep. 10)

May 21

1786
Carl Wilhelm Scheele, Swedish chemist who discovered oxygen and several other chemical elements, died in Köping, Sweden. (☞ Dec. 9)

1860
Willem Einthoven, Dutch physiologist, Nobel laureate, was born in Semarang, Java,

Dutch East Indies (now Indonesia). Einthoven graduated in medicine from the University of Utrecht and served as professor of physiology at the University of Leiden from 1886 until his death. In 1889 Einthoven saw the London physiologist Augustus Desiré Waller demonstrate his technique of recording the current of the heart, as deduced from the body surface, using a capillary electrometer. Einthoven repeated this experiment and, in 1895, while working on the construction of this instrument and developing the necessary photographic equipment, with the capillary electrometer he registered graphic reproductions of the variations of the electric charges induced by the contractions of the heart's musculature, as well as heart sounds of humans and animals. He defined the physical constants of the capillary electrometer and calculated the true curve, which he called the 'electrocardiogram.' In 1901, Einthoven invented a new galvanometer for producing electrocardiograms using a fine quartz string coated with silver, which was an improvement over the capillary electrometer. Einthoven recognised differences in the records or tracings obtained from different kinds of heart disease. From 1908 to 1913 he studied the patterns of records of normal heart activity in order to gain precision in recognising and interpreting deviations. Einthoven was awarded the 1924 Nobel Prize for Physiology or Medicine for his discovery of the mechanism of the electrocardiogram. (☞ Sep. 29)

1889

Gaston Planté, French physicist who produced the first electric storage battery, or accumulator, died in Paris, France. (☞ Apr. 22)

1921

Andrey Dmitriyevich Sakharov, Soviet nuclear physicist, Nobel laureate, was born in Moscow, Russia. Sakharov graduated from the Moscow State University in 1942 and continued his studies during World War II, taking his doctorate in physics at the P N. Lebedev Institute in 1947. From 1948 to 1956 he conducted research in controlled nuclear fusion, and with the Soviet physicist Igor Tamm, made a proposal that led to the construction of the Soviet hydrogen bomb. In 1953 he became a member of the Academy of Sciences of the Soviet Union. Sakharov also made a contribution of crucial importance to our quest to understand the evolution of our universe following its physical beginnings in the Big Bang – the disappearance of antimatter. In 1967, he provided the clue for understanding this by suggesting that there is a very tiny difference between the behaviour of matter and of antimatter, with several general postulates that had been made separately in other contexts. The most intriguing of these postulates is that the proton may in fact decay just like other forms of subnuclear matter, albeit very slowly – a hypothesis that has yet to be confirmed. By 1968, Sakharov had virtually abandoned scientific research, becoming instead a spokesman for international disarmament and nuclear weapons control. Sakharov was awarded the 1975 Nobel Peace Prize for his struggle for human rights, disarmament, and cooperation between all nations with peace as its final goal. (☞ Dec. 14)

1934
Bengt Ingemar Samuelsson, Swedish biochemist, Nobel laureate, was born in Halmstad, Sweden. Samuelsson graduated from the University of Lund, where the eminent Swedish biochemist Sune K. Bergström was one of his professors. He continued his studies at the Karolinska Institute, Stockholm, earning doctorates in biochemistry in 1960 and medicine in 1961. He remained at the Karolinska Institute as a member of the faculty, eventually succeeding Bergström as dean of the medical faculty. Samuelsson and Bergström together isolated and characterised a family of chemical messengers called prostaglandins, a family of natural compounds that influence blood pressure, body temperature, allergic reactions, and other physiological phenomena in mammals. In 1962 they became the first to determine the molecular structure of a prostaglandin. In 1964 they announced that prostaglandins are derived from arachidonic acid, an unsaturated fatty acid that is found in certain meats and vegetable oils. Samuelsson subsequently determined how arachidonic acid combines with oxygen to eventually form prostaglandins. In the 1970s he discovered several new prostaglandins, including thromboxane, which is involved in blood clotting and the contraction of blood vessels. Samuelsson was awarded the 1982 Nobel Prize for Physiology or Medicine, jointly with Bergström and the English biochemist John Robert Vane, for their discoveries concerning prostaglandins and related biologically active substances. (☞ Jan. 10, Mar. 29)

1936
Günter Blobel, German-born American cellular and molecular biologist, Nobel laureate, was born in Waltersdorf, Silesia, Germany (now Niegosławice, Poland). Blobel studied medicine, first in Frankfurt and then in Kiel, Munich and Tübingen, graduating in 1960 from the University of Tübingen. In 1967 he received a PhD in oncology from the University of Wisconsin in Madison, USA. In the same year he joined the Rockefeller University protein laboratory in New York City, and in 1976 became a professor at the university. Blobel's research mainly concerned the transport of large protein molecules across tightly sealed membranes surrounding the cell organelles. In 1971 he formulated a first version of the 'signal hypothesis'. He postulated that proteins secreted out of the cell contain an intrinsic signal that governs them to and across membranes. By 1980 Blobel had established the general principles underlying the mechanism by which proteins are targeted to specific organelles within a cell. In collaboration with other research groups, he conducted a series of experiments that showed that each protein carries an address code within its molecular structure that directs it to the proper locale inside the cell. The address code, consisting of a sequence of amino acids, specifies whether the protein will pass through the membrane of a specific organelle, become integrated into the membrane, or be exported out of the cell. Blobel also showed that proteins enter organelles through a pore-like channel that opens in the organelle's outer membrane when the correct protein arrives at the organelle. Blobel's work shed light on such hereditary diseases as cystic fibrosis and provided the basis for bioengineered drugs, including insulin. Blobel was awarded the 1999

Nobel Prize for Physiology or Medicine for the discovery that proteins have intrinsic signals that govern their transport and localisation in the cell.

1964

James Franck, German-born American physicist, Nobel laureate (Physics, 1925), who verified the quantised nature of energy transfer by excitation and ionisation of atoms by electron bombardment, died in Göttingen, Germany. (☞ Aug. 26)

May 22

1848

Louis Pasteur, French chemist and microbiologist, presented before the Paris Academy of Sciences a paper reporting a remarkable discovery he had just made – that certain chemical compounds were capable of splitting into a 'right' component and a 'left' component, one component being the mirror image of the other. He was studying the effect of tartaric acid on polarised light. Under a microscope, he discovered that there were two types of tartrate crystals, which rotated polarised light in opposite directions. His discovery became the basis of a new science – stereochemistry. (☞ Dec. 27)

1906

American inventors Wilbur Wright and Orville Wright were awarded US patent for their 'flying machine', the early version of the present day aeroplane. (☞ Apr. 16, Aug. 19, Dec. 17)

1912

Herbert Charles Brown, American chemist, Nobel laureate, was born in London, England. Brown grew up in Chicago and received his BS in 1936 and his PhD in 1938 from the University of Chicago. He remained at Chicago as an instructor until 1946, when he was appointed assistant professor at the Wayne University in Detroit. In 1947 he moved to the Purdue University, West Lafayette, Indiana, as associate professor, becoming Wetherill distinguished professor there in 1959 and emeritus in 1978. Brown's main contribution has been in developing new reagents containing boron. One of his reagents, sodium borohydride, has become the reagent of choice for reduction of carbonyl compounds. He has also modified the borohydrides into reagents for highly selective chemical transformations. In addition Brown has introduced an entirely new class of compounds, called the organoboranes, obtained by reacting diborane with olefins. Organoboranes have become the most versatile reagents ever created in organic chemistry. They can be used for reductions, rearrangements, additions and have opened up a range of new possibilities for linking carbon atoms to each other. Brown was awarded the 1979 Nobel Prize in Chemistry, jointly with German chemist Georg Wittig, for their development of the use of boron- and phosphorus-containing compounds, respectively, into important reagents in organic synthesis. (☞ Jun. 16)

1920

Thomas Gold, Austrian-born British astronomer who promulgated the 'steady-state theory' of the universe, was born in Vienna, Austria. Gold carried out research on cosmology and magnetic fields, and coined the term

'magnetosphere' for the Earth's magnetic fields. Soon after the discovery of pulsars in 1968, he correctly identified these objects as rapidly rotating neutron stars with strong magnetic fields. He is best known for developing, along with the British cosmologist Hermann Bondi and mathematician and astronomer Fred Hoyle, the steady-state theory of the universe. It assumes that we live in a world that has no beginning or end, in which matter is constantly being created. Gold, Hoyle and Bondi had developed the theory because they could not believe that all the matter in the universe was created at the initial singularity (the Big Bang). But to account for the fact that the universe is continually expanding, the theory requires that matter be continually created at the rate of a few hydrogen atoms per cubic metre per billion years. The theory initially seemed plausible, but the discovery of the cosmic microwave background in 1965 dealt it its first big blow. In 1952 Gold joined the staff of the Royal Greenwich Observatory, London. Five years later he became professor of astronomy at the Harvard University, Cambridge, Massachusetts, USA. There he worked on the maser amplifier for use with radio telescopes. In 1959 he joined the faculty of the Cornell University in Ithaca, New York, as professor of astronomy. He served as director of the Center for Radiophysics and Space Research in Ithaca from 1959 to 1981. (☞ Jun. 23, Jun. 24, Nov. 1)

1927

George A. Olah, Hungarian-born American chemist, Nobel laureate, was born in Budapest, Hungary. Olah attended school in Budapest and in 1949 received a doctorate from the Technical University of Budapest, where he taught until 1954. He worked at the Hungarian Academy of Sciences for two years before emigrating from Hungary in 1956 to join the Dow Chemical Company in Montreal, Canada as a research scientist. He was a professor at the Case Western Reserve University, Cleveland, Ohio from 1965 to 1977, before joining the faculty of the University of Southern California at Los Angeles, where he became director of the Loker Hydrocarbon Research Institute in 1980. Olah's researches mainly concerned the discovery of positively charged hydrocarbons – what chemists call 'carbocations'. In the early 1960s Olah and co-workers discovered that stable carbocations could be prepared through the use of a new type of extremely acid compounds which came to be known as 'superacids' that could be prepared, for example, from hydrogen fluoride (HF) and antimony pentafluoride (SbF_5). Olah's work also showed how basic knowledge on superacids and carbocations could be applied to the facile synthesis of new and important organic compounds and that several small organic molecules, with widespread use as starting material in many large-scale syntheses, could be produced in a simple and inexpensive way using superacids as catalysts. Olah was awarded the 1994 Nobel Prize for Chemistry for his contribution to carbocation chemistry.

1983

Albert Claude, Belgian-born American cytologist, Nobel laureate (Physiology or Medicine, 1974), who developed the principal methods of separating and analysing components of the living cell, died in Brussels, Belgium. (☞ Aug. 24)

1989
India test-fired its first intermediate-range ballistic missile, *Agni* from the Satish Dhawan Space Centre, Sriharikota, Andhra Pradesh. *Agni* is a two-stage solid fuel missile with the first stage using the first-stage solid-fuel booster motor of the SLV-3 satellite launch vehicle. The 18-metre-tall, 7.5-tonne missile has a range of up to 2,500 kilometres and is capable of delivering a 1,000-kg payload. (☞ Jan. 25, Apr. 11)

1997
Alfred D. Hershey, American biologist, Nobel laureate (Physiology or Medicine, 1969), who discovered recombination in bacteriophages, died in Syosset, New York, USA. (☞ Dec. 4)

May 23

1707
Carolus Linnaeus, Swedish botanist and explorer, who was the first to frame principles for defining genera and species of organisms and to create a uniform system for naming them, was born in Rashult, Sweden. He studied at the universities of Lund and Uppsala and received his degree in medicine from the latter. Linnaeus headed an expedition to Lapland in 1732, travelling 7,400 kilometres and crossing the Scandinavian Peninsula by foot to the Arctic Ocean. On the journey he discovered a hundred botanical species. In 1734, he mounted another expedition to central Sweden. He took his medical degree in 1735 from the University of Harderwijk in Gelre, Netherlands, and then went to the University of Leiden for further studies. In 1735, he published his classification of plants based on their sexual parts in *Systema Naturae*. His method of binomial nomenclature using genus and species names was further expounded when he published *Fundamenta Botanica* in 1736 and *Classes Plantarum* in 1738. This system used the flower and the number and arrangements of its sexual organs of stamens and pistils to group plants into twenty-four classes, which in turn are divided into orders, genera and species. In his publications, Linnaeus provided a concise, usable survey of the entire world's plants and animals as then known – about 7,700 species of plants and 4,400 species of animals. These works helped to establish and standardise the consistent binomial nomenclature for species which he introduced on a world-scale for plants in 1753, and for animals in 1758, and which is used even today. (☞ Jan. 10)

1785
The invention of bifocal glasses was documented in a letter from the American printer and publisher, author, inventor and scientist Benjamin Franklin. He was writing from France to a friend describing the solution to carrying around two pairs of glasses to see objects at different distances, with the comment, 'I have only to move my eyes up and down as I want to see far or near.' Franklin incorporated a two-part lens for each eye, each part having a different focussing power. (☞ Jan. 17, Apr. 17)

1908
John Bardeen, American physicist, Nobel laureate, was born in Madison, Wisconsin, USA. Bardeen completed his schooling at the Madison Central High School in 1923, and obtained his bachelor's degree in electrical engineering in 1928 and master's degrees in 1929 from the University of Wisconsin, Madison. He obtained his doctorate in mathematical physics from the Princeton University, New Jersey, in 1936. During World War II he served as principal physicist at the US Naval Ordnance Laboratory in Washington, DC. In 1945 Bardeen joined the Bell telephone Laboratories in Murray Hill, New Jersey, where he, together with the American engineer William B. Shockley and physicist Walter H. Brattain conducted research on the electron-conducting properties of semiconductors that led to the invention of the transistor in 1947, and ushered in the electronic revolution. In 1951 Bardeen joined the University of Illinois, Urbana-Champaign, as professor of physics and electrical engineering and resumed his research on superconductivity that he had begun in the 1930s. Working with a graduate student John R. Schrieffer and American physicist Leon N. Cooper, Bardeen developed a theory to explain the disappearance of electrical resistance in materials at temperatures close to absolute zero. The BCS theory of superconductivity (from the initials of Bardeen, Cooper, and Schrieffer) was first advanced in 1957 and became the basis for all later theoretical work in superconductivity. Bardeen was also the author of a theory explaining certain properties of semiconductors. Bardeen was awarded the 1956 Nobel Prize for Physics, jointly with Shockley and Brattain, for their researches on semiconductors and their discovery of the transistor effect. He also received the 1972 Nobel Prize for Physics, jointly with Schrieffer and Cooper, for their jointly developed theory of superconductivity, usually called the BCS theory. (☞ Jan. 30, Feb. 10, Feb. 13, Feb. 28, May 31)

1915
Clyde E. Wiegand, American physicist, was born in Long Beach, Washington, USA. Wiegand attended elementary school in Oakland and did his graduation in physics from the University of California, Berkeley, California. He went to Los Alamos, New Mexico to work on the Manhattan Project in 1943, and was with physicist Robert Oppenheimer in Alamogordo at the first detonation of an atomic bomb. Wiegand returned to the Berkeley campus in 1946 and received his PhD in 1950 under the direction of the Italian-born American physicist Emilio Segrè at the Radiation Laboratory of Ernest O. Lawrence. In the 1950s, Wiegand was part of a team that discovered the antiproton, using the bevatron particle accelerator at the Lawrence Berkeley National Laboratory. Although two other members of the research team (Owen Chamberlain and Emilio Segrè) won the 1959 Nobel Prize in Physics for this work, Wiegand's crucial contribution was not so recognised. (☞ Feb. 1, Jul. 5, Jul. 10)

1925
Joshua Lederberg, American geneticist, Nobel laureate, was born in Montclair, New Jersey, USA. Lederberg received his bachelor's degree from the Columbia University, New York, in 1944 and received his PhD in 1948 from the Yale University, New Haven, Connecticut, under the American geneticist Edward Lawrie Tatum. The two scientists showed that bacteria could also reproduce sexually, and that bacterial genetic systems are similar to

those of multicellular organisms. Scientists previously believed that bacteria only reproduced asexually; i.e., by cells splitting in two. In 1952 Lederberg revealed that certain bacteriophages (bacteria-infecting viruses) were capable of carrying a bacterial gene from one bacterium to another, a phenomenon called 'transduction.' These discoveries greatly increased the utility of bacteria as a tool in genetics research. The discovery of transduction provided the first hint that genes could be inserted into cells, thereby opening up the possibility of genetic manipulation through recombinant DNA technology. From 1947–1959, Lederberg taught at the University of Wisconsin, Madison, where he established a department of medical genetics. Lederberg was awarded the 1958 Nobel Prize for Physiology or Medicine, jointly with Tatum and the American geneticist George Wells Beadle. Lederberg received half of the prize money for his discoveries concerning genetic recombination and the organisation of the genetic material of bacteria, while Tatum and Beadle shared the other half for their discovery that genes act by regulating definite chemical events. (☞ Oct. 22, Dec. 14)

1960
Georges Claude, French engineer, chemist, and inventor of the neon light, died in Saint-Cloud, France. (☞ Sep. 24)

May 24

1543
Nicolaus Copernicus, Polish astronomer who proposed the Sun-centred model of the solar system, died in Frauenburg, East Prussia (now Frombork, Poland). (☞ Feb. 19)

1544
William Gilbert, English scientist, the founder of electrical studies, and a pioneer researcher into magnetism, was born in Colchester, Essex, England. Gilbert studied at the St. John's College, Cambridge, from where he obtained his BA degree in 1561, an MA in 1564, and finally an MD in 1569. He set up a medical practice in London in the 1570s and became a member of the Royal College of Physicians. In 1601 Gilbert was appointed physician to Queen Elizabeth I, and upon her death in 1603 was appointed physician to King James I. Although a physician by profession, Gilbert spent years investigating magnetic and electrical attractions. He coined the terms 'electric attraction', 'electric force', and 'magnetic pole'. More importantly, he made the first clear distinction between magnetic attraction and static electricity. After years of experiments he concluded that a compass needle points north–south and dips downward because the Earth acts as a bar magnet. His principal work, *De Magnete, Magneticisque Corporibus, et de Magno Magnete Tellure* ('On the magnet, magnetic bodies, and the great magnet of the Earth'), published in 1600, gives a full account of his research on magnetic bodies and electrical attractions. Like Copernicus, Gilbert believed that the Earth rotates on its axis, and that the fixed stars were not all at the same distance from the Earth. (☞ Dec. 10)

1844
The first telegraph message was sent from Washington DC. USA to his partner, Alfred Vail, in the Mount Clare station

of the Baltimore and Ohio Railroad Co., Baltimore, Maryland by the American inventor Samuel Morse, thus, formally opening America's first telegraph line. The message said, 'What hath God wrought?' (☞Apr. 27)

May 25

1865

Pieter Zeeman, Dutch physicist, Nobel laureate, was born in Zonnemaire, Netherlands. Zeeman's talent for natural science first became apparent in 1883, when, while still attending secondary school, he gave an apt description and drawing of an aurora borealis, which was published in the journal *Nature*. After schooling at Zierikzee, he studied at the Leyden University where he was a student of the Dutch physicist Hendrik A. Lorentz. In 1890 he was appointed assistant to Lorentz, which enabled him to participate in an extensive research programme. He obtained his doctor's degree in 1893. After spending a year at Strasbourg, he returned to the Leyden University in 1894 and became an 'extra-mural lecturer.' Two years later, at the suggestion of Lorentz, he investigated the effect of magnetic fields on a source of light and found that each of the lines in the spectrum of emitted light split into several lines; this became known as the 'Zeeman effect'. Lorentz had postulated that, because vibrating electrons produce light, the spectral lines should split as well as widen when a magnetic field was applied. The importance of the discovery can be judged by the fact that at one stroke the phenomenon not only confirmed Lorentz' theoretical conclusions with regard to the state of polarisation of the light emitted by flames, but also demonstrated the negative nature of the oscillating particles. The Zeeman effect also offered astronomers a means of verifying the existence of strong magnetic fields on the surface of the Sun, since these should alter spectral lines derived from the celestial body. In 1897, Zeeman joined the University of Amsterdam as a lecturer. In 1900 he was appointed professor of physics, and in 1908 given additional charge as director of the Physics Institute at the university. Zeeman was awarded the 1902 Nobel Prize for Physics, jointly with Lorentz, in recognition of the extraordinary service they rendered by their researches into the influence of magnetism upon radiation phenomena. (☞ Jul. 18, Oct. 9)

1889

Igor Ivan Sikorsky, Russian-born American pioneer in aircraft design who invented the helicopter, was born in Kiev, Russia. Sikorsky graduated from the Petrograd (now St. Petersburg) Naval College, studied engineering in Paris, France, and then joined the Mechanical Engineering College of the Polytechnical Institute in Kiev in 1907. In 1909, he went back to Paris, then the aeronautical centre of Europe, to learn what he could of the embryonic science of aeronautics. In Paris he bought a 25 h.p. Anzani engine and went back to Kiev to begin building a rotary-wing aircraft, although the initial attempts failed. His earliest successes were with fixed-wing aircraft, including his prize-winning S-6-A, which he built in 1912. Sikorsky first achieved international recognition

in 1913 when he designed and flew the first multi-engine airplane. After the Russian Revolution he emigrated to USA in 1919 and re-established himself as an aircraft designer. After a few years in USA, Sikorsky established his own company, the Sikorsky Aero Engineering Corporation. By 1929, the company was producing the S-38 twin-engine amphibians – planes which could land on land and water. In 1931, the first S-40, the *American Clipper*, rolled off the assembly line. In the following years, his planes set new records for altitude, speed, and payload. Sikorsky made his most important contributions in the area of helicopter design. His *VS-300*, first built and flown in 1939, became the world's first successful helicopter and introduced a new mode of military and commercial transportation. (☞ May 13, Sep. 14, Oct. 26)

1921

Jack Steinberger, German-born American physicist, Nobel laureate, was born in Bad Kissingen, Germany. Steinberger studied physics at the University of Chicago, from where he received his PhD degree in 1948. Two years later he joined the Columbia University, New York, as professor of physics, and from 1968 he was a physicist at the European Organization for Nuclear Research (CERN) in Geneva, Switzerland. In the early 1960s at the Columbia University, Steinberger and his two American colleagues Leon M. Lederman and Melvin Schwartz conducted a landmark experiment in particle physics using the particle accelerator at the Brookhaven National Laboratory, New York, and obtained the first laboratory-made stream of neutrinos – subatomic particles that have no electric charge and virtually no mass. In the process, they discovered a new type of neutrino called a muon neutrino. After this success, high-energy neutrino beams became a basic research tool in the study of subatomic particles and nuclear forces, making possible the study of radioactive decay processes involving the weak nuclear force, one of the four fundamental forces in nature. Steinberger was awarded the 1988 Nobel Prize for Physics, jointly with Lederman and Schwartz, for the neutrino beam method and the demonstration of the doublet structure of the leptons through the discovery of the muon neutrino. (☞ Jul. 15, Nov. 2)

1940

First test of penicillin was conducted on mice. Three Oxford scientists, Australian pathologist Howard Florey, German-born British biochemist Ernst Chain, and British biochemist Norman Heatley, inoculated eight mice with a lethal dose of *streptococci* bacteria and then injected four of them with penicillin. Next day the four mice given *streptococci* alone were dead, the four given penicillin were healthy, proving the efficacy of penicillin. The three scientists had produced enough antibiotic for testing by isolating the active ingredient from what its discoverer Alexander Fleming had called 'mould juice'. When Fleming had found that penicillin would not work in animals when given orally, his interest waned and he had not pursued further research on it. (☞ Jun. 19, Aug. 6, Sep. 24)

1973

Skylab-2, the first manned mission to the American orbiting space station *Skylab* was launched from the Kennedy Space Center, Florida, USA. The crew comprising Charles C. Conrad Jr., Paul J. Weitz, and Joseph Kerwin rendezvoused with *Skylab* on the fifth orbit. After making substantial repairs, including deployment of a parasol sunshade, which cooled the inside temperatures to 23.8°C from a sizzling 52°C, the space station was in full operation. In orbit the crew conducted solar astronomy and Earth resources experiments, medical studies, and five student experiments. (☞ Feb. 8, May 14, Jun. 22, Jun. 28, Jul. 11, Sep. 25, Nov. 16)

May 26

1906

Benjamin Peary Pal, Indian agricultural scientist, Fellow of the Royal Society, was born in Mukundpur village of Jullundhar district, Punjab, India. Pal had his early education in Burma (now Myanmar).

After doing his MSc from the Rangoon University in 1929, Pal went to Cambridge, England, where he did his PhD in plant breeding and genetics. After returning to India he was appointed as the Imperial Economic Botanist at the Imperial (now Indian) Agricultural Research Institute of which he became the Director in 1950. Pal's main research was on agricultural crops, especially wheat. Wheat was particularly susceptible to rust, a kind of fungal disease which damages a plant just as rust corrodes iron. In Pal's days, although some varieties of wheat that could give higher yields had been found, there were none that could combat rust. In 1954 Pal developed the NP (New Pusa) 809 variety of wheat, after 18 years of painstaking research, which could combat all three types of rust and was a big step forward in increasing wheat production. He laid the strong foundation of coordinated research, education and agricultural extension in India that ushered in the first phase of the Green Revolution in the country. He also had great interest in flowers, and several new flower varieties have been named after him. (☞ Sep. 24)

1922

Ernest Solvay, Belgian industrial chemist who invented the Solvay process for producing washing soda, or sodium carbonate, died in Brussels, Belgium. (☞ Apr. 16)

1969

The command module of the American spacecraft *Apollo-10*, carrying astronauts Thomas P. Stafford, John W. Young, and Eugene A. Cernan, safely splashed down in the Pacific Ocean. (☞ May 18)

1999

India's remote sensing satellite *IRS-P4* (*Oceansat*) was launched by ISRO's polar satellite launch vehicle (PSLV-C2) from the Satish Dhawan Space Centre, Sriharikota, Andhra Pradesh. Two foreign satellites – Korean *KITSAT-3* and German DLR-*TUBSAT* – were also launched during the same flight. The 1,050-kg *Oceansat* is the first satellite primarily built for ocean applications and is equipped with an ocean colour monitor (OCM) and a multi-frequency scanning microwave radiometer

(MSMR). The OCM is a solid-state camera operating in eight narrow spectral bands. It is used to collect data on chlorophyll concentration, detect and monitor phytoplankton blooms and obtain data on atmospheric aerosols and suspended sediments in the water. The MSMR, which operates in four microwave frequencies, collects data on sea surface temperature, wind speed, cloud water content and water vapour content in the atmosphere above the ocean. Issue of daily forecasts for fishing zones based on ocean colour and sea surface temperature obtained from the *Oceansat* has led to a four to five-fold increase in fish catch.

May 27

1897

John Douglas Cockcroft, British physicist, Nobel laureate, was born in Todmorden, Yorkshire, England. Cockcroft was educated at the Todmorden Secondary School and studied mathematics at the Manchester University. He took the Mathematical Tripos at St. John's College, Cambridge, in 1924, and then worked under the British physicist Ernest Rutherford in the Cavendish Laboratory, Cambridge. In 1928 he started working on the acceleration of protons by high voltages and was soon joined by the Irish physicist Ernest T. S. Walton. In 1932 they designed the Cockcroft–Walton generator, the first 'atom-smasher', and used it to disintegrate lithium atoms by bombarding them with high-energy protons. They conducted further research on the splitting of other atoms and established the importance of accelerators as a tool for nuclear research. Cockcroft was Jacksonian professor of natural philosophy at the University of Cambridge from 1939 to 1946. In 1944 he went to Canada to take charge of the Canadian Atomic Energy project and became Director of the Montreal and Chalk River Laboratories until 1946 when he returned to England as Director of the Atomic Energy Research Establishment, Harwell. From 1954 to 1959 he was scientific research member of the UK Atomic Energy Authority. Cockcroft was awarded the 1951 Nobel Prize for Physics, jointly with Walton, for their pioneer work on the transmutation of atomic nuclei by artificially accelerated atomic particles. (☞ Sep. 18, Oct. 6)

1907

Rachel Louise Carson, American biologist well known for her writings on environmental pollution and the natural history of the sea, was born in Springdale, Pennsylvania, USA. Carson graduated with honours from the Pennsylvania College for Women (now Chatham College) in 1929, earning a scholarship to continue her studies at the Johns Hopkins University in Baltimore, Maryland, from where she completed her MA in zoology in 1932. After completing her education, Carson joined the US Bureau of Fisheries as the writer of a radio show entitled *Romance under the Waters*, in which she was explored life under the seas and brought it to listeners. In 1936, after being the first woman to take and pass the civil service test, the Bureau of Fisheries hired her as a full-time junior biologist, and over the next 15 years, she rose in the ranks until she was the chief editor of all publications for the US Fish and Wildlife Service. During the

1940s, Carson began to write books on her observations of life under the sea, a world as yet unknown to the majority of people. She resigned from her government position in 1952 in order to devote all her time to writing. Disturbed by the profligate use of synthetic chemical pesticides after World War II, Carson reluctantly changed her focus in order to warn the public about the long-term effects of misusing pesticides. In her book, *Silent Spring*, published in 1962, she challenged the practices of agricultural scientists and the government, and called for a change in the way humankind viewed the natural world. (☞ Apr. 14)

1910
Robert Koch, German physician, Nobel laureate (Physiology or Medicine, 1905), who discovered the tubercle bacillus and the cholera bacillus, died in Baden-Baden, Germany. (☞ Dec. 11)

1930
Cellophane transparent adhesive tape, also know as Scotch tape, was patented by American inventor Richard G. Drew of St. Paul, Minnesota, USA. Drew got the idea after observing the problem faced by auto manufacturers while giving two-tone paint finishes on cars. He found that the auto painters lacked suitable materials to attach protective masking paper. After two years of effort, Drew came up with an effective pressure-sensitive masking tape, which could be easily peeled off after use.

1931
Auguste Piccard, Swiss-born Belgian physicist and explorer with his associate Paul Kipfer, made the first manned balloon flight into the stratosphere, lifting off from Augsberg, Germany. They set a new altitude record of 15,838 metres. For this flight, Piccard designed the first pressurised cabin, which was made of aluminium. During the flight, he gathered valuable information regarding the intensity of cosmic rays in the stratosphere and also recorded a wide range of stratospheric temperatures. (☞ Jan. 28, Mar. 24)

1987
John Howard Northrop, American biochemist, Nobel laureate (Chemistry, 1946), who isolated and crystallised enzymes for the first time, died in Wickenberg, Arizona, USA. (☞ Jul. 5)

1988
Ernst August Friedrich Ruska, German electrical engineer, Nobel laureate (Physics, 1986), who invented the electron microscope, died in Berlin, Germany. (☞ Dec. 25)

May 28

585 BCE
The first ever solar eclipse predicted in advance occurred. It was predicted by Thales of Miletus, a pre-Socratic Greek philosopher and one of the Seven Sages of Greece. This was the most famous eclipse of ancient times and it ended a five-year war between the Lydians and the Medes, two Middle Eastern armies. As the eclipse progressed, 'the day was turned into night'. The sight of this was startling enough to cause both nations to stop fighting at once.

1807
Louis Agassiz, Swiss-born American naturalist, geologist, and teacher, was born in Môtiers, Switzerland. Agassiz was educated in the medical tradition and qualified as a physician. He developed a keen interest in natural history at an early

age and attended the Universities of Zürich, in Switzerland and Heidelberg in Germany before returning to Switzerland in 1827 after suffering from a severe case of typhoid fever. He then studied at the University of Munich in Germany from 1827 to1830. In 1829 he received the PhD degree from the University of Erlangen, and a degree of doctor of medicine in 1830 from the University of Munich. Agassiz began his scientific work in Europe while studying at the University of Munich and then working as a professor in natural history in Neuchatel in Switzerland. While there he published his landmark multi-volume description and classification of fossil fish in 1829, *Selecta Genera et Species Piscium*. In 1837 Agassiz was the first to scientifically propose that the Earth had been subject to a past ice age. In 1846 Agassiz moved to USA to deliver a course of lectures on zoology at the Lowell Institute in Boston, Massachusetts. Two years later he joined the Harvard University, Cambridge, Massachusetts as professor of zoology and geology. He became an American citizen in 1861. Agassiz is best known for his revolutionary contributions to the study of natural science with landmark work on glacier activity and extinct fishes. In the last years of his life he worked to establish a permanent school where zoological science could be pursued amid the living subjects of its study. His innovative teaching methods altered the character of natural science education. (☞ Dec. 14)

1942

Stanley Ben Prusiner, American neurologist, Nobel laureate, was born in Des Moines, Iowa, USA. Prusiner studied in the Walnut Hills High School in Cincinnati and majored in chemistry from the University of Pennsylvania. He then joined the Medical School at Pennsylvania and obtained his MD in 1968. After spending four years in biochemical research, he became a resident in neurology at the University of California, San Francisco, School of Medicine in 1972. He joined the faculty there in 1974 and became a professor of neurology and biochemistry. In 1972, while a resident, Prusiner came across a patient who was exhibiting progressive loss of memory and difficulty in performing some routine tasks. The patient was supposed to be dying of a 'slow virus' infection called Creutzfeldt–Jakob disease which evoked no response from the body's defences. But no virus could be detected. The symptoms appeared similar to related diseases like kuru found in some New Guinea tribes and scrapie of sheep. In 1974 Prusiner set up a laboratory to study scrapie and in 1982 he claimed to have isolated the scrapie-causing agent, which he named 'prion'. Prions are novel infectious agents differing from all other known pathogenic agents. They are simple proteins that are much smaller than viruses. Unlike the other known pathogens, such as a virus or bacterium, prion consisted only of protein and lacked the genetic material contained within all life forms that is necessary for replication. When first published, the prion theory met with much criticism, but by the early 1990s, the existence of prions was accepted in many quarters of the scientific community, although the mechanism by which a normal prion protein is converted into the disease-causing form is still obscure. Prusiner was awarded the 1997

Nobel Prize for Physiology or Medicine for his discovery of prions – a new biological principle of infection.

1971
Soviet unmanned spacecraft *Mars-3* was launched towards Mars from the Baikanur cosmodrome in Kazakhstan. It arrived at Mars in December, when a descent module was released that entered the Martian atmosphere. Through aerodynamic

braking, parachutes, and retro-rockets, the lander achieved a soft landing, becoming the first spacecraft to land successfully on Mars. The lander failed after relaying 20 seconds of video data to the orbiter. The *Mars-3* orbiter returned data until August 1972, sending measurements of surface temperature and atmospheric composition. (☞ Dec. 2)

1991
A patent was issued for the 'production of taxol or taxol-like compounds in cell culture'. Taxol was an important breakthrough in cancer treatment. However, it only existed naturally in the bark of the Pacific Yew, *Taxus brevifolia*, found solely in the Pacific Northwest, where the number of trees is limited. Moreover, a vast number of trees needed to be felled in order to collect the large amount of bark necessary for commercial drug production. Inventors A.A. Christen, J. Bland, and G.M. Gibson developed an alternative route to produce taxol by cell culture. The patent describes the production and recovery of paclitaxel and taxane-like compounds by cell cultures of *T. brevifolia*, obviating the need for large-scale cutting of trees.

2003
Ilya Prigogine, Russian-born Belgian physical chemist, Nobel laureate (Chemistry, 1977), who worked on non-equilibrium thermodynamics and attempted to reconcile a tendency in nature for disorder to increase with so-called 'self-organisation', died in Brussels, Belgium. (☞ Jan. 25)

2003
The first cloned horse was born in a natural delivery. It was also the first cloned mammal born to its genetic mother. The foal, named Prometea, was created in the lab by fusing an adult skin cell and an enucleated egg then returning the resulting embryo to the female's womb after a few days. The cloning was accomplished by Cesare Galli, of the Laboratory of Reproductive Technologies, Cremona, Italy. It was the only successful one of 328 reconstructed embryos. DNA tests confirmed that the foal is genetically identical to her mother and is her twin.

May 29

1829
Humphry Davy, English chemist who discovered several chemical elements and compounds, and invented the miner's safety lamp, died in Geneva, Switzerland. (☞ Dec. 17)

1912
Chien-Shiung Wu, Chinese-born American physicist, was born in Shanghai, China and grew up in Liuhe in the Jiangsu

province north of the city. Wu graduated from the National Central University in Nanking, in 1934 and in 1936 went to USA to pursue graduate studies in physics at the University of California at Berkeley, California, studying under the American physicist and inventor of the cyclotron, Ernest O. Lawrence. She received the doctorate in physics in 1940. She then taught at the Smith College, Northampton, Massachusetts, and the Princeton University, New Jersey, before joining the Columbia University, New York, in 1944, where she spent 37 years of her life doing research and teaching as a senior scientist. She became associate professor in 1952, full professor in 1958, and the first Pupin Professor of Physics in 1973. During World War II she participated in the Manhattan Project and developed the process of separating uranium-235 from uranium-238 by gaseous diffusion. In 1956 she conducted a crucial experiment using radioactive cobalt-60 at near absolute zero at the National Bureau of Standards in Washington, DC., that provided the first experimental proof that the principle of parity conservation does not hold in weak subatomic interactions. Earlier, the non-conservation of parity was theoretically predicted by the Chinese-born American physicists Tsung Dao Lee and Chen Ning Yang. Lee and Yang went on to receive the 1957 Nobel Prize for Physics for their theory, but Wu was never recognised for her contribution to the project. She was the first woman President of the American Physics Society. (☞ Feb. 16)

1919
Bending of starlight by the Sun's gravitational field, as predicted by Albert Einstein's theory of relativity, was proved during a total solar eclipse. Separate expeditions of the Royal Astronomical Society travelled to Brazil and off the west coast of Africa to make measurements of the position of stars visible close to the Sun during the eclipse. The English astrophysicist Arthur Eddington made observations from the Principe Island off West Africa. These observations showed that, indeed, the light from stars was bent as it passed through the gravitational field of the Sun. The verification of the predictions of Einstein's theory, proved during the solar eclipse, was a dramatic landmark scientific event. (☞ Mar. 14)

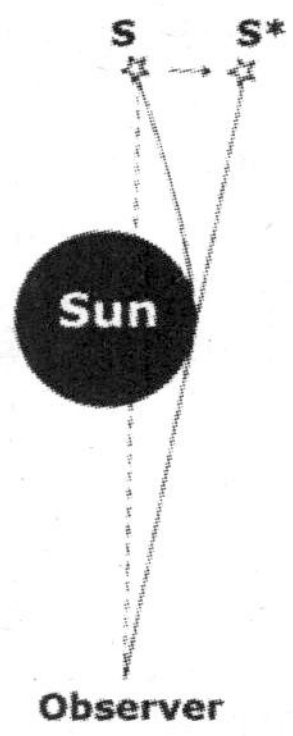

2001
Vulimiri Ramalingaswami, Indian nutritionist, Fellow of the Royal Society, who pioneered the usage of iodised salt in India, died in New Delhi, India. (☞ Aug. 8)

May 30

1876
Nikolaus August Otto German engineer was awarded a patent for a four-stroke internal combustion engine, which was described in the patent as a 'gas motor engine'. The engine included a water-jacketed

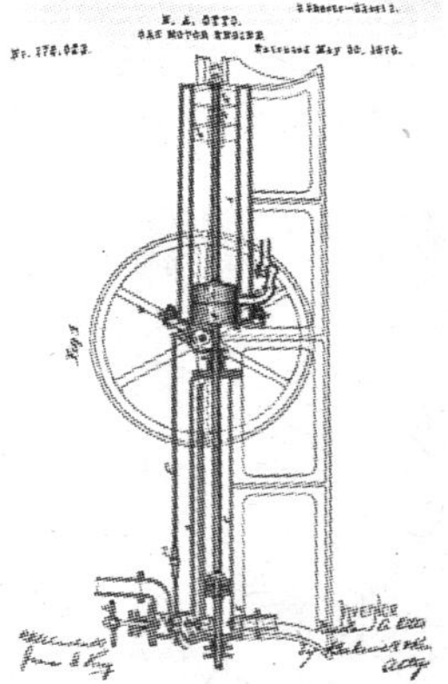

cylinder open at both ends, with a working piston and a loose piston inside it. The loose piston was connected by a rod to another piston within a hydraulic cylinder. The four strokes were admission, compression, combustion, and discharge. (☞ Jun. 10)

1898
English chemists, William Ramsay and Morris William Travers discovered the element krypton (Kr, atomic number 36). In fact, the discovery was made by chance. They were trying to isolate neon by fractional distillation of a mixture of noble gases when they noticed a bright yellow line, revealing the presence of an unknown gas. Krypton is present in the atmosphere to the extent of 1 part per million. The name derives from the Greek word for 'hidden' and it was so named due to the circumstances of its discovery.

1908
Hannes Alfvén, Swedish astrophysicist, Nobel laureate, was born in Norrköping, Sweden. After high school, Alfvén entered the University of Uppsala where he studied mathematics, experimental, and theoretical physics, received his doctorate in 1934, and was appointed lecturer in physics. In 1937 he became research physicist at the Nobel Institute for Physics in Stockholm. In 1940, at the age of only 32, Alfvén was appointed professor of electromagnetic theory and electric measurements at the Royal Institute of Technology, Stockholm. While at the Nobel Institute, Alfvén became more and more interested in the acceleration of charged particles to very high energies, and especially to the extreme energies of cosmic rays. His early attempt to develop a theory of the origin of cosmic radiation was published in the journal *Nature* in 1933. He explained the origin of the cosmic rays not by introducing any new hypotheses, but only by applying the kinetic gas theory to the conditions of world space. Alfvén's work on the cosmic ray problem led him to propose in 1937 the existence of a galactic magnetic field. Only much later was the existence of the galactic magnetic field confirmed. Alfvén was an original contributor to the potent new discipline of magneto-hydrodynamics, to which he brought a background of work in such varied and related fields as cosmic rays, fundamental electronics, aurorae, earth magnetism, sunspots, and the design of electron tubes. He also discovered a number of other fundamental concepts that are not so closely associated with his name. These include simplifications in the concepts with which we treat the behaviour of ionised gases (plasmas). Alfvén was awarded the 1970 Nobel Prize for Physics, jointly with the French physicist Louis Néel, for fundamental work and discoveries in magneto-hydrodynamics with fruitful applications in different parts of plasma physics. (☞ Apr. 2, Nov. 22)

1912
Wilbur Wright, American inventor and aviation pioneer, who with his brother Orville achieved the first powered and controlled aeroplane flight, died in Dayton, Ohio, USA. (☞ Apr. 16, Dec. 17)

1912
Julius Axelrod, American biochemist and pharmacologist, Nobel laureate, was born in New York, USA. Axelrod obtained his BSc in 1933 from the City College of New York and his MA in 1941 from the New York University. From 1935–46 he worked as chemist in the Laboratory of Industrial Hygiene. He then joined the research division of the Goldwater Memorial Hospital and began conducting

research on the chemistry of analgesic (pain-relieving) medications. In 1949, Axelrod accepted a position as a research chemist at the National Heart Institute, a part of the rapidly expanding National Institutes of Health in Bethesda, Maryland. He obtained his PhD in pharmacology from the George Washington University, Washington, DC. in 1955. In 1957 Axelrod began his most famous research project, which focussed on the activity of neurotransmitter hormones. Axelrod investigated the mechanism of formation of the hormone noradrenaline as an important nerve impulse transmitter. He discovered and isolated the enzyme he called catechol-o-methyl transferase, which degrades chemical neurotransmitters when no longer needed. In effect this resets the nerve-to-nerve connections ready to transmit the next impulse that arrives. This helped the search for treatment of nervous and mental illnesses. Axelrod's work enabled researchers during the 1970s to develop a new class of antidepressant medications, especially selective serotonin re-uptake inhibitors such as Prozac. Over the next thirty years, until his retirement in 1984, he worked on a wide array of research projects in pharmacological science. Axelrod was awarded the 1970 Nobel Prize for Physiology or Medicine, jointly with the British biophysicist Bernard Katz and Swedish physiologist Ulf von Euler, for their discoveries concerning the humoral transmitters in the nerve terminals and the mechanism for their storage, release and inactivation. (☞ Feb. 7, Mar. 26)

1934

Aleksey Leonov, Soviet cosmonaut, the first man to climb out of a spacecraft in space, was born in Listvyanka in the Altay region of Siberia, USSR. Leonov had his schooling in Kaliningrad and then joined the Soviet Air Force in 1953. He completed his flight training in 1957 and served as a fighter pilot until 1959. In 1960 he joined the cosmonaut corps when he was selected for cosmonaut training. Five years later, Leonov flew on board *Voskhod-2* along with cosmonaut Pavel Belyayev and came out of the spacecraft while it was orbiting about 177 kilometres above the Crimea. Tethered to the ship, Leonov made observations, took motion pictures, and practised maneuvering in free-fall for about 10 minutes before re-entering the spacecraft over western Siberia. The spacecraft landed after spending 26 hours in Earth orbit. In 1975, Leonov was commander of the Soviet *Soyuz* craft that linked in orbit with the US *Apollo* craft. Later in his career, he worked at the cosmonaut-training centre. He also worked on the Soviet lunar landing program and trained for the *Salyut* space station project. (☞ Mar. 18, Jul. 17)

1959

The first experimental hovercraft, the *SR-N1* made its first trip at Cowes on the Isle of Wight, England. It was designed by the British inventor Christopher Cockerell. He first considered the possibility of travel by hovercraft, a vehicle that can move across land or water on a cushion of air,

in the early 1950s. Using a can of cat food inside a coffee tin and reversed airflow from a vacuum cleaner, he proved his theory on the mud floor of his boatyard. The invention was considered initially only for military use, but was released for civilian use in 1959. The hovercraft is now used in a variety of roles, from military transportation to ferrying cars and passengers over waterways around the world. (☞ Jun. 4, Dec. 12)

1964
Leo Szilard, Hungarian-born American physicist who helped design the first nuclear reactor that sustained nuclear chain reaction, died in La Jolla, California, USA. (☞ Feb. 11)

1966
American Moon-probe *Surveyor-1* was launched towards the Moon from the Kennedy Space Center, Florida, USA. (☞ Jun. 2)

1971
American unmanned spacecraft *Mariner-9* was launched towards the planet Mars from the Kennedy Space Center, Florida, USA. The 506-kilogram spacecraft was sent on a mission to photograph the surface and analyse the atmosphere of the Red planet with infrared and ultraviolet instruments from orbit over a period of one Martian year. (☞ Nov. 13)

Mariner-9

May 31

1931
John Robert Schrieffer, American physicist, Nobel laureate, was born in Oak Park, Illinois, USA. Schrieffer graduated from the Massachusetts Institute of Technology, Cambridge, and joined the University of Illinois, Champaign-Urbana, where he worked with fellow American physicists John Bardeen and Leon N. Cooper. He received his PhD in 1957 for his thesis on the theory of superconductivity, explaining why metals lose their electrical resistance at very low temperatures. He spent the academic year 1957–58 as a National Science Foundation fellow at the University of Birmingham in England, and the Niels Bohr Institute in Copenhagen, Denmark, where he continued research in superconductivity. He taught at the University of Chicago for a year before returning to the University of Illinois in 1959 as a faculty member. In 1962 he joined the faculty of the University of Pennsylvania, Philadelphia. The theory of superconductivity developed by Bardeen, Cooper and Schrieffer, commonly known as the 'BCS theory', explains how certain metals and alloys lose all resistance to electrical current at extremely low temperatures. The insight offered by the BCS theory is that at very low temperatures, under certain conditions, electrons can form bound pairs, called 'Cooper pairs', which act as a single particle in superconductivity. Schrieffer's

later researches dealt with particle physics, metal impurities, spin fluctuations, and chemisorption. Schrieffer was awarded the 1972 Nobel Prize for Physics, jointly with Bardeen and Cooper, for their jointly developed theory of superconductivity, usually called the BCS theory. (☞ Feb. 28, May 23)

1941

Louis J. Ignarro, American pharmacologist, Nobel laureate, was born in Brooklyn, New York, USA. After receiving his schooling in the Central Grade School and the Long Beach High School, Ignarro did his graduation in pharmacology from the Columbia University in 1962, and received his PhD in pharmacology from the University of Minnesota in 1966. In 1979 he became a professor of pharmacology at the Tulane University's School of Medicine in New Orleans, Louisiana, a position he held until becoming a professor of pharmacology at the University of California, Los Angeles, in 1985. At UCLA Ignarro's research involved the study of the mechanism of action of nitroglycerine and other nitrovasodilators as well as the mechanism by which nitric oxide (NO) activates cytosolic guanylate cyclase. His work showed that nitroglycerine generates nitric oxide through chemical reactions involving thiols and formed intermediate S-nitrosothiols, which serve as carriers and donors of nitric oxide. This led to the discovery that nitric oxide relaxes vascular smooth muscle; inhibits platelet aggregation; and is a neurotransmitter mediating erectile function, thus demonstrating an entirely new mechanism by which blood vessels in the body relax and widen. This discovery opened up new applications for nitric oxide. For instance, the development of the successful anti-impotence drug sildenafil citrate, commonly known as *Viagra*, was based upon this research. Researchers suggested that nitric oxide could be a key to improved treatments for heart disease, shock, and cancer. Ignarro was awarded the 1998 Nobel Prize for Physiology or Medicine, jointly with fellow American pharmacologists Robert F. Furchgott and Ferid Murad, for their discoveries concerning nitric oxide as a signalling molecule in the cardiovascular system. (☞ Jun. 4, Sep. 14)

1976

Jacques Lucien Monod, French biochemist, Nobel laureate (Physiology or Medicine, 1965), who did much to elucidate genetic regulation of cell metabolism, died in Cannes, France. (☞ Feb. 9)

1981

India's satellite launch vehicle SLV-3 D1 was launched on its first developmental flight from Sriharikota, Andhra Pradesh carrying a *Rohini* satellite. The 35-kilogram satellite was placed in a near-Earth elliptical orbit ranging in height from 300 kilometres to 900 kilometres.

1986

Leo James Rainwater, American physicist, Nobel laureate (Physics, 1975), who discovered that not all atomic nuclei are symmetrically shaped, died in Yonkers, New York, USA. (☞ Dec. 9).

June 1

1796

Sadi Carnot, French physicist who described the Carnot cycle, relating to the theory of heat engines, was born in Paris, France. After graduating from the École Polytechnique in Paris in 1814, Carnot went to the École du Génie at Metz to take the two-year course in military engineering. In 1824 he published a short work entitled *Réflexions sur la puissance motrice du feu*, in which he attempted to clarify how to get the most work out of an engine, such as a steam engine, from a given quantity of heat. Engineers of Carnot's time had tried various mechanical means, such as high pressure of steam or using some fluid other than steam, to maximise the amount of work from engines. But Carnot argued that, if inefficiencies such as friction and heat loss were removed, efficiency could be increased by simply maximising the ratio of the input and output temperatures. He proposed the most efficient heat engine cycle, known as the Carnot cycle. Consisting of two isothermal processes and two adiabatic processes, the Carnot cycle can be thought of as the most efficient heat engine cycle allowed by physical laws. (☞ Aug. 24)

1907

Frank Whittle, English aviation engineer and pilot who was a pioneer in the field of jet propulsion, was born in Coventry, Warwickshire, England. Whittle entered the Leamington College at the age of 11 and became an apprentice in the Royal Air Force College as a pilot officer in 1928. He became interested in aircraft propulsion. He knew that even though aircraft flying at higher altitudes had the advantage of clear weather and less turbulence, the thinness of air drastically reduced the performance of piston engines. In other words, a piston engine simply could not power a plane to fly high enough in order for the plane to gain a huge increase in speed. At just 21 years of age, Whittle proposed a revolutionary alternative to the piston engine, which would solve this dilemma. Whittle's 1929 thesis, entitled *Future Developments in Aircraft Design*, would lay the foundation for the successful creation of the turbojet engine. Whittle's proposed engine drew in air through a series of compressors. The air entered a combustion chamber, where it was mixed with fuel and ignited. As the resulting hot gas blasted out of the back of the chamber, it pushed the engine forward. The exiting hot gas turned a shaft that spun a turbine – which drew in more air, creating an ingenious loop of intake and combustion. By 1930, Whittle had designed and patented a jet aircraft engine. Eleven years later, in 1941, the first aircraft powered by Whittle's engine made a historic 17-minute flight. Whittle settled in America in 1976, and was a member of the Faculty of the Naval Academy, Annapolis, Maryland, USA. (☞ Apr. 12, May 15, Aug. 8)

1917
William S. Knowles, American chemist, Nobel laureate, was born in Taunton, Massachusetts, USA. Knowles attended school at Berkshire in western Massachusetts. He recieved a PhD from the Columbia University, New York, in 1942 after which he conducted research at the Monsanto Company in St. Louis, Missouri, until his retirement in 1986. Knowles' research at Monsanto focussed on 'chirality', or 'handedness', which describes how organic molecules often appear in two forms, one the mirror image of the other, much as a person's two hands mirror each other. In nature, one form often dominates, so human cells will naturally bind to one form of a molecule over the other, while the other form is ignored or may even be harmful. In 1968 Knowles discovered that it was possible to use transition metals to make chiral catalysts for an important type of reaction called hydrogenation, thereby obtaining the desired mirror image form as the final product. His research quickly led to an industrial process for the production of the L-dopa drug, which is used in the treatment of Parkinson's disease. The Japanese biochemist Ryoji Noyori has led the further development of this process to today's general chiral catalysts for hydrogenation. Knowles was awarded the 2001 Nobel Prize for Chemistry, jointly with Noyori and the American biochemist Karl Barry Sharpless. Knowles and Noyori shared half the prize money for their work on chirally catalysed hydrogenation reactions, while Sharpless received the other half for his work on chirally catalysed oxidation reactions. (☞Apr. 28, Sep. 3)

1919
The earliest artificial transmutation of an element was performed by the English physicist Ernest Rutherford in 1911. Nitrogen-14 was bombarded with alpha particles, producing oxygen-17 and protons. This transmutation of nitrogen into oxygen was the first artificially induced nuclear reaction. (☞ Aug. 30)

1965
German-born American astrophysicist Arno Penzias and American radio astronomer Robert W. Wilson detected a 3 degree Kelvin cosmic background radiation using a horn reflector antenna built for radio astronomy. The cosmic background radiation is believed to be the remnant of the Big Bang that is assumed to have created our universe some 15,000 million years ago. The high-energy radiation produced when the universe was very young and very hot would have cooled down to microwave wavelengths as the universe expanded. The uniformity of microwave background indicates that the universe was homogeneous until it was a few million years old. The detection of the cosmic background radiation came as strong evidence in support of the Big Bang origin of the universe. (☞ Jan. 10, Apr. 26)

1979
Werner Forssmann, German surgeon, Nobel laureate (Physiology or Medicine, 1956), who developed cardiac catheterisation, died in Schopfheim, Germany. (☞ Aug. 20)

1999
Christopher Cockerell, British engineer, inventor of the hovercraft, died in Hampshire, UK. (☞ May 30, Jun. 4)

2000
The 220-MW third unit of the Rajasthan Atomic Power Project at Rawatbhata, near Kota, Rajasthan, was commissioned. It is

a pressurised heavy water reactor that uses natural uranium as fuel and heavy water as moderator and coolant. (☞ Apr. 1, Dec. 16, Dec. 23)

June 2

1896
The Italian physicist and inventor Guglielmo Marconi received a patent in England for his wireless telegraphy apparatus for 'improvements in transmitting electrical impulses and signals, and in apparatus there-for.' (☞ Apr. 25, Jul. 20)

1966
American Moon-probe *Surveyor-1* became the first spacecraft to soft-land on the Moon. The spacecraft was equipped with two television cameras, one for the approach and the other for imaging the Moon's surface. The landing was successful and the probe continued to take 11,240 images, which it transmitted back to Earth. (☞ May 30)

1975
Debendra Mohan Bose, Indian physicist, Foundation Fellow, Indian National Science Academy, died in Kolkata. (☞ Nov. 26)

1983
Soviet unmanned spacecraft *Venera-15* was launched towards Venus. *Venera-15* was part of a two spacecraft mission to Venus, along with *Venera-16*, which was launched later. Both the spacecraft were equipped with a synthetic aperture radar (SAR) and a radio altimeter to map the surface of Venus. (☞ Oct. 10)

June 3

1657
William Harvey, English physician and discoverer of the true nature of the circulation of the blood and the function of the heart as a pump, died in London, UK. (☞ Apr. 1)

1769
Famous transit of Venus across the face of the Sun was observed by British naval captain, navigator, and explorer James Cook from the South Pacific island of Tahiti, on a voyage to New Zealand. The transit was also observed by David Rittenhouse, American astronomer, instrument maker and inventor. For observing the transit, Rittenhouse constructed a high-precision pendulum clock, an astronomical quadrant, an equal altitude instrument, and an astronomical transit. (☞ Apr. 8, Jun. 8, Oct. 27, Dec. 6)

1873
Otto Loewi, German-born American physician and pharmacologist, Nobel laureate, was born in Frankfurt am Main, Germany. Loewi graduated in medicine in 1896 from the German University in Strasbourg. He then studied and taught in European universities, and in 1909 joined the Karl Franzens University in Graz, Austria, as professor of pharmacology. Loewi's neurological researches provided the first proof that chemicals were involved in the transmission of impulses from one nerve cell to another and from neuron to the responsive organ. He and his colleagues, by stimulating the nerves in the heart of a frog, slowed the heart's

rate of contraction. The fluid perfusing this heart was allowed to perfuse a second heart in which the nerves were not stimulated; the second heart slowed in rate too, indicating the presence of a reactive substance in the fluid. This substance was shown to be acetylcholine, whose physiological properties were known. In 1921 Loewi presented his theory about the humoral transduction of neural impulses by cardiac nerve fibres to the German Pharmacological Society at its annual meeting in Freiburg and identified acetylcholine as the long sought-for agent of neural transmission. In 1940 he went to the United States; he was made research professor at the School of Medicine of New York University, New York, where he remained until his death. Loewi was awarded the 1936 Nobel Prize for Physiology or Medicine, jointly with the English physiologist Henry Dale, for their discoveries relating to chemical transmission of nerve impulses. (☞ Jun. 9, Dec. 25)

1899
Georg von Békésy, Hungarian-born American physicist and physiologist, Nobel laureate, was born in Budapest, Hungary. Békésy received his early education in Munich in Germany, Constantinople in Turkey, Budapest in Hungary, and in a private school in Zurich, Switzerland. He later studied chemistry at the University of Berne in Switzerland. After a short military service he received his PhD in Physics in 1923 from the University of Budapest. Later he entered the services of the Hungarian Post Office in Budapest where he stayed until 1946. His work in the research laboratory of the Hungarian Post Office mainly concerned problems in long-distance telephone transmission. But Békésy was more interested in the study of the ear as a main component of the transmission system. He discovered the physical mechanism of stimulation within the cochlea – the snail-shaped tube in the inner ear – by which sound is analysed and communicated in the cochlea. He also developed anatomical techniques that allowed rapid, non-destructive dissection of the cochlea. In 1946 Békésy left Hungary in 1946 for Sweden, where he was a guest of the Karolinska Institute and did research at the Technical Institute in Stockholm. It was during this period that he developed a new type of audiometer that could be operated by the patient and has applications outside the field of hearing. In 1947, he went to the United States and worked in the psycho-acoustic laboratory at the Harvard University, Cambridge, Massachusetts. Békésy was awarded the 1961 Nobel Prize for Physiology or Medicine for his discoveries of the physical mechanism of stimulation within the cochlea. (☞ Jun. 13)

1924
Torsten Nils Wiesel, Swedish neuro-biologist, Nobel laureate, was born in Uppsala, Sweden. Wiesel got a degree in medicine from the Karolinska University, Stockholm in 1954 and moved to USA to take up a research position at the Johns Hopkins University, Baltimore, Maryland, where he began his work with the Canadian-born American neurobiologist, David Hunter Hubel, investigating brain function. They moved to the Harvard University, Cambridge, Massachusetts, in 1959 where they collaborated on studies of the visual

cortex, which is located in the occipital lobes of the cerebrum. Working with laboratory animals, they analysed the flow of nerve impulses from the eye to the visual cortex and were thereby able to discern many structural and functional details of that part of the brain. They noted that various forms of visual stimuli are handled by different sections of the brain. Wiesel was awarded the 1981 Nobel Prize for Physiology or Medicine, jointly with Hubel and American neurobiologist Roger Wolcott Sperry. Wiesel shared half the prize money with Hubel for their discoveries concerning information processing in the visual system, while Sperry received the other half for his discoveries concerning the functional specialisation of the cerebral hemispheres. (☞ Feb. 27, Aug. 20)

1929

Werner Arber, Swiss microbiologist, Nobel laureate, was born in Gränichen, Switzerland. Arber studied natural sciences at the Swiss Polytechnical School in Zürich. In 1953 he became a laboratory assistant at the University of Geneva, where he had access to an electron microscope. He developed preparation techniques for observation of biological specimens using the electron microscope. The electron microscope also helped him become familiar with fundamental questions of bacteriophage physiology and genetics of viruses. This work directed him to microbial genetics, in particular the generation of hybrid recombinants between the genome of a bacterial virus and that of its host, which formed the topic of his PhD thesis that he completed in 1958. Arber did his postdoctoral work at the University of Southern California in USA. In 1960 he returned to Geneva and started a research program studying the effects of radiation on micro-organisms. In the course of these studies he observed that many bacteria possess enzyme systems that enable them to identify if DNA entering a cell originates from other types of organisms than the one they are infecting. If the DNA strand is foreign, it is cut into fragments by enzymes, thereby restricting its entry. These enzymes were named restriction enzymes. Ten years later, when restriction enzymes had been isolated and purified, they were found to be very useful tools in structural and eventually functional studies of genomes. Arber was awarded the 1978 Nobel Prize for Physiology or Medicine, jointly with American microbiologists Daniel Nathans and Hamilton Othanel Smith, for the discovery of restriction enzymes and their application to problems of molecular genetics. (☞ Aug. 23, Oct. 30)

1948

The 200-inch (5.08 m) reflecting Hale telescope at the Palomar Mountain Observatory in California, USA, was dedicated. This was the first in the world with a 200-inch diameter mirror and was, at that time, the largest telescope in the world. The giant mirror of the Hale telescope was made of a particularly heat-resistant and stable glass called Pyrex. After casting, the glass was cooled slowly over an 11-month period. The resulting 20-tonne glass disc then required 11 years of careful grinding and polishing. The telescope was officially named after the American astronomer George Ellery Hale who conceived, designed and promoted this telescope, though he died before it was

completed. For almost 30 years the 200-inch Hale telescope held the honour of being the world's largest reflecting telescope, till the 6-metre reflector of the Special Astrophysical Observatory at Zelenchukskaya, in the Caucasus Mountains of Russia, became operational in 1976.

1965
Astronaut Edward H. White became the first American to space walk during NASA's *Gemini-4* mission, which was commanded by astronaut James McDivitt. White spent a total of 36 minutes floating outside the spacecraft. (☞ Jan. 27)

1977
Archibald Vivian Hill, British physiologist and biophysicist, Nobel laureate (Physiology or Medicine, 1922), who discovered the mechanism of heat production in muscles, died in Cambridge, UK. (☞ Sep. 26)

1990
Robert Norton Noyce, American engineer and co-inventor of the integrated circuit, died in Austin, Texas, USA. (☞ Dec. 12)

June 4

1783
Joseph-Michel and Jacques-Etienne Montgolfier, French brothers, made the first public demonstration of their hot air balloon at the marketplace in Annonay, France. They filled their balloon with heated air by burning straw and wool under the opening at the bottom of the bag. The balloon rose into the air about 1,000 metres, remained there some 10 minutes, and then settled to the ground about two and a half kilometres from where it rose. (☞ Jan. 6, Aug. 26, Sep. 19, Nov. 21)

1877
Heinrich Otto Wieland, German chemist, Nobel laureate, was born in Pforzheim, Germany. Wieland studied at the Universities of Munich, Berlin and Stuttgart, and then returned to the Baeyer Laboratory in Munich where, in 1901, he received his doctorate. He became professor of chemistry at the University of Munich in 1925 and held that post until his retirement in 1950. Wieland's scientific work covered a wide field in the realm of organic chemistry and biochemistry. For several decades organic nitrogenous compounds claimed his attention. He made important contributions to structural organic chemistry with his findings, in 1911, that different forms of nitrogen in organic compounds could be detected and distinguished from one another. The following year he began his research on bile acids, secretions of the liver known for the best part of a century to consist of a large number of substances. He found that the three acids he had isolated – cholic acid, deoxycholic acid, and lithocholic acid – were steroids of similar structure that were related to cholesterol. His later work led him to believe that oxidation in living tissues is a matter of removing hydrogen atoms (dehydrogenation) and not of adding oxygen. This theory proved of great importance to physiology, biochemistry, and medicine. Wieland was awarded the 1927 Nobel Prize for Chemistry for his investigations of the constitution of bile acids and related substances. (☞ Aug. 5)

1910

Christopher Cockerell, British engineer, inventor of the hovercraft, was born in Cambridge, UK. Cockerell studied engineering at the Cambridge University, and joined the Italian physicist and inventor Guglielmo Marconi as a wireless engineer in 1935. During World War II he worked on airborne navigational equipment and on radar. In 1950 he left Marconi and bought a boat building/hire business. In 1954, he performed a crucial experiment using kitchen scales, tin cans, and a vacuum cleaner to show that a stream of air could produce the required lift to keep a boat above the water surface by reducing drag considerably. Later he experimented with the first hovercraft prototype 'SR-N1', using a curtain of high-pressure air ejected from the rim of the craft to 'float' the hull. The prototype weighed 7 tonnes and was capable of moving at the rate of 60 knots. It crossed the English Channel in 1959. The hovercraft entered regular cross-channel service in 1968. (☞ May 30, Jun. 1, Dec. 12)

1916

Robert Francis Furchgott, American pharmacologist, Nobel laureate, was born in Charleston, South Carolina, USA. Furchgott graduated in chemistry in 1937 from the University of North Carolina and in 1940 received his PhD in chemistry from the Northwestern University. He joined the State University of New York-Brooklyn's department of pharmacology in 1956, a position he held until 1988, when he became an adjunct professor at the University of Miami School of Medicine in Florida. Furchgott is recognised for his numerous discoveries and original methodological concepts in the fields of cardiovascular pharmacology, peripheral neural mechanisms, and drug-receptor theory, among others. In the 1970s, many researchers noticed that the innermost cell layer of blood vessels, the endothelium, not only exhibited passive, protective properties but served as an organ, producing factors invol-ved in the control of cardiovascular dynamics. Furchgott showed that endothelium produces a factor responsible for relaxation of blood vessels and, therefore, control of blood pressure. He named this factor EDRF (endothelium-derived relaxing factor), which later turned out to be nitric oxide, produced in small quantities in the human body. These findings led to the discoveries of new, completely unexpected, metabolic pathways and revealed a new concept in biology: intracellular signalling using gaseous molecules. Furchgott's findings also led to the understanding of physiological processes like regulation of blood pressure, penile erection, and tumour formation. Furchgott was awarded the 1998 Nobel Prize for Physiology or Medicine, jointly with fellow American pharmacologists Louis J. Ignarro and Ferid Murad, for their discoveries concerning nitric oxide as a signalling molecule in the cardiovascular system. (☞ May 31, Sep. 14)

2002

A new planet-like object was found circling the Sun more than one and a half billion kilometres beyond Pluto. The discovery was made by Michael Brown of the California Institute of Technology in Pasadena, USA and colleague Chadwick Trujillo using a telescope at the Palomar Observatory in California, and was followed-up with the

Hubble Space Telescope. Known as 2002 LM60, the object measures about 1,280 kilometres; that is, about one-tenth the diameter of Earth and circles the Sun every 288 years. (☞ Nov. 14)

June 5

1819

John Couch Adams, British mathematician and astronomer, was born in Laneast, Cornwall, England. Couch graduated from the St. John's College, Cambridge, in 1843. At that time Uranus was the outermost planet known, but discrepancies were observed in its orbit that pointed to the presence of another planet beyond. Using mathematical calculations, based on the laws of Kepler and Newton, to explain discrepancies with Uranus's orbit, Adams predicted the position of the then unknown planet Neptune. But, because of the delay in England in making a telescopic search for the planet, the credit for the discovery went to a German astronomer named Johann Gottfried Galle; he was assisted by the French astronomer, Urbain Le Verrier, who had independently calculated the size and position of the unknown body. In 1858, Adams became professor of mathematics at the St. Andrews University in Scotland, but he soon returned to Cambridge, to occupy the Lowndean chair of astronomy and geometry until his death. From 1861 he was also director of the university observatory, preferring this post to that of astronomer royal, which was offered to him in 1881. He made valuable studies of the Moon's motions, of the Leonids in the great meteor shower of 1866, and of terrestrial magnetism. (☞ Jan. 21, Sep. 23)

1862

Allvar Gullstrand, Swedish ophthalmologist, Nobel laureate, was born in Landskrona, Sweden. Gullstrand passed his matriculation in 1880 and studied in Uppsala, Vienna in Austria and Stockholm. He earned a doctorate from the University of Stockholm in 1890 and was appointed as a lecturer in ophthalmology in 1891. After holding various appointments as doctor and lecturer and serving on the Swedish Medical Board, he was appointed as the first professor of ophthalmology at the Uppsala University in 1894; he was appointed professor of physiological and physical optics there in 1913. Gullstrand's research mainly concerned the structure and function of the cornea and astigmatism. He expanded the classic theory of the German physicist Hermann von Helmholtz to include the re-disposition of internal parts of the lens structure in accommodation, a mechanism by which the eye can focus for near or far vision within certain limits. Gullstrand showed that although about two-thirds of the accommodation action depends on the increase in convexity of the lens surface, the remaining one-third does not. The remaining third was produced by what Gullstrand termed the 'intracapsular mechanism' and depended on the fact that the eye was not a homogeneous medium. Gullstrand was awarded the 1911 Nobel Prize for Physiology or Medicine for his work on the dioptrics of the eye. (☞ Jul. 28)

1878
James Dewar, British chemist and physicist, exhibited liquid air at the Royal Institution, London. Dewar obtained liquid air at a temperature of −192°C, following the small-scale production of liquid air by the Swiss chemist and physicist Raoul Pierre Pictet and the French physicist Louis Paul Cailletet. In 1893, he gave six well-illustrated Christmas Lectures on 'Air: gaseous and liquid' at the Royal Institution, liquefying some of the air in the room in the presence of the audience. He also demonstrated several physical properties of liquid air. (☞ Jan. 19, Sep. 20, Dec. 22)

1900
Dennis Gabor, Hungarian-born British electrical engineer, Nobel laureate, was born in Budapest, Hungary. Gabor studied electrical engineering first in Budapest and later in Berlin. In 1934 he went to England and joined the British Thomson-Houston Co. Research Laboratory in Rugby, first on an inventor's agreement and later as a member of the research staff. He stayed there for 14 years. He initially worked on a low-pressure gas discharge lamp, which was not a success. Gabor conceived the idea of holography – a lens-less method of three-dimensional photography – in 1947. Initially he took up the task of improving the quality of the flat images seen through an electron microscope. Gabor considered the possibility of correcting the images by optical means, introducing a 'reference wave.' This idea ultimately led to the discovery of holography. Gabor first used conventional filtered-light sources, but because such sources had limitations of either too little light or too diffuse, holography was not commercially feasible until the invention of the laser in 1960. Gabor also did research on high-speed oscilloscopes, communication theory, physical optics, and television; he held more than 100 patents. Gabor was awarded the 1971 Nobel Prize for Physics for his invention and development of the holographic method. (☞ Feb. 8)

1972
The first UN conference on human environment opened in Stockholm, Sweden. The conference called upon governments and peoples to exert common efforts for the preservation and improvement of the human environment, for the benefit of all the people and for their posterity. The day is observed every year as World Environment Day.

June 6

1761
The transit of Venus was observed by German farmer and amateur astronomer, Johann Georg Palitzsch. He was the first to notice a black band linking Venus and the Sun near the beginning and end of the transit – the 'black drop effect' – and correctly interpreted this as evidence that Venus possessed an atmosphere. (☞ Jun. 3, Jun. 8, Nov. 24)

1850
Karl Ferdinand Braun, German physicist, Nobel laureate, was born in Fulda, Hesse-Kassel (now in Germany). Braun received his doctorate from the University of Berlin in 1872. After appointments at Wurzburg, Leipzig, Marburg, Karlsruhe, and Tubingen, he became director of the Physical Institute and professor of physics at the University of Strasbourg in 1895. Braun helped the Italian

physicist and inventor Guglielmo Marconi in increasing the range of his wireless transmission antenna by producing a spark-less antenna circuit (patented in 1899) that linked transmitter power to the antenna circuit inductively. He was one of the first to send electric waves in definite directions. These inventions greatly increased the broadcasting range of a transmitter and have been applied to radar, radio, and television. Braun's discovery of crystalline materials that act as rectifiers, allowing current to flow in one direction only, led to the development of crystal radio receivers. In 1897, Braun demonstrated the first oscilloscope in which a narrow stream of electrons was guided by means of alternating voltage that could trace patterns on a fluorescent screen. This invention, the forerunner of the television tube and radarscope, also became an important laboratory research instrument. Braun was awarded the 1909 Nobel Prize for Physics, jointly with Marconi, in recognition of their contributions to the development of wireless telegraphy. (☞ Apr. 20, Apr. 25)

1882

Henry W. Seeley, American inventor, was issued a patent for the electric flat iron. His design involved the use of a carbon resistance 'of such size and shape that it will heat the face of the iron sufficiently' and 'mounted to be electrically insulated from the metal of the iron'. It also had the following provision 'an adjustable resis-tance may be included in the circuit to regulate the heat', and a 'safety-catch, preferably within the plug to protect the system in case of a short circuit'. Seeley's flat iron weighed almost 7.5 kilogrammes and took a long time to warm up.

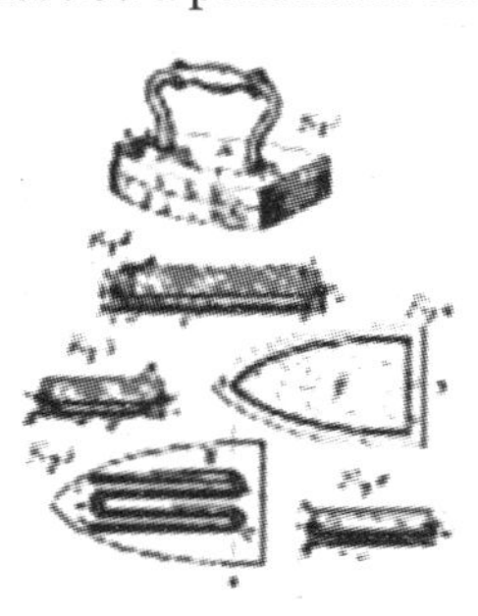

1918

Edwin Gerhard Krebs, American biochemist, Nobel laureate, was born in Lansing, Iowa, USA. Krebs completed his schooling at the Urbana High School, and did undergraduate work in organic chemistry at the University of Illinois in Urbana-Champaign. He received a medical degree from the Washington University School of Medicine, St. Louis, Missouri, in 1943. He joined the faculty of biochemistry at the University of Washington, Seattle, in 1948 and became full professor in 1957. During the 1950s, at Seattle, Krebs and fellow American biochemist Edmond Fischer began investigating the process by which muscle cells obtain energy from glycogen – the form in which the body stores sugar. The American biochemist couple Carl and Gerty Coris had previously demonstrated that cells use an enzyme called phosphorylase to release glucose from glycogen. Extending their work, Krebs and Fischer showed that phosphorylase could be converted from an inactive to an active form by the addition of a phosphate group taken from the compound adenosine triphosphate (ATP), catalysed by enzymes called protein kinases. Krebs and Fischer further showed that phosphorylase is inactivated by the removal of a phosphate group – a reaction catalysed by enzymes called phosphatases. Krebs and Fischer were jointly awarded

the 1992 Nobel Prize for Physiology or Medicine for their discoveries concerning reversible protein phosphorylation as a biological regulatory mechanism. (☞ Apr. 6)

1933

Heinrich Rohrer, Swiss physicist, Nobel laureate, was born in Buchs, St. Gallen, Switzerland. Rohrer was educated at the Swiss Federal Institute of Technology in Zurich, where he received his bachelor's degree in 1955 and his doctorate in 1960. After post-doctoral work at the Swiss Federal Institute and Rutgers University in the United States, Rohrer joined IBM's newly set up Zurich Research Laboratory in 1963 where he studied, among other things, magneto-resistance and anti-ferromagnets before turning his attention to scanning tunnelling microscopy. In 1981, Rohrer and fellow researcher, German physicist Gerd Binnig, built the first scanning tunnelling microscope (STM), in which a fine probe passes within a few angstroms of the surface of the sample. A positive voltage on the probe enables electrons to move from the sample to the probe by the tunnel effect, and the detected current can be used to keep the probe at a constant distance from the surface. As the probe moves in parallel lines, a 3-D image of the surface can be constructed. The STM can image atomic details as tiny as $^1/_{25}$th the diameter of a typical atom, which corresponds to a resolution several orders of magnitude better than the best electron microscope. Rohrer was awarded the 1986 Nobel Prize for Physics, jointly with Binnig and the German electrical engineer Ernst Ruska. Rohrer and Binnig shared half of the prize money for their design of the scanning tunnelling microscope; Ruska received the other half for his fundamental work in electron optics, and for the design of the first electron microscope. (☞ Jul. 20, Dec. 25)

1943

Richard Errett Smalley, American chemist and physicist, Nobel laureate, was born in Akron, Ohio, USA. Smalley received a bachelor's degree from the University of Michigan in 1965, and a PhD from the Princeton University in 1973. He began work at the Rice University in Texas in 1976 and became a professor in the department of physics in January 1990. In 1996, Smalley was appointed Director of the Center for Nanoscale Science and Technology at the Rice University. Smalley's research at the Rice University led to pioneering advances in the development of new experimental techniques and their application to a broad range of vital areas in chemical physics. He is widely known for the discovery and characterisation of C_{60}, a soccerball-shaped molecule dubbed 'Buckminsterfullerene', or 'buckyball', which, together with other fullerenes such as C_{70}, now constitutes the third elemental form of carbon (after graphite and diamond).

Smalley's group has been the first to generate fullerenes with metals trapped on the inside. The group have also produced continuous carbon fibres, which are essentially giant single-fullerene molecules. Just a few nanometres in width, but many centimetres in length, these fullerene fibres are expected to be the strongest fibres ever made. Smalley was awarded the 1996 Nobel Prize for Chemistry, jointly with the

American chemist Robert F. Curl, Jr., and the English chemist Harold W. Kroto, for their discovery of fullerenes. (☞ Aug. 23, Oct. 7, Oct. 28, Nov. 14)

1944
Phillip Allen Sharp, American molecular biologist, Nobel laureate, was born in Falmouth, Kentucky, USA. Sharp majored in chemistry and mathematics from the Union College, in Kentucky. In 1969, he received a doctorate in chemistry from the University of Illinois at Urbana-Champaign. In that year he began working at the California Institute of Technology, Pasadena, California, and two years later moved to the Cold Spring Harbor Laboratory in New York. In 1974 he went to the Massachusetts Institute of Technology, Cambridge, Massachusetts, where he worked on DNA transcription. In 1977 Sharp and his team discovered that the messenger RNA of an adenovirus corresponded to four separate, discontinuous segments of DNA. They found that the segments of DNA that coded for proteins, now called exons, were separated by long stretches of DNA, now called introns, which did not encode protein structure. At the same time, a team working independently under the English molecular biologist Richard J. Roberts at the Cold Spring Harbor Laboratory in New York came up with the same finding. Following the discovery of Sharp and Roberts, it was demonstrated that the discontinuous gene structure, which came to be known as 'split genes', is the most common one found in all higher organisms, including human beings. The discovery of split genes has been of fundamental importance for today's basic research in biology, as well as for research into the development of cancer and other diseases. Sharp and Roberts were jointly awarded the 1993 Nobel Prize for Physiology or Medicine for their discovery of split genes. (☞ Sep. 6)

1948
Louis Lumière, French inventor, who with his brother Auguste, were pioneer manufacturers of photographic equipment and who devised an early motion-picture camera and projector called the Cinématographe, died in Bandol, France. (☞ Oct. 5)

1961
Carl Gustav Jung, Swiss psychologist who founded analytic psychology, died in Küsnacht, Switzerland. (☞ Jul. 26)

1996
George Davis Snell, American geneticist, Nobel laureate (Physiology or Medicine, 1980), who identified the histocompatibility factor in mice, died in Bar Harbor, Maine, USA. (☞ Dec. 19)

June 7

1826
Joseph von Fraunhofer, German physicist who first studied the dark lines of the Sun's spectrum, now known as Fraunhofer lines, died in Munich, Germany. (☞ Mar. 6)

1862
Philipp Lenard, German physicist, Nobel laureate, was born in Pressburg, Hungary (now Bratislava, Slovakia). Lenard studied physics successively at Budapest, Vienna, Berlin and Heidelberg, and in 1886 took

his PhD at Heidelberg. After serving as a lecturer and as an assistant to the German physicist Heinrich Hertz at the University of Bonn in 1893, Lenard became professor of physics successively at the Universities of Breslau (1894), Aachen (1895), Heidelberg (1896), and Kiel (1898). In 1907 he returned to the University of Heidelberg as professor of physics, where he stayed until his retirement in 1931. During his stay at Bonn, Lenard performed some experiments on cathode rays with Hertz. Later, after many experiments with aluminium foil of various thicknesses he published, in 1894, his great discovery that the plate of quartz that had, until then, been used to close the discharge tube, could be replaced by a thin aluminium foil just thick enough to maintain the vacuum inside the tube, yet thin enough to allow the cathode rays to pass out. It thus became possible to study the cathode rays, and also the fluorescence they caused, outside the discharge tube. Lenard concluded from his experiments that the cathode rays could propagate through the air for distances of the order of a decimetre. The experiments also demonstrated that the power of substances to absorb the rays depends on their density and not on their chemical nature and that absorption decreases with increasing velocity of the rays. In similar experiments in 1899 he proved that cathode rays are created when light strikes metal surfaces; this phenomenon later became known as the photoelectric effect. Lenard was awarded the 1905 Nobel Prize for Physics for his work on cathode rays. (☞ May 20)

1875

Upendranath Brahmachari, Indian physician and discoverer of a cure for *kala-azar*, was born in Jamalpore, Bihar, India. Brahmachari passed his BA examination in 1893 from the Hooghly College with honours in mathematics and chemistry. In 1894 he got his MA degree in chemistry from the Presidency College, Kolkata. Brahmachari also studied medicine at the Medical College in Kolkata from where he received his LMS degree in 1899 and the MB degree in 1900. He got his MD degree in 1902 and a PhD in physiology in 1904 from the Calcutta University.

Brahmachari carried out his research on *kala-azar* – a fatal disease – at the Campbell Hospital in Kolkata. After testing several chemicals, including tartar emetic, colloidal antimony, and several compounds of antimony, he found the best results with the urea salt of *para*-amino-phenyl stibnic acid – a compound he named 'urea stibamine,' which showed clear promise as a very potent remedy in the treatment of *kala-azar*. Brahmachari's discovery soon led to an effective treatment of the disease and its almost total disappearance from the country. Brahmachari was a founder member of the National Institute of Science of India (now the Indian National Science Academy). (☞ Feb. 6)

1877

Charles Glover Barkla, British physicist, Nobel laureate, was born in Widnes, Lancashire, UK. Barkla graduated from the University College, Liverpool, in 1898 and obtained his master's degree in 1899. After brief stints at the Trinity College, Cambridge, and the King's College, London, Barkla worked successively as a demonstrator, an assistant lecturer in physics and a special lecturer in advanced electricity at the Liverpool University, from

1905 to 1909. In 1909 he joined the University of Edinburgh as professor of natural philosophy, a position he held until his death. Barkla started his experiments with X-rays in 1902. In 1906, he demonstrated that when X-rays are scattered in certain directions by a block of carbon they are polarised, showing that X-rays were transverse electromagnetic waves, of the same nature as light except for the wavelength. In 1911, he found that X-rays were scattered in the same way as alpha particles, and the results could be used to calculate the number of electrons in atoms of light elements. He noted that, for the light elements, the number of electrons was about half the atomic weight. This technique proved to be particularly useful in the study of atomic structures. Barkla also discovered that each element has its own characteristic X-ray spectrum. Barkla was awarded the 1917 Nobel Prize for Physics for his discovery of the characteristic Röntgen radiation of elements. (☞ Oct. 23)

1896

Robert Sanderson Mulliken, American chemist and physicist, Nobel laureate, was born in Newburyport, Massachusetts, USA. Mulliken took a BSc degree in 1917 at the Massachusetts Institute of Technology, Cambridge, Massachusetts, and a PhD, based on research into the separation of isotopes of mercury by evaporation, from the University of Chicago, Illinois, in 1921. In 1922, he first suggested a method of isotope separation by evaporative centrifuging. In 1923, he went to the Harvard University, Cambridge, Massachusetts, to learn spectrographic techniques and quantum theory, and in 1925 and 1927, travelled to Europe, working with outstanding spectroscopists and quantum theorists. Mulli-ken was particularly influenced by the German physicist Friedrich Hund, who had been working on quantum interpretation of band spectra of diatomic molecules, the same spectra which Mulliken had investigated at Harvard. In 1927 Mulliken worked with Hund to develop his molecular orbital theory of chemical bonding, in which electrons are assigned to states that extend over an entire molecule. Mulliken's molecular orbital method proved to be more flexible and applicable to a larger variety of types of molecules compared to the valence bond method used earlier. He also made major contributions to the theory and interpretation of molecular spectra. Mulliken was awarded the 1966 Nobel Prize for Chemistry for his fundamental work concerning chemical bonds and the electronic structure of molecules by the molecular orbital method. (☞ Oct. 31)

1954

Alan Mathison Turing, English mathematician and logician who pioneered in the field of computer theory, cryptanalysis, cognitive science, and artificial intelligence, died in Wilmslow, Cheshire, England. (☞ Jun. 23)

1978

Ronald George Wreyford Norrish, British chemist, Nobel laureate (Chemistry, 1967), who developed techniques for the study of fast chemical reactions, died in Cambridge, UK. (☞ Nov. 9)

1979

Bhaskara-I, the first Indian experimental Earth observation satellite, was launched by a Soviet Cosmos-C rocket from Kapustin Yar in Russia. Named after Bhaskara-I, the famous Indian mathe-matician of 7th century AD, the 444-kg satellite was placed in a low-Earth orbit ranging in height between 394 and 399 kilometres, and went

round the Earth once every 92.5 minutes. The satellite carried two TV cameras – one working in the visible and the other in the near-infrared band – and a 3-frequency passive microwave radiometer for remote sensing.

1983
Soviet unmanned spacecraft *Venera-16* was launched towards Venus on a mission to map its surface. The spacecraft was equipped with a synthetic aperture radar (SAR) for mapping the surface because nothing else would be able to penetrate the dense clouds of Venus. It was also equipped with a radio altimeter dish antenna, and a large communication dish, which allowed it to send images back to Earth. (☞ Oct. 14)

June 8

1625
Giovanni Domenico Cassini, Italian-born French astronomer who, among others, discovered 'Cassini's division,' the dark gap between the rings A and B of Saturn, was born in Perinaldo, Republic of Genoa (now Italy).

In 1648, Cassini started making astronomical observations from the Panzano Observatory in Bologna. In 1650, he became professor of mathematics and astronomy at the University of Bologna. In 1652–53 he observed a comet and published an account of his observations. Beginning in 1664, Cassini, observing the sky with new powerful telescopes, made a series of new discoveries. In July 1664 he measured the period of rotation of Jupiter on its axis, discovered the bands and spots on the planet, and saw that the planet was flattened at its poles. In 1666 he measured the period of rotation of Mars on its axis, getting a value within three minutes of the correct one, and observed surface features. He published detailed series of observations of the moons of Jupiter in 1668. Cassini's brilliant discoveries gave him an international reputation and led to him being invited to Paris by Louis XIV in 1668. At the Paris Observatory, Cassini continued to make revolutionary discoveries. He was the first to observe four of Saturn's moons: Iapetus, Rhea, Tethys, and Dione. In 1675, he discovered the gap in the ring system of Saturn now known as the 'Cassini division.' He also correctly proposed that the rings were composed of large numbers of tiny moons each orbiting the planet. (☞ Sep. 14, Oct. 25)

1637
The book *Discourse on Method of Rightly Conducting the Reason, and Seeking Truth in the Sciences*, by René Descartes, French mathematician and scientist was published. Regarded as a major work in science and mathematics this work of Descartes ushered in the 'scientific

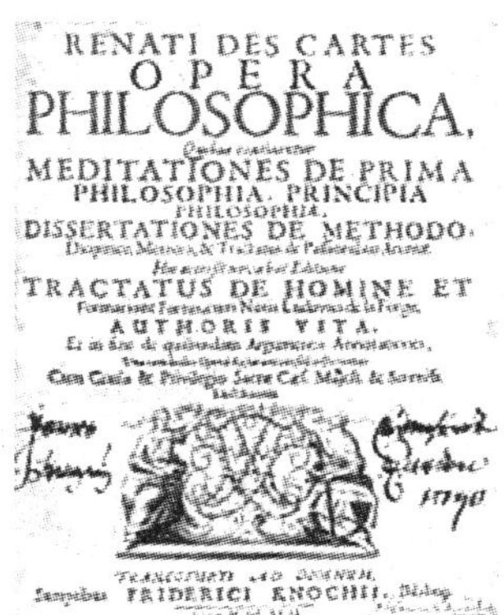
RENATI DES CARTES
OPERA
PHILOSOPHICA,
MEDITATIONES DE PRIMA
PHILOSOPHIA. PRINCIPIA
PHILOSOPHIÆ.
DISSERTATIONES DE METHODO.
TRACTATUS DE HOMINE ET
AUTHORIS VITA.

revolution' along with Galileo and Newton, and swept aside ancient and medieval traditions of philosophical methods and investigation. (☞ Mar. 31)

1695
Christian Huygens, Dutch mathematician, astronomer, and physicist, who founded the wave theory of light, and discovered the true shape of the rings of Saturn, died in The Hague, Netherlands. (☞ Apr. 14)

1869
Ives McGaffey of Chicago, Illinois, USA, patented the vacuum cleaner, a 'sweeping machine' – the first patent for a device that cleaned rugs. He built it in his basement and called it the 'Whirlwind.' It was instantly popular. Before his invention, carpets had to be beaten or shaken. (☞ Aug. 30)

1904
Biresh Chandra Guha, Indian biochemist and nutritionist, who made valuable contributions in the study of biosynthesis of vitamin C, was born in Mymensingh, Bengal (now in Bangladesh). Guha passed his matriculation examination from the Sri Krishna Pathshala and his ISc examination from the City College, both located at Calcutta (now Kolkata). In 1923 he passed the BSc examination with honours in chemistry from the St. Xavier's College, holding the first position in first class and later obtained his MSc degree in organic chemistry from the University College of Science in the city. He started his research career under the renowned India chemist PC. Ray. After working with Ray for a year, Guha went to England with a Tata Memorial Scholarship and a travelling fellowship. He earned his PhD and DSc from the London University. Guha spent most of his life working on nutrition and vitamins of which ascorbic acid (vitamin C) played the most important role. Guha's interest in vitamin C originated from his association and discussions with Albert Szent-Györgyi at Cambridge. Guha with his expertise in bioassay helped Szent-Györgyi understand the nature of the antiscorbutic property of vitamin C. On his return to India Guha joined the Indian Institute for Medical Research (now Indian Institute of Chemical Biology) as a Founder Member. In 1936 he joined Calcutta University as Ghosh Professor and took charge of the Applied Chemistry Department. (☞ Mar. 20)

1916
Francis Harry Compton Crick, British biophysicist, Nobel laureate, was born in Northampton, England. Crick was educated at the Northampton Grammar School and the Mill Hill School, London. He studied physics at the University College, London, obtaining a BSc in 1937. During World War II, Crick interrupted his education to work as a physicist in the development of magnetic mines for use in naval warfare. In 1947 he left the Admiralty to study biology at the Strangeways Research Laboratory, Cambridge University. In 1949 he joined the Medical Research Council Unit in

Cambridge, and in 1951, he started working with the American geneticist and biophysicist James D. Watson at the Cavendish Laboratory of the Cambridge University. Building upon the X-ray diffraction patterns obtained by the British chemist Rosalind Franklin, Crick and Watson together developed the proposal of the double-helical structure of deoxyribonucleic acid (DNA), which they published in 1953. DNA is the molecule ultimately responsible for hereditary control of life functions. By 1961 Crick had evidence to show that each group of three bases – a codon – on a single DNA strand designates the position of a specific amino acid on the backbone of a protein molecule. He also helped to determine which codons code for each of the 20 amino acids normally found in protein, and thus helped clarify the way in which the cell eventually uses the DNA 'message' to build proteins. Crick and Watson were awarded the 1962 Nobel Prize for Physiology or Medicine, jointly with the New Zealand-born British biophysicist Maurice Wilkins, for their discoveries concerning the molecular structure of nucleic acids and its significance for information transfer in living material. (☞ Apr. 6, Dec. 15)

1936

Kenneth Geddes Wilson, American physicist, Nobel laureate, was born in Waltham, Massachusetts, USA. Wilson graduated from the Harvard University, Cambridge, Massachusetts, in 1956. In 1961 he received a PhD from the California Institute of Technology, where he completed a dissertation under the American physicist Murray Gell-Mann. After a year at the European Council for Nuclear Research (CERN), Wilson was appointed assistant professor at the Cornell University in 1963; he was professor of physics at Cornell from 1971 to 1988. From 1988 he taught at the Ohio State University. Wilson's work at Cornell mainly concerned phase transitions – sudden and profound structural changes resulting from variations in environmental conditions – in matter, especially second order phase transitions, which occur throughout the entire volume of a material as soon as the characteristic temperature, called the critical point, is reached. One example of such a transition is the complete loss of ferromagnetic properties of certain metals when they are heated to their Curie points (about 750°C for iron). Wilson's work provided a mathematical strategy for constructing theories that could apply to physical systems near the critical point. Wilson was awarded the 1982 Nobel Prize for Physics for his theory for critical phenomena in connection with phase transitions.

1940

Creation of element 93, neptunium (symbol Np) was announced by the American nuclear physicist Edwin M. McMillan and physical chemist Philip H. Abelson, working at the Berkeley Radiation Laboratory of the University of California, Berkeley. The isotope Np-239 was produced by bombarding uranium with accelerated neutrons in a cyclotron. (☞ Sep. 18)

1947

Eric F. Wieschaus, American developmental biologist, Nobel laureate, was born in South Bend, Indiana, USA. Wieschaus graduated from the University of Notre Dame in Indiana in1969 and then, after doing his

doctorate from the Yale University, New Haven, Connecticut, in 1974, pursued postdoctoral work at the University of Zürich in Switzerland. From 1978 to 1982 he worked with the German developmental geneticist Christiane Nüsslein-Volhard at the European Molecular Biology Laboratory in Heidelberg, Germany. In 1981 he joined the faculty of the Princeton University, Princeton, New Jersey as assistant professor and later became full professor. While at Heidelberg, Wieschaus and Nüsslein-Volhard examined mutations in 40,000 fruit fly families, and discovered that about 5,000 of the fly's 20,000 genes are important to embryonic development of which about 140 are essential. Their research generated the widely accepted model of three sets of genes that control subdivision in the developing embryo, viz., gap genes, which guide general body development; pair-rule genes, which subdivide these general regions into body segments; and segment-polarity genes, which affect specific structures within these segments. The work of Wieschaus and Nüsslein-Volhard helped scientists to better understand congenital mutations in other animals, including humans. Wieschaus was awarded the 1995 Nobel Prize for Physiology or Medicine, jointly with Nüsslein-Volhard and the American geneticist Edward B. Lewis for their discoveries concerning the genetic control of early embryonic development. (☞ May 20, Oct. 20)

1955

Tim Berners-Lee, the inventor of the World Wide Web, the system with text links and multimedia capabilities that made the Internet accessible to mass audiences, was born in London, UK. Berners-Lee grew up in London in the 1960s. His parents worked with a computer firm and they taught him to think unconventionally. He got interested in electronics at an early age. Later, at Oxford, he built his own working electronic computer out of spare parts and a TV set. While doing a six-month stint as a software engineer at CERN, the European Laboratory for Particle Physics, in Geneva, in 1980 he devised a piece of software that could keep 'track of all the random associations one comes across in real life and brains are supposed to be so good at remembering but sometimes mine wouldn't'. Building on ideas that were current in software design at the time, Berners-Lee fashioned a kind of 'hypertext' notebook. Words in a document could be 'linked' to other files on his computer; he could follow a link by number and automatically pull up its related document. Berners-Lee wrote the original Web software himself in 1990 and made it available on the Internet in 1991. He also devised a relatively easy-to-learn coding system – HTML (HyperText Mark-up Language) – that has come to be the lingua franca of the Web, and devised a set of rules that permitted documents to be linked together on computers across the Internet. He called that set of rules HTTP (HyperText Transfer Protocol). He also designed an addressing scheme that gave each Web page a unique location, or URL (universal resource locator). Finally, Berners-Lee put together the World Wide Web's first browser, which allowed users anywhere to view his creation on their computer screen. The World Wide Web debuted in 1991 and has been growing since. Berners-Lee joined MIT's Laboratory for Computer Science in 1994 and remains a leading authority on Internet issues.

1975

Soviet unmanned spacecraft *Venera-9* was launched towards Venus from the Baikanur cosmodrome in Kazakhstan. *Venera-9* carried a lander, which after landing sent back data for 53 minutes. (☞ Oct. 22)

2004

A transit of Venus occurred, the first after a gap of 122 years. The transit was observed by amateur and professional astronomers and millions of people across the world. A transit of Venus across the solar disc occurs only four times every 243 years, and before the advent of modern technology the detailed observation of such transits offered the only reliable method of determining the distance between the Earth and the Sun. (☞ Jun. 3, Dec. 6)

June 9

1776

Amedeo Avogadro, Italian scientist, was born in Turin, Italy. Avogadro took a law degree and practised in the legal profession for several years. In 1800, he began studying mathematics and physics privately and was so interested that he made it his career, beginning in 1806 as a demonstrator at the Academy of Turin. In 1820, he was appointed professor of mathematical physics at the Turin University. Avogadro was the first scientist to distinguish between atoms and molecules. He is best known for his hypothesis, propounded in 1811, which states, 'equal volumes of all gases under the same conditions of temperature contain the same number of molecules'. Unfortunately his hypothesis made little impression at the time, for a variety of reasons. It was only 50 years later at the Karlsruhre Congress in Germany in 1861 that the Italian chemist Stanislao Cannizzaro was able to get large-scale agreement about the usefulness of Avogadro's hypothesis. Avogadro's constant, named after him, also known as Avogadro's number (*N*), is the exact number of molecules in one mole (molecular weight in grams) in any substance. The value of *N* is 6.0221367×10^{23}. (☞ Jul. 9)

1781

George Stephenson, English engineer and principal inventor of the railroad locomotive was born in Wylam, Northumberland, England. Stephenson grew up with a keen interest in machines. His first employment was herding cows but when he was 14 he joined his father at the Dewley Colliery. At the age of 18 he began attending evening classes where he learnt to read and write. When he was 27, Stephenson found employment as an engineman at the Killingworth Colliery. In his spare time, he took the steam engines, including those made by Thomas Newcomen and James Watt, to pieces in order to understand how they were constructed. By 1812 his knowledge of engines resulted in him being employed as the colliery's engine-wright. By 1814 he had constructed a locomotive, the *Blucher* that could pull eight loaded wagons carrying 30 tonnes of coal at 6 kilometres per hour. Not satisfied, he sought to improve his locomotive's power and introduced the 'steam blast', by which exhaust steam was redirected up the chimney, pulling air after it and increasing the draft. Stephenson continued to try and improve his locomotive and in 1815 he changed the design so that the connecting rods drove the wheels directly. These wheels were coupled together by a chain. Over the next five years Stephenson built 16 engines at Killingworth. In September 1825, railroad transportation was born when the first public passenger train,

pulled by Stephenson's *Locomotion*, ran from Darlington to Stockton in England, carrying 450 persons at 24 km/h. From then on, railroad building spread rapidly throughout Britain, Europe, and North America, and Stephenson continued as the chief guide of the revolutionary transportation medium, solving problems in roadway construction, bridge design, and locomotive and rolling-stock manufacture. (☞ Feb. 6, Aug. 12, Sep. 27)

1812
Johann Gottfried Galle, German astronomer who was the first to observe the planet Neptune, was born in Gräfenhainichen, Saxony, Germany. Galle was educated at Berlin and worked at the Berlin Observatory until 1851, when he became professor of astronomy and director of the Breslau Observatory (now in Wroclaw, Poland). In 1846, Urbain Le Verrier, director of the Paris Observatory sent to Galle, details of the position of a new planet, predicted mathematically from its apparent gravitational effect on Uranus, then the outermost known planet. He requested Galle to search for the 'new planet' at a predicted location. Within one hour of beginning their search, Galle and his colleague Heinrich d'Arrest at the Berlin Observatory located Neptune, less than 1° away from the predicted position. Galle was also the first to distinguish the Crêpe Ring, an inner ring around Saturn, in 1838. He also suggested a method of measuring the scale of the solar system by observing the parallax of minor planets, first applying his method to the minor planet Flora in 1873. (☞ Mar. 11, Jul. 10, Sep. 23)

1875
Henry Hallett Dale, English physiologist, Nobel laureate, was born in London, UK. Dale attended the Leys School, Cambridge, and in 1894 entered Trinity College with a scholarship. He graduated through the Natural Sciences Tripos, specialising in physiology and zoology. From 1898 to 1900 he studied physiology at the Trinity College, after which he gained a scholarship and entered the St. Bartholomew's Hospital, London, for the clinical part of the medical course. He earned an MD from the Cambridge University in 1909. In 1914, Dale was appointed Director of the Department of Biochemistry and Pharmacology at the National Institute for Medical Research in London, and became Director of the Institute in 1928. Dale's researches involved a painstaking investigation of the pharmacology of ergot alkaloids. He also worked on the properties of histamine and related substances, including their actions in allergic and anaphylactic conditions. In 1914, he isolated acetylcholine from certain ergot extracts and proved its role as a neurotransmitter, thereby extending the work of the German pharmacologist Otto Loewi, who had shown earlier that a substance released by electrical stimulation of the vagus nerve was responsible for effecting changes in heartbeat. Dale's work established that chemical as well as electrical stimuli are involved in nerve action. Dale and Loewi were jointly awarded the 1936 Nobel Prize for Physiology or Medicine

for their discoveries relating to chemical transmission of nerve impulses. (☞ Jun. 3, Jul. 23)

1913
Patrick Christopher Steptoe, British obstetrician and gynaecologist and a pioneer of fertility treatment, was born in Witney, Oxfordshire, England. Steptoe was educated at the King's College London and graduated from the St George's Hospital Medical School, London in 1939. In 1951 he started to work at the Oldham General Hospital, Oldham, Manchester, England, where he learned the technique of laparoscopy. In Oldham he conducted research on sterilisation and infertility and in 1967 published *Laparoscopy in Gynaecology*, concerning the use of the laparoscope, a narrow tube with a built-in fibre light. In 1966, to help women with blocked Fallopian tubes, a major cause of infertility, he teamed up with Cambridge physiologist Robert Edwards, who had developed a way to fertilise human eggs in the lab. In 1968 Steptoe and Edwards began to collaborate on the fertilisation of human eggs. Steptoe treated volunteer patients with a fertility drug to stimulate maturation of the eggs in the ovary. In 1969, Steptoe became the Director of the Centre for Human Reproduction, Oldham, where he devoted all his time to work on *in vitro* fertilisation. Using laparoscopy, he collected the ova from volunteering infertile women who saw his place as their last hope to achieve a pregnancy. In 1977 Steptoe and Edwards for the first time succeeded in implanting in the womb an egg fertilised outside the body. The first 'test-tube baby' – a baby girl named Louise Joy Brown – was born in 1978. (☞ Mar. 21, Jul. 25, Sep. 27)

1959
Adolf Windaus, German organic chemist, Nobel laureate (Chemistry, 1928), who discovered how vitamin D is produced in the body by the action of sunlight, died in Göttingen, Germany. (☞ Dec. 25)

1961
Camille Guérin, French veterinarian and co-developer, with the French bacteriologist Albert Calmette, of Bacillus Calmette–Guérin, or BCG, a vaccine that was widely used in combating tuberculosis, died in Paris, France. (☞ Dec. 22)

1989
George Wells Beadle, American geneticist, Nobel laureate (Physiology or Medicine, 1958), who helped found biochemical genetics by demonstrating that heredity genes act by regulating definite chemical events, died in Pomona, California, USA. (☞ Oct. 22)

1997
INSAT-2D, the fourth of India's second generation INSAT series of satellites, was launched by the Ariane rocket of European launch provider Arianespace from Kourou, French Guiana. Designed and built by the Indian Space Research Organisation, the 2,070-kg satellite carried 12 transponders in the normal C band and six in the extended C band. It was also fitted with three Ku band transponders providing coverage for Indian metros, and had separate transponders for mobile satellite services, besides one S band transponder to broadcast satellite services. The satellite went out of service in October 1997 after a short circuit in the power lines.

June 10

1832

Nikolaus August Otto, German engineer who developed the four-stroke internal-combustion engine, was born in Holzhausen, Germany. As a young man he began experimenting with gas engines and built his first gasoline-powered engine in 1861. In 1876 he built an internal-combustion engine using the four-stroke cycle, which offered the first practical alternative to the steam engine as a power source. In a petrol engine the cycle begins with the induction of a fuel mixture as the piston goes down on its first stroke. Going up, the piston compresses the mixture in the top of the cylinder. An electric spark ignites the mixture, and the gases produced force the piston down on its third, power stroke. On the fourth stroke the piston expels the burnt gases from the cylinder into the exhaust. Though the design for the four-stroke internal combustion engine was patented in 1862 by a French engineer Alphonse Beau de Rochas, it is commonly known as the Otto cycle since Otto was the first to build such an engine. The light, efficient Otto-cycle engines largely enabled the creation of automobiles, powerboats, motorcycles and airplanes. (☞ Jan. 26, May 30)

1836

André-Marie Ampère, French mathematician and physicist who founded and named the science of electrodynamics, now known as electromagnetism, died in Marseilles, France. (☞ Jan. 22)

1858

Robert Brown, Scottish botanist best known for his description of the natural continuous motion of minute particles in solution, commonly known as 'Brownian movement,' died in London, UK. (☞ Dec. 21)

1902

Americus F. Callahan of Chicago, Illinois was issued the first US patent for a 'window envelope'. This style of envelope saved the expense of printing or labour of writing addresses and in addition saved time in preparing the message for dispatch when the customary addresses are already on the letter paper itself.

1907

Lumière brothers, Louis and Auguste, announced the 'Autochrome Lumière' – the first practical colour photography process. It remained the only colour photography process available in the market until 1935. (☞ Oct. 5, Oct. 19)

1943

The ballpoint pen was patented by the Hungarian inventor Laslo Biro. Biro was a journalist who realised that the quick-drying ink used in the printing trade could be adapted for use in a pen. This new type of pen would write without leaving

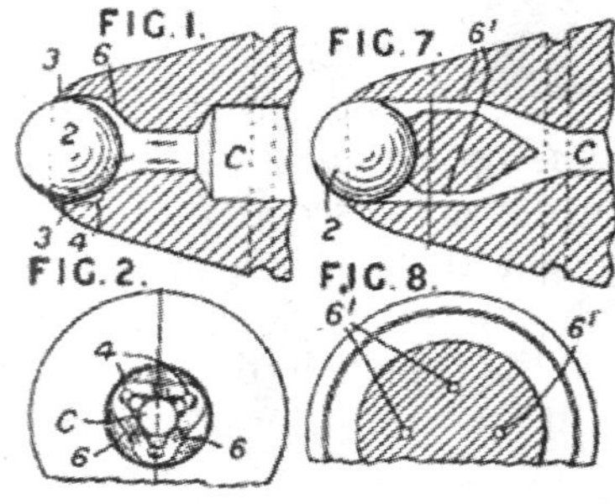

Figure illustrating the ballpoint taken from the patent

blots and would be unaffected by low or changing atmospheric pressure. The first ballpoint was manufactured from Bakelite and used a copper refill unit, which could be replaced when it got over.

2003
American robotic rover *Spirit* was launched from Cape Canaveral on a mission to Mars. The main objective was to look for geological clues about whether parts of formerly Mars had environments wet enough to be hospitable to life. (☞ Jan. 4)

June 11

1644
The invention of the barometer was described by the Italian philosopher and mathematician, Evangelista Torricelli in a letter to his friend Michelangelo Ricci in Rome. In 1643 Torricelli had discovered that the rise of a liquid in a tube, as in the barometer, is due to atmospheric pressure. Torricelli's barometer, also known as the 'torricellian tube', is made by filling a graduated glass tube about 34 inches (860 mm) long with mercury and inverting it in a cup containing mercury. The column of mercury in the tube descends until balanced by the weight of the atmosphere, and its rise or fall under varying conditions is a measure of the change in the atmospheric pressure. At sea level its ordinary height is about 30 inches (760 mm). (☞ Oct. 15, Oct. 25)

1904
Samanta Chandra Sekhar (also known as Pathani Samanta), Indian astronomer and author of *Siddhanta Darpana*, died in Puri, Orissa, India. (☞ Dec. 13)

1910
Jacques-Yves Cousteau, French naval officer, oceanographer, marine biologist and ocean explorer, was born in St. André de Cubzac, near Bordeaux, France. Cousteau always loved the water and in his early teens, he became interested in machines. After passing high school, Cousteau entered the Ecole Navale (Naval Academy) in Brest. In 1933, he joined the French Navy as a gunnery officer. It was during this time that he began his underwater explorations and started working on a breathing machine for longer dives. He was co-inventor of the aqualung, which made scuba diving possible. Cousteau was named a *capitaine de corvette* of the French navy in 1948, and two years later he became president of the French Oceanographic Campaigns. That same year, Cousteau purchased the ship *Calypso* to further his explorations. To finance his trips and increase public awareness of his undersea investigations, Cousteau produced numerous films and published many books. His films include *The Silent World* (1956) and *World Without Sun* (1966). In 1968, Cousteau was asked to make a TV series. For the next 8 years, *The Undersea World of Jacques Cousteau* introduced the public to a world of sharks, whales, dolphins, sunken treasure, and coral reefs. (☞ Jun. 25)

1985
Soviet spacecraft *Vega-1* arrived at Venus en route to Comet Halley and deployed

a 1500-kg descent module comprising a lander and an instrumented balloon to study the atmosphere of the planet. The balloon transmitted data for 56 minutes. The lander reached the surface of Venus and returned valuable data about the atmosphere of Venus and its soil composition for about an hour. After the successful Venus balloon and lander mission, the mother spacecraft continued on to a rendezvous with Comet Halley. (☞ Mar. 6, Dec. 15)

June 12

1849
Lewis P. Haslett of Louisville, Kentucky, USA was granted an American patent for his air-purifying device called 'lungs protector.' Similar in design to present-day gas masks, it used two one-way valves for inhalation and exhalation: one permitting the air to enter through a bulb-shaped filter, and the other permitting the exit of the air directly into the atmosphere. The filter materials used in the device – wool or other porous substance moistened with water – could only trap solid particulates such as dust. It could not filter out gaseous matter.

1899
Fritz Albert Lipmann, German-born American biochemist, Nobel laureate, was born in Koenigsberg, Germany (now Kaliningrad, Russia). Lipmann was educated at the Universities of Koenigsberg, Berlin, and Munich, where he studied medicine. He got his MD degree in 1924 and a PhD in 1927, from the University of Berlin. From 1927 to 1930, he conducted research in the laboratory of the biochemist Otto Meyerhof at the University of Heidelberg. In 1932 he went to Denmark to join the Biological Institute of the Carlsberg Foundation in Copenhagen, as a research associate. There he became interested in the metabolism of fibroblasts. This prompted him to investigate the 'Pasteur effect,' which led to important discoveries on the mechanism of this reaction and on the part played by glycolysis in the metabolism of the cells of embryos. In 1939 Lipmann went to USA and became research associate in the Department of Biochemistry, Cornell Medical School, New York. In 1941 he joined the research staff of the Massachusetts General Hospital in Boston, first as a research associate in the department of surgery, then heading his own group in the biochemical research laboratory of the hospital. In 1945, Lipmann found a catalytically active, heat-stable factor in pigeon liver extracts. He subsequently isolated, named, and determined the molecular structure of this factor, coenzyme A (or CoA), which is now known to be bound to acetic acid as the end product of sugar and fat breakdown in the absence of oxygen. Coenzyme A is one of the most important substances involved in cellular metabolism; it helps in the conversion of amino acids, steroids, fatty acids, and haemoglobins into energy. A new way for the transmission of energy in the cell was demonstrated by this discovery. In 1949 Lipmann became professor of biological chemistry at the Harvard Medical School, Boston. In 1957, he was appointed a Member and Professor of the Rockefeller University, New York, a post he held till his death. Lipman was awarded the 1953 Nobel Prize for Physiology or Medicine, jointly with the German-born British biochemist Hans Adolf Krebs, for his discovery of co-enzyme A and its

importance for intermediary metabolism. (☞ Jul. 24, Aug. 25)

1942

Bert Sakmann, German medical doctor and research scientist, Nobel laureate, was born in Stuttgart, Germany. Sakmann studied at the Universities of Tübingen and Munich, graduating in 1967. He obtained his PhD from the University of Göttingen in 1974. The same year he started research at the Max-Planck-Institut für Biophysikalische Chemie, along with the German physicist Erwin Neher. Together they developed the 'patch clamp' technique to record the small ionic currents (only a few trillionths of an ampere) that flow through a single ion channel in neuronal membranes. The patch clamp technique involves attaching tiny pipettes directly to a cell, making possible very precise measurements of the electrical flow. In 1976 Sakmann and Neher made the first recordings of single ion channel currents in nicotinic acetylcholine receptors from muscle. His subsequent work has led to the development of many important drugs used for diseases of the circulation and nervous system. This research identified the very sophisticated interplay of channels across the membranes of cells, which regulate the flow of sodium, potassium and calcium ions in response to chemical signals acting on the cells during nerve stimulation. Sakmann and Neher were jointly awarded the 1991 Nobel Prize for Physiology or Medicine, for their discoveries concerning the function of single ion channels in cells. (☞ Mar. 20)

1967

Soviet unmanned spacecraft *Venera-4* was launched towards the planet Venus from Baikanur Cosmodrome in Kazakhstan to carry out direct atmospheric studies. (☞ Oct. 18)

1979

The first crossing of the English Channel by an aircraft *Gossamer Albatross*, powered solely by human muscle, was completed successfully. The aircraft was powered using pedals to drive a large two-bladed propeller. Piloted by professional cyclist Bryan Allen it completed the 35.8-km crossing in 2 hr 49 min, achieving a top speed of 29 km/h and an average altitude no more than 15 metres. The *Gossamer Albatross* was constructed using plastic over a carbon fibre frame, with the structure of the wings provided with expanded polystyrene ribs. The entire structure was then wrapped in a thin, transparent plastic. The empty weight of the structure was only 32 kilogrammes, although the gross weight for the Channel flight was almost 100 kilogrammes. To maintain the craft in the air it was designed with very long tapering wings, like those of a glider, allowing some of the flight to be undertaken without constant power.

1982

Karl von Frisch, Austrian zoologist, Nobel laureate (Physiology or Medicine, 1973), whose studies of communication among bees added significantly to the knowledge of the chemical and visual sensors of insects and simple animals, died in Munich, Germany. (☞Nov. 20)

1990
INSAT-1D, the fourth of India's INSAT series of multipurpose geostationary satellites, was launch by a Thor Delta rocket from Cape Canaveral, Florida, USA. The satellite was designed by the Indian Space Research Organisation and built by Ford Aerospace, USA. The satellite carried 12 C-band transponders, 3 S-band transponders, and a very high resolution radiometer, and was meant to provide telecommunication, networking and meteorological services. The satellite was placed in the geostationary orbit at 83°E longitude.

June 13

1773
Thomas Young, English physician and physicist who reinforced the wave theory of light with his study of interference of light, was born in Milverton, Somerset, UK. After completing his schooling, Young entered St Bartholomew's Hospital, London, in 1793 to study medicine. He continued his studies at the Edinburgh University and the Göttingen University in Germany, from where he got his doctorate in physics in 1796. In 1797 he entered the Emmanuel College, Cambridge, where he turned his attention to scientific matters. As a medical student, Young had discovered how the shape of the eye's lens changes to focus on objects at different distances. In 1801, he recognised the cause of astigmatism. Using his famous 'double-slit' experiment, Young demonstrated the wave nature of light and interference fringes, and explained the colours seen in thin films such as soap bubbles. He associated wavelength with colour of light, and suggested that the retina responded to only three 'principal colours' – red, blue and green – which combined to form all the other colours. He also worked on surface tension, elasticity, and gave one of the earliest scientific definitions of energy. Young's modulus is named after his work with elasticity – he determined that if a body is put under pressure the proportional amount it compresses has a constant relationship for a given material. He was also an Egyptologist who helped decipher the Rosetta Stone – a granite stone with writing in two languages, Egyptian and Greek, and thus considered as a key to understanding hieroglyphics. Young was elected Fellow of the Royal Society in 1794. (☞ May 10)

1831
James Clerk Maxwell, Scottish physicist and mathematician, whose work united electricity and magnetism into the concept of the electromagnetic field, was born in Edinburgh, Scotland. Maxwell graduated with a degree in mathematics from the Trinity College, Cambridge, in 1854, and held professorships at the Marischal College in Aberdeen and the King's College in London. He became the first Cavendish Professor of Physics at Cambridge in 1871. In London, around 1862, Maxwell calculated that the speed of propagation of an electromagnetic field is approximately that of the speed of light. He proposed that the phenomenon of light is therefore an electromagnetic phenomenon. The four partial differential equations, now known as Maxwell's equations, first appeared in fully developed form in *Electricity and Magnetism*, published in 1873. These papers from the 1860s contain the basis of all Maxwell's electromagnetic work

– identifying light as an electromagnetic wave (rather than a wave in a material ether), and introducing the hypothetical entity, the displacement current, which made possible a theoretical explanation of the propagation of electromagnetic energy in space. Maxwell's ideas also paved the way for Einstein's special theory of relativity and the quantum theory. (☞Nov. 5)

1870

Jules Bordet, Belgian immunologist, Nobel laureate, was born in Soignies, Belgium. Bordet was educated in Brussels where he graduated as Doctor of Medicine in 1892. In 1894 he went to Paris, France, to work at the Pasteur Institute until 1901 when he returned to Belgium and founded the Pasteur Institute, Brussels of which he was the Director until his death. Bordet's early research, conducted at the Pasteur Institute in Paris between 1894 and 1901, mainly concerned substances in blood that destroy bacteria and red blood cells. In 1895 he found that two components of blood serum are responsible for the rupture of bacterial cell walls: one is a heat-stable antibody found only in vaccinated animals already immune to the bacterium; the other is a heat-sensitive substance found

in all animals and named alexin (it is now called complement). Using this observation he developed a method of diagnosing microbes by blood sera. In 1898, he discovered haemolytic sera that destroyed red blood cells and showed that the mechanism of their action on foreign blood is similar to that by which an antimicrobic serum acts on microbes. Bordet was awarded the 1919 Nobel Prize for Physiology or Medicine for his discoveries relating to immunity. (☞ Apr. 6)

1911

Luis W. Alvarez, American experimental physicist, Nobel laureate, was born in San Francisco, California, USA. Alvarez did his BSc in 1932, an MSc in 1934, and got his PhD in 1936 from the University of Chicago. Early in his scientific career, Alvarez worked concurrently in the fields of optics and cosmic rays. For several years

he concentrated his experimentation in the field of nuclear physics. In 1937 he gave the first experimental demonstration of the existence of the phenomenon of K-electron capture by nuclei. Another early development was a method for producing beams of very slow neutrons. This method subsequently led to a fundamental investigation of neutron scattering in ortho- and para-hydrogen, and to the first measurement of the magnetic moment of the neutron. During World War II, at the Massachusetts Institute of Technology, Alvarez designed a landing system for aircrafts and a radar system for locating planes. While at the Los Alamos Laboratory, Alvarez developed the detonators for setting off the plutonium bomb. Later, he helped develop the hydrogen bubble chamber for detecting subatomic particles, which led to the discovery of over 70 previously unknown elementary particles and resulted in a major revision of nuclear theories. Alvarez was responsible for the design and construction of the Berkeley 40-foot proton linear accelerator, which was completed in 1947. Since 1967 Alvarez devoted most of this time to the study of cosmic rays, using balloons and superconducting magnets. Alvarez was awarded the 1968 Nobel Prize for Physics for his decisive contributions to elementary particle physics, in particular the discovery of a large number of resonance states, made possible through his development of the technique of using hydrogen bubble chamber and data analysis. (☞ Sep. 1)

1938
Charles Édouard Guillaume, French physi-cist, Nobel laureate (Physics, 1920), who discovered iron–nickel alloys with unique properties, died in Sèvres, France. (☞ Feb. 15)

1972
Georg von Békésy, Hungarian-born American physicist and physiologist, Nobel laureate (Physiology or Medicine, 1961), who discovered the physical mechanism of hearing, died in Honolulu, Hawaii. (☞ Jun. 3)

1983
The American space probe *Pioneer-10* reached a distance beyond all the known planets, as it crossed the orbit of Neptune, the outermost planet at that time. *Pioneer-10* was powered by radioisotope thermoelectric generators, which could operate in deep space far beyond the useful range of solar cells, allowing the probe to transmit information about the interstellar space environment until 2003. *Pioneer-10* was the first spacecraft to fly-by Jupiter, the largest planet in the solar system. (☞ Jan. 22, Mar. 2, Apr. 25, Dec. 3)

June 14

1736
Charles-Augustin de Coulomb, French physicist, pioneer in electrical theory, was born in Angouleme, France. Coulomb graduated in 1761 and served as a military engineer for France in the West Indies. In 1777 he invented the torsion balance – an instrument he used with great success for the experimental investigation of the distribution of charge on surfaces and of the laws of electrical and magnetic force. With this invention, Coulomb was able to formulate the principle, now known as Coulomb's law, which states that the force between two electrical charges is proportional to the product of the charges and inversely proportional to the square of the distance between them. Coulombic force is one of the principal forces involved in atomic reactions. In 1779 he published an important investigation of the laws of friction in the treatise *Theorie des machines simples* ('Theory of simple machines'). When the French Revolution began in 1789, Coulomb was deeply involved with his scientific work. He retired to Blois, France, to continue research in magnetism, friction, and electricity. After the end of the Revolution, Coulomb came out of retirement and assisted the new government in devising a metric system of weights and measures. The unit of quantity used to measure electrical charges, the coulomb, is named after him. (☞ Aug. 23)

1864
Alois Alzheimer, German psychiatrist, who recognised the disease named after him, was born in Markbreit, Bavaria, Germany. Alzheimer was educated at medical schools in Tübingen, Berlin, and Würzburg, received his MD from the University of Würzburg, and was appointed a clinical assistant at the Irrenanstalt (asylum) in Frankfurt. In the early 1900s he became renowned for his published works on cerebro-arteriosclerosis. His early career as a professor of psychology in Germany subsequently led to his work with the German psychiatrist Franz Nissl; together they established the pathologic anatomy of mental illness. At a meeting of German psychiatrists in November 1907, Alzheimer reported on a 51-year old patient with signs of dementia, which he described as a disorder of the cerebral cortex. Later the condition of presenile dementia was designated 'Alzheimer's disease', a progressive, degenerative disorder that affects the brain and the first symptoms of which are loss of memory, inability to think and understand, and gradual behavioural changes. In 1912, Alzheimer was appointed professor of psychiatry at the University of Breslauand, and director of the Psychiatric and Neurologic Institute. (☞ Dec. 19)

1868
Karl Landsteiner, Austrian immunologist and pathologist, Nobel laureate, was born in Vienna, Austria. Landsteiner studied medicine at the University of Vienna, graduating in 1891. Well grounded in chemistry, he spent the next five years in various laboratories in Zurich in Switzerland, and Wurzburg and Munich in Germany to further his knowledge. After returning to Vienna, Landsteiner resumed his medical studies at the Vienna General Hospital. From 1898 to 1908, he was a research assistant at the University Department of Pathological Anatomy in Vienna, where he discovered basic differences in different human blood that explained the danger involved in indiscriminate transfusions of whole blood. In 1901 he was able to show that there were at least three major types of human blood that vary according to the kinds of antigens attached to the outer envelope of the red blood cells. Landsteiner labelled the three blood groups as types A, B, and O. A fourth group, possessing both antigens A and B, called type AB, was found in the following year. In 1940, he also discovered the Rhesus, or Rh factor, which explained complications in pregnancy and birth when the Rh factor of the mother and baby do not match. Such blood type analysis proved useful also in genetic and legal applications. As a result of his research in immunology and the chemistry of antigens and serological reactions, Landsteiner made valuable contributions in haemolysis and in methods of studying poliomyelitis. In 1909, Landsteiner was appointed professor of pathology at the University of Vienna. In 1922 he move to USA to join the Rockefeller Institute for Medical Research in New York City, and later became a US citizen. Landsteiner was awarded the 1930 Nobel Prize for Physiology or Medicine for his discovery of human blood groups. (☞ Jun. 26)

1924
James Whyte Black, British pharmacologist, Nobel laureate, was born in Uddingston, Scotland. Black earned a medical degree

from the University of St. Andrews in Scotland in 1946. He taught at various universities for the next 10 years and then joined the Imperial Chemical Industries as a senior pharmacologist in 1958. He became head of biological research at the Smith Kline & French Laboratories in 1964, and joined the Wellcome Research Laboratories as director of therapeutic research in 1978. From 1984 to 1993, he was professor of analytical pharmacology at the King's College Hospital Medical School, London. Black's contribution has been mainly to basic scientific and clinical knowledge in cardiology, both as a physician and as a basic scientist. He invented propranolol, a beta-blocker, which is used to treat abnormal heart rhythms, angina and high blood pressure, prevent migraine attacks, alleviate symptoms of anxiety, and makes adrenalin flow more slowly. Propanalol has revolutionised the medical management of angina pectoris, and is considered to be one of the most important contributions to clinical medicine and pharmacology of the 20th century. Black discovered another important drug, cimetidine. The discovery led to a class of drugs used to treat stomach and duodenal ulcers. He also developed a drug that could block the histamine receptors that stimulate the secretion of gastric acid in the stomach. Black was awarded the 1988 Nobel Prize for Physiology or Medicine, jointly with the American pharmacologists George Herbert Hitchings and Gertrude Belle Elion, for their discoveries of important principles for drug treatment. (☞ Jan. 23, Apr. 18)

1946
John Logie Baird, Scottish engineer and inventor, the first man to televise pictures of moving objects, died in Bexhill-on-Sea, Sussex, England. (☞ Aug. 13)

1961
Kariamanikkam Srinivasa Krishnan, Indian physicist, Fellow of the Royal Society, co-discoverer of the Raman effect, died in New Delhi, India. (☞ Dec. 4)

1975
Soviet unmanned spacecraft *Venera-10* was launched towards Venus from the Baikanur Cosmodrome in Kazakhstan on a landing mission. (☞ Oct. 25)

June 15

1667
The first fully documented human blood transfusion was administered in France by Jean Baptiste, personal physician to Louis IV. He transfused the blood of a sheep to a 15-year old boy (the boy later died).

1752
Benjamin Franklin, American inventor and scientist, proved that lightning and electricity were related while flying a kite with a key attached. He was the first to use a lightning conductor to protect his house against lightning damage. Earlier, in 1750, he had devised an experiment involving a sentry box with a pointed rod on its roof, to be erected on a hilltop or in a church steeple, with a rod attached to a Leyden jar that would collect the electrical charge, and thus prove lightning to be a form of electricity. (☞ Jan. 17, Apr. 17)

1844
Charles Goodyear, American inventor, received a patent for vulcanisation,

described as 'an improvement in India-rubber fabrics'. He had spent several years experimenting, trying to improve natural rubber, which became sticky in the summer heat and brittle in winter. He invented vulcanisation in 1839 when he accidentally dropped a mixture of natural rubber and powdered sulphur on a hot stove in his kitchen. The heat completed the process of making weather-proof rubber. Despite the value of his invention, he died in poverty in 1860. (☞ Dec. 29)

1915

Thomas Huckle Weller, American physician and virologist, Nobel laureate, was born in Ann Arbor, Michigan, USA. Weller was educated at public schools in Ann Arbor, and later at the University of Michigan from where he graduated in 1936. In 1940 he took his MD degree from the Harvard University, Cambridge, Massachusetts, and began his clinical training at the Children's Hospital in Boston. His work here was, however, interrupted by military service and he joined the Army Medical Corps during World War II. He then returned to the Children's Medical Center, Boston, for a further year of clinical training and, in 1947, joined the American virologist John Franklin Enders in the organisation of the new research division of infectious diseases at the Children's Medical Center. In 1949 he was appointed assistant director of this division and subsequently held the appointments of instructor in the department of comparative pathology and tropical medicine, and associate professor in the Harvard Medical School of Public Health. Working with Enders and the American paediatrician and virologist Frederick C. Robbins, Weller succeeded in propagating poliomyelitis virus in laboratory suspensions of human embryonic skin and muscle tissue. This made it possible to study the virus 'in the test tube' – a procedure that led to the development of polio vaccines. Weller was also the first (with the American physician Franklin Neva) to achieve the laboratory propagation of rubella (German measles) virus and to isolate chicken pox virus from human cell cultures. Weller was awarded the 1954 Nobel Prize for Physiology or Medicine, jointly with Enders and Robbins, for their discovery of the ability of poliomyelitis viruses to grow in cultures of various types of tissue. (☞ Feb. 10, Aug. 25)

1917

John B. Fenn, American chemist, Nobel laureate, was born in New York, USA. In 1928 the family moved to Berea, Kentucky, where Fenn studied in an institution known at the time as the Berea College and Allied Schools from where he graduated in 1937 with a BA degree in chemistry. Three years later he received a PhD in chemistry from the Yale University, New Haven, Connecticut. He spent the next 12 years in industry before returning to the Princeton University, Princeton, New Jersey, where he became Director of Project SQUID – a program of research in fields relating to jet propulsion, such as combustion fluid flow and heat transfer. In 1956 he became professor in the department of mechanical engineering, which subsequently merged with the department of aeronautical engineering to become a department of aerospace and mechanical engineering sciences. While at Princeton he became involved in molecular beam research, which he continued at the Yale University where he became professor of applied science and chemistry in a newly organised department of engineering and applied science. While at Yale he developed the technique known as electrospray mass spectrometry. In the method called electrospray ionisation that Fenn published in 1988, charged droplets of protein solution are produced which shrink as the water evaporates. Eventually freely hovering protein ions remain. Their masses are then determined by setting them in motion and measuring their time of flight over a known distance. Mass spectrometry is a very important analytical method used in practically all chemistry laboratories the world over. Previously only fairly small molecules could be identified, but Fenn's work made it possible to analyse biological macromolecules as well.

Fenn was awarded the 2002 Nobel Prize for Chemistry, jointly with the Japanese chemist Koichi Tanaka and Swiss chemist Kurt Wüthrich, for the development of methods for identification and structure analyses of biological macromolecules. Fenn and Tanaka shared half the prize money for their development of soft desorption ionisation methods for mass spectrometric analyses of biological macromolecules, while Wüthrich received the other half for his development of nuclear magnetic resonance spectroscopy for determining the three-dimensional structure of biological macromolecules in solution. (☞ Aug. 3, Oct. 4)

1919

The first non-stop flight across the Atlantic was completed by two British aviators, Captain John Alcock and Lieutenant Arthur Whitten Brown, in a modified *Vimy IV* biplane. They had taken off from Lester's Field, near St. Johns, Newfoundland, Canada, the previous day and landed at Clifden in Ireland after a flight lasting 16 hours and 27 minutes. For their accomplishment Alcock and Brown were presented with Lord Northcliffe's *Daily Mail* prize of £10,000 by Winston Churchill, who was then Britain's Secretary of State. A few days later both men were knighted at Buckingham Palace by King George V in recognition of their pioneering achievement.

Alcock (left) and Brown

1969

Darashaw Nosherwan Wadia, Indian geologist, Fellow of the Royal Society, an authority on Himalayan geology and founder of the Wadia Institute of Himalayan Geology, died in New Delhi, India. (☞ Oct. 25)

1971

Wendell Meredith Stanley, American biochemist, Nobel laureate (Chemistry, 1946), who purified and crystallised viruses, demonstrating their molecular structure, died in Salamanca, Spain. (☞ Aug. 16)

1985

Soviet spacecraft *Vega-2* arrived at Venus en route to Comet Halley and deployed a 1500-kg descent module comprising a lander and an instrumented balloon to study the atmosphere of the planet. The balloon transmitted data for 46.5 hours. The lander reached the surface of Venus and returned valuable data about the atmosphere of Venus and its soil composition for about one hour. After the successful Venus balloon and lander mission, the mother spacecraft continued on to rendezvous with Comet Halley. (☞ Mar. 9, Dec. 21)

June 16

1657
A patent for a pendulum clock was issued to the Dutch mathematician, physicist and astronomer Christiaan Huygens. Huygens invented the pendulum clock in 1656, based on the pendulum introduced by the Italian scientist Galileo Galilei. Huygens's pendulum clock was regulated by a mechanism with a 'natural' period of oscillation and had an error of less than 1 minute a day – the first time such accuracy had been achieved. His later refinements reduced his clock's errors to less than 10 seconds a day. (☞ Apr. 14)

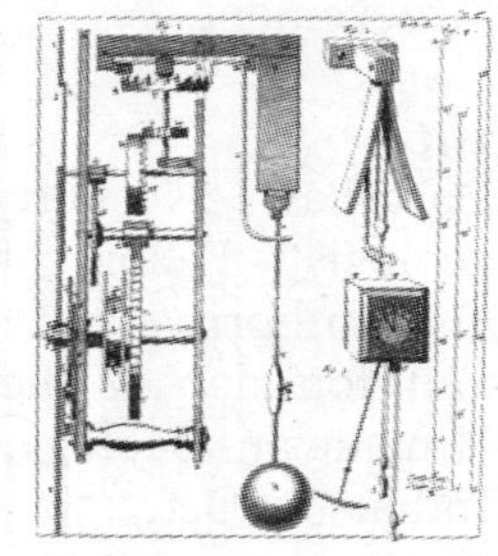

1858
John Snow, English physician, who discovered that cholera is transmitted through contaminated water, died in London, UK. (☞ Mar. 15)

1874
The Cavendish Laboratory was opened at the University of Cambridge, England. It started as a teaching laboratory with a regular course of instruction and was initially located on the New Museums Site in the centre of Cambridge. Named after the famous English physicist Henry Cavendish, the laboratory housed the department of physics of the Cambridge University from the early 1870s, at the time of the first Cavendish Professor James Clerk Maxwell, until it moved to new laboratories in West Cambridge in the early 1970s. Both Maxwell, and after him Lord Rayleigh much expanded the knowledge of physics. The third Cavendish Professor was J.J. Thomson, discoverer of the electron. The Cavendish Laboratory has had significant influence on biology, mainly through the application of X-ray crystallography to the study of structures of biological molecules. American geneticist James D. Watson and British biophysicist Francis Crick, who discovered the double-helix structure of DNA, carried out their Nobel Prize-winning work at the Cavendish Laboratory. Other areas in which the laboratory has been influential since 1950 include superconductivity, electron microscopy, and radio astronomy.

1878
Crawford Williamson Long, American physician who was the first to use ether as anaesthetic in surgery, died in Athens, Georgia, USA. (☞ Mar. 30, Nov. 1)

1897
Georg Wittig, German chemist, Nobel laureate, was born in Berlin, Germany. Wittig studied at Kassel and Marburg, and taught at Freiburg, and Tubingen, before joining the faculty of the University of Heidelberg in 1956, where he became emeritus in 1965. His main research was into organic reactions of alkali metals and elaborations of carbon-based chemistry. In 1954 he first demonstrated the chemical reaction now

known after him, using a class of organic phosphorus compounds called ylides. In Wittig reaction, a carbonyl compound (aldehyde or ketone) reacts with an ylide to form a hydrocarbon with a double bond, an olefin (alkene). The Wittig reaction allows the joining of two carbon atoms from different molecules in a predictable manner to form a double bond. It has been widely used for synthesising complex compounds such as vitamin A, vitamin D derivatives, steroids, and biological pesticides. Wittig was awarded the 1979 Nobel Prize for Chemistry, jointly with the American chemist Herbert Charles Brown, for their development of the use of boron- and phosphorus-containing compounds, respectively, into important reagents in organic synthesis. (☞ May 22, Aug. 26)

1902

Barbara McClintock, American geneticist, Nobel laureate, was born in Hartford, Connecticut, USA. McClintock studied plant genetics at the Cornell University's College of Agriculture in Ithaca, New York, from where she obtained her undergraduate degree in 1923 and a doctorate in botany in 1927. From 1931–1933 she was supported by a fellowship from the National Research Council; and from 1941 until her death she worked at the Cold Spring Harbor Laboratory in New York. While teaching at Cornell she helped show through her laboratory experiments with maize that trait-determining information is transferred between chromosomes in the crossing-over stage of cell division. By observing and experimenting with variations in the colouration of kernels of maize, she discovered that genetic information is not stationary; that genes are transposable – they can move around – on and between chromosomes. She called these mobile genetic elements 'transposons.' By tracing pigmentation changes in maize and using a microscope to examine its large chromosomes, she isolated two genes that she called 'controlling elements'. She found that the controlling elements could move along the chromosome to a different site, and that these changes affected the behaviour of neighbouring genes. She theorised that these genes controlled the genes that were actually responsible for pigmentation of maize kernel, and suggested that these transposable elements were responsible for new mutations in pigmentation or other characteristics. Though her initial discovery was made in 1944, McClintock confirmed, controlled, and extended her observations for six years, publishing at last in 1950. McClintock was awarded the 1983 Nobel Prize for Physiology or Medicine for her discovery of mobile genetic elements. (☞ Sep. 2)

1944

Prafulla Chandra Ray, Indian chemist, educationist, and nationalist visionary, died in Kolkata, India. (☞ Aug. 2)

1963

Russian cosmonaut Valentina Tereshkova was launched into space aboard *Vostok-6*, and became the first woman to travel in space. During her flight she made 48 orbits around the Earth. Tereshkova never made a second trip into space. She became an important member of the Communist Party and a representative of the Soviet government. (☞ Mar. 6, Jun. 19)

1977

Wernher von Braun, German-born American rocket engineer who played a prominent pioneering role in all aspects of rocketry and space exploration, died in Alexandria, Virginia, USA. (☞ Mar. 23)

June 17

1832

William Crookes, British chemist and physicist noted for his discovery of the element thallium and for his cathode ray studies, was born in London, UK. After attending grammar school at Chippenham, Crookes entered the Royal College of Chemistry in London. In 1854 he became superintendent of the meteorological department at the Radcliffe Observatory in Oxford, and in 1855 was appointed lecturer in chemistry at the Chester training college. After 1880 he did all his scientific work in his private laboratory at home. The scientific work of Crookes extended over the fields of both chemistry and physics. His first great discovery was that of the element thallium, announced in 1861, through the study of spectrum. By this work his reputation became firmly established, and he was elected a fellow of the Royal Society in 1863. One of Crookes' most well-known areas of study was the properties of highly rarefied gases, in the course of which he built the 'Crookes radiometer', in which a system of vanes, each blackened on one side and polished on the other, is set in rotation when exposed to radiant energy. Of more fundamental importance were his researches on the passage of the electrical discharge through rarefied gases. He found that as the gas pressure inside a sealed discharge tube was reduced to near vacuum the dark space round the negative electrode extended, while rays, now known as cathode rays, proceed from the electrode. He investigated the properties of the rays, showing that they travel in straight lines, cause phosphorescence in objects upon which they impinge, and by their impact produce great heat. He believed the rays to consist of streams of particles of ordinary molecular magnitude; today we know these rays consist of streams of negative electrons. Crookes's experimental work in this field was fundamental to the development of atomic physics. (☞ Apr. 4)

1867

Joseph Lister, British surgeon and medical scientist who was the founder of antiseptic medicine, became the first surgeon to perform surgery under antiseptic conditions at the Glasgow Royal Infirmary. He was able to announce at a British Medical Association meeting that his wards at the infirmary had remained clear of sepsis for nine months. (☞ Apr. 5)

1920

François Jacob, French biologist, Nobel laureate, was born in Nancy, France. Jacob received his MD degree in 1947 and a doctorate in science in 1954 from the Faculty of Medicine and the Faculty of Science, respectively, in Paris. In 1950, he joined the Pasteur Institute under André Lwoff. In 1956 he was appointed laboratory director, and in 1960 head of the department of cell genetics, at the Institute. In 1964 he was appointed professor of cell genetics at the Collège de France. In 1977 he became a member of the French Academy of Sciences. At the Pasteur Institute, Jacob discovered that the genes of a bacterium are arranged linearly in a ring and that the ring can be broken at almost any point. In 1958 Jacob, in collaboration with the French biochemist Jacques Monod began studies of the regulation of bacterial enzyme synthesis. One of their first major contributions

was the discovery of regulator genes, so called because they control the activities of structural genes. Jacob and Monod also proposed the existence of an RNA messenger, a partial copy of deoxyribonucleic acid (DNA) that carries genetic information to other parts of the cell. Jacob and his team found that a regulator gene produces a repressor substance that prevents an operator gene from providing messenger RNA, blocking production of protein. Jacob was awarded the 1965 Nobel Prize for Physiology or Medicine, jointly with Lwoff and Monod, for their discoveries concerning genetic control of enzyme and virus synthesis. (☞ Feb. 9, May 8)

1940

Arthur Harden, English biochemist, Nobel laureate (Chemistry, 1929), known for his study of the chemistry of the fermentation of sugar by yeast juice, died in Bourne, Buckinghamshire, UK. (☞ Oct. 12)

2001

Donald James Cram, American chemist, Nobel laureate (Chemistry, 1987), who for the first time synthesised three-dimensional cyclic molecules that mimic the chemical behaviour of molecules found in living systems, died in Palm Desert, California, USA. (☞ Apr. 22).

June 18

1799

William Lassell, amateur English astronomer, Fellow of the Royal Society, who discovered several moons of planets, was born in Bolton, Lancashire, England. Although he was in the brewery business Lassel was deeply interested in astronomy and built an observatory at Starfield near Liverpool, where in 1843 he constructed a reflecting telescope with a 24-inch-diameter speculum metal mirror. This was the first telescope with a large mirror mounted 'equatorially' to allow easy tracking of the stars. It paved the way for future developments in reflector technology, leading to the large equatorial instruments seen in many national observatories around the world. Lassel used this telescope at Malta to discover several planetary moons, including Triton, a moon of Neptune in 1846, and Ariel and Umbriel, moons of Uranus in 1851. Ten years later he erected a 48-inch reflector, which he used to observe and catalogue hundreds of new nebulae. He was elected a fellow of the Royal Society of London in 1849 and was president of the Royal Astronomical Society from 1870 to 1872. (☞ Oct. 5, Oct. 24)

1845

Alphonse Laveran, French physician, pathologist, and parasitologist, Nobel laureate, was born in Paris, France. Alphonse completed his education in Paris at the Collège Saint Baube and later at the Lycée Louis-le-Grand. After attending a course at the Public Health School in Strasbourg he was appointed as a resident medical student in the Strasbourg civil hospitals in 1866. Laveran served as an army surgeon in the Franco-German War (1870–71). In 1874 he was appointed to the Chair of Military Diseases and Epidemics at the École de Val-de-Grâce, previously occupied by his father, where he taught military medicine. In 1897, he joined the Pasteur Institute, Paris, where he spent the rest of his life as a researcher.

As a French military surgeon in Algeria in 1888, Laveran discovered the parasite that causes human malaria in the red blood cells of patients; it was a protozoan, which was later named *Plasmodium*. He founded the medical field of protozoology, doing important work on other protozoal diseases, including sleeping sickness and *kala-azar*. Laveran was awarded the 1907 Nobel Prize for Physiology or Medicine in recognition of his work on the role played by protozoa in causing diseases. (☞ May 18)

1901
Peter Cooper Hewitt, American inventor was issued a patent for a mercury vapour lamp. The energy efficiency of the lamp proved to be two or three times that of the contemporary incandescent lamp. Creating a nearly shadow-free light with less glare, the mercury lamp immediately found wide use for industrial and street lighting in the United States.

1905
Per Teodor Cleve, Swedish chemist and geologist who discovered the elements holmium and thulium, died in Uppsala, Sweden. (☞ Feb. 10)

1918
Jerome Karle, American crystallographer, Nobel laureate, was born in New York, USA. Karle graduated from the City College of New York in 1937, received his master's degree from the Harvard University, Cambridge, Massachusetts, in 1938, and his PhD in physical chemistry from the University of Michigan in 1943. He worked on the Manhattan Project at the University of Chicago during 1943–44, and then joined the Naval Research Laboratory, Washington, where his college mate crystallographer Herbert A. Hauptman joined him. They devised mathematical equations to describe the arrangements of numerous spots that appear on photographic film as a result of a crystal's diffraction of X-rays. Their equations enabled the location of atoms within the crystal's molecules to be pinpointed based upon an analysis of the intensity of the spots. This eventually led to the development of the direct methods for deducing the crystal structure of simple biological molecules such as hormones, vitamins and antibiotics, which previously took years to determine. Karle was awarded the 1985 Nobel Prize for Chemistry, jointly with Hauptman, for their outstanding achievements in the development of direct methods for the determination of crystal structures. (☞ Feb. 14)

1926
Allan Rex Sandage, American astronomer who discovered the first quasar, was born in Iowa City, Iowa, USA. Sandage was educated at the University of Illinois, Chicago, and the California Institute of Technology, Pasadena, California, where he earned his doctorate under the German-born American astronomer Walter Baade. In 1952 he joined the staff of the Mt. Wilson and Palomar Observatories, now the Observatories of the Carnegie Institution of Washington, where he began as a graduate student assistant to the American astronomer Edwin P. Hubble. His research primarily concerned stellar astronomy and observational cosmology. Following Hubble's death in 1953, Sandage continued with Hubble's passion to study the expansion of the universe. He demonstrated in the

early 1950s that the observed characteristics of the light and colour of the brightest stars in various globular clusters indicate that the clusters can be arranged in order according to their age. This information provided insight into stellar evolution and galactic structure. In 1960, Sandage made the first optical identification of a quasar. He found that there was a faint optical object with an extraordinary spectrum at the position of the compact radio source 3C 48 in the constellation Virgo. This faint object was emitting more intense radio waves and ultraviolet radiation than a typical star. This enigmatic object was called a quasi-stellar object, or 'quasar' in short. Later he discovered many more similar objects, including some that did not emit in radio wavelengths. (☞ Sep. 28, Nov. 20)

1932

Dudley R. Herschbach, American chemist and educator, Nobel laureate, was born in San Jose, California, USA. Herschbach attended the Stanford University, Stanford, California, and received the BS degree in mathematics in 1954 and the MS in chemistry in 1955. He obtained his PhD in chemical physics from the Harvard University, Cambridge, Massachusetts, in 1958 on a doctoral thesis titled 'Internal rotation and microwave spectroscopy'. He was appointed an assistant professor of chemistry in 1959 at the University of California, Berkeley, where he developed what is known as the 'crossed molecular beam technique', a technique in which beams of molecules are brought together at supersonic speed under carefully controlled conditions. This procedure enabled a detailed, molecule-by-molecule investigation of the changes that occur in chemical reactions. Herschbach's chief experiments dealt with reactions of alkali atoms with alkyl iodides. Herschbach developed the field further after he returned to Harvard as professor of chemistry. He studied a wide range of alkali reactions and found several prototype modes of reaction dynamics, which could be correlated, with the electronic structure of the target molecule. Herschbach was awarded the 1986 Nobel Prize for Chemistry, jointly with Yuan T. Lee and John C. Polanyi, for their contributions concerning the dynamics of chemical elementary processes. (☞ Jan. 23, Nov. 29)

1971

Paul Karrer, Swiss chemist, Nobel laureate (Chemistry, 1937), who determined the structure of vitamin A, died in Zürich, Switzerland. (☞ Apr. 21)

1983

American space shuttle *Challenger* was launched on its second flight from the Kennedy Space Center, Florida, USA, carrying the first woman crewmember. Sally K. Ride became America's first woman in space. The other members of the crew were Robert L. Crippen, Frederick H. Hauck, John M. Fabian, and Norman E. Thagard. During the mission, two communications satellites – *ANIK C-2* for TELESAT Canada and *PALAPA-B1* for Indonesia – were deployed. The mission also carried the Shuttle Pallet Satellite (*SPAS*-1) built by a West German aerospace firm. *SPAS-1* was launched free of the payload bay, flew alongside *Challenger* for several hours, then was grabbed by the shuttle's remote arm and brought back into the payload bay. Ten experiments mounted on *SPAS-1* performed research in forming metal alloys in microgravity and in the use of remote sensing scanner. (☞ Jan. 28, Apr. 4, Jun. 24)

June 19

1623
Blaise Pascal, French mathematician, physicist, was born in Clermont-Ferrand, France. From an early age Pascal showed an inclination toward mathematics. He studied privately, tutored mostly by his father, Etienne, who was a scientist and a government official. At the age of 16 Pascal formulated one of the basic theorems of projective geometry, known as Pascal's theorem and described it in his *Essai pour les coniques* ('Essay on conics'). in 1639. In 1642 he invented the first mechanical adding machine – forerunner of the modern digital computer – and later the hydraulic press, and the syringe. In 1648 he proved by experimentation that the level of the mercury column in a barometer is determined by an increase or decrease in the surrounding atmospheric pressure rather than by a vacuum. He performed an experiment in which he found that the height of the mercury dropped with altitude, indicating pressure decreases with altitude. Six years later, together with the French mathematician Pierre de Fermat, Pascal formulated the mathematical theory of probability, which has become important in such fields as mathematical, and social statistics and as a fundamental element in the calculations of modern theoretical physics. Around 1650 he formulated Pascal's law, which states that in a fluid at rest in a closed container a pressure change in one part is transmitted without loss to every portion of the fluid and to the walls of the container. The hydraulic press works on this principle. The unit of pressure in the SI system is named in his honour; a 'pascal' is a pressure of one newton per square metre. (☞ Aug. 19)

1897
Cyril Norman Hinshelwood, British chemist, Nobel laureate, was born in London, England. Hinshelwood was educated at the Westminster City School and the Oxford University where he received his MA and DSc degrees and became professor of chemistry in 1937. After retiring from Oxford in 1964 he became a senior research fellow at the Imperial College, London. The early researches of Hinshelwood concerned molecular kinetics and he subsequently worked on chemical changes in the bacterial cell, providing physicochemical explanations for the biological responses of bacteria to changes in environment, especially development of drug resistance. His findings proved to be of great importance in later research work on antibiotics and therapeutic agents. He also worked on reaction rates and reaction mechanisms, particularly that of the combination of hydrogen and oxygen to form water. He showed that the products of this reaction help to spread the reaction further in what is essentially a chain reaction. Hinshelwood was awarded the 1956 Nobel Prize for Chemistry, jointly with the Russian physical chemist Nikolay Semenov, for their researches into the mechanism of chemical reactions. (☞ Apr. 15, Oct. 9)

1906
Ernst Boris Chain, German-born British biochemist, Nobel laureate, was born in Berlin, Germany. Chain was educated at

the Luisengymnasium, Berlin, and attended the Friedrich-Wilhelm University, Berlin, from where he graduated in chemistry in 1930. In 1933, he emigrated to England, where he spent two years working on phospholipids at the School of Biochemistry, Cambridge. In 1935 he joined the University of Oxford, where in 1939, with the Australian pathologist Howard Walter Florey, he began a systematic study of antibacterial substances produced by micro-organisms. This led to his best known work, the re-investigation of penicillin, which had been described by the Scottish bacteriologist Alexander Fleming nine years earlier, and to the discovery of its chemotherapeutic action. Later he worked on the isolation and elucidation of the chemical structure of penicillin and other natural antibiotics. After 1948 he carried out research on the carbohydrate–amino acid relationship in nervous tissue, a study of the mode of action of insulin, fermentation technology, penicillinase-stable penicillins, lysergic acid production in submerged culture, and the isolation of new fungal metabolites. Chain served as the director of the International Research Centre for Chemical Microbiology, Superior Institute of Health, Rome, from 1948 until 1961. He then joined the faculty of the Imperial College, University of London, where he was professor of biochemistry from 1961 to 1973, and professor emeritus and senior research fellow from 1973 to 1976. Chain was awarded the 1945 Nobel Prize for Physiology or Medicine, jointly with Fleming and Florey, for the discovery of penicillin and its curative effect in various infectious diseases. (☞ Aug. 6, Aug. 12, Sep. 24)

1910

Paul J. Flory, American physical chemist, Nobel laureate, was born in Sterling, Illinois, USA. Flory graduated from the Manchester College, Indiana, in 1931 and did his PhD from the Ohio State University, Columbus, Ohio, in 1934. He then joined the Central Research Department of the DuPont Company, where he worked with the American chemist Wallace H. Carothers, inventor of nylon and neoprene, and became interested in exploration of the fundamentals of polymerisation and polymeric substances. After the death of Carothers in 1937, Flory left DuPont and joined the Basic Science Research Laboratory of the University of Cincinnati for a period of two years. With the outbreak of World War II and the urgency of research and development on synthetic rubber, he returned to the industry and researched commercially successful polymers, the processes by which polymers form, and their properties in bulk and in solution. He showed how an understanding of the sizes and shapes of these flexible molecules is important to establish relationships between their chemical structures and their physical properties. From 1961 to 1976 he was a professor of chemistry at the Stanford University, California. Flory was awarded the 1974 Nobel Prize for Chemistry for his fundamental achievements, both theoretical and experimental, in the physical chemistry of the macromolecules. (☞ Apr. 27, Sep. 9)

1922

Aage Niels Bohr, Danish physicist, Nobel laureate, son of physicist Niels Bohr, was born in Copenhagen, Denmark. Bohr was educated at the University of Copenhagen,

where he received his master's degree in 1946 and a doctorate in 1954. During the 1940s he worked as assistant to his father, Niels Bohr, on the development of the atomic bomb at Los Alamos, New Mexico. From 1946 he was associated with the Niels Bohr Institute of Theoretical Physics, founded in Copenhagen by his father, first as a research fellow and from 1956 as a professor of physics at the University of Copenhagen. He succeeded his father as director in 1963. Inspired by the theory propounded by the American physicist James Rainwater that not all atomic nuclei are spherical, Bohr, in collaboration with the American-born Danish physicist Ben R. Mottelson, conducted experiments in the early 1950s, which led to the discovery that the motion of subatomic particles can indeed distort the shape of the nucleus. The discovery helped explain many nuclear properties by showing that nuclear particles can vibrate and rotate so as to distort the shape of the nucleus from the expected spherical symmetry into an ellipsoid. It was also important for the understanding and development of nuclear fusion. Bohr was awarded the 1975 Nobel Prize for Physics, jointly with Rainwater and Mottelson, for the discovery of the connection between collective motion and particle motion in atomic nuclei and the development of the theory of the structure of the atomic nucleus based on this connection. (☞ Jul. 9, Dec. 9)

1963

Russian cosmonaut Valentina Tereshkova, who became the first woman to travel in space aboard *Vostok-6*, safely landed back on Earth. (☞ Mar. 6, Jun. 16)

1976

The American unmanned spacecraft *Viking-1* went into orbit around Mars. (☞ Jul. 20, Aug. 20)

1981

APPLE, India's first geostationary satellite, was launched by the European Space Agency's Ariane rocket from Kourou in French Guiana. Designed and built in India by the Indian Space Research Organisation, the experimental telecommunications satellite carried a C-band transponder. It was placed in the geostationary orbit at 102°E longitude and was used for the first nationwide broadcast of television programmes in India. APPLE is an acronym for *A*riane *P*assenger *P*ay*L*oad *E*xperiment. (☞ Jul. 22)

June 20

1840

Samuel F.B. Morse, American painter and inventor received a patent for telegraphy

S. F. B. Morse.
Telegraph Signs.
No 1,647.
Patented Jun. 20, 1840.

signs (Morse code) in which each letter of the alphabet is represented by a combination of short or long, on or off signals. Morse had applied for a patent in September 1837 after he had discovered how to use a relay to reinforce a feeble current after it travelled some distance, and he was able to use his machine in his apartment through a circuit of about 520 metres of wire strung back and forth across his room at the New York University. This patent was later re-issued with amendments in 1846 and 1848. (☞ Apr. 27)

1861

Frederick Gowland Hopkins, British biochemist, Nobel laureate, was born in East Sussex, England. Hopkins received his BSc degree from the University College, London in 1888 and his medical degree from the Guy's Hospital, London in 1894, and then taught physiology and toxicology in the same college for four years. In 1901 Hopkins discovered the amino acid tryptophan, isolated it from protein, and eventually showed that certain amino acids, including tryptophan, known as essential amino acids, cannot be manufactured by some animals from other nutrients but must be supplied in the diet. Hopkins later discovered that no animal can live on a mixture of pure protein, fat, and carbohydrate, even when mineral salts are added, and termed the missing factors – later called vitamins – 'accessory substances'. Hopkins then worked on the metabolic changes occurring in muscular contractions and rigor mortis and devised a new colour reaction for lactic acid. In 1921 he isolated a substance, which he named glutathione and showed that it was widely distributed in rapidly multiplying cells of plants and animals. Later he proved it to be the tripeptide of glutamic acid, glycine and cysteine. Hopkins was awarded the 1929 Nobel Prize for Physiology or Medicine, jointly with the Dutch physician and pathologist Christiaan Eijkman, for his discovery of the growth-stimulating vitamins. (☞ May 16, Aug. 11)

1917

James Mason Crafts, American chemist who, in collaboration with the French organic chemist Charles Friedel discovered the chemical process known as the Friedel–Crafts reaction, died in Cambridge, Massachusetts, USA. (☞ Mar. 8)

1958

Kurt Alder, German chemist, Nobel laureate (Chemistry, 1950), who in collaboration with the German organic chemist Otto Paul Hermann Diels developed the Diels–Alder reaction, died in Cologne, Germany. (☞ Jul. 10)

1966

Georges Lemaître, Belgian astronomer and cosmologist, who first proposed the Big Bang theory of the origin of the universe, died in Louvain, Belgium. (☞ Jul. 17)

1996

American space shuttle *Columbia*, carrying astronauts Terence T. Henricks, Kevin R. Kregel, Susan J. Helms, Richard M. Linnehan, Charles E. Brady, Jr., Jean-Jacques Favier, Robert Brent Thirsk, Pedro Duque, and Luca Urbani, was launched on its 20th flight from the Kennedy Space Center, Florida, USA. The primary payload of the mission was the *Life and Microgravity Spacelab* (*LMS*). Developed by the European Space Agency, the LMS was a re-usable laboratory designed to allow scientists to perform experiments in microgravity conditions while orbiting the Earth. More than 40 experiments grouped into the following two areas were performed: life sciences, which included human physiology and space biology, and microgravity science, which included basic fluid physics investigations, advanced semiconductor and metal alloy materials processing, and medical research in protein crystal growth. (☞ Jul. 7)

2005
Jack S. Kilby, American electrical engineer, Nobel laureate (Physics, 2000), who developed the first monolithic electronic circuit called integrated circuit, or IC, which later came to be known as the 'chip', died in Dallas, Texas, USA. (☞ Nov. 8)

June 21

1874
Anders Jonas Ångström, Swedish physicist, one of the founders of spectroscopy, after whom a unit of wavelength of light is named, died in Uppsala, Sweden. (☞ Aug. 13)

1957
Johannes Stark, German physicist, Nobel laureate (Physics, 1919), who discovered the Doppler effect in canal rays and the splitting of spectral lines in electric fields, died in Traunstein, Germany. (☞ Apr. 15)

1994
William Wilson Morgan, American astronomer, who was the first astronomer to demonstrate that the Milky Way galaxy has a spiral structure, died in Williams Bay, Wisconsin, USA. (☞ Jan. 3)

2004
SpaceShipOne became the first private manned spacecraft to reach an altitude of more than 100 kilometres during a suborbital flight. It was piloted by test pilot Mike Melvill, who became the first civilian to fly a spaceship out of the Earth's atmosphere. (☞ Sep. 29, Oct. 4)

June 22

1633
Galileo Galilei, Italian astronomer, was forced to confess the falsehood of the heliocentric theory of the Polish astronomer Nicolaus Copernicus before the Inquisition in Rome. Threatened with torture, he publicly confessed that he had been wrong to have said that the Earth moves around the Sun. Legend has it that after his confession, Galileo quietly whispered 'And yet, it moves'. As a punishment he was put under house arrest until his death in 1642. (☞ Jan. 8, Feb. 15, Oct. 31)

1675
The Royal Greenwich Observatory was established in England by the Royal Warrant of Charles II. John Flamsteed was appointed as the first Astronomer Royal. The observatory was primarily used for practical astronomy – navigation, timekeeping, and determination of star positions. In 1767 the observatory began

publishing *The Nautical Almanac*, which established the longitude of Greenwich as a baseline for time calculations. The almanac's popularity among navigators led in part to the adoption in 1884 of the Greenwich meridian as the Earth's prime meridian (0° longitude) and the international time zones.

1887

Julian Sorell Huxley, English biologist, philosopher, educator, and author, Fellow of the Royal Society, was born in London, England. Huxley was educated at Eton and the Balliol College, Oxford. His scientific researches included important work on hormones, developmental processes, ornithology, and ecology. He taught at the Rice Institute, Houston, Texas, USA from 1912 to 1916, and then became professor of zoology at the King's College, London University. He was secretary to the Zoological Society of London from 1935 to 1942. From 1946 to 1948 he served as the first director-general of the United Nations Educational, Scientific, and Cultural Organization (UNESCO). Although Huxley was one of the foremost biologists of his time, his most important contribution was not primarily in genetic evolution but in culture, and the interrelationships between the two processes. He was perhaps the first evolutionary theorist to recognise the reality and causal significance of human society and culture: a reality which materialism – by the very nature of its premises – is forced to ignore. Huxley greatly influenced the modern development of embryology, systematics, and studies of behaviour and evolution. He wrote many popular articles and essays, especially on ornithology and evolution, and co-produced several history films. He was awarded the UNESCO Kalinga Prize for popularisation of science in 1953. (☞ Feb. 14)

1973

The crew of Skylab-2, which rendezvoused with the American space station *Skylab*, splashed down safely after 28 days in space. During the mission the crew fitted a sunshade to cool the station, which had heated up due to the loss of a meteoroid shield during launch. It completed 404 orbits, and 'space walks' totalling 6 hours 20 minutes were carried out. (☞ Feb. 8, May 14, May 25, Jun. 28, Jul. 11, Sep. 25, Nov. 16)

1978

A moon of Pluto was discovered by American astronomer James W. Christy of the Naval Observatory in Flagstaff, Arizona, USA, when he obtained a photograph of Pluto that showed the orb to be distinctly elongated. The elongations appeared to change position with respect to the stars over time. The moon, which was found orbiting Pluto at a distance of about 19,000 kilometres with a period of 6.4 days, was named Charon, after the boatman in Greek mythology who took the souls of the dead across the River Styx to Pluto's underworld.

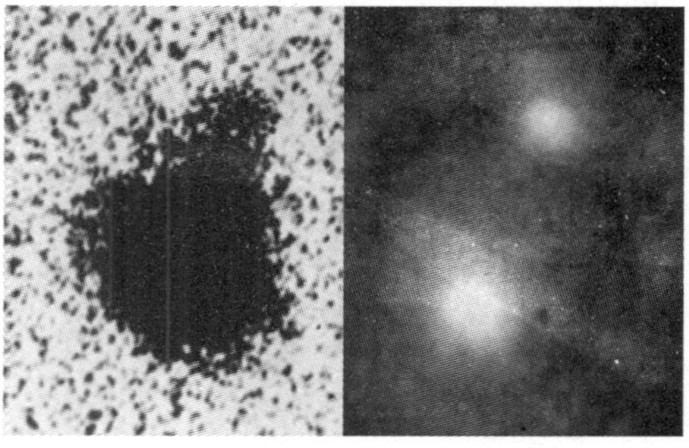

1990

Ilya Mikhaylovich Frank, Russian physicist, Nobel laureate (Physics, 1958), who, with his compatriot Igor Y. Tamm, theoretically explained the mechanism of Cherenkov radiation, died in Russia. (☞ Oct. 23)

June 23

1912

Alan Mathison Turing, English mathematician and logician, was born in London, England. Turing learnt deep mathematics and read Albert Einstein's papers on relativity while at school. He also read about quantum mechanics in the English astronomer Arthur Eddington's *The Nature of the Physical World*. Turing entered the King's College, Cambridge, in 1931 to study mathematics. In 1933, he read the Hungarian-born American John von Neumann's 1932 text on quantum mechanics. Turing graduated in 1934 and, in the spring of 1935, attended an advanced course on the foundations of mathematics. In 1936 he published 'On computable numbers, with an application to the *Entscheidungs problem* (German: decision problem).' It is in this paper that Turing introduced an abstract machine, now called a 'Turing machine' – a hypothetical computing device, which moved from one state to another using a precise finite set of rules and depending on a single symbol it read from a tape. The Turing machine could write a symbol on the tape, or delete a symbol from the tape. The most remarkable feature of Turing's work on Turing machines was that he was describing a modern computer before technology had reached the point where construction was a realistic proposition. From November 1942 until March 1943 Turing was in the United States liaising over decoding issues and also on a speech secrecy system. Turing was not directly involved with the successful breaking of the complex German codes, but his ideas proved of the greatest importance in this work. By 1951 he was working on the application of mathematical theory to biological forms. In 1952 he published the first part of his theoretical study of morphogenesis, the development of pattern and form in living organisms. Turing was elected a Fellow of the Royal Society of London in 1951, mainly for his work on Turing machines. (☞ Jun. 7)

1995

Jonas Edward Salk, American physician and medical researcher, who developed the first safe and effective vaccine for poliomyelitis, died in La Jolla, California, USA. (☞ Oct. 28)

2004

Thomas Gold, Austrian-born British astronomer who propounded the steady-state theory of the universe, died in Ithaca, New York, USA. (☞ May 22)

June 24

1881

The first generally recognised photograph of a comet was taken by the French astronomer P.J.C. Janssen and American photographer J.W. Draper.

1883

Victor Francis Hess, Austrian-born physicist, Nobel laureate, was born in Waldstein, Styria, Austria. Hess received his PhD from the University of Graz in Austria in 1906. From 1910 to 1920 he worked as an assistant at the Institute of Radium Research of the Vienna Academy of Sciences. Hess started his research on atmospheric electricity in 1911, trying to find out the source of an ionising background radiation that appeared to permeate the atmosphere. Initially it was believed that the radiation came from the Earth, but when balloons were sent up carrying devices to measure the radiation, it was found that the intensity of the radiation increased with height. The possibility of the

Sun being the source was ruled out, as the radiation remained high even during a solar eclipse. Thus Hess proved that the high-energy radiation came from outer space. In 1925, the American physicist Robert Andrews Millikan confirmed Hess's theory and called the radiation 'cosmic rays'. Cosmic-ray research soon emerged as an important branch of physics and led to the discovery of several new fundamental particles – including the positron. Hess was awarded the 1936 Nobel Prize for Physics, jointly with American physicist Carl David Anderson, for his discovery of cosmic radiation. (☞ Sep. 3, Dec. 17)

1915

Fred Hoyle, English astronomer, Fellow of the Royal Society who coined the term 'Big Bang' mockingly, in rejection of the theory, was born in Bingley, Yorkshire, England. Hoyle was educated in mathematics and theoretical physics at the University of Cambridge. During World War II he worked on the development of radar and returned to Cambridge in 1945 as a lecturer in mathematics. His early work on stellar evolution led to his famous collaboration with the English astronomer Margaret Burbidge, her husband Geoffrey Burbidge, and the American astro-physicist William Alfred Fowler on the synthesis of the elements beyond helium in stars. He also did theoretical work on the formation of – in older, hotter stars – carbon, oxygen, and eventually elements up to iron as a result of fusion of helium nuclei. In 1948 Hoyle, together with mathematician Hermann Bondi, and the Australian-born British astronomer Thomas Gold, developed the 'steady state cosmological model', according to which the universe has always remained as it is today. Hoyle provided a mathematical theory of the model consistent with the general theory of relativity. He also worked on numerous other problems in theoretical astrophysics and cosmology. In 1967 he founded the Institute of Theoretical Astronomy at Cambridge and served as its first director. Hoyle made several long-term visits to the California Institute of Technology and the Mt. Wilson and Palomar Observatories in USA. In 2000, he along with Geoffrey Burbidge and the Indian theoretical astrophysicist Jayant V. Narlikar proposed a quasi-steady state

theory. Hoyle was a leading populariser of science and became Britain's best-known astronomer in 1950 with his broadcast lectures on the nature of the universe. He was the author or co-author of several science fiction books, a play, and more than twenty non-fiction books. (☞ May 22, Jul. 19, Aug. 20, Nov. 1)

1927

Martin Lewis Perl, American physicist, Nobel laureate, was born in New York, USA. Perl graduated from the Brooklyn Polytechnic Institute with a degree in chemical engineering in 1948. After working

as a chemical engineer for two years, he studied nuclear physics at the Columbia University in New York from where he received his PhD in 1955. Thereafter he worked as an instructor and associate professor at the University of Michigan before joining the faculty of the Stanford University in California in 1963. Between 1974 and 1977, Perl and his colleagues discovered, through a series of experiments, at the Stanford Linear Accelerator Center (SLAC), a national laboratory operated by the Stanford University, that the collisions between electrons and positrons formed heavy leptons, later called 'tau' particles, that decay in less than a trillionth of a second into neutrinos and either an electron or a muon. He also discovered the 'anti-tau' particle, which decays into neutrinos and either a positron or an anti-muon. Perl's discovery of the tau was the first sign that a third 'family of fundamental building blocks' existed (the first family included the electron and the neutrino; the second family included the muon and the muon neutrino). Some years later, a further fundamental particle was discovered – one of the family's two quarks, the bottom quark. Its other quark, the top quark, was discovered almost 18 years later. Perl was awarded the 1995 Nobel Prize for Physics, jointly with the American physicist Frederick Reines, for pioneering experimental contributions to lepton physics. Perl received half the prize money for the discovery of the tau lepton, while Frederick received the other half for the detection of the neutrino. (☞ Mar. 16)

1984

American space shuttle *Challenger* carrying astronauts Sally K. Ride, Robert L. Crippen, Frederick H. Hauck, John M. Fabian, and Norman E. Thagard landed at the Edwards Air Force Base, California, USA, after completing its second flight during which three satellites were deployed. (☞ Apr. 4, Jun. 18)

1999

The *Cassini–Huygens* spacecraft made the second fly-by of Venus at a distance of 600 kilometres, en route to Saturn. (☞ Jan. 14, Apr. 26, Jul. 1, Aug. 18, Oct. 15, Dec. 24, Dec. 30)

June 25

1864

Walther Hermann Nernst, German scientist, Nobel laureate, was born in Briesen, Prussia (now in Germany). Nernst spent his early school years at Graudentz, and subsequently went to the Universities of Zürich in Switzerland, Berlin in Germany, and Graz in Austria, studying physics and mathematics. He graduated from the Wurzburg University in 1887 with a thesis on electromotive forces produced by magnetism in heated metal plates. In 1894 he joined the Göttingen University as professor of physical chemistry. He founded the Institute for Physical Chemistry and Electrochemistry and became its director. In 1905 he was appointed professor of chemistry and later of physics, in the University of Berlin. Nernst's early studies were in electrochemistry and in 1889 he elucidated the theory of galvanic cells by assuming an 'electrolytic pressure of dissolution' which forces ions from electrodes into solution and which was opposed

to the osmotic pressure of the dissolved ions. In 1906 Nernst announced his heat theorem, or 'third law of thermodynamics', in which he postulated that entropy has an absolute value such that crystalline materials have zero entropy at absolute zero (−273° C) temperature. In practical terms, this theorem implies the impossibility of attaining absolute zero, since as a system approaches absolute zero, the further extraction of energy from that system becomes more and more difficult. Nernst was awarded the 1920 Nobel Prize for Chemistry in recognition of his work in thermochemistry. (☞ Nov. 18)

1894

Hermann Oberth, German scientist who is considered to be one of the founders of modern astronautics, was born in Nagyszeben, Austria-Hungary (now Sibiu, Romania). Oberth's interest in rocketry was sparked at the age of 11. His mother gave him a copy of Jules Verne's *From the Earth to the Moon*, a book which he later recalled he read 'at least five or six times and, finally, knew by heart'. It was a young Oberth, who discovered that many of Verne's calculations were not simply fiction, and that the very notion of interplanetary travel was not as fantastic as had been assumed by the scientific community. Oberth enrolled in the University of Munich, Germany, to study medicine, but his studies were interrupted with the outbreak of World War I when he served in the Austro-Hungarian army. After the war, he remained in Germany – his home city having become part of Romania – and resumed his studies at the Munich University, but this time studying physics, especially astronautics. In 1922, his PhD degree dissertation based on his liquid-propellant rocket design was rejected by the University of Heidelberg. He then had the 92-page work privately published as the book *Die Rakete zu den Planetenräumen* (The Rocket into Interplanetary Space) in 1923. The book, which mathematically explained how rockets could achieve a speed that would allow them to escape Earth's gravitational pull, received widespread recognition. In 1931 Oberth received a patent for a liquid-propellant rocket from the Romanian Patent Office, and launched his first rocket in May 1931 near Berlin. In 1938 Oberth joined the faculty of the Technical University of Vienna, Austria. He became a German citizen in 1940 and in 1941 moved to the German rocket development centre at Peenemünde, where he worked with Wernher von Braun, his former assistant. After World War II he spent a year in Switzerland as a rocket consultant, and in 1950 moved to Italy, where he worked on solid-propellant anti-aircraft rockets for the Italian navy. In 1955 von Braun, by this time the head of a US army ballistic missile effort at Huntsville, Alabama, invited Oberth to work for him on his program. Oberth worked for a short time on these efforts, but in 1959 he retired and returned to Feucht, where he lived the rest of his life. (☞ Dec. 29)

1907

Johannes Hans Jensen, German physicist, Nobel laureate, was born in Hamburg, Germany. Jensen studied physics, mathematics, physical chemistry and philosophy at the Universities of Hamburg and Freiburg. After obtaining his PhD in physics in 1932 from the Hamburg University he became a scientific assistant at the Institute for Theoretical Physics of the University of Hamburg from where he obtained a DSc in 1936. In 1949 he was

appointed professor at the University of Heidelberg. Jensen's main research work concerns the shell model of the nucleus, which he proposed independently of the German-born American physicist Maria Goeppert-Mayer in 1949. According to the shell nuclear model, an atomic nucleus is not a random aggregation of neutrons and protons but rather a structure of shells, or spherical layers, of differing radii, each of which is filled with neutrons and protons. Jensen was awarded the 1963 Nobel Prize for Physics, jointly with Goeppert-Mayer and the Hungarian-born American physicist Eugene Paul Wigner. Jensen and Goeppert-Mayer shared half the prize money for their discoveries concerning the nuclear shell structure, while Wigner received the other half for his contributions to the theory of the atomic nucleus and the elementary particles, particularly through the discovery and application of fundamental symmetry principles. (☞ Feb. 11, Jun. 28, Nov. 17)

1911

William Howard Stein, American biochemist, Nobel laureate, was born in New York, USA. Stein had his early education at the Lincoln School of Teachers College of the Columbia University in New York. He received his PhD degree from the Columbia College of Physicians and Surgeons, New York, in 1938 and joined the staff of the Rockefeller Institute for Medical Research (now the Rockefeller University), also in New York where he became professor in 1954. At Rockefeller, Stein and fellow American biochemist Stanford Moore were assigned to devise accurate analytical methods for the determination of the amino acid composition of proteins. They started to separate amino acids first on columns of potato starch and then on columns of ion exchange resins and worked on amino acid analysis, and later on structural analysis of proteins and the structure of ribonuclease. Between 1949 and 1963 they deciphered how ribonuclease catalyses the digestion of food. Stein was awarded the 1972 Nobel Prize for Chemistry, jointly with Moore and Christian B. Anfinsen. Stein and Moore shared half the prize money for their contribution to the understanding of the connection between chemical structure and catalytic activity of the active centre of the ribonuclease molecule, while Anfinsen received the other half for his work on ribonuclease, especially concerning the connection between the amino acid sequence and the biologically active conformation. (☞ Feb. 2, Mar. 26, Sep. 4)

1928

Alexei A. Abrikosov, Russian theoretical physicist, Nobel laureate, was born in Moscow, USSR (now Russia). Abrikosov completed high school in 1943 and was accepted as a student at the Institute for Power Engineers. In 1945 he joined the Physics Department of the Moscow State University, from where he obtained an MSc degree in 1948. He received his PhD in 1951 from the Institute for Physical Problems (now the P.L. Kapitsa Institute) and joined the same institute as a junior scientist. In 1955 he was awarded a degree of Doctor of Physical and Mathematical Sciences for a thesis on quantum electrodynamics at high energies. After that, from 1965 to 1988 he worked in the Landau Institute for Theoretical Physics (USSR Academy of Sciences). Since 1987 he has been an Academician of the USSR Academy of

Sciences. Abrikosov's main contribution has been in the field of condensed matter physics. In 1952 Abrikosov discovered the way in which magnetic flux can penetrate a superconductor. The phenomenon is known as type-II superconductivity, and the accompanying arrangement of magnetic flux lines has come to be known as the 'Abrikosov vortex' lattice. Since 1991 he has been working in the Materials Science Division at the Argonne National Laboratory in Illinois, USA. He is a citizen of both Russia and USA. Abrikosov was awarded the 2003 Nobel Prize for Physics, jointly with the Russian physicist Vitaly L. Ginzburg and American physicist Anthony J. Leggett, for pioneering contributions to the theory of superconductors and superfluids. (☞ Mar. 26. Oct. 4)

1971
John Boyd Orr, Scottish scientist and authority on nutrition, Nobel laureate (Peace, 1949), who did extensive studies on the nutritional status and campaigned for healthy diets for all peoples everywhere, died in Edzell, Angus, Scotland. (☞ Sep. 23)

1995
Ernest Thomas Sinton Walton, Irish physicist, Nobel laureate (Physics, 1951), who co-developed the first nuclear particle accelerator and achieved transmutation of atomic nuclei, died in Belfast, Northern Ireland. (☞ Oct. 6)

1997
Jacques-Yves Cousteau, French naval officer, oceanographer, marine biologist and ocean explorer, who co-invented the 'aqualung,' died in Paris, France. (☞ Jun. 11)

1997
In the first ever space accident of its kind, the Russian space station *Mir* suffered a near-fatal mishap when a *Progress* ferry craft accidentally rammed into the *Spektr* science module, punching a hole in the pressure vessel and damaging its solar arrays beyond use. The craft was being docked with *Mir* via remote control by Russian cosmonaut Vasily Tsibliyev. To salvage the station, which consisted of a core, a connecting node, and five science modules, crew members severed electrical and data connections between the *Spektr* module and the rest of the station and then sealed off the module. They saved the station but lost about half of their electrical power. (☞ Feb. 20, Mar. 23)

June 26

1730
Charles Messier, French astronomer, was born in Lorraine, France. The appearance of a great 6-tailed comet sparked Messier's interest in astronomy when he was 14 years old. An annular solar eclipse visible from his hometown in July 1748 further stimulated this interest. In 1751 he went to Paris, and was employed by the astronomer of the Navy, Joseph Nicolas Delisle. His first job was copying a large map of China. Besides these activities, he was instructed to keep careful records of his observations at the observatory on Hôtel de Cluny in the city. Messier's first documented observation was that of the Mercury transit of May 6, 1753. In 1754, he was regularly employed as a depot clerk of the Navy. Messier's life was dedicated to the discovery of comets, and in this he was eminently successful. When

searching the sky obscure comet-like fuzzy patches of light often hampered his observation, so he set about compiling a list of each of these nebulae and star clusters so that he and others could ignore them and reduce the risk of any confusion with comets. It was while searching the night sky for the return of Comet Halley in 1758 that Messier came across a comet-like fuzzy patch in the constellation of Taurus. He measured its position again some days later and it became the first entry, M1, in his famous catalogue. This object later turned out to be one of the most interesting objects in the sky, the remnant of the supernova 1054, now commonly called the Crab nebula. (☞ Apr. 11)

1796
David Rittenhouse, American astronomer, instrument maker and inventor who was an early observer of the transit of Venus, died in Philadelphia, Pennsylvania, USA. (☞ Apr. 8)

1810
Joseph-Michel Montgolfier, French inventor, who, with his brother Jacques-Etienne Montgolfier, pioneered hot air ballooning, died in Balaruc-les-Bains, France. (☞ Jan. 6, Aug. 26)

1824
Lord Kelvin, Scottish physicist, mathematician and engineer, Fellow of the Royal Society, was born as William Thomson, in Belfast, Ireland. Thomson attended the Glasgow University from the early age of 10, and began university level work in 1838 when he was 14 years old. He graduated from the University of Cambridge in 1845. The following year he was appointed professor of natural philosophy at the University of Glasgow and remained there for the rest of his career. Here he set up the first physics laboratory at a British university. One of his early projects was to calculate the age of the Earth, based on the rate of cooling of the planet, but his value was far short of the current estimate of 4.5 billion years. Thomson was also interested in the improvement of physical instrumentation, and designed and built many new devices, including the mirror-galvanometer that was used in the first successful sustained telegraph transmissions in transatlantic submarine cable, as a detector for extremely feeble signals.

His thermodynamics studies led to his proposal in 1848 of an absolute scale of temperature. He calculated that molecular motion stops at −273°C. He called this temperature absolute zero, the lowest possible temperature. The Kelvin absolute temperature scale, developed later, is named after him. In 1852, Thomson observed what is now called the Joule–Thomson effect – the decrease in temperature of a gas when it expands in a vacuum, which makes refrigerators work. Thomson published more than 600 scientific papers and filed a total of 70 patents. He was the president of the Royal Society from 1890 to 1895. (☞ Dec. 17)

1886
Henri Moissan, French chemist, became the first to successfully isolate the highly reactive element fluorine. He did this through the electrolysis of potassium fluoride and hydrofluoric acid. His apparatus consisted of two platinum–iridium electrodes sealed into a platinum

U-tube containing an electrolyte solution of dry potassium acid fluoride in anhydrous hydrofluoric acid chilled with methylene chloride to a temperature of –23°. The ends were closed with fluorspar screw caps covered with a layer of gum lac. Electrolysis produced a gas at the anode. When Moissan tested it with silicon, it immediately burst into flame, which he regarded as a test for fluorine gas. Two days later, his discovery was announced at the Academy of Science, Paris. (☞ Feb. 20, Sep. 28)

1919
Mikhail Semyonovich Tsvet, Russian botanist who invented adsorption chromatography, died in Voronezh, Russia. (☞ May 14)

1937
Robert C. Richardson, American physicist, Nobel laureate, was born in Washington, DC, USA. Richardson attended the Arlington County public schools and graduated from Washington-Lee High School in 1954. He obtained his BA degree in physics in 1954 and MS degree in physics in 1960 from the Virginia Polytechnic Institute. After a brief stint in the US Army he returned to graduate school in physics at the Duke University, Durham, North Carolina, where he received his PhD degree in 1966 for his thesis on NMR studies of solid helium-3. In 1966 he joined the Cornell University, Ithaca, New York, to work with the American physicist David Lee on a project to observe the nuclear magnetic phase transition in solid helium-3. He became director of the laboratory of atomic and solid-state physics there in 1990. In collaboration with Douglas Osheroff, a student who joined the group in 1967, Richardson and Lee worked on cooling techniques and NMR instrumentation for studying low temperature helium liquids and solids. In 1971, they made the accidental discovery of the transition of liquid helium-3 to a superfluid state. Richardson was awarded the 1996 Nobel Prize for Physics, jointly with Lee and Osheroff, for their discovery of superfluidity in helium-3. (☞ Jan. 20, Aug. 1)

1943
Karl Landsteiner, Austrian immunologist and pathologist, Nobel laureate (Physiology or Medicine, 1930), who discovered human blood groups, died in New York, USA. (☞ Jun. 14)

June 27

1869
Hans Spemann, German embryologist, Nobel laureate, was born in Stuttgart, Germany. From 1878 to 1888 he attended the Eberhard-Ludwig School at Stuttgart and afterwards spent a year in his father's publishing business. In 1891 he entered the University of Heidelberg to study medicine. In 1894 he joined the Zoological Institute at the University of Würzburg from where he took his degree in zoology, botany, and physics in 1895. In 1898 he qualified as a lecturer in zoology at the University of Würzburg, and in 1908 became professor of zoology and comparative anatomy at Rostock. In 1914 he became associate director of the Kaiser Wilhelm Institute of Biology at Berlin-Dahlem. In 1919 he was appointed professor of zoology at the University of Freiburg-im-Breisgau, a post which he held until he retired and became Emeritus Professor in 1935. Spemann specialised in experimental embryology. While working with the relatively large eggs

of newts he discovered in 1924 that, in the earliest stages, the fate of the embryonic parts is not decided in advance. If a piece of presumptive skin tissue is excised and transplanted into an area of presumptive nervous tissue, it will form nervous tissue, not skin. He identified areas in embryo called 'organiser centres', which he said, 'organised' those transformations. Spemann was awarded the 1935 Nobel Prize for Physiology or Medicine, for his discovery of the organiser effect in embryonic development. (☞ Sep. 12)

1931
Martinus J.G. Veltman, Dutch physicist, Nobel laureate, was born in Waalwijk, Netherlands. Veltman attended high school in Waalwijk and then received his PhD in physics from the University of Utrecht in 1963. He joined the same university as professor in 1966. In 1981 he moved to the United States to teach at the University of Michigan, Ann Arbor, where he became professor emeritus in 1997. When Veltman was at Utrecht the fundamental theory of particle physics, termed the standard model, was incomplete in that it failed to provide for detailed calculations of physical quantities. While the theory of electro-weak interaction developed in the 1960s was a great step forward, the research community at first found it difficult to accept. When they tried to use the theory for calculating in more detail the properties of the new W and Z particles (and many other physical quantities) it gave unreasonable results. But the situation changed after Gerardus 't Hooft, a research student, joined him in 1969. They looked for a method of re-normalising non-abelian gauge theories. In 1971 they published two articles that represented an important breakthrough in the research programme. It was only through the work of Veltman and 't Hooft that more precise prediction of physical quantities involving properties of W and Z could start, including the mass of the top quark, which was directly observed in 1995. Veltman was awarded the 1999 Nobel Prize for Physics, jointly with 't Hooft, for elucidating the quantum structure of electro-weak interactions in physics. (☞ Jul. 5)

1954
The world's first atomic power station was commissioned at Obninsk, near Moscow, Russia. It had a generation capacity of 5,000 kW.

1982
American space shuttle *Columbia* was launched on its fourth flight, carrying astronauts Thomas K. Mattingly, and Henry W. Hartsfield Jr. During the mission the crew took data for two medical experiments on themselves, operated a remote manipulator arm to swing the induced environment contamination monitor around the orbiter, and took photos of lightning activity in the Earth's atmosphere. (☞ Jul. 4)

1995
American space shuttle *Atlantis*, carrying astronauts Robert Gibson, Charles J. Precourt, Bonnie Dubbar, Ellen S. Baker, and Gregory J. Harbaugh, was launched from the Kennedy Space Center, Florida, USA, on its 14th flight. (☞ Jun. 29, Jul. 7)

June 28

1873
Alexis Carrel, French scientist, surgeon, biologist, Nobel laureate, was born in Sainte-Foy-lès-Lyon, France. Carrel received his BSc degree in 1889 and his PhD in 1900 from the University of Lyons. He then continued his medical work at the Lyons Hospital and also taught anatomy and operative surgery at the University. During World War I, Carrel served as a Major in the French Army Medical Corps and at this time he helped to devise the well-known and widely used Carrel–Dakin method of treating war wounds with antiseptic fluids in order to prevent infection. Carrel's researches mainly concerned experimental surgery and the transplantation of tissues and whole organs. As early as 1902 he published, in the *Lyons Medical*, a technique for the end-to-end surgical joining of blood vessels with minimum number of stitches, and in 1910 he demonstrated that blood vessels could be kept for long periods in cold storage before they were used as transplants in surgery. In 1904 he went to the USA, and in 1905 worked in the department of physiology in the University of Chicago. In 1906 he was attached to the Rockefeller Institute for Medical Research (now the Rockefeller University), New York, where he carried out experiments on joining and transplantation of blood vessels. He also did notable work on the problem of keeping tissue alive after removal from a living organism. The most famous example was a piece of tissue from the heart of a chicken embryo, which was kept alive from 1912 to 1946, by circulating blood through it. Techniques developed by Carrel have made possible the surgical transplantation of blood vessels and body organs. Carrel was awarded the 1912 Nobel Prize for Physiology or Medicine in recognition of his work on vascular suture and the transplantation of blood vessels and organs. (☞ Nov. 5)

1906
Maria Goeppert-Mayer, German-born American physicist, Nobel laureate, was born in Kattowitz, Germany (now Katowice, Poland). Maria Goeppert went to private and public schools in Göttingen. She studied physics at the University of Göttingen from where she received her PhD in 1930 for her thesis on the decay of excited states by the simultaneous emission of two quanta. In 1930 she married the American chemical physicist Joseph E. Mayer, and a short time later they moved to USA and joined the Johns Hopkins University in Baltimore, Maryland. During World War II, she worked on uranium isotope separation, under the Manhattan Project that developed the atom bomb. After the war, she moved to Chicago, where she worked at the Institute for Nuclear Studies at the University of Chicago and at the Argonne National Laboratory. In 1948 she started to work on the magic numbers (2, 8, 20, 28, 50, 82 and 126) that ultimately led to the nuclear shell model. She developed a theory that the nucleus consists of several shells, or orbital levels, and that the distribution of protons and neutrons among these shells produces the characteristic degree of stability of each

species of nucleus. In 1959 she joined the new University of California campus at San Diego as professor. Goeppert-Mayer was awarded the 1963 Nobel Prize for Physics, jointly with the German physicist Johannes Hans Jensen and the Hungarian-born American physicist Eugene Paul Wigner. Goeppert-Mayer and Jensen shared half the prize money for their discoveries concerning the nuclear shell structure, while Wigner received the other half for his contributions to the theory of the atomic nucleus and the elementary particles, particularly through the discovery and application of fundamental symmetry principles. (☞ Feb. 20, Jun. 25, Nov. 17)

1927

Frank Sherwood Rowland, American chemist, Nobel laureate, was born in Delaware, Ohio, USA. Rowland was educated in public schools in Delaware and received his BA degree from the Ohio Wesleyan University in the city in 1948. He did his MS in 1951, and PhD in 1952 from the University of Chicago in Chicago, Illinois. He held academic posts at the Princeton University, Princeton, New Jersey during 1952–56, and at the University of Kansas, Lawrence, Kansas during 1956–64, before becoming a professor of chemistry at the University of California, Irvine, in 1964. At Irvine in the early 1970s he began working with the Mexican-born American chemist Mario Molina. The two conducted experiments on pollutants in the atmosphere, discovering that man-made substances such as chlorofluorocarbons (CFCs) used in spray cans and refrigerators rise into the stratosphere, where ultraviolet radiation breaks them into their component elements of chlorine, fluorine, and carbon. Their researches showed that CFCs could harm the fragile ozone layer that protects the world from the dangerous ultraviolet radiation of the Sun. The discoveries of Rowland, Molina, and the German chemist Paul Crutzen, that some industrially manufactured gases deplete the ozone layer, led to the signing of the Montreal Protocol in 1987 to limit the widespread use of CFCs. Rowland was awarded the 1995 Nobel Prize for Chemistry, jointly with Molina and Crutzen, for their work in atmospheric chemistry, particularly concerning the formation and decomposition of ozone. (☞ Mar. 19, Dec. 3)

1943

Klaus von Klitzing, German physicist, Nobel laureate, was born in Sroda, Poland (formerly Schroda, Germany). At the end of World War II, Klitzing's parents moved to West Germany where he attended the Technical University of Brunswick, Lower Saxony, and graduated in 1969. He received a doctorate in physics from the University of Würzburg in 1972 for his thesis on 'Galvanomagnetic properties of tellurium in strong magnetic fields'. In 1980 he became a professor at the Technical University of Munich, and in 1985 director of the Max Planck Institute for Solid State Physics in Stuttgart, Germany. In 1980 Klitzing discovered precise steps in the behaviour of electrons under certain applications in semiconductor electronics. He demonstrated that electrical resistance varies by discrete steps rather than smoothly and continuously – a phenomenon that came to be known as 'quantised Hall effect'. The Hall effect denotes the voltage that develops between the edges of a thin current-carrying ribbon placed between the poles of a strong magnet. The ratio of this voltage to the current is called the Hall resistance. When the magnetic field

is very strong and the temperature very low, the Hall resistance varies only in discrete jumps, as was observed by Klitzing. For his experiments Klitzing used a magnetic field roughly 400,000 times stronger than that of the Earth's surface magnetism. His discovery made the extremely precise measurement of electrical resistance possible – a field of importance in computer science and other present-day technology. Klitzing was awarded the 1985 Nobel Prize for Physics for the discovery of the quantised Hall effect.

1965
Commercial TV transmission was started via *Early Bird*, the world's first commercial communications satellite. Positioned 36,000 kilometres above the equator, *Early Bird* provided line-of-sight communications between Europe and North America. (☞ Apr. 6)

1972
Prasanta Chandra Mahalanobis, Indian statistician, Fellow of the Royal Society, who did pioneering work on anthropometric variation in India, died in Kolkata, West Bengal, India. (☞ Jun. 29)

1973
'Skylab-3', the second manned mission to the American orbiting space station *Skylab* was launched from the Kennedy Space Center, Florida, USA. The crew comprising astronauts Alan L. Bean, Jack R. Lousma, and Owen K. Garriott continued maintenance of the space station and carried out extensive scientific and medical experiments. They installed a twin-pole solar shield to replace the meteoroid shield lost during lift off. During the mission the crew carried out a total of 13 hours, 43 minutes of space walk. (☞ Feb. 8, May 14, May 25, Jun. 22, Jul. 11, Sep. 25, Nov. 16)

June 29

1868
George Ellery Hale, American astronomer known for his development of important astronomical instru-ments, including the 200-inch (508-cm) Hale telescope on Mount Palomar near Pasadena, California, was born in Chicago, Illinois, USA. Hale developed an interest in astronomy at a young age. He graduated from the Massachusetts Institute of Technology, Cambridge, Massachusetts, in 1890 with a bachelor's degree in physics. By 1891 Hale was effectively equipped with his own private solar astronomical laboratory and his scientific reputation was already well established. Although he never completed a doctorate, in 1892 he was appointed professor of astronomy at the University of Chicago when he launched a life-long campaign to fund and build ever better astronomical observatories. Over the years he organised the founding of three world-class astronomical observatories – University of Chicago's Yerkes Observatory, in nearby William's Bay, Wisconsin; a solar observatory on Mt. Wilson in California; and the observatory on Mt. Palomar, California, with a 5-metre telescope. In 1895 Hale co-founded, and edited for nearly 30 years, *The Astrophysical Journal*. The journal was originally envisioned as an international forum for the publication of astronomically relevant papers in the field of spectroscopy, but

the journal rapidly expanded its scope to become the world's leading research journal in the field of astrophysics. Hale's most acclaimed scientific work was his demonstration that sunspots are the seat of strong magnetic fields, and that their polarity reveals striking spatial and temporal regularities that betray the presence of a well-organised, large-scale magnetic field in the solar interior. (☞ Feb. 21, Jun. 3)

1893

Prasanta Chandra Mahalanobis, Indian statistician, Fellow of the Royal Society, was born in Calcutta (now Kolkata), India. Mahalanobis did his schooling at the Brahmo Boys' School, and then obtained his BSc in physics from the Presidency College, Kolkata. He then went to England and studied mathematics and physics at the Cambridge University. After his return to India, Mahalanobis was offered a position to teach at the Presidency College which he was hard put to turn down since it allowed him to pursue his increasing interest in statistical problems. Statistics ended up overtaking his first love of physics, and so, even though he continued to teach physics until 1948, he was devoting more and more of his free time and career studies to statistics. Initially he worked on analysing university exam results, anthropometric measurements on Anglo-Indians of Kolkata, and some metrological problems. His most important contributions are related to large-scale sample surveys. He introduced the concept of pilot surveys and advocated the usefulness of sampling methods. The National Sample Survey was started in 1950, and it grew quickly into an agency noted for its use of continuing sample surveys for the collection of socioeconomic and demographic data that covered the whole country. Mahalanobis founded the Indian Statistical Institute in Kolkata in 1931. He also started a new journal in statistics called *Sankhya*. Mahalanobis was the author of India's First Five-year Plan, which was based on the two-sector growth model, and called the 'Mahalanobis Model'. It showed that an increase in the allocation of investment in capital goods vis-à-vis consumer goods results in a correspondingly higher rate of savings ultimately accelerating the GNP growth. Mahalanobis made three notable contributions in the field of statistics. He formulated the concept of optimum design survey in terms of the variance and cost functions. He showed how pilot surveys could be profitably used to estimate the parameters of variance and cost functions. He proposed various techniques for assessment and control of errors, which were inevitably, creeping into the data. Other areas where he made contributions include meteorological statistics, operations research, and errors in field experimentation. (☞ Jun. 28)

1895

Thomas Henry Huxley, English biologist, educator and advocate of Darwinism, died in Eastbourne, Sussex, UK. (☞ May 4)

1966

Damodar Dharmanand Kosambi, Indian scientist, mathematician, historian, and Sanskrit scholar, died. (☞ Jul. 31)

1968

The Wadia Institute of Himalayan Geology was founded. An autonomous research institute under the Department of the Science & Technology, Govt. of India it started as a small nucleus under the name Institute of Himalayan Geology in two rooms of the botany department, Delhi University, and later shifted to Dehra Dun, Uttar Pradesh (now Uttaranchal) in April 1976. It was re-named as the Wadia Institute of Himalayan Geology in memory of its founder – the late D. N. Wadia (FRS and National Professor), in recognition of his contributions to geology of the Himalayas. During the last quarter century the Institute has grown into a centre of excellence in Himalayan geology and is recognised as a national laboratory of international repute with well equipped laboratories and other infrastructural facilities for undertaking an advanceed level of research in the country. The institute carries out research towards the development of new concepts and models concerning geodynamic evolution of the Himalayas through an integrated interdisciplinary approach. It also operates three field research stations in the remote areas of the Himalayan region. (☞ Oct. 25)

1995
American space shuttle *Atlantis*, on its 14th flight, docked with the Russian orbiting space station *Mir*, forming the largest structure in orbit at that time. The docking took place 258 kilometres above Central Asia near the Russian/Mongolian border. On board *Atlantis* were Commander Robert Gibson; pilot Charles J. Precourt, and mission specialists Bonnie Dubbar, Ellen S. Baker and Gregory J. Harbaugh. They were joined by the two-man replacement crew for *Mir*, cosmonauts Anatoly Solovyev and Nikolia Budarin. (☞ Feb. 20, Jun. 27, Jul. 7)

June 30

1908
An enormous aerial explosion flattened approximately 2,000 square kilometres

of pine forest near the Podkamennaya Tunguska River in central Siberia in Russia. The energy of the explosion, which is believed to have been caused by a comet fragment colliding with the Earth, was equivalent to that of 10 to 15 megatons of TNT. The remote site of the explosion was first investigated from 1927 to 1930 in expeditions led by Russian scientist Leonid Alekseyevich Kulik. The incident has come to be known as the 'Tunguska event'.

1916
An electric public address system was first used at the Madison Square Garden in New York, USA. The system was developed by AT&T and used new amplifiers, microphones, and loudspeakers developed by the Bell Laboratories, New Jersey.

1919
Lord Rayleigh (John William Strutt), English physical scientist, Nobel laureate (Physics, 1904), who discovered and isolated argon, one of the rare gases of the atmosphere, died in Witham, Essex, England. (☞ Nov. 12)

1926
Paul Berg, American biochemist, Nobel laureate, was born in New York, USA. Berg graduated from high school in 1943 and enrolled at New York's City College to study chemical engineering. Later he

enlisted in the US Navy to be a pilot, and while waiting to be called for pre-flight training, enrolled at the Pennsylvania State University, Philadelphia, to study biochemistry from where he graduated in 1948. He then took a doctorate from the Western Reserve University, Cleveland, Ohio, in 1952, and pursued further studies at the Institute of Cytophysiology in Copenhagen, Denmark, and at the Washington University in St. Louis, Missouri, USA, where he remained as assistant professor of microbiology until 1959. From 1959 he was associated with the Stanford University's Medical Centre in California. While at Stanford, he conceived the notion of using the SV40 tumour virus as a means of introducing new genes into mammalian cells much in the way that bacteriophage transduces cellular DNA among infected cells. Berg and his colleagues succeeded in developing a general way to join two DNAs together *in vitro*, which ultimately led to the emergence of the recombinant DNA technology thereby providing a major tool for analysing mammalian gene structure and function. Berg was awarded the 1980 Nobel Prize for Chemistry, jointly with the American molecular biologist Walter Gilbert and English biochemist Frederick Sanger. Berg received half the prize money for his fundamental studies of the biochemistry of nucleic acids, with particular regard to recombinant DNA, while Gilbert and Sanger shared the other half for their contributions concerning the determination of base sequences in nucleic acids. (☞ Mar. 21, Aug. 13)

1934

Chintamani Nagesa Ramachandra Rao, India physical chemist, Fellow of the Royal Society of London, was born in Bangalore, Karnataka, India. Rao graduated in science from the Mysore University in 1951 and obtained his MSc from the Banaras Hindu University in 1953. Later he did his PhD in 1958 from the Purdue University, USA, and DSc in 1961 from the Mysore University. Rao was lecturer at the Indian Institute of Science, Bangalore from 1959 to 1963 and then moved on to the Indian Institute of Technology Kanpur to take over as professor and dean. In 1976 he returned to IISc, Bangalore as professor of chemistry and became its director in 1984. He has been the Director of the Jawaharlal Nehru Centre for Advanced Scientific Research in Bangalore since 1989. He was conferred the first India Science Award in 2006. Rao started his research career working on spectra and structure of molecules and moved to solid state and materials chemistry, surface science and related areas. His interests in materials chemistry related mostly to phenomena and properties exhibited by transition metal oxide systems that include high temperature superconductivity, colossal magneto resistance and metal-insulation transitions. He has made outstanding contributions in synthesis and characterisation of designer solids with novel structures and properties. He has also contributed widely to several aspects of spectroscopy and structure by employing vibrational, electronic and high-energy spectroscopy, and study of fullerenes and metal clusters.

1961

Lee De Forest, American inventor of the audion vacuum tube, which made possible live radio broadcasting and became the key component of all radio, telephone, radar, television, and computer systems before the invention of the transistor, died in Hollywood, Los Angeles, California, USA. (☞ Aug. 26)

1972

One extra second, called a 'leap second', was added to the world's time in order to keep the super-accurate atomic clocks in step with the Earth's rotation. Leap seconds are inserted at the end of June or December as an additional second after 23:59:59 Coordinated Universal Time (UTC). The additional second is the 61st second of the last minute of the month, and it is written as 23:59:60 (or 11:59:60 PM in 12-hour format). The second is inserted at the same time all over the world – the actual local time would therefore depend on the time zone. Since the adoption of this system in 1972, firstly due to the initial choice of the value of the second (1/86,400 mean solar day of the year 1900) and secondly to the general slowing down of the Earth's rotation, it has been necessary to add over 20 seconds to UTC.

1973

A total solar eclipse, predicted as the longest for 1,000 years, was observed by British, French and American scientists aboard the French Concorde supersonic aircraft on a flight from Las Palmas, Canaries Islands to Fort Lamy, Chad. The path of totality crossed the Atlantic, the Sahara Desert and East Africa. The Moon's shadow travelled at over 3,000 km per hour. Flying at 16,800 metres, the jet's speed made possible a continuous view of the solar eclipse for 74 minutes, ten times longer than could be seen by an observer on the ground. During each millennium there are typically fewer than 10 total solar eclipses exceeding 7 minutes. The next one will be on 25 June 2150. The longest total solar eclipse during the 8,000-year period from 3000 BCE to 5000 CE will occur on 16 July 2186, when totality will last 7 min 29 s.

July 1

1646

Gottfried Wilhelm Leibniz, German philosopher, mathematician, and logician who is probably most well known for having invented the differential and integral calculus independently of Isaac Newton, was born in Leipzig, Germany. Leibniz was educated in the Nicolai School but was largely self-taught in the library of his father, who had died in 1652. In 1661, he entered the University of Leipzig as a law student and there he came into contact with the writings and thoughts of men who had revolutionised science and philosophy – men such as Galileo, Francis Bacon, and René Descartes. As early as 1666 he wrote *De Arte Combinatoria* (On the Art of Combination), in which he laid the modern foundation of the movement from decimal to binary, laying out a method for reducing all logic to exact statements. After completing his legal studies in 1666, Leibniz applied for the degree of doctor of law, but was refused because of his age. He consequently left Leipzig and moved to Altdorf – the university town of the free city of Nuremberg. There he got his doctor's degree for his dissertation *De Casibus Per plexis* (On perplexing cases) and was immediately offered a professor's chair, which he however declined. Leibniz played a major role in the European politics and diplomacy of his day. He also occupies an equally large place in both the history of philosophy and the history of mathematics. In 1675 Leibniz made his most important scientific discovery – the differential and integral calculus – which became the basis for modern mathematics. The discovery resulted in a controversy with Isaac Newton over whether he or Newton was the inventor. Nowadays it is generally agreed that they both discovered the basic foundations of calculus independently. After spending time in London and Paris in diplomatic appointments Leibniz became, in 1676, librarian to the Duke of Brunswick at Hanover, Germany, a position which he retained until his death. In March 1679 he perfected the binary system of numeration (i.e., using two as a base), the foundation of virtually all modern computer architectures. At the end of the same year he proposed the basis for analysis situs (geometry of situation), now known as general topology, a branch of mathematics that deals with selected properties of collections of related physical or abstract elements. He was, along with René Descartes and Baruch Spinoza, one of the three great 17th century rationalists. Leibniz also made major contributions to physics and technology, and anticipated notions that surfaced much later in biology, medicine, geology, probability theory, psychology, knowledge engineering, and information science. By the turn of the 17th century Leibniz's fame as a philosopher and scientist had spread all over Europe; he was named a foreign member by the Academy of Sciences of Paris in 1700 and was in correspondence with most of the important European scholars of the day. In his *Monadologie*, published in 1714, Leibniz maintained that the divine order of the

universe is reflected in each of its parts, each part being called a monad, a term that means 'unit' or 'unity' – or a counterpart of atom. Leibniz had a lifelong interest in and pursuit of the idea that the principles of reasoning could be reduced to a formal symbolic system, an algebra or calculus of thought, in which controversy would be settled by calculations. (☞ Nov. 14)

1796

The first experiment with smallpox inoculation was made by Edward Jenner of Berkeley Gloucestershire, England. He made two cuts in the arm of eight-year-old James Phipps and worked a small amount of smallpox matter into them. Jenner had previously vaccinated the boy with cowpox in May. Now, he was not harmed by the otherwise deadly smallpox material, proving the efficacy of vaccination, a term derived from the medical name for cowpox, *vaccinia*. (☞ May 14)

1858

In the first public presentation of his theory of evolution, the English naturalist Charles Darwin's paper on the 'Perpetuation of varieties and species by natural means of selection' was read to the Linnean Society in London, jointly with the British naturalist and geographer Alfred Russel Wallace's paper 'On the tendency of species to form varieties'. Neither scientist was present at the reading. Darwin had gradually developed his ideas on evolution by natural selection since his voyage aboard *HMS Beagle*. His paper contained extracts from a book that he was writing; the book would later be *On the Origin of Species*. (☞ Jan. 8, Feb. 12, Oct. 2, Nov. 24, Dec. 27)

1860

Charles Goodyear, American inventor of the vulcanisation process that made possible the commercial use of rubber, died in New York, USA. (☞ June 15, Dec. 29)

1905

The paper on 'Special theory of relativity' by the German-born physicist Albert Einstein was published in the journal *Annalen der Physik*. Publication of the paper turned the scientific ideas of the English physicist Isaac Newton upside down. The most striking implication of the theory is that mass is a form of energy, and that a small amount of mass would be equivalent to a huge amount of energy. (☞ Mar. 14)

1929

Gerald Maurice Edelman, American biochemist, Nobel laureate, was born in New York, USA. Edelman studied at the University of Pennsylvania Medical School from where he got his MD in 1954. After two years in the Army Medical Corps in Paris, France, he joined the Rockefeller University as a graduate fellow in 1957. After receiving his PhD degree in 1960, he remained at the Rockefeller University as assistant dean of graduate studies and started work in his own laboratory. In 1963, he became associate dean of graduate studies, a position from which he retired in 1966. At Rockefeller, Edelman studied the structure of human immunoglobulin. He first split this large molecule into smaller portions by reducing and splitting the disulphide bonds. He proposed that the molecule contained two kinds of polypeptide chains: light and heavy. In 1969, Edelman and his team succeeded in constructing a precise model of an entire antibody molecule, immunoglobulin G, which was found to be a four-chain structure consisting of more than 1,300 amino acids. They were then able to identify the precise locations on the molecule where antigenic binding occurs. His subsequent research focussed on morphogenesis, i.e., the formation and

differentiation of tissues and organs. As director of the Neurosciences Institute at the Rockefeller University from 1981, Edelman tried to construct a general theory of neural development and brain function, which he discussed in his book *Neural Darwinism: The Theory of Neuronal Group Selection* (1987). In place of the traditional view of the human nervous system as a fixed and immutable structure, Edelman argued that neural systems change and evolve. Those connections used frequently tend to be maintained, while those that are not decay or are used for other purposes. Edelman was awarded the 1972 Nobel Prize for Physiology or Medicine, jointly with the British biochemist Rodney Porter, for their discoveries concerning the chemical structure of antibodies. (☞ Oct. 8)

1941

Alfred Goodman Gilman, American pharmacologist, Nobel laureate, was born in New Haven, Connecticut, USA. Gilman received his BS in biochemistry in 1962 from the Yale University, New Haven, and his MD and PhD in pharmacology in 1969 from the Case Western Reserve University, Cleveland, Ohio, where he worked on cyclic AMP (adenosine monophosphate) in the thyroid gland. He received further training as a pharmacology research associate in the laboratory of biochemical genetics at the National Institutes of Health during 1969–71. In 1971 Gilman began a ten-year stay at the University of Virginia in Charlottesville. He became chairman of the department of pharmacology at the University of Texas Southwestern Medical Center at Dallas in 1981. In 1980, Gilman and his team characterised and purified a set of guanine nucleotide-binding regulatory proteins termed G-proteins, which are intermediaries in the multi-step pathway cells use to react to an incoming signal, such as a hormone or a neurotransmitter. G-protein is so named because it becomes activated when bound to a molecule called guanosine triphosphate (GTP). Abnormally functioning G-proteins can disrupt the normal signal transduction process and play a role in diseases such as cholera, cancer, and diabetes. Gilman's observations provided for the first time a firm molecular basis for understanding certain signal transduction processes present throughout nature. Gilman was awarded the 1994 Nobel Prize for Physiology or Medicine, jointly with the American biochemist Martin Rodbell, for their discovery of G-proteins and the role of these proteins in signal transduction in cells. (☞ Dec. 1)

1961

Kalpana Chawla, Indian-born American astronaut, was born in Karnal, Haryana, India. Chawla did her schooling from the Tagore Bal Niketan School in Karnal. Her interest in flight was inspired by the veteran Indian aviation pioneer and industrialist, J R D Tata. She earned her BSc degree in aeronautical engineering from the Punjab Engineering College in Chandigarh, India in 1982. She then moved to the USA and obtained an MSc degree in aerospace engineering from the University of Texas at Arlington, Texas in 1984. She earned a second Master's degree in 1986 and a PhD in aerospace engineering from the University of Colorado in 1988. Later that same year she began working for NASA Ames Research Center in Moffett Field,

California. In 1993 she joined Overset Methods Inc., Los Altos, California, as Vice President and Research Scientist to form a team with other researchers specialising in simulation of moving multiple body problems. Chawla joined the NASA astronaut corps in March 1995 and was selected for her first flight in 1996. She went on her first space mission in November 1997 when she was a member of the six-astronaut crew that flew the space shuttle *Columbia* on its 24th flight. Chawla was the first Indian-born woman and the second person of Indian origin to fly in space, after cosmonaut Rakesh Sharma who flew in 1984 in a Soviet spacecraft. In 2000 Chawla was selected for her second flight as part of the crew of the ill-fated *Columbia* on its 28th flight in January 2003. She was a mission specialist in charge of the microgravity experiments. (☞ Jan. 16, Feb. 1)

1971
William Lawrence Bragg, Australian-born British physicist and X-ray crystallographer, Nobel laureate (Physics, 1915), who formulated the law of X-ray diffraction, died in Ipswich, Suffolk, England. (☞ Mar. 31)

1983
Richard Buckminster Fuller, American engineer and architect who developed the geodesic dome, the only large dome that can be set directly on the ground as a complete structure, died in Los Angeles, USA. (☞ Jul. 12)

1992
The 220-MW second unit of the Narora Atomic Power Project at Narora, Uttar Pradesh, India, was commissioned. It is a pressurised heavy water reactor that uses natural uranium as fuel and heavy water as moderator and coolant. (☞ Jan. 1)

2004
The *Cassini–Huygens* spacecraft arrived at Saturn. After a nearly seven-year journey covering 3.5 billion kilometres, the six-tonne spacecraft swung into a highly elliptical orbit around the giant ringed planet, after an on-board engine fired for 96 minutes, slowing it down for orbit insertion. *Cassini* approached Saturn from below the planet's ring plane, crossing through the large gap between the F-ring and the G-ring. On its journey from Earth, the *Cassini–Huygens* made two fly-bys of Venus and one fly-by each of Earth and Jupiter to gain speed by gravity assist. The spacecraft will perform scientific investigations of Saturn, its moons and it magnificent rings until 2008. Along with an array of science instruments, *Cassini* carried the European Space Agency's *Huygens* probe, which was dropped into the thick atmosphere of Titan – Saturn's largest moon – in January 2005 to study its composition. (☞ Jan. 14, Apr. 26, Jun. 24, Aug. 18, Oct. 15, Dec. 24, Dec. 30)

July 2

1862
William Henry Bragg, British physicist, Nobel laureate, was born in Wigton, Cumberland, UK. Bragg attended the Market Harborough Grammar School and the King William's College in the Isle of Man. In 1882 he was granted a scholarship at the Trinity College, Cambridge; and two years later he obtained third place in the Mathematical Tripos. In 1885, at the

age of 23, he was appointed professor of mathematics and physics at the University of Adelaide, in Southern Australia. While at Adelaide he made important discoveries about the ranges of alpha particles and carried on research involving the properties of X-rays. In 1909, on his return to England, he was appointed Cavendish Professor at the University of Leeds. Working with his son William Lawrence Bragg, who was studying physics at Cambridge, he developed a technique of using X-rays to study crystals. By examining the pattern of X-rays diffracted by various crystals, Bragg and his son were able to establish some fundamental mathematical relationships between an atomic crystal structure and its diffraction pattern, which could be used to determine the molecular structure of crystals. He also constructed an X-ray spectrometer for measuring the wavelengths of X-rays. In the 1920s, while director of the Royal Institution in London, he initiated X-ray diffraction studies of organic molecules. William Henry Bragg and his son William Laurence Bragg were jointly awarded the 1915 Nobel Prize for Physics for their services in the analysis of crystal structure by means of X-rays. (☞ Mar. 12, Mar. 31)

1900

The zeppelin, a rigid gas-filled airship designed and built by the retired German army officer Ferdinand Graf von Zeppelin and named after him, was flown for the first time on Lake Constance, near Friedrichshafen, Germany. The huge 127-metre-long cigar-shaped airship was driven by two 16-horsepower internal combustion engines geared to two propellers, each

fitted to external cars suspended beneath the rigid frame that contained the gas-filled bags. It remained in the air for 18 minutes. More than 100 zeppelins were used for military operations in World War I and the first zeppelin passenger service was started in 1910. (☞ May 6)

1906

Hans Albrecht Bethe, German-born American theoretical physicist, Nobel laureate, was born in Strassburg, Germany (now Strasbourg, France). Bethe attended the Gymnasium in Frankfurt from 1915 to 1924, and then studied at the Universities of Frankfurt and at Munich, taking his PhD in theoretical physics in 1928. He immigrated to England in 1933 where he held a temporary position as lecturer at the University of Manchester and a fellowship at the University of Bristol. In 1935 he was appointed assistant professor at the Cornell University, Ithaca, New York, USA. He was a professor there from 1937 to 1975, when he became professor emeritus. In 1939 Bethe calculated the Sun's energy production, which results from the fusion of four hydrogen atoms into one helium atom. Bethe showed that the energy is produced not by direct fusion but through a four-step 'carbon–nitrogen cycle'. A carbon isotope of mass 12 reacts successively with three hydrogen nuclei (protons) to form the nitrogen isotope of mass 15; energy is produced through the fusion of a fourth hydrogen nucleus to

release a helium nucleus (alpha particle) and the original carbon isotope. Bethe became a US citizen in 1941, and during World War II, he worked at the Radiation Laboratory at the Massachusetts Institute of Technology, working on microwave radar, and then at the Los Alamos Scientific Laboratory, which was engaged in assembling the first atomic bomb. Bethe was awarded the 1967 Nobel Prize for Physics for his contributions to the theory of nuclear reactions, especially his discoveries concerning the energy production in stars. (☞ Mar. 6)

1931
Stephen Moulton Babcock, American agricultural research chemist, who developed a simple method of measuring the butterfat content of milk known as the 'Babcock test', died in Madison, Wisconsin, USA. (☞ Oct. 22)

1940
Enrico Fermi, the Italian-born American physicist and his team was issued a US patent for a process of producing radioisotopes. The invention related to the 'production of isotopes of elements from other isotopes of the same or different elements by reaction with neutrons, and especially to the production of artificial radioactivity by formation of unstable isotopes'. (☞ Sep. 29)

1946
Richard Axel, American biologist, Nobel laureate, was born in New York, USA. Axel graduated from the Stuyvesant High School in New York in 1963, and received his AB degree in 1967 from the Columbia University in the same city. He received his MD in 1970 from the Johns Hopkins University School of Medicine, Baltimore, Maryland. In 1978 he joined the Columbia University as professor of pathology and biochemistry. Since 1984 he has been investigator at the Howard Hughes Institute, and professor at the Columbia University. In his earlier work, Axel and his colleagues developed gene transfer techniques, which permit the introduction of virtually any gene into any cell, allowing the analysis of gene function *in vivo*. Axel's primary research interest has been on how the brain interprets the sense of smell, specifically mapping the parts of the brain that are sensitive to specific olfactory receptors. This work has provided insight into how the recognition of odours is translated into an internal representation of sensory quality in the brain. In 1991, Axel and fellow American biologist Linda B. Buck cloned for the first time olfactory receptors, showing that they belong to the family of 'G-protein-coupled receptors'. By analysing rat DNA they estimated that there were approximately one thousand different genes for olfactory receptors in the mammalian genome. This research opened the door to the genetic and molecular analysis of the mechanisms of olfaction. Axel was awarded the 2004 Nobel Prize for Physiology or Medicine, jointly with Buck, for their discoveries of odorant receptors and the organisation of the olfactory system. (☞ Jan. 29)

1985
European Space Agency's *Giotto* spacecraft was launched on a mission to observe Comet Halley from close range. The spacecraft was named after Giotto di Bondone, Italian painter and architect, who was considered as the first genius of art in the

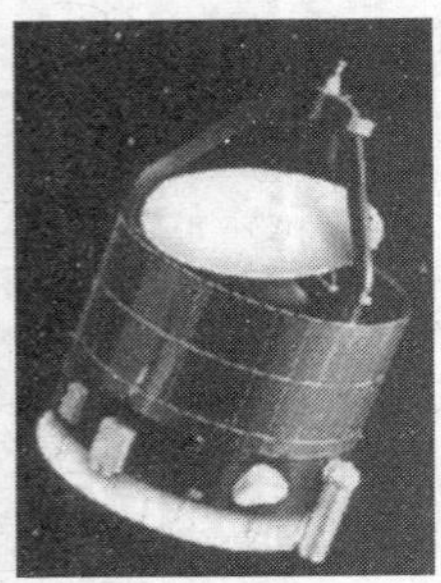

Italian Renaissance. The major objectives of the mission were to: (1) obtain colour photographs of the comet's nucleus; (2) determine the elemental and isotopic composition of volatile components in the cometary coma; (3) characterise the physical and chemical processes that occur in the cometary atmosphere and ionosphere; (4) determine the elemental and isotopic composition of dust particles; (5) measure the total gas-production rate and derive the dust-to-gas ratio; and (6) investigate the macroscopic systems of plasma flows resulting from the cometary–solar wind interaction. (☞ Mar. 14)

July 3

1886
German mechanical engineer Karl Benz drove the world's first practical automobile to be powered by an internal combustion engine – a three-wheeler – in Mannheim, Germany, reaching a top speed of 16 km/h. (☞ Jan. 29, Nov. 25)

1928
Scottish engineer and inventor John Logie Baird demonstrated the world's first colour television transmission at his studios in Long Acre, London, England. The low-definition system was based on a Nipkow scanning disc with red, blue and green filters. The transmission showed red and blue scarves, a policeman's helmet, a man sticking his tongue out, the glowing end of a cigarette, and a bunch of flowers. The British journal *Nature* reported: 'Delphiniums and carnations appeared in their natural colours and a basket of strawberries showed the red fruit very clearly'. (☞ Jan. 27, Feb. 9, Feb. 17, Aug. 13)

1929
Foam rubber was developed at the Dunlop Latex Development Laboratories in Birmingham, UK. British scientist E A Murphy whipped up the first batch, using an ordinary kitchen mixer to froth natural latex rubber, and then set it. His colleagues were initially unimpressed, but when they sat on it they were pleasantly surprised. Within five years foam rubber found application almost everywhere – motorcycle seats, bus seats, theatre seats, and eventually in mattresses.

July 4

1054
One of the most powerful supernovas – a violently exploding star – was observed by Chinese and other astronomers. The supernova was so bright that it was visible in daylight for 23 days and at night for almost 2 years. The Crab Nebula in the constellation Taurus is believed to be the remnant of this supernova, which is the earliest such object on record. Rock paintings in North America suggest that

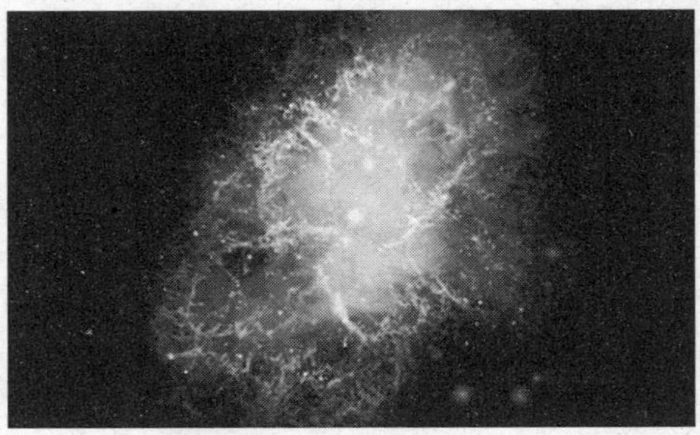

American Indians in Arizona and New Mexico saw it. There are no European or Indian records of the event.

1790
George Everest, British surveyor and geographer, Fellow of the Royal Society, was born in Gwernvale, Wales, England. After schooling at Marlow, Everest went to the military academy at Woolwich, after which he was sent to India in 1806 as a cadet in the Bengal Artillery. After being employed in various engineering works throughout India, he was appointed in 1818 as an assistant to Colonel William Lambton, the founder of the great trigonometrical survey of India. In 1823, on Colonel Lambton's death, he succeeded to the post of superintendent of the survey. Everest was largely responsible for completion of the trigonometric survey of India along the meridian arc from the south of India extending north to Nepal – a distance of approximately 2,400 kilometres. The survey was started in 1806 and lasted several decades, but Everest never climbed Mount Everest or even entered Nepal. However, the completion of the Indian survey allowed the subsequent survey of Everest and calculation of its summit height, establishing it as the world's highest mountain. It was later re-named in his honour. In 1830 Everest was appointed by the East India Company as Surveyor-General of India. After his retirement from service in 1843 he went back to England, where he became an active member of several other scientific associations. In 1862 he was chosen vice-president of the Royal Geographical Society. (☞ Dec. 1)

1910
Giovanni Virginio Schiaparelli, Italian astronomer who is remembered for his observations of Mars and mapping of Martian 'canals', died in Milan, Italy. (☞ Mar. 14)

1933
American physical chemist William David Coolidge obtained a patent for the X-ray tube – popularly called the 'Coolidge tube' in which the cathode (the source of electrons) is an electrically heated filament. The Coolidge tube completely revolutionised the use of X-rays and it still remains the model that X-ray tubes for medical purposes are patterned after. (☞ Oct. 23, Dec. 30)

1934
Marie Curie (*nee* Maria Sklodowska), Polish-born French chemist and physicist, Nobel laureate (Physics, 1903; Chemistry, 1911), who discovered radium and polonium, died near Sallanches, France. (☞ Nov. 7)

1982
American space shuttle *Columbia*, carrying astronauts Thomas K. Mattingly, and Henry W. Hartsfield Jr., landed at the Edward Air Force Base, California, USA. (☞ Apr. 12, Jun. 27)

1997
American space probe *Mars Pathfinder* landed on Mars at a place called Chryse Planitia, at the mouth of a large flood channel. As the spacecraft descended through the Martian atmosphere, it was slowed successively by a heat shield, a parachute, and rockets. Its impact on the surface was cushioned by an enveloping cluster of air bags, on which it bounced to rest – the first time such a landing technique had been tried. The spacecraft consisted of

two parts – a 370-kg lander and a 10.6-kg rover. Once on the surface, the lander was formally named the Carl Sagan Memorial Station after the American astronomer and writer Carl Sagan. The lander carried a stereo camera system that sent back more than 16,500 images. The rover was named *Sojourner* in honour of the 19th-century African American civil rights advocate Sojourner Truth. *Sojourner* carried two black-and-white cameras, one colour camera, and an alpha proton X-ray spectrometer for determining the composition of rocks and soil. It sent back 550 images to Earth through the lander. (☞ Sep. 27, Dec. 4)

2005
An impactor released from NASA's *Deep Impact* spacecraft hit Comet Tempel-1 as planned and carved out a large crater on its surface. *Deep Impact* watched the explosion from a safe distance of 500

kilometres, and collected data about the comet's composition, which may throw new light on the origin of our solar system. (☞ Jan. 12)

July 5

1833
Joseph Nicéphore Niepce, French chemist and inventor who was the first to make a permanent photographic image, died in Chalon-sur-Saône, France. (☞ Mar. 7)

1888
Herbert Spencer Gasser, American physiologist, Nobel laureate, was born in Platteville, Wisconsin, USA. After attending school in Platteville, Gasser attended the University of Wisconsin at Madison, taking some medical school courses. He received a bachelor's degree in 1910 and a master's degree in 1911. As a student at Wisconsin, Gasser was involved in several studies on the effects of oxygen stress and on the responses of the heart to exercise. He went to the Johns Hopkins Medical School for his clinical studies, obtaining his MD in 1915. After a year in pharmacology at Wisconsin he went to Washington University, St. Louis, where he worked with the American physiologist Joseph Erlanger, and later became professor of pharmacology in 1921. From 1931 to 1935 he was professor of physiology at the medical college of the Cornell University, and from 1935 to 1953, he was director of the Rockefeller Institute for Medical Research. While at St. Louis, Gasser and Erlanger began studying ways in which the then newly developed field of electronics could be applied to physiological investigations. They studied the barely detectable electrical impulses carried by isolated mammalian nerve

fibres. By 1924, they had succeeded in adapting the oscillograph to physiological research, enabling them to visualise amplified nerve impulses on a fluorescent screen. Using this device, they demonstrated that different nerve fibres exist for the transmission of specific kinds of impulses, such as those of pain, cold, or heat. Their work also made it possible to construct improved recording machines to diagnose brain and nervous disorders. Gasser and Erlanger were jointly awarded the 1944 Nobel Prize for Physiology or Medicine, for their discoveries relating to the highly differentiated functions of single nerve fibres. (☞ Jan. 5, May 11)

1891

John Howard Northrop, American biochemist, Nobel laureate, was born in Yonkers, New York, USA. After public school education, Northrop entered the Columbia University in 1908 to study zoology and chemistry, completing his BSc in 1912, MA in 1913, and finally his PhD in chemistry in 1915. In 1916 he joined the staff of the Rockefeller Institute. In 1949 he was appointed professor of bacteriology at the University of California, and later, professor of biophysics. Northrop's researches at Columbia were chiefly concerned with carbohydrates and his early work at the Rockefeller Institute was connected with theories of duration of life. During World War I, Northrop conducted research for the US Army Chemical Warfare Service on the production of acetone and ethanol through fermentation. This work led to the study of enzymes. In 1929 he isolated and crystallised pepsin, a digestive enzyme present in gastric juice, and determined that it was a protein. In 1938, using the same chemical methods, he isolated the first bacterial virus (bacteriophage), and found it to be a nucleoprotein. He also isolated and crystallised trypsin, chymotrypsin, and carboxypeptidase. Northrop was awarded the 1946 Nobel Prize for Chemistry, jointly with the American biochemists James B. Sumner and Wendell M. Stanley. Northrop and Stanley shared half of the prize money for their preparation of enzymes and virus proteins in a pure form, while Sumner received the other half for his discovery that enzymes can be crystallised. (☞ May 27, Aug. 16, Nov. 19)

1927

Albrecht Kossel, German biochemist, Nobel laureate (Physiology or Medicine, 1910), who discovered nucleic acids – the bases in the DNA molecule, died in Heidelberg, Germany. (☞ Sep. 16)

1946

Gerardus 't Hooft, Dutch physicist, Nobel laureate, was born in Den Helder, Netherlands. 't Hooft completed his schooling in the Hague and studied physics and mathematics at the University of Utrecht. His research interests were mostly in high-energy physics and in 1969 he joined fellow Dutch physicist Martinus J G Veltman at Utrecht as a doctoral student to help in the search for a method of re-normalising non-abelian gauge theories. 't Hooft succeeded beyond all expectation and in 1971 published two articles that represented an important breakthrough in the research programme. While the theory of electro-weak interaction developed in the 1960s was a great step forward, the research community at first found it difficult to accept. When they tried to use the theory for calculating the properties of the new W and Z particles (and many other physical quantities) in more detail, it gave unreasonable results. It was only through 't Hooft's and Veltman's work that more precise prediction of physical quantities involving properties of W and Z particles could start, including the mass of the top quark, which was directly observed in 1995. 't Hooft earned his doctorate in physics from the University of Utrecht in 1972 and five years later became a professor there. 't Hooft was awarded the 1999 Nobel Prize for Physics, jointly with Veltman, for elucidating the quantum structure of electro-weak interactions in physics. (☞ Jun. 27)

1966
George Charles de Hevesy, Hungarian chemist, Nobel laureate (Chemistry, 1943), whose development of isotopic tracer techniques enabled understanding of the chemical paths of life processes, died in Freiburg im Breisgau, Germany. (☞ Aug. 1)

1996
Dolly, a cloned sheep, was born at the Roslin Institute, Edinburgh, Scotland. Dolly was the first mammal to be cloned from a somatic cell. Scientists Ian Wilmut and his colleagues at Roslin Institute had removed cells from the udder of a six-year-old ewe and introduced the entire nucleus, containing the full complement of DNA, from this cell into an egg cell taken from another ewe from which the nucleus had been removed. The manipulation was done using microscopic needles, a method pioneered in human fertility treatments in the 1970s. The resulting embryo was implanted into the womb of a third, surrogate sheep and Dolly was the result. The birth of Dolly was announced in 1997. (☞ Feb. 14)

Dolly (left)

1996
Clyde E. Wiegand, American physicist who worked on the Manhattan Project, which produced the atomic bomb, and later, in the 1950s, was part of a team that discovered the antiproton, using the bevatron particle accelerator at the Lawrence Berkeley National Laboratory, Berkeley, died in Oakland, California, USA. (☞ May 23)

July 6

1854
Georg Simon Ohm, German physicist who showed by experiment that all conductors have some resistance and discovered the law named after him, died in Munich, Germany. (☞ Mar. 16)

1885
Louis Pasteur, French microbiologist, and his colleagues injected the first of 14 daily doses of anti-rabies vaccine, made up of rabbit spinal cord suspensions containing inactivated rabies virus, into

9-year-old Joseph Meister, who had been severely bitten by a rabid dog two days before. The immunisation was successful and the child did not develop rabies. This was the beginning of the modern era of immunisation, which had been presaged by Edward Jenner nearly 100 years earlier. Pasteur's rabies immunisation procedure was rapidly adopted throughout the world. Meister grew up and became director of the Pasteur Institute. (☞ Sep. 28, Dec. 27)

1903

Axel Hugo Teodor Theorell, Swedish biochemist, Nobel laureate, was born in Linköping, Sweden. Theorell studied at the Karolinska Institute in Stockholm and received his medical degree in 1930. During 1932–33 he was assistant professor of biochemistry at the University of Uppsala. He became research director at the Karolinska Institute in 1937. Theorell was interested in biochemistry while still a student; in 1926 he discovered and described the lipoproteins in blood plasma. Six years later he became the first person to crystallise and describe some of the properties of myoglobin, an oxygen-carrying protein found in red muscle. He also purified and crystallised other haemoproteins, including cytochrome C and lactoperoxidase. In 1934 he isolated from yeast a pure sample of the so-called 'yellow enzyme', which is instrumental in the oxidative interconversion of sugars by the cell. His study of enzymes that facilitate oxidation reactions in living cells contributed to the understanding of enzyme action and led to the discovery of the ways in which nutrients are used by organisms in the presence of oxygen to produce usable energy. Theorell was awarded the 1955 Nobel Prize for Physiology for his discoveries concerning the nature and mode of action of oxidation enzymes. (☞ Aug. 15)

1959

The first open heart surgery in India was performed on 12-year-old Beulah in the Department of Thoracic & Cardiovascular Surgery of Christian Medical College and Hospital, Vellore, Tamil Nadu by Dr. Reeve H. Betts, a missionary from USA.

July 7

1843

Camillo Golgi, Italian physician and cytologist, Nobel laureate, was born in Corteno, Italy. Golgi received his degree in medicine from the University of Pavia. Then, after serving as a physician in a Pavia hospital for seven years, he became chief physician at a home for incurables in a village named Abbiategrasso, where he worked out a method of silver nitrate staining nerve tissue, an invaluable tool in subsequent nerve studies. In 1880 he was appointed professor of histology at the University of Pavia. In 1883 he discovered the presence in nerve cells of an irregular network of fibrils (small fibres), cavities, and granules, now known as the Golgi complex or Golgi apparatus. The Golgi complex is found in all cells except bacteria and plays an important role in the modification and transport of proteins within the cell. Apart from his research on cell structures, Golgi also made important observations on malaria. He distinguished between different

varieties of the disease and showed that the severity of a malarial attack depends upon the number of parasites in the blood. Golgi was awarded the 1906 Nobel Prize for Physiology or Medicine, jointly with the Spanish histologist Santiago Ramón y Cajal, in recognition of their work on the structure of the nervous system. (☞ Jan. 21, May 1)

1981
Solar Challenger, the first solar-powered aircraft, crossed the English Channel. Weighing 108 kilograms, with a 14.3-metre wingspan, the *Solar Challenger* flew 262 kilometres from France to England at

3,400 metres in 5 hours and 22 minutes. Flying on less power than 2 hairdryers, it remains the only manned aircraft ever to fly on solar energy alone.

1995
American space shuttle *Atlantis* landed at the Kennedy Space Center, Florida, USA, after completing its 14th flight, during which it docked with the Russian orbiting space station *Mir*, forming the largest structure in orbit at that time. (☞ Feb. 20, Jun. 27, Jun. 29)

1996
American space shuttle *Columbia*, carrying astronauts Terence T. Henricks, Kevin R. Kregel, Susan J. Helms, Richard M. Linnehan, Charles E. Brady, Jr., Jean-Jacques Favier, Robert Brent Thirsk, Pedro Duque, and Luca Urbani, landed at the Kennedy Space Center, Florida, USA after completing its 20th flight. The shuttle carried the *Life and Microgravity Spacelab* (*LMS*) orbital laboratory on the mission. (☞ Jun. 20)

2003
American robotic rover *Opportunity* was launched from Cape Canaveral on a mission to Mars. The main objective was to look for geological clues about whether parts of Mars formerly had environments wet enough to be hospitable to life. (☞ Jan. 25)

July 8

1695
Christiaan Huygens, Dutch mathematician, astronomer, and physicist, who patented the first pendulum clock, founded the wave theory of light, and discovered a moon of Saturn, died in The Hague, Netherlands. (☞ Mar. 25, Apr. 14)

1894
Pyotr Leonidovich Kapitsa, Russian physicist, Nobel laureate, was born in Kronstadt, near Leningrad, Russia. Educated at the Petrograd Polytechnical Institute, Kapitsa remained there as a lecturer until 1921. He went to England to study at the University of Cambridge. There he worked with the British physicist Ernest Rutherford and became assistant director of magnetic research at the Cavendish Laboratory in 1924, designing apparatus that achieved an extremely high magnetic field of 500,000 gauss. In his last years in Cambridge, Kapitsa turned

to low-temperature research and developed a new and original apparatus for the liquefaction of helium based on the adiabatic principle. In 1934 he returned to Moscow and organised the Institute for Physical Problems where he continued his research on strong magnetic fields, low-temperature physics and cryogenics and began a series of experiments to study the properties of liquid helium that led to discovery of the superfluidity of helium in 1937. Late in the 1940s Kapitsa turned his attention to a totally new range of physical problems. He invented high-power microwave generators – planotron and nigotron and discovered a new kind of continuous high pressure plasma discharge with electron temperatures over a million K. For his achievements in science during the 1930s and 1940s, Kapitsa was given the title of 'Hero of Socialist Labour', the Soviet Union's highest civilian award, in 1945. Kapitsa's research on high-power microwave generators in the late 1950s turned his interests to controlled thermonuclear fusion, on which he published a series of papers beginning in 1969. Kapitsa was awarded the 1978 Nobel Prize for Physics, jointly with the German-born American astrophysicist Arno Allan Penzias and American radio astronomer Robert Woodrow Wilson. Kapitsa received half of the prize money for his basic inventions and discoveries in the area of low-temperature physics, while Penzias and Wilson shared the other half for their discovery of cosmic microwave background radiation. (☞ Jan. 10, Apr. 8, Apr. 26)

1895

Igor Yevgenyevich Tamm, Soviet physicist, Nobel laureate, was born in Vladivostok, Siberia, Russia. Tamm graduated from the Moscow State University in 1918 and taught there from 1924 to 1937. Tamm's early researches were devoted to crystallo-optics and the quantum theory of diffused light in solid bodies. He later turned his attention to the theory of relativity and quantum mechanics and evolved a method for interpreting the interaction of nuclear particles. In 1934 the Soviet physicist Pavel A.Cherenkov had discovered that light is emitted when gamma rays pass through a liquid medium. In 1936 Tamm and the Soviet physicist Ilya Mikhaylovich Frank explained this phenomenon as the emission of light waves by electrically charged particles moving faster in a medium than the speed of light. Tamm spent the latter decades of his career at the P N Lebedev Physical Institute in Moscow, where he worked on methods for controlling thermonuclear fusion reactions and proposed the first tokamak in 1951. Tamm was awarded the1958 Nobel Prize for Physics, jointly with Cherenkov and Frank for the discovery and the interpretation of the Cherenkov effect. (☞ Apr. 12, July 28, Oct. 23)

1979

Robert Burns Woodward, American chemist, Nobel laureate (Chemistry, 1965), best known for his syntheses of complex organic substances, including quinine, cholesterol and cortisone, and vitamin B_{12}, died in Cambridge, Massachusetts, USA. (☞ Apr. 10)

1979
Shin'ichiro Tomonaga, Japanese physicist, Nobel laureate (Physics, 1965), who developed a quantum field theory that was consistent with the theory of special relativity, died in Tokyo, Japan. (☞ Mar. 31)

1988
The first patent was issued on an animal life form by the US Patent Office to Harvard University scientists for a genetically engineered mouse. The genetically modified mouse, called 'Harvard oncomouse', was genetically altered by inserting a cancer-causing gene which made it highly susceptible to breast cancer. The mouse is specially bred for use in cancer research.

July 9

1856
Nikola Tesla, Serbian-born American inventor, was born on the stroke of midnight in Smiljan, Croatia. Tesla studied electrical engineering at the Austria Politechnic in Graz, Austria. While in Graz, he studied the uses of alternating current. He also developed a telephone repeater (or amplifier). Tesla was one of most prolific inventors of his time. His most famous contribution was the theory of polyphase alternating current electricity, which he used to build the first induction motor in 1883. In 1884, Tesla moved to USA to accept a job with the Edison Company in New York. Since alternating current could be transmitted over greater distances than direct current, in 1885, Tesla could persuade George Westinghouse, head of the Westinghouse Electric Company in Pittsburgh, USA, to buy the patent rights to his polyphase system of alternating current dynamos, transformers, and motors. Westinghouse used Tesla's system to light the World's Columbian Exposition at Chicago in 1893, and installed the first power machinery at Niagara Falls, that carried power to the city of Buffalo, New York in 1896. Tesla's theory and many of his patents form the basis for modern electric power systems. Tesla soon established his own laboratory and experimented with shadowgraphs similar to those that would later be used by the German physicist Wilhelm Röntgen when he discovered X-rays in 1895. Tesla's countless experiments included work on a carbon button lamp, on the power of electrical resonance, and on various types of lighting including the first neon light. While staying at Colorado Springs, Colorado, Tesla made what he regarded as his most important discovery – terrestrial stationary waves. By this discovery he proved that the Earth could be used as a conductor and would be as responsive as a tuning fork to electrical vibrations of a certain frequency. Tesla is also noted for inventing the Tesla coil. (☞ Jan. 7)

1856
Amedeo Avogadro, Italian scientist who was the first scientist to distinguish between atoms and molecules and propounded the law of chemistry known after him, died in Turin, Italy. (☞ Jun. 9)

1926
Benjamin Roy Mottelson, American-born Danish physicist, Nobel laureate, was born in Chicago, Illinois, USA. Mottelson

graduated from the Lyons Township High School in 1944 and did his BSc from the Purdue University, Layfayette, Indiana. After completing his doctorate in theoretical physics from the Harvard University, Cambridge, Massachusetts, in 1950, he got a fellowship at the Niels Bohr Institute of Theoretical Physics in Copenhagen, Denmark, and then joined the faculty of the Nordic Institute for Theoretical Nuclear Physics there as professor in 1959. Since 1980, Mottelson has been a physicist for the Bohr Institute in Copenhagen. Mottelson began collaborating with the Danish physicist Aage N. Bohr in 1951 to develop a structural model of the atomic nucleus. From experiments conducted in the early 1950s, they discovered that the motion of subatomic particles could distort the shape of the nucleus, thus challenging the widely accepted theory that all nuclei are perfectly spherical. Subsequently it was discovered that such asymmetries occur in atoms of all elements. Mottelson was awarded the 1975 Nobel Prize for Physics, jointly with Bohr and the American physicist James Rainwater, for the discovery of the connection between collective motion and particle motion in atomic nuclei and the development of the theory of the structure of the atomic nucleus based on this connection. (☞ Jun. 19, Dec. 9)

1932

King Camp Gillette, American inventor of the safety razor and disposable blade, died in Los Angeles, California, USA. (☞ Jan. 5)

1957

Discovery of element 102, nobelium (No) was announced by a group of scientists working at the Nobel Institute of Physics in Stockholm, Sweden. They produced this new element by bombarding a target of curium-244 with ions of carbon-13 using a device called a cyclotron. They obtained an isotope with mass number between 251 and 255. The isotope they created had a half-life of 10 minutes. They named the element nobelium after the Swedish inventor Alfred Nobel who invented dynamite. In 1958, another group of scientists, Albert Ghiorso, Glenn T. Seaborg, Torbørn Sikkeland and John R. Walton, working at the Lawrence Radiation Laboratory in Berkeley, California, USA, attempted to confirm the Nobel Institute's discovery. They were unable to produce any isotope of nobelium with a half-life of 10 minutes, but were able to produce nobelium-254, with a half-life of three seconds, by bombarding curium-246 with carbon-12. Ten isotopes of nobelium are known to exist, with the isotope No-259 having the longest half-life of about 58 minutes.

1979

The American unmanned spacecraft *Voyager-2* flew-by Jupiter. It was one of a pair of unmanned interplanetary probes launched by the American space agency NASA in 1977. Although it was launched before *Voyager-1*, *Voyager-2* was designed to travel more slowly, and fly-by all the giant planets. *Voyager-2* flew by Saturn in August 1981, but continued on to Uranus in January 1986 and Neptune in August 1989. Data from the two probes included photographs of Jupiter showing a variety of cloud forms around Jupiter, and volcanic activity on its moon, Io. (☞ Jan. 24, Aug. 20, Aug. 25)

July 10

1851

Louis Jacques Mande Daguerre, French artist, chemist, and photographer, who invented the Daguerreotype process of photography, died in Bry-sur-Marne, France. (☞ Jan. 9, Nov. 18)

1902

Kurt Alder, German chemist, Nobel laureate, was born in Königshütte, Upper Silesia (now in Poland). Alder studied chemistry at the universities of Berlin and Kiel, where he received his doctorate in 1926. From 1930 to 1936 Alder taught chemistry at the Kiel University after which he took up the appointment as head of department in the science laboratories of the I.G. Farben-Industrie, at their works in Leverkusen, where he worked on the preparation and constitution of a synthetic rubber called 'Buna'. In 1940 Alder was appointed to the chair for experimental chemistry and chemical technology at the University of Cologne and also became director of the Chemical Institute. As early as 1927–1928, while working for his doctorate at Kiel, Alder had studied problems of systematic organic chemistry in collaboration with his teacher, the German organic chemist Otto Paul Hermann Diels, and this led to their joint discovery of the principle of the diene-synthesis, which they investigated and determined in all its aspects. The Diels–Alder reaction involves the adding of an organic compound that has two double bonds separated by a single bond (called a conjugated diene) to a compound with only one, activated double bond (termed a dienophile). The diene synthesis stimulated and made easier the understanding of this important group of natural products. The reaction proceeds in the mildest conditions, is of general application, and hence of great utility in synthesis. It is used in the synthesis of natural products, such as sterols, vitamin K, cantharides, and synthetic polymers. Alder and Diels were jointly awarded the 1950 Nobel Prize for Chemistry for their discovery and development of the diene synthesis. (☞ Jan. 23, Jun. 20)

1908

Helium was liquefied for the first time at a temperature of 4.2 K (about −269°C) by the Dutch physicist Heike Kamerlingh Onnes. He had worked for many years to liquefy this element, which until then had persisted as a gas even to the lowest temperature. The gas was finally liquefied by compressing it, cooling it below the inversion temperature and then allowing it to expand, which caused further cooling resulting in the liquefaction of some of the gas. Kamerlingh Onnes used liquid air to produce liquid hydrogen and then the liquid hydrogen to jacket the liquefaction apparatus, to produce about 60 millilitres of liquid helium. At his cryogenic laboratory, he had liquefied air in large quantities in 1892, and built a large hydrogen liquefier in 1906. Kamerlingh Onnes received the 1913 Nobel Prize for Physics for his work in low temperature physics. (☞ Sep. 21)

1910

Johann Gottfried Galle, German astronomer who was the first to observe the planet Neptune, on the basis of calculations made by the French astronomer Urbain Le Verrier, died in Potsdam, Germany. (☞ Jun. 9)

1920

Owen Chamberlain, American physicist, Nobel laureate, was born in San Francisco,

California, USA. After obtaining his bachelor's degree at the Dartmouth College, Hanover, New Hampshire, in 1941, Chamberlain entered graduate school in physics at the University of California, but his studies were interrupted when the United States joined World War II. From 1942 to 1946, he joined the Manhattan Project that led to the development of the atom bomb. Working under the Italian-born American physicist Emilio Segrè, he investigated the spontaneous fission of heavy elements. Later, while completing his PhD at the University of Chicago, he worked at the Argonne National Laboratory, in Illinois. In 1948 he joined the faculty of the University of California at Berkeley, where he became a professor in 1958. There he conducted research on alpha particle decay, neutron diffraction in liquids, and high-energy nuclear particle reactions. In 1955, in a series of proton scattering experiments he and Segrè discovered the antiproton, a particle exactly like a proton but with a negative charge. The following year they confirmed the existence of the antineutron. Chamberlain and Segrè were awarded the 1959 Nobel Prize for Physics, for their discovery of the antiproton. (☞ Feb. 1)

1960

CIRUS, the 40-MW research reactor of the Atomic Energy Centre (re-named in 1967 as the Bhabha Atomic Research Centre) at Trombay, near Bombay (now Mumbai) attained criticality. Originally known as the Canada–India Reactor, or *CIR* (re-named *CIRUS* in 1965), it was the first major atomic energy project undertaken with international assistance. *CIRUS* is a powerful source of neutrons and has been extensively used for investigating and characterising materials.

1962

Telstar-1, the world's first telecommunication satellite, was launched from Cape Canaveral, Florida, USA, to relay TV and telephone signals between the United States and Europe. It was placed in an elliptical orbit ranging in height from 945 km to 5,643 km. *Telstar* made possible the first live television transmission across the Atlantic. Viewers in France and England saw President Kennedy conduct a press conference, and audiences in the

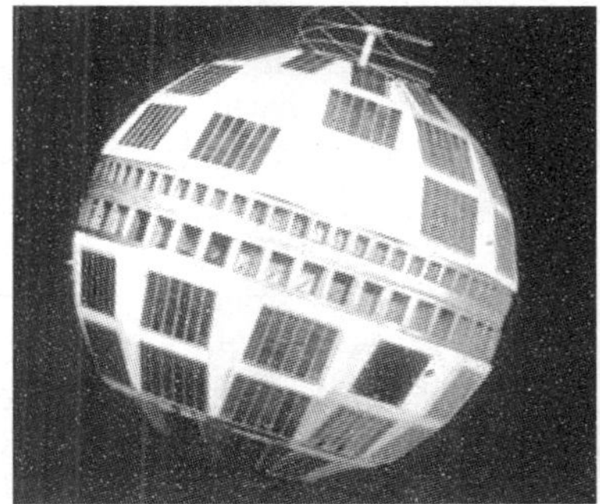

United States watched French singer Yves Montand and the changing of the guard at England's Buckingham Palace. (☞ May 7)

1992

INSAT-2A, the first of India's second generation INSAT series of satellites, was launched by the European Space Agency's Ariane rocket from Kourou in French Guiana. Designed and built by the Indian Space Research Organisation, the multipurpose satellite carried two S-band transponders, 18 C-band transponders, and two very high resolution radiometers, and was meant to provide telecommunication, networking and meteorological services. The satellite was placed in the geostationary orbit at 74°E longitude.

July 11

1916

Aleksandr Mikhaylovich Prokhorov, Russian physicist, Nobel laureate, was born in Atherton, Queensland, Australia. Prokhorov graduated from the Leningrad State University in 1939 and did his post-graduation from the P.N. Lebedev Physical Institute in Moscow, where he worked on the problems of propagation of radio waves. Starting in 1947, Prochorov carried out a study of the coherent radiation of electrons in the synchrotron in the region of centimetre waves. In 1951, he received a PhD for his thesis on 'Coherent radiation of electrons in the synchrotron accelerator'. In 1952 he and fellow physicist Nikolay Gennadiyevich Basov jointly suggested the maser principle of amplifying and emitting coherent electromagnetic waves that are all in phase and of the same wavelength. Maser stands for 'microwave amplification by stimulated emission of radiation'. By the time they published their suggestion in 1954, the American physicist Charles Hard Townes had already built the first working maser. In 1954 Prokhorov became head of the institute's Oscillation Laboratory and later professor at the Moscow M.V. Lomonosov State University. Prokhorov was awarded the 1964 Nobel Prize for Physics, jointly with Basov and Townes, for fundamental work in the field of quantum electronics, which led to the construction of oscillators and amplifiers based on the maser laser principle. (☞ Jul. 28, Dec. 14)

1924

César Mansueto Giulio Lattes, Brazilian physicist who, with American physicist Eugene Gardner, discovered the pi meson, was born in Curitiba, Brazil. In the early 1940s, Lattes studied at the University of São Paulo with the Italian physicist Giuseppe Occhialini. In 1946 Lattes and Occhialini joined the University of Bristol in UK, to work with the British physicist Cecil Powell. In 1947 they discovered tracks of particles on photographic film of two kinds of mesons. They called the first a pi meson, or pion, and the second a mu meson, muon. In 1935 the Japanese physicist Hideki Yukawa had suggested that a meson (a charged particle with a mass intermediate between those of the electron and the proton) might be exchanged between nucleons, binding them together in an atomic nucleus. The pion turned out to be a particle about 1.3 times more massive than the muon. In 1948, together with the American physicist Eugene Gardner, at the University of California, Berkeley, Lattes confirmed the existence of the heavy and light mesons formed during the bombardment of carbon nuclei with alpha particles.

1927

Theodore Harold Maiman, American physicist who developed and patented the laser, was born in Los Angeles, California, USA. Mainman attended the University of Colorado and received a BS in engineering physics in 1949. He then went on to do graduate work at the Stanford University, where he received an MS in electrical engineering in 1951 and a PhD in physics in 1955. In 1960 Maiman

built the first functioning laser in the world while working at the Hughes Research Laboratories at Malibu, California. His invention was influenced by Arthur L. Schawlow and Charles H. Townes' article, 'Infrared and optical masers', which appeared in the journal *The Physics Review* in 1958. In their article the two scientists laid out a theoretical basis for the laser construction and outlined some of its problematic aspects. Maiman used the three-level principle for maser worked out by the Dutch-born American physicist Nicolaas Bloembergen and constructed a ruby cylinder with the ends polished flat and parallel and silvered. Energy was fed in using a flash lamp. He got an output of monochromatic coherent beam of radiation. By way of analogy with the maser, Maiman named his device laser, an acronym for 'light amplification by stimulated emission of radiation'. (☞ May 16)

1979

The first American space station *Skylab* re-entered Earth's atmosphere and disintegrated over Western Australia and the Indian Ocean, casting large pieces of debris in populated areas. Before re-entry the *Skylab* orbited the Earth 2,476 times during the 171 days and 13 hours of its occupation during the three manned Skylab missions. Astronauts performed ten space walks totalling 42 hours 16 minutes from it. Many of the experiments performed on it were on the astronauts' adaptation to extended periods of microgravity, and each Skylab mission set a record for the duration of time astronauts spent in space. (☞ Feb. 8, May 14, May 25, Jun. 22, Jun. 28, Sep. 25, Nov. 16)

1998

Discovery of the tau neutrino was announced by researchers at the Fermi National Accelerator Laboratory (Fermilab) in Batavia, Illinois, USA. The tau neutrino is a mass-less (or nearly mass-less) particle that carries no electric charge and barely interacts with surrounding matter. The tau is the third and last neutrino to be confirmed on the Standard Model of elementary particles, a theoretical description of the basic constituents of matter and the fundamental forces of nature. The first two neutrinos to be discovered were the electron and the muon. The electron neutrino was discovered in 1956 by Frederick Reines and Clyde Cowan of the Los Alamos National Laboratory using the Hanford reactor in Washington and the Savannah River reactor in South Carolina, USA. The muon neutrino was discovered in 1962 by Leon Lederman, Melvin Schwartz, and Jack Steinberger at the Brookhaven National Laboratory, Long Island, New York, USA.

July 12

1854

George Eastman, American inventor, who invented the photographic roll film that revolutionised amateur photography, was born in Waterville, New York, USA. Eastman was a self-educated man who contributed greatly to the field of photography. In 1877, he embarked upon the intricate tasks of preparing the necessary emulsions, coating the 'wet plates' on which most pictures were then

taken, and developing the prints. He pursued eagerly all available literature on the subject and was attracted by a formula for a 'dry plate' emulsion that appeared in an English almanac. The formula suggested the possibility of reducing the size and weight of outdoor photographic equipment. The first commercial film, put into production a year later, was cut in narrow strips and wound on a roller device patented by Eastman. Film rolls sufficient for 100 exposures were mounted in a small box camera – the Kodak camera – the first camera designed specifically for roll film, which was introduced in 1888 priced at $25. In 1892 he established the Eastman Kodak Company, at Rochester, New York, one of the first firms to mass-produce standardised photography equipment. This company also manufactured the flexible transparent film, which proved vital to the subsequent development of the motion picture industry. (☞ Mar. 14, Sep. 4, Oct. 14)

1863

Albert Calmette, French bacteriologist and co-developer of the tuberculosis vaccine Bacillus Calmette-Guérin (BCG), was born in Nice, France. Calmette studied medicine in Paris and graduated in 1886. In 1891 he founded the Pasteur Institute at Saigon (now Ho Chi Minh City), Indochina (present Vietnam), where he discovered a protective serum against snake venom. He returned to France in 1896 and became founder director of the Pasteur Institute at Lille. There in 1908 he discovered that he could turn virulent bovine tubercle bacilli into a less virulent strain if he used a bile-containing culture medium. These attenuated bacilli were able to confer a certain amount of immunity against infection with either bovine or human tubercle bacilli. This avirulent strain was used to make BCG vaccine, which was introduced about 15 years later for the vaccination of children against tuberculosis. Although widely used in continental Europe, the vaccine was not adopted in the United States and Great Britain until after American studies in 1940 and a trial sponsored by the British Medical Research Council indicated BCG's substantial protective action against tuberculosis. Calmette also described a diagnostic test for tuberculosis, known as Calmette's reaction. (☞ Oct. 29, Dec. 22)

1870

John Wesley Hyatt, and his brother, Isaiah S. Hyatt of Albany, New York, USA were issued a US patent for a process for making celluloid, the first synthetic plastic. While looking for a substitute material for ivory billiard balls they developed a process of producing a horn-like material using cellulose nitrate and camphor, which was named 'Celluloid'. Unfortunately, it was inflammable. It was used for production of photographic film, and many other plastic products.

1895

Richard Buckminster Fuller, American engineer and architect who developed the geodesic dome, the only large dome that can be set directly on the ground as a complete structure, was born in Milton, Massachusetts, USA. Fuller studied in the Milton Academy. He was later expelled from the Harvard University during freshman year for 'irresponsibility and lack of interest'. He served in the US Navy in World War I. In 1927 Fuller designed the Dymaxion House, which was intended to house the technology needed for people to live. With its wire construction it was in no way intended to be aesthetic but designed to be a totally practical machine to live in. For many years Fuller worked on ideas for buildings, eventually arriving at the structure for which he is most famous. He examined a vectorial system of geometry, based on the tetrahedron, which provides maximum strength with minimum structure. This led to his patent of a geodesic dome in 1947, a building the strength of which increases as the log of its size, and the only practical kind of building that has no limiting dimensions. Another of Fuller's projects, which he began in 1928, involved the design of a highly unconventional car. It was called the Dymaxion car, a three-wheeled machine capable of carrying 12 passengers, of crossing fields as well as having a top speed of nearly 200 km/h. With a manoeuvrability, which allowed it to turn within its own length, this was a remarkable concept, and development continued until 1943, but it was never launched as a commercial venture. Fuller also invented a wide range of other machines and structural systems. He was especially interested in high-strength-to-weight designs, with a maximum utility for minimum material. His designs and engineering philosophy are part of the foundation of contemporary high-tech design aesthetics. He held over 2,000 patents. (☞ Jul. 1)

1913

Willis Eugene Lamb, Jr., American physicist, Nobel laureate, was born in Los Angeles, California, USA. Lamb was educated in the public schools of Los Angeles. In 1930 he entered the University of California at Berkeley and received a BS degree in chemistry in 1934. His graduate work in theoretical physics at the same university led to the PhD degree in 1938; his thesis was on the electromagnetic properties of nuclear systems. He went to the Columbia University in New York as instructor in physics in 1938, and became a professor in 1948. From 1943 to 1951, he was associated also with the Columbia Radiation Laboratory where he carried out studies on the fine structure of the hydrogen spectrum. In 1951 he went to the Stanford University in California as a professor of physics. While at Stanford he devised microwave techniques for examining the hyperfine structure of the spectral lines of helium. He was professor of theoretical physics at the University of Oxford, UK until 1962, when he was appointed professor of physics at the Yale University, New Haven, Connecticut. In 1974 he became professor of physics and optical sciences at the University of Arizona, Tucson, Arizona. Lamb was awarded the 1955 Nobel Prize for Physics, jointly with the German-born American physicist Polykarp Kusch, for his discoveries concerning the fine structure of the hydrogen spectrum. (☞ Jan. 26)

1928

Elias James Corey, American organic chemist, Nobel laureate, was born in Methuen, Massachusetts, USA. After schooling in Lawrence, a city near Methuen, Corey entered the Massachusetts Institute of Technology in 1945. He earned his BS in 1948 and a PhD in chemistry in 1951 from MIT. From 1951–59 he taught at the University of Illinois in Champaign-Urbana and then joined the Harvard University, Cambridge, Massachusetts, as professor of

chemistry. At Harvard, Corey and his students synthesised some 100 molecules hitherto found only in nature. Corey also pioneered the synthesis of prostaglandins – hormone-like compounds used to treat infertility and induce labour. Another substance first made by Corey is ginkgolide B, an active chemical originally found in the ginkgo tree, which is used to treat asthma and circulatory problems. Beginning in the 1960s, Corey coined the term, and developed the concept of 'retrosynthetic analysis'. Starting from the structure of the molecule he wanted to produce – the target molecule – he established rules for how it should be dissected into smaller parts, and what strategic bonds should be broken. In this way, less complicated building blocks were obtained, which could later be assembled in the process of synthesis. These building blocks were then analysed in the same way until simple compounds had been reached, whose synthesis was already described in literature, or which were commercially available. The identified smaller units could then be re-assembled into the target molecule in the least possible number of steps using the simplest possible reactions, in which computer programming could be of great help. The technique makes the synthesis of even very complex molecules faster, cheaper, and more efficient. Corey was awarded the 1990 Nobel Prize for Chemistry for his development of the theory and methodology of organic synthesis.

1957
Leroy Burney, US Surgeon General, issued a report, which linked smoking and lung cancer. He was the first government official to publicly acknowledge the connection between smoking and lung cancer. His report said, 'It is clear that there is an increasing and consistent body of evidence that excessive cigarette smoking is one of the causative factors in lung cancer'.

July 13

1826
Stanislao Cannizzaro, Italian chemist, teacher, and legislator who recognised the distinction between atomic and molecular weights, was born in Palermo, Sicily, Italy. Cannizzaro joined the medical school in the Palermo University when he was 15, but did not get his degree; he was more interested in chemistry. In 1845 he joined the University of Pisa to learn chemistry, but was condemned to death for his part in the Sicilian revolution in 1848. Cannizzaro escaped to Marseille, France, and arrived in Paris in1849 where he helped prepare cyanamide in 1851. The same year he became professor of chemistry and physics at the Technical Institute of Alessandria, Piedmont, Italy, where in 1853 he discovered that treatment of benzaldehyde with a strong alcoholic hydroxide produced equal amounts of benzyl alcohol and the salt of benzoic acid. The reaction came to be known as 'Cannizzaro reaction'. Cannizzaro laid the foundations of modern chemistry with his experimental atomic weight determinations. As professor of chemistry at the University of Genoa, he showed in 1858 that the atomic weights of the elements in the molecules of a volatile compound can be calculated by applying Avogadro's law concerning gases, and in the case of a non-volatile compound by a

measurement of specific heat. He suggested that since hydrogen is the lightest of all elements, the weight of half a hydrogen molecule should be used as the standard to which all other weights were compared. From 1861 he was professor at the Palermo University and from 1871 at the University of Rome, where he was also a member of the senate. (☞ May 10)

1896
Friedrich August Kekulé, German chemist who laid the groundwork for the modern structural theory in organic chemistry, died in Bonn, Germany. (☞ Sep. 7)

1921
Gabriel Lippmann, French physicist, Nobel laureate (Physics, 1908), who produced the first colour photographic plate, died at sea, en route from Canada to France. (☞ Aug. 16)

1974
Patrick Maynard Stuart Blackett, English physicist, Nobel laureate (Physics, 1948), whose studies explained the disintegration of stable atomic nuclei in cosmic radiation, died in London, UK. (☞ Nov. 18)

1995
The American spacecraft *Galileo* released a probe that plunged into the thick clouds of Jupiter, becoming the first space probe ever to penetrate the atmosphere of any of the outer gas giants. The mission's scientific objectives included measurement of the temperature and pressure structure of Jupiter's atmosphere and its chemical composition. Measurements were made of the amount of helium relative to hydrogen on Jupiter, and winds in the atmosphere. The probe could also detect lightning and energetic protons and electrons trapped in Jupiter's magnetic field. (☞ Sep. 21, Oct. 18, Dec. 7)

July 14

1867
Dynamite was first demonstrated by the Swedish chemist and inventor Alfred Bernhard Nobel at a quarry in Redhill, Surrey, England. Nobel had produced dynamite as a safe and manageable form of nitroglycerine. Following an explosion at a dynamite plant that killed Nobel's younger brother and four other workers, Nobel came up with a safer explosive called gelignite. Other inventions followed including ballistite, a form of smoke-less powder, artificial gutta-percha and mild steel for armour plating. (☞ Oct. 21, Nov. 25, Dec. 10)

1907
William Henry Perkin, English chemist and inventor, who discovered the first artificial dye, died in Sudbury, near Harrow, Middlesex, England. (☞ Mar. 12)

1914
Robert Hutchings Goddard, American professor and inventor, was granted the first patent for a liquid-fuelled rocket design. His first rocket reached a height of 12.5 metres. (☞ Mar. 16, Oct. 5)

1921
Geoffrey Wilkinson, British chemist, Nobel laureate, was born in Todmorden, Yorkshire, England. Wilkinson graduated from the Imperial College of Science and Technology, University of London, UK, in 1941 and worked with the Atomic Energy Project in Canada from 1943 to 1946. He then taught at the University of California at Berkeley, the Massachusetts Institute of Technology, and the Harvard University in USA before returning in 1956 to the Imperial College in London,

where he became professor emeritus in 1988. While working with the Atomic Energy Project in Canada, Wilkinson discovered many new isotopes in the products of atomic fission reactions. In 1951 after reading about a puzzling, newly synthesised compound called dicyclopentadienyl-iron (now called ferrocene) he deduced its structure to be consisting of a single iron atom sandwiched between two five-sided carbon rings. Wilkinson went on to synthesise a number of other 'sandwich' compounds, or metallocenes and also studied the chemistry of ruthenium, rhodium and rhenium, in compounds of unsaturated hydrocarbons and with metal-to-hydrogen bonds. The latter led to work on homogeneous catalytic reactions such as hydrogenation and hydroformylation of olefins. His research on metal-to-hydrogen bonding, particularly his discovery of Wilkinson's catalyst, a homogeneous hydrogenation catalyst for alkenes, proved to have important industrial applications. Wilkinson was awarded the 1973 Nobel Prize for Chemistry, jointly with the German theoretical chemist Ernst Otto Fischer, for their pioneering work, performed independently, on the chemistry of the organometallic, so-called sandwich, compounds. (☞ Sep. 26, Nov. 10)

1965
The first close-up photographs of Mars were sent back by the American spacecraft *Mariner-4*. The pictures, taken from a distance of about 9,850 kilometres, depicted the regions on Mars known as Cebrenia, Arcadia, and Amazonis. *Mariner-4* sent back 22 television pictures covering about 1% of the planet. The photographs revealed Mars to have a cratered, rust-coloured surface, with signs on some parts of the planet that liquid water had once etched its way into the soil. (☞ Nov. 28)

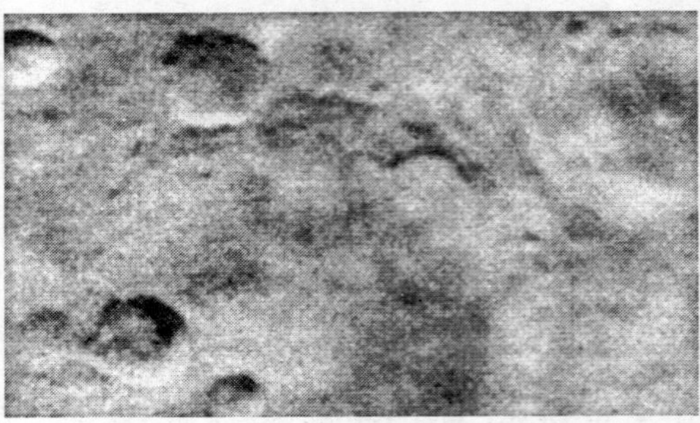

1989
The large electron positron collider (LEP) was inaugurated at the CERN research centre in Switzerland. The largest accelerator built till 2005, the LEP lies in a tunnel, 3.8 metres wide and contains 4,600 magnets to guide the particle beams. The LEP is used to collide together bunches of electrons with bunches of positrons, as they travel in opposite directions round a ring 27 kilometres in circumference, at velocities close to the speed of light. When the bunches of particles meet, some electrons and positrons annihilate, creating, for a fraction of a second, bursts of high energy which echo the state of the early Universe.

July 15

1918
Bertram N. Brockhouse, Canadian physicist, Nobel laureate, was born in Lethbridge, Alberta, Canada. Brockhouse graduated from the University of British Columbia in 1947 and obtained his master's degree from the University of Toronto in 1948.

During World War II, he served as an on-shore electronic technician in the Royal Canadian Navy. After the war he obtained his PhD degree in 1950 from the University of Toronto. In the same year he joined the staff of the Atomic Energy Project of the National Research Council of Canada at the Chalk River Nuclear Laboratory where he began his work on developing new neutron scattering techniques – a method of 'seeing' the atomic structure of solid materials by bombarding them with neutrons. Brockhouse used inelastic neutron scattering in his pioneering examination of phonons – units of the lattice vibrational energy expended by the scattered neutrons. He also developed the neutron spectrometer and was one of the first to measure the phonon dispersion curve of a solid. His triple-axis crystal spectrometer is now widely used to investigate atomic structures, even including virus and DNA molecules. He was a professor at the McMaster University in Hamilton, Ontario, from 1962 until his retirement in 1984. Brockhouse was awarded the 1994 Nobel Prize for Physics, jointly with the American physicist Clifford G. Shull, for pioneering contributions to the development of neutron scattering techniques for studies of condensed matter. (☞ Sep. 23, Oct.13)

1919

Emil Fischer, German organic chemist, Nobel laureate (Chemistry, 1902), who determined the structures of uric acid, xanthine, caffeine, theobromine, and other related compounds, and showed that they are all derivatives of a single compound that he named purine, died in Berlin, Germany. (☞ Oct. 9)

1921

Robert Bruce Merrifield, American biochemist and educator, Nobel laureate, was born in Fort Worth, Texas, USA. Merrifield graduated from the University of California at Los Angeles in 1943 and obtained his PhD in biochemistry there in 1949. In the same year he joined the staff of the Rockefeller Institute for Medical Research (now the Rockefeller University) in New York where he worked on a dinucleotide growth factor he had discovered in graduate school and on peptide growth factors. These studies led to the need for peptide synthesis and, eventually, to the idea of synthesising chains of amino acids, or polypeptides, in any predetermined order on a solid matrix. His innovative method, developed during the 1950s and 60s, grew from his idea that the key to the synthesis of polypeptides was the anchoring of the first amino acid to an insoluble solid. Other amino acids could then be joined, one by one, to the fixed terminus. At the end of the sequence of steps, the completed chain could be detached from the solid. The process could be carried out by machine and proved highly efficient and of great significance for the commercial manufacture of drugs like insulin and substances like interferon. The technique also made possible the development of highly specific antibody probes for proteins and provided strategies for studies of structure–activity relationships in proteins and peptides. Merrifield was awarded the 1984 Nobel Prize for Chemistry for his development of methodology for chemical synthesis on a solid matrix.

1922

Leon Max Lederman, American physicist, Nobel laureate, was born in New York, USA. Lederman obtained his BS degree from the City College of New York in 1943, and received his PhD in physics from the Columbia University, New York, in 1951. He joined the faculty at the Columbia University the same year and became a full professor in 1958. From 1960 to 1962, Lederman, together with his fellow Columbia University researchers Melvin Schwartz and Jack Steinberger, collaborated in an important experiment at the Brookhaven National Laboratory on Long Island, New York in which they used a particle accelerator to produce the first laboratory-made beam of neutrinos – elusive subatomic particles that have no detectable mass and no electric charge and that travel at the speed of light. It was known that when neutrinos interact with matter, either electrons or electron-like particles known as muons (mu mesons) are created. The three scientists showed that the neutrinos that produced muons were a distinct and previously unknown type of neutrino, one that the scientists named 'muon neutrinos.' The discovery of muon neutrinos was followed by discoveries by other scientists of a number of different 'families' of subatomic particle. Together, they now form a 'standard model,' a scheme that has been used to classify all known elementary particles. In 1979 Lederman became director of the Fermi National Accelerator Laboratory in Batavia, Illinois, where he supervised the construction and utilisation of the first superconducting synchrotron. Lederman was awarded the 1988 Nobel Prize for Physics, jointly with Schwartz and Steinberger, for the neutrino beam method and the demonstration of the doublet structure of the leptons through the discovery of the muon neutrino. (☞ May 25, Nov. 2)

1956

Ashoke Sen, Indian theoretical physicist, Fellow of the Royal Society, who works on string theory, was born in Calcutta (now Kolkata), West Bengal. Sen had his schooling in the Shailendra Sircar Vidyalaya, and did his graduation from the Presidency College in Kolkata in 1975. He obtained his Master's degree from the Indian Institute of Technology Kanpur in 1978 and went to USA to do his graduate studies at the State University of New York, Stonybrook. He did his post-doctoral work at the Fermilab, Batavia, and the Stanford Linear Accelerator Center, Stanford, USA. He returned to India in 1988 and joined the Tata Institute of Fundamental Research, Mumbai, Maharashtra. Since 1995 he has been a theoretical physicist at the Harish-Chandra Research Institute in Allahabad, Uttar Pradesh. Sen's work mostly concerns the duality of the String Theory. String theorists believe one-dimensional strings vibrate at sub-atomic scales like miniature harps in a symphony. The vibrations produce fundamental tones whose 'notes' represent all the known forces. To cosmologists, however, string theory's cold, dainty fingers could not have orchestrated the titanic processes that caused the Big Bang and brought the universe into existence. The apparent contradiction was resolved when Sen established that super-symmetric theories (every fundamental fermion has a bosonic

super-partner and vice versa) would require the existence of certain solitons (self-reinforcing solitary wave), with both electrical and magnetic charges. He co-discovered S-duality (strong–weak duality) and proposed a successful explanation of open string tachyon (a hypothetical particle that travels faster than light) condensation. He has also co-authored many papers on string field theory.

1975
American manned mission *Apollo-18* carrying astronauts Thomas P. Stafford, Vance D. Brand, and Donald K. Slayton was launched from the Kennedy Space Center in Florida, USA. This was the final flight of the Apollo spacecraft and it was aimed primarily at docking with the Russian spacecraft *Soyuz-19*, the first docking of spacecraft built by different nations. This presaged the era of cooperation between Russians and Americans. (☞ Jul. 17)

1975
Russian spacecraft *Soyuz-19* carrying cosmonauts Alexei Leonov and Valeri Kubasov was launched from Baikanur, USSR (now Kazakhstan). Two days later the spacecraft docked with the American spacecraft *Apollo-18* marking the first international link-up in space and presaging the era of cooperation between Russians and Americans. (☞ Jul. 17)

July 16

1746
Giuseppe Piazzi, Italian astronomer, was born in Ponte di Valtellina, Italy. Piazzi's early education was in philosophy, and only later in life did he turn to the study of mathematics and astronomy. While working as a professor at the University of Palermo in 1780, he established an observatory at Palermo and mapped with great accuracy the positions of 7,646 stars. He also discovered that the star 61 Cygni had a large proper motion, which led the German astronomer Friedrich Wilhelm Bessel to choose it as the object of his parallax studies. Piazzi discovered 'Ceres' – the first minor planet discovered – in 1801. (Ceres has now been designated as a dwarf planet.) This discovery led other astronomers to search for other objects in the same area of the sky where Ceres was found; three more minor planets were discovered during Piazzi's lifetime. Today more than 1,600 minor planets are known. The thousandth minor planet discovered was named Piazzia in his honour. (☞ Jan. 1, Jul. 22)

1867
Reinforced concrete was patented by F. Joseph Monier, a gardener in Paris, to reinforce garden tubs, beams and posts. The French inventor had found that the tensile weakness of plain concrete could be overcome if steel rods were embedded in a concrete member. The new composite material, called reinforced cement concrete, combined the high compressive strength of concrete with the high tensile strength of steel which made the structures extraordinarily strong.

1888
Frits Zernike, Dutch scientist, Nobel laureate, was born in Amsterdam, the Netherlands. Frits inherited his passion for physics from his father; as a boy he already possessed an arsenal of pots, crucibles, tubes, which he scraped together with his

own pocket money, or was awarded as gifts from understanding manufacturers. At secondary school he excelled in all scientific subjects. He entered the University of Amsterdam in 1905, studying chemistry, with physics and mathematics as minor subjects. In 1915 he was appointed as a lecturer in mathematical physics at Groningen, where he was made a full professor in 1920. From 1930 on he turned to optics – developing his phase contrast method (the introduction of a filter into an imaging system to implement a phase contrast for an intensity mapping of a pure phase image), studied imaging errors of the concave grating and worked on partial coherence. In addition to its capacity to render colourless and transparent objects visible in the microscope, the use of the phase contrast method also enables one to detect slight flaws in mirrors, telescope lenses, and other instruments indispensable for research. Although discovered in 1930, the concept of phase contrast did not immediately receive attention and it was not until 1941 that the first phase-contrast microscopes were manufactured. Zernike was awarded the 1953 Nobel Prize for Physics for his demonstration of the phase contrast method, especially for his invention of the phase contrast microscope. (☞ Mar. 10)

1899

Robert Bunsen, German chemist who, with his compatriot Gustav Kirchhoff, observed that each element emits a light of characteristic wavelength, and who is better known for the laboratory gas burner named after him, died in Heidelberg, Germany. (☞ Mar. 31)

1916

Ilya Ilyich Mechnikov, Russian-born French zoologist and microbiologist, Nobel laureate (Physiology or Medicine, 1908), who propounded a theory of phagocytosis as a basis for the theory of immunity, died in Paris, France. (☞ May 16)

1926

Irwin Rose, American biochemist, Nobel laureate, was born in Brooklyn, New York, USA. Rose attended the Hebrew School and grew up in Spokane, Washington. He later studied at the Washington State College and then served in the US Navy as a radio technician near the end of World War II. He completed his undergraduate course in 1949 at the University of Chicago and went on to earn his doctorate in biochemistry there. He served on the faculty of the Yale Medical School's biochemistry department from 1954 to 1963. In 1963, he became a senior member of the Fox Chase Cancer Center in Philadelphia, where he remained until his retirement in 1995. Rose is currently a specialist at the department of physiology and biophysics at the University of California Irvine College of Medicine in California. Starting in the late 1970s, Rose, together with the Israeli biochemists Avram Hershko and Aaron J. Ciechanover, did a series of epoch-making biochemical studies in his laboratory at the Fox Chase Cancer Center, which ultimately led to the discovery of a molecule called ubiquitin that plays a key role in protein breakdown within cells. Their work showed that the breakdown is not indiscriminate but takes place through a process that is controlled in detail so that the proteins to be broken down at any given moment are given a molecular label by the ubiquitin molecule. The labelled proteins are then fed into the cells' 'waste disposers', the so-called proteasomes, where they are chopped into small pieces and destroyed. The

ubiquitin–proteasome pathway has a critical role in maintaining the homeostasis of cells and is believed to be involved in the development and progression of diseases such as cancer, muscular and neurological diseases, immune and inflammatory responses, and cystic fibrosis. Rose was awarded the 2004 Nobel Prize for Chemistry, jointly with Hershko and Ciechanover, for their discovery of ubiquitin-mediated protein degradation. (☞ Oct. 1, Dec. 31)

1929

The Indian Council of Agricultural Research (ICAR) was set up on the recommendation of the Royal Commission on Agriculture. With its headquarters in New Delhi, ICAR is an autonomous apex national organisation registered as a society, which plans, conducts and promotes research, education, training and transfer of technology for advancement of agriculture and allied sciences. Over the years the Council has developed a large research and training infrastructure in order to meet the agricultural research and education needs of the country. It also works on the production and other emerging problems confronted in agriculture to meet the growing demands for food, fodder, fibre and fuel. It operates through 46 Central Research Institutes, five National Bureaux, ten Project Directorates, 30 National Research Centres, 90 All-India Coordinated Research Projects, 261 Krishi Vigyan Kendras, and eight Trainers Training Centres. ICAR acts as a repository of information and provides consultancy on agriculture, horticulture, resource management, animal sciences, agricultural engineering, fisheries, agricultural extension, agricultural education, home science and agricultural communication. It has the mandate to coordinate agricultural research and development programmes and develop linkages at the national and international level with related organisations to enhance the quality of life of the farming community.

1945

The first atomic bomb was exploded at Los Alamos, New Mexico, USA. The atomic bomb was produced under the 'Manhattan Project' which was launched by the USA to pre-empt a possible German bomb during the World War II. What makes the atomic bomb so powerful and destructive is the conversion of mass into energy during the fission of uranium nuclei in a chain reaction. Fission occurs when the central part of a uranium atom, the nucleus, breaks up into two equal fragments. Once a neutron breaks up the uranium atom, the fragments release other neutrons that break up more atoms and so on. In an atomic bomb this chain reaction takes place in a millionth of a second. During break up of the nucleus some mass is lost which appears as energy, as given by Einstein's famous equation $E = mc^2$, where E is the energy, m the mass lost and c the velocity of light. The most complicated issue in making an atomic bomb at that time was the production of ample amounts of 'enriched' uranium to sustain a chain reaction. At the time, uranium-235 was very hard to extract, and the final refined uranium extracted from the ore contained less than one per cent of the uranium-235 required for making the bomb, the rest being non-fissile uranium-238. No ordinary chemical extraction method could separate them; only mechanical methods could work. A massive enrichment plant was constructed at Oak Ridge, Tennessee to separate the two isotopes. Next, a gas centrifuge was

used to further separate the lighter uranium-235 from the heavier uranium-238. Over the course of six years, from 1939 to 1945, more than US$2 billion was spent during the history of the Manhattan Project. Chief among the people who unleashed the power of the atom was J. Robert Oppenheimer, who oversaw the project from conception to completion. Upon witnessing the explosion, Oppenheimer quoted a fragment from the *Bhagavad Gita*. 'I am become Death', he said, 'the destroyer of worlds'. (☞ Aug. 6, Aug. 9)

1952
The Central Road Research Institute (CRRI), a constituent national laboratory of the Council of Scientific and Industrial Research (CSIR), was inaugurated. Located in New Delhi, CRRI carries out R&D relating to pavement design and performance, road condition monitoring, maintenance planning and management, and improved transportation planning for emerging urban needs. Besides, applied research on planning and engineering aspects of rural roads, material characterisation, pavement evaluation, highway instrumentation, deterioration and rehabilitation of bridges, transportation planning, traffic engineering, road safety and environmental problems forms an integral part of the overall R&D programme of the institute. CRRI has developed many new techniques for road construction, especially for stabilised soil in rural roads, mechanically stabilised soil sub-base and thin sand asphalt surfacing. Other techniques developed pertain to strengthening and rehabilitation of existing thin cement concrete pavements which show signs of distress, by providing suitable flexible and rigid overlays. The institute has pioneered a number of control techniques and preventive measures to avert landslides. Notable among these are the techniques of horizontal drilling to provide drains for quick ingress of water, spreading coir netting to promote vegetation on the hill slopes, geogrids to control rock falls and geotextiles to improve the strength and load bearing capacity of weak strata in many situations.

1969
American manned lunar mission *Apollo-11*, carrying astronauts Neil A. Armstrong, Edwin E. Aldrin, and Michael Collins, was launched by a Saturn-V rocket from the Kennedy Space Center in Florida, USA, on the first manned mission for landing on the Moon. (☞ Jul. 21, Jul. 24)

1994
Julian Seymour Schwinger, American physicist, Nobel laureate (Physics, 1965) who developed equations that served to unite the electromagnetic theory with quantum mechanics to form the new field of quantum electrodynamics, died in Los Angeles, California, USA. (☞ Feb. 12)

July 17

1790
The first patent for a complete machine for sewing was issued to the English inventor and cabinetmaker, Thomas Saint. The patent described an awl that punched a hole in leather and passed a needle through the hole. It is not known if Saint actually built a working prototype of his invention. (☞ Aug. 12, Sep. 9)

1894

Georges Lemaître, Belgian astronomer and cosmologist, was born in Charleroi, Belgium. Lemaître studied humanities at a Jesuit school and entered the civil engineering school of the Catholic University of Louvain, when World War I interrupted his studies. After the war, he undertook studies in physics and mathematics and began to prepare for priesthood. He obtained his doctorate in 1920. In 1923, he visited the University of Cambridge in England where the British astronomer Arthur Eddington introduced him to modern stellar astronomy and numerical analysis. He spent the following year at the Harvard College Observatory in Cambridge, Massachusetts, USA, with the American astronomer Harlow Shapley, who had just gained a name for himself for his work on nebulae. In 1925, on his return to Belgium, he became a part-time lecturer at the University of Louvain and in 1927 was appointed professor. Lemaître laid out the basis of his theory about the expanding universe in a note dating from 1927. In 1931 Eddington published the English translation of this note along with a lengthy commentary. Lemaître was invited by the British Association to come to London. There he proposed his original theory of the expanding universe and launched his idea about the origin of the universe. He published an extensive account of his theory in the Monthly Notices of the Royal Astronomical Society, explaining that the further one goes back in time, the closer the galaxies become, eventually compressing the universe into a single 'primeval atom'. The explosion of this gigantic primitive atom causes the universe to expand.

His proposition came to be known as the 'Big Bang' theory. This theory was later developed by the Russian-born American physicist George Gamow and others. (☞ Mar. 4, Jun. 20)

1911

Rustam Jal Vakil, Indian cardiologist, was born in Bombay (now Mumbai), India. Vakil studied at the Elphinstone College and the Royal Institute of Medicine in Mumbai and went to England to study at the St Thomas Hospital Medical School in London. He returned to India in 1938 with an MRCP and an MD degree of the London University, and built up a large private practice. Vakil specialised in the treatment of heart ailments and became the most sought-after cardiologist in India. His main contribution to medical science was his firm establishment of the therapeutic value of reserpine, extracted from the roots of the plant *Rauwolfia serpentina* and an important drug used in Ayurveda for hundreds of years, for the control of hypertension. In 1949 Vakil published a paper in *British Heart Journal* that summed up 10 years of his own conscientious work and the opinion of 50 other physicians and brought the first proven remedy for high blood pressure into Western medicine. Vakil is also credited with opening a new epoch in the therapy of mental disorder using reserpine. Vakil thus introduced ancient folk remedies of the East into Western medicine and 'inspired new and successful lines of treating both heart and mental or emotional ailments'. Vakil was long associated with the KEM Hospital, Mumbai. (☞ Nov. 20)

1975
The first international link-up of two spacecraft in orbit was achieved when the American spacecraft *Apollo-18* docked with the Russian spacecraft *Soyuz-19*. The two spacecraft were launched within seven and a half hours of one another, and docked two

days later. Three hours after they docked, astronaut Thomas Stafford and cosmonaut Alexei Leonov met in the middle and shook hands in orbit, exchanging flags and gifts, including the seeds of trees that were later planted in each others' countries. (☞ Jul. 15)

1994
The first of 21 major fragments of the comet Shoemaker-Levy 9, which broke up two years earlier, hit Jupiter, creating a 1,900-km wide and 950-km high fireball, giving scientists their first chance to observe

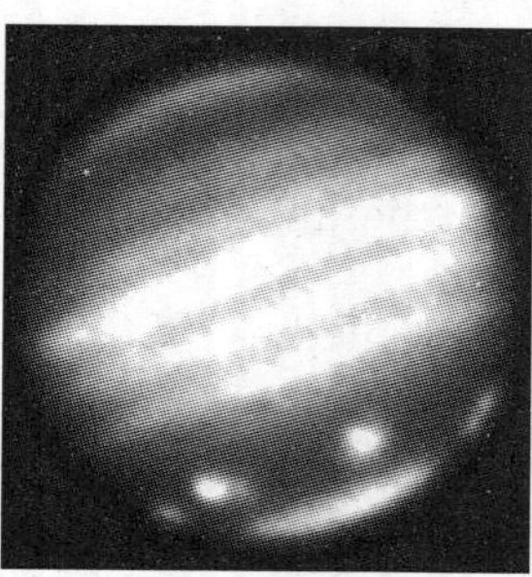

Jupiter after the impact of fragments from Shoemaker-Levy 9

such a collision as it happened. The larger fragments had diameters up to two kilometres. The collisions continued for six days and left dark patches on Jupiter's cloud layer. (☞ Mar. 24)

July 18

1635
Robert Hooke, English physicist who discovered the law of elasticity known as Hooke's law, was born in Freshwater, Isle of Wight, UK. Hooke was educated at the Westminster College and the Christ Church, Oxford. In 1655 he became an assistant to the Anglo-Irish chemist Robert Boyle and helped him improve the air pump. In 1662 he was appointed curator of the newly founded Royal Society of which he was elected a fellow in 1663 and served as its secretary from 1677 to 1683. Hooke was a virtuoso scientist who did research in a remarkable variety of fields ranging from physics and astronomy, to chemistry, biology, and geology, to architecture and naval technology. He was the first to build a Gregorian reflecting telescope and observe the rotation of Jupiter. He was also the first to draw maps of the surface of Mars. He invented the balance spring for clocks; invented or improved meteorological instruments such as the barometer, anemometer, and hygrometer; served as Chief Surveyor, and helped rebuild London after the Great Fire of 1666. In 1678 Hooke stated that the force tending to restore a spring or any elastic system to its equilibrium position is proportional to the distance by which it is displaced from that position. This came to be known as

'Hooke's law'. According to Hooke's law, elastic behaviour of solids can be explained by the fact that small displacements of their constituent molecules, atoms, or ions from normal positions are proportional to the force that causes the displacement. Hooke is also remembered for discovering the porous structure of cork and coining the term 'cell' as units of living things. (☞ Mar. 3, Nov. 3)

1853

Hendrik Antoon Lorentz, Dutch physicist, Nobel laureate, was born in Arnhem, Netherlands. After finishing his schooling in Arnhem, Lorentz obtained his BSc degree in mathematics and physics from the University of Leiden in 1871. He obtained a doctorate from the same university in 1875, for his thesis on the reflection and refraction of light in which he refined the electromagnetic theory of the British physicist James Clerk Maxwell, thereby laying the grounds for Albert Einstein's special theory of relativity. His fundamental work in the fields of optics and electricity revolutionised conceptions of the nature of matter. Lorentz's view that the electron plays a role during electromagnetic phenomena in matter made it possible to apply the molecular theory to the theory of electricity, and to explain the behaviour of light waves passing through moving, transparent bodies. According to Maxwell's theory, electromagnetic radiation is produced by the oscillation of electric charges, but the charges that produce light were unknown. Since it was generally believed that an electric current was made up of charged particles, Lorentz theorised that the atoms of matter might also consist of charged particles and suggested that the oscillations of these charged particles inside the atom were the source of light. If this were true, then a strong magnetic field ought to have an effect on the oscillations and therefore on the wavelength of the light thus produced. The phenomenon was demonstrated in 1896 by his pupil Pieter Zeeman, and came to be known as the 'Zeeman effect.' Lorentz was awarded the 1902 Nobel Prize for Physics, jointly with Zeeman, in recognition of the extraordinary service they rendered by their researches into the influence of magnetism upon radiation phenomena. (☞ Feb. 4, May 25)

1898

Polonium (at. no. 84) was discovered by the French chemist Marie Curie and her husband physicist Pierre, while seeking the cause of radioactivity of the mineral pitchblende. It required several tonnes of pitchblende to produce very small amounts of polonium. (☞ May 15, Nov. 7)

1931

Oskar Minkowski, German physiologist and pathologist who introduced the concept that diabetes results from the suppression of a pancreatic substance (later identified as insulin), died in Fürstenberg an der Havel, Germany. (☞ Jan. 13)

1937

Roald Hoffmann, Polish-born American chemist, Nobel laureate, was born in Zloczów, Poland. Hoffmann immigrated to USA with his family in 1949 and graduated from the Columbia University, New York. He received his PhD from the Harvard University, Cambridge, Massachusetts, in 1962 and worked with the American chemist Robert B. Woodward for three years before joining the Cornell University faculty in 1965. His work was mainly concerned with theoretically anticipating the course of chemical reactions. At Harvard, while studying the synthesis of vitamin B_{12}, Hoffmann and Woodward discovered that many reactions involving the formation or breaking of rings of atoms could be explained on the basis of quantum mechanics, which attempts to explain how atoms behave. Orbital interaction and symmetry relations between molecules or parts of molecules are fundamental to this theory of conservation of orbital symmetry in chemical reactions. Their theory, expressed in a set of statements now called the 'Woodward–Hoffmann rules', could account for the failure of certain cyclic compounds to form from apparently appropriate starting materials, though others are readily produced. It could also clarify the geometric arrangement of the atoms in the products formed when the rings in cyclic compounds are broken. Hoffmann was awarded the 1981 Nobel Prize for Chemistry, jointly with the Japanese chemist Fukui Kenichi, for their theories, developed independently, concerning the course of chemical reactions. (☞ Oct. 4)

1948
Hartmut Michel, German biochemist, Nobel laureate, was born in Ludwigsburg, Germany. Michel studied biochemistry at the University of Tübingen and earned his doctorate from the University of Wurzburg in 1977. In 1979 he joined the staff of the Max Planck Institute for Biochemistry in Martinsried, Germany, where he worked with fellow German biochemists Johann Deisenhofer and Robert Huber to determine the three-dimensional structure of a four-protein complex found in certain bacteria. The protein complex – called the photosynthetic reaction centre – was taken from a bacterium which, like green plants and algae, uses light energy from the Sun to synthesise organic substances. Michel crystallised the membrane-bound protein complex to a pure crystalline form, thus making it possible to determine the protein's structure atom-by-atom by means of X-ray diffraction techniques. He became head of the Max Planck Institute for Biophysics in Frankfurt am Main in 1987. Michel was awarded the 1988 Nobel Prize for Chemistry, jointly with Deisenhofer and Huber, for the determination of the three-dimensional structure of a photosynthetic reaction centre. (☞ Feb. 20, Sep. 30)

1968
Corneille Heymans, Belgian physiologist, Nobel laureate (Physiology or Medicine, 1938), who discovered the role played by the sinus and aortic mechanisms in the regulation of respiration, died in Knokke, West Flanders, Belgium. (☞ Mar. 28)

1980
India's satellite launch vehicle SLV-3 carrying a *Rohini* satellite was launched on its first successful flight from Sriharikota, Andhra Pradesh. The 40-kilogram satellite was placed in a near-Earth elliptical orbit ranging in height from 307 km to 921 km.

July 19

1921
Rosalyn S. Yalow, American medical physicist, Nobel laureate, was born in New

York, USA. Yalow graduated with honours from the Hunter College of the City University of New York in 1941 and four years later received her PhD in nuclear physics from the University of Illinois at Urbana-Champaign. In 1945 she joined the Federal Telecommunications Laboratory as an engineer, and then in 1946 took up a full-time teaching position at her alma mater, Hunter College. In 1947, while still on the faculty at Hunter, Yalow became a consultant in nuclear physics at the Veterans Administration Hospital in the Bronx, New York, where research was being carried out on medical applications of radioactive materials. In 1950, Yalow left Hunter and accepted a full-time position with the Veterans Administration Hospital where she, together with fellow American physician Solomon A. Berson, carried out investigations on the use radioisotopes to obtain more accurate estimates of blood volume. But their first major contribution was not in this field but in the study of how the thyroid gland and kidneys remove iodine from the blood. They developed a method of discerning the quantity of blood cleared of iodine by the thyroid gland per unit of time. Using radioisotopes, Yalow and Berson could readily ascertain clearance rates in a 35-min sitting, providing a quick determination of thyroid activity. They called their method radioimmunoassay – the first technique to use radioisotopes for the study of the primary reaction of antigen with antibody. Yalow and her co-workers also pioneered new directions in diabetes research with the aid of radioimmunoassay and were able to elucidate the physiology of the peptide hormones insulin, adrenocorticotrophic hormone, and growth hormone; they were able to throw light upon the pathogenesis of diseases caused by abnormal secretion of these hormones. Yalow was awarded the 1977 Nobel Prize for Physiology or Medicine, jointly with the Polish-born American endocrinologist Andrew V. Schally and French-born American physiologist Roger Charles Louis Guillemin. Yalow received half of the prize money for the development of radioimmunoassays of peptide hormones, while Schally and Guillemin shared the other half for their discoveries concerning the peptide hormone production of the brain. (☞ Jan. 11, Nov. 30)

1938

Jayant V. Narlikar, Indian theoretical astrophysicist, was born in Kolhapur, Maharashtra, India, in 1938. Narlikar received his early education at the Banaras Hindu University, Varanasi, Uttar Pradesh, and his PhD in 1960 and ScD in 1976 from the Cambridge University, England. After spending several years at Cambridge as a fellow of the King's College and as staff member of the Institute of Theoretical Astronomy, he returned to India in 1972 as a professor at the Tata Institute of Fundamental Research in Mumbai. He was later invited to Pune to set up the Inter-University Centre for Astronomy and Astrophysics, of which he was the founder director from 1988 to 2003. A well-known cosmologist, Narlikar worked with the British-born American astronomer Geoffrey Burbidge and the British mathematician and astronomer Fred Hoyle on a variant of steady-state cosmology called 'quasi-steady state theory' that would nevertheless account for all the observations – the radio source counts, the quasar

red-shifts, the microwave background and the abundances of helium and light isotopes. Their work was summarised in their 1999 book *A Different Approach to Cosmology*. Narlikar is a well-known popular science and science-fiction writer and is co-recipient of the 1996 UNESCO's Kalinga Prize for the Popularisation of Science.

July 20

1897

Tadeus Reichstein, Swiss chemist, Nobel laureate, was born in Wloclawek, Poland. Reichstein was educated in Zurich, Switzerland, and held posts in the department of organic chemistry at the Federal Institute of Technology, Zurich, from 1930. In 1938 he was appointed professor in pharmaceutical chemistry, and director of the Pharmaceutical Institute in the University of Basel. From 1946 he worked as a professor of organic chemistry at the same university. Between the years 1948–1952 he supervised the building and equipment of a new Institute of Organic Chemistry, of which he became director in 1960. Reichstein and his colleagues accomplished the first isolation of four active hormones from the adrenal cortex, including cortisone – a hormone of the adrenal cortex and an anti-inflammatory agent useful in the treatment of arthritis. They also accomplished the first synthesis of one of them, and provided the proof of the steroid nature of these hormones. In 1933 Reichstein succeeded in synthesising vitamin C. Reichstein was awarded the 1950 Nobel Prize for Physiology or Medicine, jointly with the American physician Philip Showalter Hench and the American chemist Edward Calvin Kendall, for their discoveries relating to the hormones of the adrenal cortex, their structure and biological effects. (☞ Feb. 28, Mar. 8)

1933

Roddam Narasimha, Indian aerospace scientist, Fellow of the Royal Society, was born in Bangalore, Karnataka, India. Narasimha did his bachelor's degree in engineering from the Mysore University, and obtained his master's degree in aeronautical engineering from the Indian Institute of Science, Bangalore. He obtained his PhD from the California Institute of Technology, Pasadena, California, USA, in 1961. From 1962 to 1967 he was assistant professor in IISc, Bangalore, and from 1967 to 1970 the chief project coordinator with Hindustan Aeronautics Limited. He returned to IISc as dean of the engineering faculty in 1980 and was professor of aerospace engineering during 1982–85. He was appointed director of the National Aeronautical Laboratory (now National Aerospace Laboratory), Bangalore, in 1984. After his retirement from NAL in 1993 he was ISRO Distinguished Professor at the Centre for Atmospheric and Oceanic Sciences at IISc and director of the National Institute of Advanced Studies and Jawaharlal Nehru Centre for Advanced Scientific Research in Bangalore. Narasimha is recognised for his outstanding contributions to fluid mechanics and related fields such as boundary layer transition, fully developed turbulence, the structure of shocks, meteorology, aircraft airworthiness and design. His research in the field of dynamics has led to a better understanding of flows in transition from laminar to turbulent and vice versa.

1937
Guglielmo Marconi, Italian physicist and inventor, Nobel laureate (Physics, 1909), who developed wireless telegraphy, died in Rome, Italy. (☞ Apr. 25, Dec. 12)

1947
Gerd Binnig, German physicist, Nobel laureate, was born in Frankfurt, Germany. Binnig's interest in physics started at the early age of ten. He began his study of physics in Frankfurt where he attended two colleges. After graduating from the Johann Wolfgang Goethe University in Frankfurt he received a doctorate from the University of Frankfurt in 1978 and then joined a physics research team at the IBM Research Laboratory in Zürich, Switzerland, where he met Heinrich Rohrer. Together they designed and built the first scanning tunnelling microscope (STM) that could reveal matter at the atomic level. The STM is not a true microscope (i.e., an instrument that gives a direct image of an object) since it is based on the principle that the structure of a surface can be studied using a stylus that scans the surface at a fixed distance from it. Vertical adjustment of the stylus is controlled by means of what is termed the tunnel effect – hence the name of the instrument. The tip of the STM's stylus is only about one angstrom (1Å) wide (1 Å = one ten-billionth of a metre, or about the width of an atom), and the distance between it and the surface being studied is only about 5 or 10 Å. An STM can produce images of the surfaces of conducting or semi-conducting materials in such fine detail that individual atoms can be clearly identified. Binnig was awarded the 1986 Nobel Prize for Physics, jointly with Rohrer and Ernst Ruska. Binnig and Rohrer shared half the prize money for their design of the scanning tunnelling microscope, while Ruska received the other half for his fundamental work in electron optics, and for the design of the first electron microscope. (☞ Jun. 6, Dec. 25)

1976
The *Viking-1* lander touched down on the western slope of Chryse Planitia on Mars. The lander was equipped with a biology instrument, gas chromatograph/ mass spectrometer, X-ray fluorescence spectrometer, seismometer, meteorology instrument, and stereo colour cameras. Besides taking photographs and collecting other science data on the Martian surface, the lander conducted three biology experiments designed to look for possible signs of life in Martian soil. These experiments discovered unexpected and enigmatic chemical activity

in the Martian soil, but provided no clear evidence for the presence of living micro-organisms in soil near the landing sites. (☞ Jun. 19, Aug. 20)

July 21

1891
The first successful flight of a human-carrying glider was completed by German

pioneer of human aviation Otto von Lilienthal near the small village of Derwitz, outside Potsdam near Berlin. In 1894, he had a conical hill, about 15 metres high and about 70 metres in diameter, constructed at Gross-Lichterfelde near Berlin from which he could glide in any direction. He continued experimenting with various gliders of his own design and undertook as many as 2,000 flights until he died after a crash in 1896.

1923

Rudolph A. Marcus, Canadian-born American chemist, Nobel laureate, was born in Montreal, Quebec, Canada. Marcus was educated at the Baron Byng High School and received his doctorate from the McGill University, Montreal, in 1946. He then joined the new post-doctoral program at the National Research Council of Canada in Ottawa. From 1951 he worked at the Polytechnic Institute of Brooklyn in New York, USA, where he began studying electron-transfer reactions. In a series of papers published between 1956 and 1965, he investigated the role of solvent molecules in determining the rate of redox reactions – oxidation and reduction reactions in which the reactants exchange electrons – in solution. He found that subtle changes occur in the molecular structure of the reactants and the solvent molecules around them, which influence the ability of electrons to move between the molecules. He further established that the relationship between the driving force of an electron transfer reaction and the reaction's rate is described by a parabola; the rate at first increases but then begins to decrease. His theory could explain diverse and fundamental phenomena such as the fixation of light energy by green plants during photosynthesis, cell metabolism, photochemical production of fuel, chemiluminescence, the conductivity of electrically conducting polymers, and simple corrosion. In 1964 Marcus joined the faculty of the University of Illinois, and in 1978 went to the California Institute of Technology as professor of chemistry. Marcus was awarded the 1992 Nobel Prize for Chemistry for his contributions to the theory of electron transfer reactions in chemical systems.

1969

Apollo-11 astronauts Neil A. Armstrong and Edwin E. Aldrin became the first men to walk on the Moon after their lunar module *Eagle* landed in the Mare Tranquillitatis (sea of tranquillity). Armstrong and Aldrin established the Tranquillity Base while Michael Collins orbited above in the command module *Columbia*. After stepping on the lunar surface Armstrong proclaimed, 'That's one small step for a man, one giant leap for mankind'. After spending about two hours gathering rock samples, taking photographs, and setting up scientific equipment for tests, the astronauts concluded their excursion on the surface. Armstrong and Aldrin later piloted the lunar module to a successful rendezvous with Collins

and the command module in lunar orbit, which later returned to Earth. (☞ Jul. 16, Jul. 24)

1988

INSAT-1C, the third of India's INSAT series of multipurpose geostationary satellites, was launched by the European Space Agency's Ariane rocket from Kourou in French Guiana. The satellite was designed by the Indian Space Research Organisation and built by Ford Aerospace, USA. The satellite carried 12 C-band transponders, 3 S-band transponders, and a very high resolution radiometer. Its main purpose was to provide telecommunication, networking and meteorological services. It was placed in the geostationary orbit at 93.5°E longitude, but after a power failure, only half its payload was functional. The satellite lost its Earth lock in November 1989 and had to be abandoned.

2000

Discovery of the subatomic particle 'tau neutrino' was announced by an international collaboration of scientists at the US Department of Energy's Fermi National Accelerator Laboratory in Batavia near Chicago, Illinois, USA. The tau neutrino is the third neutrino – a group of particles with zero mass and zero charge – of the Standard Model of elementary particles, a theoretical description that groups all particles into three generations. The first generation electron neutrinos were created in 1956 and the second generation muon neutrinos in 1962.

2004

Edward B. Lewis, Nobel laureate (Physiology or Medicine, 1995), who discovered how genes regulate the development of specific regions of the body, died in Pasadena, California, USA. (☞ May 20)

July 22

1784

Friedrich Wilhelm Bessel, German astronomer, was born in Minden, Brandenburg (now in Germany). Bessel came from a poor family and at the age of 15, entered an export–import firm as an apprentice. During his apprenticeship, he studied languages, geography, the habits of distant peoples, and the principles of navigation, which led him to the study of astronomy and mathematics. In 1804 he calculated the orbit of the comet Halley from observations made in 1607 and sent a paper to the astronomer Wilhelm Olbers, who was so impressed that he arranged its publication in a German journal. On Olbers' suggestion Bessel joined the Lilienthal Observatory as an assistant. After he had spent just four years at Lilienthal, the then Prussian government gave him the responsibility of constructing the first major German observatory at Königsberg in Prussia (now Kaliningrad, Russia). He directed the observatory from the date of its completion in 1813 until the end of his life. In 1810 he was appointed professor of astronomy at the University of Königsberg, where he worked hard on the reconstruction of the science of astronomical observations. During 1831–32 he directed geodetical measurements of meridian arcs in East Prussia, and in 1841 he deduced the value of the Earth's ellipticity – the amount of elliptical distortion in the Earth's shape – as 1/299. He was the first to make effective use of the heliometer, an instrument designed for measuring the apparent diameter of the Sun. Having established exact positions for thousands of individual stars at his observatory in Königsberg, Bessel was ready to observe the stars' exceedingly small but highly

significant motions, relative to one another. In 1838, choosing '61 Cygni', a star barely visible to the naked eye, Bessel showed that every year, the star apparently moved in an ellipse in the sky. He said that this back and forth motion, called parallax, could only be interpreted as being caused by the motion of the Earth around the Sun. Calculation indicated the distance from Earth to 61 Cygni to be of 10.3 light years. This was the first measurement of the distance of a star other than the Sun, using the method of parallax. Bessel was honoured for this achievement by the Royal Astronomical Society of London and others. (☞ Mar. 17, Oct. 11)

1822

Gregor Mendel, Austrian botanist and plant experimenter who is often called the father of genetics for his study of the inheritance of traits in pea plants, was born in Heinzendorf, Austria. Mendel worked as a gardener in his childhood and later attended the Olmutz Philosophical Institute. In 1843 he entered an Augustinian monastery in Brno (now in Czech Republic) and later joined the University of Vienna in Austria to study. Mendel experimented with the pea plant in his monastery's experimental garden. Between 1856 and 1863 he cultivated and tested some 28,000 pea plants. He self-pollinated the plants, wrapping them to guard against cross-pollination. He noticed that characteristics of offspring, such as height, shape of pods, etc., exhibited recessive and dominant behaviour. His experiments brought forth two generalisations, which later became known as Mendel's Laws of Inheritance. He published his results in the early 1860s in the *Proceedings of the Natural Science Society of Brno*, but his paper was ignored until rediscovered by the Dutch botanist Hugo de Vries, German botanist Carl Erich Correns, and Austrian botanist Erich Tschermark von Seysenegg in 1900, leading to the development of modern genetics. (☞ Jan. 6)

1826

Giuseppe Piazzi, Italian astronomer, who discovered the first dwarf planet Ceres, died in Naples, Italy. (☞ Jul. 16)

1887

Gustav Ludwig Hertz, German physicist, Nobel laureate, was born in Hamburg, Germany. Hertz attended the Johanneum School in Hamburg before commencing his university education at Göttingen in 1906; he subsequently studied at the Universities of Munich and Berlin, graduating in 1911. He was appointed as a research assistant at the Physics Institute of the Berlin University in 1913, where he began to work with the German-born American physicist James Franck. Together with Franck he began his studies on electron impact and spent much patient work on the study and measurement of ionisation potentials in various gases. He later demonstrated the quantitative relations between the series of spectral lines and the energy losses of electrons in collision with atoms corresponding to the stationary energy states of the atoms. The Franck–Hertz experiment confirmed

that energy can be absorbed by an atom only in definite amounts and provided an important confirmation of the Bohr atomic model. In 1925 Hertz was appointed professor of physics at the University of Halle and in 1928 as professor of physics at the Technische Hochschule in Berlin. In 1932 he devised a method of separating the isotopes of neon. Hertz was awarded the 1925 Nobel Prize for Physics, jointly with Franck, for their discovery of the laws governing the impact of an electron upon an atom. (☞ Aug. 26, Oct. 30)

1888

Selman Abraham Waksman, Ukrainian-born American biochemist, Nobel laureate, was born in Priluka, Ukraine. Waksman received his BSc degree in Agriculture in 1915 and his MSc in 1916 from the Rutgers College, New Jersey, USA. In the same year, he became a naturalised American citizen and was appointed a research fellow at the University of California where he received his PhD in biochemistry in 1918. He was appointed associate professor at the Rutgers University in 1925 and became professor in 1930. When the University's department of microbiology was organised in 1940, he became professor of microbiology and head of the department. In 1949, he was appointed director of the Institute of Microbiology (now the Waksman Institute). Waksman's research mainly encompassed the microbiological population of the soil, micro-organisms and soil fertility; production and nature of antibiotic substances; taxonomy, physiology, and biochemistry of the actinomycetes – filamentous, bacteria-like micro-organisms found in the soil. He coined the term antibiotic, meaning 'against life'. During his extensive study of the actinomycetes, he extracted from them antibiotics valuable for their killing effect not only on gram-positive bacteria, against which penicillin is effective, but also on gram-negative bacteria, of which the tubercle bacillus (*Mycobacterium tuberculosis*) is one. He isolated, together with his students and associates, a number of new antibiotics, including actinomycin, streptothricin, streptomycin, neomycin, and others. Two of these – streptomycin and neomycin – have found extensive application in the treatment of numerous infectious diseases of men, animals and plants. Waksman was awarded the 1952 Nobel Prize for Physiology or Medicine for his discovery of streptomycin, the first antibiotic effective against tuberculosis. (☞ Aug. 16)

1972

Soviet unmanned spacecraft *Venera-8* landed on Venus. It transmitted data during the descent and continued to send back data for 50 minutes after landing. The probe confirmed the earlier data on the high Venus surface temperature and pressure returned by *Venera-7*, and also measured the light level as being suitable for surface photography, finding it to be similar to the amount of light on Earth on an overcast day. (☞ Mar. 26)

1981

APPLE, India's first geostationary satellite, became operational. It was used for the first nationwide television broadcast on Independence Day. (☞ Jun. 19)

July 23

1906

Vladimir Prelog, Yugoslavian-born Swiss chemist, Nobel laureate, was born in

Sarajevo, Bosnia, Austria-Hungary (now in Bosnia and Herzegovina). Prelog attended school in Zagreb and Osijek, and from 1924 studied chemistry at the Czech Institute of Technology in Prague, Czechoslovakia (now Czech Republic) from where he received his PhD in 1929. He then spent several years in a commercial laboratory, and in 1935 began teaching at the University of Zagreb, first as a lecturer and later as professor of organic chemistry. In 1942 he joined the faculty of the Federal Institute of Technology in Zürich, Switzerland, where he served as head of the laboratory of organic chemistry from 1957 to 1965. Prelog's main interests were natural products. While working in this field he became interested in problems of stereochemistry – the study of the three-dimensional arrangements of atoms within molecules. He performed wide-ranging research on the stereochemistry of alkaloids, antibiotics, enzymes, and other natural compounds. His research has contributed to the explanation of the structure of steroids, tri-terpene, quinine, strychnine, solanine and other alkaloids introducing the so-called 'Prelog's regulation', which defines the conformational relations between reactants and products. In particular he contributed to the understanding of stereoisomerism, in which two compounds of identical chemical composition have different, mirror-image configurations. He also devised systematic rules for naming molecules and their mirror-image versions. He used X-ray diffraction and elucidated the structure of several antibiotics. Prelog was awarded the 1975 Nobel Prize for Chemistry, jointly with the British chemist John W. Cornforth, for his work on the stereochemistry of enzyme-catalysed reactions. (☞ Jan. 7, Sep. 7)

1916
William Ramsay, Scottish chemist, Nobel laureate (Chemistry, 1904), who discovered the 'inert gases', neon, krypton and xenon, and co-discovered argon, radon, calcium and barium, died in Buckinghamshire, England. (☞ Aug. 13, Oct. 2)

1937
The isolation of the pituitary hormone was announced by scientists of the Yale University, Killingworth, Connecticut, USA. The pituitary gland is a small gland the size of a peanut at the base of the brain and is considered to be the master endocrine gland in vertebrate animals. The hormones secreted by the pituitary stimulate and control the functioning of almost all the other endocrine glands in the body. Pituitary hormones also promote growth and control the water balance of the body.

1968
Henry Hallett Dale, English physiologist, Nobel laureate (Physiology or Medicine, 1936), who discovered the neurotransmitter action of acetylcholine, died in Cambridge, UK. (☞ Jun. 9)

1993
INSAT-2B, the second of India's second generation INSAT series of satellites, was launched by the European Space Agency's Ariane rocket from Kourou in French Guiana. Designed and built by the Indian Space Research Organisation, the multipurpose satellite carried two S-band transponders, 18 C-band transponders, and two very high resolution radiometers, and was meant to provide telecommunication, networking and meteorological services.

The satellite was placed in the geostationary orbit at 93.5° E longitude.

1998
An international team of scientists, lead by Ryuzo Yanagimachi, of the University of Hawaii, announced in a paper published in the journal *Nature* the first reproducible cloning of a mammal from adult cells to produce three generations of cloned mice, more than 50 identical sisters in all. Their 'Honolulu technique' of cloning was said to be more reliable than the one used to create Dolly the sheep by the British embryologist Ian Wilmut.

1999
The *Chandra X-ray Observatory* was deployed from the American space shuttle *Columbia*. The observatory is named in honour of Indian-American physicist Subrahmanyan Chandrasekhar who is

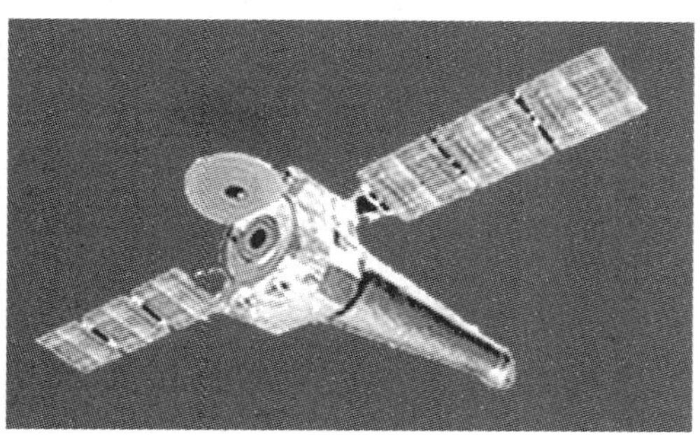

known for determining the mass limit for white dwarf stars to become neutron stars. Since the Earth's atmosphere absorbs most of the X-rays coming from space, they are not detectable from Earth-based telescopes. Observation of these X-rays would require a space-based telescope. The data gathered by *Chandra* since its launch has greatly advanced the field of X-ray astronomy. (☞ Oct. 19)

July 24

1853
Henri-Alexandre Deslandres, French physicist and astrophysicist who invented the spectroheliograph, was born in Paris, France. Deslandres graduated from the École Polytechnique in 1874 and spent seven years in the army. Later he worked in the laboratories of the École Polytechnique and the Sorbonne. From 1886 to 1891 he studied the spectra of radiation emitted by molecules. He joined the Paris Observatory in 1889, where he carried out studies of the molecular spectra and spectra of planets, the Sun, and other stars. He continued his work at the Meudon Observatory and in 1908 was appointed its director. After the Paris and Meudon Observatories were merged in 1926, Deslandres remained in charge of them until his retirement in 1929. During his career, Deslandres was elected to several scientific societies, including the Académie des Sciences, the Royal Astronomical Society and the Royal Society of Britain, and the National Academy of Sciences of the United States. (☞ Jan. 15)

1911
The Indian Institute of Science (IISc) started functioning at Bangalore, Karnataka, with the first batch of students in the departments of general and applied chemistry and electrotechnology. With the establishment of the University Grants Commission in 1956, the Institute came under its purview as a deemed university. IISc is neither a national laboratory, which concentrates solely on research and applied

work, nor a conventional university, which is mainly involved with teaching – the Institute concerns itself with research in frontier areas and education in current technologically important areas. This is also the first Institute in the country to introduce innovative integrated PhD programmes in biological, chemical and physical sciences for science graduates. Besides formal education and research, the Institute has been playing an active part in offering short-term courses to scientists and technologists in service. The Continuing Education Programme covers a wide range of topics and over 1,500 working scientists and engineers go through such courses every year. In keeping with its aims and objects, the Institute has organised a Centre for Scientific and Industrial Consultancy through which the know-how generated in the Institute percolates to industries via industry-sponsored projects. The Jawaharlal Nehru Centre for Advanced Scientific Research with organic links with the Institute has been functioning in the Institute's campus since 1989. The Council of the Institute confers Honorary Fellowships on eminent scholars and scientists and on those who have made noteworthy and lasting contributions to the cause of science and industry in India.

1956
A patent for oral penicillin was obtained by Austrian researchers Ernst Brandl and Hans Margreiter. Penicillin – derived from the *Penicillium* mould – was one of the first and still one of the most widely used antibiotic agents. In 1952 Brandl had discovered acid-stable penicillin, which could not be decomposed by gastric acids. Thus he found a way to give the active substance also as a tablet or syrup. One month later his colleague Hans Margreiter succeeded in isolating the acid-stable penicillin in pure form. This new penicillin, called 'Penicillin-V' was first produced in 1953 under the name 'Ospen' and was a world sensation. (☞ Aug. 6)

1974
James Chadwick, English physicist, Nobel laureate (Physics, 1935), who discovered the neutron and later worked on the Manhattan Project to develop the atom bomb, died in Cambridge, England. (☞ Oct. 20)

1969
The command module of *Apollo-11* carrying astronauts Neil Armstrong and Edwin Aldrin, who were the first men to set foot on the Moon, and Michael Collins, who remained in orbit, splashed down safely in the Pacific Ocean, about 812 nautical miles southwest of Hawaii. (☞ Jul. 16, Jul. 21)

1986
Fritz Albert Lipmann, German-born American biochemist, Nobel laureate (Physiology or Medicine, 1953), who discovered coenzyme A, an important catalytic substance involved in the cellular conversion of food into energy, died in Poughkeepsie, New York, USA. (☞ Jun. 12)

July 25

1783
French aristocrat Marquis de Jouffroy, made the first steamboat run on the Saone River in Lyon, France. His 50-metre-long boat used a Newcomen-style steam engine for power. For 15 minutes, the boat chuffed past cheering crowds who had lined up along the riverbank.

1814
Joseph von Fraunhofer, German physicist, plotted more than 500 dark lines in the solar spectrum and discovered that the relative positions of the lines remained constant for each element. The dark lines came to be known as 'Fraunhofer lines'. (☞ Mar. 6)

1843
Charles Macintosh, Scottish chemist and manufacturer, who invented rubberised waterproof cloth for raincoats, died near Glasgow, Scotland. (☞ Dec. 29)

1920
Rosalind Franklin, British scientist whose X-ray diffraction photographs of deoxyribonucleic acid (DNA) were crucial for the determination of its molecular structure by the American geneticist James D. Watson and British biophysicist Francis Crick, was born in London, England. Franklin graduated from the Cambridge University and in 1951 went to work as a research associate at the King's College. A chemist by training, Franklin made original and essential contributions to the understanding of the structure of graphite and other carbon compounds. Franklin used the technique of X-ray crystallography to study a variety of molecules. She applied her chemist's expertise to the unwieldy DNA molecule. After complicated analysis, she was the first to state that the sugar–phosphate backbone of DNA lies on the outside of the molecule. She also elucidated the basic helical structure of the molecule. X-ray diffraction photographs of DNA taken by her were shown to Watson and Crick without her knowledge by Maurice Wilkins, a colleague of hers. Without Franklin's data and X-ray diffraction patterns of DNA, Watson and Crick may not have succeeded in elucidating the molecular structure of DNA. (☞ Apr. 5, Apr. 16, Jun. 8, Dec. 15)

1946
The United States carried out the first underground nuclear test at Bikini Atoll in the Pacific. The test was made to record the effect of a nuclear explosion on naval vessels. The targets brought into Bikini Lagoon were outdated US Navy ships and German and Japanese vessels captured during World War II. When the bomb was detonated, a massive column of steam and water erupted from the lagoon, and the explosion created a series of huge waves that sunk all the target ships.

1978
The world's first 'test-tube' baby, Louise Joy Brown, was born in Oldham, England. She had been conceived through the technique of *in vitro* fertilisation (IVF). The technique of IVF was pioneered by Robert Edwards of Cambridge University, who was keen to extend already existing work on animals to humans and treat women with blocked Fallopian tubes. These blocked tubes prevent eggs travelling from the ovaries to the womb, where they can be fertilised. In 1968 Edwards succeeded in fertilising the human ova outside the womb. He collaborated with Patrick Steptoe, a gynaecologist in Oldham, who he met while lecturing at the Royal Society of Medicine. (☞ Jun. 9, Sep. 27)

1984
Russian cosmonaut Svetlana Savitskaya became the first woman to walk in space. Along with fellow cosmonauts, Vladimir

Dzhanibekov and Igor Petrovich Volk, she had travelled to the Russian orbiting space station *Salyut-7* to conduct experiments. Savitskaya, along with Dzhanibekov did the space walk that lasted 3.58 hours and was part of the *Soyuz T-12* mission.

1997
A team of researchers from the Johns Hopkins University, Baltimore, Maryland, USA, led by John D. Gearhart, a professor of gynaecology and obstetrics, announced the successful cloning of human stem cells taken from aborted human embryos. Stem cells are the basic, unspecialised cells from which all other cells in the body develop during the growth of a baby in the womb. The announcement was made at an international symposium on the ethics of human cloning and stem cells 'to spark discussion about establishing guidelines for the ethical use of such cells'.

2000
A *Concorde* supersonic airliner bound for New York crashed in a ball of fire shortly after taking off from the Paris Charles de Gaulle airport in France, killing 113 people. This was the first and only crash of a *Concorde* since it was first put into passenger service in January 1976. Later investigations showed that a metal strip on the runway had punctured a tyre and started the catastrophic chain of events. Debris from the puncture pierced the under-wing fuel tanks and started the fire that brought the plane down. (☞ Jan. 9, Jan. 21, Mar. 2, Oct. 1, Oct. 24)

July 26

1875
Carl Gustav Jung, Swiss psychologist who founded analytic psychology, was born in Kesswil, Switzerland. Jung was a rather solitary adolescent, who didn't care much for school, and especially couldn't take competition. He went to boarding school in Basel, Switzerland, where he found himself the object of a lot of jealous harassment. He began to use sickness as an excuse, developing an embarrassing tendency to faint under pressure. Although his first career choice was archaeology, he went on to study medicine at the University of Basel. He obtained his MD from the University of Zürich, Switzerland, in 1902. While working under the famous neurologist Krafft-Ebing, he settled on psychiatry as his career. After graduating, he took a position at the Burghoeltzli Mental Hospital in Zurich. He also taught classes at the University of Zurich and had a private practice. It was at this time that he invented word association. He met and collaborated with Sigmund Freud in Vienna from 1907 to 1912, but then developed his own theories, to distinguish them from Freud's psychoanalysis. He proposed and developed the concepts of the extroverted and introverted personality, archetypes, and the collective unconscious. In Jung's terms, archetypes are instinctive patterns, having a universal character, expressed in behaviour and images. His work has been influential in psychiatry and in the study of religion. (☞ Jun. 6)

1971
American manned lunar mission *Apollo-15*, carrying astronauts David R. Scott, Alfred M. Worden, and James B. Irwin, was launched towards the Moon by a Saturn-V rocket from the Kennedy Space Center, Florida, USA. It carried the first lunar rover to the Moon. (☞ Jul. 30, Jul. 31, Aug. 7)

2005
The space shuttle *Discovery*, carrying a seven-member crew comprising Eileen Collins, Jim Kelly, Charlie Camarda, Wendy Lawrence, Soichi Noguchi, Steve Robinson, and Andy Thomas, was launched from NASA's Kennedy Space Center, Florida, USA, ending a two-and-a-half year wait for the historic return to flight mission after the *Columbia* accident of February 2003. The mission included breathtaking in-orbit manoeuvres, tests of new equipment and procedures, and a first-of-its-kind space walking repair. The *Discovery* crew members' primary objectives were to test and evaluate new safety procedures and conduct assembly and maintenance tasks on the *International Space Station*. The crew also carried out the first ever on-orbit repair of the shuttle heat shield. Two crew members, Robinson and Noguchi, ventured outside the shuttle three times on space walks. The first demonstrated repair techniques on the shuttle's protective tiles. During the second space walk, they replaced a failed control moment gyroscope, which helps keep the *International Space Station* oriented properly. During the third walk, they installed the external stowage platform, a sort of space shelf for holding spare parts during *Station* construction. (☞ Feb. 1, Aug. 9)

July 27

1844
John Dalton, English meteorologist and chemist who proposed the atomic theory of matter, died in Manchester, England. (☞ Sep. 6)

1881
Hans Fischer, German biochemist, Nobel laureate, was born in Hoechst, Germany. Fischer went to a primary school in Stuttgart, and matriculated in 1899 from the "humanistische Gymnasium" in Wiesbaden. He studied chemistry and medicine simultaneously, first at the University of Lausanne and obtained his chemistry degree at Marburg in 1904. In 1908, after he qualified for his MD, Fischer spent his first working years at the Second Medical Clinic in Munich and at the First Berlin Chemical Institute. He returned to Munich in 1911 and qualified as lecturer on internal medicine one year later. In 1913 he was appointed at the Physiological Institute in Munich, where he became a lecturer in physiology. From April 1921 until his death Fischer held the position of professor of organic chemistry at the Technische Hochschule (Technical University) in Munich. Fischer's scientific work was mostly concerned with the investigation of the constitutive properties of the pigments in blood, bile, and also leaves, as well as with the chemistry of nitrogen-containing organic compounds called pyrrole. Of special importance was his synthesis of the bile pigment bilirubin.

Fischer was awarded the 1930 Nobel Prize for Chemistry for his researches into the constitution of haemin and chlorophyll and especially for his synthesis of haemin. (☞ Mar. 31)

1917
Theodor Emil Kocher, Swiss surgeon, Nobel laureate (Physiology or Medicine, 1909), who did pioneering work on the physiology, pathology and surgery of the thyroid gland, died in Bern, Switzerland. (☞ Aug. 25)

1949
The British De Havilland *Comet*, the world's first jet-propelled airliner, made its maiden flight in England, ushering in the jet age. The new aircraft could carry 36 passengers at a cruising speed of 720 km/h over a distance of 4,000 km. Earlier commercial airliners all used piston engine driven propellers that made them slow and also incapable of flying at high altitudes. Flying was made difficult by the bad weather at these low cruising levels. The development of jet engines for warplanes during World War II led to a new milestone in air transport that reduced the time of long-distance flights drastically. (☞ Apr. 8, May 2)

1987
Sálim Moizuddin Abdul Ali, better known as Salim Ali, the pre-eminent ornithologist of India, died in Mumbai. (☞ Nov. 12)

July 28

1851
The first astronomical photograph of a total solar eclipse, a Daguerrotype, was taken by astronomer Berkowski at the Königsberg observatory (then in Prussia, now Kalinigrad in Russia). The photograph, taken with an exposure of 24 seconds using a heliometer, showed prominences (large-scale gaseous formation above the surface of the Sun) and corona (the upper atmosphere of the Sun which appears as a halo around the Sun during the eclipse).

1858
William Herschel, Chief Magistrate of the Hooghly district in Jungipoor (now in West Bengal), India, first used fingerprints as a means of identification of native contracts. Herschel had Rajyadhar Konai, a local businessman, impress his handprint on the back of a contract. The idea was merely 'to frighten (him) out of all thought of repudiating his signature'. Herschel made a habit of requiring palm prints – and later, simply the prints of the right index and middle fingers – on every contract made with the locals. Personal contact with the document, they believed, made the contract more binding than if they simply signed it. Thus, the first wide-scale, modern-day use of fingerprints was predicated, not upon scientific evidence, but upon superstitious beliefs.

1904
Pavel Alekseyevich Cherenkov, Soviet physicist, Nobel laureate, was born in Novaya Chigla, Russia. Cherenkov

graduated from the Voronezh State University in 1928; he later became a research student at the PN Lebedev Institute of Physics in the USSR Academy of Sciences. In 1940 he was awarded the degree of doctor in physico-mathematical sciences. It was in 1934 while doing his dissertation research that Cherenkov observed a faint blue glow when electrons passed through a transparent liquid at high velocity. This 'Cherenkov Effect', associated with charged atomic particles moving at velocities higher than the speed of light, proved to be of great importance in subsequent experimental work in nuclear physics and for the study of cosmic rays. This phenomenon, which was interpreted by the Russian physicists Igor Y. Tamm and Ilya M. Frank, led to the development of the Cherenkov counter, or 'Cherenkov detector,' which was later used extensively in experimental nuclear and particle physics. Cherenkov continued to do research in nuclear physics at the Lebedev Institute, where he became a full professor in 1959. Cherenkov was awarded the 1958 Nobel Prize for Physics, jointly with Tamm and Frank, for the discovery and interpretation of the Cherenkov Effect. (☞ Jan. 6, Jul. 8, Oct. 23)

1915

Charles Hard Townes, American physicist, Nobel laureate, was born in Greenville, South Carolina, USA. Townes attended the Greenville public schools and received his BA and BS degrees from the Furman University in Greenville in 1935. He received his MA degree from the Duke University, Durham, New Carolina, in 1937, and a PhD from the California Institute of Technology, Pasadena, California, in 1939, with a thesis on isotope separation and nuclear spins. He then joined the technical staff of the Bell Telephone Laboratories, Inc., where he worked until 1947. Townes worked extensively during World War II in designing radar-bombing systems and received a number of patents in related technology. From this he turned his attention to applying the microwave technique of wartime radar research to spectroscopy. In 1948 he joined the faculty of the Columbia University, New York, where he continued research in microwave physics, particularly in studying the interactions between microwaves and molecules, and using microwave spectra for the study of the structure of molecules, atoms, and nuclei. In 1951, Townes conceived the idea of the maser, and a few months later he and his associates began working on a device using ammonia gas as the active medium and completed the first such device in December 1953. They gave it the name 'maser', an acronym for 'microwave amplification by stimulated emission of radiation'. In 1958 Townes and fellow American physicist Arthur L. Schawlow showed that it was possible to construct a similar device using light, which came to be known as the laser. Townes was awarded the 1964 Nobel Prize for Physics, jointly with the Soviet physicists Aleksandr M. Prokhorov and Nikolay G. Basov, for fundamental work in the field of quantum electronics, which led to the construction of oscillators and amplifiers based on the maser–laser principle. (☞ Mar. 23, Jul. 11, Dec. 14)

1925

Baruch S. Blumberg, American research physician, Nobel laureate, was born in

New York, USA. Blumberg attended the Far Rockaway High School and then joined the US Navy in 1943. He obtained his undergraduate degree in physics at the Union College in New York, and in 1946 began graduate work in mathematics at the Columbia University. He entered the College of Physicians and Surgeons of Columbia University in 1947 from where he received his MD degree. He received his PhD degree in biochemistry from the Oxford University, England, in 1957. In 1960 he became chief of the geographic medicine and genetics section of the US National Institutes for Health, in Maryland. In 1964 he was appointed associate director for clinical research at the Institute for Cancer Research, Philadelphia, and also served as professor of medicine, human genetics, and anthropology at the University of Pennsylvania. In 1963, while examiningbloodsamplesfromwidelydiverse populations in an attempt to determine why the members of different ethnic groups vary widely in their susceptibility to disease, he discovered in the blood serum of an Australian aborigine an antigen that he later found was part of a virus that causes hepatitis B, the most severe form of hepatitis. The discovery of this so-called Australian antigen, which causes the body to produce antibody responses to the virus, made it possible to screen blood donors for possible hepatitis B transmission. It also led to the development by other researchers of a successful vaccine against the disease. Blumberg was awarded the 1976 Nobel Prize for Physiology or Medicine, jointly with the American physician Daniel Carleton Gajdusek, for their discoveries concerning new mechanisms for the origin and dissemination of infectious diseases. (☞ Sep. 9)

1930
Allvar Gullstrand, Swedish ophthalmologist, Nobel laureate (Physiology or Medicine, 1911), who discovered how the eye produces images on the retina, died in Stockholm, Sweden. (☞ Jun. 5)

1964
American space probe *Ranger-7* was launched towards the Moon. (☞ Jul. 31)

1968
Otto Hahn, German chemist, Nobel laureate (Chemistry, 1944), who co-discovered nuclear fission, died in Göttingen, Germany. (☞ Mar. 8)

2002
Archer J.P. Martin, British biochemist, Nobel laureate (Chemistry, 1952), who developed the technique of paper partition chromatography, a quick and economical analytical technique permitting extensive advances in chemical, medical, and biological research, died in Llangarron, Herefordshire, England. (☞ Mar. 1)

2004
Francis Harry Compton Crick, British biophysicist, Nobel laureate (Physiology or Medicine, 1962), who co-discovered the three-dimensional molecular structure of deoxyribonucleic acid (DNA), died in San Diego, California, USA. (☞ Jun. 8)

July 29

1876
The Indian Association for the Cultivation of Science (IACS) was inaugurated in

Calcutta (now Kolkata). It is the oldest institute in India devoted to the pursuit of fundamental research in the frontier areas of basic sciences. Originally located at 210, Bowbazar Street, IACS is a national institution for higher learning whose primary purpose is to foster high-quality fundamental research in frontier disciplines of the basic sciences. Founded by Dr. Mahendra Lal Sircar, a philanthropist, the activities at the Association in the very early years were supported by generous public contributions. The original objective of the Association, which continues even today, was to cultivate science in all its departments both with a view to its advancement by original research and to its varied applications to the arts and comforts of life. Many distinguished scientists of modern India had carried out research here. Notable among them was Sir C V Raman who did his monumental work on physical optics leading to the discovery of the celebrated Effect that bears his name and won for him and India the first Nobel Prize in Science. The present campus of the Association is located in Jadavpur to the south of Kolkata. (☞ Feb. 28, Nov. 2)

1898
Isidor Isaac Rabi, American physicist, Nobel laureate, was born in Rymanów, Austria-Hungary (now in Poland). Rabi was brought to USA by his family, in 1899, and had his early education in New York City. In 1919 he received his BS in Chemistry at the Cornell University, New York. After two years in non-scientific occupation, he started postgraduate studies in physics at Cornell in 1921, which he later continued at the Columbia University, New York. In 1927 he received his PhD degree for work on the magnetic properties of crystals. He spent two years in Europe, working at different times with Arnold Sommerfeld, Niels Bohr, Wolfgang Pauli, Otto Stern, and Werner Heisenberg. On his return in 1929 he was appointed lecturer in theoretical physics at the Columbia University, and became professor in 1937. He spent most of his life (1929–67) at the Columbia University, where he performed most of his pioneering research in radar and the magnetic moment associated with electron spin in the 1930s and 1940s. Rabi's most important scientific work was his development of a method for measuring the magnetic properties of atoms, atomic nuclei, and molecules. The method is based on measuring the spin of the protons in the atom's core. With the application of his magnetic resonance method, several mechanical and magnetic properties, as well as the shape, of an atomic nucleus can be deduced. The precise measurements yielded by this method made possible such subsequent applications as the atomic clock, the maser, and the laser, as well as the nuclear magnetic resonance imaging used in diagnostic medicine. Rabi was awarded the 1944 Nobel Prize for Physics for his resonance method used in recording the magnetic properties of atomic nuclei. (☞ Jan. 11)

1982
Vladimir Kosma Zworykin, Russian-born American electronic engineer and a pioneer of modern television technology and co-inventor of the electron microscope, died in Princeton, New Jersey, USA. (☞ Apr. 20, Jul. 30)

1985
American space shuttle *Challenger*, carrying astronauts Gordon Fullerton, Roy D. Bridges, F. Story Musgrave, Anthony W. England, Karl G. Henize, Loren W. Acton

and John-David F. Bartoe, was launched on its 8th flight from the Kennedy Space Center, Florida, California, USA. The main payload was the second Spacelab mission, *Spacelab-2*, jointly sponsored by the European Space Agency and the American space agency NASA to conduct advanced research in space. *Spacelab-2* carried several instruments for making observations of the Sun and also equipment to carry out studies in life sciences, plasma physics, and atmospheric physics. (☞ Apr. 4, Aug. 6)

1994
Dorothy Crowfoot Hodgkin, English chemist, Nobel laureate (Chemistry, 1964), who determined the structure of important biochemical substances like insulin and vitamin B_{12}, died in Shipston-on-Stour, Warwickshire, England. (☞ May 12)

2005
The discovery of the tenth planet (now designated as a dwarf planet) of the solar system was announced by planetary scientist Michael Brown of the California Institute of Technology, USA. The planet-like object was first spotted in October 2003 with the Samuel Oschin Telescope at the Palomar Observatory near San Diego, California. Known by the catalogue number 2003UB313, the new object is 2,683 kilometres in diameter and orbits the Sun in an elliptical orbit tilted some 45 degrees to the plane of the solar system's eight major planets, taking 560 years to complete one trip around the Sun. At its most distant, the dwarf planet is 97 times farther from the Sun than Earth. At its closest, it passes inside the orbit of Pluto at a distance of some 36 astronomical units. It is currently the most distant object known in the solar system. The object has been designated as a dwarf planet by the International Astronomical Union. (☞ Aug. 24)

July 30

1863
Henry Ford, American industrialist who introduced assembly line manufacturing for mass production, was born near Dearborn, Michigan, USA. Ford enjoyed a childhood typical of the rural nineteenth century, spending days in a one-room school and doing farm chores. At an early age, he showed an interest in mechanical things and a dislike for farm work. In 1879, Ford left home for the nearby city of Detroit to work as an apprentice machinist. He remained an apprentice for three years and then returned to Dearborn. During the next few years, he divided his time between operating or repairing steam engines, finding occasional work in a Detroit factory, and over-hauling his father's farm implements. In 1891, Ford became an engineer with the Edison Illuminating Company in Detroit. His promotion as chief engineer in 1893 gave him enough time and money to devote attention to his personal experiments on internal combustion engines, which culminated in 1896 with the completion of his own self-propelled vehicle – the Quadricycle. After two unsuccessful attempts to establish a company to manufacture automobiles, the Ford Motor Company was incorporated in 1903 with Henry Ford as vice-president and chief engineer. The infant company produced only a few cars a day at the Ford factory on Mack Avenue in Detroit. Groups of two or three men worked on each car from components made to order by other companies. Ford realised his dream of producing an automobile that was reasonably priced, reliable, and efficient with the introduction of the 'Model T' in 1908. For this he introduced the assembly line manufacturing process for the first time. The Model T initiated a new era in

personal transportation. It was easy to operate, maintain, and handle on rough roads, immediately becoming a huge success. Ford not only revolutionised industrial production in the United States and Europe, but also had such influence over the 20th century economy and society that his combination of mass production, high wages and low prices is called 'Fordism'. (☞ Jan. 13, Apr. 7, Oct. 1, Dec. 1)

1889

Vladimir Kosma Zworykin, Russian-born American electronic engineer and a pioneer of modern television technology, was born in Murom, Russia. Zworykin studied at the St. Petersburg Institute of Technology in St. Petersburg and the Collegè de France, in Paris, France. He emigrated to USA in 1919 and in 1920 joined the Westinghouse Electric Corporation in Pittsburgh, Pennsylvania, where he developed an entirely new, fully electronic television system. In 1923 he filed a patent application for the 'iconoscope', or television transmission tube, which produced pictures by electronically scanning images. In 1924 he filed a patent application for the 'kinescope', the modern TV picture tube that projects pictures it receives onto a screen. Zworykin's television system provided the impetus for the development of modern television as an entertainment and education medium. Zworykin was director of electronics research laboratory of the Radio Corporation of America (RCA), at Camden, New Jersey, from 1929 to 1942 where he helped develop an early version of the electron microscope. (☞ Apr. 20, Jul. 29)

1913

John Milne, English seismologist and geologist who invented the seismograph and promoted the establishment of seismological stations, died in Shide, Isle of Wight, England. (☞ Dec. 30)

1971

Apollo-15 astronauts David R. Scott and James B. Irwin on board lunar module *Falcon* touched down on Moon near the sinuous channel known as Hadley Rille. Alfred Worden remained in orbit in the command module *Endeavour*. During their stay on the Moon, Scott and Irwin rode more than 27.36 km in the lunar rover and collected a sample of ancient lunar crust nicknamed the 'Genesis Rock' due to its extreme formation age. (☞ Jul. 26, Jul. 31, Aug. 7)

July 31

1800

Friedrich Karl Wöhler, German chemist who was the first to synthesise an organic compound from an inorganic substance, was born in Escherheim, Germany. Wöhler was educated at the Universities of Marburg and Heidelberg. While studying medicine at Heidelberg, he became interested in chemistry and went to Stockholm, Sweden, to study with the Swedish chemist Jons Jakob Berzelius. In 1827 he devised a new method for isolating aluminium. In 1828, he succeeded in the isolation of beryllium as a black-grey powder; he also isolated yttrium and crystalline silicon. He taught chemistry from 1825 to 1831 at the Polytechnic School in Berlin. A pioneer in the field of organic chemistry, Wöhler is famous for his synthesis of the organic compound urea from the inorganic salt ammonium cyanate in 1828. By this synthesis he proved, contrary to scientific thinking of the time, that a product of the living processes of animals could be made in the laboratory from inorganic materials. Until that time it was believed that organic substances could only be formed under the influence of the vital force in the bodies of animals and plants. Wöhler also conducted important research on uric acid and the oil of bitter almonds, in collaboration with the German chemist Justus von Liebig. In 1862, he discovered calcium carbide and prepared acetylene from it. In 1836 Wöhler became professor of chemistry at the University of Göttingen where he remained till his death. (☞ Feb. 22, Sep. 23)

1907

Damodar Dharmanand Kosambi, Indian scientist, mathematician, historian, and Sanskrit scholar, was born in Goa, India. Kosambi went to USA in 1918 with his father and graduated from the Harvard University, Cambridge, Massachusetts, in 1926 in mathematics, history and languages. On his return to India he taught at the Banares Hindu University in Banares (now Varanasi), the Aligarh Muslim University, Aligarh, United Provinces (now Uttar Pradesh), and then at the Ferguson College, Pune, and the Tata Institute of Fundamental Research, Bombay (now Mumbai), Maharashtra. Kosambi was one of the few Indian scientists who had grasped the nature of twentieth century science and technology and its implications for humanity. He showed genuine awareness of the interaction between science and social processes particularly in the context of under-developed countries. His approach to science and its application was always an integrated one and not purely technical. He devoted a great deal of his time to the peace movement and the campaign against nuclear weapons. He made valuable contributions in the field of pure and applied mathematics, and theoretical and applied statistics. (☞ Jun. 29)

1918

Paul D. Boyer, American biochemist, Nobel laureate, was born in Provo, Utah, USA. Boyer received a BS degree in chemistry from the Brigham Young University in Provo in 1939 and a PhD in biochemistry from the University of Wisconsin, Madison, Wisconsin, in 1943. He held a number of teaching positions before joining the faculty of the University of California at Los Angeles in 1963. There he served as professor (1963–89) and director of the Molecular Biology Institute (1965–83) before becoming professor emeritus in 1990. In the early 1950s Boyer began his independent research career at the University of Minnesota and introduced kinetic, isotopic, and chemical methods for investigating enzyme mechanisms to find out how cells form the energy-storage molecule adenosine triphosphate (ATP), a process that occurs in animal cells in a structure called a mitochondrion. He followed up the work of the British chemist Peter D. Mitchell and revealed more specifically what is involved in ATP synthesis. His work focussed on the enzyme ATP synthase, and he demonstrated how the enzyme harnesses the energy produced by the hydrogen flow to form ATP out of adenosine diphosphate (ADP) and inorganic phosphate. The British chemist John E. Walker and his co-workers established the structure of the enzyme and verified the mechanism proposed by

Boyer. Boyer was awarded the 1997 Nobel Prize for Chemistry, jointly with Walker and the Danish biophysicist Jens C. Skou. Boyer and Walker shared half the prize money for their elucidation of the enzymatic mechanism underlying the synthesis of adenosine triphosphate (ATP), while Skou received the other half for the first discovery of an ion-transporting enzyme, Na^+, K^+-ATPase. (☞ Jan. 7, Oct. 8)

1964
American space probe *Ranger-7* crashed on the Moon in the Mare Nubium. It sent back 4,308 photographs of the lunar surface before impact. (☞ Jul. 28)

1969
The American unmanned spacecraft *Mariner-6* flew-by Mars at a distance of 3,550 kilometres and sent back close-up photographs and data on the Martian atmosphere and its chemical composition. *Mariner-6* and its twin spacecraft *Mariner-7*, which flew by the Red Planet a few days later, together sent back to Earth a combined total of over 200 images of the Martian surface. Some of the photos showed Phobos, the inner and larger of Mars's two tiny moons, and the northern and southern polar caps. The photographs also revealed geologic features – including cratered deserts, huge concentrically terraced impact regions, collapsed ridges and craterless depressions on Mars. (☞ Feb. 24, Aug. 4)

1971
American astronaut David R. Scott became the first person to drive a vehicle on the Moon. The lunar roving vehicle (LRV) was carried on board *Apollo-15*, the first by an Apollo mission. Weighing 210 kg, the Lunar Rover was driven a total of 27.8 km in 3 hours, 2 minutes of driving time during the mission. Power was provided by two 36-volt silver–zinc potassium hydroxide non-rechargeable batteries, which were used to power the drive and steering motors and also a 36-volt utility

David R. Scott on the lunar rover

outlet mounted on front of the LRV to power the communications relay unit or the TV camera. The next two missions – *Apollo-16* and *Apollo-17* – also carried Lunar Rovers. (☞ Jul. 26, Jul. 30, Aug. 7)

August 1

1744

Jean-Baptiste Lamarck, French biologist who is best known for his idea that acquired traits are inheritable, was born in Bazentin-le-Petit, Picardy, France. Lamarck was sent to a Jesuit school at Amiens, but after his father died he enlisted and served in an infantry regiment for several years. While stationed on the French Riviera, he became interested in plants and, after resigning from the army, embarked upon a study of medicine and botany. He soon devoted himself entirely to botany and recognition of his skill followed upon publication of the 3-volume work *Flore Française* in 1778. He was elected to the French Academy of Sciences, and in 1781 appointed botanist to the French king. From 1793 he was professor of invertebrate zoology at the Museum of Natural History, Paris. His ideas concerning the origin of species were first made public in his *Système des animaux sansvertèbres, ou table général des classes* ('System of Invertebrate Animals, or General Table of Classes') published in 1801. He introduced the terms 'biology' and 'Invertebrata' and suggested the invertebrate classes Infusoria, Annelida, Crustacea, Arachnida, and Tunicata. He is also considered the founder of invertebrate palaeontology. Lamarck's theory of evolution, or Lamarckism, asserts that all life forms have arisen by a continuous process of gradual modification throughout geologic history. To explain this process he cited the then generally accepted theory of acquired characteristics, which held that new traits in an organism develop because of a need created by the environment and that the traits are transmitted to the organism's offspring. Although the latter hypothesis was disputed during Lamarck's lifetime and rejected altogether as the principles of heredity were established, Lamarck's theory of evolution was an important forerunner of the work of the British naturalist Charles Darwin. (☞ Dec. 18)

1774

Joseph Priestley, English clergyman and chemist, identified a gas, which he called 'dephlogisticated air' – later known as oxygen. Priestley found that mercury heated in air became coated with a 'red rust of mercury', which, when heated separately, converted back to mercury with 'air' being released. He found that this 'air' made candles burn very brightly. Also, when a mouse was kept in a sealed vessel with this air it could survive much longer than in ordinary air. (☞ Feb. 6, Mar. 8, Mar. 13)

1793

France became the first country to use the metric system of weights and measures. The metre was introduced as the unit of length and gramme as the unit of mass. The metre was defined as the distance between two lines on a standard bar of an alloy of 90% platinum and 10% iridium, which was equal to 10000/254 inches, or approximately 39.37 inches. In 1960, the definition of the metre changed to

1,650,763.73 wavelengths of the orange-red radiation of krypton-86. According to the current definition, the metre is the length of the path travelled by light in absolute vacuum during a time interval of 1/299,792,458 of a second. For the metric unit of mass, the gramme was defined as the mass of one cubic centimetre of pure water at a given temperature. (☞ Mar. 30)

1861

The first weather 'forecast' was published in newspapers. The forecast was prepared and sent by Admiral Robert Fitzroy, who was the Captain of the ship *HMS Beagle* and had travelled with the British naturalist Charles Darwin between 1831 and 1836. In 1859, Fitzroy was given the job of setting up a service to warn of nasty weather. He began by collecting observations from England and Ireland, which were then relayed to Lloyds of London, England, where the ship's captains would look at them. But publication of weather forecasts was stopped by the Royal Society of London in 1865 after Fitzroy committed suicide, demoralised by the criticism of his forecasts. Fitzroy's dream of weather forecasting finally came true in January 1921 when the radio station at the University of Wisconsin in USA started the first public daily weather forecast.

1885

George Charles de Hevesy, Hungarian chemist, Nobel laureate, was born in Budapest, Hungary. Hevesy did his matriculation at the Gymnasium of the Piarist Order in 1903. He then studied at the Budapest University and the Berlin Technical University. He obtained his doctor's degree from the University of Freiburg im Breisgau in Germany in 1908. He worked for two years as an assistant at the Institute of Physical Chemistry, Technical University of Switzerland. De Hevesy developed an interest in isotopes while working in the British physicist Ernest Rutherford's laboratory in Manchester, England between 1910 and 1913. He carried out the first radioactive tracer research, jointly with Frederic Paneth, at the Vienna Institute of Radium Research from 1913 to 1915. In 1920 de Hevesy joined Niels Bohr's Institute in Copenhagen, Denmark. Six years later he moved to the University of Freiburg in Germany as professor of physical chemistry. In 1930 he was appointed Baker Lecturer at the Cornell University, Ithaca, USA, and returned to the Niels Bohr's Institute four years later. In Copenhagen, de Hevesy's researches were initially concerned with isotopic separations and in 1923, together with the Danish chemist Dirk Coster he discovered the element hafnium (Hf). He was responsible for pioneer work in the use of isotopic indicators both in inorganic and life sciences. Later, in Freiburg, he was involved in the first clinical use of isotopes. On his return to Copenhagen, he demonstrated the formation of new artificially radioactive isotopes and subsequently introduced a method of activation analysis based on neutron bombardment of the element to be investigated. After preparing a radioactive isotope of phosphorus, in 1934, he began to study life processes using these atoms as tracers attached to molecules. Thus labelled, the movement and changes of these molecules could be followed in the body's life processes. De Hevesy was awarded the 1943 Nobel Prize for Chemistry for his work on the use of isotopes as tracers in the study of chemical processes. (☞ Jul. 5)

1895
Gopal Chandra Bhattacharya, Indian naturalist, researcher, and science populariser, was born in Lonsing, a remote village in Faridpur district, Bengal, India (now in Bangladesh). Bhattacharya passed the matriculation examination with a first division, but his financial position compelled him to forgo an academic career and take up the job of a schoolteacher to maintain his family. Bhattacharya was a keen observer of nature right from his childhood. He wrote an article on bioluminescence, which was published in *Prabashi*, a famous Bengali periodical and drew the attention of the Indian plant physiologist and physicist Jagadis Chandra Bose, who offered him a job in his Bose Institute, an event that marked a turning point in Bhattacharya's career. He joined the institute as an assistant and had to do odd jobs like instrument repair, making drawings, etc. But soon, he was allowed to start his own research.

Bhattacharya started observing the behaviour of various insects, spiders and tadpoles, and photographing them, as he was an expert photographer. He persuaded an Indian variety of ants known as Occophylia to make nests of transparent cellophane so that he could silently watch and photograph their activities. He noticed that only a special food, newly sprouted leaves and buds, induced the formation of queens. This remarkable finding was published in the *Transactions of the Bose Institute* of Kolkata. Based on his observations, Bhattacharya also wrote popular articles in his native tongue Bengali, which were widely read. Despite his scholarly work exploring the natural world, Bhattacharya did not get much recognition, except an Honorary DSc from the Calcutta University in 1981, a few months before his death. (☞ Apr. 8)

1924
Georges Charpak, Polish-born French physicist, Nobel laureate, was born in Dabrovica, Poland. Charpack did his BSc in mining from the Ecole des Mines in Paris in 1948 and received a PhD in physics in 1955 from the College de France in Paris, where he worked in the laboratory of the French physicist Frédéric Joliot-Curie. In 1959 he joined the staff of CERN (European Organization for Nuclear Research) in Geneva and in 1984 became a Joliot–Curie professor at the School of Advanced Studies in Physics and Chemistry, Paris. He was made a member of the French Academy of Science in 1985. Charpak's main contribution has been in the field of particle detectors.

Earlier detectors of charged particles, such as the bubble chamber, could record photographically the tracks left by particles at the rate of only one or two per second. Charpack built a new kind of detector called the 'multi-wire proportional chamber' in 1968, which could record up to one million tracks per second and send the data directly to a computer for analysis. The speed and precision of the multi-wire chamber and its descendants, the drift chamber and the time projection chamber, revolutionised high-energy physics. By the 1990s such detectors were at the heart of almost every experiment in particle physics. Charpack was awarded the 1992 Nobel Prize for Physics for his invention and development of particle detectors, in particular the multi-wire proportional chamber.

1945

Douglas D. Osheroff, American physicist, Nobel laureate, was born in Aberdeen, Washington, USA. Osheroff received his bachelor's degree in 1967 from the California Institute of Technology in Pasadena, and a doctorate in 1973 from the Cornell University in Ithaca, New York. In 1972, while working as a graduate student, investigating the properties of the isotope helium-3 under temperatures of just a few thousandths of a degree above absolute zero (−273°C) with the American physicists David Lee and Robert C. Richardson, Osheroff noticed minute jumps in the internal pressure of the sample of helium-3 under investigation. The deviations were eventually traced to a phase transition of the helium-3 to a superfluid state, in which a liquid's atoms lose their randomness and move about in a coordinated manner and even flow out of a container spontaneously. Such a substance lacks all internal friction, flows without resistance, and behaves according to quantum mechanical laws rather than to those of classical fluid mechanics. The discovery of superfluidity in helium-3 enabled scientists to study directly the quantum mechanical effects in macroscopic systems that had previously been studied only indirectly in molecules, atoms, and subatomic particles. From 1972 to 1982 Osheroff conducted research at the Bell Telephone Laboratories in New Jersey and headed solid-state and low-temperature research there from 1982 to 1987. He became a professor at the Stanford University in California in 1987. Osheroff was awarded the 1996 Nobel Prize for Physics, jointly with Lee and Richardson, for their discovery of superfluidity in helium-3. (☞ Jan. 20, Jun. 26)

1967

Richard Kuhn, German biochemist, Nobel laureate (Chemistry, 1938), who discovered at least eight carotenoids, prepared them in pure form, and determined their constitution, died in Heidelberg, Germany. (☞ Dec. 3)

1967

American spacecraft *Lunar Orbiter-5* was launched on a photographic mission to the Moon. Between August 6 and 18 *Lunar Orbiter-5* completed photographing the far side of the Moon and acquired medium (20 m) and high resolution (2 m) images of 36 pre-selected areas. In all, five *Lunar Orbiter* missions were sent and all were successful, photographing 99% of the Moon's surface with a resolution of 60 m or better.

1970

Otto Heinrich Warburg, German biochemist, Nobel laureate (Physiology or Medicine, 1931), who discovered the nature and mode of action of the respiratory enzyme, died in Berlin, Germany. (☞ Oct. 8)

August 2

1799

Jacques-Étienne Montgolfier, French inventor, who, with his brother Josef-Michel Montgolfier, pioneered hot air ballooning, died en route from Lyon to Annonay in France. (☞ Jan. 6, Aug. 26)

1820

John Tyndal, British physicist who demonstrated why the sky is blue, was born in Leighlinbridge, County Carlow, Ireland. Tyndall left school at the age of 17, and in 1839 joined the Irish Ordnance

Survey where he spent three years becoming an accomplished practical surveyor and draftsman. When the Irish survey was finished in 1842 he was transferred to the English survey. In 1848 he arrived in Marburg, Germany, to work with the German chemist Robert Bunsen. With intense effort and Bunsen's inspiration, Tyndall completed all the work required for his doctoral degree in less than two years. By his second year he decided to concentrate on physics and began a series of studies on diamagnetism and magnetic optical properties of crystals. This was to be his major research for nearly six years. He also carried out research on radiant heat, studied spontaneous generation and the germ theory of disease, glacier motion, sound, the diffusion of light in the atmosphere and a host of related topics. He showed that ozone was an oxygen cluster rather than a hydrogen compound. His experiments also showed that molecules of water vapour, carbon dioxide and ozone are the best absorbers of heat radiation and that even in small quantities these gases absorb much more strongly than the atmosphere itself. One of his most important inventions, the light pipe, led to the development of fibre optics. In the course of his study on light beams he discovered in 1869 the Tyndall effect – the diffusion of light by large molecules and dust. His suggestion that the sky is blue due to the scattering of the Sun's rays by molecules in the atmosphere was later explained theoretically by the English physical scientist Lord Rayleigh. (☞ Dec. 4)

1861

Prafulla Chandra Ray, Indian scientist, educationist, and nationalist visionary, was born in Raruli, a village in Jessore, India (now in Bangladesh). After early schooling in his village, Ray moved to Calcutta (now Kolkata) and passed the entrance examination of the Albert School in 1879. He then joined the Metropolitan Institute (now Vidyasagar College). He went to England in 1882 and received his BSc degree in 1885, and a DSc degree in 1887 from the University of Edinburgh. After returning to India in 1888, Ray spent almost a year working with his friend the Indian physicist Jagadis Chandra Bose in his laboratory. In 1889 he entered the Bengal Education Service and was appointed Assistant Professor of Chemistry in the Presidency College. To encourage indigenous development of the chemical industry in India, he himself set an example by establishing a chemical firm called the Bengal Chemical and Pharmaceutical Works in 1901. In 1902, he published his famed *History of Hindu Chemistry*, which came out in a revised and enlarged form in 1904. As a scientist Ray was particularly esteemed among his peers for his seminal contributions in the preparation of mercurous nitrite and its derivatives. In 1916 Ray left the Presidency College and joined the newly founded Calcutta University College of Science as a Professor of Chemistry. Ray was an exponent of the cultivation of arts and sciences through one's mother tongue. He himself wrote a lot in his native tongue Bangla and encouraged others to write in the mother tongue. (☞ Jun. 16)

1922

Alexander Graham Bell, Scottish-born American audiologist and inventor of the telephone, died in Cape Breton Island, Nova Scotia, Canada. (☞ Mar. 3)

August 3

1811
Elisha Otis, American inventor of the automatic safety brake for elevators, was born in Halifax, Vermont, USA. From a young age Otis worked on several jobs and invented several labour-saving machines. In 1852 his employer sent him to Yonkers, New York, to operate a new factory and to install its machinery. There he designed and installed what he called the 'safety hoist', the first elevator, or lift, equipped with an automatic safety device to prevent it from falling if the lifting chain or rope broke. Before this invention, elevators of his time were extremely dangerous. In May 1854, he demonstrated his elevator at the Crystal Palace in New York City. In front of a large crowd, Otis ascended in his new elevator. He called for the elevator's cable to be cut with an axe, but the elevator platform did not fall. The brake he invented used toothed guide rails in the elevator shaft and a spring-loaded bar that automatically caught in the toothed rail if the elevator car cable failed. (☞ Jan. 15, Apr. 8)

1942
Richard Willstätter, German chemist, Nobel laureate (Chemistry, 1915), who discovered the constitution of the plant pigment chlorophyll and blood pigment haemoglobin, died in Locarno, Switzerland. (☞ Aug. 13)

1958
The American nuclear-powered submarine *USS Nautilus* became the first vessel to sail under the North Pole, 90°N. It had 116 men on board with Commander William R. Anderson at the helm. (☞ Jan. 21, Mar. 3)

1959
Koichi Tanaka, Japanese scientist, Nobel laureate, was born in Toyama Prefecture, Japan. After completing his schooling in Toyama, Tanaka joined the Faculty of Technology, Tohoku University in 1978 to study electrical engineering. In 1983, he graduated with a bachelor's degree in engineering and joined the Shimadzu Corporation, where he was engaged in the development of mass spectrometers. He also developed a technique for mass spectroscopic analysis of macromolecules, such as proteins. For mass spectroscopic analyses of macromolecules, the sample must be ionised and vapourised by laser irradiation, which breaks up the sample into tiny fragments destroying its structure. In February 1985, Tanaka found that by using a mixture of ultra-fine metal powder and glycerol as a matrix, a macromolecule sample can be ionised without disrupting its structure. His work was reported for the first time at the Annual Conference of the Mass Spectrometry Society of Japan held in Kyoto, Japan, May 1987. This phenomenon, referred to as soft laser desorption (SLD), provided the basis for 'matrix-assisted laser desorption ionisation time-of-flight mass spectrometry', or MALDI-TOF/MS in short, which is indispensable for research in life sciences today, especially in the field of proteomics. Tanaka was awarded the 2002 Nobel Prize for Physics, jointly with the American chemist John B. Fenn, and Swiss chemist Kurt Wüthrich, for the development of methods for identification

and structure analyses of biological macromolecules. Tanaka and Fenn shared half the prize money for their development of soft desorption ionisation methods for mass spectrometric analyses of biological macromolecules, while Wüthrich received the other half for his development of nuclear magnetic resonance spectroscopy for determining the three-dimensional structure of biological macromolecules in solution. (☞ Jun. 15, Oct. 4)

1994
The first human heart transplant in India was performed at the All India Institute of Medical Sciences, New Delhi by cardiologist P. Venugopal on Devi Ram, a mechanic.

August 4

1956
Apsara, India's first 1 MW nuclear reactor became critical at the Atomic Energy Establishment (re-named Bhabha Atomic Research Centre), Trombay, Maharashtra, India. (☞ Jan. 20)

1969
The American unmanned spacecraft *Mariner-7* flew by Mars at a distance of 3,550 kilometres and sent back close-up photographs and data on the Martian atmosphere and its chemical composition. *Mariner-7* and its twin spacecraft *Mariner-6*, which flew by the Red Planet a few days earlier, together sent back to Earth a combined total of over 200 images of the Martian surface. Some of the photos showed Phobos, the inner and larger of Mars's two tiny moons, and the northern and southern polar caps. The photographs also revealed geologic features – including cratered deserts, huge concentrically terraced impact regions, collapsed ridges and craterless depressions on Mars. Most famously, *Mariner-7* took the first close-up images of the deep valley, which now bears its name, Valles Marineris. (☞ Mar. 27, Jul. 31)

1977
Edgar Douglas Adrian, English electrophysiologist, Nobel laureate (Physiology or Medicine, 1932), one of the founders of modern neurophysiology, died in London, UK. (☞ Nov. 30)

2003
Frederick Chapman Robbins, American paediatrician and virologist, Nobel laureate (Physiology or Medicine, 1954), who developed a technique for cultivating the poliovirus in tissue culture, died in Cleveland, Ohio, USA. (☞ Aug. 25)

August 5

1729
Thomas Newcomen, British inventor of the atmospheric steam engine, died in London, England. (☞ Feb. 24)

1862
'Elephant Man' Joseph Carey Merrick was born in Leicester, England. Merrick began growing disfiguring tumours before the age of two and his condition rapidly worsened, rendering one of his arms completely useless. Nevertheless, he was described as a wonderfully imaginative and intelligent boy. He was dubbed the 'Elephant Man' because of the roughness of his

skin and a facial deformity that was said to resemble a trunk. He became the best-known resident patient of London Hospital from 1886 until his death. According to the modern diagnosis Merrick suffered from a very rare disorder called 'Proteus Syndrome'. (☞ Apr. 11)

1914
The world's first traffic lights were installed at each corner of the intersection of 105th Street and Euclid Avenue in Cleveland, Ohio, USA. Installed by the American Traffic Signal Company, it had only red and green lights and no amber light.

1930
Neil Alden Armstrong, American astronaut and the first man to walk on the Moon, was born in Wapakoneta, Ohio, USA. Armstrong became a licensed pilot on his 16th birthday and a naval air cadet in 1947. His graduated in aeronautical engineering from the Purdue University, West Lafayette, Indiana, and in 1955 became a civilian research pilot for the National Advisory Committee for Aeronautics, which later became the National Aeronautics and Space Administration (NASA). He flew more than 1,100 hours, testing various supersonic fighters as well as the X-15 rocket plane. In 1962, he became the first civilian to enter the astronaut-training program. He was command pilot of the *Gemini-8* mission, which accomplished the first docking of two orbiting spacecraft. Later he was commander of the *Apollo-11* lunar mission and became the first man to land on the Moon. In 1971 he became professor of aerospace engineering at the University of Cincinnati. (☞ Mar. 16, Jul. 21)

1957
Heinrich Otto Wieland, German chemist, Nobel laureate (Chemistry, 1927), who determined the constitution of bile acids, died in Munich, Germany. (☞ Jun. 4)

August 6

1181
A supernova was seen in the constellation of Cassiopeia. First seen from southern China, and independently observed one day later from Japan, this 'guest star' remained visible for 185 days or more than 6 months.

1881
Alexander Fleming, Scottish bacteriologist, Nobel laureate, was born in Lochfield, Scotland. Fleming attended the Louden Moor School, Darvel School, and Kilmarnock Academy before moving to London where he attended the Polytechnic. He worked as a clerk in London until 1901 when he began to study medicine. He proved to be a good researcher and so he joined the research department of St Mary's Hospital, London in 1906. He was to spend his entire working life here. In 1928, while working on influenza virus, he observed that mould had developed accidentally on a staphylococcus culture plate and that the mould had created a bacteria-free circle around itself. He experimented further and found that a mould culture prevented growth of staphylococci, even when diluted 800 times. The active substance, which he named penicillin, initiated the highly effective practice of antibiotic therapy for

infectious diseases. He was elected Fellow of the Royal Society in 1943. Fleming was awarded the 1945 Nobel Prize for Physiology or Medicine, jointly with German-born British biochemist Ernst Boris Chain and Australian pathologist Howard Walter Florey for the discovery of penicillin and its curative effect in various infectious diseases. (☞ Mar. 11, Jun. 19, Jul. 24, Sep. 24)

1930

Sivaramakrishna Chandrasekhar, Indian liquid-crystal physicist, Fellow of the Royal Society, was born in Calcutta (now Kolkata), West Bengal. Chandrasekhar received his MSc degree in physics from the Nagpur University, Nagpur, Maharashtra in 1951. Later, he joined the Raman Research Institute in Bangalore, Karnataka and did his doctoral work on optical rotatory dispersion measurements on crystals under the guidance of his maternal uncle, physicist Chandrasekhara Venkata Raman. After receiving his DSc degree from the Nagpur University in 1954 he went to the Cavendish Laboratory in Cambridge, England as an 1851 exhibition scholar and obtained a second doctorate degree from the Cambridge University. After returning to India in 1961, Chandrasekhar joined the Mysore University at Mysore, Karnataka as the first head of the department of physics when he turned his attention to liquid crystals. Liquid crystals made of rod-like molecules had been discovered in 1888, and many compounds had been synthesised in the 1920s and 30s, but the subject later languished. Chandrasekhar and his co-workers made contributions to the application of the dynamical theory of reflections to study the fascinating optical properties of cholesteric liquid crystals which have a helical structure, and to the extensions of the molecular theory of nematic liquid crystals. In 1971 Chandrasekhar joined the Raman Research Institute (RRI) in Bangalore and set up a liquid-crystal research laboratory which became one of the leading centres of research in the world. Chandrasekhar's greatest achievement was the discovery in 1977 of a new type of liquid crystal made of molecules in the shape of discs rather than the well-studied rods. He predicted that these simple disc-like molecules can form column-like structures that show properties of translation, as in a solid, in two dimensions, and liquid-like disorder in the third dimension. The physics of this class of liquid crystals is different in many respects from the classical liquid crystals of rod-like molecules. After retiring from RRI in 1990, Chandrasekhar set up the Centre for Liquid Crystal Research in Bangalore of which he was the founder-director. Chandrasekhar was largely responsible for placing India on the international map in the field of liquid crystals. (☞ Mar. 8)

1945

The atom bomb was used in warfare for the first time. At 8:15 AM, a B29 bomber named 'Enola Gay' of the US Air Force dropped an atom bomb named 'Little Boy' on Hiroshima, Japan. The devise exploded approximately 580 metres above ground,

Hiroshima after bombing

not far from the T-shaped bridges that was its target. The destruction was incredible. In an instant, the explosion flattened virtually every structure in a 5-km radius; a City centre was literally vaporised. The ensuing fireball spread and engulfed several kilometres more of the city in fire. It has been estimated that over 80,000 people were killed at once, and more than 60,000 died later as a result of exposure to radiation from this bomb. (☞ Jul. 16, Aug. 9)

1961
The Russian cosmonaut Gherman Titov became the first person to spend a full day in orbit. After launch from the Baikanur cosmodrome, Kazakhstan, he spent 25 hours, 18 minutes in orbit on board the spacecraft *Vostok-2*, completing 17 orbits of the Earth.

1979
Feodor Lynen, German biochemist, Nobel laureate (Physiology or Medicine, 1964), died in Munich, Germany. (☞ Apr. 6)

1985
American space shuttle *Challenger*, carrying astronauts Gordon Fullerton, Roy D. Bridges, F. Story Musgrave, Anthony W. England, Karl G. Henize, Loren W. Acton and John-David F. Bartoe, landed at the Edwards Air Force Base, California, USA, after completing its 8th flight. The shuttle carried *Spacelab-2*, an orbital laboratory during the mission. (☞ Jan. 28, Apr. 4, Jul. 29)

August 7

1802
Germain Henri Hess, Swiss-born Russian chemist whose studies of heat in chemical reactions formed the foundation of thermochemistry, was born in Geneva, Switzerland. Hess studied at the University of Dorpat (now Tartu, Estonia), and in Stockholm, Sweden, under chemist Jöns Jacob Berzelius. He then took part in a geological expedition to the Urals before setting up a medical practice in Irkutsk, Russia. After practising medicine for several years in Irkutsk, Hess became professor of chemistry in 1830 at the Technological Institute, University of St. Petersburg. His early investigations concerned minerals and the natural gas found near Baku. Hess formulated an empirical law called the 'law of constant heat summation' (also known as the Hess' Law) in 1840, which states that the heat evolved or absorbed in a chemical process is the same whether the process takes place in one or in several steps. It is explained by thermodynamic theory, which holds that enthalpy is a state function. Chemists have made great use of the Hess' Law in establishing the heats of formation of compounds that are not easily formed from their constituent elements. In 1842 Hess proposed his second law, the 'law of thermoneutrality', which states that in exchange reactions of neutral salts in aqueous solution, no heat effect is observed. (☞ Nov. 30)

1848
Jöns Jacob Berzelius, Swedish scientist, one of the founders of modern chemistry, especially noted for his determination of atomic weights, the development of modern chemical symbols and his electrochemical theory, died in Stockholm, Sweden. (☞ Aug. 20)

1903
Louis Seymour Bazett Leakey, Kenyan archaeologist and anthropologist, was born in Kabete, Kenya, of English missionaries parents. Leakey was educated at the University of Cambridge in England. He began his archaeological research and led fossil-hunting expeditions to eastern Africa in 1924. In 1931 he began his research at

Olduvai Gorge in Tanzania, which became the site of his most famous discoveries. Leakey's discoveries formed the basis for the most important subsequent research into the earliest origins of human life. Leakey was largely responsible for convincing scientists that Africa, rather than Java or China, was the most significant area to search for evidence of human origins. In 1964, on an expedition to the Olduvai Gorge in Tanzania, he found fossil remains of, what he believed to be the earliest member of the genus of human beings. He named the species *Homo habilis*. Leakey's discoveries formed the basis for the most important subsequent research into the earliest origins of human life. (☞ Oct. 1)

1925
Monkombu Sambasivan Swaminathan, Indian agricultural scientist, Fellow of the Royal Society, was born in Kumbakonam, Tamil Nadu, India. Swaminathan holds two bachelors of science degrees from the Travancore University and the Madras University as well as a PhD from the School of Agriculture, University of Cambridge, England. He became the director of IARI in 1966 and the director-general of the Indian Council of Agricultural Research in 1972. Swaminathan brought into India seeds developed in Mexico by American agricultural scientist Norman Borlaug and, after crossbreeding them with local species, created a wheat plant that yielded much more grain than traditional types, thus starting the Green Revolution across Southeast Asia.

Swaminathan has been acclaimed by *Time* magazine as one of the twenty most influential Asians of the 20th century. He has also been in the forefront of the gene conservation campaign and was instrumental in building up the wheat and rice germplasm collections at IARI, New Delhi. He was awarded the Ramon Magsaysay Award for Community Leadership in 1971. In 1982, Swaminathan was appointed director-general of the International Rice Research Centre, Manila, Philippines. In July 1988, he established the M.S. Swaminathan Research Foundation in Chennai, as a non-profit organisation with the goals of preserving regional ecosystems, developing innovative technology to help realise sustainable agriculture, and promoting the application of ecotechnology in agricultural communities.

1946
John Cromwell Mather, American astrophysicist and cosmologist, Nobel laureate, was born in Roanoke, Virginia, USA. Mather did his BA in Physics in 1968 from the Swarthmore College, Philadelphia and received his PhD from the University of California, Berkeley in 1974. For the next three years he was NRC Postdoctoral Fellow at the Columbia University Goddard Institute for Space Studies in New York, during which he led the proposal efforts on the Cosmic Background Explorer (*COBE*) project. The success of *COBE*

was the outcome of prodigious teamwork involving more than 1,000 researchers, engineers and other participants. Mather coordinated the entire process and also had primary responsibility for the experiment that revealed the blackbody form of the microwave background radiation measured by *COBE*. University of California physics professor, George Smoot had the main responsibility for measuring the small variationsinthetemperatureoftheradiation.

Mather is at present a senior astrophysicist in the Observational Cosmology Laboratory at NASA's Goddard Space Flight Center (GSFC) in Maryland. His research centres on infrared astronomy and cosmology. He came to GSFC in 1976 to be the study scientist. In 1988 he became the project scientist and also the principal investigator for the far infrared absolute spectrophotometer (FIRAS) on *COBE*. He showed that the cosmic microwave background radiation has a blackbody spectrum within 50 ppm. As senior project scientist for the James Webb Space Telescope, he leads the science team, and represents scientific interests within the project management. He has served on advisory and working groups for the National Academy of Sciences, NASA, and the National Science Foundation (for the ALMA, the Atacama Large Millimetre Array, and for the CARA, the Centre for Astrophysical Research in the Antarctic). He is a member of the Astrophysics Subcommittee of the NASA Advisory Committee and of the Standing Review Board for the Kepler project. Mather was awarded the 2006 Nobel Prize for Physics, jointly with George F. Smoot, for their discovery of the blackbody form and anisotropy of the cosmic microwave background radiation. (☞ Feb. 20, Nov. 18)

1947
The balsa raft *Kon Tiki* with the Norwegian ethnologist and adventurer Thor Heyerdahl and five companions on board smashed into the reef at Raroia in the Tuamotu Islands in Polynesia after sailing from Peru across the Pacific Ocean for 101 days, covering almost 6,900 kilometres. The voyage demonstrated that the ancient Polynesians could in the same way have originated from South America. Heyerdahl and his companions had modern equipment such as radio and military expedition food, but found they could live off the fruits of the ocean, as the raft attracted lots of marine life. (☞ Apr. 28)

1971
The command module of *Apollo-15*, carrying astronauts Alfred M. Worden, James B. Irwin, and David R. Scott, who became the first person to drive a vehicle on the Moon, splashed down safely in the Pacific Ocean. (☞ Jul. 26, Jul. 30, Jul. 31)

1976
Amerian unmanned spacecraft *Viking-2* went into orbit around Mars. (☞ Sep. 3, Sep. 9)

August 8

1576
Tycho Brahe, Danish astronomer, began work in his observatory at Uraniborg, the first purpose-built observatory ever constructed. The observatory was built on the little island of Hven off the coast of southern Sweden, near Copenhagen.

It was the first of the modern observatories in that it was completely supported by the state and produced the first organised, extensive array of dependable data in astronomical history, including a catalogue of more than 1,000 stars. Brahe designed and built new instruments, calibrated them. He instituted nightly observations and trained a generation of young astronomers there in the art of observing. (☞ Oct. 24, Dec. 14)

1876

Thomas Alva Edison, American inventor, was awarded a patent for his mimeograph machine, or stencil duplicator. It was a document-copying machine that was a precursor to the modern photocopy machine. It used waxed paper 'stencils', which were placed in a typewriter to create the original with the typewriter cutting through the paper. The stencil was wrapped around the drum of the (manual or electrical) duplicating machine, which forced ink out through the cut marks on the stencil, making copies on paper. (☞ Jan. 27, Feb. 11, Feb. 19, Apr. 15, Aug. 9, Aug. 27, Oct. 18, Oct. 21, Dec. 6)

1897

Viktor Meyer, German organic chemist best known for the method of determining vapour densities named after him, died in Heidelberg, Germany. (☞Sep. 8)

1901

Ernest Orlando Lawrence, American physicist, Nobel laureate, was born in Canton, South Dakota, USA. Lawrence received his BA degree in chemistry from the University of South Dakota, Vermillion, South Dakota, in 1922. The following year he received his MA from the University of Minnesota. He spent a year at the Chicago University doing physics and was awarded a PhD from the Yale University in 1925. In 1928 he was appointed associate professor of physics at the University of California, Berkeley, and two years later he became professor. In 1936 he became director of the University's Radiation Laboratory as well, remaining in these posts until his death. Lawrence's research centred on nuclear physics. His early work was on ionisation phenomena and the measurement of ionisation potentials of metal vapours. In 1929 he invented the cyclotron, a device for accelerating nuclear particles to very high velocities without the use of high voltages, by making them move in spiralling paths under a strong magnetic field. The swiftly moving particles were used to bombard atoms of various elements, disintegrating the atoms to form, in some cases, completely new elements. Hundreds of radioactive isotopes of the known elements were also prepared using the cyclotron. Lawrence was awarded the 1939 Nobel Prize for Physics for the invention and development of the cyclotron and for results obtained with it, especially with regard to artificial radioactive elements. (☞ Aug. 27)

1902

Paul A. M. Dirac, English theoretical physicist, Nobel laureate (Physics, 1933), was born in Bristol, Gloucestershire, England. Dirac was educated at the Merchant Venturer's Secondary School, Bristol, and received the BSc (Engineering) degree from the Bristol University in 1921. He then studied mathematics for two years

at the Bristol University, later going on to St. John's College, Cambridge, from where he received his PhD in mathematics in 1926. In 1932, he was appointed Lucasian Professor of Mathematics at Cambridge. Dirac's work has been mainly concerned with the mathematical and theoretical aspects of quantum mechanics. He began work in the new field of quantum mechanics as soon as it was introduced by the German physicist Werner Heisenberg in 1928 and wrote a series of papers on the subject, published mainly in the *Proceedings* of the Royal Society. Around the 1930s, in a paper on 'the theory of holes', he proposed the existence of a positive particle having the same mass and charge as the negatively charged electron. The American physicist Carl David Anderson discovered the existence of this positive electron, called 'positron', experimentally in 1932. The importance of Dirac's work lies essentially in his famous wave equation, which introduced special relativity into Schrödinger's equation. Taking into account the fact that, mathematically speaking, the theory of relativity and quantum theory are not only distinct from each other, but also oppose each other, Dirac's work could be considered a fruitful reconciliation between the two theories. Dirac was awarded the 1933 Nobel Prize for Physics, jointly with the Austrian physicist Erwin Schrödinger, for the discovery of new productive forms of atomic theory. (☞ Aug. 12, Oct. 20)

1921

Vulimiri Ramalingaswami, Indian nutritionist, Fellow of the Royal Society, who pioneered the usage of iodised salt in India, was born in Srikakulam, Andhra Pradesh, India. Ramalingaswami did his MBBS and MD from the Andhra Medical College, Visakhapatnam, and went to England to do his DPhil and DSc from the Oxford University. Beginning his career in 1947 as a pathologist in the Indian Council of Medical Research he went on to become its director-general in 1979. He was deputy director, and then director and professor of pathology at the All India Institute of Medical Sciences, New Delhi from 1969 to 1979. He has also served as adviser to the executive director of UNICEF and to the director-general of WHO. Ramalingaswami sought to answer the fundamental questions underlying the high incidence of goitre in South Asia. As a young scientist, he traversed long distances in the Himalayan foothills and was appalled at finding that in Bangladesh and India, the proportion of malnourished children was significantly higher than in even the poorest countries of the sub-Sahara. Ramalingaswami's research showed that iodine is necessary for the production of thyroxin – a hormone, which plays a pivotal role in foetal growth and development and in childhood and adolescence. Varying degrees of thyroxin deficiency at the foetal and neo-natal stage, he found, could thwart mental growth, ranging from loss of intellectual potential to extreme forms of irreversible brain and body defects. In recognition of his research, Ramalingaswami was conferred the title of the National Research Professor of India by the Government of India. (☞ May 29)

1931

Roger Penrose, British mathematician and relativist who in the 1960s calculated

many of the basic features of black holes, was born in Colchester, Essex, England. Penrose graduated from the University College, London, with honours in mathematics and decided to go to Cambridge to undertake research in pure mathematics. He began research in algebraic geometry at St John's College and was awarded a PhD for his work in algebra and geometry from the University of Cambridge in 1957. Beginning in 1959, Penrose published a series of important papers on cosmology. In 1964 he was appointed as a reader at Birkbeck College, London, and two years later he became professor of applied mathematics there. One of Penrose's major breakthroughs was his introduction of 'twistor theory' – a remarkable mathematical theory combining powerful algebraic and geometric methods – in an attempt to unite relativity and quantum theory. Penrose is well known for his 1974 discovery of 'Penrose tilings', which are formed from two tiles that can only tile the plane aperiodically. In 1984, similar patterns were found in the arrangement of atoms in quasi-crystals. Penrose has written books such as *The Emperor's New Mind* where he argues that known laws of physics do not constitute a complete system, and that true artificial intelligence is impossible. In this controversial book, he proves the former based on claims that humans can do things outside the power of formal logic systems, such as realising the truth of unprovable statements, or solving the insoluble halting problem – deciding, given a description of a programme and finite inputs, if the programme runs forever or finishes running. Later on, Penrose followed up *The Emperor's New Mind* with *Shadows of the Mind* and *The Large, the Small and the Human Mind*, further updating and explaining his theories.

1985

Dhruva, the 100-MW research reactor at the Bhabha Atomic Research Centre in Trombay, Maharashtra, attained criticality. The indigenously designed *Dhruva* is a natural uranium-fuelled reactor, which uses heavy water both as a coolant and as a moderator. But unlike the reactors used to generate power in atomic power stations, *Dhruva* is a research reactor, which produces very little heat but is a source of high-intensity beams of neutrons that has been widely used by researchers for various kinds of studies.

1996

Frank Whittle, English aviation engineer and pilot who was a pioneer in the field of jet propulsion, died in Columbia, Maryland, USA. (☞ Apr. 12, Jun. 1)

1996

Nevill Francis Mott, English physicist, Nobel laureate (Physics, 1977), who did independent researches on the magnetic and electrical properties of amorphous semiconductors, died in Milton Keynes, Buckinghamshire, England. (☞ Sep. 30)

2001

American spacecraft *Genesis* was launched from the Cape Canaveral Air Force Station, Florida, California, USA, on an ambitious three-year mission to catch a piece of the Sun and safely return the sample to Earth. *Genesis* was designed to collect pristine material tossed off the turbulent Sun as particles of solar wind on more than 200 ceramic tiles,

before swinging back to Earth for recovery of the sample through a returning capsule. Flying some 1.5 million kilometres away from the Earth, the spacecraft spent two years cruising outside the influence of our planet's magnetic field, sweeping through pure solar wind with a set of sophisticated collectors. Scientists hoped to recover from *Genesis*, samples of the same type of matter that combined eons ago to form the Sun, eight planets and their concomitant moons as well as comets and asteroids. The Sun is thought to contain more than 99 per cent of the various matter found throughout the solar system, and material blasted off its surface is constantly carried through interplanetary space by the solar wind. The *Genesis* spacecraft was equipped with collector arrays that were designed to gather solar matter being continuously blown toward the Earth. (☞ Feb. 22)

August 9

1892

Thomas Alva Edison, American inventor, received a patent for a two-way electric telegraph that could simultaneously receive and send signals. (☞ Jan. 27, Feb. 11, Feb. 19, Apr. 15, Aug. 8, Aug. 27, Oct. 18, Oct. 21, Dec. 6)

1899

Edward Frankland, English chemist who was one of the first investigators in the field of structural chemistry died in Golaa, Norway. (☞ Jan. 18)

1910

The electric-powered washing machine was patented by Alva J. Fisher of Chicago, Illinois, USA. The machine was of a drum type with a galvanised tub and an electric motor.

1911

William Alfred Fowler, American nuclear astrophysicist, Nobel laureate, was born in Pittsburgh, Pennsylvania, USA. Fowler attended school in Lima, Ohio and studied at the Ohio State University from where he received his BS degree in 1933. In 1936 he received a PhD from the California Institute of Technology, Pasadena, California, where he became a professor in 1939. In the 1950s he developed, together with the English astrophysicists Fred Hoyle, Margaret Burbidge, and Geoffrey Burbidge, a theory of nucleosynthesis of elements in stars. They suggested that in stars, elements are synthesised progressively from lighter elements to heavy ones, by nuclear reactions that also produce light and heat. With the collapse of more massive stars, the explosive rebound known as supernova occurs; according to theory, this phase makes possible the synthesis of the heaviest elements. After 1964 Fowler worked on theoretical problems involving supernovae, gravitational collapse, neutrinos, and formation of light elements. Fowler was awarded the 1983 Nobel Prize for Physics, jointly with the Indian-born American astrophysicist Subrama-nyan Chandra-sekhar. Fowler received half the prize money for his theoretical and experimental studies of the nuclear reactions of importance in the formation of the chemical elements in the universe, while Chandrasekhar received the other half for his theoretical studies of the physical processes of importance to the structure and evolution of the stars. (☞ Mar. 14, Oct. 19)

1945

The first plutonium fission bomb, nicknamed the 'Fat Man' was dropped by US Air Force on the Japanese city of Nagasaki, which killed between 35,000 and 40,000 people, injured a like number and flattened every structure over an area of about three square kilometres. This was the last time an atom bomb was used in war. (☞ Aug. 6)

1948
Yellapragada SubbaRow, Indian-born American biochemist, who contributed to the development of synthetic folic acid, tetracycline, and anti-malarial drugs, died in New York, USA. (☞ Jan. 12)

1969
Cecil Frank Powell, English educator, physicist, Nobel laureate (Physics, 1950), who developed the photographic method of studying nuclear processes and discovered the pi-meson, died near Milan, Italy. (☞ Dec. 5)

2005
Space shuttle *Discovery*, carrying a seven-member crew comprising Eileen Collins, Jim Kelly, Charlie Camarda, Wendy Lawrence, Soichi Noguchi, Steve Robinson, and Andy Thomas, landed at the Edwards Air Force Base in California, USA, after successfully completing the historic return-to-flight mission since the loss of space shuttle *Columbia* in February 2003. (☞ Feb. 1, Jul. 26)

2006
James Alfred Van Allen, American physicist who discovered the zones of radiation encircling the Earth named after him, died in Iowa City, Iowa, USA. (☞ Sep. 7)

August 10

1897
Aspirin was produced for the first time by Felix Hoffmann, German chemist and an employee of the German dyestuffs company Bayer, by acetylating the phenol group of a compound called salicylic acid. Chemically known as acetylsalicylic acid, aspirin was the world's first truly synthetic drug and probably the most successful medicine in history that paved the way for the modern pharmaceuticals industry.

1902
Arne Wilhelm Kaurin Tiselius, Swedish biochemist, Nobel laureate, was born in Stockholm, Sweden. Tiselius studied at the University of Uppsala, specialising in chemistry. He became research assistant to the Swedish chemist Theodore Svedberg in 1925 when he developed the use of electrophoresis for the delicate task of separating proteins in suspension on the basis of their electrical charge. In electrophoresis an electrical field is applied to the sample, and particles with different sizes migrate at different rates to the pole of opposite charge, enabling them to be detected and identified. After obtaining his doctorate in 1930, Tiselius was appointed assistant professor in chemistry at the University of Uppsala. In 1934,

with support of a Rockefeller Foundation Fellowship, he went to USA to conduct research at the Institute for Advanced Study at Princeton, New Jersey. After his return to Uppsala in 1937, he became a professor of biochemistry and resumed his interest in proteins, and the application of physical methods to biochemical problems in general. This led among other things to a much-improved method of electrophoretic analysis. Using the technique on blood serum Tiselius confirmed the existence of four different groups of proteins – albumins and alpha, beta, and gamma globulins. He also conducted work on chromatography, partition and gel filtration. Tiselius was awarded the 1948 Nobel Prize for Chemistry for his research on electrophoresis and adsorption analysis, especially for his discoveries concerning the complex nature of serum proteins. (☞ Oct. 29)

1910

American pilot James McCurdy flies the first radio-equipped aircraft and sends the first aircraft-to-ground wireless message in New York, USA.

1913

Wolfgang Paul, German physicist, Nobel laureate, was born in Lorenzkirch, Germany. Paul studied at technological institutes in Munich and Berlin and received a doctorate in physics from the Technical University in Berlin in 1939. From 1944 onwards, he taught for a few years at the University of Göttingen. In 1952 he was appointed professor of experimental physics at the University of Bonn and director of the Physics Institute. From 1965 to1967 he was director of the division of nuclear physics at the European Organization for Nuclear Research (CERN). While in Bonn, Paul and his students started activities in new fields that included molecular beam physics, mass spectrometry and high-energy electron physics. They also developed focussing methods in molecular beam physics with quadrupole and sextupole lenses. It was during this period that the quadrupole mass spectrometer and the ion trap, also known as the 'Paul trap', were conceived and studied. The Paul trap is an electromagnetic device that captures ions and holds them long enough for study and for precise measurement of their properties, which in turn allows physicists to study atomic properties and test physical theories with high degrees of precision. Paul also invented a way of separating ions of different masses and storing them in the ion trap, using a principle that was subsequently widely applied in modern spectrometers. Paul was awarded the 1989 Nobel Prize for Physics, jointly with Hans G. Dehmelt and Norman F. Ramsey. Paul and Dehmelt shared half of the prize money for the development of the ion trap technique, while Ramsey received the other half for the invention of the separated oscillatory field's method and its use in the hydrogen maser and other atomic clocks. (☞ Aug. 27, Sep. 9, Dec. 6)

1915

Henry Gwyn Jeffreys Moseley, English physicist who firmly established the relationship between atomic number and the charge of the atomic nucleus, was killed in battle at Gallipoli, Turkey. (☞ Nov. 23)

1927

Manali Kallat Vainu Bappu, Indian astrophysicist, and the main architect of the revival of astronomical studies in India, was born in Madras (now Chennai), Tamilnadu, India. Young Bappu inherited from his father, who was an astronomer at the Nizamiah

Observatory, Hyderabad, a deep fascination for the wide and mysterious cosmos. Vainu Bappu graduated from the Nizamia College, Hyderabad in 1946, from where he also obtained his Master's degree in physics in 1949. His research contributions started right during this period, and *Current Science* published his first papers on visual observations of variable stars in 1946. He also obtained at the young age of nineteen, spectra of the night sky requiring exposures of several nights, using a spectrograph of his own design. After obtaining his MSc, Bappu earned a state scholarship for higher studies in astrophysics at the Harvard University, Cambridge, Massachusetts in USA. He got his PhD from Harvard in 1952. He then spent a year at Mount Wilson and Palomar Observatories as a Carnegie Post-Doctoral Fellow, during which he jointly discovered an important phenomenon in stellar chromospheres – what came to be known as the 'Wilson–Bappu Effect.' Bappu returned to India in 1953 to lead a team of scientists in establishing a modern observatory at Nainital, Uttar Pradesh (now in Uttaranchal). He left Nainital in 1960 to take over as director of the Kodaikanal Observatory, which he transformed into an active centre of astronomical research. Bappu was elected vice-president of the International Astronomical Union in 1967, and became its president in 1979. (☞ Jan. 7, Aug. 19)

1945

Robert Hutchings Goddard, American professor, physicist and inventor, who pioneered contemporary space exploration, died in Baltimore, Maryland, USA. (☞ Oct. 5)

1966

American spacecraft *Lunar Orbiter-1* was launched on a mission to photograph smooth areas of the lunar surface in order to select safe landing sites for the *Apollo* missions. The spacecraft was also equipped to collect data on radiation intensity and

micrometeoroid impact on the Moon. The spacecraft photographed the Moon from August 18 to 29. In all, five *Lunar Orbiter* missions were sent and all were successful, photographing 99% of the Moon's surface with a resolution of 60 m or better. (☞ Aug. 23)

1990

The *Magellan* spacecraft, named after the sixteenth-century Portuguese explorer whose expedition first circumnavigated the Earth, went into orbit around Venus. The spacecraft carried a sophisticated imaging radar, which was used to make the most highly detailed map of Venus ever captured during its four years in orbit around Venus from 1990 to 1994. After concluding its radar mapping, *Magellan* also made global maps of Venus's gravity field. (☞ May 4, Sep. 15)

August 11

1854

Macedonio Melloni, Italian physicist, who was the first to show the similarity between infrared radiation and light, died in Naples, Italy. (☞ Apr. 11)

1858

Christiaan Eijkman, Dutch scientist, physician, Nobel laureate, was born in Nijkerk, Netherlands. Eijkman received his early education at Zaandam, and in 1875, after undergoing preliminary examinations, he became a student at the Military Medical School of the University of Amsterdam. Here he was trained as a medical officer for the Netherlands Indies Army, passing through all his examinations with honours. He received his doctorate from the University of Amsterdam in 1883 and went to Indonesia as a medical officer of health at Tjilatjap, a small village on the south coast of Java. He caught malaria, which later so impaired his health that in 1885 he had to return to Europe on sick-leave. He then worked with the German physician Robert Koch in Berlin on bacteriological research and in 1886 returned to Java to investigate the cause of beriberi (Sinhalese for 'extreme weakness'). The main symptoms of beriberi include tingling, or loss of sensation in hands and feet, wasting of muscle with loss of function or paralysis of the lower extremities, and potential brain damage and death. Eijkman accidentally discovered that beriberi is basically produced by a diet of polished rice, which lacked some vital substance. Experimenting with polished rice he found that the vital substance is found in the thin coating of unpolished rice, which is removed during polishing. The essential substance was later identified as vitamin B_1 (thiamine). Eijkman was awarded the 1929 Nobel Prize for Physiology or Medicine, jointly with the British biochemist Frederick Gowland Hopkins, for his discovery of the anti-neuritic vitamin. (☞ Jun. 20, Nov. 5)

1877

Asaph Hall, American astronomer, discovered a moon of Mars, which he named Deimos (Greek for 'flight') after the attendant of Mars mentioned in Homer's *Iliad*. The moon is composed of carbon-rich rock like C-type minor planets and ice. But its density is so low that it cannot be pure rock. The moon is heavily cratered. Deimos is probably a minor planet perturbed by Jupiter into an orbit that allowed it to be captured by Mars. (☞ Aug. 16, Oct. 15)

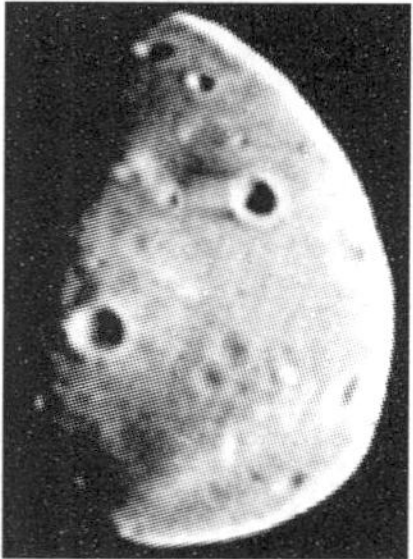

1926

Aaron Klug, British chemist, Nobel laureate, was born in Zelvas, Lithuania. Klug studied at the Durban High School in Durban, South Africa, where the family had emigrated soon after his birth. He graduated from the University of Witwatersrand at Johannesburg with a science degree and did his MSc from the University of Cape Town. In 1949, he went to Trinity College, Cambridge, England, on a fellowship and completed his doctorate in 1953. He then joined the Birkbeck College of the University of London on a research fellowship and took up the study of the structure of tobacco mosaic virus and other viruses using a technique he developed in which a series of electron micrographs, taken of two-dimensional crystals from different

angles, could be combined to produce three-dimensional images of particles. In four short years, Klug and his team were able to map out the general outline of the three-dimensional structure of tobacco mosaic virus. In 1958 Klug became director of the Virus Structure Research Group at Birkbeck College. In 1962, at the invitation of the British biophysicist Francis Crick, Klug returned to Cambridge as a staff member of the Medical Research Council. Klug was awarded the 1982 Nobel Prize for Chemistry for his development of crystallographic electron microscopy and his structural elucidation of biologically important nucleic acid–protein complexes.

1941
Madabusi Santanam Raghunathan, Indian mathematician, Fellow of the Royal Society, was born in Anantapur, Andhra Pradesh, India. Raghunathan completed his schooling in Madras (now Chennai) and completed his Intermediate education at the St. Joseph's College, Bangalore, Karnataka during 1955–57. He then returned to Chennai and joined the Vivekananda College to study BA (Hons) in mathematics. After completing his BA (Hons) in 1960, he joined the Tata Institute of Fundamental Research as a research assistant at the age of 19. After initial training during 1960–62, he worked on a research problem suggested by M.S. Narasimhan, on 'Deformations of linear connections and Riemannian metrics', and solved it by the summer of 1963. Raghunathan completed his PhD from the University of Bombay in 1966 and then spent a year at the Institute for Advanced Study, Princeton, New Jersey, USA, for post-doctoral research. He was promoted to the faculty position of associate professor at TIFR at the young age of 25, and rose to the highest academic position of professor of eminence in 1987. Discrete subgroups of Lie groups have been the central objects of his researches. He is well-known for the conjecture named after him which has been influential in the study of dynamics of a class of flows and their applications in various areas.

1972
Max Theiler, South African-born American microbiologist, Nobel laureate (Physiology or Medicine, 1951), whose research on yellow fever led to the development of the first attenuated strain of the virus, died in New Haven, Connecticut, USA. (☞ Jan. 30)

1978
Scientists at the Center of Disease Control in Atlanta, Georgia, USA successfully isolated the bacterium *Legionella pneumophila* that causes legionnaire's disease, a severe form of infection which includes pneumonia. The name of the disease (and of the bacterium) is derived from a 1976 state convention of the American Legion, a US war veteran's organisation, at a Philadelphia hotel where 182 Legionnaires contracted the disease, 29 of them fatally.

August 12

1848
George Stephenson, English engineer and principal inventor of the railroad locomotive, died in Chesterfield, Derbyshire, England. (☞ Jun. 9)

1851
Isaac Merrit Singer, American inventor was issued a patent for the first practical domestic

sewing machine. However, because Singer had used the basic eye-pointed needle and the lock switch that were developed by fellow American inventor Elias Howe, the patent was challenged and Howe was awarded the name of the inventor of the sewing machine. (☞ Sep. 9)

1887

Erwin Schrödinger, Austrian theoretical physicist, Nobel laureate, was born in Vienna, Austria. Schrödinger studied at the University of Vienna from 1906 and received his doctorate in physics in 1910. It was during this period that he acquired a mastery of eigenvalue problems in the physics of continuous media, thus laying the foundation for his future great work. Beginning in 1920 he took up teaching positions at the Universities of Stuttgart, Breslau, and Zurich, where he settled for six years and produced his most valuable papers that form the foundations of quantum wave mechanics. In those papers he described his partial differential equation that is the basic equation of quantum mechanics. Adopting a proposal made by the French physicist Louis de Broglie in 1924 that particles of matter have a dual nature and in some situations act like waves, Schrödinger introduced a theory describing the behaviour of such a system by a wave equation that is now known as the Schrödinger's equation. The solutions to Schrodinger's equation are wave functions that can only be related to the probable occurrence of physical events, thus bringing in the more abstract notion of probability. In 1927 Schrödinger moved to Berlin, Germany, but left in 1933 to join the faculty of the University of Graz in Austria. In 1938 he moved to the newly created Institute for Advanced Studies in Dublin, Ireland, where he became director of the School for Theoretical Physics. He remained in Dublin until his retirement in 1955. In 1956 he returned to Vienna as professor emeritus at the university. Schrödinger was awarded the 1933 Nobel Prize for Physics, jointly with the British theoretical physicist Paul Adrien Maurice Dirac, for the discovery of new productive forms of atomic theory. (☞ Jan. 4, Aug. 8)

1919

Vikram Ambalal Sarabhai, Indian physicist, space scientist, was born in Ahmedabad, Gujarat, India. Sarabhai had his early education at a private school run by his parents. After passing Inter-Science from the Gujarat College in Ahmedabad, he joined the St. John's College, Cambridge, England, where he took his Tripos in natural sciences in 1939. On his return to India, Sarabhai took up cosmic ray research under the guidance of the Indian physicist C V. Raman at the Indian Institute of Science in Bangalore. He built his own equipment with which he took measurements at Bangalore, Poona and the Himalayas. He returned to Cambridge in 1945 and received his PhD in 1947 for his thesis, 'Cosmic ray investigations in tropical latitudes'. Sarabhai founded the Physical Research Laboratory in Ahmedabad in 1947 where he continued his scientific activities. In 1962, when the Department of Atomic Energy created the Indian National Committee for Space Research (INCOSPAR), Sarabhai was appointed its Chairman. He directed the setting up of the country's first rocket launching station at Thumba near

Thiruvananthapuram, Kerala, and initiated a programme to fabricate sounding rockets in India. In 1966, he was appointed the Chairman of the Atomic Energy Commission. Sarabhai was associated with India's first satellite TV experiment called the 'Satellite Instructional Television Experiment' (SITE), carried out during 1975–76, in which specially tailored television programmes were directly beamed to 2,400 specially augmented TV sets in six village clusters across the country. Sarabhai took keen interest in encouraging scientific temperament in the young and established the Community Science Centre in Ahmedabad for popularising science among children. (☞ Dec. 31)

1948
Harry Brearley, English metallurgist who invented stainless steel, died in Torquay, England. (☞ Feb. 18)

1955
James Batcheller Sumner, American biochemist, Nobel laureate (Chemistry, 1946), who first crystallised enzymes, showing their protein nature, died in Buffalo, New York. (☞ Nov. 19)

1960
The first telecommunications satellite, *Echo-1*, was launched by the National Aeronautics and Space Administration (NASA) of USA from Cape Canaveral, Florida. The satellite was a large metallised balloon designed to act as a passive communications reflector to bounce communi-cation signals transmitted from one point on Earth to another. The satellite was successfully used to re-direct transcontinental and intercontinental telephone, radio, and television signals.

1973
Walter Rudolf Hess, Swiss physiologist, Nobel laureate (Physiology or Medicine, 1949), who developed techniques for studying the brain using implanted electrodes, died in Locarno, Switzerland. (☞ Mar. 17)

1973
Karl Ziegler, German chemist, Nobel laureate (Chemistry, 1963), who developed organometallic catalysts for synthesis of high polymers, died in Mülheim, Germany. (☞ Nov. 26)

1979
Ernst Boris Chain, German-born British biochemist, Nobel laureate (Physiology or Medicine, 1945), who re-discovered penicillin and studied its chemotherapeutic action, died in Ireland. (☞ Jun. 19)

1989
William B. Shockley, English-born American engineer and teacher, Nobel laureate (Physics, 1956), co-inventor of the transistor, died in Palo Alto, California, USA. (☞ Feb. 13)

2005
The American space Agency, NASA launched *Mars Reconnaissance Orbiter* on a mission to Mars. The main objective of the mission is to search for evidence that water persisted on the surface of Mars for a long period of time. While other Mars missions have shown that water

flowed across the surface in Mars' history, it remains a mystery whether water was ever around long enough to provide a habitat for life. After a seven-month cruise to Mars and six months of aerobraking, *Mars Reconnaissance Orbiter* reached its desired orbit in 2006. With its sophisticated science instruments, it will seek to find out the history of water on Mars. These studies would help determine if there are deposits of minerals that form in water over long periods of time, detect any shorelines of ancient seas and lakes, and analyse deposits placed in layers over time by flowing water. (☞ Mar. 10)

August 13

1814

Anders Jonas Ångström, Swedish physicist, one of the founders of spectroscopy, was born in Lögdö, Sweden. Ångström was educated at the University of Uppsala, where he became a lecturer in 1839. He was appointed keeper of the Uppsala Astronomical Observatory in 1843 and became professor of physics there in 1858. He used a combination of spectroscopy with photography to prove that the Sun contains hydrogen. In 1867 he was the first to examine the spectrum of aurora borealis (northern lights). He published his extensive research on the solar spectrum including detailed measurements of more than 1000 spectral lines in the classic work *Recherches sur le spectre solaire* in 1868. In a paper presented to the Stockholm Academy in 1853, he pointed out that an electric spark yields two superimposed spectra, one from the metal of the electrode and the other from the gas in which it passes, and deduced that an incandescent gas emits luminous rays of the same refractive capacity as those which it can absorb. He also published works on thermal theory and carried out geomagnetic measurements in different places around Sweden. The Angstrom unit (1 Å = 10^{-9} m), used in spectroscopy and for measurements on atomic scales, is named in his honour. (☞ Jun. 21)

1826

René Théophile Hyacinthe Laënnec, French physician who invented the stethoscope and is generally considered the father of chest medicine, died at Kerlouanec, France. (☞ Feb. 17)

1872

Richard Willstätter, German chemist, Nobel laureate, was born in Karlsruhe, Germany. Willstätter attended the technical high school in Munich and obtained his doctorate from the University of Munich, on a dissertation on the structure of cocaine, in 1894. He taught at the University of Munich and the Technishe Hochschule in Zürich in Switzerland, before becoming director of the Kaider Wilhelm Institute in Berlin in 1912, where his investigations revealed the structure of many of the pigments of flowers and fruits. In 1916 he returned to Munich and taught at the university till he resigned in 1925. Willstätter made his most important research contributions between the years 1905 and 1912 when he was a young professor at Zürich, working on the green plant pigment chlorophyll. He discovered the constituents of

chlorophyll to be two closely related substances containing magnesium, pyrrole residues, and a long-chain alcohol, phytol. He also showed that the blood pigment haeme bears a structural resemblance to the porphyrin compound found in chlorophyll, with the magnesium atom in chlorophyll occupying a similar position as the iron atom in haeme, the coloured component of haemoglobin. Willstätter was awarded the 1915 Nobel Prize for Chemistry for his researches on plant pigments, especially chlorophyll. (☞ Aug. 3)

1888
John Logie Baird, Scottish engineer, the first man to televise pictures of moving objects, was born in Helensburgh, Dunbarton, Scotland. Baird showed early signs of ingenuity – rigging up a telephone exchange to connect his bedroom to those of his friends across the street. His studies at the Glasgow University were interrupted by the outbreak of war in 1914. Later he tried his hand at business, setting up a jam factory and a soap project in Trinidad, which made little headway. After returning to Britain in 1922, he applied himself to creating a television. His first crude apparatus included a motor set on a tea chest, a projection lamp housed in a biscuit tin, and scanning discs cut out of cardboard; he also utilised cheap cycle lenses. Scrap-wood, darning needles, string, and sealing wax held the apparatus together. By 1924 Baird managed to transmit the flickering image of a Maltese cross across a few metres and in January 1926, before an audience of some fifty scientists, he gave the world's first demonstration of a true television in his attic workshop. (☞ Jan. 27, Feb. 9, Feb. 17, Jun. 14, Jul. 3)

1889
The first coin operated public telephone, invented by William Gray, was installed at a bank in Hartford, Connecticut, USA. He got a patent for his invention in 1891.

1894
The discovery of the rare gas argon was announced at a meeting of the British Association in London, England, by the English physical scientist Lord Rayleigh and Scottish chemist William Ramsay. Rayleigh had asked chemists to explain the difference between the atomic weight of nitrogen found in chemical compounds and the heavier, pure nitrogen found in the atmosphere. Ramsay and Rayleigh exchanged information about their individual experimental results daily to satisfy their curiosity of the unknown gas. Rayleigh determined that a litre of pure nitrogen weighs 1.2505 grams, but a litre of nitrogen gas extracted from air, by removing the oxygen, the carbon dioxide, and the water vapour, was slightly heavier and weighed 1.2572 grams. This led to Ramsay's prediction that there must be another unknown element. The two scientists devised a chemical reaction in which nitrogen extracted from air was repeatedly passed over heated magnesium. This removed all of the real nitrogen in the form of solid magnesium nitride, leaving behind the unknown gas. They named the remaining gas, which was approximately 1% of air, Argon from the Greek word for lazy or inactive. (☞ Oct. 2, Nov. 12)

1902
Felix Wankel, German engineer and inventor of the Wankel rotary engine, was born in Lahr, Germany. Wankel did not receive any formal school education. From 1921 to 1924, he worked at the University bookshop in Heidelberg, dealing mainly in scientific publications. Suffering a kidney

complaint during his time there, he was transferred from the printing room away from the lead type, and positioned in the storeroom where he quickly devised a new and efficient method to stack books very high without risk of the pile collapsing. He did this using an unconventional arrangement by which the stability of the stack was maximised. In 1924, he began his work on the rotary engine while still in the employ of the university bookshop. He obtained his first patent on the rotary engine in 1929. Between the years 1926 and 1945, he worked as a sealing specialist at the German Aeronautical Research Establishment attaining the development work on rotary valves. In 1951, he began working with the research department of a German motor manufacturer, NSU Motorenwerk AG at Neckarsulm. In 1954, after much experimentation, he completed the design of a rotary–piston engine – the first rotary internal combustion engine. The year after, he successfully produced a model with a curved equilateral triangular rotor in a figure-of-eight-shaped chamber. The advantages of the Wankel engine include lightweight, few moving parts and compactness. By 1957, the first unit was tested. Mazda, a Japanese automobile company, produced and developed the Wankel engine. Despite its many advantages, the Wankel engine has not been adopted on a large-scale – it has a few shortcomings such as improper sealing of the rotor. In recognition of his contributions to engineering science, Wankel was awarded an honorary doctorate from the Technical University of Munich in 1969. (☞ Oct. 9)

Wankel rotary engine

1903

The journal *Nature* reported that helium gas is produced by the radioactive decay of radium. This key discovery by the British chemists William Ramsay and Frederick Soddy helped to reveal the structure of atoms. As early as 1908, the British physicist Ernest Rutherford had confirmed that alpha rays and these emanations from radium were one and the same: the nuclei of helium atoms, bearing a positive electrical charge.

1912

Salvador E. Luria, Italian-born American biologist, Nobel laureate, was born in Turin, Italy. Luria obtained his MD from the University of Turin in 1935. From 1938 to 1940 he was a research fellow at the Radium Institute in Paris, France. In 1940 he went to USA as a research assistant in surgical bacteriology at the Columbia University, New York. In 1950 he was appointed professor of microbiology at the University of Illinois, Urbana-Campaign, from where he moved to the Massachusetts Institute of Technology in 1959 to become professor of microbiology. Soon after his arrival in USA, Luria met the German-born American biologist Max Delbrück, through whom he became involved in solving problems of viral self-replication. In 1942 he obtained one of the electron micrographs of phage particles, which confirmed earlier descriptions of them as consisting of a round head and a thin tail. In 1943 Luria and Delbrück published a paper showing that, contrary to

the current view, viruses undergo permanent changes in their hereditary material. They also devised the 'fluctuation test', which provided experimental evidence that phage-resistant bacteria were the result of spontaneous mutations rather than a direct response to changes in the environment. Luria was awarded the 1969 Nobel Prize for Physiology or Medicine, jointly with Delbrück and the American biologist Alfred Day Hershey, for their discoveries concerning the replication mechanism and the genetic structure of viruses. (☞ Sep. 4, Dec. 4)

1917
Eduard Buchner, German biochemist, Nobel laureate (Chemistry, 1907), who explained cell-free alcoholic fermentation of sugar, died in Focsani, Romania. (☞ May 20)

1918
Frederick Sanger, English biochemist, Nobel laureate, was born in Rendcombe, Gloucestershire, England. Sanger was educated at Bryanston School and at St. John's College, Cambridge, where he took his BA degree in natural sciences in 1939. He obtained a PhD from the Cambridge University in 1943 for his thesis on the metabolism of the amino acid lysine. Since 1943 his work has been largely concerned with problems related to the determination of the structure of proteins. He worked continuously on biochemical research at Cambridge until 1951, when he began working under the auspices of the Medical Research Council. He spent 10 years elucidating the structure of the bovine insulin molecule, and by 1955 he had determined the exact order of all that molecule's amino acids. These studies resulted in the determination of the complete structure of insulin, which was an essential requirement for the laboratory synthesis of insulin. The laboratory techniques that he developed for determining the order in which amino acids are linked in proteins also opened the way toward the determination of the structure of many other complex proteins. Sanger later turned his attention to determining the sequence of nitrogenous bases in molecules of DNA and RNA, and by 1977 he had succeeded in elucidating the sequences of nucleotides in the DNA molecule of a bacteriophage (a virus that infects bacteria) called phi-X 174. To achieve this feat, he developed a technique for splitting DNA into fragments whose base sequences can then be determined. Sanger was awarded the 1958 Nobel Prize for Chemistry for his work on the structure of proteins, especially that of insulin. He also received the 1980 Nobel Prize for Chemistry, jointly with the American chemists Paul Berg and Walter Gilbert. Sanger shared half of the prize money with Gilbert for their contributions concerning the determination of base sequences in nucleic acids, while Berg received the other half for his fundamental studies of the biochemistry of nucleic acids, with particular regard to recombinant DNA. (☞ Mar. 21, Jun. 30)

1963
Sisir Kumar Mitra, Indian radio physicist, Fellow of the Royal Society, pioneer of ionospheric research in India, died in Kolkata, West Bengal, India. (☞ Oct. 24)

August 14

1777
Hans Christian Ørsted, Danish physicist and chemist who discovered that electric

current in a wire can deflect a magnetised compass needle and founded the science of electromagnetism, was born in Rudköbing, Denmark. Øersted was educated at the University of Copenhagen from where he received a degree in pharmacy in 1797. After completing his doctorate in 1799 he worked for awhile as a chemical assistant for the medical staff of the same university and then spent several years travelling in Europe. In 1806 he became professor of physics at the University of Copenhagen. In 1819, while conducting an experiment for his students, he noticed that when a compass was brought near a wire carrying electric current, the compass needle pointed in a direction at right angles to the current. He published his discovery in 1820, initiating the development of the electromagnetic theory. His contribution in the field of chemistry included the discovery of piperine, one of the pungent components of pepper, in 1920 and the preparation of metallic aluminium in 1925. In 1932 the unit of magnetic field strength was named the oersted in his honour. (☞ Mar. 9)

1807

Robert Fulton, American inventor, engineer, and artist, who invented the first practical steamboat, made his first successful steamboat trip from New York to Albany in USA. He made the trip in thirty-two hours and successfully demonstrated to the world the possibilities of steam navigation. (☞ Feb. 9, Nov. 14)

1933

Richard R. Ernst, Swiss researcher and teacher, Nobel laureate, was born in Winterthur, Switzerland. Ernst studied chemistry at the Swiss Federal Institute of Technology in Zürich and received his BA degree in chemistry in 1957 and his PhD in physical chemistry in 1962 from the same institute. From 1963 to 1968 he worked as a research chemist with Varian Associates in Palo Alto, California, USA, where his research mainly concerned nuclear magnetic resonance spectroscopy – primarily designing and building advanced electronic equipment for improved NMR spectrometers. He was also involved in developing the theoretical background for the optimum performance of these instruments. He designed and constructed high sensitivity radio frequency pre-amplifiers and in particular high sensitivity probe assemblies, initially for a 25 MHz, and later for a 75 MHz proton resonance spectrometer. In 1966, working with an American colleague, Ernst discovered that the sensitivity of NMR techniques could be dramatically increased by replacing the slow, sweeping radio waves traditionally used in NMR spectroscopy with short, intense pulses. His discovery enabled analysis of many more types of nuclei and smaller amounts of materials. His second major contribution to the field of NMR spectroscopy was a technique that enabled a high-resolution, 'two-dimensional' study of larger molecules, which enabled scientists to determine the three-dimensional structure of organic and inorganic compounds and of biological macromolecules such as proteins. Ernst returned to Zürich in 1968 to take up a teaching position at the Swiss Federal Institute of Technology, where he became assistant professor in 1970 and full professor in 1976. Ernst was awarded

the 1991 Nobel Prize for Chemistry for his contributions to the development of the methodology of high-resolution nuclear magnetic resonance (NMR) spectroscopy.

1941
Tiruppattur Venkatachalamurti Ramakrishnan, Indian physicist, Fellow of the Royal Society, was born in Madras (now Chennai). Ramakrishnan did his BSc in 1959 and MSc in 1961 from the Banaras Hindu University, Varanasi, Uttar Pradesh. He obtained his PhD from the Columbia University, New York, USA in 1966 and joined the Indian Institute of Technology, Kanpur, as lecturer, later to become professor of physics. He was professor of physics at the Indian Institute of Science, Bangalore from 1981 to 1984 and DAE Homi Bhabha Professor at the Benaras Hindu University from 1984 to 1986. He was consultant to the Bell Laboratories in USA during 1980–81. Ramakrishnan has made crucial contributions to the field of condensed matter physics. His research work helped in starting two new subfields in condensed matter physics, namely the onset of electronic localisation in disordered systems and the modern theory of liquid–solid phase transitions. His contributions to the theory of localisation and mixed valence in rare earth metals have generated a great deal of activity in the field and are of fundamental significance. He has also been involved in research on high temperature superconductivity.

1941
Paul Sabatier, French organic chemist, Nobel laureate (Chemistry, 1912), who developed a method of hydrogenating organic compounds in the presence of finely divided metals, died in Toulouse, France. (☞ Nov. 5)

1958
Frédéric Joliot-Curie, French physical chemist, Nobel laureate (Chemistry, 1935), who with his wife Irène Joliot-Curie discovered several artificially prepared, radioactive isotopes of new elements, died in Arcouest, France. (☞Mar. 19)

August 15

1892
Louis-Victor de Broglie, French physicist, Nobel laureate, was born in Dieppe, France. Broglie studied theoretical physics at the Sorbonne in Paris, but he was also earning a degree in history, to advance a career in the diplomatic service, according to family tradition. He took his degree in history in 1910 and the degree in science in 1913. But he gave up history and chose for his PhD thesis a subject in physics. In this thesis, submitted in 1924, Broglie developed his revolutionary theory of electron waves, which he had published earlier in scientific journals. He argued that since light could be seen to behave under some conditions as particles (photoelectric effect) and other times as waves (diffraction), matter should be considered as having the same ambiguity of possessing both particle and wave properties. The ideas set out in his work, which first gave rise to astonishment owing to their novelty, were subsequently fully confirmed by the American physicists Clinton Joseph Davisson and Lester

Halbert Germer's discovery of electron diffraction by crystals in 1927. Broglie's work also served as the basis for developing the general theory, nowadays known by the name of 'wave mechanics', a theory that has utterly transformed our knowledge of physical phenomena on the atomic scale. Between 1930 and 1950, Broglie's work has been chiefly devoted to the study of the various extensions of wave mechanics: Paul Dirac's electron theory, the new theory of light, the general theory of spin particles, applications of wave mechanics to nuclear physics, etc. Broglie was awarded the 1929 Nobel Prize for Physics for his discovery of the wave nature of electrons. (☞ Mar. 19)

1896

Gerty Theresa Cori, Czech-born American biochemist, Nobel laureate, was born in Prague, Czechoslovakia (now Czech Republic). Cori received her primary education at home before entering a Lyceum for girls from where she graduated in 1912. She received her Doctorate in Medicine from the Medical School of the German University of Prague in 1920. She then spent two years at the Carolinen Children's Hospital in Vienna, Austria, before emigrating to America with her husband, Carl, whom she married in 1920. They worked together in the State Institute for the Study of Malignant Diseases in Buffalo, New York, and when Carl moved to the Washington University Medical School in St. Louis, Missouri, she joined him as a research associate. Gerty Cori was made professor of biochemistry in 1947. The Coris first studied the fate of sugar in the animal body and the effects of insulin and epinephrine. In 1936, they isolated glucose-1-phosphate, known as the 'Cori ester', and traced its presence to the activity of the phosphorylase, which catalyses the breakdown and synthesis of polysaccharides: this discovery made possible the enzymatic synthesis of glycogen and starch *in vitro*. Six years later they isolated and purified the enzyme responsible for catalysing the glycogen–Cori ester reaction, and with it they achieved the test-tube synthesis of glycogen in 1943. Gerty Cori was awarded the 1947 Nobel Prize for Physiology or Medicine, jointly with his husband Carl and the Argentinean physiologist Bernardo Houssay, for their discovery of the course of the catalytic conversion of glycogen. (☞ Apr. 10, Oct. 26, Dec. 5)

1969

Indian Space Research Organisation (ISRO) was formed under the Department of Atomic Energy. ISRO is India's premier agency that is entrusted with organising and executing all national space activities, including the development, launch and operation of rockets and satellites and their applications. It operates through a network of more than half-a-dozen centres spread over the country, which take care of all its technological and operational needs. The Vikram Sarabhai Space Centre (VSSC) in Thiruvananthapuram, Kerala, is the lead centre for all rocket and launch vehicle development. The ISRO Satellite Centre (ISAC) in Bangalore, Karnataka, is the centre where all Indian satellites are designed and built. The tracking and control of all Indian satellites except the *INSAT*s are carried out by the ISRO Telemetry, Tracking and Command Network (ISTRAC) in Bangalore, while the *INSAT* Master Control Facility (MCF) at Hasan, Karnataka, controls and operates the *INSAT* satellites. Research and development on application of space technology are carried out at ISRO's Space Applications Centre (SAC), Ahmedabad, Gujarat. Propellants for satellite launching rockets are manufactured at the Propellant Fuel Complex at Thumba, Kerala, and at the Liquid Propulsion Systems Centre (LPSC) at Thiruvananthapuram. Liquid propulsion test facilities are located at Mahendragiri, Tamilnadu. ISRO's main satellite launch facility is the Satish Dhawan Space Centre at Sriharikota, Andhra Pradesh. The Thumba Equatorial Rocket Launching Station (TERLS) in Thumba, Kerala, is an international facility mainly used for launching rockets for scientific research.

1982
Axel Hugo Teodor Theorell, Swedish biochemist, Nobel laureate (Physiology or Medicine, 1955), who discovered the nature and mode of oxidation enzymes, died in Stockholm, Sweden. (☞ Jul. 6)

2006
Voyager-1, launched by NASA in September 1977, reached 100 astronomical units from the Sun; that is, 100 times the distance between Earth and the Sun. In more common terms, the spacecraft was about 15 billion kilometres from the Sun. *Voyager-1* is now at the outer edge of our solar system, in an area called the 'heliosheath', the zone where the Sun's influence wanes. This region is the outer layer of the 'bubble' surrounding the Sun – no one knows how big this bubble actually is. Travelling at a speed of about one and a half million kilometres per day, *Voyager-1* could cross into interstellar space within the next 10 years. (☞ Mar. 7, Nov. 5, Sep.5, Nov. 12)

August 16

1705
Jacob Jacques Bernoulli, Swiss mathematicians who introduced the first principles of the calculus of variation, died in Basel, Switzerland. (☞ Jan. 6)

1845
Gabriel Lippmann, French physicist, Nobel laureate, was born in Hollerich, Luxembourg. Lippmann grew up in Paris and despite the fact that he never qualified as a teacher, in 1883, he was appointed professor of mathematical physics at the Sorbonne. Three years later he was appointed head of the Sorbonne's Laboratories of Physical Research. Lippmann made many valuable fundamental contributions to many different branches of physics, especially electricity, thermodynamics, optics and photochemistry. He studied the relationship between electrical and capillary phenomena: this led to the development, amongst other instruments, of his extraordinarily sensitive capillary electrometer.

In 1891 Lippmann revealed a revolutionary colour photography process, later called the Lippmann process, which utilised the natural colours of light wavelengths instead of using dyes and pigments. He placed a reflecting coat of mercury behind the emulsion of a panchromatic plate. The mercury reflected light rays back through the emulsion to interfere with the incident rays, forming a latent image that varied in depth according to each ray's colour. The development process then reproduced the image, and the result, when viewed, was brilliantly accurate. But Lippmann's direct method of colour photography was slow and tedious because of necessarily long exposure times, and no copies of the original could be made. It never achieved popularity, but it was an important step in the development of colour photography. His other inventions include the 'coleostat' for steady, long-exposure sky photography. Lippmann was awarded the 1908 Nobel Prize for Physics for his method of reproducing colours photographically based on the phenomenon of interference. (☞ Jul. 13)

1877

American astronomer Asaph Hall discovered a second moon of Mars, which he named Phobos, Greek for 'fear'. The moon is composed of carbon-rich rock like C-type minor planets and ice. But its density is so low that it cannot be pure rock. The moon is heavily cratered. Phobos is probably a minor planet perturbed by Jupiter into an orbit that allowed it to be captured by Mars. (☞ Aug. 11, Oct. 15)

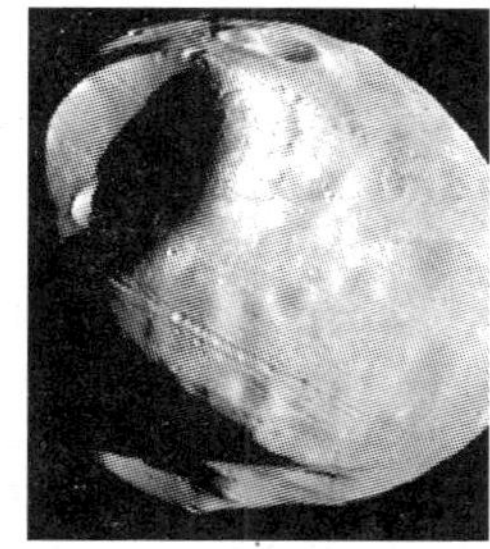

1899

Robert Wilhelm Bunsen, German chemist, who is better known for the laboratory gas burner named after him, died in Heidelberg, Baden, Germany. (☞ Mar. 31)

1904

Wendell Meredith Stanley, American biochemist, Nobel laureate, was born in Ridgeville, Indiana, USA. Stanley graduated in science from the Earlham College in 1926 and entered the University of Illinois, from where he did his MSc in 1927 and PhD in chemistry in 1929. In 1930, Stanley went to Munich, Germany as a National Research Council Fellow and returned to USA in 1931 to take up a post as assistant at the Rockefeller Institute. He remained with the Institute until 1948, when he was appointed professor of biochemistry and director of the Virus Laboratory, University of California, Berkeley. In 1935 Stanley crystallised the tobacco mosaic virus, which causes disease in tobacco plants, and showed that it is a rod-shaped aggregate of protein and nucleic acid molecules. His work enabled other scientists, utilising methods of X-ray diffraction, to ascertain the precise molecular structures and the modes of propagation of several viruses including the virus that causes influenza in humans. Stanley was awarded the 1946 Nobel Prize for Chemistry, jointly with fellow American biochemists John Northrop and James Sumner. Stanley and Northrop shared half the prize money for their preparation of enzymes and virus proteins in a pure form, while Sumner received the other half for his discovery that enzymes can be crystallised. (☞ Jun. 15, Jul. 5, Nov. 19)

1957

Irving Langmuir, American physical chemist, Nobel laureate (Chemistry, 1932), whose studies of molecular films on solid and liquid surfaces opened new fields in colloid research and biochemistry, died in Falmouth, Massachusetts, USA. (☞ Jan. 31)

1973

Selman Abraham Waksman, Ukrainian-born American biochemist, Nobel laureate (Physiology or Medicine, 1952), discoverer of streptomycin and other antibiotics, died in Hyannis, Massachusetts, USA. (☞ Jul. 22)

2003

The man-made element 110 was given the official name of 'darmstadtium', with symbol Ds, by the International Union of Pure and

Applied Chemistry (IUPAC) Council at its 42nd General Assembly in Ottawa, Canada. This name continues the long-established tradition of naming an element after the place of its discovery. (☞ Nov. 9)

August 17

1601
Pierre de Fermat, French lawyer and mathematician, often called the founder of the modern theory of numbers, was born in Beaumont-de-Lomagne, France. Fermat attended the University of Toulouse before moving to Bordeaux in the second half of the 1620s. From Bordeaux he went to Orléans where he studied law at the University and received a degree in civil law. In Bordeaux, he began his first serious mathematical researches. He anticipated differential calculus with his method of finding the greatest and least ordinates of curved lines. Fermat published nothing in his lifetime, and gave no systematic exposition of his methods. Some of the most striking of his results were found after his death on loose sheets of paper or written in the margins of works which he had read and annotated, and are unaccompanied by any proof. He proposed the famous Fermat's Last Theorem while studying the work of the ancient Greek mathematician Diophantus. He wrote in pencil in the margin, 'I have discovered a truly remarkable proof which this margin is too small to contain, that when the Pythagorean Theorem is altered to read $a^n + b^n = c^n$, the new equation cannot be solved in integers for any value of n greater than 2'. (☞ Jan. 12)

1915
The first practical automobile self-starter was patented by Charles F. Kettering of Detroit, Michigan, USA. Clyde J. Coleman of New York City, USA had originally patented the self-starter in 1903, but it was found to be impractical. Kettering, who worked at General Motors in Detroit, modified the self-starter and made it practical. It was first installed in Cadillac cars in 1911. By 1920, nearly every car had a self-starter.

1958
John Hubert Marshall, English archaeologist, former Director General of the Archaeological Survey of India, who was responsible for the excavation of ancient sites at Ṭaxila, Vaisali, Nalanda, Rajagriha and Sarnath, died in Guildford, Surrey, England. (☞ Mar. 19)

1969
Otto Stern, German-born American physicist, Nobel laureate (Physics, 1943), who developed the molecular beam as a tool for studying the characteristics of molecules and measured the magnetic moment of the proton, died in Berkeley, California, USA. (☞ Feb. 17)

1970
Soviet unmanned spacecraft *Venera-7* was launched towards Venus from the Baikanur cosmodrome in Kazakhstan, to study the Venusian atmosphere and other phenomena of the planet. (☞ Dec. 15)

August 18

1868
Helium was first discovered on the Sun by the British astronomer Joseph Norman Lockyer.

While studying the solar spectrum taken during a total eclipse of the Sun, he noticed a yellow line that could not be accounted for by elements known on the Earth. It was known that each element produces a characteristic spectral pattern of bright lines, but the yellow line was not due to any element known at that time. Lockyer concluded that the unknown spectral line must be due to an unknown element, which he named helium after the Greek word *Helios*, meaning Sun. Helium was discovered on Earth by the British chemist William Ramsay in 1895, almost 27 years after it was discovered on the Sun. (☞ Oct. 2)

1951
The Indian Institute of Technology, India's first, was inaugurated at Kharagpur, West Bengal. Initially, the IIT started functioning from 5, Esplanade East, Calcutta (now Kolkata) in 1950 and very soon shifted to the old Hijli Detention Camp in Kharagpur where some of India's great freedom fighters toiled and sacrificed their lives for the country's independence. The Institute started its academic programme with only 224 freshers and 42 teachers in ten departments. Its first director was the eminent Indian chemist Jnan Chandra Ghosh. Today, out of the seven IITs in India, IIT Kharagpur offers courses in the largest number of engineering disciplines. They range from aerospace engineering to agricultural & food engineering to reliability engineering. It also offers courses in geology & geophysics, humanities & social sciences, mathematics, physics & meteorology, rubber technology, rural development, and management.

1986
William Schroeder, retired American Government quality-control inspector, who was implanted with a *Jarvik-7* artificial heart, died at the Humana Audubon Hospital in Louisville, Kentucky, USA, after living for a record 620 days with the artificial heart. (☞ Nov. 25)

1994
Richard Laurence Millington Synge, British biochemist, Nobel laureate (Chemistry, 1952), who co-developed partition chromatography, died in Norwich, Norfolk, England. (☞ Oct. 28)

1998
The *Cassini–Huygens* spacecraft swung by Earth at a distance of 1,171 kilometres, en route to Saturn. (☞ Jan. 14, Apr. 26, Jun. 24, Jul. 1, Oct. 15, Dec. 24, Dec. 30)

August 19

1646
John Flamsteed, English astronomer who established the Greenwich Observatory, was born in Denby, Derbyshire, England. Between 1662 and 1669 Flamsteed studied astronomy on his own, after poor health forced him to leave school. He later continued his education at the University of Cambridge, receiving his MA degree in 1674. In 1677 he was elected to the Royal Society. He was one of a group of scientists who convinced King Charles II to build a national observatory. The Royal Observatory at Greenwich was built for him and although he had to procure his own instruments, he began using it in 1676. He was the first 'Astronomer Royal', though that title was not yet formally attached to his post. His task was to run the new Royal Observatory and he remained in

charge there until his death in 1719. His main work at the observatory was collecting improved observations and position measurements for stars for use in navigation. Flamsteed published his star catalogue *Historia Coelestis Britannica* in 1725, which contained data on 3,000 stars. It listed more stars and gave their positions considerably more accurately than any other previous publication had done. He also worked on the motions of the Sun and Moon, tidal tables, and was one of the only astronomers to maintain that the comets of 1680 and 1681 were the same, viewed before and after passing around the Sun. (☞ Dec. 31)

1662
Blaise Pascal, French mathematician, physicist, who formulated the Pascal's law, invented the hydraulic press, mechanical calculator, and injection syringe, died in Paris, France. (☞ Jun. 19)

1839
The French Government made details of Louis Daguerre's invention, which it had purchased from the inventor, public 'as a gift to the world from France'. It also granted pensions to both Daguerre and the late Joseph Nicephore Niépce's son Isidor 'for the glory of endowing the world of science with a new mechanical pictorialism'. The day is designated as World Photography Day. (☞ Jan. 2, Jan. 9, Nov. 18)

1871
Orville Wright, American inventor who with his brother, Wilbur, achieved the first powered, sustained, and controlled airplane flight, was born in Dayton, Ohio, USA. Even as a child, mechanics fascinated him. He became interested in bicycles, and in 1893, the two brothers opened the Wright Cycle Company, a bicycle sales and repair business in their hometown. After reading about the death of pioneer glider pilot Otto Lilienthal in 1896, they became interested in flying. By 1896, they had made up their mind to build a flying machine. They began seriously reading on the subject in 1899, and soon obtained all the scientific knowledge of aeronautics then available. Orville helped Wilbur built and test their biplanes made of wood and cloth. Orville rode on their first ever manned powered flight at Kitty Hawk in December 1903. (☞ Jan. 30, Dec. 17)

1960
The Russian spacecraft *Sputnik-5*, carrying two dogs, Belka and Strelka ('Squirrel' and 'Little Arrow' in Russian), was launched into orbit around the Earth. The animals were later safely returned to Earth, becoming the first living beings to return from space. (☞ Aug. 20)

1964
Syncom-3, the world's first geostationary satellitewaslaunchedfromtheKennedySpace Centre, Florida, USA. It was an experimental geosynchronous communications satellite placed over the equator at 180 degrees longitude over the Pacific Ocean. The satellite provided live television coverage of the 1964 Olympic games in Tokyo, Japan and was used for various communications tests.

1968
George Gamow, Russian-born American nuclear physicist, cosmologist and writer

who was one of the foremost advocates of the Big Bang theory, died in Boulder, Colorado, USA. (☞ Mar. 4)

1982
M K Vainu Bappu, Indian astrophysicist, and the main architect of the revival of astronomical studies in India, died in Munich, Germany. (☞ Aug. 10)

1994
Linus Carl Pauling, American chemist, Nobel laureate (Chemistry, 1954; Peace, 1962), who worked on chemical bonding, touted the benefits of vitamin C, and campaigned for nuclear peace, died in Big Sur, California, USA. (☞ Feb. 28)

August 20

1779
Jöns Jacob Berzelius, Swedish scientist, one of the founders of modern chemistry, especially noted for his determination of atomic weights, the development of modern chemical symbols and his electrochemical theory, was born near Linköping, Sweden. Berzelius received his MD from the Uppsala University in 1802. In 1807 he became a professor at the Stockholm University and began his analysis of the composition of chemical compounds; he studied about 2,000 compounds over a 10-year period. While conducting experiments, he discovered the 'law of constant proportions,' which states that inorganic substances are composed of different elements in constant proportions by weight. By 1828, Berzelius compiled, from these studies, a table of the relative atomic weights (with oxygen as a reference) of all elements then known. Taken together, this work was a strong confirmation of the atomic hypothesis that inorganic chemical compounds were composed of atoms combining in whole number amounts. By carrying out experiments on the electrolysis of various solutions, Berzelius developed his dualistic electrochemical theory that compounds are made up of two electrically different (positively and negatively charged) components. Berzelius also developed a system of chemical notation in which the elements were given simple written labels – such as O for oxygen, or Fe for iron – and proportions were noted with numbers. This is the same basic system as is used today. Berzelius discovered the elements cerium (Ce), selenium (Se), and thorium (Th), and isolated silicon, zirconium, and titanium. Known as the father of gravimetric analysis, he introduced the use of the water bath, desiccator, wash bottle, filter paper, rubber tubing, and improved the blowpipe technique. (☞ Aug. 7)

1897
Ronald Ross, British bacteriologist, Nobel laureate, discovered the malarial parasite in the stomach of the Anopheles mosquito while working at an army hospital in Secunderabad (now in Andhra Pradesh), India. While dissecting the stomach tissue of an anopheline mosquito which fed four days previously on a patient with malaria fever, he found the malaria parasite and went on to prove the role of Anopheles mosquitoes in the transmission of malaria parasites in humans. Ross wrote a paper, 'On some peculiar pigmented cells found in two mosquitoes fed on malarial blood', and sent it immediately to the *British Medical Journal*, which took three months to publish it. He continued his research into malaria at the Presidency General

Hospital (now SSKM Hospital) in Calcutta (now Kolkata), where he worked out the life history of the malaria parasite in birds. By July 1898, he had demonstrated that mosquitoes could serve as intermediate hosts for bird malaria. After feeding mosquitoes on infected birds, he found that the malaria parasites could develop in the mosquitoes and migrate to the insects' salivary glands, allowing the mosquitoes to infect other birds during subsequent blood meals. (☞ May 13, Sep. 16)

1904

Werner Forssmann, German surgeon, Nobel laureate, was born in Berlin, Germany. Forssmann was educated at the Askanische Gymnasium in Berlin and went to the University of Berlin to study medicine, passing his State Examination in 1929. For clinical instruction in surgery he went to the August Victoria Home at Eberswalde near Berlin, where he was the first to develop a technique for the catheterisation of the heart. This he did by inserting a cannula into his own antecubital vein, through which he passed a catheter for 65 cm and then walked to the X-ray department, where a photograph was taken of the catheter lying in his right auricle, proving that a catheter could be inserted safely into a human heart. Subsequently he worked at the Charité, Berlin, and the City Hospital at Mainz, and then went to the Rudolf Virchow Hospital in Berlin for specialist training in urology. He was later appointed Chief of the Surgical Clinic of the City Hospital at Dresden-Friedrichstadt and at the Robert Koch Hospital, Berlin. Forssmann's procedure, with slight modifications, was put into practice in 1941 by the American physiologist Dickinson W. Richards and French-born American physiologist André F. Cournand, and has since become an extremely valuable tool in diagnosis and research. It has made possible, among other things, precise measurement of intracardiac pressure and blood flow, injection into the heart of drugs and of opaque material visible on X-ray photographs, and insertion of electrodes for the regulation of the heartbeat. Forssmann was awarded the 1956 Nobel Prize for Physiology or Medicine, jointly with Cournand and Richards, for their discoveries concerning heart catheterisation and pathological changes in the circulatory system. (☞ Jun. 1, Sep. 24, Oct. 30)

1913

Roger Wolcott Sperry, American neurobiologist, Nobel laureate, was born in Hartford, Connecticut, USA. Sperry earned a bachelor's degree in English literature and a master's degree in psychology from the Oberlin College, Ohio, and a doctorate in zoology from the University of Chicago in 1941. In 1946 he joined the faculty of the University of Chicago and in 1954 moved to the California Institute of Technology in Pasadena as a professor of psychobiology. As early as 1941, Sperry had suggested that chemical gradients were responsible for the orderly re-growth of nerve fibre connections. This came to be known as the chemoaffinity theory; Sperry later developed it in more detail. Sperry was best known for his research on 'split brain' patients, demonstrating how the two halves of the brain functioned. He was responsible for overturning the widespread belief that the left brain is dominant by showing that several

cognitive abilities were localised in the right brain. His studies demonstrated that the left side of the brain is normally dominant for analytical and verbal tasks, while the right hemisphere assumes dominance in spatial tasks, music, and certain other areas. Sperry's experiments with cats and later with monkeys paved the way for cutting the corpus callosum that connects the two hemispheres of the human brain, as a treatment for severe epilepsy. After the patients recovered from the operation they no longer had seizures. The surgical techniques developed by Sperry laid the groundwork for much more specialised explorations of the mental functions carried out in different areas of the brain. Sperry was awarded the 1981 Nobel Prize for Physiology or Medicine, jointly with the Canadian-born American neurobiologist David Hunter Hubel and the Swedish neurobiologist Torsten Nils Wiesel, for his discoveries concerning the functional specialisation of the cerebral hemispheres. (☞ Feb. 27, Apr. 17, Jun. 3)

1915
Paul Ehrlich, German medical scientist, Nobel laureate (Physiology or Medicine, 1908), known for his pioneering work in haematology, and chemotherapy and for his discovery of the first effective treatment for syphilis, died in Bad Homburg vor der Höhe, Germany. (☞ Mar. 14)

1917
Adolf von Baeyer, German chemist, Nobel laureate (Chemistry, 1905), who synthesised indigo and formulated its structure, died in Starnberg, near Munich, Germany. (☞ Oct. 31)

1936
Hideki Shirakawa, Japanese chemist, Nobel laureate, was born in Tokyo, Japan. Shirakawa earned both his BS and PhD in chemical engineering from the Tokyo Institute of Technology. In 1966, having just received his doctorate, he joined the Chemical Resources Laboratory in Tokyo as a research assistant. Around 1966 he began to synthesise polymers, notably polyacetylene. Initially his goal was to determine the mechanism for the polymerisation of acetylene using Ziegler–Natta catalysts, but during an experiment, the amount of catalyst added went wrong and a 'ragged film' rather than the usual black powder form of polyacetylene was obtained. More importantly, the film also had a metallic lustre to it. Shirakawa's laboratory began to study the unusual properties of this film. When Alan G. MacDiarmid, a New Zealand-born American chemistry professor at the University of Pennsylvania, USA, heard of this work he invited Shirakawa to USA to collaborate. In 1977 they, along with the American chemist Alan J. Heeger, published their work on conductive polymers in *Chemical Communications*. The discovery opened up the possibilities of a wide variety of commercial applications of these unique polymers, such as anti-static substances for photographic film, and for 'smart' windows that can exclude sunlight. In addition, semi-conductive polymers have recently been developed into light-emitting diodes, solar cells and as displays in mobile telephones and mini-format television screens. After his return to Japan, Shirakawa joined the Institute of Materials Science at the University of Tsukuba as associate professor, being promoted to full professor in 1982. He retired as professor emeritus in 2000. Shirakawa was awarded the 2000 Nobel Prize for Chemistry, jointly with MacDiarmid and Heeger, for the discovery and development of conductive polymers. (☞ Jan. 22, Apr. 14)

1960
The Russian spacecraft *Sputnik-5* carrying two dogs Belka and Strelka ('Squirrel' and 'Little Arrow' in Russian) returned from space. After a day in orbit, the spacecraft's retrorocket was fired and the landing capsule and the dogs were safely recovered. Belka and Strelka were the first living beings to return alive from

space. Just seven months later, Russian cosmonaut Yuri Gagarin became the first man in space. (☞ Aug. 19)

1961
Percy Williams Bridgman, American physicist, Nobel laureate (Physics, 1946), who developed techniques for studying matter under extremely high pressures, died in Randolph, New Hampshire, USA. (☞ Apr. 21)

1975
The American space probe *Viking-1* was launched from the Kennedy Space Center,

Florida, USA. The probe carried a lander equipped with a complete biology lab to test for signs of life in Martian soil. (☞ Jun. 19, Jul. 20)

1977
The American unmanned spacecraft *Voyager-2* was launched from the Kennedy Space Center, Florida, USA, on a mission to the outer planets. Although it was launched before its twin spacecraft *Voyager-1*, *Voyager-2* was designed to travel more slowly, and fly-by all the outer giant planets. *Voyager-2* flew by Saturn in August 1981, but

continued on to Uranus in January 1986 and Neptune in August 1989. Data from the two probes included photographs of Jupiter showing a variety of cloud forms around Jupiter, and volcanic activity on its moon, Io. (☞ Jan. 24, Jul. 9, Aug. 24, Aug. 25)

2001
Fred Hoyle, English astronomer, Fellow of the Royal Society who coined the term 'Big Bang' and was one of the propounders of the 'steady state theory' of the universe, died in Bournemouth, England. (☞ Jun. 24)

August 21

1814
Benjamin Thompson Rumford, American-born British physicist, and a founder of the Royal Institution of Great Britain, London, died in Auteuil, France. (☞ Mar. 26)

1986
Carbon dioxide eruption from Lake Nyos, a volcanic lake in the West African nation of Cameroon killed more than 1,700 people. Carbon dioxide being denser than air hugged the ground and flowed down valleys killing humans and animals. The cloud travelled as far as 25 km from the lake. In addition to the human deaths caused by suffocation, 845 people were hospitalised and 3,000 cattle died.

1995
Subrahmanyan Chandrasekhar, Indian-born American astrophysicist, Nobel laureate (Physics, 1983), who discovered the Chandrasekhar limit, died in Chicago, USA. (☞ Oct. 19)

August 22

1647
Denis Papin, French-born British physicist, Fellow of the Royal Society, who invented the pressure cooker, was born in Blois, France. Papin attended a Jesuit school in Blois, and in 1661 he began his studies at the University of Angers from where he graduated with a medical degree in 1669. He assisted the Dutch physicist Christiaan Huygens with air-pump experiments, and went to London in 1675 to work with the English physicist Robert Boyle. In 1679 he invented his steam digester (a type of pressure cooker). This device consisted of a vessel provided with a tightly fitting lid, so that under pressure its contents could be raised to a high temperature for quick cooking; a safety valve was used, for the first time, to guard against an excessive rise in the pressure. Papin conceived of using steam to drive a piston in a cylinder after he observed that the pressure of the steam in his cooker tended to raise the lid. But he never built an engine of his own; his idea was improved by others and led to the development of the steam engine, which brought about the Industrial Revolution. Papin is known to have died in London in 1712, but there is no record of the actual date of his death.

1935
Gurdev Singh Khush, Indian botanist, Fellow of the Royal Society, was born in the village Rurkee in Jallandhar district of Punjab, India. Khush did his BSc in agriculture from the Punjab University in 1955 and earned his PhD in genetics from the University of California, Davis, California, USA in 1960. He stayed on at UC Davis until 1967 doing groundbreaking research on the cytogenetics of tomatoes and writing an authoritative reference on cytogenetics. In 1967, Khush joined the International Rice Research Institute in Manila, Philippines, and became its principal plant breeder and head of the plant breeding, genetics, and biochemistry division, developing 300 new rice varieties and triggering the Green Revolution in rice in Asia. Prior to the beginning of the green revolution, varieties of rice took 6–7 months to mature and yielded about 1–2 tonnes per hectare. Khush modified the plant by reducing its height, shortening maturing time, and increasing response to fertilisers. Under optimal conditions, these plants can yield up to 10 tonnes per hectare. In 1994, Khush announced a new type of 'super rice', which has the potential to increase yields by

25 per cent. Khush managed to do all these with traditional plant breeding methods without once using the modern genetic engineering techniques. The super rice is designed to yield up to 12 tonnes per hectare in irrigated tropical conditions, but can adjust its genetic characteristics to match tastes and environment conditions. The improved varieties of rice developed by Khush and his team are now planted on approximately 70 per cent of the world's rice-growing land and have more than doubled rice production worldwide. Khush was the recipient of the 1996 World Food Prize for developing improved rice varieties. After retiring from IRRI in 2000 he has been with the University of California, Davis, USA, as adjunct professor.

1939

Julian Seth Kahn of New York, USA, was issued a patent for dispensing liquids under pressure from a disposable container, precursor of the aerosol spray. Kahn's device, described as an 'apparatus for mixing a liquid with gas', was capable of dispensing liquids that had been stored under pressure. It included a valve mechanism that was both inexpensive and disposable. Kahn intended his invention to be used for whipping cream by putting the cream in a glass soda bottle and adding gas in a controlled manner. The device was later refined into an aerosol spray can pressurised by fluorocarbons for spraying mosquitoes that carried malaria. Its use was extended to applications such as dispensing paints, pharmaceuticals, and body sprays. (☞ Nov. 23)

1967

Gregory Pincus, American endocrinologist whose work on the anti-fertility properties of steroids led to the development of the first effective oral contraceptive, died in Boston, Massachusetts, USA. (☞ Apr. 9)

1974

Jacob Bronowski, Polish-born British mathematician who wrote extensively on the humanistic aspects of science, died in East Hampton, New York, USA. (☞ Jan. 18)

1989

The first complete ring around Neptune was discovered by the American space probe *Voyager-2*. Earlier Earth-based observations showed only faint arcs instead of complete rings. (☞ Aug. 20, Aug. 24)

2004

India's first homemade civil passenger aircraft, *Saras*, made its inaugural flight in Bangalore, Karnataka. Named after the Indian Crane, the twin turbo-prop *Saras* was designed by the National Aerospace Laboratory (NAL), a constituent laboratory of India's Council of Scientific and Industrial Research (CSIR). It is a multi-role small aircraft, designed to offer

maximum operational flexibility. It can operate as a passenger carrier, executive craft, air ambulance, air rescue craft or coast guard aeroplane. In its primary role as a regional commuter craft, the *Saras* provides space for 14 to 18 passengers.

August 23

1806

Charles-Augustin de Coulomb, French physicist, pioneer in electrical theory, best known for the formulation of Coulomb's law, died in Paris, France. (☞ Jun. 14)

1931
Hamilton Othanel Smith, American microbiologist, Nobel laureate, was born in New York, USA. Smith matriculated from the University of Illinois, Urbana-Champaign, majoring in mathematics, and graduated from the University of California, Berkeley in 1952. After receiving his MD degree from the Johns Hopkins University Medical School, Baltimore, Maryland in 1956, he joined the Barnes Hospital in St. Louis, Missouri, for a medical internship. In 1962 he joined the Department of Human Genetics at the University of Michigan, Ann Arbor. In 1967 he returned to Johns Hopkins, becoming professor of microbiology in 1973. In 1968, while studying the mechanism of DNA uptake from the phage virus P22 by the bacterium *Haemophilus influenzae*, Smith and his colleagues discovered what came to be called type II restriction enzymes that could recognise specific sequences of nucleotides in a molecule of DNA and cut the molecule at that particular point. This predictable behaviour made type II restriction enzymes valuable tools for DNA analysis. Smith was awarded the 1978 Nobel Prize for Physiology or Medicine, jointly with Swiss microbiologist Werner Arber and American microbiologist Daniel Nathans, for the discovery of restriction enzymes and their application to problems of molecular genetics. (☞ Jun. 3, Oct. 30)

1933
Robert F. Curl, Jr., American chemist, Nobel laureate, was born in Alice, Texas, USA. Curl received his BA in 1954, from the Rice University, Houston, Texas, and obtained his PhD from the University of California, Berkeley, in 1957. Curl has remained at the Rice University as professor of chemistry, pursuing his research career from 1958. Curl's research interests cover several areas of physical chemistry, a great deal of which involves collaborating with others both at Rice and elsewhere. In September 1985, Curl met with chemist Harold W. Kroto of the University of Sussex, England, and American chemist Richard E. Smalley, a colleague at Rice, and in just 11 days of research, they discovered fullerenes, a class of closed, hollow, aromatic carbon compounds that are made up of 12 pentagonal and differing numbers of hexagonal faces. The three scientists used a laser to vaporise graphite rods in an atmosphere of helium gas to obtain C_{60}, the first fullerene to be discovered. The molecule was named after the American architect R. Buckminster Fuller, whose geodesic dome design is similar to the molecular structure of C_{60}. Fullerenes are the third form of pure carbon known to exist, after diamond and graphite. Curl was awarded the 1996 Nobel Prize for Chemistry, jointly with Kroto and Smalley, for their discovery of fullerenes. (☞ Jun. 6, Oct. 7, Nov. 14)

1966
The American spacecraft *Lunar Orbiter-1* sent back the first photograph of the Earth from the Moon. (☞ Aug. 10)

1977
American pilot Bryan Allen pedalled the *Gossamer Condor* for more than two kilometres at the Shafter Airport,

California, USA, to win the Kremer Prize of US$95,000 for the first human-powered flight. The plane, made of thin aluminium tubes covered with Mylar plastic and braced with stainless steel wires, was airborne for a total duration of 7 minutes, 27.5 seconds and flew at a speed of between 16 and 18 km/h.

1982
Stanford Moore, American biochemist, Nobel laureate (Chemistry, 1972), who contributed to the development of the first automatic amino-acid analyser, a machine that greatly facilitated the analysis of the amino acid sequences of proteins, died in New York, USA. (☞ Sep. 4)

1997
John Cowdery Kendrew, British biochemist, Nobel laureate (Chemistry, 1962), who determined the structure of the muscle protein myoglobin, died in Cambridge, England. (☞ Mar. 24)

August 24

1832
Sadi Carnot, French physicist who described the Carnot cycle, relating to the theory of heat engines, died in Paris, France. (☞ Jun. 1)

1888
Rudolf Clausius, German mathematical physicist who formulated the second law of thermodynamics and is credited with making thermodynamics a science, died in Bonn, Germany. (☞ Jan. 2)

1898
Albert Claude, Belgian-born American cytologist, Nobel laureate, was born in Longlier, Belgium. Claude obtained his medical degree from the Université de Liège, Liège, Belgium, in 1928. After spending a year in Berlin, first at the Institute für Krebsforschung, and then at the Kaiser Wilhelm Institute, Dahlem, to study tissue culture, he went to USA to join the Rockefeller Institute (now the Rockefeller University) in New York, where he trained in the culture of cancerous cells. While attempting to isolate the Rous sarcoma virus from chicken tumours, he discovered that centrifugal separation of the cell components made possible a biochemical analysis of them that confirmed that the separated particles consisted of distinct organelles. Such analysis enabled Claude to discover the endoplasmic reticulum – a membranous network within cells – and to establish the function of the mitochondria as the centres of respiratory activity. Use of the electron microscope also helped Claude to isolate and take photos of a cancerous virus. His demonstration of the instrument's usefulness in this regard eventually helped scientists to correlate the biological activity of each cellular component with its structure and its place in the cell. In 1949 he returned to Belgium and became director of the Tumour Centre in the Institut Bordet,

and professor emeritus in the Faculty of Medicine, at the University of Brussels, Belgium. Later, he held professorships at the Rockefeller University, New York, USA, and the Université Libre in Brussels, Belgium. Claude was awarded the 1974 Nobel Prize for Physiology or Medicine, jointly with George Palade, and Christian de Duve, for their discoveries concerning the structural and functional organisation of the cell. (☞ May 22, Oct. 2, Nov. 19)

1989
The American unmanned spacecraft *Voyager-2* made its closest approach to Neptune, passing about 4,950 kilometres above Neptune's North Pole. Five hours later, *Voyager-2* passed about 40,000 kilometres from Neptune's largest moon, Triton, before turning onto a course that will take it to interstellar space. (☞ Jan. 24, Jul. 9, Aug. 20, Aug. 25)

2006
The International Astronomical Union (IAU), at its 26th General Assembly in Prague, Czeck Republic, adopted a resolution defining a 'planet', according to which Pluto does not qualify to be a planet. As a consequence, the Solar System would henceforth consist of eight 'planets' – Mercury, Venus, Earth, Mars, Jupiter, Saturn, Uranus and Neptune. A new distinct class of objects called 'dwarf planets' was also decided. It was agreed that 'planets' and 'dwarf planets' are two distinct classes of objects. The first members of the 'dwarf planet' category are Ceres, Pluto and 2003 UB313 (temporary name). More 'dwarf planets' are expected to be announced by the IAU in the coming months and years. Currently a dozen candidate 'dwarf planets', including the objects '2002 LM60' and '2003 VB12' are listed on IAU's 'dwarf planet' watch list, which keeps changing as new objects are found and the physics of the existing candidates becomes better known.

August 25

1819
James Watt, Scottish instrument maker and inventor whose steam engine contributed substantially to the Industrial Revolution, died near Birmingham, Warwick, England. (☞ Jan. 19)

1822
William Herschel, German-born British astronomer, who discovered the planet Uranus, died in Slough, Buckinghamshire, England. (☞ Nov. 15)

1841
Theodor Emil Kocher, Swiss surgeon, Nobel laureate, was born in Bern, Switzerland. Kocher received his medical degree from the University of Bern and went on to study in Berlin, London, Paris and Vienna. In 1872 he returned to the University of Bern as professor of clinical surgery, a post he held for 45 years. A staunch supporter of the British surgeon Joseph Lister, who had pioneered antiseptic surgery, Kocher performed about 5,000 operations for goitre under scrupulously aseptic conditions, by 1912. From his study on the effects of the removal of the thyroid gland, he discovered that complete removal of the gland leads to a characteristic disease pattern, including serious disturbances in nutrition. Metabolism is significantly diminished; growth ceases; the skin and the subcutaneous tissues are the site of mucous infiltration;

degenerative processes occur in internal organs; serious disturbances make their appearance in the functions of the nervous system and muscles. His work was instrumental in coordinating studies on thyroid malfunction that ultimately led to the introduction of a thyroxin hormone for treatment of thyroid disorders. Kocher was awarded the 1909 Nobel Prize for Physiology or Medicine for his work on the physiology, pathology and surgery of the thyroid gland. (☞ Jul. 27)

1867
Michael Faraday, English chemist, physicist and inventor, who built the first dynamo, and stated the basic laws of electrolysis, died in Hampton Court, Surrey, England. (☞ Sep. 22)

1900
Hans Adolf Krebs, German-born British biochemist, Nobel laureate, was born in Hildesheim, Germany. Krebs was educated at the Gymnasium Andreanum at Hildesheim, and between the years 1918 and 1923 he studied medicine at the Universities of Göttingen, Freiburg-im-Breisgau, and Berlin. He received his MD degree in 1925 from the University of Hamburg and then spent one year studying chemistry at Berlin. In 1934, he was appointed demonstrator of biochemistry in the University of Cambridge, England. In 1935, he was appointed lecturer in pharmacology at the University of Sheffield in England and in 1938 he became lecturer-in-charge of the newly founded Department of Biochemistry there. In 1954 he was appointed Whitley Professor of Biochemistry in the University of Oxford. In 1937, in an elegant series of experiments, Krebs demonstrated the existence of a cycle of chemical reactions that converts – in the presence of oxygen – substances that are formed by the breakdown of sugars, fats, and protein components to carbon dioxide, water, and energy-rich compounds. The cycle, which came to be known as the Krebs cycle or citric acid cycle, regenerates oxaloacetic acid through a series of intermediate compounds while liberating carbon dioxide and electrons that are immediately utilised to form high-energy phosphate bonds in the form of adenosine triphosphate (ATP) – the chemical-energy reservoir of the cell. The Krebs cycle explains two simultaneous processes: the degradation reactions, which yield energy, and the building-up processes, which use up energy. This metabolic cycle occurs in most plants, animals, fungi, and many bacteria. Krebs was awarded the 1953 Nobel Prize for Physiology or Medicine, jointly with the German-born American biochemist Fritz Lipmann, for his discovery of the citric acid cycle. (☞ Jun. 12, Nov. 22)

1908
Antoine-Henri Becquerel, French physicist, Nobel laureate (Physics, 1903), who discovered radioactivity through his investigations of uranium and other substances, died in Le Croisic, France. (☞ Dec. 15)

1916
Frederick Chapman Robbins, American paediatrician and virologist, Nobel laureate, was born in Auburn, Alabama, USA. Robbins was educated at the University of Missouri, where he took the BA degree in 1936 and the BS in 1938. In 1940 he graduated from the Harvard Medical School and was appointed as resident physician in bacteriology at the Children's Hospital Medical Center in

Boston, Massachusetts. He continued his training there until 1942 when he joined the US Army and served in Italy, and North Africa during World War II as chief of the US Army's 15th medical general laboratory virus and rickettsia section, where he investigated epidemics of infectious hepatitis, typhus, and Q fever. After the war, in 1948, he joined American virologists John Enders and Thomas Weller at the Children's Hospital, Boston, where they helped solve the difficult problem of propagating viruses. Robbins chiefly studied the cultivation of poliomyelitis virus in tissue culture and the application of this technique. This accomplishment made possible the production of polio vaccines and the development of sophisticated diagnostic methods. He also investigated the viruses of mumps, herpes simplex and vaccinia. In 1952, he moved to Cleveland, Ohio, and served as director of the department of paediatrics and contagious diseases at the Cleveland Metropolitan General Hospital and later as professor of paediatrics and dean at the Case Western Reserve University School of Medicine, Cleveland, Ohio. Robbins was awarded the 1954 Nobel Prize for Physiology or Medicine, jointly with Enders and Weller, for their discovery of the ability of poliomyelitis viruses to grow in cultures of various types of tissue. (☞ Feb. 10, Jun. 15, Aug. 4)

1928

Herbert Kroemer, German physicist, Nobel laureate, was born in Weimar, Germany. Kroemer attended school in Weimer and joined the University of Jena to study physics. After receiving a PhD from the Georg August University in Göttingen in 1952, Kroemer joined the RCA Laboratories in the United States in 1954. In 1957 he carried out theoretical calculations showing that a heterostructure transistor – a tranistor made of a combination of different materials as against the conventional transistors that are made from only one kind of material – would be superior to a conventional transistor, especially for certain high-frequency uses and other applications. Scientists later showed that heterostructure transistors can operate at frequencies 100 times higher than conventional transistors, and they also work better as amplifiers. Kroemer's work was used by the Russian physicist Zhores Ivanovich Alferov, to develop the first practical heterostructure electronic device in 1966. The heterostructure laser, proposed independently by Kroemer and Alferov in 1963 and first developed by Alferov, has made fibre-optic communications possible. They are used in numerous everyday products, including computers and video players. Kroemer was awarded the 2000 Nobel Prize, jointly with Alferov and the American electrical engineer Jack S. Kilby, for basic work on information and communication technology. Kroemer and Alferov shared half of the prize money for developing semiconductor heterostructures used in high-speed- and opto-electronics, while Kilby received the other half for his part in the invention of the integrated circuit. (☞ Mar. 15, Nov. 8)

1981

The American unmanned spacecraft *Voyager-2* flew by Saturn at a distance of 100, 800 kilometres, sending back pictures that showed not a few, but thousands of rings. *Voyager-2* also took a closer look at some of the ring features found by *Voyager-1*. (☞ Jan. 24, Jul. 9, Aug. 20, Aug. 24)

2003
The *Spitzer Space Telescope* (formerly SIRTF, the Space Infrared Telescope Facility) was launched by the American space agency NASA from Cape Canaveral, Florida. Carrying an 85-centimetre telescope kept at a temperature of 5 degrees above absolute zero (−268°C), and three cryogenically cooled science instruments, *Spitzer* is the largest infrared telescope ever launched into space. During its 2.5-year mission, *Spitzer* obtained images and spectra by detecting the infrared energy, or heat, radiated by objects in space between wavelengths of 3 and 180 microns (1 micron is one-millionth of a metre). Most of this infrared radiation from space is blocked by the Earth's atmosphere and cannot be observed from the ground. The telescope has already sent back dazzling images of a dusty, spiral galaxy; a planet-forming disc; a glowing, stellar nursery and a young, buried star, demonstrating the power of its infrared eyes to spy hidden objects.

August 26

1723
Antonie van Leeuwenhoek, Dutch scientist, microscopist and zoologist who was the first to observe bacteria and protozoa, died in Delft, Netherlands. (☞ Oct. 24)

1740
Joseph-Michel Montgolfier, French inventor, who, with his brother Jacques-Etienne Montgolfier, pioneered hot-air ballooning, was born in Annonay, France. His father Pierre Montgolfier had prosperous paper factories in the small town of Vidalon, near Annonay. While carrying on their father's paper business, he and his brother Jacques-Étienne maintained their interest in scientific experimentation. In 1782 Joseph-Michel and his brother discovered that heated air, when collected inside a large lightweight paper or fabric bag, caused the bag to rise into the air. They made the first public demonstration of this discovery in June 1783 at the marketplace in Annonay. They filled their balloon with heated air by burning straw and wool under the opening at the bottom of the bag. The balloon rose into the air about 1,000 metres, remained there some 10 minutes, and then settled to the ground more than two kilometres from where it rose. He and his brother were honoured by the French Académie des Sciences. (☞ Jun. 26, Sep. 19, Nov. 21)

1743
Antoine-Laurent Lavoisier, French scientist, usually regarded as the founder of modern chemistry, was born in Paris, France. Lavoisier's attended the College Mazarin in Paris from 1754 to 1761, studying chemistry, botany, astronomy, and mathematics. His first chemical publication appeared in 1764. Beginning in 1775 he served on the Royal Gunpowder Administration, where his work led to improvements in the production of gunpowder and the use of agricultural chemistry. In 1778, he found that air

consists of a mixture of two gases, which he called oxygen and nitrogen. He disproved the phlogiston theory – all flammable materials contain phlogiston, a substance without colour, odour, taste or weight that is liberated in burning – by demonstrating that oxygen is required for combustion, rusting and respiration. Lavoisier is best known for his synthesis of chemical knowledge in his *Traité elémentaire de chimie*, published in 1789, in which for the first time the modern system of elements was laid out systematically, replacing the three or four elements of classical chemistry. Lavoisier also worked out reactions in chemical equations that respect the conservation of mass, showing that the mass of the reactants had to equal the mass of the products. He was a member of the commission appointed to secure uniformity of weights and measures throughout France. The report of the commission led to the adoption of the metric system. As an important public official during the reign of Louis XVI (including the early Revolutionary period), Lavoisier was branded a traitor by the revolutionists and guillotined. (☞ May 8, Nov. 1)

1850

Charles Richet, French physiologist, Nobel laureate, was born in Paris, France. Richet received his MD in 1869 and a DSc in 1878 from the University of Paris. From 1887 to 1927 Richet was a professor at the University of Paris, Sorbonne. While there he discovered 'anaphylaxis', a term he coined for the sometimes-fatal allergic reaction by a sensitised individual to a second injection of an antigen. His research helped to elucidate problems of hay fever, asthma, and other allergic reactions. For 24 years, from 1878 to 1902, Richet edited the journal *Revue Scientifique*. In physiology, he worked out the mechanism of thermoregulation in homoeothermic animals. Before his researches on polypnoea and shivering due to temperature, carried out between 1885 and 1895, little was known about the methods by which animals deprived of cutaneous transpiration can guard against overheating and how chilled animals can warm themselves again. He is also known for his investigations into the physiology of respiration and digestion, as well as epilepsy, the regulation of body heat, and a wide array of other subjects, including parapsychology. Richet was awarded the 1913 Nobel Prize for Physiology or Medicine in recognition of his work on anaphylaxis. (☞ Dec. 4)

1873

Lee De Forest, American inventor of the audion vacuum tube, was born in Council Bluffs, Iowa, USA. De Forest obtained his bachelor's degree and a PhD in physics from the Sheffield School of Science at the Yale University, New Haven, Connecticut in 1899. He then worked briefly for several companies, Western Electric Company in Chicago among them. He was a charter member of the Institute of Radio Engineers, the predecessor of the IEEE. But most of his life was spent as an independent inventor of gadgets for which he received over 300 patents. He invented the audion tube – the forerunner of a triode – in 1906 by introducing

a third electrode (De Forest used a piece of wire bent into the shape of a grid-iron, hence it was called the grid) in the centre of a vacuum tube, between the anode and the cathode. This simple modification allowed the audion tube to be used as an amplifier, as applying a small voltage to the grid controlled the amount of a larger current flowing across the tube. The audion tube allowed for voice amplification for radio reception. De Forest also discovered, in 1912, that by feeding part of the output of his triode vacuum tube back into its grid, he could cause a self-regenerating oscillation in the circuit. The signal from this circuit, when fed to an antenna system, was far more powerful and effective than that of the crude transmitters then generally employed. In 1913, the American company AT&T installed audions to boost voice signals as they crossed the US continent. Soon the audion was being used in radios as well. De Forest's audion tube and its later developments made live radio broadcasting possible and became the key component of all radio, telephone, radar, television, and computer systems before the invention of the transistor in 1947. During the 1930s De Forest developed Audion-diathermy machines for medical applications and, during World War II, conducted military research for Bell Telephone Laboratories. (☞ Jun. 30, Oct. 20)

1882
James Franck, German-born American physicist, Nobel laureate, was born in Hamburg, Germany. After attending the Wilhelm Gymnasium in Hamburg, Franck studied chemistry for a year at the University of Heidelberg, and then studied physics at the University of Berlin, from where he received his PhD in 1906. After World War I, Franck was appointed member and head of the physics division in the Kaiser Wilhelm Institute for Physical Chemistry at Berlin. In 1920, he became professor of experimental physics and director of the Second Institute for Experimental Physics at the University of Göttingen. Here he and fellow German physicist Gustav Ludwig Hertz carried out studies of the changes of energy that occur when atoms and electrons collide. By bombarding gases with electrons of different energies, they showed that an electron needs to have a certain minimum energy level to ionise an atom, thereby verifying the quantised nature of energy transfer. Franck immigrated to USA in 1935 and became a professor at the John Hopkins University, Baltimore, Maryland. In 1938 he joined the University of Chicago as professor of physical chemistry. Franck was awarded the 1925 Nobel Prize for Physics, jointly with Hertz, for their discovery of the laws governing the impact of an electron upon an atom. (☞ May 21, Jul. 22)

1906
Albert Bruce Sabin, Polish-born American physician and microbiologist who developed the first oral polio vaccine, was born in Bialystok, Poland. Sabin emigrated with his parents to the USA in 1921 and later became an American citizen. He received his MD from the New York University in 1931, where he began research on human poliomyelitis. In 1935 he joined the staff of the Rockefeller Institute for Medical Research (now the Rockefeller University) in New York, where he was the first researcher

to demonstrate the growth of poliovirus in human nervous tissue outside the body. After the outbreak of World War II, he joined the US Army and went on special missions in the Middle East, Africa, Sicily, Okinawa and the Philippines. During this period, he isolated viruses for sand fly fever, found a vaccine for dengue fever and developed a vaccine against Japanese encephalitis. He returned to polio research in 1939 when he joined the University of Cincinnati College of Medicine as associate professor of paediatrics and chief of the division of infectious diseases at the Children's Hospital Research Foundation of the college. Sabin postulated that live, weakened virus, administered orally, would provide immunity over a longer period of time than killed, injected virus. By 1957 he had isolated strains of each of the three types of poliovirus that were not strong enough to produce the disease itself but were capable of stimulating the production of antibodies. In 1960, after extensive trials, the Sabin oral polio vaccine was approved for use in USA. Mass immunisations began in early 1960s with his live virus vaccine, which was administered orally, on a lump of sugar. Sabin's oral vaccine became more popular than the injectable vaccine developed by the American physician Jonas Edward Salk. Sabin became professor emeritus at the University of Cincinnati in 1971, and from 1974 to 1982 he was a research professor at the University of South Carolina, Charleston. (☞ Mar. 3, Oct. 28)

1950

The Central Glass and Ceramics Research Institute (CGCRI), a constituent national laboratory of the Council of Scientific and Industrial Research (CSIR), was inaugurated. Located at Jadavpur on the outskirts of Kolkata, CGCRI initially conducted research directed towards identifying mineral resources within the country and their suitability for specific product development. Later, the development of various types of optical glasses brought CGCRI into the limelight in the international arena. The institute's pilot plant with an annual capacity of 10 tonnes went into production in 1961. Several varieties of optical glass representing the principal requirements of the consumers are still being produced in this plant. Besides the development of optical glass, work on introduction of modern techniques in glass and ceramic, basic studies on generation of colour in glasses, and evaluation of different minerals of the country such as clay and mica for their suitability in specific uses are some of the important activities taken up by the institute. Work in the field of optical fibre for telecommunications, sol-gel processing of glass and ceramic materials, production of glass fibre-based composites and application of ceramic materials in electronics have placed CGCRI in the global map. CGCRI has also started work in three strategic sectors, viz., atomic energy, defence and space, which have been facing sanctions from the developed countries.

1987

Georg Wittig, German chemist, Nobel laureate (Chemistry, 1979), who discovered the Wittig reaction, died in Heidelberg, Germany. (☞ Jun. 16)

1998

Frederick Reines, American physicist, Nobel laureate (Physics, 1995), who discovered the tau lepton, died in Orange, California, USA. (☞ Mar. 16)

August 27

1874

Carl Bosch, German Industrial chemist, Nobel laureate, was born in Cologne, Germany. From 1894 to 1896 Bosch studied metallurgy and mechanical engineering at the Technische Hochschule in Charlottenburg, but started reading chemistry at the Leipzig University and graduated from there in 1898. He joined the Badische-Anilin-und Sodafabrik, Ludwigshafen, Rhine as a chemist in 1899 and participated actively in the development of the then new industry of synthetic indigo. Here, in 1909, he undertook the task of improving a process of high-pressure synthesis of ammonia directly from hydrogen and nitrogen, which had been developed by the German physical chemist Fritz Haber at the Technische Hochschule in Karlsruhe. The development of the Haber–Bosch process involved more than 20,000 experiments and led to the substitution of the carbon–steel reaction container with one of alloy steel, and the replacement of osmium and uranium catalysts with cheaper and more easily available pure iron with certain additives. The Haber–Bosch process became important in the manufacture of explosives and fertilisers, thus providing practically every country in the world with sufficient fertilisers for agricultural purposes. Bosch also developed industrial processes for the syn-thesis of methanol, hydro-genation and commercial production of hydrogen. Bosch was awarded the 1931 Nobel Prize for Chemistry, jointly with fellow German chemist Friedrich Bergius, in recognition of their contributions to the invention and development of chemical high-pressure methods. (☞ Apr. 26, Oct. 11)

1875

French chemist Paul-Émile Lecoq de Boisbaudran discovered the element 31, which he named gallium (Ga) in honour of France (Gallia). He extracted it from large quantities of zinc blende. His first spectroscopic analysis of the tiny amount (about 0.01 mg) of the prepared sample showed a previously unknown violet line at 417.0 Å, indicating a new element. Metallic gallium was prepared for the first time in November 1875 by electrolysis of an ammoniacal solution of its sulphate. A rare, soft silvery metallic poor metal, gallium is a brittle solid at low temperatures but liquefies slightly above room temperature and indeed will melt in the hand. It occurs in trace amounts in bauxite and zinc ores. Gallium filled up a long-standing vacant space in the Periodic Table.

1910

American inventor Thomas Alva Edison demonstrated his kinetophone, which made talkies a reality. It was a machine that combined a kinetoscope – the forerunner of the motion-picture film projector – and a phonograph synchronised so as to reproduce a scene and its accompanying sounds simultaneously. In the early version, the viewer would look into the peepholes of the kinetoscope to watch the motion picture while listening to the accompanying phonograph through two rubber ear tubes connected to the machine. The picture and sound were made somewhat synchronous by connecting the two with a belt. In 1913, a different version of the kinetophone was introduced to the public. This time, the sound was made to synchronise with a motion picture projected onto a screen. However, his attempt to combine the phonograph and motion pictures failed commercially and was superseded by better recording techniques. (☞ Jan. 27, Feb. 11, Feb. 19, Apr. 15, Aug. 8, Aug. 9, Oct. 18, Oct. 21, Dec. 6)

1915

Norman Foster Ramsey, American physicist, Nobel laureate, was born in Washington, D.C. Ramsey graduated from the Columbia University, New York, in 1935 and received his PhD degree in physics there in 1940. He then went to Washington, D.C. as a Carnegie Institution Fellow, where he studied neutron–proton and proton–helium scattering. In 1943 he went to Los Alamos, New Mexico, to work on the Manhattan Project. After the war he returned to the Columbia University as a professor and research scientist and started measurements of nuclear magnetic dipole and electric quadrupole moments. In 1946, when the Brookhaven National Laboratory was established on Long Island, New York, he became the first head of the physics department. In 1947 he moved to the Harvard University, Cambridge, Massachusetts, where he established a molecular beam laboratory with the intent of doing accurate molecular beam magnetic resonance experiments and invented the 'separated oscillatory field' method to study the structure of atoms. He used the technique to measure several molecular and nuclear properties including nuclear spins, nuclear magnetic dipole and electric quadrupole moments, spin–rotational interactions, spin–spin interactions, electron distributions in molecules, etc., and also applied it to atomic clocks. In an effort to attain even greater accuracy and to do so with atomic hydrogen, Ramsey and his students invented the atomic hydrogen maser, which they used for accurate measurements of the hyperfine separations of atomic hydrogen, deuterium and tritium and for determining the extent to which the hyperfine structure was modified by the application of external electric and magnetic fields. Ramsey was awarded the 1989 Nobel Prize for Physics, jointly with the German physicist Wolfgang Paul and German-born American physicist Hans Georg Dehmelt. Ramsey received half the prize money for the invention of the separated oscillatory field method and its use in the hydrogen maser and other atomic clocks. Paul and Dehmelt shared the other half for the development of the ion trap technique. (☞ Aug. 10, Sep. 9)

1939

The first jet-propelled aircraft *He-178* was flown in Germany. Designed by the German aircraft designer and builder Ernst Heinkel, the *He-178* was powered by an improved 'HeS-3B' turbojet engine with a 300-kilogram thrust. The plane attained a maximum speed of 700 kmph during the flight which lasted about five minutes.

1958

Ernest Orlando Lawrence, American physicist, Nobel laureate (Physics, 1939), who invented the cyclotron, died in Palo Alto, California, USA. (☞ Aug. 8)

1962
The American space probe *Mariner-2* was launched towards Venus. *Mariner-2* was the second of a series of spacecraft used for planetary exploration in the fly-by mode and the first spacecraft to successfully encounter another planet. The objective of the *Mariner-2* mission was to fly-by Venus and return data on the planet's

atmosphere, magnetic field, charged particle environment, and mass. It also made measurements of the interplanetary medium during its cruise to Venus and after the fly-by. (☞ Dec. 14)

2003
Mars and Earth made their closest approach in 60,000 years – a 'mere' 56 million kilometres apart.

August 28

1789
Saturn's moon Enceladus was discovered by the German-born British astronomer William Herschel. (☞ Nov. 15)

1878
George Hoyt Whipple, American pathologist, Nobel laureate, was born in Ashland, New Hampshire, USA. Whipple was educated at the Andover Academy in Andover, Massachusetts, and then went to the Yale University, New Haven, Connecticut, where he took his AB degree in 1900. He then went to the Johns Hopkins University in Baltimore, Maryland where he took his MD degree in 1905 and was appointed assistant in pathology. In 1914 he was appointed professor of research medicine at the University of California Medical School, San Francisco, California, and director of the Hooper Foundation for Medical Research at the same university. In 1921 he was appointed professor of pathology and dean of the School of Medicine and Dentistry at the University of Rochester, New York. Whipple's main researches were concerned with anaemia and the physiology and pathology of the liver. In 1908 he began a study of bile pigments. This led to his interest in the body's manufacture of the oxygen-carrying haemoglobin, which is also an important constituent in the production of bile pigments. In 1920 he demonstrated that liver as a dietary factor greatly enhances haemoglobin regeneration in dogs. During 1923–25, he also carried out experiments in artificial anaemia, which established iron as the most potent inorganic factor involved in the formation of red blood cells. Whipple was awarded the 1934 Nobel Prize for Physiology or Medicine, jointly with the American physicians George Richards Minot and William Parry Murphy, for their discoveries concerning liver therapy in cases of anaemia. (☞ Feb. 1, Feb. 6, Dec. 2)

1919
Godfrey Newbold Hounsfield, English electrical engineer, Nobel laureate, was born in Newark, Nottinghamshire, England. After schooling at the Magnus Grammar School in Newark, Hounsfield joined the Royal Air Force as a volunteer reservist and worked as a radar expert during World War II. He joined the staff of EMI Ltd., an international electronics and entertainment corporation, in Middlesex in 1951, where he worked for a while on radar and guided weapons and later ran a small design laboratory. During this time he became particularly interested in computers. While working at the EMI Central Research Laboratories at Hayes near London, he developed 'the EMI-Scanner' and the technique of computed tomography. In 1972 he built the prototype for the first CT scan machine, which produced detailed images of cross-sections of the human body. In a CT scan, a high-resolution X-ray picture of an imaginary slice through the body (or head) is built up from information taken from detectors rotating around the patient, which allows delineation of very small changes in tissue density. Introduced in 1973, early CT scan machines were used for the diagnosis of diseases of the brain, but the technique has since been extended to the whole body. Although he had no formal university education, Hounsfield was granted an honorary doctorate in medicine by the City University, London in 1975. Hounsfield was awarded the 1979 Nobel Prize for Physiology or Medicine, jointly with the South African-born American physicist Allan Cormack, for the development of computer-assisted tomography. (☞ Feb. 23)

1928
Mambillikalathil Govind Kumar Menon, Indian physicist, Fellow of the Royal Society, was born in Mangalore, Karnataka, India. Menon did his BSc in 1946 from the Jaswant College, Jodhpur, and his MSc in 1949 from the Royal Institute of Science, Mumbai. He did his PhD from the University of Bristol in England. In 1955, Menon joined the Tata Institute of Fundamental Research in Mumbai where he became director in 1966. His research mainly concerned cosmic ray studies and in particular the study of high-energy interactions of elementary particles using nuclear emulsion techniques. He was instrumental in starting the high altitude balloon programme at TIFR for studying primary cosmic rays using a variety of detector systems. Menon has held many prominent positions including Secretary, Department of Science and Technology; Secretary, Department of Electronics, Government of India; Member, Planning Commission; and Scientific Adviser to the Prime Minister of India. He served as President of the International Council of Scientific Unions from 1988–1993.

1993
The first moon of a minor planet was discovered. The photograph of the minor planet Ida and its newly discovered moon, Dactyl, was taken by NASA's *Galileo* spacecraft, about 14 minutes before its closest approach to the minor planet. (☞ Mar. 26, Oct. 18)

August 29

1831

Electromagnetic induction was discovered by the English physicist Michael Faraday. He wound one side of a thick iron ring with insulated wire that was connected to a battery. He then wound the opposite side with wire connected to a galvanometer. He found that by closing and opening the battery circuit, the galvanometer showed deflection in opposite directions – the first example of electromagnetic induction, which forms the basis of all electric motors, generators and transformers. Faraday demonstrated the first electric transformer at the Royal Institute, London, UK. (☞ Sep. 22)

1893

Whitcomb Judson of Chicago, Illinois, USA was issued a US patent for the zip fastener. The patent described it as a 'clasp locker or unlocker for shoes'. The new fastener was exhibited at the World's Columbian Exposition in Chicago, USA, in 1893.

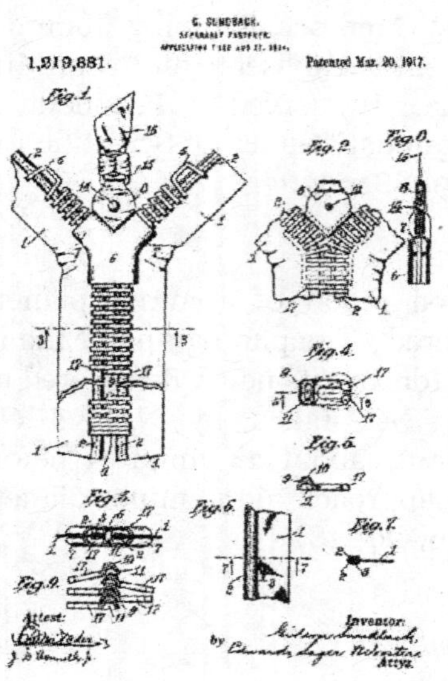

1965

American astronaut Gordon Cooper in orbit 160 kilometres above the Earth aboard the spacecraft *Gemini-V* chatted directly with aquanaut M. Scott Carpenter in the submersible *Sealab-II*, which was at that time 63 metres below the surface of the Pacific Ocean. It was the first time an astronaut in space spoke with an aquanaut under the sea. Carpenter was the second American to go into orbital flight around the Earth in May 1962. A project of the U.S. Navy and the University of California's Scripps Institution of Oceanography, *Sealab-II* was USA's most ambitious effort of that time to explore and eventually exploit the ocean's great store of food, oil and mineral resources. *Gemini-V* splashed down later in the day.

Sealab-II

1982

An atom of element 109 was made at the Heavy Ion Research Laboratory, Darmstadt, West Germany, by bombing a target of Bi-209 with accelerated nuclei of Fe-58. The team of physicists who created the new element confirmed the existence of element 109 by four independent measurements. It has been given the name of meitnerium, symbol Mt, after the Austrian-born British physicist Lise Meitner. The experiment demonstrated the feasibility of using fusion techniques as a method of making new, heavy nuclei.

1991
IRS-1B, the second satellite in India's IRS series was launched by a Semiorka rocket from Baikanur in Khazakistan. Designed and built by the Indian Space Research Organisation, the satellite carried two linear imaging and self scanner (LISS) cameras for Earth imaging and was placed in a Sun-synchronous polar orbit at an inclination of 99 degrees at an average altitude of 904 km. It took 103 minutes to complete one orbit around the Earth, making 14 orbits per day. The satellite returned periodically over the same geographical area, giving a complete coverage of the globe in 22 days.

August 30

1852
Jacobus Henricus van't Hoff, Dutch physical chemist, Nobel laureate, was born in Rotterdam, Netherlands. van't Hoff obtained his technology diploma from the Polytechnic School at Delft in 1871. He then worked briefly in Bonn, Germany and Paris, France. In 1874 he postulated that the four valences of a carbon atom are directed towards the corners of a regular tetrahedron. This allows the carbon atom to be asymmetric, i.e., connected to four different atoms or groups, in certain compounds, and it is these compounds that exhibit optical activity. van't Hoff ascribed to the asymmetric carbon atom in the molecule, the ability to rotate the plane of polarised light, and showed that optical isomers are left- and right-handed forms – mirror images – of the same molecule. In 1878 van't Hoff joined the University of Amsterdam as professor of chemistry, mineralogy, and geology. In 1884 he published *Etudes de Dynamique chimique* ('Studies in Chemical Dynamics'), which contained the principles of chemical kinetics, described a new method for determining the order of a reaction, and applied the laws of thermodynamics to chemical equilibriums. He also introduced the modern concept of chemical affinity. In 1886 he showed a similarity between the behaviour of dilute solutions and gases and demonstrated that the 'osmotic pressure' in solutions which are sufficiently dilute is proportionate to the concentration and the absolute temperature so that this pressure can be represented by a formula which only deviates from the formula for gas pressure by a coefficient *i*. He also determined the value of *i* by various methods. Thus he was able to prove that thermodynamic laws are not only valid for gases, but also for dilute solutions. van't Hoff was awarded the first Nobel Prize for Chemistry in 1901 in recognition of the extraordinary services he has rendered by the discovery of the laws of chemical dynamics and osmotic pressure in solutions. (☞ Mar. 1)

1871
Ernest Rutherford, New Zealand-born British physicist, Nobel laureate, was born in Spring Grove, New Zealand. Rutherford received his early education in Government schools. He received his master's degree from the Canterbury University College, Christchurch, with a double first in mathematics and physical science in 1893 and a BSc degree in geology and chemistry from the same college in 1894. He entered the Trinity College, Cambridge, England, on a '1851 Exhibition Science Scholarship' to do research at the Cavendish Laboratory under the English physicist J.J. Thomson. In 1898 he joined the faculty of the McGill University, Montreal, Canada, and returned

to England in 1907 where he joined the Manchester University as Langworthy Professor of Physics. In 1919 he succeeded Joseph John Thomson as Cavendish Professor of Physics at Cambridge. Rutherford's researches laid the groundwork for the development of nuclear physics. He developed several ingenious techniques to study the mechanism whereby normally insulating gases become electrical conductors when a high voltage is applied across them. When X-rays were discovered a few months later, he used them to initiate electrical conduction in gases. He repeated the same with rays from radioactive atoms when they were discovered in 1896. His interest soon switched to understanding radioactivity itself. While at the McGill University, Rutherford discovered radon, a chemically unreactive but radioactive gas. Later, with the help the English chemist Frederick Soddy, he unravelled the mysteries of radioactivity and announced that radioactive decay involves a series of transformations. He also discovered that two quite separate types of emissions came from radioactive atoms and he named them alpha and beta rays. In 1908 he showed that the alpha particle was a helium atom stripped of its electrons. Beta rays were soon shown to be high-speed electrons. With the German physicist Hans Geiger, he devised the alpha-particle scattering experiment that led in 1911 to the discovery of the atomic nucleus. In 1919 he produced the first artificial disintegration of an element when he bombarded nitrogen with alpha particles and transmuted it into an isotope of oxygen. Rutherford was awarded the 1908 Nobel Prize for Chemistry for his investigations into the disintegration of the elements, and the chemistry of radioactive substances. (☞ Oct. 19)

1884

Theodor H.E. Svedberg, Swedish chemist, Nobel laureate, was born in Fleräng, near Gävle, Sweden. Svedberg obtained his BA degree from the Uppsala University in 1905. He received his master's degree in 1907 and his PhD in 1908 from the same university. In 1909 he joined the faculty of the Uppsala University as lecturer and demonstrator of physical chemistry and became professor of physical chemistry in 1912. Svedberg's work has mainly been concerned with colloids and macromolecular compounds. In his doctoral thesis, he described a new method of producing colloid particles and gave convincing evidence of the existence of molecules as manifested by Brownian motion. Svedberg also studied the physical properties of colloids, such as their diffusion, light absorption, and sedimentation, from which could be concluded that the gas laws could be applied to disperse systems. For the study of sedimentation he constructed the ultracentrifuge, in which large molecules in solution, such as proteins, carbohydrates, and high polymers could be put under the influence of centrifugal forces up to about 10^6 times gravity. He introduced the ultracentrifuge as a means for investigating the molecular weights of very large molecules. He showed that the molecules of certain pure proteins are all of one size and demonstrated the use of the ultracentrifuge to show the presence of contaminants. He also contributed to the development of the cyclotron and the electrophoresis process of chemical analysis. Svedberg was awarded the 1926 Nobel Prize for Chemistry for his work on disperse systems. (☞ Feb. 25)

1901
Hubert Cecil Booth, English mechanical and civil engineer was issued a patent for a dust removing 'suction cleaner'. It was the predecessor of the modern vacuum cleaner. Earlier vacuum cleaners used compressed air to blow the dust away, but Booth came up with the idea of sucking up dust, instead of blowing it away. Built on a horse-drawn cart, Booth's vacuum machine had an engine driving a pump provided with a long hose to extend into a house to be cleaned. One of its earliest tasks was to clean the great blue coronation carpet in Westminster Abbey for the coronation of Edward VII. The first commercial bag-on-a-stick upright vacuum cleaner was produced by William Henry Hoover in Ohio, USA, in 1907. (☞ Jun. 8)

1912
Edward Mills Purcell, American physicist, Nobel laureate, was born in Taylorville, Illinois, USA. Purcell was educated in public schools in Taylorville and in Mattoon, Illinois, and graduated in electrical engineering from the Purdue University in Indiana in 1933. He received his PhD from the Harvard University, Cambridge, Massachusetts, in 1938. After serving two years as an instructor in physics at Harvard, he joined the Radiation Laboratory, Massachusetts Institute of Technology, which was organised in 1940 for military research and development of microwave radar. He became head of the Fundamental Developments Group in the Radiation Laboratory, which was concerned with the exploration of new frequency bands and the development of new microwave techniques. In 1946 he developed his nuclear magnetic resonance (NMR) detection method, which was extremely accurate. NMR, which detects and measures the magnetic fields of atomic nuclei, has become widely used to study the molecular structure of pure materials and the composition of mixtures. In 1949 Purcell returned to Harvard as associate professor of physics and continued to work in the field of nuclear magnetism, with particular interest in relaxation phenomena, related problems of molecular structure, measurement of atomic constants, and nuclear magnetic behaviour at low temperatures. Purcel was awarded the 1952 Nobel Prize for Physics, jointly with the Swiss-born American physicist Felix Bloch, for their development of new methods for nuclear magnetic precision measurements and discoveries in connection therewith. (☞ Mar. 7, Oct. 23)

1928
Wilhelm Wien, German physicist, Nobel laureate (Physics, 1911), who formulated the displacement law concerning the radiation emitted by the perfectly efficient blackbody, died in Munich, Germany. (☞ Jan. 13)

1940
Joseph John Thomson, English physicist, Nobel laureate (Physics, 1906), who helped revolutionise the knowledge of atomic structure by his discovery of the electron, died in Cambridge, England. (☞Apr. 30, Dec. 18)

1984
American space shuttle *Discovery*, carrying a six-man crew of Henry W. Hartsfield, Michael L. Coats, Judith A. Resnik, Steven A. Hawley, Richard M. Mullane, and Charles D. Walker, was launched on its first flight from the Kennedy Space Center, Florida, USA. It carried aloft

three communications satellites – *SBS-D*, *Syncom IV-2* and *Telstar* – for deployment by its crew. (☞ Sep. 5)

August 31

1821

Hermann von Helmholtz, German scientist who made fundamental contributions to physiology, optics, electrodynamics, and mathematics, was born in Potsdam, Prussia (now Germany). In 1838, after attending the gymnasium, Helmholtz entered the Friedrich Wilhelm Medical Institute in Berlin, to study medicine. At the institute he attended physics lectures, and worked his way through the standard textbooks of higher mathematics. After graduation from the medical school in 1843, Helmholtz was assigned to a regiment at Potsdam. Because his army duties were few, he did experiments in a makeshift laboratory which he set up in the barracks. His first important scientific achievement, an 1847 physics treatise on the conservation of energy was written in context of his medical studies and philosophical background. He had discovered the principle of conservation of energy while studying muscle metabolism. He tried to prove that no energy is lost in muscle movement, and that there were no 'vital forces' necessary to move a muscle. In 1851 he invented the ophthalmoscope, an instrument that can be used to look into the human eye. His work on the physiology of the senses also provided empirical theories on spatial and colour vision. In 1855 he became professor of anatomy and physiology at the University of Bonn, moving in 1858 to Heidelberg. During these years his scientific interests progressed from physiology to physics and he started studies in electromagnetism. He predicted electromagnetic radiation from Maxwell's equations, and the wave equation now carries his name. His growing scientific stature was further recognised in 1871 by the offer of the professorship of physics at the University of Berlin. In 1888, he was appointed as first director of the Physico-Technical Institute at Berlin, the post that he held for the rest of his life. (☞ Sep. 8)

1909

Chemotherapy was first used by the German Nobel laureate Paul Ehrlich and his colleague Hata Sahachiro to treat a rabbit infected with syphilis. The infected rabbit was injected with what was dubbed as 'Preparation 606' – marking the 606th chemical devised and tested by Ehrlich's team at his Frankfurt laboratory. The compound was so successful that the sores on the rabbit promptly healed. The term 'chemotherapy' was coined by Erhlich. (☞ Mar. 14)

1913

Bernard Lovell, English radio astronomer, was born in Gloucestershire, Eng-land. Lovell studied physics at the University of Bristol and after graduating in 1934, went on to do research, obtaining his PhD in 1936. He then taught physics at the University of Manchester for a year before becoming a member of the cosmic ray research team at the university. He worked in this capacity until the outbreak of World War II in 1939. During the war, Lovell worked for the Air Ministry conducting valuable research to develop a radar system to improve

the aim of bombers in night raids. After World War II, using an ex-army mobile radar unit he showed that radar echoes could be obtained from daytime meteor showers, which were invisible using optical astronomical techniques. Lovell was the leader of the team that built, at Jodrell Bank, near Manchester, what was then the largest steerable radio telescope, with a dish diameter of 76 metres. Completed in 1957, just in time to track the first artificial satellite, *Sputnik-I*, the telescope has been instrumental in pinpointing the exact locations of manned and unmanned satellites and space probes. Lovell was appointed to the posts of senior lecturer in the University of Manchester in 1947; he became reader in 1949 and then professor of radio astronomy in 1951, a position he held until 1980. From 1951 to 1981, he was also director of the Jodrell Bank Experimental Station (now the Nuffield Radio Astronomy Laboratories). In 1950, Lovell discovered that galactic radio sources emitted at a constant wavelength and that the fluctuations recorded on the Earth's surface were introduced only as the radio waves met and crossed the ionosphere. In 1958 the British Broadcasting Corporation invited Lovell to give a series of radio talks known as the 'Reith Lectures,' which were published a year later under the title of *The Individual and the Universe*.

1949

Hugh David Politzer, American theoretical physicist, Nobel laureate, was born in New York City, New York, USA. Politzer graduated from the Bronx High School of Science in New York 1966, received his bachelor's degree from the University of Michigan in 1969, and his PhD in 1974 from the Harvard University, Cambridge, Massachusetts. After graduating from Harvard, Politzer moved to the California Institute of Technology in Pasadena, California, where he is currently professor of theoretical physics. In 1973, Politzer published his first paper in which he proposed that in the atomic nucleus, the closer the quarks are to each other, the weaker the strong interaction, given by the colour charge, will be between them. This meant that when quarks are extremely close together, the nuclear force between them is so weak that they behave almost like free particles. This result, which came to be known as 'asymptotic freedom', was independently discovered at around the same time by the American physicists David J. Gross and Frank Wilczek at the Princeton University. The concept of asymptotic freedom has turned out to be extremely important in the development of quantum chromodynamics, the theory of strong nuclear interactions. Politzer was awarded the 2004 Nobel Prize for Physics, jointly with Gross and Wilczek, for the discovery of asymptotic freedom in the theory of the strong interaction. (☞ Feb. 19, May 15)

1955

The first Sun-powered automobile was publicly demonstrated in Chicago, Illinois, USA. The car, built by William G. Cobb of the General Motors Corporation used an array of 12 selenium photovoltaic cells that turned sunlight into electricity, which was then used to power the vehicle.

1971

The world's first videocassette recorder was launched by the Dutch electronics company Philips at the Internavex exhibition in London, England. The Philips video compact cassette (VCC) was the first practical home video cassette recorder system; it used square cassettes with a recording time of one hour. The machine was equipped with crude timers that used rotary dials, but it was expensive and the system never caught on. It was not until the late 1970s, when European and Japanese companies developed more technically advanced machines with more accurate electronic timers and greater tape duration that the VCR started to become a mass market consumer product.

1983
INSAT-1B, the second of India's INSAT series of multipurpose geostationary satellites, was deployed in orbit from the American space shuttle *Challenger* during its 3rd flight. The satellite was designed by the Indian Space Research Organisation and built by Ford Aerospace, USA. The satellite carried 12 C-band transponders and 3 S-band transponders, and a very high resolution radiometer, and was meant to provide telecommunication, networking and meteorological services. The satellite was placed in the geostationary orbit at 74°E longitude and was commissioned in November 1983.

1985
Frank Macfarlane Burnet, Australian physician and virologist, Nobel laureate (Physiology or Medicine, 1960), who did pioneering work in explaining the immune process, died in Melbourne, Australia. (☞ Sep. 3)

2002
George Porter, English chemist, Nobel laureate (Chemistry, 1967), who co-developed the technique of flash photolysis for studying extremely fast chemical reactions, died in Canterbury, England. (☞ Dec. 6)

September 1

1877
Francis William Aston, British physicist and chemist, Nobel laureate, was born in Harborne, Birmingham, England. Aston attended the Birmingham University to study chemistry and physics. In 1898 he was awarded the Forster Scholarship, which allowed him to conduct research on the optical properties of tartaric acid derivatives; the results of this work were published in 1901. He began working as an assistant at the Cavendish Laboratory in Cambridge in 1909 and conducted research on the properties of the 'Crookes dark space' in discharge tubes. Within a short time he had discovered the phenomenon, which is known as the 'Aston dark space.' In 1910 he became an assistant to the British physicist Joseph John Thomson at Cambridge, who was investigating positively charged rays emanating from gaseous discharges. Together they obtained, from experiments with neon, the first evidence of isotopes. In 1919, Aston made his most significant contribution to atomic science with the invention of the mass spectrograph – a device that separates atoms or molecular fragments of different mass and measures those masses with remarkable accuracy. Using the mass spectrograph, Aston successfully identified 212 naturally occurring isotopes. The mass spectrograph enabled him to discover that elements are composed of atoms of varying mass, and that the atomic weight of an element is an average of the atoms comprising it. The mass spectrograph also prompted him to devise his famous 'whole-number rule', which states that, 'the mass of the oxygen isotope being defined, all the other isotopes have masses that are very nearly whole numbers.' The rule became crucial to later developments in nuclear energy technology. Aston was awarded the 1922 Nobel Prize for Chemistry for his discovery, by means of his mass spectrograph, of isotopes, in a large number of non-radioactive elements, and for his enunciation of the whole-number rule. (☞ Nov. 20)

1939
Polythene was commercially produced for the first time by the Imperial Chemical Industries Ltd, at their Wallerscote plant in Cheshire, England. Polythene had a major impact on the war effort. The long chain, low density white polymer could be extruded as film and coated around wires and cables to give greatly enhanced performance. Polythene was a superb insulator with high dielectric and low loss factor and was thus in great demand for submarine cables, radar applications, etc.

1979
American unmanned spacecraft *Pioneer-11* flew to within 21,000 kilometres of Saturn and took the first close-up pictures of the planet. The spacecraft's instruments located two previously undiscovered small moons and an additional ring, and charted Saturn's magnetosphere and magnetic field. It also found the planet-size moon, Titan, to be too

cold for life. Hurtling underneath the ring plane, *Pioneer-11* sent back amazing pictures of Saturn's rings. The rings, which normally seem bright when observed from Earth, appeared dark in the *Pioneer* pictures, and the dark gaps in the rings seen from Earth appeared as bright rings. (☞ Apr. 5, Dec. 2)

1985
The wreckage of the luxury ship *Titanic* was located on the ocean floor, 3,978 metres below the North Atlantic Ocean, by the Woods Hole Oceanographic Institution's robot submarine *Argo*, operated from the US Navy research ship *Knorr*. The *Titanic* had sunk in 1912 off

the coast of Newfoundland, Canada, on her maiden voyage from Southampton in England to New York, USA, with a loss of 1,522 lives. *Argo* was equipped with an array of cameras (looking forward and down), strobes and incandescent lighting to illuminate the ocean floor.

1988
Luis W. Alvarez, American experimental physicist, Nobel laureate (Physics, 1968), who discovered more than 70 elementary particles and developed the technique of using the hydrogen bubble chamber and data analysis, died in Berkeley, California, USA. (☞ Jun. 13)

1990
A four-year old girl named Ashanthi DeSilva became the first to receive gene therapy treatment at the National Institutes of Health, Bethesda, Maryland, USA. She suffered from a genetic disease which led to the deficiency of an enzyme called adenosine deaminase (ADA) and left her defenceless against infections. White blood cells were taken from her, and the normal genes for making adenosine deaminase were inserted into them. They were then re-injected into her blood. Doctors W. French Anderson and his colleagues R. Michael Blaese and Kenneth Culver, all then working at the NIH, used a virus to deliver the correct ADA gene to Ashanthi. She received repeated treatments over a period of two years. After the treatment, Ashanthi recovered fully and attended school like any normal child.

1995
The 220-MW second unit of the Kakrapar Atomic Power Station at Kakrapar, near Surat, Gujarat, India, was commissioned. It is a pressurised heavy water reactor that uses natural uranium as fuel and heavy water as moderator and coolant. (☞ May 6)

September 2

1853
Friedrich Wilhelm Ostwald, German chemist, Nobel laureate, was born in Riga, Latvia. Ostwald took his doctorate from the University of Dorpat (now Tartu State University, Tartu, Estonia) in 1878 and taught at Riga before joining the University of Leipzig in Germany as professor of physical chemistry in 1887. Ostwald started his experimental work in 1875, with an investigation on the law of mass action of water in relation to the problems of chemical affinity, with special emphasis on electrochemistry and chemical dynamics.

His pioneering work especially in the field of electrochemistry, led to the discovery of the law of dilution named after him. Ostwald became one of the founders of classical physical chemistry. In 1894 he gave the first modern definition of a catalyst and turned his attention to catalytic reactions. His process for the conversion of ammonia to nitric acid, the 'Ostwald process', patented in 1902, became of great industrial importance. Ostwald was awarded the 1909 Nobel Prize for Chemistry in recognition of his work on catalysis and for his investigations into the fundamental principles governing chemical equillibria and rates of reaction. (☞ Apr. 4)

1877
Frederick Soddy, English chemist and physicist, Nobel laureate, was born in Eastbourne, Sussex, England. Soddy was educated at the Eastbourne College and the University College of Wales, Aberystwyth, Wales. He graduated from the Oxford University in 1898. After two years of research at Oxford he went to Canada and from 1900 to 1902 was demonstrator in the chemistry department of the McGill University, Montreal. Here he worked with the British physicist Ernest Rutherford on problems of radioactivity. Together they published a series of papers on radioactivity and concluded that it was a phenomenon involving atomic disintegration with the formation of new kinds of matter. They also investigated the gaseous emanation of radium. Soddy then worked with the British chemist William Ramsay at the University College, London where he continued the study of radium emanation. Soddy and Ramsay were able to demonstrate, by spectroscopic means, that the element helium was produced in the radioactive decay of radium bromide. From 1904 to 1914 Soddy taught at the University of Glasgow and also did practical chemical work on radioactive materials. During this period he evolved the so-called 'Displacement Law', which stated that emission of an alpha particle from an element causes that element to move back two places in the Periodic Table. He suggested that different elements produced in different radioactive transformations were capable of occupying the same place on the Periodic Table, and was the first to name such species as 'isotopes' from the Greek words meaning 'same place.' In 1919 he joined the Oxford University as professor of chemistry, a post he held until 1937. Soddy was awarded the 1921 Nobel Prize for Chemistry for his contributions to our knowledge of the chemistry of radioactive substances, and his investigations into the origin and nature of isotopes. (☞ Feb. 18, Sep. 22)

1980
William Howard Stein, American biochemist, Nobel laureate (Chemistry, 1972), who contributed to the understanding of the connection between chemical structure and catalytic activity of the active centre of the ribonuclease molecule, died in New York, USA. (☞ Jun. 25)

1992
Barbara McClintock, American geneticist, Nobel laureate (Physiology or Medicine, 1983), who discovered transposons, or mobile genetic elements, died in Huntington, New York, USA. (☞ Jun. 16)

2001
Christiaan Barnard, South African surgeon who performed the first human heart transplant operation, died in Paphos, Cyprus. (☞ Nov. 8, Dec. 3)

September 3

1869
Fritz Pregl, Austrian chemist, Nobel laureate, was born in Laibach, Austria. Pregl attended the local Gymnasium (grammar school), and studied medicine at the University of Graz from where he received his MD in 1894. After a brief stay in Germany he returned to Graz and joined the Medico-Chemical Institute. He was appointed as a forensic chemist for the Graz circuit in 1907. His work often involved analysis of bile acids and other substances. But getting sufficient starting materials was a big problem. This fact forced him to look for methods requiring smaller amounts when making quantitative analyses of elements in compounds. By 1912 he had refined his micro-analytical technique enough to make reliable measurements of carbon, hydrogen, nitrogen, and sulphur with only 5–13 milligrams of starting material, with results as accurate as those obtained by macro-analysis. Later he perfected his techniques so that as little as 2.5 mg were adequate for an accurate analysis. Pregl also developed several micro-methods for measuring atomic groups and developed a series of apparatus, including a sensitive micro-balance, necessary for his work. His techniques are of immense importance in organic chemistry. Pregl was awarded the 1923 Nobel Prize for Chemistry for his invention of the method of micro-analysis of organic substances. (☞ Dec. 13)

1887
Leopold Stephen Ruzicka, Swiss chemist, Nobel laureate, was born in Vukovar, Croatia. Ruzicka attended the primary school in Osijek and, although there was no chemistry in the curriculum, he decided to study this subject out of his interest in the composition of natural products. He began his chemical studies in 1906 at the Technische Hochschule at Karlsrohe, Germany and completed his laboratory courses in $1^3/_4$ years. He then immediately started his doctoral work on ketenes with the German chemist Hermann Staudinger. While working as an assistant to Staudinger, he investigated the composition of pyrethrins – a plant product, toxic to insects and other cold-blooded animals. They thus opened a new chapter of alicyclic chemistry. He accompanied Staudinger to the Federal Institute of Technology in Zürich, Switzerland. He became a Swiss citizen and lectured at the institute. Ruzicka began his studies of fragrant natural compounds in 1916, with a series of investigations on monoterpenes, culminating in the discovery that the molecules of muskone and civetone, used in perfume industry, contain rings of 15 and 17 carbon atoms, respectively. He also showed that the carbon skeletons of terpenes and many other large organic molecules are constructed from multiple units of isoprene. In 1926 he became professor of organic chemistry at the University of Utrecht in the Netherlands, and three years later he returned to Switzerland

to become professor of chemistry at the Federal Institute of Technology. In the mid-1930s, Ruzicka discovered the molecular structure of several male sex hormones, notably testosterone and androsterone, and subsequently synthesised them. Ruzicka was awarded the 1939 Nobel Prize for Chemistry, jointly with the German biochemist Adolf Butenandt, for his work on polymethylenes and higher terpenes. (☞ Mar. 24, Sep. 26)

1899

Frank Macfarlane Burnet, Australian physician and virologist, Nobel laureate, was born in Traralgon, Australia. Burnet was educated at the Victoria State Schools and at Geelong College, completing his medical course at the University of Melbourne, where he received his MB, BS degree in 1922, and his MD in 1923. During 1926–27, he was a research fellow at the Lister Institute of Preventive Medicine in London, UK. He became assistant director of the Walter and Eliza Hall Institute of Medical Research at Royal Melbourne Hospital in 1928 and later was its director and professor of experimental medicine at the University of Melbourne. Burnet's experimental work on bacteriophages (viruses that attack bacteria) and animal viruses, especially influenza virus, resulted in major discoveries concerning their nature and replication. In 1935 he isolated a strain of 'influenza A' virus in Australia, and subsequently did much work on serological variations of the influenza virus and on Australian strains of the swine influenza. He made pioneering use of fertile hen's eggs as hosts for virus multiplication. He proposed two concepts in immunology – acquired immunological tolerance and the clonal selection theory of antibody production – which proved to be of critical importance in stimulating research and led to a more complete understanding of immune processes. He developed the notion of immunological tolerance to explain why humans do not form antibodies to their own bodily constituents. The concepts he developed are part of the basis for current theories about viruses as cancer-causing agents. Burnet was awarded the 1960 Nobel Prize for Physiology or Medicine, jointly with the British zoologist Peter Brian Medawar, for discovery of acquired immunological tolerance. (☞ Feb. 28, Aug. 31)

1905

Carl David Anderson, American physicist, Nobel laureate, was born in New York, USA. Anderson graduated from the California Institute of Technology in 1927 with a BSc degree in physics and engineering. He was awarded his PhD degree by the same institute in 1930. His early researches were in the field of X-rays. For his doctoral thesis, he studied the space distribution of photoelectrons ejected from various gases by X-rays. In 1930, using a modified cloud chamber he studied the tracks left by cosmic rays – energetic particles reaching Earth from outer space – in a strong magnetic field. He found that besides the curved paths of negative electrons, the photographs also contained paths deviating in the opposite direction, corresponding to positively charged particles – yet having the same mass as an electron.

This particle was named 'positron'; its existence had been theoretically predicted earlier by the English theoretical physicist Paul A.M. Dirac. Anderson was awarded the 1936 Nobel Prize for

Physics, jointly with the Austrian physicist Victor Francis Hess, for his discovery of the positron. (☞ Jan.11, Jun. 24)

1938

Ryoji Noyori, Japanese chemist, Nobel laureate, was born in Kobe, Japan. Noyori had his schooling in Kobe and earned his Bachelor's degree in 1961 and Master's degree in 1963 from the Kyoto University. He received his PhD from the Kyoto University in 1967 and the following year joined the faculty at the Nagoya University, where he became professor in 1972. In 2000 he assumed the directorship of the university's Research Center for Materials Science. Noyori's research mainly concerned asymmetric hydrogenation and chiral synthesis. Organic molecules often appear in two 'chiral' forms, one the mirror image of the other, much as a person's two hands mirror each other. In nature, one form often dominates, so human cells will naturally bind to one form of a molecule over the other, while the other form is ignored or may even be harmful. Noyori's work has led to further development of chiral catalysts for hydrogenation. Noyoti was awarded the 2001 Nobel Prize for Chemistry, jointly with the American biochemist William S. Knowles and American biochemist Karl Barry Sharpless. Knowles and Noyori shared half the prize money for their work on chirally catalysed hydrogenation reactions, while Sharpless received the other half for his work on chirally catalysed oxidation reactions. (☞ Apr. 28, Jun. 1)

1976

The American unmanned spacecraft *Viking-2* landed on a gently sloping hillside in Utopia Planitia on Mars. Like its twin *Viking-1*, *Viking-2* carried three full-fledged biology laboratories packed into a space of just one cubic foot (28.3 litres). It carried instruments that examined the physical and magnetic properties of the soil; and analysed the atmosphere and weather patterns of Mars. The lander was also equipped with a robotic arm that could scoop out Martian soil for analysis in the lander's biology labs. But chemical tests on the Martian soil samples failed to bring up any evidence of life on the planet. (☞ Aug. 7, Aug. 20, Sep. 9)

2006

European Space Agency's *SMART-1* spacecraft crashed on the Moon after completing its mission. It crashed close to the planned landing site – a volcanic plain called 'Lake of Excellence', producing a brilliant infrared flash that was observed by astronomers at the 3.6-metre Canada–France–Hawaii Telescope (CFHT) on Mauna Kea, Hawaii. The spacecraft hit the Moon at about 7,200 kilometres per hour and gouged out a crater 5 to 10 metres wide. Scientists hope the impact observations will reveal the chemical make-up of the rock and dust ejected by the impact. *SMART-1* completed more than 2,000 orbits of the Moon and mapped the mineralogy of the lunar surface. The data sent back by the spacecraft would help scientists piece together the history of the Moon. One of its most important discoveries was a 'Peak of Eternal Light', a mountain top near the Moon's north pole that remains in constant, year-round sunlight and could provide an ideal site for solar-powered Moon bases. But the orbiter's main purpose was to test new technologies, including an innovative ion engine that nudged the spacecraft through space by expelling xenon ions. (☞ Sep. 28)

September 4

1888

The American inventor George Eastman was granted a patent for a roll-film camera that made photography accessible to the common man. Eastman's 'Kodak' camera used flexible rolls of film that could take as many as 100 photographs. (☞ Jul. 12, Oct. 14)

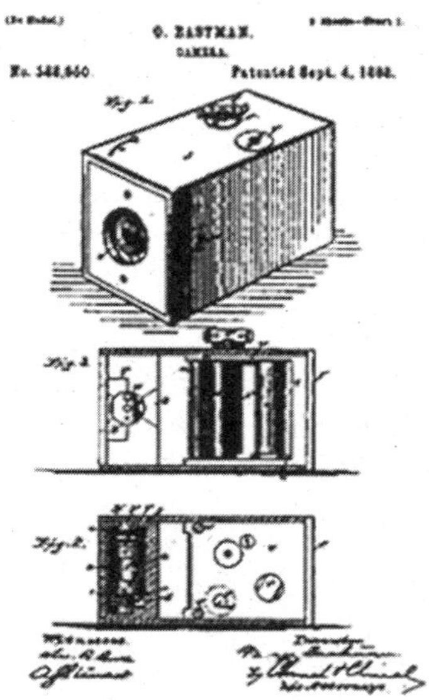

1893

Jnan Chandra Ghosh, Indian chemist and educationist, was born in Purulia, West Bengal. After completing his schooling, he joined the Presidency College, Calcutta (now Kolkata) in 1909. He obtained his MSc degree from the Calcutta University in 1915. He was awarded the Palit Scholarship to do his doctorate at the University College, London, England, during 1919–1921. At the University College, he took up investigation problems of photochemistry and placed the theory of strong electrolytes on a permanent footing. It was only after Ghosh's postulate that the concept of strong electrolytes came to be modified by the German chemist Peter Debye and German physicist Eric Hueckel, which finally gave shape to a better theory. Ghosh's career in the field of science and engineering education was spread over pre-eminent institutions in the country: He taught at the universities of Calcutta and Dacca (now Dhaka in Bangladesh), where he built up an active research school devoted to photochemistry and gas reactions. In 1939 he left Dacca for Bangalore where he joined the Indian Institute of Science as Director, succeeding C.V. Raman. In 1950 he became the first Director of the newly opened Indian Institute of Technology at Kharagpur, West Bengal, and spearheaded the development of the Institute in the field of engineering education and creating new departments – Electrical Communication Engineering, Internal Combustion Engineering, Metallurgy, and Power Engineering. He left IIT Kharagpur in 1954 to become the vice-chancellor of Calcutta University. In 1955, he joined the Planning Commission as a Member where he remained till his death. (☞ Jan. 21, Aug. 18)

1906

Max Delbrück, German-born American biologist, Nobel laureate, was born in Berlin, Germany. Delbrück studied astrophysics, and theoretical physics, at the University of Göttingen. After receiving his PhD in physics from Göttingen in 1930, he travelled through England, Denmark, and Switzerland during which he met the Austrian physicist Wolfgang Pauli and the Danish physicist Niels Bohr, who got him interested in biology. Delbrück went back to Berlin in 1932 as a research assistant at the Kaiser Wilhelm Institute for Chemistry. It was here that he got interested in bacteriophages – viruses that infect bacteria. In 1937 he moved to the USA to join the faculty of the California Institute of Technology, Pasadena where he started research on the genetics of the fruit fly, *Drosophila melanogaster*. In 1939 he joined the Vanderbilt University in Nashville, Tennessee as an instructor in physics, but returned to Caltech in 1947 as a professor of biology. While at Nashville, Delbrück developed a method for purifying viruses in order to study them and discovered that a single virus invades a single bacterial cell. In 1942, he and the Italian-born American biologist Salvador Luria demonstrated that bacterial resistance to virus infection is caused by random mutation and not adaptive change. These findings enabled others to develop vaccines against such virus-caused diseases as mumps, polio, and German measles. Delbrück was awarded the 1969 Nobel Prize for Physiology or Medicine, jointly with Luria and the American biologist Alfred Day Hershey, for their discoveries concerning the replication mechanism and the genetic structure of viruses. (☞ Mar. 9, Aug. 13, Dec. 4)

1913

Stanford Moore, American biochemist, Nobel laureate, was born in Chicago, Illinois, USA. Moore graduated from the Vanderbilt University, Nashville, Tennessee with a major in chemistry in 1935, and received his PhD in organic chemistry from the University of Wisconsin in 1938. He joined the staff of the Rockefeller University in New York in 1939, and became a professor in 1952. At Rockefeller, Moore and fellow American biochemist William Howard Stein began a collaboration that led to the development of quantitative chromatographic methods for amino acid analysis and their automation. In 1958 they developed the first automatic amino acid analyser, which drastically cut down the time required for the analysis of the amino acid sequences of proteins. In 1959 they determined for the first time, the complete chemical structure of the enzyme, ribonuclease, using their new machine. Moore was awarded the 1972 Nobel Prize for Chemistry, jointly with Stein and fellow American biochemist Christian Boehmer Anfinsen. Moore and Stein shared half the prize money for their contribution to the understanding of the connection between the chemical structure and catalytic activity of the active centre of the ribonuclease molecule, while Anfinsen received the other half for his work on ribonuclease, especially concerning the connection between the amino acid sequence and the biologically active conformation. (☞ Mar. 26, Jun. 25, Aug. 23)

1974

The Institute of Physics, an autonomous research institution of the Department of Atomic Energy, was founded. Located in Bhubaneswar, Orissa, the Institute had its beginnings in 1972 as the All India Institute of Research and Advanced Studies established by the Government of Orissa with the objective of promoting fundamental and applied research in all branches of physics and allied disciplines, and training manpower. The institute moved to the present campus on Nandankanan Road in 1981. It was taken over by the Department of Atomic Energy, India in 1985. Research programmes at the Institute covers the frontline areas of high-energy physics, condensed matter, atomic and nuclear physics and accelerator-based applications. In high energy and particle physics research, the institute has made fundamental contributions in field theories, phase transitions in the early universe, cosmology, string theory and high-energy phenomenology. Experimental research at the institute is primarily centred on a 3-MV tandem pelletron accelerator, which provides a variety of ion beams, ranging from protons to heavy ions. Accelerator mass spectrometry is applied to measure the age of samples using carbon-14 and beryllium-10 nuclides. Radiocarbon dating has applications in a vast number of areas such as geology, archaeology, oceanography, and environmental science. A compact 'molecular beam epitaxy' (MBE) facility has been installed at the institute for the study of Si and Ge multi-layers in atomic dimensions. A low energy cluster beam deposition (LECBD) facility for synthesising cluster-assembled nanomaterials has been developed.

1989
Aditya, India's first indigenously designed and built tokamak, was commissioned at the Institute for Plasma Research in Gandhinagar, Gujarat. A tokamak is a toroidal or doughnut-shaped plasma containment device used for research in nuclear fusion. *Aditya* is a medium-sized tokamak that produces a plasma stream with a major radius of 0.75 m and a minor radius of 0.25 m. The magnetic field is generated with the help of 20 toroidal field coils spaced symmetrically in the toroidal direction. *Aditya* is being used for experiments on edge plasma fluctuations, turbulence and other related fields. Research conducted with *Aditya* is primarily aimed at developing nuclear fusion as a futuristic energy source through experimental and theoretical research in plasma physics. Since its original construction, *Aditya* has been gradually upgraded with the integration of various discharge cleaning systems and online diagnostics.

September 5

1906
Ludwig Eduard Boltzmann, Austrian physicist who founded statistical mechanics, which explains and predicts how the properties of atoms determine the visible properties of matter, died in Duino, Italy. (☞ Feb. 20)

1916
Harry Brearly, English metallurgist, was awarded a patent for stainless steel. The patent related to 'new and useful improvements in cutlery or other hardened and polished articles of manufacture where non-staining properties are desired'. Brearley produced stainless steel by adding 'an amount of chromium anywhere between 9% and 16%, and also an amount of carbon not greater than 0.7%'. (☞ Feb. 18)

1939
Susumu Tonegawa, Japanese molecular biologist and immunologist, Nobel laureate, was born in Nagoya, Japan. Tonegawa earned a BS degree from the University of Kyoto in 1963 and a PhD in molecular biology from the University of California at San Diego, USA, in 1969. In 1971 he joined the newly established Basel Institute for

Immunology in Basel, Switzerland, to do research on immunology. Between 1974 and 1981, Tonegawa proved in a series of experiments that approximately 1,000 pieces of genetic material in the antibody-manufacturing portion of B lymphocytes (the cells that produce antibodies) can be shuffled or recombined into different

sequences, the resulting variations enabling the production of as many as one billion different types of antibodies, each specific to a different antigen. Prior to Tonegawa's work it was not known how the B lymphocytes could, with their limited number of genes, produce the millions of differently structured antibodies that are specific to a comparable diversity of antigens. In 1981, Tonegawa moved to the USA to become a professor of biology at the Center for Cancer Research at the Massachusetts Institute of Technology where, in addition to conducting immunological investigations, he was involved in studies of molecular and cellular aspects of neurobiology. In 1994 he became director of MIT's Center for Learning and Memory. Tonegawa was awarded the 1987 Nobel Prize for Physiology or Medicine for his discovery of the genetic principle for generation of antibody diversity.

1977
The American unmanned spacecraft *Voyager-1* was launched from the Kennedy Space Center, Florida, USA, on a mission to Jupiter and Saturn. Originally planned as *Mariner-11* of the Mariner program, *Voyager-1* was equipped with plutonium oxide radioisotope thermoelectric generators, which provided power during the mission. The spacecraft made a

number of important discoveries about Jupiter and Saturn and their moons as it flew by the giant planets. (☞ Mar. 7, Nov. 5, Nov. 12)

1984
American space shuttle *Discovery*, carrying a six-man crew of Henry W. Hartsfield, Michael L. Coats, Judith A. Resnik, Steven A. Hawley, Richard M. Mullane, and Charles D. Walker, landed at the Edwards Air Force Base, California, USA after its first flight. (☞ Aug. 30)

September 6

1766
John Dalton, English meteorologist and chemist who proposed the atomic theory, was born in Eaglesfield, Cumberland, England. Dalton taught in a school in Cumberland before becoming a teacher of mathematics and natural philosophy at New College in Manchester. He resigned this position in 1800 to become secretary of the Manchester Literary and Philosophical Society and served as a public and private teacher of mathematics and chemistry. In 1817 he became president of the Philosophical Society, an honorary office that he held until his death. For 57

years, beginning 1787, Dalton kept a daily record of the weather. His meteorological diary contained more than 200,000 observations. He was the first to confirm the theory that rain is caused not by any alteration in atmospheric pressure but by a lowering of temperature. His interest in atmospheric gases led him to investigate gases in general. In 1801 he developed the law of partial pressures, which states that the total pressure of a mixture of gases equals the sum of the pressures of the gases in the mixture, each gas acting independently. Dalton believed all matter to be composed of tiny, indivisible particles that he called 'atoms', and maintained that the atoms of different elements differ because of the difference in their masses, or weights. In 1803 he prepared the first table of atomic weights and propounded the law of multiple proportions, which states that a chemical combination of different elements occurs in simple numerical ratios by weight. He set the atomic weight of hydrogen to be identically equal to one and developed the table of atomic weights for the other elements. Dalton was the first to make a systematic study of colour blindness, which is also known as 'Daltonism.' (☞ Jul. 27)

1876

John James R. Macleod, Scottish physiologist, Nobel laureate, was born in New Clunie, Tayside, Scotland. Macleod attended the Grammar School in Aberdeen and took his medical degree with honours from the Marischal College of the University of Aberdeen in 1898. In 1899 he was appointed as demonstrator of physiology at the London Hospital Medical School, and in 1902 he became lecturer in biochemistry at the same College. In 1903, he was appointed professor of physiology at the Western Reserve University at Cleveland, Ohio, USA. In 1918 he was elected professor of physiology at the University of Toronto, Canada. Here he was associate dean of the Faculty of Medicine and director of the physiological laboratory, where Frederick Banting and Charles H. Best began investigating the secretions of the pancreas. Eventually Macleod, jointly with Banting and Best, succeeded in isolating and preparing insulin in 1921. Before this discovery, Macleod had been interested in carbohydrate metabolism and especially in diabetes since 1905. He was a very successful teacher and director of research. In 1928 he was appointed Regius Professor of Physiology at the University of Aberdeen, a post that he held until his death. Macleod was awarded the 1923 Nobel Prize for Physiology or Medicine, jointly with Banting and Best, for the discovery of insulin. (☞ Feb. 27, Mar. 16, Apr. 15, Nov. 14)

1892

Edward Victor Appleton, English physicist, Nobel laureate, was born in Bradford, UK. Appleton received a first class degree in Natural Sciences from the Cambridge University in 1913. After the outbreak of the war in August 1914, he joined the Royal Engineers, where he was trained in radio technology. After the war, he returned to Cambridge and started work under the English physicist J.J. Thomson, the discoverer of the electron, before becoming an assistant demonstrator of physics. His interest centred upon the way radio waves propagated through the atmosphere

and on interruptions of radio signals. In 1924, while carrying out an experiment on the strength of the radio signals received at Cambridge from a broadcasting station in London, Appleton discovered that the strength of the signal was constant during the day but varied during the night, rising and falling in an almost regular manner. He conjectured that, at night, the receiver at Cambridge was receiving not one but two waves, one travelling directly and the other being reflected by the atmosphere. The fading of the signal was a result of the interference. Through further experiments, Appleton was able to prove that there was a reflecting layer in the atmosphere at altitudes greater than 160 km, which reflected radio waves causing the interference and fading of signals. Appleton realised that this was the layer – now called the 'Appleton layer' – which reflected short wave radio waves round the world, making radio communication across the globe possible. Appleton was awarded the 1947 Nobel Prize for Physics for his investigations of the physics of the upper atmosphere especially for the discovery of the so-called Appleton layer. (☞ Apr. 21)

1906

Luis Federico Leloir, Argentinean biochemist, Nobel laureate, was born in Paris, France. Leloir obtained a medical degree from the University of Buenos Aires in 1932 and started his scientific career as an assistant at the Institute of Physiology, University of Buenos Aires. In 1936 he worked at the Biochemical Laboratory of Cambridge, England, for a year and returned to the Institute of Physiology, where he undertook investigations of the oxidation of fatty acids. In 1947 he set up the Institute for Biochemical Research, Buenos Aires, where he began research on the formation and breakdown of lactose, or milk sugar, in the body. In 1949 he published the discovery that became the foundation for a remarkable development. He found that in certain reactions involving the transformation of one sugar to another sugar, the participation of a so far unidentified substance containing a sugar moiety bound to a nucleotide was essential. Compounds of this type are now called 'sugar nucleotides,' which are key elements in the processes by which sugars stored in the body are converted into energy. Today more than one hundred sugar nucleotides are known and well characterised. Leloir's work demonstrated that the polysaccharide biosynthesis is not a reversal of the biological breakdown, as had been assumed earlier. Leloir also investigated the formation and utilisation of glycogen and discovered certain liver enzymes that are involved in its synthesis from glucose. Leloir was awarded the 1970 Nobel Prize for Physiology or Medicine for his discovery of sugar nucleotides and their role in the biosynthesis of carbohydrates. (☞ Dec. 2)

1943

Richard J. Roberts, English molecular biologist, Nobel laureate, was born in Derby, England. Roberts graduated from the Sheffield University in 1965 with honours in science and received a PhD in organic chemistry from the same university in 1968. After his postdoctoral research at the Harvard University, Cambridge, Massachusetts, USA, he joined Cold Spring Harbor Laboratory in New York in

1972. In 1977, working with a common cold-causing virus, called adenovirus, whose genes display important similarities to those in higher organisms, Roberts discovered that an individual gene can comprise not only one but several DNA segments separated by irrelevant DNA. Such discontinuous genes exist in organisms more complex than those studied earlier. This discovery was contrary to the general concept prevailing during the mid-1970s regarding the hereditary material and its function according to which a gene exists as a continuous stretch within a long, double-stranded DNA molecule. When the gene is activated, its information is copied into a single-stranded RNA molecule, called messenger RNA, which translates the information into a protein. Roberts' discovery changed our view on how genes in higher organisms develop during evolution. The discovery also led to the prediction of a new genetic process, namely that of splicing, which is essential for expressing genetic information. The discovery of 'split genes' has been of fundamental importance for basic research in biology, as well as for more medically-oriented research concerning the development of cancer and other diseases. Roberts was awarded the 1993 Nobel Prize for Physiology or Medicine, jointly with the American molecular biologist Phillip A. Sharp, for their discoveries of split genes. (☞ Jun. 6)

1970
Arthur William Sidney Herrington, American engineer and manufacturer who developed a series of military vehicles, the best known of which was the jeep, died in Indianapolis, Indiana, USA. (☞ Mar. 30)

1985
Rodney Robert Porter, British biochemist, Nobel laureate (Physiology or Medicine, 1972), who determined the chemical structure of antibodies, died near Winchester, Hampshire, England. (☞ Oct. 8)

September 7

1829
Friedrich August Kekulé, German chemist who laid the groundwork for the modern structural theory in organic chemistry, was born in Darmstadt, Germany. After early schooling in his native town, where he studied drawing and mathematics, Kekulé went to the University of Giessen and specialised in architecture. But later he came under the influence of the German chemist Justus von Liebig and switched to chemistry. Between 1854 and 1856 he worked as an assistant to the chemist, J. Stenhouse at the St Bartholomew's Hospital, London, where he brought out his first publication, *Valence in Organic Compounds* in English. He taught at the Heidelberg University for two years till he was called upon to be professor of chemistry at the University of Ghent in Belgium in 1858. After working there till 1867, he returned to Germany to become professor of chemistry at the University of Bonn, where he remained till his death in 1896. Kekulé made his most outstanding contribution to chemistry in 1858 when he postulated that (i) the valency of carbon is always four; (ii) carbon atoms can combine with one another to form chains of any length and complexity; (iii) the study of reaction products can give

information about the structure. Kekulé told of a 'waking dream' on a London bus, in which he had seen the atoms grouping themselves in space. The ring structure for benzene is also said to have come to him in his dream in 1865 when he saw a snake biting its tail while in a whirling motion. (☞ Jul. 13)

1914

James Alfred Van Allen, American physicist who discovered the zones of radiation encircling the Earth named after him, was born in Mt. Pleasant, Iowa, USA. Van Allen graduated from the Iowa Wesleyan College in 1935 and received his MS in 1936 and PhD in 1939 from the University of Iowa. During World War II he was a naval officer and helped develop the radio proximity fuse for naval artillery shells. In 1946 he was placed in charge of high altitude research at the Applied Physics Laboratory of the Johns Hopkins University, Silver Spring, Maryland. Van Allen's career took an important turn in 1955 when he and several other American scientists developed proposals for the launch of a scientific satellite as part of the research program conducted during the International Geophysical Year (IGY) of 1957–1958. Van Allen's *Explorer-1* spacecraft, which carried a Geiger counter designed to observe cosmic rays, was launched in January 1958, and returned enormously important scientific data, which led to the discovery of the radiation belts circling the Earth. The Earth actually has two radiation belts of different origins. The inner belt, the one discovered by Van Allen, occupies a compact region that lies about 6,400 km above the equator and is a by-product of cosmic radiation. It consists of very energetic protons. The outer radiation belt is a larger region containing ions and electrons of much lower energy – mostly plasma trapped in the magnetosphere. Van Allen was closely associated with several other American space probes including the first *Pioneers*, several of the *Mariner* probes. (☞ Jan. 31, May 1, Aug. 9)

1917

John W. Cornforth, British chemist, Nobel laureate, was born in Sydney, Australia. Despite a hearing problem, which rendered him gradually but completely deaf, Cornforth attended the Sydney Boy's High School and graduated from the Sydney University in 1937 with first class honours and a University medal. In 1939, he was awarded an Oxford University scholarship in England where he worked on the antibiotic penicillin under the British chemist Robert Robinson, receiving his PhD in 1941. He continued his collaboration with Robinson even after he joined the staff of the National Institute for Medical Research in London in 1946, and by 1951 they were able to complete, simultaneously with Robert Burns Woodward in USA, the first total synthesis of a non-aromatic steroid. While working at the Milstead Laboratory of Chemical Enzymology set up by the Shell Research Ltd., Cornforth made a study of the stereochemistry of enzymic reactions by means of asymmetry artificially introduced by isotopic substitution. Cornforth's methods helped clarify the stereochemistry – the connection between the course of chemical reactions and the geometric form of the participating molecules – of several biologically significant reactions including the synthesis of steroids from mevalonic

acid via squalene and the condensations of acetate with glyoxylic acid to malic acid and with oxalylacetic acid to citric acid. Cornforth was awarded the 1975 Nobel Prize for Chemistry, jointly with the Swiss chemist Vladimir Prelog, for his work on the stereochemistry of enzyme-catalysed reactions. (☞ Jul. 23)

1991
Edwin Mattison McMillan, American nuclear physicist, Nobel laureate (Chemistry, 1951), who co-discovered neptunium and plutonium, died in El Cerrito, California, USA. (☞ Sep. 18)

September 8

1848
Viktor Meyer, German organic chemist best known for the method of determining vapour densities named after him, was born in Berlin, Germany. Meyer was educated in Berlin, Heidelberg and Württemberg, and received his PhD from the University of Heidelberg in 1867. He was appointed professor at the Stuttgart Polytechnic in 1870 at the young age of 22. Later he became professor at the Universities of Göttingen and Heidelberg. In 1871, Meyer experimentally proved Avogadro's hypothesis by measuring the vapour densities of volatile substances (Vapour density is half the molecular weight, or relative molecular mass, of a volatile substance.). He went on to determine the vapour densities of inorganic substances at high temperatures and found that diatomic molecules of iodine and bromine dissociate into atoms upon heating. In 1876 he began to work in the field of physical chemistry and devised a new method for the determination of gas density. From benzene obtained from petroleum, Meyer in 1883 isolated thiophene, a heterocyclic compound containing sulphur, which much later was to become an important component of various synthetic drugs. Originator of the term 'stereochemistry' – the study of molecules identical in chemical structure but possessing different spatial configurations – Meyer discovered the organic compounds called oximes and demonstrated their stereoisomerism. He also coined the term 'steric hindrance' to signify the energy barrier to rotation of different parts of an organic molecule brought about by the presence in the molecule of bulky side groups. In 1888, Meyer succeeded the German chemist, Robert Bunsen as professor of chemistry at the University of Hiedelberg and remained there till his death. (☞ Aug. 8)

1894
Hermann von Helmholtz, German scientist who made fundamental contributions to physiology, optics, electrodynamics, and mathematics, discovered the principle of conservation of energy, and invented the ophthalmoscope, died in Berlin, Germany. (☞ Aug. 31)

1918
Derek Harold Richard Barton, British chemist, Nobel laureate, was born in Gravesend, Kent, England. Barton obtained his BSc in 1940 and his PhD in 1942 from the Imperial College, University of London. Barton's research mainly concerned conformation of organic molecules. In 1950, in a brief paper in *Experientia* entitled 'The conformation of the steroid nucleus', Barton showed

that organic molecules in general and steroid molecules in particular could be assigned a preferred conformation; that is, a 3-D geometric structure, based upon data gathered by chemical physicists. It was further demonstrated that the chemical and physical properties of a molecule could be interpreted in terms of that preferred conformation. Conformational analysis helps in the elucidation of configuration, in the planning of organic synthesis, and in the analysis of reaction mechanisms. It is fundamental to a complete understanding of enzymatic processes. Barton was awarded the 1969 Nobel Prize for Chemistry, jointly with Norwegian physical chemist Odd Hassel, for their contributions to the development of the concept of conformation and its application in chemistry. (☞ Mar. 16, May 17)

1965
Hermann Staudinger, German chemist, Nobel laureate (Chemistry, 1953), who demonstrated that polymers are long-chain molecules, died in Freiburg im Breisgau, Germany. (☞ Mar. 23)

1967
American Moon probe *Surveyor-5* was launched towards Moon from the Kennedy Space Center, Florida, USA. (☞ Sep. 11)

1980
Willard Frank Libby, American chemist, Nobel laureate (Chemistry, 1960), whose technique of radiocarbon dating provided an extremely valuable tool for archaeologists, anthropologists, and earth scientists, died in Los Angeles, California, USA. (☞ Dec. 17)

1981
Hideki Yukawa, Japanese physicist, Nobel laureate (Physics, 1949), who predicted the existence of mesons, died in Kyoto, Japan. (☞ Jan. 23)

1985
John Franklin Enders, American virologist and microbiologist, who discovered the ability of poliomyelitis viruses to grow in cultures of various types of tissue, died in Waterford, Connecticut, USA. (☞ Feb. 10)

September 9

1737
Luigi Galvani, Italian physicist and physician, was born in Bologna, Italy. Galvani graduated from the University of Bologna in 1759. After obtaining a doctorate in medicine in 1762, he was appointed lecturer in anatomy at the University of Bologna and professor of obstetrics at the Institute of Arts and Sciences. Galvani's early research was in comparative anatomy – such as the structure of renal tubules, nasal mucosa, and the middle ear – with a tendency toward physiology, a direction appropriate to the later work for which he is noted. His work in comparative anatomy and physiology included a study of the kidneys of birds and of their sense of hearing. His developing interest was indicated by his lectures on the anatomy of the frog in 1773 and in electrophysiology in the late 1770s, when, after acquiring an electrostatic machine and a Leyden jar, he began to experiment with muscular stimulation by electrical means. His notebooks indicate that, from the early 1780s, animal electricity remained his major field of investigation. He elicited

twitching of a dead frog's legs without the aid of the electrostatic machine by pressing a brass hook into the frog's spinal cord and hanging the hook on an iron railing. The twitching occurred only when the brass hook touched the iron railing. Galvani chose to interpret his results in terms of 'animal electricity', but his discoveries led the Italian physicist Alessandro Volta to invent the voltaic pile, a kind of battery that makes possible a constant source of current electricity. (☞ Feb. 18, Dec. 4)

1846

Elias Howe of Spencer, Massachusetts, USA, American inventor, received a patent for his hand-cranked sewing machine. At 250 stitches a minute, his lockstitch mechanism was faster than the output of five hand sewers with a reputation for speed. The first mechanical sewing machines were used in the garment factory production lines. It was not until 1889 that a sewing machine for use in the home was designed and marketed. By 1905, the electrically powered sewing machine was in wide use. (☞ Jul. 17, Aug. 12)

1896

The first successful surgery of the human heart was performed in Frankfurt, Germany by surgeon Ludwig Rehn on a 22-year-old man who was stabbed in the heart and had collapsed. Rehn repaired the 1.5-cm-wide stab wound to the right ventricle using a small intestinal needle and silk suture. The patient survived.

1922

Hans Georg Dehmelt, German-born American physicist, Nobel laureate, was born in Görlitz, Germany. Dehmelt completed his schooling at the Gymnasium zum Grauen Kloster, Berlin in 1940 and then served in the German army until the American forces captured him in the Battle of the Bulge in 1945. Having studied physics during the war at the Universität Breslau under an army technical program, he resumed his studies after the War at the University of Göttingen, and earned a doctorate in physics in 1950. He then went to the USA and began teaching at the University of Washington in Seattle, where he became a full professor in 1961, the year in which he also became an American citizen. Dehmelt developed his 'Penning trap' – a kind of ion trap – in 1955 at Seattle, for confining electrons and ions in a small space for long periods of time in relative isolation. He and his co-workers used ion-trap spectroscopy mainly to study electrons. In 1973 Dehmelt succeeded for the first time in observing a single electron in a trap, and two years later he introduced a method for 'cooling' the electron – two inventions which opened the way for the precise measurement of key properties of electrons and improved accuracy considerably. Dehmelt and his colleagues went on to develop methods for measuring atomic frequencies and individual quantum jumps (the transitions between atomic energy levels) with unprecedented precision. Later in the 1970s, Dehmelt succeeded in observing a single ion in a trap, opening the way to a new kind of spectroscopy, single-ion spectroscopy, which has been further refined. He used his trap to measure an electron's magnetic moment to an accuracy of four parts in a trillion, the most precise measurement of that quantity at the time.

Dehmelt was awarded the 1989 Nobel Prize for Physics, jointly with the German physicist Wolfgang Paul and the American physicist Norman F. Ramsey. Dehmelt and Paul shared half the prize money for the development of the ion-trap technique; Ramsey received the other half for the invention of the separated oscillatory fields method and its use in the hydrogen maser and other atomic clocks. (☞ Aug. 10, Aug. 27)

1923
Daniel Carleton Gajdusek, American physician and medical researcher, Nobel laureate, was born in Yonkers, New York, USA. Gajdusek graduated from the University of Rochester, New York, in 1943 and received his MD from the Harvard University Medical School, Boston, Massachusetts in 1946. He did his postdoctoral research at both the California Institute of Technology and the Harvard University before being drafted for military service at the Walter Reed Army Medical Service Graduate School in Washington, D.C. as a research virologist. From 1952 to 1953, he held a position at the Institut Pasteur in Tehran where he worked on rabies, plague, arbovirus infections, scurvy and other epidemic diseases in Iran, Afghanistan and Turkey. In 1954, he went to work as a visiting investigator at the Walter and Eliza Institute of Medical Research in Melbourne, Australia where he started work on a remarkable neurological disorder, called Kuru (meaning 'trembling' in the Fore language). This disease was rampant among the South Fore people of New Guinea in the 1950s and 1960s. Gajdusek lived among the Fore, studying their language and culture. He performed autopsies on Kuru victims and concluded that the disease was transmitted by the ritualistic eating of the brains of deceased relatives, which was practised by the Fore people. With elimination of this practice, Kuru disappeared among the South Fore people within a generation. Though Gajdusek was not able to identify the infective agent that spread Kuru, further research led to the identification of rogue proteins called 'prions' as the cause of Kuru. Gajdusek was awarded the 1976 Nobel Prize for Physiology or Medicine, jointly with the American physician Baruch S. Blumberg, for their discoveries concerning new mechanisms for the origin and dissemination of infectious diseases. (☞ Jul. 28)

1975
Amerian space probe *Viking-2* was launched from the Kennedy Space Center, Florida, USA. The probe carried a lander equipped with a complete biology lab to test for signs of life in Martian soil. (☞ Aug. 7, Aug. 20, Sep. 3)

1978
Soviet unmanned spacecraft *Venera-11* was launched on a mission to Venus from the Baikanur Cosmodrome in Kazakhstan. *Venera-11* was part of a two-spacecraft mission to study Venus and the interplanetary medium. Each of the two spacecraft, *Venera-11* and *Venera-12*, carried a lander probe. The spacecraft was equipped with instruments to study solar wind composition, gamma-ray bursts, ultraviolet radiation, and the electron density of the ionosphere of Venus. The lander carried eleven instruments that studied the detailed chemical composition of the atmosphere, the nature of the clouds, and heat exchange within the atmosphere. (☞ Sep. 14, Dec. 25)

1985
Paul J. Flory, American physical chemist, Nobel laureate (Chemistry, 1974), who made significant contributions to our understanding of the physical chemistry of macromolecules, died in Big Sur, California, USA. (☞ Jun. 19)

2003
Edward Teller, Hungarian-born American nuclear physicist, known as 'the father of the H-bomb', died in Stanford, California, USA. (☞ Jan. 15)

September 10

1892
Arthur Holly Compton, American physicist and engineer, Nobel laureate, was born in Wooster, Ohio, USA. Compton was educated at the Wooster College, obtaining his BSc in 1913. He spent three years in postgraduate study at the Princeton University, New Jersey, receiving his MA degree in 1914 and his PhD in 1916. In 1920, he was appointed professor of physics and

head of the department of physics at the Washington University, St. Louis, Missouri. In 1923 he moved to the University of Chicago as professor of physics. Compton returned to St. Louis as Chancellor in 1945 and from 1954 until his retirement in 1961 he was Distinguished Service Professor of Natural Philosophy at the Washington University. In his early days at Princeton, Compton began his studies in the field of X-rays. He developed a theory that made use of the intensity of X-ray reflection from crystals as a means of studying the arrangement of electrons and atoms in the crystal and in 1918 started a study of X-ray scattering. This led, in 1922, to his discovery of the increase of wavelength of X-rays due to scattering of the incident radiation by free electrons. This effect, known as the 'Compton Effect', clearly illustrates the particle concept of electromagnetic radiation, the proof of which was provided by the Scottish physicist C. T. R. Wilson who, in his cloud chamber, could show the presence of the tracks of the recoil electrons. Compton was awarded the 1927 Nobel Prize for Physics, jointly with Wilson, for his discovery of the effect named after him. (☞ Feb. 14, Mar. 15)

1920
Calyampudi Radhakrishna Rao, Indian statistician, Fellow of the Royal Society, was born. Rao earned his PhD and ScD degrees from the Cambridge University, UK. He held several important positions, as the Director of the Indian Statistical Institute, Jawaharlal Nehru Professor and National Professor in India, University Professor at the University of Pittsburgh and Eberly Professor of statistics at the Pennsylvania State University in USA. He discovered the Cramér–Rao and the Rao–Blackwell inequality both related to quality of estimators. His work not only had a profound influence in theory and applications of statistics, but also had far-reaching implications in fields as varied as economics, genetics, anthropology, geology, national planning, demography, biometry, and medicine. In 2002, Rao was honoured by the President of the USA with the prestigious National Medal of Science, 'for his pioneering contributions to the foundations of statistical theory and

multivariate statistical methodology and their applications, enriching the physical biological, mathematical, economic and engineering sciences'.

1941
Stephen Jay Gould, American palaeontologist, evolutionary biologist, and science writer, was born in New York, USA. Gould graduated from the Antioch College, Yellow Springs, Ohio, and received his PhD from the Columbia University, New York in 1967. Since then he has been professor of geology and zoology at the Harvard University, Cambridge, Massachusetts. For his doctoral thesis, he investigated variation and evolution in an obscure Bermudan land snail, anchoring his later theorising in intense scrutiny of a single group of organisms. At one point, he hoped to find a correlation between variation and different ecologies within the creature's range, but the snails' sizes, colours and shell shapes seemed to vary quite independently of local environment. Gould was a great believer in evolution's unpredictability. His best-known contribution to evolutionary theory is called punctuated equilibrium, which states that most evolution occurs in short bursts, interspersed with long periods of stasis. He criticised the belief that evolution comes up with the best or only, solution to a biological problem. He emphasised evolution's tendency to cobble together from available materials or to put something that evolved for one purpose to a different use. Gould's powers as a populariser made him famous beyond science. He authored more than 20 books – nine are collections of essays, culled from the monthly column in the *Natural History* magazine that he wrote for 25 years. (☞ May 20)

1975
George Paget Thomson, English physicist, Nobel laureate (Physics, 1937), who proved the wave nature of electrons by a diffraction experiment, died in Cambridge, UK. (☞ May 3)

1983
Felix Bloch, Swiss-born American physicist, Nobel laureate (Physics, 1952), who helped develop the nuclear magnetic resonance method of measuring the magnetic field of atomic nuclei, died in Zurich, Switzerland. (☞ Oct. 23)

September 11

1841
John Goffe Rand, an artist, was issued a US patent for collapsible metal tubes, as an 'improvement in the construction of vessels or apparatus for preserving paint, etc'. He designed the tube as a mode of preserving paints, and other fluids, by confining them in close metallic vessels so constructed as to collapse with slight pressure, and thus force out the paint or fluid confined therein through proper openings for that purpose. The tubes, moulded of lead and used to hold oil colours, were provided with caps to keep them airtight. Before Rand made the screw-cap tube, pigments ground in oil were tied in small parcels of animal bladder or tin foil. The parcels were punctured with pins to squeeze out the colour; then the pins were replaced in the holes to seal them. Rand's first tubes were manufactured in England and later exported to the USA. The idea was later used in 1892 for the commercial packaging of toothpaste. The first collapsible polythene tubes were produced for skin-tanning lotion in 1953.

1869
Thomas Graham, Scottish chemist, Fellow of the Royal Society, who propounded the law relating to the diffusion of gases known after him and is often referred to as 'the father of colloid chemistry', died in London, England. (☞Dec. 20)

1877
James Hopwood Jeans, English mathematician, physicist, and astronomer, Fellow of the Royal Society, was born in London, England. Jeans was educated in Merchant Taylor's School in London and went to the Trinity College, Cambridge, where he obtained a first class degree in the Mathematical Tripos of 1900. While he was still an undergraduate, Jeans had gained experience in experimental physics having worked in the Cavendish Laboratory during the academic year 1899–1900. He was very active in research and published work on a variety of topics in applied mathematics, physics and astronomy from 1901 onwards. Jeans was appointed a lecturer in mathematics at Cambridge in 1904. He then went to the USA and was appointed professor of applied mechanics at the Princeton University in New Jersey, from 1905 until 1909. In 1909 Jeans returned to England and the following year he was appointed Stokes Lecturer in Applied Mathematics at Cambridge. In 1923 he became a research associate at the Mt. Wilson Observatory, Pasadena, California, where he remained until 1944. In 1928 he was the first to propose the 'steady-state hypothesis', which states that the universe is constantly expanding and maintaining a constant average density through the continuous creation of new matter. The hypothesis was later developed by the British mathematician and astronomer Fred Hoyle, Austrian-born British astronomer Thomas Gold and the British cosmologist Herman Bondi. Jeans proposed a 'catastrophic' or tidal theory for the origin of the solar system, negating the nebular hypothesis of the French astronomer and physicist Pierre-Simon, Marquis de Laplace. But the current paradigm for the solar system origin suggests that its formation began with the collapse of part of an interstellar cloud of gas and dust, as postulated by Laplace. Jeans wrote extensively on astronomy. His most popular books on the subject include *The Universe around Us*, *The Mysterious Universe*, *The Stars in their Course*, and *Through Space and Time*. (☞ Sep. 16)

1952
The world's first implantation of an artificial valve in the human heart was performed by Charles A. Hufnagel, professor of surgical research at the Georgetown University Medical Center, Washington D.C., USA. The artificial 'caged ball' valve was made of a plastic called methyl methacrylate, also known as Perspex, Lucite or Plexiglass. It was inserted into the descending thoracic aorta in a patient with aortic insufficiency. It demonstrated, for the first time, that a prosthetic valve device could be successfully placed in the human circulatory system.

1967
American Moon-probe *Surveyor-5* landed on Moon in Mare Tranquillitatis, 25 km from the later *Apollo-11* landing site. It sent back 18,000 photographs and scientific data. (☞ Sep. 8)

1997
The American space probe *Mars Global Surveyor* went into an elliptical orbit

around Mars. It employed a technique known as aerobraking – using the drag of the Martian upper atmosphere on the spacecraft to slow it down gradually – to achieve a final 400-km circular polar orbit in which it circled Mars 12 times a day. This orbital configuration allowed the spacecraft to collect data from the entire Martian surface once about every seven days as Mars rotated beneath it. The spacecraft completed the primary mission in January 2001, after having observed Mars during an entire Martian year (687 Earth days), but it continued in an extended mission phase. During the mission, *Mars Global Surveyor* sent back pictures of gullies and debris flow features that suggest that there may be current sources of liquid water, similar to an aquifer, at or near the surface of the planet. The spacecraft's magnetometer data showed that the planet's magnetic field is not globally generated in the planet's core, but is localised in particular areas of the crust. New temperature data and close-up images of the Martian moon Phobos showed its surface to be composed of powdery material at least one metre thick, caused by millions of years of meteoroid impacts. Data from the spacecraft's laser altimeter have given scientists their first 3-D views of Mars' north polar ice cap. In its first three years of operation, *Mars Global Surveyor* returned more data about Mars than all prior Mars missions combined. (☞ Nov. 7)

September 12

1897

Irène Joliot-Curie, French chemist, Nobel laureate, was born in Paris, France. Irene began her formal education at a school on the Rue Cassini, near the Observatory and finished her high school education at the College of Sevigne in Paris. She entered the Sorbonne in 1914 to prepare for a degree in mathematics and physics, but her education was interrupted by World War I. In 1916, she left the Sorbonne to assist her mother Marie Curie as a nurse radiographer during the war. In 1918, Irene joined the staff of the Curie Institute as her mother's assistant. In 1925 she obtained a doctorate in science, having prepared a thesis on the alpha-rays of polonium. With her husband Frédéric, she did important work on natural and artificial radioactivity, transmutation of elements, and nuclear physics. In 1938 her research on the action of neutrons on heavy elements was an important step in the discovery of uranium fission. She was appointed lecturer in 1932, and became professor in the Faculty of Science in Paris in 1937, and afterwards, in 1946, Director of the Radium Institute. Irène was awarded the 1935 Nobel Prize for Chemistry, jointly with her husband Frédéric, in recognition of their synthesis of new radioactive elements. (☞ Mar. 17, Mar. 19)

1941

Hans Spemann, German embryologist, Nobel laureate (Physiology or Medicine, 1935), who discovered the effect now known as embryonic induction, the influence exercised by various parts of the embryo that directs the development of groups of cells into particular tissues and organs, died in Freiburg im Breisgau, Germany. (☞ Jun. 27)

1959
The Russian Moon-probe *Luna-2* was launched from Baikanur Cosmodrome in USSR (now Kazakhstan). It was the first spacecraft to crash land on the Moon. (☞ Sep. 14)

1970
The Soviet Moon-probe *Luna-16* was launched towards the Moon from the Baikanur Cosmodrome in USSR (now Kazakhstan). It was the first unmanned spacecraft to land on the Moon and carry lunar soil samples back to Earth. (☞ Sep. 20, Sep. 24)

2002
India's first exclusive meteorological satellite *Kalpana-1*, originally named *METSAT* (*MET*eorological *SAT*ellite), was launched by ISRO's polar satellite launch vehicle (PSLV) from the Satish Dhawan Space Centre at Sriharikota, Andhra Pradesh. The 1,060-kg satellite is named after the late Indian-born astronaut Kalpana Chawla, who died in the *Columbia* space shuttle accident. *Kalpana-1* is the first satellite launched by PSLV and placed in the geostationary orbit. It is positioned at 74°E longitude. For meteorological observation, *Kalpana-1* carries a very high resolution radiometer (VHRR) capable of imaging the Earth in the visible, thermal infrared and water vapour bands. It also carries a data relay transponder (DRT) for collecting data from unattended meteorological platforms. The satellite relays the data sent by these platforms to the Meteorological Data Utilisation Centre at New Delhi to aid in better weather forecasting. (☞ Feb. 1)

September 13

1886
Robert Robinson, British chemist, Nobel laureate, was born near Chesterfield, Derbyshire, England. Robinson was educated at the Chesterfield Grammar School, Fulneck School, near Leeds, and at the Manchester University where he obtained his BSc in 1906 and DSc in 1910. In 1912, he was appointed professor of pure and applied organic chemistry in the University of Sydney, Australia. He returned to England in 1915 and taught at various British universities before being appointed in 1930 to the Waynflete chair of chemistry at the University of Oxford, where he remained until his retirement in 1955. Robinson's extensive researches in organic chemistry dealt not only with the structure and synthesis of organic molecules, but also with the electrochemical mechanism of organic reactions. His interest in the chemical constitution of plant dyestuffs, mainly anthocyanins, soon extended to alkaloids – complex, naturally occurring, nitrogen-containing organic compounds that can have profound effects on living things. His efforts to determine the chemical reactions that form alkaloids in plants led him to discover the structures of morphine

and strychnine. These discoveries led to the successful production of penicillin and certain anti-malarial drugs. Robinson was awarded the 1947 Nobel Prize for Chemistry for his investigations on plant products of biological importance, especially the alkaloids. (☞ Feb. 8)

1949
Schack August Steenberg Krogh, Danish physiologist, Nobel laureate (Physiology or Medicine, 1920), who discovered the motor-regulating mechanism of capillaries, died in Copenhagen, Denmark. (☞ Nov. 15)

1955
George de Mestral, Swiss engineer and inventor, was issued a patent for Velcro – the hook and loop fasteners, named for the French words *velour* meaning velvet, and *crochet* meaning hook. Mestral got the idea from cockleburs sticking to his clothes while walking in the mountains. He discovered the natural hook-like shape of the burrs when he observed them under a microscope. After a few years of development and testing, he applied for a patent for a product in which one surface was covered with tiny hooks and the other with flexible loops. When the two surfaces were pressed together they remained stuck till pulled apart. Velcro uses two nylon tapes, one with stiff 'hooks' like the burrs, which clings to the second tape with soft 'loops' like the fabric of Mestral's clothes. After considerable experimentation, looms and proprietary hook cutting systems were developed, which set the standards for hook and loop fastener performance, consistency and quality. The product was released for the public in April 1978. (☞ Apr. 2)

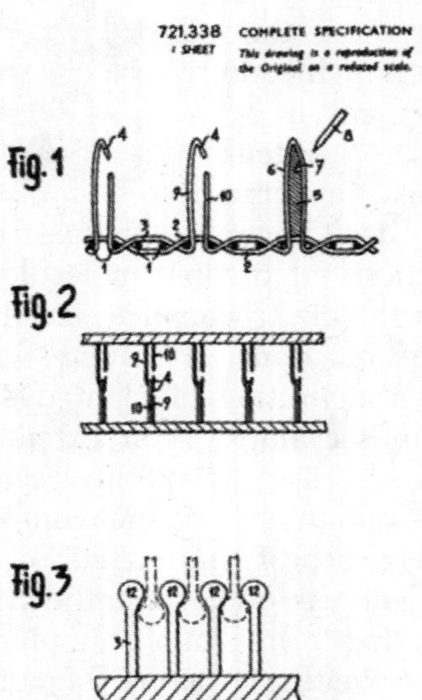

September 14

1712
Giovanni Domenico Cassini, Italian-born French astronomer who, among others, discovered 'Cassini's division,' the dark gap between the rings A and B of Saturn, died in Paris, France. (☞ Jun. 8)

1849
Ivan Petrovich Pavlov, Russian physiologist, Nobel laureate, was born in Ryazan, Russia. Pavlov was educated first at the church school in Ryazan and then at the theological seminary there. Although educated to enter priesthood, he elected to be a scientist. In 1870 he entered the University of St. Petersburg, where he studied chemistry and physiology. After receiving his MD at the Imperial Medical Academy in St. Petersburg in 1883, he studied in Germany under the direction of the German cardiovascular physiologist Carl Ludwig in Leipzig and the gastrointestinal physiologist Rudolf Heidenhain. Having worked with

Ludwig, Pavlov's first independent research was on the physiology of the circulatory system. From 1888 to 1890, at the Institute of Experimental Medicine in St. Petersburg, he investigated cardiac physiology and the regulation of blood pressure. He became so skilful a surgeon that he was able to introduce a catheter into the femoral artery of a dog almost painlessly without anaesthesia and to record the influence on blood pressure of various pharmacological and emotional stimuli. Pavlov did the bulk of his research on the physiology of digestion at St. Petersburg during the years 1891–1900. He was investigating the gastric function of dogs, by tapping a salivary gland so he could collect, measure, and analyse the saliva produced in response to food under different conditions. He noticed that the dogs tended to salivate before food was actually delivered to their mouths, by simple auditory stimulation, and set out to investigate this 'psychic secretion', as he called it. This discovery opened a new era in the development of physiology. He carried out a long series of experiments in which he manipulated the stimuli occurring before the presentation of food. In 1903 Pavlov published his results calling this a 'conditional reflex', different from an innate reflex, such as yanking a hand back from a flame, in that it only occurred conditional to specific previous experiences of the animal. Pavlov was awarded the 1904 Nobel Prize for Physiology or Medicine in recognition of his work on the physiology of digestion, through which knowledge on vital aspects of the subject has been transformed and enlarged. (☞ Feb. 27)

1936

Ferid Murad, American pharmacologist, Nobel laureate, was born in Whiting, Indiana, USA. Murad received his MD degree in 1965 from the Case Western Reserve University in Cleveland, Ohio, from which he also received a PhD in pharmacology that same year. He then joined the University of Virginia, Charlottesville, Virginia, where he was made professor in 1970, before moving to the Stanford University, California, in 1981. Murad's key work was in showing that nitroglycerine and related drugs worked by releasing nitric oxide into the body, with the gas somehow acting as a signalling molecule in the cardiovascular system. The colourless, odourless gas signals blood vessels to relax and dilate which lowers blood pressure and relieves the pain of angina pectoris. Nitric oxide was earlier known mainly as a harmful air pollutant, released into the atmosphere from automobile engines and other combustion sources. Murad's discovery opened up new applications for the molecule. For instance, the development of the successful anti-impotence drug sildenafil citrate, commonly known as *Viagra*, was based upon this research. It is now believed that in minute quantities nitric oxide acts as the body's most important signalling molecule and is profoundly involved in blood pressure, heart function, infections, lung problems, and the defence of the body against tumours, as well as having the potential to treat disease. Researchers suggest that nitric oxide could be a key to improved treatments for heart disease, shock, and cancer. Murad was awarded the 1998 Nobel Prize for Physiology or

Medicine, jointly with fellow American pharmacologists Robert F. Furchgott and Louis J. Ignarro, for their discoveries concerning nitric oxide as a signalling molecule in the cardiovascular system. (☞ May 31, Jun. 4)

1939
The first flight of a helicopter was made. The Russian inventor Igor Sikorsky made the first flight in his *VS-300* helicopter, which was constructed of an open welded tubular steel frame with three-wheel main landing gear. This machine had a single three-bladed lifting rotor and a small vertical rotor at the tail to offset twisting effects and to supply directional control. A four-cylinder 75-horsepower air-cooled engine powered both rotors. During the first flight, Sikorsky lifted the helicopter a few metres off the ground and set it down several times. During the flight the aircraft remained tethered to the ground. (☞ May 13, May 25)

1959
The Russian Moon-probe *Luna-2* crashed on the Moon east of the Sea of Serenity. It was the first man-made object to reach a celestial body. (☞ Sep. 12)

1978
Soviet unmanned spacecraft *Venera-12* was launched on a mission to Venus from the Baikanur Cosmodrome in Kazakhstan. *Venera-12* was part of a two-spacecraft mission to study Venus and the interplanetary medium. Each of the two spacecraft, *Venera-11* and *Venera-12*, carried a lander probe. The spacecraft was equipped with instruments to study solar wind composition, gamma-ray bursts, ultraviolet radiation, and the electron density of the ionosphere of Venus. The lander carried eleven instruments that studied the detailed chemical composition of the atmosphere, the nature of the clouds, and heat exchange within the atmosphere. (☞ Sep. 9, Dec. 21)

September 15

1929
Murray Gell-Mann, American theoretical physicist, Nobel laureate, was born in New York, USA. Gell-Man obtained his BSc in physics from the Yale University, New Haven, Connecticut in 1948, and his PhD from the Massachusetts Institute of Technology, Cambridge, Massachusetts, in 1951. In 1952 he joined the Institute for Nuclear Studies at the University of Chicago. His first major contribution to high-energy physics was made in 1953, when he demonstrated how some puzzling features of hadrons (particles responsive to the strong force) could be explained by a new quantum number, which he called 'strangeness'. As defined by Gell-Mann, strangeness is conserved when any subatomic particle interacts via the strong force, i.e., the force that binds the components of the atomic nucleus. In 1961 Gell-Mann and Yuval Ne'eman, an Israeli theoretical physicist, independently proposed the eightfold way (after Buddha's Eightfold Path to Enlightenment and bliss) to define the structure of particles. The scheme grouped mesons and baryons (e.g., protons and neutrons) into

multiplets of 1, 8, 10, or 27 members on the basis of various properties. This led to Gell-Mann's postulate of the quark, a name he coined. Gell-Mann joined the faculty of the California Institute of Technology, Pasadena, in 1955 and was appointed Millikan professor of theoretical physics in 1967. Gell-Man was awarded the 1969 Nobel Prize for Physics for his contributions and discoveries concerning the classification of elementary particles and their interactions.

1964
Alfred Blalock, American surgeon, who together with Vivien T. Thomas and Helen B. Taussig, developed a technique to treat infants born with the condition known as the 'blue baby' syndrome, died in Baltimore, Maryland, USA. (☞ Apr. 5)

1971
John Desmond Bernal, Irish physicist and X-ray crystallographer who did pioneering work in X-ray crystallography, died in London, England. (☞ May 10)

1977
India's first variable energy cyclotron was commissioned at the Variable Energy Cyclotron Centre (VECC) in Kolkata, West Bengal. Operated by the Department of Atomic Energy, the VECC is a national centre for accelerator-based research. The cyclotron is used for research in nuclear physics and radiochemistry, and for producing radioisotopes for various applications. Energetic heavy ion beams are now available from the cyclotron for experiments. The Centre is setting up a superconducting cyclotron in conjunction with the Bhabha Atomic Research Centre (BARC), and a radioactive ion beam facility. (☞ Jan. 20)

1990
The *Magellan* spacecraft, named after the sixteenth-century Portuguese explorer whose expedition first circumnavigated the Earth, started radar photography of the surface of Venus, using wavelengths in the centimetre range. It used synthetic aperture radar to reveal details as small as 120 metres across increasing previous resolution by a factor of 10. The four-year Magellan project involved the mapping of the whole surface of Venus. (☞ May 4, Aug. 10)

Venus mountain revealed by Magellan

September 16

1736
Gabriel Fahrenheit, German physicist, Fellow of the Royal Society, who invented the Fahrenheit scale mercury thermometer, died in The Hague, Netherlands. (☞ May 14)

1853
Albrecht Kossel, German biochemist, Nobel laureate, was born in Rostock, Germany. Kossel was educated at Rostock and at the University of Strasbourg, from where he earned his medical degree in 1878. He then joined the Physiological Institute in Berlin to carry out research on 'nuclein', a vaguely defined cell substance that had been isolated a decade earlier. By a careful series of experiments he was able to show that nuclein contains a protein portion, and a non-protein portion – the nucleic acid – which was different from any natural product known before. The breakdown products of the nucleic acids included nitrogen-containing cyclic molecules called purines and pyrimidines. Kossel isolated two purines – adenine and guanine – and three different pyrimidines – thymine, cytosine, and uracil. This work was of crucial importance for the determination of the structure of DNA several decades later. In 1895 Kossel became professor of physiology and director of the Physiological Institute at Marburg; six years later he accepted the corresponding post at the University of Heidelberg, from where he retired in 1924. Kossel was awarded the 1910 Nobel Prize for Physiology or Medicine in recognition of the contributions to our knowledge of cell chemistry made through his work on proteins, including the nucleic substances. (☞ Jul. 5)

1893
Albert Szent-Györgyi, Hungarian biochemist, Nobel laureate, was born in Budapest, Hungary. Szent-Györgyi matriculated in 1911 and joined his uncle's laboratory in the department of anatomy at the University of Budapest when World War I broke out. After the war he obtained a medical degree from the University of Budapest in 1917. Then, for more than 10 years he toured the world, and obtained his PhD from the Cambridge University, England. While working at the Cambridge University, from 1927 to 1929, and at the Mayo Foundation, Rochester, Minnesota, USA, in 1928, Szent-Györgyi found and isolated ascorbic acid and prepared it from paprika. Between 1930 and 1936, while a professor at Szeged University in Hungary, and working on the chemistry of cell respiration, he proved that hexuronic acid, also known as ascorbic acid, which he had previously isolated from plants such as oranges and lemons, is identical with vitamin C, which had been discovered in 1907 by the Norwegian hygienist and bacteriologist Axel Holst and the Norwegian paediatrician Theodor Christian Brun Frölich. Szent-Györgyi was professor at the university in Budapest from 1945 to 1947, when he emigrated to the United States. He was immediately appointed director of the Institute for Muscle Research, Woods Hole, Massachusetts, where for many years he headed the research on the causes of cellular division. While studying the biochemistry of muscular action, he discovered in the muscles a protein he termed 'actin', and demonstrated that this, in combination with the muscle protein myosin, is responsible for muscular contraction, and demonstrated that the compound adenosine triphosphate (ATP) is the primary source of energy in muscular contraction. Fumaric acid is involved in the production of energy in the form of ATP from

food. Szent-Györgyi was awarded the 1937 Nobel Prize for Physiology or Medicine for his discoveries in connection with the biological combustion processes, with special reference to vitamin C and the catalysis of fumaric acid. (☞ Oct. 22)

1932
Ronald Ross, British bacteriologist, Nobel laureate (Physiology or Medicine, 1902), who discovered that malaria is transmitted by *Anopheles* mosquitoes, died in London, England. (☞ May 13, Aug. 20)

1946
James Hopwood Jeans, English mathematician, physicist, and astronomer, Fellow of the Royal Society, who was the first to propose that matter is continuously created throughout the universe, died in Dorking, Surrey, England. (☞ Sep. 11)

1987
The 'Montreal Protocol on Substances that Deplete the Ozone Layer' was signed in Montreal, Canada. The protocol called for phasing out the production and consumption of most ozone depleting substances such as chlorofluorocarbons (CFCs), which destroy ozone in the stratosphere, by 2000. It was further adjusted and amended at subsequent meetings of the parties between 1990 and 1997. Because of their relatively high ozone depletion potential, several man-made compounds in addition to CFCs, namely, carbon tetrachloride, methyl chloroform, and halons were targeted as the first to be phased out. About 190 nations have become party to the Protocol and since it came to effect, the atmospheric concentrations of many CFCs and related hydrocarbons have either levelled off or decreased. Therefore the Montreal Protocol has often been called the most successful international environmental agreement to date.

September 17

1764
John Goodricke, English astronomer, Fellow of the Royal Society, who was the first to notice that some variable stars were periodic, was born in Groningen, Netherlands. Found to be deaf at the age of one, possibly due to a fever, Goodricke was from an early age sent to a college for the deaf and dumb in Edinburgh, Scotland. There he learned the three 'r's', and to speak and read lips. In 1778, after acquiring his basic education, he was enrolled at the Warrington Academy, in Warrington, where he excelled in mathematics and got interested in astronomy. Goodricke was the first scientist to investigate a short period variable star. He is also given credit for discovering two variable stars, beta Lyrae and delta Cephei, and with starting a whole new field of observation in astronomy. In 1781 Goodricke started making his own astronomical observations, and in November 1782 he noticed that the brightness of the variable star known as Algol varied over a period of 68 hours and 50 minutes, during which the star changed the degree of its brightness by more than a magnitude as seen from Earth. He also put forth the theory that Algol was an eclipsing binary, which was not proved until over one hundred years later. (☞ Apr. 20)

1857
Konstantin Eduardovich Tsiolkovsky, Russian research scientist in aeronautics and astronautics, was born in Ijevskoe, Ryasan Province, Russia. At the age of 10, Tsiolkovsky lost his hearing as the result of

scarlet fever. After that he could not attend school. Not receiving any formal education, he taught himself by reading books. His extraordinary success at this venture was recognised by his family who then sent him to Moscow to complete his education. Because of his proficiency in mathematics and the sciences, he eventually won a teaching post at Kaluga where he continued his research in astronautics and aeronautics. Tsiolkovsky theorised many aspects of space travel and rocket propulsion, and is considered the father of astronautics and rocket dynamics. His most famous work, *The Exploration of Cosmic Space by Means of Reaction Motors*, was published in 1903 and was the first academic treatise on rocketry. In this work, he discussed in mathematical terms the problems involved in overcoming the Earth's gravitational pull by means of rockets. In 1929, Tsiolkovsky presented a design for a multistage rocket, which he called a rocket train. He also proposed the construction of artificial Earth satellites, including manned space platforms to be used as way stations in interplanetary travel. Unfortunately his ideas never made it out of Russia, and progress in this field lagged until other scientists independently made the same calculations decades later. (☞ Sep. 19)

1877
William Fox Talbot, English mathematician, physicist and chemist who invented the negative-positive photographic process, died near Chippenham, Wiltshire, England. (☞ Feb. 11)

1901
Peter Cooper Hewitt of New York, an American electrical engineer, was issued the first US patents for a mercury vapour lamp. The eight patents covered the design of an elongated vacuum tube with a little mercury inside that used series-wired carbon filament lamps as the ballast. This converted the resistive by-product, heat, into light and the resultant configuration was the forerunner of today's self-ballasted or mercury fluorescent lamp. Light was produced when an electric current passed through the mercury vapour. However, it was a garish blue-green colour, lacking any red light.

1918
Elmer Sperry, American inventor and engineer, received a patent for the 'gyrocompass', essential to modern ship navigation. Sperry began work on gyroscopes in 1896, combining electrical and mechanical elements into gyroscopic compasses and stabilisers for ships and airplanes. His gyroscope-guided autopilot became known as a 'Metal Mike,' the mechanical helmsman. Until Sperry's work, the gyroscope was little more than a scientific curiosity. But it was known that a properly mounted gyroscope would always turn to match its plane of rotation with that of the Earth – just as a compass turns to match the Earth's magnetic field.

September 18

1819
Jean-Bernard-Léon Foucault, French physicist who introduced and helped develop a technique for measuring the absolute velocity of light with extreme

accuracy, was born in Paris, France. Foucault received his early schooling at home and after obtaining his Bachelor of Arts degree, he entered medical school but soon abandoned study when his interests turned to experimental physics. Considered one of the most versatile experimentalists of the 19th century, Foucault is well known for his experiments in optics and for proving, with the use of a pendulum, that the Earth rotates on its axis. Foucault met the French physicist Armand Fizeau in 1839 and they worked in close collaboration for almost a decade. Together they took the first detailed pictures of the Sun's surface in 1849. In 1850 Foucault developed a more precise way to measure the absolute velocity of light in the laboratory using a rapidly rotating mirror. He also proved independently that the velocity of light in air is greater than it is in water and that the velocity of light in different media varies inversely as the refractive indices of the media. While employing a pendulum to create a clock used for controlling telescopes, he stumbled upon the idea of applying the technique to prove that the Earth rotates on its axis. Foucault demonstrated his proof for the first time at the 1851 World Fair in the Pantheon in Paris, using a pendulum with a heavy bob that came to be known as the 'Foucault pendulum'. He showed that although the pendulum seemed to change its path during the day, it was actually the floor that was rotating underneath the pendulum. Because the floor is attached to the Earth, it must be the Earth that is rotating. The invention of the gyroscope followed from this. In 1855 Foucault was appointed physicist of the Paris Observatory and was one of the first to demonstrate the existence of eddy currents generated by magnetic fields. (☞ Jan. 8, Feb. 11)

1896

Armand Hippolyte-Louis Fizeau, French physicist, who was the first to measure the velocity of light, died in Nanteuil-le-Haudouin, France. (☞ Sep. 23)

1907

Edwin Mattison McMillan, American nuclear physicist, Nobel laureate, was born in Redondo Beach, California, USA. McMillan spent his early years in Pasadena, California, and received his education there. He obtained his BSc degree in 1928 and MSc degree in 1929 from the California Institute of Technology. He received his PhD from Princeton University, New Jersey, in 1932. The same year he entered the University of California at Berkeley as a National Research Fellow under the physicist Ernest O. Lawrence, and started measurement of the magnetic moment of the proton by the molecular beam method. In 1934 he became a staff member of the Radiation Laboratory, studying nuclear reactions and their products, and helping in the design and construction of cyclotrons and other equipment. He was appointed instructor in the Department of Physics at Berkeley in 1935, and became assistant professor in 1936, associate professor in 1941, and professor in 1946. During World War II McMillan also did research on radar and sonar and worked on the first atomic bomb. He was appointed director of the Lawrence Radiation Laboratory (now the Lawrence Berkeley National Laboratory) in 1958. In 1940, with the American physical chemist Philip H. Abelson he discovered neptunium (element 93), the first element heavier than uranium and with nuclear chemist Glenn T. Seaborg and others, plutonium (element 94). It was during 1945 that McMillan had the idea of

'phase stability' which led to the development of the synchrotron and synchrocyclotron; these machines have already extended the energies of artificially accelerated particles into the region of hundreds of MeV and have made possible many important researches. McMillan was awarded the 1951 Nobel Prize for Chemistry, jointly with Seaborg, for their discoveries in the chemistry of the transuranium elements. (☞ Apr. 19, Sep. 7)

1967
John Douglas Cockcroft, British physicist, Nobel laureate (Physics, 1951), who pioneered the use of particle accelerators in studying the atomic nucleus, died in Cambridge, England. (☞ May 27)

September 19

1783
The French brothers, Joseph-Michel and Jacques-Etienne Montgolfier gave the first public demonstration of their hot air balloon carrying living creatures at Versailles, France. The balloon carried a sheep, a rooster, and a duck aloft as passengers. The balloon floated for about eight minutes and landed safely more than three kilometres from the place it took off. (☞ Jan. 6, Jun. 4, Aug. 26, Nov. 21)

1848
Hyperion, moon of Saturn, was discovered by the American astronomer William Cranch Bond. With a dimension of 360 km at its widest diameter and a pockmarked body, it is the largest irregularly shaped satellite ever observed. It orbits Saturn at an average distance of 1.48 million km.

1926
Masatoshi Koshiba, Japanese physicist, Nobel laureate, was born in Toyohashi city, Aichi Prefecture, Japan. Koshiba graduated from University of Tokyo with physics major in 1951 and then joined the Graduate School at the University of Tokyo. He received his PhD in physics from the University of Rochester, Rochester, New York, USA for his thesis on 'Ultra-high-energy phenomena in cosmic rays'. From 1955 to 1958, he conducted research in the department of physics at the University of Chicago as a research associate. In 1958 he returned to the University of Tokyo as an associate professor at the Institute of Nuclear Study. From 1959 to 1962, while on leave from the University of Tokyo, he was a senior research associate and acting director of the Laboratory of High Energy Physics and Cosmic Radiation at the University of Chicago in USA. Koshiba returned to the University of Tokyo in 1963 and remained there until 1987. Thereafter he taught at the Tokai University until 1997. In 1980s, using a gigantic detector, called 'Kamiokande' at the University of Tokyo, Koshiba led a group of researchers to confirm the results of the American physicist Raymond Davis Jr. who had earlier detected the elusive subatomic particles called neutrinos. Koshiba's group captured 12 of the total of 10^{16} neutrinos that passed through the detector. They were also able to detect for the first time neutrinos from a distant supernova explosion that was observed in February 1987. Koshiba was

awarded the 2002 Nobel Prize for Physics, jointly with the American physicist Raymond Davis Jr., and the Italian-born American physicist Riccardo Giacconi, for pioneering contributions to astrophysics, in particular for the detection of cosmic neutrinos. (☞ Feb. 24, Oct. 6, Oct. 14)

1935
Konstantin Eduardovich Tsiolkovsky, Russian research scientist in aeronautics and astronautics, who is considered the father of astronautics and rocket dynamics, died in Kaluga, Russia. (☞ Sep. 17)

1957
The United States conducted its first 'fully contained' underground nuclear test, with no release of radioactivity, in the Nevada desert in Arizona. The underground nuclear detonation named the 'Rainier event' was conducted in a horizontal tunnel, about 47 metres into the mesa and 274 metres beneath the top of the mesa.

1968
Chester Floyd Carlson, American physicist who invented the dry copying process known as xerography, died in New York, USA. (☞ Feb. 8)

September 20

1842
James Dewar, British chemist and physicist, was born in Kincardine-on-Forth, Scotland. Dewar was educated at the University of Edinburgh. In 1869 he became a lecturer at the Royal Veterinary College, and it was here that he began his life's work on cryogenics. He was professor of experimental natural philosophy at the University of Cambridge, England, in 1875, and in 1877 he was appointed to the Fullerian Chair at the Royal Institute in London, where his best work – specifically cryogenics – was completed. By 1891 he had designed and erected at the Royal Institution, an apparatus that could produce several litres of liquid oxygen. He was the first person to liquefy, and later solidify, hydrogen in 1899, and show that metals had increased ability to conduct electricity at low temperatures. Later he investigated the gas-absorbing powers of charcoal when cooled to low temperatures, and applied them to the production of high vacuum and to gas analysis. Dewar invented the Dewar flask in 1892 for storing liquid gases. The Dewar flask or vacuum flask has double walls with a vacuum between and silvering on the inner walls to help reflect heat. The vacuum greatly reduces the transfer of heat, preventing a temperature change. The whole fragile flask rests on a shock-absorbing spring within a metal or plastic container. The common thermos flask is an adaptation of the Dewar flask. Dewar also invented cordite, the first smokeless powder – an explosive that was to prove vital in World War I. (☞ Jan. 19, Mar. 27)

1848
The American Association for the Advancement of Science (AAAS) was founded in Washington D.C., USA. The AAAS, is an international non-profit organisation dedicated to advancing science around the world by serving as an educator, leader, spokesperson and professional association. In addition to organising membership activities, AAAS publishes the journal *Science*, as well as many scientific newsletters, books and reports, and spearheads programmes that

raise the bar of understanding for science worldwide. The non-profit AAAS is open to all and fulfils its mission to 'advance science and serve society' through initiatives in science policy; international programs; science education; and more. It seeks to 'advance science and innovation throughout the world for the benefit of all people'.

1938
Wallace Hume Carothers, a research chemist for DuPont was issued a US patent for 'synthetic fibre' (nylon) which he had invented. The patent covered synthetic linear condensation polymers called polyamides, capable of being drawn into strong pliable fibres, as well as the process for making them. One of the first consumer uses of this new wonder plastic was replacing hog bristles in toothbrushes. (☞ Feb. 24, Apr. 27)

1954
The first FORTRAN computer program was run. A high-level programming language for mathematical and scientific purposes, FORTRAN stands for FORmula TRANSlation. John Backus at IBM supervised the development of the programming language that would allow users to express their problems in commonly understood mathematical formulae. FORTRAN was the first high-level programming language to be developed, and the concepts have been copied numerous times in the development of other programming languages. By 1958 the language was expanded to FORTRAN II, which included subroutines, functions and common blocks, and in 1962 IBM introduced the extended FORTRAN IV.

1970
Russian Moon-probe *Luna-16* landed in Mare Fecunditatis on the Moon. This was the first landing made on the dark side of the Moon, as the Sun had set about 60 hours earlier. The spacecraft was equipped with a television camera, radiation and temperature monitors, telecommunications equipment, and an extendable arm with a drilling rig for the collection of a lunar soil sample. The drill was deployed to collect lunar rocks and soil in a sample container. After 26 hours and 25 minutes on the lunar surface, the ascent stage, with the hermetically sealed soil sample container, lifted off from the Moon carrying 101 grams of collected material and returned to Earth. The lower stage of *Luna-16* remained on the lunar surface and continued transmission of lunar temperature and radiation data. (☞ Sep. 12, Sep. 24)

2004
India's exclusively educational satellite *Edusat* was launched by the geostationary satellite launch vehicle, GSLV-F01 from the Satish Dhawan Space Centre at Sriharikota, Andhra Pradesh. The satellite, built by ISRO, weighs 1,950 kilograms and carries 6 Ku-band and 6 C-band

transponders. The satellite is positioned at 74 degrees east longitude and will be used to train teachers and provide primary and secondary education by linking classrooms across India. The launch was the third for the GSLV and its first operational flight.

September 21

1853

Heike Kamerlingh Onnes, Dutch physicist, Nobel laureate, was born in Groningen, Netherlands. After attending the secondary school in his native town, Kamerlingh Onnes entered the University of Groningen, obtaining his BSc in 1871. After spending two years in Heidelberg, Germany, as a student of the German chemist Robert Bunsen and physicist Gustav Robert Kirchhoff, Kamerlingh Onnes returned to Groningen and obtained his MSc in 1878 and a PhD in 1879 from the same university. In 1882, he was appointed professor of experimental physics and meteorology at the Leiden University where he re-organised the Physical Laboratory (now known as the Kamerlingh Onnes Laboratory) in a way to suit his own programme. In particular, he had in mind the establishment of a cryogenic laboratory, which would enable him to verify his compatriot Johannes D. Van der Waals' law of corresponding states over a large range of temperatures. His efforts to reach extremely low temperatures culmi-nated in the liquefaction of helium in 1908. He brought the temperature of the helium down to 0.9, the nearest to absolute zero then achieved. In 1911 Kamerlingh Onnes made a momentous discovery – that of the superconductivity of pure metals such as mercury, tin and lead at very low temperatures – when electrical resistance of these metals disappeared totally. Kamerlingh Onnes was awarded the 1913 Nobel Prize for Physics for his investigations on the properties of matter at low temperatures, which led, *inter alia*, to the production of liquid helium. (☞ Feb. 21, Jul. 10)

1866

Charles-Jules-Henri Nicolle, French bacteriologist, Nobel laureate, was born in Rouen, France. Nicolle received early tuition in biology from his father and, after education at the Lycée Corneille de Rouen, he entered the local medical school where he studied for three years before joining the Pasteur Institute in Paris from where he obtained his MD in 1893. He returned to Rouen to become a member of the medical faculty and in 1896 he was appointed Director of the Bacteriological Laboratory. He continued in this capacity until 1903 when he was appointed Director of the Pasteur Institute in Tunis, a position he held until his death in 1936. In Tunis, Nicolle was immediately brought into contact with the scientific and practical problems that typhus epidemics had created in that country. He discovered that although typhus patients continued to spread infection before entering the hospital, they were no longer a threat after being bathed and put in new clothing, which pointed to the presence of some parasite on the patient's body and in his clothing. He subsequently identified the carrier to be the body louse (*Pediculus humanus humanus*), which he proved to be the culprit in 1909 by a series of experiments involving monkeys. Nicolle was awarded the 1928 Nobel Prize for Physiology or Medicine for his work on typhus. (☞ Feb. 28)

1926

Donald A. Glaser, American physicist, Nobel laureate, was born in Cleveland, Ohio, USA. Glaser received his early education in the public schools of Cleveland Heights, Ohio, and took his BSc degree in physics and mathematics at the Case Institute of Technology in 1946. In the same year he began his graduate studies at the California Institute of Technology, Pasadena, California, obtaining his PhD in 1949. His doctoral thesis dealt with an experimental study of the momentum spectrum of high-energy cosmic rays and mesons at sea level. Glaser began his career of full-time teaching and research in the Physics Department of the University of Michigan in 1949, becoming a professor in 1957. In 1959 he became professor of physics at the University of California, Berkeley. Glaser's main research interest was the elementary particles, particularly the strange particles. He examined various experimental techniques that seemed useful in his research and constructed a number of diffusion cloud chambers and parallel-plate spark counters before finally beginning to develop the ideas that led to the invention of the bubble chamber in 1952. Particle tracks in Glaser's chamber are composed of small gas bubbles in a superheated liquid. Glaser's idea was that an atomic particle passing through the liquid would be able to provoke boiling by means of the ions, which the atomic particle produces along its path. A flash picture records the particle's path. The bubble chamber has become a widely used instrument because it allows precise measurement of the paths of subatomic particles. Glaser was awarded the 1960 Nobel Prize for Physics for the invention of the bubble chamber.

1971
Bernardo Alberto Houssay, Argentine physiologist, Nobel laureate (Physiology or Medicine, 1947), who discovered the role of the anterior pituitary hormones in the metabolism of sugar, died in Buenos Aires, Argentina. (☞ Apr. 10)

1974
The American unmanned spacecraft *Mariner-10* makes the second close approach to Mercury. During three fly-bys at intervals of six months, *Mariner-10* could cover only half of the surface of Mercury, as the planet's slow rotation left the other half always in the dark when *Mariner-10* returned. (☞ Feb. 5, Mar. 16, Mar. 29, Nov. 3)

2003
The American spacecraft *Galileo* ended its eight-year mission to Jupiter as planned. The Jet Propulsion Laboratory in Pasadena, California directed the craft into Jupiter's atmosphere to burn up, totally vaporizing its structure. This prevented the possibility of any later uncontrolled fall onto a Jovian moon causing contamination with bacterial life from Earth. *Galileo* changed the way we look at our solar system. The spacecraft was the first to fly past a minor planet and the first to discover a moon of a minor planet. It provided the only direct observations of a comet colliding with a planet. It found evidence of subsurface saltwater on Europa, Ganymede and Callisto and revealed the intensity of volcanic activity on Io. *Galileo* was first conceived in 1976 as a mission to Jupiter and its moons, Ganymede, Callisto, Io, and Europa. (☞ Jul. 13, Aug. 28, Oct. 18)

September 22

1791

Michael Faraday, English chemist, physicist and inventor, whose work led to the practical use of electricity, was born in Newington, Surrey, England. Faraday had little formal education and was apprenticed to a bookbinder at the age of 14. There he became interested in the physical and chemical works of the time. After attending a lecture by the English chemist Humphry Davy, he sent Davy the notes he had made of his lectures. As a result Faraday was appointed, at the age of 21, as an assistant to Davy in the laboratory of the Royal Institution in London. In 1833 he was appointed professor of chemistry at the Royal Institution. Initially, Faraday occupied himself mainly with chemical problems. He discovered two new chlorides of carbon and succeeded in liquefying chlorine and other gases. He isolated benzene in 1825, the year in which he was appointed director of the laboratory. Faraday's research into electricity and electrolysis was guided by the belief that electricity is only one of the many manifestations of the unified forces of nature. His ideas about conservation of energy led him to believe that since an electric current could cause a magnetic field, a magnetic field should be able to produce an electric current. In 1831 he demonstrated this principle of electromagnetic induction, which was a landmark in applied science, for it made possible the dynamo, or generator, which produces electricity by mechanical means. Faraday also produced the first dynamo, stated the basic laws of electrolysis, and discovered that a strong magnetic field will rotate the plane of polarisation of light, which came to be known as the Faraday effect. (☞ Aug. 25, Aug. 29)

1901

Charles Brenton Huggins, Canadian-born American surgeon and urologist, Nobel laureate, was born in Halifax, Nova Scotia, Canada. Huggins received his bachelor's degree from the Acadia University, Wolfville, Nova Scotia, in 1920, and received his MD from the Harvard University Medical School, Boston, Massachusetts. USA in 1924. He then interned at the University of Michigan Hospital for further training in surgery, and in 1927 joined the faculty of the University of Chicago, where he served as director of the Ben May Laboratory for Cancer Research from 1951 to 1969. While working on prostate cancer in the early 1940s, Huggins found that he could retard its growth by blocking the action of the patient's male hormones with doses of the female hormone oestrogen. This clearly demonstrated that some cancer cells, like normal body cells, depend on hormonal signals to survive and grow and if those signals were blocked, the growth of tumours could be slowed down, at least temporarily. In 1951 Huggins showed that removing the ovaries and adrenal glands, which are the source of oestrogen, could also reverse the growth of breast cancer in some patients. Huggins's discovery of the hormone dependence of normal and neoplastic cells in experimental animals and their immediate practical application to the treatment of human prostate and breast cancer leading to the development of drugs that block the body's production of oestrogen have given many years of active and useful life to patients with advanced cancer over the entire

civilised world. Huggins was awarded the 1966 Nobel Prize for Physiology or Medicine, jointly with the American pathologist Peyton Rous, for his discoveries concerning hormonal treatment of prostate cancer. (☞ Jan. 12, Oct. 5)

1922
Chen Ning Yang, Chinese-born American theoretical physicist, Nobel laureate, was born in Hofei, Anhwei, China. Yang grew up in the campus of the Tsinghua University, just outside of Beijing, where his father was a professor of mathematics. After obtaining his bachelor's degree in 1942 at the National Southwest University in Kunming, he entered the Tsinghua University, obtaining his master's degree in 1944. He took his PhD in nuclear physics from the University of Chicago in USA in 1948 and served for a year at the same university as an instructor. In 1949, Yang joined the Institute for Advanced Study in Princeton, New Jersey, where he became a professor in 1955. While at Chicago, Yang met Tsung-Dao Lee, who had also come to Chicago on a fellowship, and the two men made a fundamental theoretical breakthrough in the field of non-conservation of parity. By 1953 it was recognised that there was a fundamental paradox in the decay of the so-called K-meson, which seemed to exhibit decay modes into configurations of differing parity. Since it was believed that parity had to be conserved, this led to a severe paradox. In 1956 Yang and Lee showed that parity – the symmetry between physical phenomena occurring in right-handed and left-handed coordinate systems – is violated in the so-called weak interactions when certain elementary particles decay. Until this discovery, it had been assumed by physicists that parity symmetry is as universal a law as the conservation of energy or electric charge. Yang was awarded the 1957 Nobel Prize for Physics, jointly with Lee, for their penetrating investigation of the so-called parity laws, which has led to important discoveries regarding the elementary particles. (☞ Nov. 25)

1956
Frederick Soddy, English chemist and physicist, Nobel laureate (Chemistry, 1921), who investigated radioactive transformations and coined the term 'isotope', died in Brighton, Sussex, England. (☞ Sep. 2)

1979
Otto Robert Frisch, Austrian-born British physicist who first described the fission of uranium nuclei under neutron bombardment, coining the term 'fission' to describe the splitting of a nucleus, died in Cambridge, England. (☞ Oct. 1)

September 23

1710
Ole Christensen Rømer, Danish astronomerwho demonstrated conclusively that light travels at a finite speed, died in Copenhagen, Denmark. (☞ Sep. 25)

1819
Armand Hippolyte-Louis Fizeau, French physicist, was born in Paris, France. Fizeau studied at the College de France and during the early period of his research collaborated with Jean-Bernard-Leon Foucault, a fellow medical student who decided on a career in physics. They met while Fizeau was working on improving the daguerreotype process. They were the first to obtain a detailed

photographic image of the Sun in 1845. In 1848 Fizeau gave an explanation of the shift in wavelength in light coming from a star (now known as the Doppler shift) and showed how it could be used to measure the relative velocities of stars that lie in the same line of sight. In 1849, almost simultaneously with Foucault, he measured the velocity of light by a laboratory experiment. He sent a narrow beam of light between gear teeth on the edge of a rotating wheel. The beam then travelled to a mirror 8 km away and returned to the wheel where, if the spin were fast enough, a tooth would block the light. Knowing this time from the rotational speed of the wheel, and the mirror's distance, Fizeau directly measured the speed of light. His figure of about 3,15,000 km/s was not as accurate as that produced by astronomical calculations, but he showed the practicability of this approach. He also found that light travels faster in air than in water, which confirmed the wave theory of light. Fizeau became a member of the French Academy in 1860 and was appointed superintendent of physics at the École Polytechnique, Paris, in 1863. (☞ Apr. 2, Sep. 18)

1846

The planet Neptune was first observed by Johann Gottfried Galle, German astronomer, from the Berlin Observatory. Galle was guided by the calculations of the French astronomer and director of the Paris Observatory, Urbain Le Verrier, who had worked out the location of the new planet as beyond Uranus, the outermost known planet at that time. (☞ Jun. 9)

1861

Robert Bosch, German engineer and industrialist who invented the spark plug and magneto for automobiles, was born in Albeck, near Ulm, Württemberg, Germany. Bosch worked as a precision mechanic for several years in renowned mechanical companies in Germany, Great Britain and the USA. In 1886 he founded his own company, the 'Workshop for precision mechanics and electro-technology', in Stuttgart. Bosch's company undertook to repair all types of electrical equipment, provide a service for the installation of telephones, house telegraphs and lightning conductors, and undertook all kinds of work in precision mechanics. Even in his first year of business, there were clear indications of his future relationship with the automobile. Bosch built a low-voltage magneto for a stationary gas motor. In 1897 Bosch had improved the low-voltage magneto for use in a motor vehicle. In 1902, a magneto ignition device with a 'spark plug' was designed which was most competitive in terms of its ignition cycling rate, precision and price. This promoted the development of high-speed petrol engines. The Bosch ignition system was exported all over the world and as early as 1914 Bosch was delivering almost 90% of his production abroad. Nearly all the larger automobile companies were his customers. With the growing importance and economic power of his enterprise, he became one of the major industrialists and philanthropists in the German economy. He encouraged general education, medicine and general welfare, civility, international understanding and the unification of Europe. (☞ Jan. 7, Mar. 9)

1877
Urbain Le Verrier, French astronomer, who predicted the existence of the planet Neptune, died in Paris, France. (☞ Mar. 11)

1880
John Boyd Orr, Scottish scientist and authority on nutrition, Nobel laureate, was born in Kilmaurs, Ayrshire, Scotland. Boyd Orr grew up in Kilmaurs, a rural community, where he studied at the local school, then studying to become a teacher at a teachers' college and Glasgow University, where he studied arts. After three years of teaching, he returned to the university to study medicine and biology, and after short stints of practice he became a researcher. In 1914 he became director of the Institute of Animal Nutrition at the University of Aberdeen and in 1929 founded the Imperial Bureau of Animal Nutrition there. During the 1920s, Boyd Orr's own research was devoted mainly to animal nutrition. During the 1930s, however, after extensive experiments with milk in the diet of mothers, children, and the underprivileged, and after large-scale surveys of nutritional problems in many nations throughout the world, his interests swung to human nutrition, not only as a researcher but also as a propagandist for healthy diets for all peoples everywhere. His report of 1936 entitled *Food, Health and Income*, revealed the 'appalling amount of malnutrition' among the people of England regardless of their economic status and became the basis for the later British policy on food during World War II. After the war, he was appointed director-general of the Food and Agriculture Organisation (FAO) of the United Nations. In 1946, under the aegis of the FAO, he set up an International Emergency Food Council, with thirty-four member nations, to meet the post-war food crisis. Boyd Orr was awarded the 1949 Nobel Prize for Peace for his efforts to eliminate world hunger. (☞ Jun. 25)

1882
Friedrich Karl Wöhler, German chemist who was the first to synthesise an organic compound from an inorganic substance, and isolated aluminium and beryllium, died in Göttingen, Germany. (☞ Jul. 31)

1915
Clifford G. Shull, American physicist, Nobel laureate, was born in Pittsburgh, Pennsylvania, USA. Shull received his BS degree from the Carnegie Institute of Technology (now the Carnegie Mellon University) in Pittsburgh in 1937 and received a doctorate in physics from New York University in 1941. Shull's most important work was done at the Oak Ridge National Laboratory in Tennessee from 1946–51. At Oak Ridge, Shull and his colleague, the cosmic ray physicist Ernest O. Wollan, systematically investigated the funda-mental principles of elastic neutron scatter-ing, leading to the development of neutron scattering techniques to analyse condensed matter. Neutron scattering has proved to be the most important single technique for elucidating the structure and dynamics of solids and fluids. Physicists earlier used X-rays to probe materials, but neutron scattering was the first technique able to detect the position of hydrogen atoms and to measure magnetic fields around atoms. Shull left Oak Ridge in 1955 to become a professor at

the Massachusetts Institute of Technology, Cambridge, Massachusetts, and retired in 1986. Shull was awarded the 1994 Nobel Prize for Physics, jointly with Canadian physicist Bertram N. Brockhouse, for the development of the neutron diffraction technique. (☞ Mar. 31, Jul. 15)

1929
Richard Zsigmondy, Austrian-born German chemist, Nobel laureate (Chemistry, 1925), who demonstrated the heterogeneous nature of colloid solutions, died in Göttingen, Germany. (☞ Apr. 1)

1930
The first commercially available photoflash bulb was patented by the German inventor Johannes Ostermeier. These bulbs were called the 'Vacublitz' and burned a thin tin foil inside a glass bulb filled with low-pressure oxygen.

1939
Sigmund Freud, Austrian neurologist, founder of psychoanalysis, died in London, England. (☞ May 6)

September 24

1541
Paracelsus (full name Theophrastus Phillippus Aureolus Bombastus von Hohenheim), German-Swiss physician and alchemist who condemned medical teaching that was not based on observation and experience, died in Salzburg, Austria. (☞ Nov. 11)

1852
The first powered and controlled flight in an airship, also known as a dirigible, was made by the French engineer and inventor Henri Giffard. He travelled a distance of 27 kilometres from the Paris racecourse to Trappes in France in a steam-powered airship at a speed of 10 km/h. His cigar-shaped, non-rigid bag was 44 metres long and could hold 3,200 cu m of hydrogen gas. A 3-horsepower steam engine drove its propeller, and it had a rudimentary vertical rudder. The gondola was suspended from a pole that hung from a net surrounding the balloon. The engine weighed 113 kilograms. The airship also had to carry a 45.4-kg boiler and the coke required to fire it. (☞ Jul. 2)

1870
Georges Claude, French engineer, chemist, and inventor of the neon light, was born in Paris. In 1897 Claude discovered that acetylene gas, which was highly explosive, could be transported safely by dissolving it in acetone. His method led to a wide expansion of the acetylene industry. In 1902 Claude developed a process for producing liquefied air, independently of the German chemist Carl von Linde. He proposed the use of liquid oxygen in iron smelting as early as 1910, but his suggestion was not adopted until after World War II. While studying the inert gases, Claude showed that when an electric current passed through certain gases such as neon, it produced a soft, coloured light. In 1910 he developed the neon lamp for use in lighting and signs. With the introduction of fluorescent coatings inside the tube, the fluorescent light was developed and began to replace the incandescent lamp in industrial and certain home-lighting uses. (☞ Jan. 15, May 23, Dec. 11)

1895

André F. Cournand, French-born American physician and physiologist, Nobel laureate, was born in Paris, France. Cournand received his early education up to secondary school level at the Lycée Condorcet, and obtained his bachelor's degree at the Faculté des Lettres of the Sorbonne in 1913. His medical studies, begun in 1914, were interrupted by his volunteering for service in the French Army. On leaving the army at the end of World War I, he resumed his medical studies and received his MD degree from the University of Paris in 1930. He then went to the USA and joined the Bellevue Hospital in New York to conduct, under the guidance of American physiologist Dickson W. Richards, research on the physiology and physiopathology of respiration. Together they collaborated in clinical lung and heart research and perfected cardiac catheterisation, developed by German surgeon Werner Forssmann, whereby a tube is passed into the heart from a vein at the elbow. He became an American citizen in 1941. The procedure developed by Cournand and Richards made it possible to study the functioning of the diseased human heart and to make more accurate diagnoses of the underlying anatomic defects. Cournand and Richards also used the catheter to examine the pulmonary artery, thus enabling improvement in the diagnosis of lung diseases as well. Cournand was appointed professor of medicine at the Columbia University College of Physicians and Surgeons in New York in 1951. He was the Chairman of the Cardiovascular Study Section of the National Heart Institute of the Public Health Service from 1956 to 1959. Cournand was awarded the 1956 Nobel Prize for Physiology or Medicine, jointly with Richards and Forssmann, for their discoveries concerning heart catheterisation and pathological changes in the circulatory system. (☞ Feb. 19, Aug. 20, Oct. 30)

1898

Howard Walter Florey, Australian pathologist, Nobel laureate, was born in Adelaide, Australia. Florey studied medicine at the University of Adelaide and received his MBBS degree in 1921. He continued his studies at the Magdalen College, Oxford University, England, as a Rhodes Scholar. In 1925 he visited the USA on a Rockefeller Travelling Fellowship for a year, returning in 1926 to a Fellowship at Gonville and Caius College, Cambridge, England, from where he received his PhD in 1927. In 1935 he became professor of pathology and a fellow of the Lincoln College, Oxford. Florey's best-known work dates from his collaboration with the German-born British biochemist Ernst Boris Chain, which began in 1938 when they conducted a systematic investigation of the properties of naturally occurring antibacterial substances. Working with Chain and the British biochemist Norman Heatley, he read Scottish bacteriologist Alexander Fleming's paper discussing the antibacterial effects of *Penicillium notatum* mould. His research team investigated the large-scale production of the mould and efficient extraction of the active ingredient, which led to the manufacture of sufficient quantities of the drug to be used in World War II to treat war wounds. Florey was elected President of the Royal Society in 1960. He was appointed chancellor of the

Australian National University, Canberra in 1965, a position he held until his death. Florey was awarded the 1945 Nobel Prize for Physiology or Medicine, jointly with Chain and Fleming, for the discovery of penicillin and its curative effect in various infectious diseases. (☞ Feb. 21, Jun. 19, Jul. 24, Aug. 6)

1904
Niels Ryberg Finsen, Danish physician, Nobel laureate (Physiology or Medicine, 1903), who made important discoveries regarding the use of light waves in the treatment of disease, died in Copenhagen, Denmark. (☞ Dec. 15)

1905
Severo Ochoa, Spanish-born American biochemist and molecular biologist, Nobel laureate, was born in Luarca, Spain. Ochoa was educated at the Málaga College, where he took his BA degree in 1921. He received his MD degree from the Medical School of the University of Madrid, in 1929. He then spent two years studying the biochemistry and physiology of muscle under the German biochemist Otto Meyerhof at the University of Heidelberg, Germany. In 1931, Ochoa was appointed Lecturer in Physiology at the University of Madrid, a post he held until 1935. In 1936, after the outbreak of the Spanish Civil War, Ochoa moved to Heidelberg, Germany where he continued the work he had started in Spain. He then joined the biochemical laboratory of the Oxford University in England in 1938, where he studied the role of vitamin B_1 in pyruvate metabolism. At Oxford he discovered the coupling of phosphorylation with oxidation of pyruvic acid in the brain. In

1942 he went to the USA and became a research associate in medicine at the New York University, New York, where he became professor of biochemistry and chairman of the department in 1954. In 1955, Ochoa discovered an enzyme named polynucleotide phosphorylase while conducting research on high-energy phosphates. It was subsequently found that the enzyme's function is to degrade RNA, not synthesise it; under test-tube conditions, however, it runs its natural reaction in reverse. The enzyme has been singularly valuable in enabling scientists to understand and re-create the process whereby the hereditary information contained in genes is translated, through RNA intermediaries, into enzymes that determine the functions and character of each cell. Ochoa continued research on protein synthesis and replication of RNA viruses until 1985, when he returned to Spain. Ochoa was awarded the 1959 Nobel Prize for Physiology or Medicine, jointly with the American biochemist Arthur Kornberg, for their discovery of the mechanisms in the biological synthesis of ribonucleic acid and deoxyribonucleic acid. (☞ Mar. 3, Nov. 1)

1925
Autar Singh Paintal, Indian physiologist, Fellow of the Royal Society, was born in Mogok in Burma (now Myanmar). Paintal received his MBBS degree from the Lucknow Medical College, Lucknow, Uttar Pradesh, and went to the University of Edinburgh, UK as a Rockefeller Fellow and obtained his PhD degree in 1953. He worked at the Vallabhbhai Patel Chest Institute in New Delhi as Assistant Director from 1954 to 1956. He also held Visiting Professorship at the Albert Einstein College of Medicine, New York, USA, and at the Universities of Utah in USA and Goettingen in Germany. After serving as Professor of Physiology at

the All India Institute of Medical Sciences in New Delhi for six years, he joined as Director of the VP Chest Institute in 1964. Paintal began his research career by introducing a new index for evaluation of galvanic skin responses in humans, new techniques for recording impulses from single active fibre nerves and providing their conduction velocities, and a method for locating visceral receptors. He is best known for his discovery in 1955 of type 'J-receptors' which cause breathlessness. These receptors produce inhibition of the somatic muscles. They produce an irritating sensation in the throat and cause a dry cough. Paintal's discoveries have had a great impact on our understanding of the behaviour of several visceral receptors and their relation to health and disease. The knowledge of the J-receptors and their various central and reflex effects discovered by him is now routinely used by clinicians in the assessment of their patients suffering from a variety of heart and lung diseases. (☞ Dec. 22)

1945
Hans Geiger, German physicist who introduced the first successful detector of individual alpha-particles and other ionising radiations, died in Potsdam, Germany. (☞ Sep. 30)

1960
The world's first nuclear-powered aircraft carrier, *USS Enterprise* was launched. The ship was equipped with four aircraft elevators and four catapults and can carry 85 planes. The power to drive its four propellers is provided by eight nuclear reactors. The ship was formally inducted into the US Navy in November 1961. With a displacement of approximately 93,500 tonnes and a speed of 35 knots, *USS Enterprise* remained the biggest and fastest warship in the world from 1961 to 1972.

1970
The re-entry capsule of the Russian Moon-probe *Luna-16*, carrying lunar soil samples, returned to Earth and parachuted down near the city of Dzhezkazgan in Kazakhstan. (☞ Sep. 12, Sep. 20)

1989
Benjamin Peary Pal, Indian agricultural scientist, Fellow of the Royal Society, died in New Delhi, India. (☞ May 26)

September 25

1644
Ole Christensen Rømer, Danish astronomer who demonstrated conclusively that light travels at a finite speed, was born in Aarhus, Denmark. Rømer was educated in Copenhagen and joined the observatory of Uraniborg on the island of Hven, near Copenhagen, in 1671. Over a period of several months, he observed about 140 eclipses of Jupiter's moon Io. In Paris, France, the Italian astronomer Giovanni Cassini also observed the same eclipses. By

comparing the times of the eclipses, the difference in longitude between Paris and Uraniborg was calculated. In 1672 Rømer went to Paris and continued observing the satellites of Jupiter as Cassini's assistant. He observed that the times between eclipses of Io got shorter as Earth approached Jupiter, and longer as Earth moved farther away. He hypothesised that this variation was due to the time it took for light to travel the lesser or greater distance, and estimated that the time for light to travel the diameter of Earth's orbit, a distance of two astronomical units, was 22 minutes. This is somewhat greater than the currently accepted value, which is about 16 minutes and 40 seconds. In 1681, he returned to Denmark and was appointed professor of astronomy at the Copenhagen University. He was also active as an observer, both at the University Observatory at the Round Tower and in his home, using improved instruments of his own construction. (☞ Sep. 23)

1866

Thomas Hunt Morgan, American zoologist and geneticist, Nobel laureate, was born in Lexington, Kentucky, USA. Morgan was educated at the University of Kentucky, where he took his BS degree in 1886. He did his postgraduate work at the Johns Hopkins University, Baltimore, Maryland, where he studied morphology and received his PhD in 1890. In 1891 he became associate professor of biology at the Bryn Mawr College for Women, Bryn Mawr, Pennsylvania, where he stayed until 1904, when he became professor of experimental zoology at the Columbia University, New York. During the next 24 years, he conducted most of his important research in heredity by which he established the chromosome theory of heredity. By carefully breeding the common fruit fly *Drosophila melanogaster*, and studying the characteristics inherited by mutants, he showed that genes are linked in a series on chromosomes and are responsible for identifiable, hereditary traits. His study of the characteristics inherited by mutants ultimately enabled him to determine the precise behaviour and exact localisation of genes. Morgan and his colleagues produced the first chromosome maps in 1911. His work played a key role in establishing the field of genetics. In 1928 Morgan was invited to organise the division of biology at the California Institute of Technology. He was also instrumental in establishing the Marine Laboratory on Corona del Mar, California, as an integral part of Caltech's biology training program. In subsequent years, Morgan and his co-workers, including a number of postdoctoral and graduate students, continued to elaborate on the many features of the chromosome theory of heredity. Morgan was awarded the 1933 Nobel Prize for Physiology or Medicine, for his discoveries concerning the role played by the chromosome in heredity. (☞ Dec. 4)

1973

The crew of *Skylab-3*, which rendezvoused with the American space station *Skylab*, splashed down safely after 59 days in space. During the mission, the crew installed a twin-pole solar shield to replace the meteoroid shield lost during lift off. The mission completed 858 Earth orbits and carried out space walks totalling 13 hours, 43 minutes. (☞ Feb. 8, May 14, May 25, Jun. 22, Jun. 28, Jul. 11, Nov. 16)

1985
William Cumming Rose, American biochemist who researched the role of amino acids in nutrition determining which were essential, and who calculated the minimum daily requirement for each of them, died in Urbana, Illinois, USA. (☞ Apr. 4)

1986
Nikolay Nikolaevich Semenov, Russian physical chemist, Nobel laureate (Chemistry, 1956), who was the first to show that chain reactions are the norm in chemical transformations of matter, died in Moscow, Russia. (☞ Apr. 15)

2004
Raja Ramanna, Indian nuclear physicist and the leading architect of the India's nuclear programme, died in Mumbai, India. (☞ Jan. 28)

September 26

1886
Archibald Vivian Hill, British physiologist and biophysicist, Nobel laureate, was born in Gloucestershire, England. Hill studied at Trinity and Kings Colleges in Cambridge and began his investigations of the physiological thermodynamics of muscle and nerve tissue at the University of Cambridge in 1911. His early experiments concerned the effects of electrical stimulation on nerve function, the mechanical efficiency of muscle, energy processes in muscle during recovery, and the interaction between oxygen and haemoglobin. Working with a strap-like thigh muscle in the frog, he was able to demonstrate that oxygen is needed only for the recovery, not the contractile phase of muscular activity, laying the foundation for the discovery of the series of biochemical reactions carried out in muscle cells that results in contraction. Later, he devised mathematical models describing heat production in muscle, and applied kinetic analysis to explain the time course of oxygen uptake during both exercise and recovery. His research helped establish the origin of muscular force in the breakdown of carbohydrates with the formation of lactic acid in the absence of oxygen. Hill was awarded the 1922 Nobel Prize for Physiology or Medicine, jointly with the German biochemist Otto Meyerhof. Hill received half the prize money for his discovery relating to the production of heat in the muscle, while Meyerhof received the other half for his discovery of the fixed relationship between the consumption of oxygen and the metabolism of lactic acid in the muscle. (☞ Apr. 12, Jun. 3)

1887
Emile Berliner, a German immigrant working in Washington, D.C., USA was granted a patent for a system for recording sound on a flat disc, or record. The system, which Berliner called the 'gramophone', involved a method for mapping out sounds in a spiralling groove etched into a flat disc (first of glass, then of zinc, then of plastic). The recording system included a device driven by a weight box and controlled by a paddle-wheeled governor that recorded four minutes of sound on an 11-inch (279 mm) glass disc at 30 rpm. After recording, the sounds could be 'read' by a needle, which transmitted the pattern of vibrations to a diaphragm, which then reproduced the original sounds. Berliner's gramophone differed from Edison's phonograph in that it used a horizontal lateral movement for

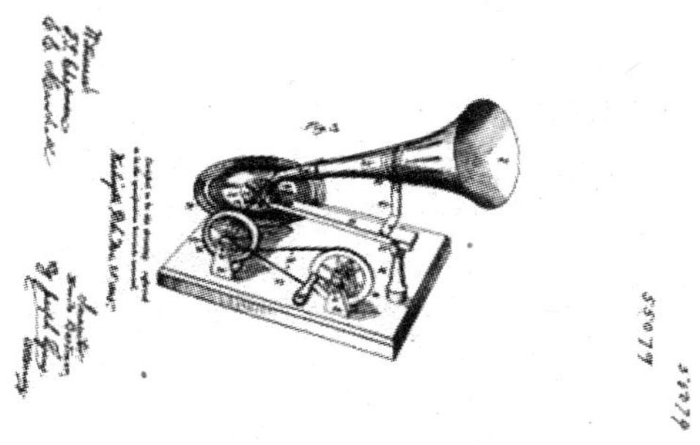

the needle, as opposed to the 'hill-and dale' principle, used by Edison for recording on cylinders. Berliner's technique allowed mass-production of records, which was not possible with the cylinder phonograph invented by Edison. (☞ May 16)

1942
The Council of Scientific and Industrial Research (CSIR) was constituted by a resolution of the Department of Commerce, Government of India. The primary objectives of the Council were to 'promote, guide and coordinate scientific and industrial research in India including the institution and financing of specific researches'. The Council was also to take up 'establishment, maintenance and management of laboratories, workshops, institutes and organisations to further scientific and industrial research and to utilise and exploit for purposes of experiment or otherwise any discovery or invention likely to be of use to Indian industries'. CSIR has since grown into India's largest network of government research laboratories, with 38 laboratories and 80 field-extension centres spread all over the country. CSIR has build, over the years, an impressive laboratory infrastructure with excellent manpower capabilities, which in most cases have been nationally and internationally acclaimed. Some areas in which CSIR has won international recognition are aerospace engineering, bio-science and bio-technology, chemistry and chemicals technology, drugs and pharmaceuticals, earth and ocean sciences, environmental technology, botanicals, toxicology and microbiology, food technology, leather, medicinal and aromatic plants and petroleum. CSIR laboratories have responded creditably to the challenges of competition following the globalisation and liberalisation of the Indian economy. It now generates a substantial portion of the income needed for its activities from sponsored work, consultancy and services and from licensing its technologies for industrial production. The notable achievements of CSIR include *Centchroman* – the world's first non-steroidal contraceptive for women, several petroleum refining technologies, eco-friendly technologies for the leather industry, *Saras* – India's first indigenous all-composite light aircraft for civilian use, and a supercomputer named *Flosolver*.

1976
Leopold Stephen Ruzicka, Swiss chemist, Nobel laureate (Chemistry, 1939), who discovered the molecular structure of several fragrant natural compounds, died in Zürich, Switzerland. (☞ Sep. 3)

1978
Karl Manne Georg Siegbahn, Swedish physicist, Nobel laureate (Physics, 1924), who made considerable improvements in the study of characteristic X-radiations of elements, died in Stockholm, Sweden. (☞ Dec. 3)

1991
Four men and four women began a two-year stay inside *Biosphere-2*, a sealed-off structure nestled in the foothills of Santa Catalina Mountains about 50 kilometres north of Tucson, Arizona, USA. Built in the late 1980s with a $150 million funding from the Texas oil magnate Edward Bass,

Biosphere-2 was designed as an airtight replica of Earth's environment. This 204,000-cubic-metre sealed glass and space-frame structure contains 5 biomes, including a 4-million-litre ocean, a rain forest, a desert, agricultural areas and a human habitat. The first *Biosphere-2* crew remained inside for two years despite various problems, including limited agricultural productivity, and emerged on September 26, 1993. After a 6-month transition period,

a second crew of seven biospherians (five men and two women) entered *Biosphere-2*. Unfortunately, after a number of physical and social problems developed, the project soon suffered scientific disdain and public ridicule. The experiments were suspended in 1994. Since then, there have been no resident crew living inside *Biosphere-2*, and no future human habitation is planned.

1996
Geoffrey Wilkinson, British chemist, Nobel laureate (Chemistry, 1973), who was the first to synthesise sandwich compounds, or metallocenes, died in London, UK. (☞ Jul. 14)

September 27

1825
The first locomotive to haul a passenger train was operated by George Stephenson's

Stockton & Darlington's line in England. The engine *Locomotion No. 1* pulled 34 wagons and one solitary coach on its journey of about 34 kilometres from Shildon, via Darlington to Stockton in County Durham. (☞ Feb. 6, Jun. 9)

1918
Martin Ryle, British radio astronomer, Nobel laureate, was born in Brighton, Sussex, England. Ryle was educated at the Bradfield College and graduated from the Oxford University in 1939. During World War II he worked on the development of radar and other radio systems for the Royal Air Force. In 1948 he was appointed a lecturer in physics and in 1949 was elected to a Fellowship at the Trinity College, Cambridge. The British astrophysicist Antony Hewish joined him at this time and together they developed a technique called aperture synthesis, which provided a tremendously increased resolving power for radio telescopes. Using interferometry, he and his team located radio-emitting regions on the Sun and pinpointed other radio sources so that they could be studied in visible light. Ryle guided the Cambridge radio astronomy group in the production of radio source catalogues, which later helped in the discovery of the first quasar in 1967. His catalogues of radio sources also led to the discovery of numerous radio galaxies

and quasars. He was Astronomer Royal from 1972 to 1982. Ryle was awarded the 1974 Nobel Prize for Physics, jointly with Hewish, for their pioneering research in radio astrophysics: Ryle for his observations and inventions, in particular of the aperture synthesis technique, and Hewish for his decisive role in the discovery of pulsars. (☞ May 11, Oct. 14)

1925
Robert Edwards, British physiologist, who together with fellow British obstetrician and gynaecologist Patrick Steptoe pioneered *in vitro* fertilisation as a means to overcome human infertility, was born in Yorkshire, England. Edwards attended the Universities of Wales and Edinburgh, and has held academic positions in Scotland and California, USA. From 1958 to 1962 he worked at the National Institute of Medical Research, London. In 1963 he moved to the department of physiology at Cambridge University. In the 1950s Edwards successfully replanted mouse embryos into the uterus of a mouse and wondered if the same process could be applied to humans. In 1965 he attempted, for the first time, the *in vitro* fertilisation of human eggs, but did not succeed. In 1966, to help women with blocked Fallopian tubes, a major cause of infertility, he teamed up with fellow British obstetrician and gynaecologist Patrick Steptoe. In 1968 Edwards and Steptoe began to collaborate on the fertilisation of human eggs and succeeded in fertilising the human ova outside the womb. In 1977 Edwards and Steptoe for the first time succeeded in implanting in the womb, an egg fertilised outside the body leading to the birth of the first 'test-tube baby' – a baby girl named Louise Joy Brown – in 1978. Their work at the Centre for Human Reproduction in Oldham, Manchester, England, resulted in the birth of more than 1,000 babies. (☞ Jun. 9, Jul. 25)

1940
Julius Wagner-Jauregg, Austrian psychiatrist and neurologist, Nobel laureate (Physiology or Medicine, 1927), who developed fever therapy and shock therapy for several mental disorders, died in Vienna, Austria. (☞ Mar. 7)

1975
Tiruvenkata Rajendra Seshadri, Indian chemist, Fellow of the Royal Society, died in New Delhi, India. (☞ Feb. 3)

1997
American space probe *Mars Pathfinder* sent back data for the last time from Mars. (☞ Jul. 4, Dec. 4)

September 28

1852
Henri Moissan, French chemist, Nobel laureate, was born in Paris, France. Moissan attended the Collège de Meaux, and later the Museum of Natural History and the School of Pharmacy in Paris. After getting his PhD degree in 1880, he became professor of toxicology in 1886 and of inorganic chemistry in 1889 at the School of Pharmacy. In 1900 he was appointed professor of inorganic chemistry at the Sorbonne in Paris. Moissan's first researches concerned the interchange of

oxygen and carbon dioxide in the leaves of plants. He soon left biology for the field of inorganic chemistry, where his early work was on the oxides of iron-group metals and chromium and a study of chromous salts. In 1884 he turned his attention to fluorine chemistry, preparing some organic and phosphorus derivatives of that element. Two years later, by electrolysing a solution of potassium fluoride in hydrofluoric acid, he prepared the highly reactive gas fluorine and made a full study of the properties of the element and its reactions with other elements. In 1892, Moisssan theorised that diamonds could be synthesised by crystallising carbon under pressure from molten iron. He designed and developed the electric-arc furnace that could attain temperatures up to 3,500°C, to assist him in work which led to the production of tiny artificial diamonds. He subsequently used the furnace to volatilise many substances that had been regarded as infusible and to prepare many new compounds, particularly carbides, silicides and borides. In 1891 he discovered carborundum, one of the hardest substances known. Moissan was awarded the 1906 Nobel Prize for Chemistry in recognition of the great services rendered by him in his investigation and isolation of the element fluorine, and for the adoption in the service of science of the electric furnace called after him. (☞ Feb. 20, Jun. 26)

1895
Louis Pasteur, French chemist and microbiologist whose contributions were among the most varied and valuable in the history of science and industry and whose work became the foundation for the science of microbiology, and a cornerstone of modern medicine, died in Saint-Cloud, near Paris, France. (☞ May 22, Dec. 27)

1953
Edwin Powell Hubble, American astronomer, who is considered the founder of extragalactic astronomy and who gave the proof of an expanding universe, died in San Marino, California, USA. (☞ Nov. 20)

2003
INSAT-3E, the fourth of India's third generation INSAT series of satellites, was launched by the Ariane-5 launch vehicle of the European launch provider Arianespace from Kourou, French Guyana. Designed and built by the Indian Space Research Organisation, the 2,775-kg communication satellite carries 24 C-band transponders and 12 upper extended C-band transponders to further augment telecom and television services in the country. *INSAT-3E* is positioned at 55° East longitude in the geostationary orbit.

2003
SMART-1, the first spacecraft using a solar–electric propulsion system (ion engine) was launched by the European Space Agency (ESA) on an exploratory mission to the Moon. The spacecraft was launched along with India's *INSAT-3E* by an Ariane-5 rocket from the Guiana Space Centre at Kourou in French Guiana. *SMART-1* is a small spacecraft, measuring only about one metre across, and weighing 287 kg. It is a part of ESA's 'Small Missions for Advanced Research in Technology' which are designed to test new spacecraft technologies. The primary objective of *SMART-1* was to test the solar-powered ion thrusters. It also tested the miniaturised scientific instruments on board the spacecraft, which are thought to be more efficient. After arrival at the Moon,

it made the first comprehensive inventory of the key chemical elements in the lunar surface from orbit. The spacecraft crashed on the Moon after completing its mission on September 3, 2006. The data sent back by the spacecraft would help scientists investigate the theory that the Moon was formed following the violent collision of a smaller planet with Earth, 4,500 million years ago. (☞ Sep. 3)

September 29

1901

Enrico Fermi, Italian-born American physicist who was one of the chief architects of the nuclear age, was born in Rome, Italy. Fermi attended the University of Pisa from where he received his doctorate in physics in 1922. In 1923, he was awarded a scholarship from the Italian Government and spent some months with the German physicist Max Born in Göttingen, Germany. In 1924, he returned to Italy to occupy the post of lecturer in mathematical physics and mechanics at the University of Florence. In 1926 he discovered the statistical laws, nowadays known as the 'Fermi statistics' that govern the particles subject to Pauli's exclusion principle (now referred to as 'fermions', in contrast with 'bosons' which obey the Bose-Einstein statistics). In 1927 he was appointed professor of theoretical physics at the University of Rome, where in 1934 he began experiments involving bombardment of a variety of elements with neutrons. He discovered that slow moving neutrons were especially effective in producing radioactive atoms. Not realising that he had split the atom, Fermi announced the discovery of what he thought were elements beyond uranium. Fermi settled in USA in 1939, and became professor of physics at the Columbia University in New York, where he continued to conduct nuclear fission experiments. In 1940, Fermi's team confirmed that absorption of a neutron by a uranium nucleus can cause the nucleus to split into two nearly equal parts, releasing several neutrons and enormous amounts of energy. Fermi's team developed the first atomic pile and produced the first nuclear chain reaction at the University of Chicago in 1942. After World War II, Fermi continued his pioneering research on high-energy particles. (☞ Nov. 28, Dec. 2)

1913

Rudolf Diesel, German thermal engineer who invented the high-compression internal-combustion engine that bears his name, died at sea in the English Channel. (☞Feb. 29, Mar. 18)

1920

Peter Dennis Mitchell, British chemist, Nobel laureate, was born in Mitcham,

Surrey, England. Mitchell was educated at the Queens College, Taunton, and at the Jesus College, Cambridge. He received his PhD from the University of Cambridge in 1950 for work on the mode of action of penicillin, and held the post of demonstrator at the Department of Biochemistry, Cambridge, from 1950 to 1955. In 1955 he set up and directed a biochemical research unit, called the Chemical Biology Unit, in the Department of Zoology, Edinburgh University, where he was appointed a senior lecturer in 1961, and a reader in 1962. In 1964 he joined the Glynn Research Laboratories as director of research, where in 1965 he embarked on the programme of research on chemiosmotic reactions and reaction systems which are involved in the transfer of energy in biological systems. Mitchell was able to clarify how adenosine diphosphate (ADP) is converted into the energy-carrying compound adenosine triphosphate (ATP) in the mitochondria of living cells. He showed how the arrangement of enzymes facilitates their use of hydrogen ions as an energy source in the conversion of ADP to ATP. Mitchell was awarded the 1978 Nobel Prize for Chemistry for his contribution to the understanding of biological energy transfer through the formulation of the chemiosmotic theory. (☞ Apr. 10)

1927
Willem Einthoven, Dutch physiologist, Nobel laureate (Physiology or Medicine, 1924), who discovered the electrical properties of the heart through the electrocardiograph, died in Leiden, Netherlands. (☞ May 21)

1931
James Watson Cronin, American particle physicist, Nobel laureate, was born in Chicago. Illinois, USA.

Cronin graduated from the Southern Methodist University at Dallas, Texas, in 1951 and received his PhD from the University of Chicago in 1955. He then joined the staff of the Brookhaven National Laboratory, Upton, New York, and in 1958 became a professor at the Princeton University, New Jersey. At Princeton, Cronin was involved in the development of the spark chamber as a practical research tool. Using this chamber, he and his students studied the decay of heavy particles called hyperons. Then Cronin joined with compatriot Val Logsdon Fitch to study neutral K-meson decays. In experiments conducted at the Brookhaven National Laboratory in 1964, Cronin and Fitch showed that the decay of subatomic particles called K-mesons could violate the general conservation law for weak interactions known as CP symmetry, which implied that reversing the direction of time would not precisely reverse the course of certain reactions of subatomic particles. In 1971 Cronin was appointed professor of physics at the University of Chicago. Cronin was awarded the 1980 Nobel Prize for Physics, jointly with Fitch, for the discovery of violations of fundamental symmetry principles in the decay of neutral K-mesons. (☞ Mar. 10)

1988
The American space shuttle *Discovery*, carrying a five-man crew of Frederick H. Hauck, Richard O. Covey, John M. Lounge, George D. Nelson, and David C. Hilmers,

was launched on its 7th flight from the Kennedy Space Center, Florida, USA, marking the resumption of shuttle flights after the *Challenger* disaster. During the four-day mission, the *Tracking and Data Relay Satellite-3* of NASA was deployed in space. (☞ Jan. 28, Aug. 30, Oct. 3)

1996
India's remote sensing satellite *IRS-1D*, was launched by the polar satellite launch vehicle, PSLV-C1, from the Satish Dhawan Space Centre, Sriharikota, Andhra Pradesh. *IRS-1D* carried a panchromatic camera, a linear imaging self scanner (LISS-III), and a wide field sensor (WiFS) camera. It provided enhanced coverage through a programmable on-board tape recorder.

2004
World's first private manned spacecraft, *SpaceShipOne*, piloted by Mike Melvill, coasted above the 100-km altitude point on its second suborbital flight and successfully completed the first of two Ansari X-Prize flights. The peak altitude reached was 103 kilometres. (☞ Jun. 21, Oct. 4)

September 30

1870
Jean-Baptiste Perrin, French physicist and Nobel laureate, was born in Lille, France. Perrin was educated at the École Normal Supérieure in Lille, becoming an assistant in physics in 1894, when he began his researches on cathode rays and X-rays. He received the doctorate in science in 1897 for a thesis on cathode and Röntgen rays, and was appointed in the same year as a reader in physical chemistry at the Sorbonne, University of Paris. He became professor there in 1910, a post which he held till 1940. In 1941 he immigrated to USA. Perrin is best known for his study of colloids and, in particular, the so-called Brownian movement. His experimental results on Brownian movement were able to confirm Einstein's theoretical prediction that colloidal particles should obey the gas laws. Using an emulsion of gum resin, Perrin was able to determine one of the most important physical constants, Avogadro's number – the number of molecules of a substance needed to make up the a mass equal to its molecular weight in grammes. The value obtained by Perrin corresponded, within the limits of error, to that given by the kinetic theory of gases. Perrin was awarded the 1926 Nobel Prize for Physics for his work on the discontinuous structure of matter, and especially for his discovery of sedimentation equilibrium. (☞ Apr. 17)

1882
Hans Geiger, German physicist who introduced the first successful detector of individual alpha-particles and other ionising radiations, was born in Neustadt-an-der-Haardt, Germany. Geiger studied physics at the University in Munich and received his PhD from the University of Erlangen in 1906. Geiger worked in Manchester, England, with the English physicist Ernest Rutherford from 1906 to 1912. Eventually, in 1911, they devised the first version of the Geiger counter to count the number of alpha-particles and other ionising radiation. With the aid of other radiation detectors, he used his counter in early experiments to identify alpha-particles as the nucleus of the helium atom. Geiger and Rutherford also demonstrated

that alpha-particles had two units of charge. Without Geiger's work on particle detection, the discovery of radioactivity may not have been possible, and without the Geiger counter, most nuclear or atomic physics would have been hindered in some way. Geiger returned to Germany in 1914. During World War I, he served as an artillery officer in the German army. In 1924, at the University of Kiel, he and the German-born American physicist Walther Müller improved the sensitivity, performance and durability of the particle counter which Geiger made before. Named the Geiger–Müller counter, the device is a gas-filled metal tube with a wire through its axis. A high voltage is applied to the wire. As particles enter the tube, they create a large avalanche of ionisation in the gas, which then discharges, creating a brief electric pulse. The tube produces the same large output pulse for virtually every charged particle that passes through the gas and so is useful for detecting individual particles. It can therefore indicate lower levels of radiation than is possible with other types of detectors. Later in life, Geiger investigated cosmic rays, artificial radioactivity, and nuclear fission. (☞ Sep. 24)

1902

William H. Walker, Arthur D. Little, and Harry S. Mork, American chemists were granted a patent for 'artificial silk' made of cellulose acetate fibre. The material was invented around 1855 by the Swiss chemist Georges Audemars, and the process refined in 1864 by the French chemist and industrialist Hilaire Bernigaud de Chardonnet. The new fibre was re-named 'rayon' in 1924. Rayon is a cellulose-based fibre that is made from wood pulp or cotton waste and is used as a substitute for silk. It was the first synthetic material to be patented. (☞ Dec. 19)

1905

Nevill Francis Mott, English physicist, Nobel laureate, was born in Leeds, England. Mott was educated at the Clifton College, Bristol and St. John's College, Cambridge, where he studied mathematics and theoretical physics. After obtaining his bachelor's degree in 1927 and master's degree in 1930 from the University of Cambridge, he became a professor of theoretical physics at the University of Bristol in 1933, where he carried out research on a theory of transition metals, on the properties of metals and semiconductors, hardness of alloys, and on the photographic latent image. After a period of military research in London during the war, he became head of the Bristol physics department, and published papers on low-temperature oxidation and the metal–insulator transition. In 1954 Mott became Cavendish professor of experimental physics at the University of Cambridge, from where he retired in 1971. Mott's studies of electrical conduction in various metals led him in the 1960s to explore the conductivity potential of amorphous materials, which are so called because their atomic structures are irregular or unstructured. He devised formulas describing the transitions that glass and other amorphous substances can make between electrically conductive states and insulating states, thereby enabling them to function as semiconductors. These amorphous substances, which are relatively

simple and cheap to produce, could replace the more expensive crystalline semiconductors in many electronic switching and memory devices, thereby leading to more affordable personal computers, pocket calculators, copying machines, and other electronic devices. Mott was awarded the 1977 Nobel Prize for Physics, jointly with the American physicists Philip W. Anderson and John Hasbrouck Van Vleck, for their fundamental theoretical investigations of the electronic structure of magnetic and disordered systems. (☞ Mar. 13, Aug. 8, Dec. 13)

1939

Jean-Marie Lehn, French chemist, Nobel laureate, was born in Rosheim, France. Lehn had his schooling in the towns of Rosheim and Obernai and studied chemical and natural sciences at the University of Strasbourg, from where he obtained his bachelor's degree in 1960 and PhD in 1963. He then spent a year in the laboratory of the American chemist, Robert Burns Woodward at the Harvard University in Cambridge, Massachusetts, USA, where he took part in the total synthesis of vitamin B_{12}. In 1966 he was appointed assistant professor at the chemistry department of the University of Strasbourg. In 1967, the American chemist, Charles J. Pedersen had published two works, describing methods of synthesising cyclic polyethers, which he named 'crown ethers'. These compounds have remarkable properties of binding even the alkali metal ions of lithium, sodium, potassium, rubidium and caesium into complexes. Expanding on Pedersen's work, Lehn, in 1969 developed bicyclic compounds of the crown ether type, which he called 'cryptands' and which show even higher selectivity when forming complexes. In the course of his efforts to synthesise three-dimensional molecules that would possess similar reactive characteristics, Lehn created a molecule that combined with the chemical acetylcholine, which is an important neurotransmitter in the brain. In 1970 he became a professor of chemistry at the Louis Pasteur University in Strasbourg. From 1979 he was also a professor at the Collège de France in Paris. Lehn was awarded the 1987 Nobel Prize for Chemistry, jointly with Pedersen and American chemist Donald J. Cram, for their development and use of molecules with structure-specific interactions of high selectivity. (☞ Apr. 22, Oct. 3)

1943

Johann Deisenhofer, German biochemist, Nobel laureate, was born in Zusamaltheim, Germany. Deisenhofer had his elementary schooling in Zusamaltheim, and studied physics at the Technical University in Munich. He received his PhD from the Max Planck Institute for Biochemistry in Martinsried, in 1974 and conducted research there under the German biochemist Robert Huber until 1987, when he moved to USA to join the scientific staff at the Howard Hughes Medical Institute in Dallas, Texas. In 1987, Deisenhofer, together with Huber and compatriot Hartmut Michel, who had joined them in 1982, were able to determine the complete three-dimensional structure of a four-protein complex called the 'photosynthetic reaction centre' from the purple bacterium *Rhodopseudomonas viridis*, using X-ray crystallography. The photosynthetic reaction centre is a biomolecule that provides the energy on which almost all life on Earth depends, namely it allows plants and some (so-called photosynthetic) bacteria to harness the

Sun's energy and convert it into energy that the biological organisms can use. The research of the three scientists helped in better understanding the mechanisms of photosynthesis and revealed similarities between the photosynthetic processes of plants and bacteria. Deisenhofer was awarded the 1988 Nobel Prize for Chemistry, jointly with Huber and Michel, for the determination of the three-dimensional structure of a photosynthetic reaction centre. (☞ Feb. 20, Jul. 18)

1985

Charles Francis Richter, American seismologist, who developed the Richter scale for measuring the intensity of earthquakes, died in Pasadena, California, USA. (☞ Apr. 26)

1994

André Michel Lwoff, French biologist, Nobel laureate (Physiology or Medicine, 1965), who contributed to the understanding of lysogeny, in which a bacterial virus, or bacteriophage, infects bacteria and is transmitted to subsequent bacterial generations solely through the cell division of its host, died in Paris, France. (☞ May 8)

October 1

1904
Otto Robert Frisch, Austrian-born British nuclear physicist, was born in Vienna, Austria. Frisch received his doctorate degree in 1926 from the University of Vienna and continued to conduct research at the universities in Berlin and Hamburg in Germany. During the early 1930s, he moved to Copenhagen, Denmark, where his aunt, Austrian-born physicist Lise Meitner informed him of a discovery by the German chemist Otto Hahn that would lead physicists to produce the first splitting of neutron-bombarded uranium into lighter elements. Frisch described the process in detail and also calculated details such as the critical mass needed to produce a chain reaction. He named the process 'nuclear fission', borrowing a term from biology. After settling in Birmingham, England in 1939, Frisch expanded on this research with the Austrian-born British physicist Rudolf Peierls and demonstrated that the fissioning of uranium had the potential to create a volatile chain reaction which, using uranium-235, could be used to develop an extremely destructive weapon. During 1943–45, he worked on the Manhattan Project at Los Alamos, New Mexico, USA that produced the first atom bomb. After World War II, he returned to England to work for the Atomic Energy Research Establishment at Harwell and was professor of natural philosophy at Cambridge University from 1947 to 1979. (☞ Sep. 22)

1908
The Ford 'Model T' car, the first car to be mass-produced on an assembly line, was introduced at a price of $825. It was an immediate sensation. Before long, it was the largest selling car in USA. As the volume of production rose, costs came down and the car sold for as low as $260. (☞ Jul. 30)

1947
Aaron J. Ciechanover, Israeli biochemist, Nobel laureate, was born in Haifa, Israel. Ciechanover received his Master of Science degree in 1971 and his MD in 1974 from the Hadassah Medical School of the Hebrew University in Jerusalem. During 1974–77, he served in the Israel Defense Forces. In 1977 he joined Avram Hershko at the Technion (Israel Institute of Technology), Haifa, as a graduate student, receiving his PhD degree in 1981. In 1982 he went to USA as a postdoctoral fellow at the Department of Biology at the Massachusetts Institute of Technology, Cambridge, Massachusetts where he studied compounds known as sialoglycoproteins and transferrin receptors. He is at present a Distinguished

Professor at the Rappaport Family Institute for Research in Medical Sciences at the Technion in Haifa. Starting in the late 1970s, during a series of sabbatical leaves, Ciechanover together with the Israeli biologist Avram Hershko and American biochemist Irwin A. Rose did a series of epoch-making biochemical studies in Rose's laboratory at the Fox Chase Cancer Center in Philadelphia, USA, which ultimately led to the discovery of a molecule called ubiquitin. Ubiquitin plays a key role in protein breakdown within cells. Their work showed that the breakdown is not indiscriminate but takes place through a process that is controlled so that the proteins to be broken down at any given moment are given a molecular label by the ubiquitin molecule. The labelled proteins are then fed into the cells' 'waste disposers', the so-called proteasomes, where they are chopped into small pieces and destroyed. The ubiquitin–proteasome pathway has a critical role in maintaining the homeostasis of cells and is believed to be involved in the development and progression of diseases such as cancer, muscular and neurological diseases, immune and inflammatory responses, and cystic fibrosis. Ciechanover was awarded the 2004 Nobel Prize for Chemistry, jointly with Hershko and Rose, for the discovery of ubiquitin-mediated protein degradation. (☞ Jul. 16, Dec. 31)

1958

The American space agency National Aeronautics and Space Administration (NASA) was created by the US Congress and the President of the United States by an Act 'to provide for research into the problems of flight within and outside the Earth's atmosphere, and for other purposes'. It took over the activities of the earlier National Advisory Committee for Aeronautics, three major research laboratories – Langley Aeronautical Laboratory, Ames Aeronautical Laboratory, and Lewis Flight Propulsion Laboratory – and two smaller test facilities. NASA quickly incorporated other organisations into the new agency, notably the space science group of the Naval Research Laboratory in Maryland, the Jet Propulsion Laboratory managed by the California Institute of Technology for the army, and the Army Ballistic Missile Agency in Huntsville, Alabama, where the German-born American space scientist Wernher von Braun's team of engineers were engaged in the development of large rockets. Eventually NASA created other centres and today it has ten located all around USA. NASA began to conduct space missions within months of its creation, and during its first four decades conducted several major programmes, including landing men on the Moon, sending space probes to the distant planets of the solar system, and exploration of Mars using remote-controlled robots. NASA has also been the only space agency to successfully utilise re-usable space shuttles.

1969

The Anglo-French supersonic airliner *Concorde* broke the sound barrier for the first time. It was the first plane in the world to be entirely controlled by a computer. As the only supersonic passenger aircraft, *Concorde* remained a brilliant technological achievement, but its impact on international air travel has been limited, mainly due to its high cost of operation. (☞ Jan. 9, Jan. 21, Mar. 2, Jul. 25, Oct. 24)

1972

Louis Seymour Bazett Leakey, Kenyan archaeologist and anthropologist, who found the fossil remains of one of the earliest members of the genus of human beings named *Homo habilis*, died in London, England. (☞ Aug. 7)

October 2

1608
The first refracting telescope was demonstrated to the Netherlands States General by Dutch spectacle maker Hans Lippershey.

1836
The British naturalist Charles Darwin returned from his voyage on the *HMS Beagle* to the Pacific. The voyage of the *HMS Beagle* and Darwin's collections, observations and publications related to its five-year journey have perhaps affected human thinking in ways that no other scientific discoveries have. It would be 23 years before Darwin published his *Origin of Species*. (☞ Feb. 12, Apr. 19)

1852
William Ramsay, Scottish chemist, Nobel laureate, was born in Glasgow, Scotland. Until 1870 Ramsay studied in his native town. He then moved to Tübingen, Germany to work with the German organic chemist Rudolf Fittig and received his PhD for a thesis on orthotoluic acid and its derivatives. In 1880 he was appointed principal and professor of chemistry at the University College, Bristol, and moved on in 1887 to the University College, London, as professor of inorganic chemistry, a post that he held until his retirement in 1913. Ramsay's earliest works were in the field of organic chemistry but in the early 1880s he was more active in physical chemistry, his many contributions to this branch of chemistry being mostly in stoichiometry and thermodynamics. It was however in inorganic chemistry that his most celebrated discoveries were made. As early as 1885–1890 he published several notable papers on the oxides of nitrogen and followed those up with the discovery of the inert gases argon, helium, neon, krypton, and xenon. Ramsay was awarded the 1904 Nobel Prize for Chemistry in recognition of his services in the discovery of the inert gaseous elements in air, and his determination of their place in the periodic system. (☞ Jul. 23, Aug. 13)

1907
Alexander Robertus Todd, British biochemist, Nobel laureate, was born in Glasgow, Scotland. Todd was educated at the Allan Glen's School and Glasgow University, where he took his BSc degree in 1928. He obtained his PhD from the University of Frankfurt-on-Maine in Germany on a thesis on the chemistry of bile acids. He returned to England and worked with the British chemist Robert Robinson on anthocyanins and other colouring matters, and took a further PhD degree at the Oxford University in 1933. Todd went back to Scotland in 1934 when he joined the staff of the Edinburgh University. In 1936 he moved to the Lister Institute of Preventive Medicine, Chelsea, near London, and became reader in biochemistry at the University of London in 1937. In 1938 he was appointed as professor of chemistry and director of the Chemical Laboratories of the University of Manchester, which position he held until 1944, when he moved to the Cambridge University as professor of organic chemistry. Todd's research was in the chemistry of natural products of biological importance, nucleotide and nucleotide coenzyme studies,

the chemistry of vitamins B_1, E and B_{12}, the constituents of the *Cannabis* species, insect colouring matters, factors influencing obligate parasitism, and various mould products. Nucleotides – found in the chromosomes of cells – are linked with the units of heredity. In combination with proteins, they constitute the virus molecules. Many coenzymes are nucleotides of low molecular weight but with a special structure. Todd's researches showed how they are connected to each other. While trying to determine the chemical reactions that form alkaloids in plants, Todd discovered the structures of morphine and strychnine. He also formulated a qualitative electronic theory of the structure of organic molecules. Todd was awarded the 1957 Nobel Prize for Chemistry for his work on nucleotides and nucleotide coenzymes. (☞ Jan. 10)

1917

Christian René de Duve, Belgian cytologist and biochemist, Nobel laureate, was born in Thames Ditton, Surrey, England. De Duve received his MD in 1941 and a master's degree in chemistry in 1946 from the Catholic University of Louvain, Belgium. In 1946–1947, he spent 18 months at the Medical Nobel Institute in Stockholm, Norway, and then six months as a Rockefeller Foundation fellow at the Washington University, St. Louis, Missouri, USA, to train in biochemistry. He returned to the Catholic University of Louvain in March 1947 to take over the teaching of physiological chemistry at the medical faculty, becoming full professor

in 1951. It was at Louvain, while studying the so-called 'latency' of release of an enzyme called acid phosphatase during the metabolism of carbohydrates by the liver, that de Duve stumbled upon a new cell part, which he named the 'lysosome'. Lysosomes have now been shown to be engaged in a series of cellular activities during which biological material is degraded. The lysosomes are used in defence mechanisms against bacteria, during resorption and secretion. They can also be used for a controlled degradation of the cell in which they are contained, e.g., to remove worn out components. De Duve's discovery of lysosomes answered the question of how the powerful enzymes used by cells to digest nutrients are kept separate from other cell components. Subsequently, De Duve and his team also developed improved instrumental, technical and conceptual tools in relation to the separation and analysis of cell components, and to apply them to an increasing variety of problems of biological and medical interest. In 1962, de Duve was appointed a professor at the Rockefeller Institute (now the Rockefeller University), in New York, USA. De Duve was awarded the 1974 Nobel Prize for Physiology or Medicine, jointly with the Belgian-born American cytologist Albert Claude and the Romanian-born American biochemist George Palade, for their discoveries concerning the structural and functional organisation of the cell. (☞ Aug. 24, Nov. 19)

1927

Svante Arrhenius, Swedish physical chemist, Nobel laureate (Chemistry, 1903), best known for his theory that electrolytes in solution are dissociated into electrically charged particles, or ions, even when there is no current flowing through the solution, died in Stockholm, Sweden. (☞ Feb. 19)

1987
Peter B. Medawar, Brazilian-born British zoologist, Nobel laureate (Physiology or Medicine, 1960), who discovered acquired immunological tolerance, died in London, England. (☞ Feb. 28)

2000
ANDi, the first genetically modified primate – a rhesus monkey – was born at the Oregon Regional Primate Research Center, in Beaverton, Oregon, USA. The name is derived from "inserted DNA" written backwards. The gene inserted into ANDi came from a jellyfish and was described as 'GFP' which stands for 'green fluorescent protein'. To create ANDi, Anthony Chan and his colleagues at the Research Center injected unfertilised rhesus eggs with a virus carrying the GFP gene and later fertilised the eggs artificially. ANDi was born out of one of the genetically modified and fertilised eggs.

The work demonstrated that a foreign gene can be delivered and inserted into a primate chromosome, which could lead to primate models of human diseases – like Alzheimer's, diabetes, heart disease and obesity – that could offer a more robust testing ground for new drugs, gene therapy and modified stem cells.

2002
Cracking of the genetic codes of the malaria-spreading mosquito *Anopheles gambiae*, and the malarial parasite *Plasmodium falciparum* was announced. The genome of *A. gambiae*, which is the main transmitter of malaria in Africa and is sometimes known as the 'malaria machine', was found to have about 13,000 genes, a similar number to that of the fruit fly. The sequencing of the mosquito genome was carried out by more than 150 international scientists, led by the US company, Celera Genomics and the French national sequencing centre, Genoscope. The genome of the parasite, *P. falciparum*, is about a hundredth the size of the human genome. It contains 5,300 genes, more than half of which have no known function and are new to biology.

October 3

1844
Patrick Manson, Scottish parasitologist, was born in Old Meldrum, Aberdeen, Scotland. Manson abandoned a potentially lucrative career in industry for medicine, graduating from the Aberdeen University in 1865. The following year he left for China where he was to spend 24 years studying such diseases as tinea, Calabar swelling, and blackwater fever. He took an interest in medical education, and founded the first western-style medical school in China and another in Hong Kong after moving there in 1883. In 1878, in China, he observed that the filarial parasite, the worms that cause elephantiasis in man, pass part of their life cycle in the Culex mosquito and that the embryonic filariae only appear in the patient's peripheral blood stream at night when mosquitoes feed.
He thus led the way in the study of the transmission of diseases caused by parasites. In 1894 he made the deduction that the parasite of malaria passes part of its life cycle in the mosquito, a theory that the British

bacteriologist Ronald Ross was to verify three years later. Manson also did valuable research on sleeping sickness and beriberi, and helped introduce vaccination to the Chinese. A founder of two schools devoted to the study of tropical diseases, one at Hong Kong and the other at London, Manson is often described as the father of tropical medicine. (☞ Apr. 9)

1904

Charles J. Pedersen, Korean-born American chemist, Nobel laureate, was born in Pusan, Korea. Pedersen obtained his bachelor's degree in chemical engineering at the University of Dayton, Ohio, USA, and his master's degree in organic chemistry from the Massachusetts Institute of Technology, Cambridge, Massachusetts. In 1927, he joined the Du Pont Company in Wilmington, Delaware, as a researcher and remained there for his entire 42-year career as a chemist. For several years Pedersen worked on the effects of various ligands on the catalytic properties of copper and the transition elements in general. During the late 1940s and 1950s he worked on the photochemistry of some new phthalocyanine adducts and of quinoneimine dioxides and developed polymerisation initiators and even made some novel polymers. In 1960 he decided to study the effects of bi- and multidentate phenolic ligands on the catalytic properties of the vanadyl group, VO. In the course of these investigations one of his experiments yielded an unexpected small quantity of unknown white crystals which was eventually identified as dibenzo18-crown-6, the first 'crown ether' – a group of organic compounds with structure-specific interactions of high selectivity when reacting with other atoms and molecules. Pedersen was awarded the 1987 Nobel Prize for Chemistry, jointly with the French chemist Jean-Marie Lehn and American chemist Donald J. Cram, for their development and use of molecules with structure-specific interactions of high selectivity. (☞ Apr. 22, Sep. 30, Oct. 26)

1942

The *V2* rocket makes first test flight in Germany. The *V2* was developed by the German rocket pioneers Wernher von Braun, Walter Dornberger and Hermann Oberth at the rocket research station at Peenemunde, on the Baltic coast. It was a 14-metre long, liquid-fuelled rocket and was capable of carrying a one-tonne warhead at supersonic speed, flying at an altitude of over 80 kilometres. The *V2* was first used in war in September 1944. (☞ Mar. 23)

1985

American space shuttle *Atlantis* was launched on its maiden flight from the Kennedy Space Center, Florida, USA. It carried a five-person crew of Karol J. Bobko, Ronald J. Grabe, David C. Hilmers, Robert L. Stewart, and William A. Pailes. During this flight, only classified military activities were conducted. *Atlantis* was the fourth shuttle to be inducted into NASA's shuttle fleet. In 1989, *Atlantis* deployed two planetary probes, *Magellan* and *Galileo*, and in 1991 it deployed the

Compton Gamma Ray Observatory. Beginning in 1995, *Atlantis* made seven straight flights to the Russian space station *Mir*. On the second *Mir* flight, it delivered a docking module and on subsequent flights it conducted astronaut exchanges. In October 2002, *Atlantis* and its six-person crew completed an 11-day mission to the *International Space Station* that involved three space walks. (☞ Oct. 7)

1988
The American space shuttle *Discovery*, carrying a five-man crew of Frederick H. Hauck, Richard O. Covey, John M. Lounge, George D. Nelson, and David C. Hilmers, landed at the Edwards Air Force Base, California, USA after completing its 7th flight. (☞ Aug. 30, Sep.29)

October 4

1916
Vitaly L. Ginzburg, Russian theoretical physicist, Nobel laureate, was born in Moscow, USSR (now Russia). Ginzburg graduated from the Physics Faculty of the Moscow State University in 1938, and received his PhD in 1940. Since 1940 he has been working at the P N. Lebedev Physical Institute of the Russian Academy of Sciences. He has been a part time professor at the Gorky State University and at the Moscow Institute for Physics and Technology. In the 1950s, Ginzburg together with the Russian physicist Lev Landau formulated a theory that could describe in more detail than before, how superconductivity disappears at certain 'critical' values of electrical current and magnetic fields. They introduced a measure for the order among electrons, which they called the superconducting order parameter. Guided by a deep physical intuition they went on to formulate mathematical equations whose solution determines the order in a superconductor. They found a close correspondence with what had been measured for superconductors known at the time. Although these theories were formulated in the 1950s, they have gained renewed importance in the rapid development of materials with completely new properties. Materials can now be made superconductive at increasingly high temperatures and strong magnetic fields. Ginzberg was awarded the 2003 Nobel Prize for Physics, jointly with the Russian-born American physicist Alexei A. Abrikosov and British-born American physicist Anthony J. Leggett, for pioneering contributions to the theory of superconductors and superfluids. (☞ Mar. 26, Jun. 25)

1918
Fukui Kenichi, Japanese theoretical chemist, Nobel laureate, was born in Nara, Japan. Fukui graduated from the Kyoto University in 1941, and was then engaged in experimental research on synthetic fuel chemistry in the Army Fuel Laboratory. He joined the Fuel Chemistry Department of the Kyoto Imperial University as lecturer in 1943, and became assistant professor in 1945 and professor in 1951. He obtained his PhD in chemical engineering from the Kyoto University in 1948. As a professor at the Kyoto University, Fukui developed the theory that during chemical reactions molecules share loosely bonded electrons, which occupy so-called frontier orbitals. He showed that certain properties of the orbits of the most loosely bound electrons and of the 'most easily accessible' unoccupied

electronic orbits had unexpectedly great significance for the chemical reactivity of molecules. He called these orbits 'frontier orbitals'. His theory advanced the understanding of the mechanism of chemical reactions, especially in the production of organic compounds. This success led his theoretical group to the chemical reactivity theory, extending more and more widely the range of compound and reactions that were discussed. Fukui was also interested in formulating the path of chemical reactions. The idea of electron delocalisation between the frontier orbitals of reactant species served to provide information on the geometrical shape of reacting molecules, and he was able to make the role of the frontier orbitals in chemical reactions more distinct through visualisation, by drawing their diagrams. Fukui was awarded the 1981 Nobel Prize for Chemistry, jointly with the Polish-born American chemist Roald Hoffmann, for their theories, developed independently, concerning the course of chemical reactions. (☞ Jan. 9, Jul. 18)

1938

Kurt Wüthrich, Swiss scientist, Nobel laureate, was born in Aarberg, Switzerland. Wüthrich studied chemistry, physics and mathematics at the University of Bern from 1957 to 1962, and obtained a PhD in inorganic chemistry from the University of Basel in 1964. Subsequently, he worked as a postdoctoral fellow at the University of California, Berkeley, USA and at the Bell Telephone Laboratories, New Jersey, USA. In 1969 he moved to the Swiss Federal Institute of Technology (ETH), Zürich, where he became professor of molecular biophysics in 1972. Since 2001 he shares his time between the ETH Zürich and The Scripps Research Institute, La Jolla, California, USA, where he is currently the Cecil H. and Ida M. Green Visiting Professor of Structural Biology. Wüthrich's research interests mainly concern molecular structural biology, and structural and functional genomics. He developed a method of determining the conformations of proteins, nucleic acids and other bio-macromolecules as such in solutions or biomembranes, where they exhibit their function. He developed a general method of systematically assigning certain fixed points in the protein molecule, and also a principle for determining the distances between these. Using the distances, he was able to calculate the three-dimensional structure of the protein. This achievement is considered an unprecedented contribution to structural and molecular biology, representing a great advance in biotechnology.

Wüthrich was awarded the 2002 Nobel Prize for Chemistry, jointly with the American chemist John B. Fenn and Japanese scientist Koichi Tanaka, for the development of methods for identification and structure analyses of biological macromolecules. Wüthrich received half the prize money for his development of nuclear magnetic resonance spectroscopy for determining the three-dimensional structure of biological macromolecules in solution, while Tanaka and Fenn shared the other half for their development of soft desorption ionisation methods for mass spectrometric analyses of biological macromolecules. (☞ Jun. 15, Aug. 3)

1947

Max Planck, German theoretical physicist, Nobel laureate (Physics, 1918), who originated quantum theory, died in Göttingen, Germany. (☞ Apr. 23)

1957

The first man-made satellite *Sputnik-1* was launched into orbit around the Earth from Kazakhstan in the erstwhile Soviet Union, ushering in the Space Age. The 92-kg satellite circled the Earth every 95 minutes

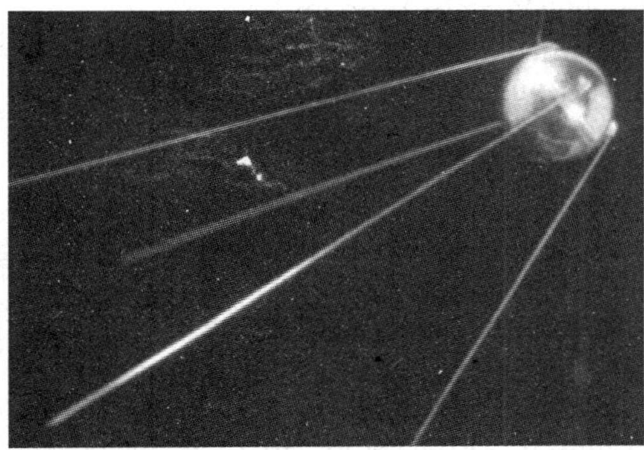

at almost 3,200 km/h at a height of 800 kilometres, and transmitted a radio signal which was picked up all around the world. The *Sputnik* (meaning 'companion') stayed in orbit for about three months.

1959

The Soviet lunar-probe *Luna-3* was launched from the Baikanur Cosmodrome in USSR (now Kazakhstan). It made the

first circumnavigation of the Moon and returned the first photographs of the Moon's far side. (☞ Oct. 7)

2000

Michael Smith, British-born Canadian biochemist, Nobel laureate (Chemistry, 1993), who developed oligonucleotide-based, site-directed mutagenesis for the study of proteins, died in Vancouver, British Columbia, Canada. (☞ Apr. 26)

2004

SpaceShipOne became the first private manned spacecraft to exceed an altitude of 100 kilometres twice within a span of a 14-day period, claiming the US$10,000,000 Ansari X-Prize. *SpaceShipOne* was designed by the American aircraft designer Burt Rutan (who made aviation news in 1986 by developing the *Voyager*, the first aircraft to fly non-stop around the world without refuelling) and his research team at the California-based aerospace company, Scaled Composites. It was piloted by test pilot Brian Binnie. To reach space, a carrier aircraft, the *White Knight*, lifted *SpaceShipOne* from the runway, and an hour later, after climbing to an altitude of approximately 15 kilometres, the carrier aircraft released the spaceship into a glide. The spaceship pilot Binnie

then fired his rocket motor for about 80 seconds, reaching Mach 3 in a vertical climb. *SpaceShipOne* then coasted up to an altitude of 112 km before falling back to Earth. During the flight, the pilot experienced a weightless environment for more than three minutes. Although flown by a single pilot so far, *SpaceShipOne* is equipped with three seats and is designed for missions that include a pilot and two passengers. (☞ Jun. 21, Sep. 29)

October 5

1864
Louis Lumière, French inventor, who invented the cinématographe, was born in Besançon, France. Louis and his brother Auguste were brought up in Lyon. Their father ran a photographic firm and both brothers worked for him – Louis as a physicist and Auguste, a manager. Louis had made some improvements to the still photograph process but it was not until their father retired in 1892 that the brothers really set to work on creating moving pictures. They patented a number of significant processes – most notably the creation of sprocket holes in the filmstrip as a means of getting the film through the camera and projector. Louis designed his Cinématographe, building on previous inventions such as the kinetoscope, invented by the American inventor Thomas Alva Edison, but with the key addition of a mechanism that allowed the intermittent motion of the film in the camera, based on the sewing machine. The small, twin-function device, which worked both as a camera and a projector and weighed about 13 kg, was patented in February 1895. Louis organised the first public screening of his film at the Grand Café on the boulevard des Capucines in Paris in December 1895, which brought wide public acclaim and marked the beginning of cinema history. (☞ Mar. 22, Jun. 6, Dec. 28)

1879
Peyton Rous, American pathologist, Nobel laureate, was born in Baltimore, Maryland, USA. Rous graduated from the Johns Hopkins Medical School, Baltimore, in 1905 and became an instructor in pathology at the University of Michigan. He joined the Rockefeller Institute for Medical Research (now the Rockefeller University) in New York in 1909 and remained there throughout his career. In 1910 Rous found that sarcomas – a type of cancer – in hens could be transmitted to birds of the same inbred stock not only by grafting tumour cells but also by injecting an agent extractable from them. This discovery showed that viruses could cause cancer. Although his research was derided at the time, subsequent experiments vindicated his thesis. Apart from cancer research, Rous did investigations on liver and gall bladder physiology, and also worked on the development of blood-preserving techniques that made the first blood banks possible. Around 1934, returning to cancer research, Rous discovered that viruses also caused benign tumours called warts, which later often developed into cancer. Rous was awarded the 1966 Nobel Prize for Physiology or Medicine, jointly with the Canadian-born American surgeon Charles Brenton Huggins, for his discovery of tumour-inducing viruses. (☞ Feb. 16, Sep. 22)

1880
William Lassell, amateur English astronomer, Fellow of the Royal Society, who discovered several moons of various planets, died in Maidenhead, Berkshire, England. (☞Jun. 18)

1882
Robert Hutchings Goddard, American professor, physicist and inventor, was born in Worcester, Massachusetts, USA.

Goddard was educated at the Worcester Polytechnic Institute and the Clark University in Worcester. From 1909 to 1943 he taught physics at various institutions, including the Worcester Polytechnic Institute and Princeton and Clark universities. His interest in rocketry began in childhood, and in 1919 he published a short book, *A Method of Reaching Extreme Altitudes*, in which he proposed a rocket that might reach the Moon. Goddard not only developed the mathematical theory of rocket propulsion, but also proved that rocket engines could produce thrust in vacuum. In March 1926, in Auburn, Massachusetts, Goddard successfully tested the first rocket using liquid fuel – a mixture of gasoline and liquid oxygen. Although the rocket flew for just 2.5 seconds and rose to a height of only 12.5 metres, it proved that liquid fuel rockets worked. In 1929 he sent up the first instrument-carrying rocket, which carried a barometer, a thermometer, and a small camera. From 1930 to 1942, with the aid of a Guggenheim Foundation grant, he worked in New Mexico. His experiments included the construction of rockets that reached a velocity of 885 km/hr and heights of up to 2 km. He accumulated more than 200 patents related to rocketry. His work virtually laid the foundation for contemporary space exploration. (☞ Mar. 16, Aug. 10)

1976

Lars Onsager, Norwegian-born American chemist, Nobel laureate (Chemistry, 1968), who developed a general theory of irreversible chemical processes, died in Coral Gables, Florida, USA. (☞ Nov. 27)

October 6

1807

Humphry Davy, British chemist, discovered and isolated potassium by electrolysis of slightly damp fused potash. When a current was passed, globules of silvery matter collected at the negative pole; most of these caught fire, but some could be collected. These globules were pure potassium. Its chemical symbol is derived from the Latin word *kalium*, which means 'alkali'. Its English name is from potash, which is the common name for a compound containing it. (☞ Dec. 17)

1893

Meghnad Saha, Indian astrophysicist, Fellow of the Royal Society, was born in Sheoratali, a village in the District of Dacca, now in Bangladesh. Saha passed the entrance examination of the Calcutta University in 1909, obtaining the highest marks in languages and mathematics. He graduated from the Presidency College, Calcutta, in 1913 and ranked first in the MSc examinations in applied mathematics in 1915. Saha decided to do research in physics and applied mathematics, but there was no proper infrastructure in the Calcutta University for higher research. In 1920, Saha developed his famous thermal ionisation equation, which, in the form perfected by the British astrophysicist Edward A. Milne, has remained fundamental in all work on stellar atmospheres. The Saha equation links the composition and appearance of the spectrum with the temperature of the light source and can thus be used to determine either the temperature of the star or the relative abundance of the chemical elements in the star. Saha became professor of physics at the University of

Allahabad in 1923, but returned to the Science College, Calcutta in 1938. During this time Saha got attracted to nuclear physics. In 1947, he established the Institute of Nuclear Physics in Calcutta, which later was named after him as the Saha Institute of Nuclear Physics. In his later years Saha increasingly turned his attention to the social responsibilities of science. He joined politics and was elected as a member of the Indian Parliament in 1951. (☞ Feb. 16)

1903

Ernest Thomas Sinton Walton, Irish physicist, Nobel laureate, was born in Dungarvan, Ireland. Walton attended day schools in Banbridge and Cookstown and in 1915 he was sent to the Methodist College, Belfast, where he excelled in mathematics and science. In 1922 he entered the Trinity College, Dublin, from where received his BSc degree in 1926 and his MSc degree in 1927. The same year he went to the Cambridge University on a scholarship to work in the Cavendish Laboratory under Lord Rutherford. He received his PhD from Cambridge in 1931. Walton returned to the Trinity College, Dublin, in 1934, where he remained as a fellow for the next 40 years and a fellow emeritus thereafter. He was Erasmus Smith professor of natural and experimental philosophy from 1946 to 1974. At the Cavendish Laboratory, Walton worked on indirect methods for producing fast particles, working on the linear accelerator and on what was later to become known as the betatron. He followed this with work on the direct method of producing fast particles by the use of high voltages, jointly with the British physicist John Douglas Cockcroft. They built the first nuclear particle accelerator, known as the Cockcroft–Walton generator, and succeeded in disintegrating the nucleus of the lithium atom by bombarding it with accelerated protons, and identifying the products as alpha particles (helium nuclei). Walton was awarded the 1951 Nobel Prize for Physics, jointly with Cockcroft, for their pioneer work on the transmutation of atomic nuclei by artificially accelerated atomic particles. (☞ May 27, Jun. 25)

1914

Thor Heyerdahl, Norwegian ethnologist and adventurer, who organised and led the famous *Kon-Tiki* and *Ra* trans-oceanic scientific expeditions, was born in Larvik, Norway. Heyerdahl studied zoology and geography at university and in 1936 travelled to the Marquesas Islands in French Polynesia in the Pacific. He spent a year there, living off the land and studying the local flora and fauna of this remote island group. Knowing that the Pacific currents ran from east to west and that many local plants were identical to those of South America, he became more interested in finding out how Polynesia had been originally populated. To prove that ancient cultures could have sailed to and populated the South Pacific, in 1947 he sailed in a tiny balsawood raft, *Kon Tiki*, on a 101-day, 6,000-km drifting voyage from Peru to Polynesia. In 1970, he crossed the Atlantic in a papyrus craft, *Ra II*, and reached the West Indies, demonstrating that the Spanish explorer Christopher Columbus may not have been the first to cross the Atlantic in a boat. Eight years

later, Heyerdahl sailed in another boat, the *Tigris*, on a journey from the Euphrates and Tigris rivers, down the Persian Gulf to Oman, Pakistan, and across the Indian Ocean to Djibouti on the Horn of Africa. The five-month journey was meant to show how the ancient Sumerians could have travelled widely. Heyerdahl's expeditions fostered a close understanding of the global environment. He voiced his concern at the increasing problem of pollution, which he had encountered even in the middle of the world's oceans. (☞ Apr. 18, Apr. 28, Aug. 7)

1927

The first talking movie, *The Jazz Singer*, was released at the Warner Theatre in New York, USA. The movie, a mixture of the silent and talkie format, was the story of

Scenes from 'The Jazz Singer'

a young Jewish man who was from a long line of cantors. Yet, he decided to break with the family tradition and become a Jazz singer. The movie was full of songs some of which became famous.

1931

Riccardo Giacconi, Italian-born American physicist, Nobel laureate, was born in Genoa, Italy. Giacconi attended school in Milan and received his doctorate from the University of Milan in 1954. After teaching at the same university for two years Giacconi went to USA on a Fulbright fellowship and joined the Indiana University, Bloomington, Indiana as research associate. In 1958 he moved to the Princeton University Princeton, New Jersey as a research associate in the Cosmic Ray Laboratory. In 1959 he joined the American Science & Engineering, Inc., Cambridge, Massachusetts as senior scientist, becoming its executive vice president in 1969. From 1973–1981 he was associate director at the High Energy Astrophysics Division, Harvard-Smithsonian Center for Astrophysics, Cambridge, Massachusetts. Giacconi's main contribution has been in detection of X-rays from space. For example, the Sun and all other stars emit electromagnetic radiation at different wavelengths, both visible and invisible light, e.g. X-rays. In order to investigate cosmic X-ray radiation, which is absorbed by the Earth's atmosphere, it is necessary to place instruments in space. Giacconi constructed such instruments and detected for the first time a source of X-rays outside our solar system. He was the first to prove that the universe contains background radiation of X-ray wavelengths. He also detected sources of X-rays that most astronomers now consider to be black holes. Giacconi constructed the first X-ray telescopes, which have provided us with completely new – and sharp – images of the universe. His contributions laid the foundations of X-ray astronomy. Giacconi was awarded the 2002 Nobel Prize for Physics, jointly with the American physicist Raymond Davis Jr. and the Japanese physicist Masatoshi Koshiba, for pioneering contributions to astrophysics, which have led to the discovery of cosmic

X-ray sources. Giacconi received half the prize money while Davis Jr. and Koshiba shared the other half. (☞ Sep. 19, Oct. 14)

1951

Otto Fritz Meyerhof, German biochemist, Nobel laureate (Physiology or Medicine, 1922), who discovered the fixed relationship between the consumption of oxygen and the metabolism of lactic acid in the muscle, died in Philadelphia, USA. (☞ Apr. 12)

1990

The European Space Agency's unmanned spacecraft *Ulysses* was deployed in space on a mission to study the Sun during the 11th flight of the American space shuttle *Discovery*. The spacecraft is equipped with instruments to measure particles, magnetic fields, and electromagnetic radiation from radio wavelengths to gamma rays. The primary goal of the *Ulysses* mission, a joint venture of ESA and NASA, was to explore for the first time the regions over the Sun's north and south poles. Since the energy needed to launch a space probe into a polar orbit around the Sun is much larger than that needed to put a satellite in Earth's polar orbit, *Ulysses* was sent on a trajectory that took it first to Jupiter for gravity assist that would give it enough power to reach a highly elliptical polar orbit around the Sun. It passed over the Sun's south pole in 1994 and 2000 and over its north pole in 1995 and 2001. After more than 10 years of close study of the Sun's north and south polar regions, *Ulysses* has returned a wealth of data that has led to a much broader understanding of the global structure of the Sun's environment – the heliosphere.

1995

Discovery of the first planet around a star outside the solar system was announced by two Swiss astronomers from the University of Geneva, Michael Mayor and Didier Queloz in the journal *Nature*. By observing the motion of the star, the scientists calculated the orbit of the extrasolar planet around 51 Pegasi, a Sun-like star in the constellation of Pegasus 47.9 light-years from Earth, with an orbital period of 4.2 days. After its discovery, many teams have confirmed its existence and obtained more observations of its properties, including the fact that it is about half the mass of Jupiter and orbits very close to the star, which raises its temperature up to around 1000°C. Since then, dozens of planets – mostly gas giants – have been detected around other stars. (☞ Jan. 27)

October 7

1806

Ralph Wedgwood, an Englishman, was granted the first patent for carbon paper. He called his product a 'stylographic writer'. In its original form, Wedgwood's carbon paper was intended to help the blind write through the use of a metal stylus instead of a quill. A piece of paper soaked in printer's ink and dried, was placed between two sheets of writing paper in order to transfer a copy onto the bottom sheet. Horizontal metal wires on the writing-board acted as feeler-guides for the stylus and presumably helped the blind to write. A few years later, Wedgwood developed the idea of making copies of private or business letters and other documents using carbon paper.

1885
Niels Henrik David Bohr, Danish physicist, Nobel laureate, was born in Copenhagen, Denmark. Bohr matriculated from the Gammelholm Grammar School in Copenhagen in 1903, took his master's degree in physics in 1909 and his doctor's degree in 1911 from the Copenhagen University. He was awarded a gold medal from the Royal Danish Academy of Sciences and Letters for his theoretical analysis of and precise experiments on the vibrations of water jets as a way of determining surface tension. In 1912 Bohr went to Manchester, England, and joined Ernest Rutherford's group studying the structure of the atom. Based on Rutherford's theories, Bohr published his model of the atomic structure in 1913, introducing the theory of electrons moving in orbits around the atomic nucleus, the chemical properties of the element being largely determined by the number of electrons in the outer orbits. Bohr also introduced the idea that an electron could drop from a higher-energy orbit to a lower one, emitting a photon (light quantum) of discrete energy. This became the basis for quantum theory. He also formulated the liquid drop model of the nucleus. In 1916, Bohr became a professor at the University of Copenhagen, and director of the newly set up Institute of Theoretical Physics in 1920. Bohr was awarded the 1922 Nobel Prize for Physics for his services in the investigation of the structure of atoms and of the radiation emanating from them. (☞ Nov. 18)

1939
Harold W. Kroto, English chemist, Nobel laureate, was born in Wisbech, Cambridgeshire, England. Kroto was brought up in Bolton, Lancashire. He graduated in chemistry from the University of Sheffield in 1961 and received his PhD there in 1964. After two years of postdoctoral research in electronic and microwave spectroscopy at the National Research Council in Ottawa, Canada, he spent one year at the Bell Laboratories New Jersey, USA, studying liquid phase interactions by Raman spectroscopy. He joined the faculty of the University of Sussex in 1967 and became a professor of chemistry there in 1985. In the course of his research, Kroto used microwave spectroscopy to discover long, chain-like carbon molecules in the atmospheres of stars and gas clouds. He wanted to study the vapourisation of carbon in the laboratory in order to find out how these carbon chains formed. For this he went to the Rice University in Houston, Texas, USA, where the American chemist Richard Smalley had designed an instrument, the laser-supersonic cluster beam apparatus, which could vaporise almost any known material and then be used to study the resulting clusters of atoms or molecules. Using Smalley's apparatus, Kroto carried out experiments that simulated the chemical reactions in the shells of red giant carbon stars. It was during these experiments that the existence of C_{60}, popularly known as 'Buckminsterfullerene', or simply fullerene, was discovered. These new forms of the element carbon contain 60 or more atoms arranged in closed shells. In follow-up investigations of this original discovery, the molecule was isolated independently at Sussex and structurally characterised. Kroto was awarded the 1996 Nobel Prize

for Chemistry, jointly with Smalley and the American chemist Robert F. Curl, for their discovery of fullerenes. (☞ Jun. 6, Aug. 23, Nov. 14)

1952
Norman Woodland, American inventor, was awarded a US patent for 'bar codes', which contains data used for identifying products and can be read by scanners.

1959
The far side of the Moon was photographed for the first time and pictures relayed back to the Earth by the Russian spacecraft *Luna-3*. The photos, taken over a period of 40 minutes, covered 70% of Moon's far side, which was lit by the Sun. Despite the poor quality, they provided the first view ever of the part of the Moon that always

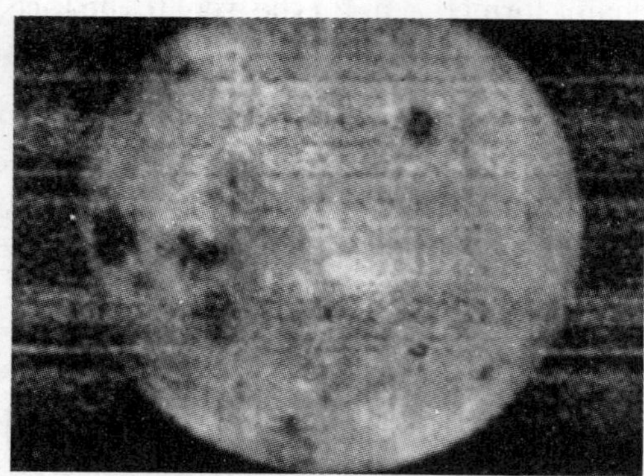

remains hidden from viewers on Earth. The photographs showed mountains and two dark areas which were named Mare Moscovrae (Sea of Moscow) and Mare Desiderii, (Sea of Dreams). (☞ Oct. 4)

1985
American space shuttle *Atlantis*, carrying astronauts Karol J. Bobko, Ronald J. Grabe, David C. Hilmers, Robert L. Stewart, and William A. Pailes, landed at the Edwards Air Force Base, California, USA, after completing its maiden flight during which the crew conducted only classified military activities. (☞ Oct. 3)

1994
Niels K. Jerne, Danish immunologist, Nobel laureate (Physiology or Medicine, 1984), who made significant contributions to explain various aspects of how the immune system defends the body against disease, died in Castillon-du-Gard, France. (☞Dec. 23)

October 8

1604
The supernova (an exploding star) now called 'Kepler's nova' was first sighted in the constellation of Ophiuchus, the Serpent Bearer. Jan Brunowski, Johannes Kepler's assistant, was the first to observe it and Kepler continued observing it until early 1606, when the supernova was no longer visible to the unaided eye. At its greatest apparent magnitude (about −2.5), the exploding star was brighter than Jupiter. It encouraged him to write *De Stella Nova* ('Concerning the New Star') in 1606. (☞ Dec. 27)

1808
Ardaseer Cursetjee, Indian engineer and ship builder, first Indian Fellow of the Royal Society, was born in Bombay (now Mumbai). Following the family tradition, Cursetjee joined the Bombay dockyard at the age of 14. In 1828, after six years of training, he was placed in charge of the supplementary shipyard at Mazagaon, where he designed and supervised the construction of several fine vessels. But more than building ships, Cursetjee was interested in steam machinery. He built a one-horsepower engine and installed

it at his premises for pumping water. This was the first steam engine built in India. In 1833, he installed a 10-horsepower marine engine in a vessel named *Indus* – the second steamer built at Bombay. Cursetjee went to England in 1839, where he made arrangements to work at a few engineering firms, and visited the Royal Dockyards of Great Britain and private foundries. After returning to India in 1841, he was appointed Chief Engineer and Inspector of Machinery in the East India Company's steam factory and foundry at Bombay. He resigned his post in 1863 and went to England, settling down at Richmond. (☞ Nov. 16)

1883

Otto Heinrich Warburg, German biochemist, Nobel laureate, was born in Freiburg im Breisgau, Germany. Otto obtained his PhD in chemistry from the Berlin University in 1906 and an MD from the Heidelberg University in 1911. In 1918 he was appointed professor at the Kaiser Wilhelm Institute for Biology in Berlin-Dahlem. In 1931 he became Director of the Kaiser Wilhelm Institute for Cell Physiology, there. Warburg's early researches were in the polypeptide field. In the early 1920s, while investigating the process by which oxygen is consumed in the cells of living organisms, he introduced the use of manometry (the measurement of changes in gas pressure) for studying the rates at which slices of living tissue take up oxygen. He devised a highly sensitive manometer for this research, enabling him to study the action of respiratory enzymes and poisons in detail. At Heidelberg, he worked on the process of oxidation. His special interest in the investigation of vital processes by physical and chemical methods led to attempts to relate these processes to phenomena of the inorganic world. His methods involved detailed studies on the assimilation of carbon dioxide in plants, the metabolism of tumours, and the chemical constituent of the oxygen-transferring respiratory enzyme. His later researches at the Kaiser Wilhelm Institute led to the discovery that flavins and nicotinamides were the active groups of the hydrogen-transferring enzymes. This, together with the iron-oxygenase discovered earlier, has given a complete account of the oxidations and reductions in the living world. Warburg was awarded the 1931 Nobel Prize for Physiology or Medicine for his discovery of the nature and mode of action of the respiratory enzyme. (☞ Aug. 1)

1917

Rodney Robert Porter, British biochemist, Nobel laureate, was born in Lancashire, England. Porter was educated at the Ashton-in-Makerfield Grammar School and obtained his BSc (Hons) in biochemistry in 1939 from the University of Liverpool and his PhD from the University of Cambridge in 1948. After a year's postdoctoral work at Cambridge, Porter joined the scientific staff of the National Institute of Medical Research at Mill Hill in 1949 and was there until 1960 when he joined the St. Mary's Hospital Medical School, London University, as a professor of immunology. In 1967, he was appointed professor of biochemistry in the University of

Oxford and Fellow of Trinity College, Oxford. Porter started his studies on the structure of antibodies in 1948 and devised a method of breaking large molecules using the enzyme papain. Using this technique, he was able to determine the structure of the smaller fragments and hence of the whole molecule. By 1969, Porter and his team were able to determine the complete structure of the immunoglobulin molecule, which turned out to be made up of four chains comprising more than 1,300 amino acids. Porter was awarded the 1972 Nobel Prize in Physiology or Medicine, jointly with the American physician and physical chemist Gerald M. Edelman, for their discoveries concerning the chemical structure of antibodies. (☞ Jul. 1, Sep. 6)

1918

Jens C. Skou, Danish biophysicist, Nobel laureate, was born in Lemvig, Denmark. Skou received his medical degree from the University of Copenhagen in 1944, and in 1954 earned a doctorate degree from the Aarhus University, where he later taught. In the late 1950s, Skou proposed that an enzyme is responsible for the transport of molecules through a cell's membrane. His work with finely ground crab nerve membranes led to the discovery of the enzyme called sodium–potassium-activated adenosine triphosphatase (Na^+/K^+-ATPase) that works with adenosine triphosphate (ATP) to regulate levels of potassium and sodium in cells. Bound to a cell membrane, Na^+/K^+-ATPase is activated by external potassium and internal sodium. The enzyme pumps sodium out of the cell and potassium into it, thereby maintaining a high intracellular concentration of potassium and a low concentration of sodium relative to the surrounding external environment. Skou was the first to describe an enzyme that can promote directed transport of substances through a cell membrane, a fundamental property of all living cells. Numerous enzymes have since been demonstrated to have essentially similar functions. Skou was awarded the 1997 Nobel Prize for Chemistry, jointly with the American biochemist Paul D. Boyer and British chemist John E. Walker, for the first discovery of an ion-transporting enzyme, Na^+/K^+-ATPase. (☞ Jan. 7, Jul. 31)

1922

Gopalasamudram Narayana Iyer Ramachandran, Indian biophysicist, Fellow of the Royal Society, was born in Ernakulam, Kerala, India. Ramachandran did his intermediate from the Maharaja's College, Ernakulam, his BSc (Hons) in Physics from St. Joseph's College, Tiruchi, and received a master's degree in physics from the Madras University in 1942, after which he moved to the Indian Institute of Science (IISc), Bangalore to work with the Nobel laureate physicist Chandrasekhar Venkata Raman. He obtained a DSc degree from the Madras University in 1947 and later a PhD from the Cambridge University, England. At IISc, Ramachandran worked on X-ray diffraction and X-ray crystallography, techniques that at that time were keys to studying the structures of biomolecules. In 1952 he moved to the Madras University, where a major portion of his research in crystallography and biophysics was performed. His interest shifted to the structure of biological macromolecules, and in 1954, Ramachandran and his associate Gopinath Kartha described the structure of collagen, the most abundant protein of connective tissue. The structure, an intertwined three-chain triple helix, elegantly accounted for the limited experimental data then available and also provided a compelling rationale for the unique amino acid composition of collagen. The proposed structure of collagen was not without controversy. To settle the controversy, Ramachandran decided to develop a good yardstick that could be used to examine and assess any structure in general, but peptides in particular. It involved marathon calculations, but the result that emerged from these calculations in 1962, now commonly known as the 'Ramachandran map', was published in the *Journal of Molecular Biology* in 1963 and has become a household name in the field of protein conformation. At the time of its publication, the crystal structure of proteins had not yet been obtained, and the map was expected to be valuable for studies of peptide and polypeptide structures. The application of the Ramachandran map and its uses slowly began to be felt in the sixties and seventies as the number of protein structures solved steadily increased. Protein crystallographers also used it as a tool for examining their structures, even at the preliminary stage of structure determination. In 1970 Ramachandran returned to IISc and founded the Molecular Biophysics Unit to supplement the various facets of his theoretical work with support from the experimental side in the field of biopolymer conformation. Ramachandran travelled regularly to USA and was a visiting professor at the University of Michigan, from 1965 to 1966, and at the University of Chicago from 1967 to 1977. During that time, he did some exciting work on three-dimensional image reconstruction from radiographs and electron micrographs, which became applicable to computer-assisted tomography (CT scan). (☞ Apr. 7)

1926

César Milstein, Argentine immunologist, Nobel laureate, was born in Bahía Blanca, Argentina. Milstein attended the universities of Buenos Aires and Cambridge, England, from where he received his PhD in 1960. After heading the division of molecular biology at the National Institute of Microbiology in Buenos Aires, he joined the Medical Research Council Laboratory of Molecular Biology in Cambridge, England in 1963. In 1983 he became head, Protein and Nucleic Acid, Chemistry Division at the Cambridge University. The major part of Milstein's research career was devoted to studying the structure of antibodies and the mechanism by which antibody diversity is generated. His early work on antibodies focussed on the nature of their diversity at the amino acid level as well as on the disulphide bonds by which they were held together. In 1975, together with Georges Köhler, a German postdoctoral fellow in his laboratory, he developed the hybridoma technique for the production of monoclonal antibodies. By fusing antibody-producing B lymphocyte cells with tumour cells that are 'immortal', they were able to produce a 'hybridoma', which could continuously synthesise antibodies. All of the antibodies produced by this single clone type of hybridoma cell were identical (monoclonal), the same as those produced by the B cell before it was fused. Monoclonal antibody production was one of the principal driving forces in the creation of the biotechnology industry. It opened the way for the commercial development of new types of drugs and diagnostic tests, in fields as diverse as cancer, the prevention of transplant rejection, pregnancy testing and treatment of

arthritis. Milstein was awarded the 1984 Nobel Prize for Physiology or Medicine, jointly with Köhler and the Danish immunologist Niels K. Jerne, for theories concerning the specificity in development and control of the immune system and the discovery of the principle for production of monoclonal antibodies. (☞ Mar. 24, Apr. 17, Dec. 23)

October 9

1852

Emil Fischer, German organic chemist, Nobel laureate, was born in Euskirchen, Prussia (now in Germany). Fischer finished high school in Bonn. After obtaining his doctorate from the University of Strasbourg in 1874, he held the post of professor at the Universities of Erlangen, Wurzburg and Berlin. In 1875 Fisher published his discovery of the organic derivatives of hydrazine, a compound of nitrogen and hydrogen. He started his research on purines in 1881 and showed that uric acid, xanthine, caffeine, theobromine, and other related compounds were all derivatives of purine, a cyclic molecule containing nitrogen. Nearly all the facts about the purine group, which he named, can be attributed to Fischer. His work had a far-reaching impact on the study of biology because purines were later found to be an important part of the nucleic acid group, including DNA – the key molecule of life. Between 1899 and 1908, Fischer made his greatest contributions to the knowledge of proteins. He developed effective methods of separating and identifying the individual amino acids, discovering a new type of them, the cyclic amino acids: proline and oxyproline. He also studied the synthesis of proteins by obtaining the various amino acids in an optically active form in order to unite them. He was able to establish the type of bond that would connect them together in chains, namely, the peptide bond, and by means of this, he obtained the dipeptides and later the tripeptides and polypeptides. His study of sugars led him to investigate the reactions and substances involved in fermentation, leading to his investigations of how enzymes break down sugars. Thus, he was instrumental in laying the foundation of enzyme chemistry. Fischer was awarded the 1902 Nobel Prize for Chemistry in recognition of the extraordinary services he has rendered by his work on sugar and purine syntheses. (☞ Jul. 15)

1879

Max von Laue, German physicist, Nobel laureate, was born in Pfaffendorf, Germany. von Laue studied mathematics, physics and chemistry at the University of Strassburg, followed by periods at the University of Göttingen and the University of Berlin. In Berlin, he worked under the physicist Max Planck in the Institute of Theoretical Physics. In 1909 he took up a teaching post without pay at the University of Munich, where he lectured on optics, thermodynamics and the theory of relativity. In 1912 he became Professor of Physics at the University of Zürich in Switzerland. In 1914 he moved, as Professor of Physics, to Frankfurt on Main in Germany, and from 1916 he was engaged in war work at the University of Würzburg on high vacuum tubes used for telephony and wireless communication. von Laue began his pioneering work on X-ray diffraction in 1912. He measured the wavelength of X-rays by their diffraction through the

closely spaced atoms in a crystal. He also worked out the mathematical formulation of it and published the results. This work led to the techniques of X-ray spectroscopy, which enabled scientists to study the structure of crystals and hence marked the origin of solid-state physics, an important field in the development of modern electronics. von Laue was awarded the 1914 Nobel Prize for Physics for his discovery of the diffraction of X-rays by crystals. (☞ Apr. 23)

1933
Peter Mansfield, British physicist, Nobel laureate, was born in Lambeth, London, UK. Mansfield graduated with honours in physics from the Queen Mary College, London in 1959 and received his PhD from the same college in 1962. After spending two years as research associate at the Physics Department, University of Illinois, Urbana, Illinois, USA, he returned to England to join the University of Nottingham as lecturer in physics in 1964 and became professor of physics in 1979. He was Medical Research Council Professorial Research Fellow during 1983–1988, after which he returned to the University of Nottingham in 1988. Mansfield's research mainly concerned improvement in magnetic resonance imaging. He utilised gradients in the magnetic field in order to show more precisely the differences in resonance. He showed how the detected signals could be analysed rapidly and effectively and transformed to

an image. This was an essential step in obtaining a practical method of imaging. Mansfield also showed how extremely rapid imaging could be achieved by very fast gradient variations (so-called echo-planar scanning). This technique became useful in clinical practice a decade later. Mansfield was awarded the 2003 Nobel Prize for Physiology or Medicine, jointly with the American chemist Paul C. Lauterbur, for their discoveries concerning magnetic resonance imaging. (☞ May 6)

1943
Pieter Zeeman, Dutch physicist, Nobel laureate (Physics, 1902), who discovered the splitting of spectral line under magnetic field, known as Zeeman effect, died in Amsterdam, Netherlands. (☞ May 25)

1950
Willis Haviland Carrier, American engineer and inventor of modern air conditioning, died in Buffalo, New York, USA. (☞ Nov. 26)

1967
Cyril Norman Hinshelwood, British chemist, Nobel laureate (Chemistry, 1956), who made significant contributions to the study of reaction rates and reaction mechanisms, died in London, England. (☞ Jun. 19)

1987
William Parry Murphy, American physician, Nobel laureate (Physiology or Medicine, 1934), who co-discovered liver therapy for treatment of anaemia, died in Brookline, Massachusetts, USA. (☞ Feb. 6)

1988
Felix Wankel, German engineer who invented the first rotary internal combustion engine, died in Lindau, Germany. (☞ Aug. 13)

October 10

1731

Henry Cavendish, English physicist and chemist, Fellow of the Royal Society, was born in Nice, France. Cavendish attended the St. Peter's College in Cambridge from 1749 to 1753 but left without taking a degree. He inherited a large fortune, which enabled him to pursue his scientific studies in diverse fields, most of which remained unpublished during his lifetime. Cavendish is generally credited with having discovered hydrogen, because he had described the density of 'inflammable air', which formed water on combustion, in the paper 'Factitious airs' that appeared in *Philosophical Transactions* of the Royal Society in 1766. The French chemist Antoine Lavoisier later reproduced his experiment and gave the element its name. Cavendish is also credited with one of the earliest accurate calculations of the mass of the Earth. He used a torsion balance to measure the gravitational attraction between lead spheres in 1798, from which he calculated the universal gravitational constant G, which he used to calculate the Earth's mass. His method has come to be known as the 'Cavendish experiment'. His other discoveries include the determination of the specific gravity of carbon dioxide, and the demonstration that it could stop a flame burning in common air. Cavendish discovered nitric acid (HNO_3), and proved that water was not an element but made up of gases. He left a large estate on his death, which was used to endow the Cavendish Laboratory at Cambridge University in 1871. (☞ Feb. 24)

1846

Neptune's moon, Triton, was discovered by the amateur English astronomer William Lassell while he was observing the newly discovered planet Neptune. Using his 24-inch (61-cm) reflecting telescope, he was attempting to confirm his observation of the previous week that Neptune had a ring. Instead he discovered that Neptune had a satellite, which was named Triton. Lassell soon proved that the ring he thought he had seen was a product of his new telescope's distortion.

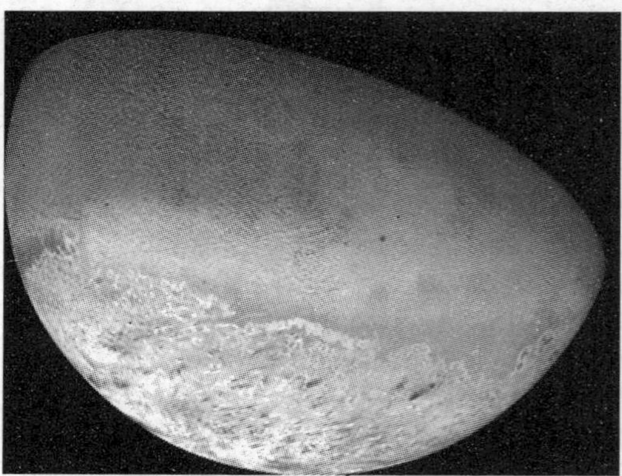

Voyager-2 image of Triton

1933

Waldo L. Semon, American chemical engineer and inventor, was issued a US patent for his method of making plasticised PVC, now known simply as vinyl. The patent was titled 'Synthetic rubber-like composition and method of making same'. Unplasticised polyvinyl chloride, commonly known as PVC, is a polymer that is hard and difficult to form into useful products. Semon invented a way to make it in a rubber-like form. The patent listed uses such as waterproof boots or shoes, insulating coatings and resilient flooring material.

1983

Soviet unmanned spacecraft *Venera-15* arrived at Venus and went round the planet in a polar orbit once in 24 hours. Its high-resolution imaging system produced images at 1–2 kilometres in resolution and found several hot spots, possibly caused from volcanic activity. (☞ Jun. 2)

October 11

1758

Heinrich Wilhelm Olbers, German astronomer and physician who discovered the asteroids Pallas and Vesta as well as five comets, was born in Arbergen, near Bremen, Germany. Olbers studied to be a physician at Göttingen. After his graduation in 1780, he began practising medicine in Bremen, Germany. At night he devoted his time to astronomical observation. In 1797 he developed the first satisfactory method for calculating the orbits of comets, but despite the fame it brought him, he remained an amateur astronomer and continued his research on comets. He was the first to detect the periodic comet of 1815, which has been named after him. He also discovered two asteroids, Pallas in 1802, and Vesta in 1807. Considering their orbits and those of the other asteroids then known, Olbers concluded that they are fragments of a disrupted planet that had formerly revolved around the Sun. He is best remembered for Olbers' paradox: 'Why is the sky dark at night?' Assuming that space is infinite and filled with stars, he suggested that the entire sky should be as bright as the surface of the Sun. The question had originally been raised by Kepler. Others offered one of several incorrect resolutions to the paradox, suggesting that absorbing clouds obscured our view of the distant stars. The correct explanation is that our universe is finite both in time and place, and the total amount of matter and energy is far too small to light up the night sky. (☞ Mar. 2, Mar. 28, Mar. 29)

1838

Friedrich Bessel, German astronomer, for the first time determined the distance of a star other than the Sun, using the method of parallax. He determined the distance of the star 61 Cygni to be 10.3 light years from Earth. (☞ Jul. 22)

1838

R. Games Slayter of the American glass manufacturing company Owens Corning was issue a patent for glass wool, as a material to be used as insulation. It was marketed under the trade name 'Fiberglas'. Fibreglass, as it is known today, is widely used for making lightweight composites.

1884

Friedrich Bergius, German chemist, Nobel laureate, was born in Goldschmieden, near Breslau, Silesia (now Wroclaw, Poland). Bergius was educated in Breslau and whilst still at school took great interest in his father's chemical factory where he was able to study various working methods under the guidance of his father and thus became acquainted with chemico-technical processes. Before entering university, Bergius was sent to the Ruhr for six months by his father, where he studied the practical aspects of a large metallurgical plant and profited greatly by this experience. In 1903 he entered the Breslau University to study chemistry and after doing one year's military service he proceeded to the Leipzig University in 1905 to write his thesis 'Über absolute Schwefelsäure als Lösungsmittel' ('On absolute sulphuric acid as a solvent'). He received his degree

at Leipzig in 1907. In 1909, Bergius went to Karlsruhe to study under the German physical chemist Fritz Haber for one term and to work on the chemical equilibrium in gas reactions, in particular on the synthesis of ammonia. This experience prompted Bergius to start his own research in Hanover in 1909. He started with a detailed investigation of the dissociation of calcium peroxide and developed a practical method for laboratory work at pressures up to 300 atmospheres. The equipment available in the laboratories of the Technische Hochschule in Hanover soon proved insufficient and, in 1910, Bergius established his own private laboratory in Hanover, where he employed a number of collaborators. During 1912 and 1913, the most important result of his research was the development of a chemical high-pressure hydrogenation method for treating a mixture of pulverised coal in oil with the gas under high pressure. Bergius succeeded in forcing hydrogen under high pressure to combine chemically with the coal, transforming more carbon from the coal into oils than is possible with conventional distillation. Bergius was awarded the 1931 Nobel Prize for Chemistry, jointly with the German industrial chemist Carl Bosch, in recognition of their contributions to the invention and development of chemical high-pressure methods. (☞ Mar. 30, Aug. 27)

1889

James Prescott Joule, English physicist who established that the various forms of energy are basically the same and can be changed, one into another, died in Sale, Cheshire, England. (☞ Dec. 24)

1923

Harish-Chandra, Indian mathematician, Fellow of the Royal Society, was born in Kanpur, Uttar Pradesh, India. After completing his schooling in Kanpur, Harish-Chandra studied theoretical physics at the University of Allahabad from where he received his MSc in 1943. He did his PhD from Cambridge, England under the supervision of the English theoretical physicist Paul Dirac, and in 1947 went to USA. During his stay in Cambridge, Harish-Chandra moved away from physics and became more interested in mathematics. In USA, Harish-Chandra worked as assistant to Dirac at the Princeton University during the latter's visit for a year. The period 1950 to 1963 was his most productive and he spent these years at the Columbia University, New York. During this time he worked on representations of semi-simple 'Lie groups'. His scientific work, being a synthesis of analysis, algebra and geometry, is still of lasting influence. Also during this period, he had close contact with the French mathematician André Weil. Harish-Chandra worked at the Institute for Advanced Study at Princeton from 1963. He was appointed IBM-von Neumann Professor in 1968. (☞ Oct. 16)

1957

The Jodrell Bank radio telescope in England, the world's first giant radio telescope with a 76-m diameter dish antenna, began operation. The telescope was designed by the English physicist and astronomer Bernard Lovell. Though the telescope is popularly known for tracking and communicating with man-made satellites, its prime function is the study

of the universe by means of radio waves emitted by distant stars. Although now superseded by larger radio telescopes, it still remains one of the biggest and most powerful radio telescopes in the world, spending most of its time investigating cosmic phenomena, which were undreamed of when it was conceived. (☞ Aug. 31)

1968
American spacecraft *Apollo-7*, carrying a crew of Walter Schirra, Donn Eisele, and Walter Cunningham, was launched from the Kennedy Space Center, Florida, USA. It was an eleven-day Earth-orbital mission, and the first manned mission in the Apollo program to be launched. It was also the first American space mission carrying a crew of three. The mission objectives included an in-orbit docking manoeuvre and the first live television broadcast from an American spacecraft. (☞ Oct. 22)

October 12

1865
Arthur Harden, English biochemist, Nobel laureate, was born in Manchester, England. Harden was educated at a private school in Victoria Park and at the Tettenhall College, Staffordshire. He graduated from the Owens College in the University of Manchester in 1885 with first class honours in chemistry. In 1887 he joined the University of Manchester as a lecturer and demonstrator, and remained there until 1897 when he was appointed chemist to the newly founded British Institute of Preventive Medicine in London, which later became the Lister Institute. In 1907 he was appointed head of the biochemistry department, a position which he held until his retirement in 1930, and in 1912 he was appointed emeritus professor of chemistry at the University of London. At Lister Institute, Harden carried out the investigation of various biological phenomena, including the chemical action of bacteria and alcoholic fermentation. He studied the breakdown products of glucose but was soon drawn to the problem of the chemistry of the yeast cell. His classic study of the chemistry of the fermentation of sugar by yeast juice lasted for more than 20 years and provided a stepping-stone for many research workers in allied fields. Harden also contributed to the vitamin field with a series of papers on anti-scorbutic and anti-neuritic vitamins and their occurrence in food and drinks. He also pioneered in studies of bacterial enzymes and metabolism. Harden was awarded the 1929 Nobel Prize for Chemistry, jointly with the Swedish biochemist Hans von Euler-Chelpin, for their investigations on the fermentation of sugar and fermentative enzymes. (☞ Feb. 15, Jun. 17)

1928
The artificial respirator, usually called an 'iron lung', was first demonstrated at the Children's Hospital in Boston, Massachusetts, USA. Invented by a Harvard professor named Philip Drinker, the device is a large machine that enables a person

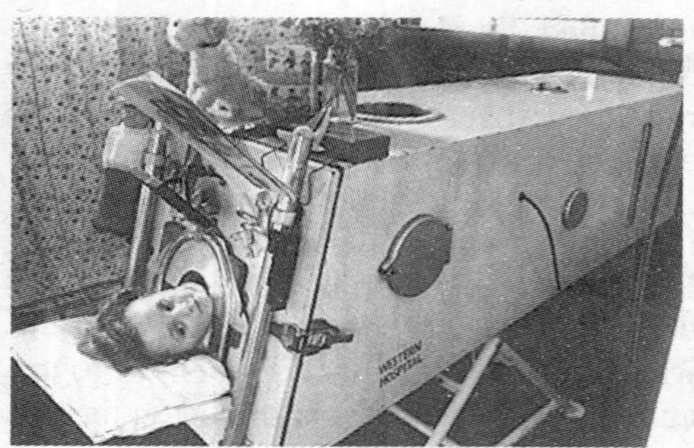

who otherwise could not breathe. Nearly the whole body, except for the head, fits into the machine, which is airtight and creates a negative pressure inside which increases the volume of the chest cavity and pulls in air through the mouth. Then the pressure inside the iron lung increases, causing the air to be expelled again. The device is used to maintain artificial respiration over an extended period of time. Iron lungs were used in the past for victims of polio, and in cases where the respiration control centre of the brain was damaged due to injury. The use of iron lungs has now almost completely ceased. Today, patients with paralysis of the breathing muscles use mechanical ventilators that push air into the airway with positive pressure.

1965
Paul Hermann Müller, Swiss chemist, Nobel laureate (Physiology or Medicine, 1948), who discovered the potent toxic effects of DDT on insects, died in Basel, Switzerland. (☞ Jan. 12)

October 13

1807
Humphry Davy, British chemist, discovered and isolated sodium by electrolysis of very dry molten caustic soda (sodium hydroxide). When a current was passed, globules of silvery matter collected at the negative pole. Most of these caught fire, but some solidified later and could be collected. These solid particles were pure sodium. Its chemical symbol is derived from the Latin word *natrium*, which means 'soda'. Its English name is from soda, which is the common name for caustic soda. In its pure form, sodium is a silvery white metal, which is soft enough to cut with a knife. It is the sixth most abundant element in the Earth's crust, occurring in large amounts in both seawater and soil in various mineral compounds, the most common of which is sodium chloride. (☞ Dec. 17)

1884
Greenwich, England, was adopted as the universal meridian at the International Meridian Conference – a meeting of delegates from 25 nations held in Washington, DC, USA. At the conference, it was decided that a single world meridian passing through the principal Transit Instrument at the Observatory in Greenwich would be adopted and that all longitude would be calculated both east and west from this meridian up to 180°. Greenwich lies on the River Thames, a few kilometres from central London.

1987
Walter Houser Brattain, American physicist, Nobel laureate (Physics, 1956), co-inventor of the transistor, died in Seattle, Washington, USA. (☞ Feb. 10)

2003
Bertram N. Brockhouse, Canadian physicist, Nobel laureate (Physics, 1994),

who independently developed the neutron-scattering techniques, died in Hamilton, Canada. (☞ Jul. 15)

October 14

1884

George Eastman, American inventor, was granted a patent for the photographic roll film. Eastman's invention revolutionised photography by using coated paper and rollers rather than heavy glass dry plates to reproduce images. This invention allowed him to mass-produce a small hand-held box camera filled with rolls of film with 100 exposures. (☞ Jul. 12, Sep. 4)

1914

Raymond Davis Jr, American physicist, Nobel laureate, was born in Washington, DC USA. Davis Jr was educated in Washington public schools and graduated from the University of Maryland in 1938 with a degree in chemistry. After working for the Dow Chemical Company in Midland, Michigan for a year, he returned to the University of Maryland to take a master's degree. He received a doctorate in physical chemistry from the Yale University, New Haven, Connecticut in 1942. After working in the Army for three years, Davis Jr joined the Monsanto Chemical Company's Mound Laboratory, in Miamisburg, Ohio, working on applied radiochemistry which was of interest to the Atomic Energy Commission. In 1948, he joined the newly created Brookhaven National Laboratory, which was dedicated to finding peaceful uses for atomic energy. At Brookhaven he started work on detection of the elusive subatomic particles called neutrinos. Neutrinos are extremely difficult to detect because they hardly interact with matter at all. For example, thousands of billions of neutrinos pass through us every second without our noticing them. It was also suspected, though not proved, that the Sun's energy was due to a fusion reaction which releases countless neutrinos. To detect these solar neutrinos, Davis Jr constructed a completely new detector – a gigantic tank filled with 600 tonnes of perchloroethylene fluid – which was placed in a mine. When neutrinos from the Sun collided with a chlorine atom of the fluid, the atom gets converted into a radioactive isotope of argon, which can then be collected and counted. Over a period of 30 years, David Jr succeeded in capturing a total of 2,000 neutrinos from the Sun and was thus able to prove that the energy of the Sun was caused by fusion. Davis Jr. was awarded the 2002 Nobel Prize for Physics, jointly with the Japanese physicist Masatoshi Koshiba and Italian-born American physicist Riccardo Giacconi, for pioneering contributions to astrophysics, in particular for the detection of cosmic neutrinos. (☞ Sep. 19, Oct. 6)

1947

American test pilot Charles E. Yeager became the first human to fly faster than the speed of sound, breaking through the sound barrier in a rocket-powered X-1 fighter plane over Murac Dry Lake, California,

USA. The bullet-shaped X-1 was carried aloft by a B-29 bomber and then released, when Yeager fired its rockets. At 11,300 metres, the X-1 flew nicely. Although it began to buffet as it approached the sound barrier, Yeager succeeded in breaking the sound barrier.

1983
Soviet unmanned spacecraft *Venera-16* went into orbit around Venus. *Venera-16* was part of a two-spacecraft mission to Venus, along with *Venera-15*, which was launched earlier. Both the spacecraft were equipped with synthetic aperture radar (SAR) for mapping the surface because nothing else could penetrate the dense clouds of Venus. Both were also equipped with a radio altimeter dish antenna, and a large communication dish, which allowed them to send images back to Earth. (☞ Jun. 7)

1984
Martin Ryle, British radio astronomer, Nobel laureate (Physics, 1974), who developed revolutionary radio telescope systems and used them to accurately locate weak radio sources, died in Cambridge, England. (☞ Sep. 27)

October 15

1564
Andreas Vesalius, Belgian anatomist and physician whose dissections of the human body and descriptions of his findings helped to correct misconceptions prevailing since ancient times, died in the island of Zacynthus, Republic of Venice (now in Greece). (☞ Dec. 31)

1608
Evangelista Torricelli, Italian physicist and mathematician, was born in Faenza, Italy. Torricelli studied mathematics and philoso-phy at a Jesuit college in Faenza from 1624 to 1626 and then went to Rome to study at the Collegio Romano. After completing his studies, he served as Galileo Galilei's secretary and succeeded Galileo as grand-ducal mathematician and professor of mathematics in the Florentine academy. He applied mathematics to fluid flow and projectile motion, and developed calculus-like procedures for calculating arc length and finding infinitesimals. He also made telescopes and microscopes by designing and grinding fine lenses. In 1643, he discovered the principle of the barometer; using a glass tube filled with mercury inverted into a dish of mercury, he discovered that the rise of a liquid in a tube, as in the barometer, is due to atmospheric pressure. The height of the mercury column inside the tube gives a measure of the atmospheric pressure, because such a column compensates the weight of the air at the exterior. Torricelli anticipated that in the case of water, the corresponding column would be about 10 metres in height, a result that was later verified by the French mathematician Blaise Pascal. The 'torr', a unit of pressure is named after him. (☞ Oct. 25)

1829
Asaph Hall, American astronomer, who discovered and named the two moons of Mars and calculated their orbits, was born in Goshen, Connecticut, USA. Hall was apprenticed as a carpenter at the age of sixteen and later enrolled at the Central College in McGrawville, New York. In 1856, he took a job at the Harvard College Observatory in Cambridge, Massachusetts, and became an expert computer of orbits.

He became assistant astronomer at the US Naval Observatory in Washington DC 1862, and within a year of his arrival was made professor of mathematics. Using the Observatory's 66-cm refracting telescope, then the largest of its kind in the world, he made his discoveries in August 1877. He named the moons Phobos and Deimos. Deimos is Greek for 'panic'; Phobos is Greek for 'fear'. These moons are very small – Phobos is only 28 kilometres and Deimos only 16 kilometres at the widest – and are composed of ice and carbon-rich rock like the minor planets. Hall also investigated stellar parallaxes and the position of the stars in the Pleiades cluster. He worked at the Observatory for 30 years until 1891. From 1896 to 1901 he was professor of astronomy at the Harvard University, Cambridge, Massachusetts. (☞ Aug. 11, Aug. 16, Nov. 22)

1867

Alfred Nobel, Swedish chemist and inventor, was granted a patent for dynamite, a special type of nitroglycerine that was safe to handle. Dynamite was made by mixing nitroglycerine with kieselguhr, an absorbent sand, making a dough-like material that could be handled safely. Nobel called it dynamite after the Greek word '*dynamis*', meaning power. (☞ Oct. 21)

1931

Avul Pakir Jainulabdeen Abdul Kalam, space scientist and the main driving force behind India's missile development program, was born in Rameshwaram, Tamil Nadu, India. After graduating in science from the St. Joseph's College in Tiruchi, Kalam specialised in aero-engineering from the Madras Institute of Technology. He joined the Defence Research and Development Orga-nisation (DRDO) in 1958 and then the Indian Space Research Organisation (ISRO) in 1963. At ISRO, he initiated fibre-reinforced plastics activities. He later joined the satellite launch vehicle team at Thumba, near Thiruvananthapuram, Kerala, and became the project director for SLV-3, India's first satellite launch vehicle. Kalam rejoined DRDO in 1982 and was instrumental in the resurgence of the Defence Research & Develop-ment Laboratory (DRDL), Hyderabad. Out of his initiative was born the Integrated Guided Missile Development Programme (IGMDP). He also headed other prestigious projects, such as the main battle tank *Arjun* and the light combat aircraft (LCA) projects. In 1997, Kalam was awarded the 'Bharat Ratna' – India's highest civilian honour. In June 2002, he became the 11th President of India.

1940

Peter C. Doherty, Australian immunologist and pathologist, Nobel laureate, was born in Brisbane, Queensland, Australia. Doherty received his bachelor's degree in veterinary science in 1962 and master's in veterinary science in 1966 from the University of Queensland. He then obtained his PhD in pathology from the University of Edinburgh, Scotland in 1970. During 1972–75, while conducting research at the John Curtin School of Medical Research in Canberra, Doherty began collaborating with Swiss immunologist Rolf Zinkernagel to study how the immune system, and particularly T-lymphocytes, in mice could protect the animals against infection from a virus able

to cause meningitis. Infected mice developed killer T-lymphocytes, which in a test-tube could kill virus-infected cells. But there was an unexpected discovery: the T-lymphocytes, even though they were reactive against that very virus, were not able to kill virus-infected cells from another strain of mice. Further research showed that T-lymphocytes must recognise two separate signals on an infected cell before they destroy it. One signal is a fragment of the invading virus that the cell displays on its surface; the other is a self-identifying tag from the cell's major histo-compatibility complex, which identify the cell as belonging to one's own body. This concept of simultaneous recognition of both self and foreign molecules formed the basis for a new understanding of the general mechanisms used by the immune system at the cellular level. Doherty and Zinkernagel's findings had an immediate impact on immunological research. Doherty taught at the Wistar Institute in Philadelphia, USA, from 1975 to 1982 and then headed the department of pathology at the Curtin School in Canberra, Australia. In 1988 he was appointed chairman of the Department of Immunology at St. Jude Children's Research Hospital, Memphis, Tennessee, USA, and a professor in the University of Tennessee Medical School. Doherty was awarded the 1996 Nobel Prize for Physiology or Medicine, jointly with Zinkernagel, for their discoveries concerning the specificity of the cell-mediated immune defence. (☞ Jan. 6)

1994

IRS-P2, the second indigenously launched satellite in India's IRS series was launched by a PSLV rocket from Sriharikota, Andhra Pradesh. Designed and built by the Indian Space Research Organisation, the satellite carried two linear imaging and self scanner (LISS) cameras for Earth imaging and was placed in a Sun-synchronous polar orbit at an inclination of 98.7 degrees with an average altitude of 815 km. It took 103 minutes to complete one orbit around the Earth, making 14 orbits per day. The satellite returned periodically over the same geographical area, giving a complete coverage of the globe in 22 days.

1997

The *Cassini–Huygens* spacecraft was launched towards Saturn from the Kennedy Space Center, Florida, USA. A joint mission of NASA, the European Space Agency and the Italian Space Agency the *Casini–Huygens* spacecraft comprised two units – the *Cassini* orbiter and the *Huygens* probe that would land on Saturn's largest moon Titan. It is the most highly instrumented and scientifically capable planetary spacecraft ever flown. It has 12 instruments on NASA's *Cassini* orbiter and six more on the ESA-built *Huygens* probe. The mission represents the best technical efforts of 260 scientists from USA and 17 European nations. The *Cassini–Huygens* mission was to study Saturn's rings, icy satellites, magnetosphere and Titan, the planet's largest moon. It entered into orbit

around Saturn in 2004 and the *Huygens* probe landed on Titan in 2005. (☞ Jan. 14, Apr. 26, Jun. 24, Jul. 1, Aug. 18, Dec. 24, Dec. 30).

2000
Konrad E. Bloch, German-born American biochemist, Nobel laureate (Physiology or Medicine, 1964), who identified the chemical process by which the body turns acetic acid into cholesterol, died in Burlington, Massachusetts, USA. (☞ Jan. 21)

October 16

1846
The first public demonstration of the use of ether as an anaesthetic was given by American dentist, William Thomas Green Morton at the Massachusetts General Hospital in Boston, USA. The patient, 20-year-old Gilbert Abbott inhaled from a blown glass flask, during surgical removal of a small superficial tumour from beneath his left lower jaw. After the success of the public demonstration, the use of ether for painless operations spread quickly to other countries. Four years, an American physician named Crawford W. Long, had used ether as an anaesthetic, but Long's success went unrecognised for years, till his work was reported in 1849 in the *Southern Medical Journal* and is now widely considered to represent the discovery of surgical anaesthesia. (☞ Mar. 30)

1982
Comet Halley was first detected on its 30th recorded visit to Earth, using the 5-m (200-inch) Hale Telescope at the Mount Palomar Observatory in California, USA by a team of astronomers led by David Jewett and G. Edward Danielson. They found the comet beyond the orbit of Saturn, about 1.6 billion km from the Sun. Being 50 million times fainter than the faintest objects our eyes can see, they needed to use not only the largest American telescope but also special electronic equipment developed for the Hubble Space Telescope. (☞ Feb. 9, Mar. 30, Apr. 11, May 19, Nov. 8, Dec. 25)

1983
Harish-Chandra, Indian mathematician, Fellow of the Royal Society, who did fundamental work in 'representation theory', died in Princeton, New Jersey, USA. (☞ Oct. 11)

October 17

1855
Henry Bessemer, a British inventor and metallurgist was granted a patent for a steel-making process. His patent was for a method of making steel by blasting compressed air through molten iron to remove impurities and excess carbon. The 'Bessemer process', made it possible to mass-produce steel inexpensively. In the course of his life, Bessemer earned more than 100 patents. (☞ Jan. 19, Mar. 15)

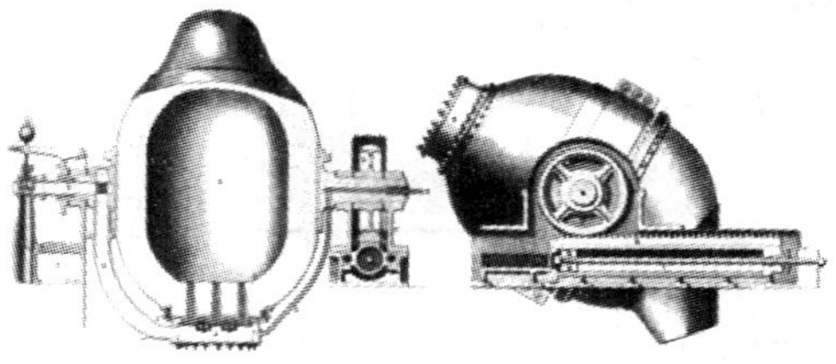

Bessemer process

1887
Gustav Robert Kirchhoff, German physicist who, with the chemist Robert Bunsen, established the theory of spectrum analysis and used it to determine the composition of the Sun, died in Berlin, Germany. (☞ Mar. 12)

1934

Santiago Ramón y Cajal, Spanish histologist, Nobel laureate (Physiology or Medicine, 1906), who established the neuron, or nerve cell, as the basic unit of nervous structure, died in Madrid, Spain. (☞ May 1)

2003

India's remote sensing satellite *RESOURCESAT-1* (*IRS-P6*) was launched by ISRO's polar satellite launch vehicle, PSLV-C5 from the Satish Dhawan Space Centre, Sriharikota, Andhra Pradesh. The satellite carried three cameras on board – a multi-spectral high spatial resolution camera, a linear imaging self scanner-4 (LISS-4), a multi-spectral linear imaging self scanner-3 (LISS-3), and a multi-spectral advanced wide field sensor (AWiFS). *RESOURCESAT-1* is designed to enhance remote sensing applications, especially, in the fields of agriculture, land and water resources management, and disaster management.

October 18

1871

Charles Babbage, British inventor of the adding machine, who is also known as the 'Father of Computing' for his contributions to the basic design of the computer through his analytical machine, died in London, England. (☞ Dec. 26)

1931

Thomas Alva Edison, American inventor, who invented the first phonograph and the prototype of the practical incandescent electric light bulb, died in West Orange, New Jersey, USA. (☞ Jan. 27, Feb. 11, Feb. 19, Apr. 15, Aug. 8, Aug. 9, Aug. 27, Oct. 21, Dec. 6)

1960

Craig Cameron Mello, American molecular biologist, Nobel laureate, was born in New Haven, Connecticut, USA. Mello attended the Fairfax High School in Virginia. He received his BS from the Brown University in 1982 and a PhD degree in cellular and developmental biology from the Harvard University in 1990. After graduation he received a postgraduate fellowship to work at the Fred Hutchinson Cancer Research Center in Seattle, Washington. In 1994 Mello joined the faculty of the University of Massachusetts Medical School (UMMS) in Worcester, Massachusetts, where he became professor of molecular medicine. It was at UMMS that Mello along with American geneticist Andrew Zachary Fire discovered the process now known as RNA interference (RNAi), or gene silencing. Ribonucleic acid (RNA) is a molecule in body cells that carries instructions from genes for the manufacture of proteins. Mello and Fire discovered that bits of double-stranded RNA can degrade these instructions and shut down the activity of genes – a process that has come to be known as 'gene silencing'. Mello and Fire carried out their RNA research on the roundworm *C. elegans*, but later discovered that the same RNAi process operates in higher organisms too, even human beings. The ability to use RNAi to turn off specific genes has proved extremely useful in medical research. By inactivating a gene, scientists can gain a better understanding of how it functions and how it may be involved in causing disease at the cellular level, including diseases such as cancer. RNAi is also important in the body's defenses against viruses, and researchers have begun to design therapies based on RNAi techniques. Mello was awarded

the 2006 Nobel Prize for Physiology or Medicine, jointly with American geneticist Andrew Zachary Fire, for their discovery of RNA interference. (☞ Apr. 27).

1967
Soviet unmanned spacecraft *Venera-4* entered the Venusian atmosphere and took measurements of the abundance of hydrogen and oxygen. During descent, on-board instruments took measurements of temperature, pressure and density. A radio altimeter also kept an eye on the instruments' height above the approaching surface. Contact was lost after 94 minutes of descent at an altitude of 24.96 km from the surface. *Venera-4* was the first spacecraft to enter the atmosphere of another planet. (☞ Jun. 12)

1985
India's Fast Breeder Test Reactor at the Indira Gandhi Atomic Research Centre in Kalpakkam, Tamil Nadu, attained criticality. The 15-MW experimental reactor was entirely designed and fabricated by Indian atomic scientists. The fuel, a mixture of 70% plutonium carbide and 30% uranium carbide, was developed locally and used for the first time in the world. All the major components were manufactured by Indian industries with very high standards of quality. The reactor does not use any moderator. It uses liquid sodium as coolant and can be used to produce plutonium fuel from uranium-238. The reactor has been supplying power, off and on, to the main grid on an experimental basis. During more than two decades of operation, the FBTR has also provided valuable experience in liquid metal fast breeder reactor technology to prepare the ground for a 500-MW prototype fast breeder reactor, planned to be constructed at Kalpakkam.

1989
The American spacecraft *Galileo* was launched on a mission to Jupiter during the 5th flight of American space shuttle *Atlantis*. It studied Jupiter's atmosphere with a descent probe and conducted long-term observations of the Jovian system from orbit. The trajectory that the 2,400-kilogram spacecraft followed was called a VEEGA (Venus–Earth–Earth gravity assist), travelling first toward the Sun for a gravity assist from Venus before

encountering the Earth two times, spaced two years apart. These encounters with Venus and the Earth allowed the spacecraft to gain enough velocity to get it out to Jupiter. The descent probe was released after *Galileo* went into orbit around Jupiter to study the composition of the planet's atmosphere. (☞ Jul. 13, Aug. 28, Sep. 21, Oct. 3, Oct. 23, Dec. 7)

October 19

1862
Auguste Lumière, Frenchman who, with his brother Louis, invented and pioneered the manufacturing of photographic equipment, was born in Besançon, France. Both the brothers attended La Martiniere, the largest technical school in Lyon, and joined their father in his photography business.

The Lumière brothers are best known for having produced an early motion-picture camera and projector called the cinématographe ('cinema' is derived from this name). Lumière brothers' cinematograph camera, developed in 1895, used a claw move-ment for advancing the film, a principle that still applies in motion photography today. The Lumières' technical innovations allowed motion pictures to venture into the world outside a studio, permitting any object in reality to become a subject of interest for the camera. The Lumière brothers also produced the Autochrome plate in 1907, the first practical colour photography process. (☞ Mar. 22, Apr. 10, Oct. 5, Dec. 28)

1875
Charles Wheatstone, English physicist, Fellow of the Royal Society, who popularised the Wheatstone bridge, a device that accurately measured electrical resistance and became widely used in laboratories, died in Paris, France. (☞ Feb. 6)

1910
Subrahmanyan Chandrasekhar, Indian-born American astrophysicist, Nobel laureate, was born in Lahore, Punjab (now in Pakistan). Until the age of 12, Chandrasekhar was taught at home. Afterwards he attended the Hindu High School. After earning a bachelor's degree in physics from the Presidency College, Madras (now Chennai), he received a scholarship to pursue graduate studies at the Cambridge University in England. He also studied for one year in Copenhagen, Denmark, at the Institut for Teoretisk Fysik prior to receiving his PhD from Cambridge in 1933. In 1937, Chandrasekhar immigrated to USA to join the faculty of the University of Chicago where he remained until his death in 1995. Chandrasekhar was one of the foremost astrophysicists of the 20th century. Around 1930, using Albert Einstein's special theory of relativity and the principles of quantum physics, Chandrasekhar showed that it is impossible for a white dwarf star – the last stage in the evolution of a star such as the Sun – to be more massive than about 1.44 times the solar mass. This became known as the 'Chandrasekhar limit'. This discovery is the basis for much of modern astrophysics, since it shows that stars much more massive than the Sun must either explode or form black holes. He also developed theories on star atmospheres, black holes, illumination of the sunlit sky, star structures, and star mass. Chandrasekhar was awarded the 1983 Nobel Prize for Physics, jointly with the American nuclear astrophysicist William Alfred Fowler. Chandrasekhar received half the prize money for his theoretical studies of the physical processes of importance to the structure and evolution of stars, while Fowler received the other half for his theoretical and experimental studies of the nuclear reactions of importance in the formation of the chemical elements in the universe. In 1999, the American space agency NASA named its most advanced X-ray telescope after Chandrasekhar. (☞ Jul. 23, Aug. 9, Aug. 21)

1916
Jean-Baptiste-Gabriel-Joachim Dausset, French haematologist and immunologist, Nobel laureate, was born in Toulouse, France. Dausset pursued his secondary

studies at the Lycée Michelet in Paris and then studied medicine. But his medical studies were interrupted by World War II. After the war, Dausset resumed his interrupted medical studies and took his degree from the University of Paris in 1945. In 1948, he was sent, as a French

trainee, to the Children's Hospital in Boston, Massachusetts, USA, where he worked in one of the Harvard Medical School laboratories. On his return to France, he took up work at the Regional Blood Transfusion Centre at Hôpital Saint-Antoine, where he immediately became interested in the new immuno-haematology techniques for red blood cells. In investigating the immunological reactions of patients who had received many blood transfusions, Dausset correctly hypothesised that a specific genetic variation among people accounted for the different levels of reaction. These antibody reactions were stimulated by certain antigens located on the surface of foreign white blood cells, which were later called human leukocyte antigens. This discovery led to the use of simple blood tests to determine whether a prospective donor is a good match to the patient for an organ transplant. In 1958, while head of the immuno-haematology laboratory at the National Blood Transfusion Centre, he described the first leucocyte antigen, called the major histo-compatibility complex, which helps the immune system of all vertebrates to distinguish between its own cells and foreign substances that gain entry to the body. In 1977 he was appointed professor of experimental medicine at the Collège de France. Dausset was awarded the 1980 Nobel Prize for Physiology or Medicine, jointly with the American immuno-geneticist George Snell and Venezuelan-born American pathologist Baruj Benacerraf, for their discoveries concerning genetically determined structures on the cell surface that regulate immunological reactions. (☞ Oct. 29, Dec. 19)

1937

Ernest Rutherford, New Zealand-born British physicist, Nobel laureate (Chemistry, 1908), who laid the groundwork for the development of nuclear physics and produced the first artificial disintegration of an element, died in Cambridge, England. (☞ Aug. 30)

October 20

1891

James Chadwick, English physicist, Nobel laureate, was born in Manchester, England. Chadwick attended the Manchester High School and graduated in physics from the Manchester University in 1911. He spent the next two years under Ernest Rutherford in the Physical Laboratory in Manchester, where he worked on various radioactivity problems, getting his MSc degree in 1913. Thereafter he went to Berlin on a scholarship to study under the German physicist Hans Geiger at the Technische Hochschule. After World War I he resumed work under Rutherford, who in the meantime had moved to the Cavendish Laboratory in Cambridge, and together they accomplished the trans-mutation of light elements by bombardment with alpha particles, and studied the properties and structure of atomic nuclei. He was elected a Fellow of the Royal Society in 1927. In 1932 Chadwick observed that beryllium, when exposed to bombardment by alpha particles, released an

unknown radiation that in turn ejected protons from the nuclei of various substances. He interpreted this radiation as being composed of particles of mass approximately equal to that of the proton, but without electrical charge – particles that were named neutrons. Chadwick remained at Cambridge until 1935 when he was elected to the Lyon Jones Chair of Physics in the University of Liverpool. From 1943 to 1946 he worked in USA as head of the British Mission attached to the Manhattan Project for the development of the atomic bomb. Chadwick was awarded the 1935 Nobel Prize for Physics for the discovery of the neutron. (☞ Feb. 27, Jul. 24)

1906

Lee De Forest, American inventor, announced his three-element audion vacuum tube (now known as a triode) to a meeting of the American Institute of Electrical Engineers. He had discovered that when a mesh, or grid, of wire was placed between the filament and collector plate in a diode tube, a large voltage-amplifying effect could be produced. When the signal from the wireless telegraph antenna was applied to the grid instead of the filament, the sensitivity of the detector could be increased substantially. De Forest patented this vacuum tube in 1907. The ability of this tube to amplify weak signals was an invention as great as radio itself, because it made long-distance communication possible. (☞ Aug. 26)

1942

Christiane Nüsslein-Volhard, German developmental biologist, Nobel laureate, was born in Magdeburg, Germany. Nüsslein-Volhard studied biology, physics and chemistry at the Johann-Wolfgang-Goethe University, Frankfurt am Main. She did her doctorate in genetics from the University of Tübingen in 1973. In 1978 she was offered a post at the new European Molecular Biology Laboratory in Heidelberg, which was led by the English molecular biologist, John Kendrew. There she shared a laboratory with the American developmental biologist Eric Wieschaus. In 1980 she co-authored a paper in *Nature*, with Wieschaus, entitled 'Mutations affecting segment number and polarity in *Drosophila*', that revolutionised the field of developmental genetics. They had systematically searched for mutant genes that affect the formation of segments in the eggs of the fruit fly *Drosophila melanogaster*, to identi-fy all of the genes of this type and understand the processes that govern development in the fruit fly embryo. They categorised the mutants as three different types of genes, which they believed controlled an increasingly complex organisation of the organism. They also found that the genes were arranged in the same order on the chromosomes as the body segments they controlled. Although the fruit fly was used as an experimental system, the principles also apply to higher animals and man. Furthermore, genes analogous to those in the fruit fly have been found in higher animals. With this research, a breakthrough was achieved that also helped to explain congenital malformations in man. Nüsslein-Volhard moved back to Tübingen in 1981 where in 1986 she became director of the Genetics Division at the Max Planck Institute for Developmental Biology. Nüsslein-Volhard was awarded the 1995 Nobel Prize

for Physiology or Medicine, jointly with Wieschaus and American developmental geneticist Edward Lewis, for their discoveries concerning the genetic control of early embryonic development. (☞ May 20, Jun. 8)

1972
Harlow Shapley, American astronomer who discovered the Sun's position in the Milky Way galaxy, died in Boulder, Colorado, USA. (☞ Nov. 2)

1984
Paul A. M. Dirac, English theoretical physicist, Nobel laureate (Physics, 1933), who introduced special relativity into Schrödinger's equation and predicted the existence of the positron, died in Tallahassee, Florida, USA. (☞ Aug. 8)

1984
Carl Ferdinand Cori, American biochemist, Nobel laureate (Physiology or Medicine, 1947), who with his wife Gerty co-discovered a phosphate-containing form of the simple sugar glucose, and its universal importance to carbohydrate metabolism, died in Cambridge, Massachusetts, USA. (☞ Dec. 5)

October 21

1824
Joseph Aspdin, a stonemason, in Yorkshire, England, was granted a patent for Portland cement, made by burning finely pulverised lime and clay at high temperatures in kilns. He named his invention 'Portland cement' because concrete made from his new cement resembled a highly prized building stone quarried on the Isle of Portland off the British coast.

1833
Alfred Bernhard Nobel, Swedish chemist and inventor of dynamite, was born in Stockholm, Sweden. At an early age, Nobel went with his family to St. Petersburg in Russia, where his father started a torpedo works. He left Russia in 1850 to spend a year in Paris, France, studying chemistry and then spent four years in USA working under the direction of John Ericsson, the builder of the ironclad warship *Monitor*. Upon his return to St. Petersburg, Nobel worked in his father's factory, which made military equipment during the Crimean War. After the war ended in 1856, the company went bankrupt. Nobel and his parents returned to Stockholm. He soon began experimenting with explosives in a small laboratory on his father's estate. At the time, the only dependable explosive for use in mines was a form of gunpowder. A recently discovered liquid compound, nitroglycerine, was a much more powerful explosive, but it was so volatile that it could not be handled safely. In 1867, by chance, he discovered that nitroglycerine was absorbed to dryness by kieselguhr, a porous siliceous earth, and the resulting mixture was much safer to use and easier to handle than nitroglycerine alone. Nobel named the new product dynamite (from the Greek word '*dynamis*', meaning power) and patented it. Dynamite established Nobel's fame worldwide and was soon put to use in blasting tunnels, cutting canals, and building railways and roads. In the 1870s and '80s Nobel built a network of factories throughout Europe to manufacture dynamite, and a web of corporations to produce and market his explosives. He also continued to experiment in search of better ones, and in 1875 he invented a more powerful form of dynamite, blasting gelatine, which he patented the following year. Besides explosives, Nobel made many other inventions, such as

artificial silk and leather. Altogether he registered more than 350 patents in various countries. Nobel amassed a huge fortune, much of which he left in a fund to endow the annual prizes that bear his name. First awarded in 1901, these prizes were for achievements in the areas of physics, chemistry, physiology or medicine, literature, and peace. The sixth prize, for economics, was instituted in his honour in 1969. (☞ Dec. 10)

1879

Thomas Alva Edison, the American inventor, successfully demonstrated the first durable and commercially practical electric light bulb at his laboratory in Menlo Park, New Jersey, USA. This model lasted 40 hours before burning out. The main difficulty in making an electric light bulb was finding a suitable material for the filament. Edison tested over 6,000 vegetable fibres including flax and bamboo as filament material and finally succeeded with a filament of carbonised sewing thread after one and a half years of work, performing 1,200 experiments. (☞ Jan. 27, Feb. 11, Feb. 19, Apr. 15, Aug. 8, Aug. 9, Aug. 27, Dec. 6)

1950

The Central Food Technological Research Institute (CFTRI), a constituent national laboratory of the Council of Scientific and Industrial Research (CSIR), was inaugurated. Located in Mysore, Karnataka, CFTRI carries out research in various areas of food technology – food processing, novel food products, and post-harvest technology. CFTRI was the first in the world to formulate a baby food from buffalo's milk, which was till then considered unsuitable for easy digestion by infants. The achievement carried the Indian dairy industry to remarkable heights. Equally acclaimed and pioneering was the Institute's development of economically cheap but nutritionally rich foods such as weaning foods, based on protein isolated from edible oilseed meals, which were till then being used as cattle feed or just discarded. Novel protein foods based on this achievement went on to create a revolution among the medically and economically vulnerable sections of India and other developing nations. Protein foods of this kind developed by CFTRI have been extensively employed in nutrition intervention, disaster relief and social welfare projects across the world. CFTRI has also developed post-harvest technology presently pursued in India and elsewhere. These technologies have indeed heralded a whole new era in the history of food processing and agro-based industries of India.

1957

Wolfgang Ketterle, German physicist, Nobel laureate, was born in Heidelberg, Germany. Ketterle received a master's degree from the Technical University of Munich in 1982, and a PhD in physics from the University of Munich in 1986. After postdoctoral work at the Max-Planck Institute for Quantum Optics in Garching, at the University of Heidelberg, and at the Massachusetts Institute of Technology, Cambridge, Massachusetts, USA, he joined the physics faculty at MIT in 1993, where he is now a professor. Ketterle's major achievements have been the observation of Bose–Einstein condensation (BEC) in an atomic gas, the first realisation of an atom laser, the development of important tools to manipulate and study Bose–Einstein condensates, and several seminal studies of the properties of Bose–Einstein

condensates. During the early 1990s, he and his collaborators developed techniques, which were crucial for the first observations of BEC. These observations led to a new subfield of atomic physics. Ketterle was awarded the 2001 Nobel Prize for Physics, jointly with the American physicists Eric A. Cornell and Carl E. Wieman, for the achievement of Bose–Einstein condensation in dilute gases of alkali atoms, and for early fundamental studies of the properties of the condensates. (☞ Mar. 26, Dec. 9)

October 22

1797
The first manned free parachute descent took place over Paris, France, when French inventor of the parachute André Jacques Garnerin descended unhurt over 920 metres after he jumped from the basket hung from a hot-air balloon, using a large parachute made of a white canvas canopy 7 metres in diameter. Garnerin continued his experiments over the next few years; his subsequent endeavours included a jump from over 2,400 metres in England in 1802.

1843
Stephen Moulton Babcock, American agricultural chemist who developed the 'Babcock test' for measuring fat content of milk, was born near Bridgewater, New York, USA. Babcock graduated from the Tufts College in Melford, Massachusetts, in 1866 and received a PhD from the University of Göttingen, Germany, in 1879. After working as a teacher and chemist in New York, he joined the staff of the University of Wisconsin, where he was professor of agricultural chemistry from 1887 to 1913. He was also chief chemist of the Wisconsin Agricultural Experiment Station. Babcock developed the test named after him in 1890. It involved centrifuging a mixture of equal parts of milk and sulphuric acid. The acid dissolved the casein and liberated the fat, which was separated by centrifuging. The test discouraged milk adulteration and provided, for the first time, an adequate standard by which fair payment for milk could be determined. It stimulated improvement of dairy production, and aided in factory manufacture of cheese and butter. Babcock did not patent his invention as he felt that he should derive no personal gain from his testing device that might benefit humanity. He worked for 43 years at the University of Wisconsin, where he established a laboratory in which he carried out pioneering research in nutrition and in the chemistry of vitamins. (☞ Jul. 2)

1881
Clinton Joseph Davisson, American experimental physicist, Nobel laureate, was born in Bloomington, Illinois, USA. Davisson attended Bloomington public schools, and in 1902 entered the University of Chicago. Unable for financial reasons to continue studying at Chicago, he found employment with a telephone company in his hometown, the following year. In January 1904 he was appointed assistant in physics at the Purdue University, West Lafayette, Indiana. He returned to Chicago University in June 1904 and obtained his

BS in 1908. He received his PhD from the Princeton University, New Jersey in 1911. From 1911 until 1917 he was an instructor in the Department of Physics at the Carnegie Institute of Technology, Pittsburgh, Pennsylvania. In 1919 he started a series of investigations which led to the discovery of electron diffraction. In 1925, Davisson and the American physicist Lester H. Germer found that a beam of electrons, when reflected from a metallic crystal, shows diffraction patterns similar to those of X-rays and light waves, which proved the wave nature of electrons, as postulated by the French physicist Louis de Broglie. This discovery verified quantum mechanical understanding of the dual nature of subatomic particles and proved to be useful in the study of nuclear, atomic, and molecular structure. Davisson was awarded the 1937 Nobel Prize for Physics, jointly with the English physicist George Paget Thomson, for their experimental discovery of the diffraction of electrons by crystals. (☞ Feb. 1, May 3)

1903

George Wells Beadle, American geneticist, Nobel laureate, was born in Wahoo, Nebraska, USA. Beadle was educated at the Wahoo High School and went to the College of Agriculture at Lincoln, Nebraska. In 1926 he took his BSc degree, and in 1927 his MSc degree at the University of Nebraska. After getting his PhD degree from the Cornell University in 1931, Beadle was awarded a National Research Council Fellowship at the California Institute of Technology at Pasadena, where he joined the American geneticist Thomas Hunt Morgan's *Drosophila* group, working on the mechanisms of gene transmission. In 1935 Beadle visited Paris for six months to study the development of eye pigment in *Drosophila*, which later led to the work on the biochemistry of the genetics of the fungus *Neurospora*, but with a different methodology. Instead of starting with a mutation and then searching for the affected chemical reaction, Beadle and the American biochemist Edward Tatum decided to search for mutations affecting known chemical reactions, and were able to isolate a number of nutritional-deficiency mutants and identify the responsible genes. Beadle and Tatum succeeded in demonstrating that body substances are synthesised in the individual cell step-by-step in long chains of chemical reactions, and that genes control these processes by individually regulating definite steps in the synthesis chain. This led them to suggest the famous 'one gene–one enzyme' hypothesis, which meant that each gene specifies the production of a single enzyme. Beadle and Tatum were awarded the 1958 Nobel Prize for Physiology or Medicine, jointly with Joshua Lederberg. Beadle and Tatum shared half the prize money for their discovery that genes act by regulating definite chemical events; Lederberg received the other half for his discoveries concerning genetic recombination and the organisation of the genetic material of bacteria. (☞ May 23, Jun. 9, Dec. 14)

1905

Karl Guthe Jansky, American electrical engineer who discovered cosmic radio emissions, was born in Norman, Oklahoma, USA. Jansky attended the University of Wisconsin. After graduation in 1928, he joined the Bell Telephone Laboratories in

New Jersey, where he worked on problems of the short-wave radiotelephone. In 1931 he was assigned the task of tracking down the crackling static noises that plagued overseas telephone reception. For tracking the source of the noise he constructed a large directional antenna system and recorded two well-known kinds of atmospheric static: crashes from local thunderstorms and noise from distant thunderstorms reflected from the ionosphere. From his records, he later discovered a weak third kind of static, which sounded like a steady hissing through headphones. At first Jansky thought that the interference came from the Sun, but after a year of careful measurement he concluded that the radio waves came from a specific region on the sky every 23 hours and 56 minutes. Suspecting that the radiation was coming from an astronomical source, he attempted to trace its origin and discovered that it came from the direction of the constellation of Sagittarius toward the centre of the Milky Way. Jansky's serendipitous discovery gave birth to a new branch of astronomy, radio astronomy. The unit of flux density of radio sources 'jansky' (symbol, Jy) is named in his honour. (☞ Feb. 14)

1938
Physicist Chester F. Carlson, an employee at a New York electronics firm, made the first successful copy of a document using an electrostatic process called xerography, from the Greek word meaning 'dry writing'. The image was the words '10-22-38 Astoria'. The process came to be popularly known as 'Xerox'. (☞ Feb. 8, Sep. 19)

1968
The command module of *Apollo-7*, carrying astronauts Walter Schirra, Donn Eisele, and Walter Cunningham, splashed down safely in the Pacific Ocean after 11 days in Earth orbit. (☞ Oct. 11)

1975
Soviet unmanned spacecraft *Venera-9* landed on Venus. It transmitted data during the descent and continued to send back pictures and data for 53 minutes after landing. It also sent back the first picture from the surface of the planet. Instruments on board the lander showed the surface temperature to be 485°C and light levels comparable to those at Earth mid-latitudes on a cloudy summer day. (☞ Jun. 8)

1986
Albert Szent-Györgyi, Hungarian biochemist, Nobel laureate (Physiology or Medicine, 1937), who discovered the roles played by certain organic compounds, especially vitamin C, in the oxidation of nutrients by the cell, died in Woods Hole, Massachusetts, USA. (☞ Sep. 16)

1992
The second laser geodynamics satellite (*Lageos-2*) was launched from the Kennedy Space Center, Florida, USA, on board space shuttle *Columbia*. It was placed in a very stable circular orbit at an altitude of 5,900 kilometres. Like its predecessor *Lageos-1*, launched in 1976, *Lageos-2* is a passive satellite that carries no electronic equipment or moving parts. The spherical satellite takes on a golf-ball appearance due to the 'retrocollectors', three-dimensional prisms that reflect light back to the source regardless of the angle of the light received by the satellite. It is primarily used as an aid in accurate measurement of the distances between points on the Earth's

surface through precise laser ranging, which help in precision Earth-dynamics measurements such as crustal motions, regional strains, fault motions, polar motion and Earth-rotation variations, solid Earth tides, and other parameters associated with earthquake assessment and alleviation. (☞ May 4)

2001
Indian polar satellite launch vehicle, PSLV-C3, was launched from the Satish Dhawan Space Centre, Sriharikota, Andhra Pradesh. It placed three satellites – *Technology Experiment Satellite* (*TES*) of ISRO, *BIRD* of Germany, and *PROBA* of Belgium – into their intended orbits.

October 23

1873
William David Coolidge, American engineer and physical chemist whose improvement of tungsten filaments was essential in the development of the modern incandescent lamp bulb and the X-ray tube, was born in Hudson, Massachusetts, USA. Coolidge graduated in electrical engineering from the Massachusetts Institute of Technology in 1896 and obtained a doctorate from the University of Leipzig in Germany in 1899. He then joined the newly opened physics department at MIT and in the following year became a research assistant in the chemistry department, where he remained for five years. In 1905 he joined the General Electric Research Laboratory at Schenectady, New York, where he was given the task of replacing the fragile carbon filaments in electric light bulbs with tungsten filaments. Ordinarily tungsten is a difficult metal to work with because of its high softening temperature, but Coolidge was able to develop a process by means of which tungsten could be made sufficiently ductile at room temperatures to permit drawing through diamond dies. Coolidge also made improvements in the X-ray tube by replacing the platinum filament used in the earlier gas-filled cold-cathode tube with a heated tungsten filament cathode and a tungsten anode in a completely evacuated tube, which came to be known as the 'Coolidge tube'. This device made the use of X-rays for medical diagnosis safe and convenient. Coolidge even invented a portable model for use during World War I. Despite subsequent advances, Coolidge's basic design has never been superseded. (☞ Feb. 3, Jul. 4, Dec. 30)

1875
Gilbert Newton Lewis, American chemist, was born in Weymouth, Massachusetts, USA. Lewis was educated at home until age 13, when he entered the preparatory school of the University of Nebraska. He obtained his BA in 1896 and PhD in 1899 from the Harvard University, Cambridge, Massachusetts, where he stayed as an instructor for a year before taking a travelling fellowship, studying under the German physical chemist Wilhelm Ostwald at Leipzig, and Walter Nernst at Göttingen, Germany. He then returned to Harvard as an instructor for three more years, and in 1904 left to become superintendent of weights and measures at the Bureau of Science of the Philippines in Manila. In 1905 he took up a faculty position at the Massachusetts Institute of Technology in Cambridge, Mass-achusetts, becoming a professor in 1911. He left MIT to become professor of physical chemistry and dean of

the College of Chemistry at the University of California, Berkeley in 1912. Lewis is best known for his octet theory and the electron-pair bonding diagrams, and for the generalised theory of acids and bases, which bears his name. In 1916, he formulated the idea that a covalent bond consisted of a shared pair of electrons. Seven years later, he formulated the electron-pair theory of acid–base reactions. In the so-called Lewis theory of acids and bases, an acid is an electron-pair acceptor and a base is an electron-pair donor. Lewis spent 25 years determining free energies of various substances. His most famous book, *Thermodynamics and the Free Energy of Chemical Substances*, written with his compatriot, chemist Merle Randall, was published in 1923. In 1926, he coined the term 'photon' for the smallest unit of radiant energy. Lewis was the first to isolate deuterium (heavy hydrogen) and produce the first sample of pure deuterium oxide (heavy water) in 1933. (☞ Mar. 23)

1905

Felix Bloch, Swiss-born American physicist, Nobel laureate, was born in Zurich, Switzerland. Bloch studied physics and mathematics at the Federal Institute of Technology in Zurich and obtained his PhD from the University of Leipzig in 1928. He taught at the University of Leipzig till 1933 and then emigrated to USA where he joined the faculty of the Stanford University, Palo Alto, California. In 1936, he published a paper in which he proposed an experiment to obtain the magnetic moment of free neutrons through the observation of scattering in iron, which would lead to the production and observation of polarised neutron beams. In 1939 he carried out the experiment in collaboration with the American physicist Luis W. Alvarez at the Berkeley cyclotron, in which they could determine the magnetic moment of the neutron with an accuracy of about one per cent. During World War II, Bloch was also engaged in the early stages of the work on atomic energy at the Stanford University and Los Alamos, New Mexico and later in counter-measures against radar at the Harvard University in Cambridge, Massachusetts. After his return to Stanford in 1945, he developed a new method of nuclear induction, a purely electromagnetic procedure for the study of nuclear moments in solids, liquids, or gases, which helped establish the relationship between nuclear magnetic fields and the crystalline and magnetic properties of various materials. It later became useful in determining the composition and structure of molecules. Nuclear magnetic resonance techniques have become increasingly important in diagnostic medicine. In 1954, Bloch was appointed the first director-general of the European Organization for Nuclear Research (CERN) in Geneva. Bloch was awarded the 1952 Nobel Prize for Physics, jointly with fellow American physicist E.M. Purcell, for their development of new methods for nuclear magnetic precision measurements and discoveries in connection therewith. (☞ Aug. 30, Sep. 10)

1908

Ilya Mikhaylovich Frank, Russian physicist, Nobel laureate, was born in St. Petersburg, Russia. Frank graduated from the Moscow State University in 1930 and in 1931 became a senior scientific officer at the State Optical

Institute in Leningrad. In 1934 he joined the P.N. Lebedev Institute of Physics of the U.S.S.R. Academy of Sciences as a scientific officer. He received his doctorate in physical and mathematical sci-ences in 1935. In 1941, he became officer in charge of the Atomic Nucleus Laboratory and in 1944 a professor at the Moscow University. In 1957 he was appointed director of the Neutron Physics Laboratory of the Joint Institute of Nuclear Research in Dubna, Russia. Frank's first investigations were in the field of photoluminescence and in photochemistry. At the Lebedev Institute of Physics Frank worked with his compatriots Pavel A. Cherenkov and Igor Y. Tamm. In 1934, Cherenkov discovered that a peculiar blue light is emitted by charged particles travelling at very high speeds through water. In 1937 Frank and Tamm provided a detailed mathematical explanation of how the radiation originates and its characteristics. This discovery led to the development of the Cherenkov counter, and methods for detecting and measuring the velocity of high-speed particles which are of great importance for research in nuclear physics. Frank was awarded the 1958 Nobel Prize for Physics, jointly with Cherenkov and Tamm, for the discovery and the interpretation of the Cherenkov effect. (☞ Jun. 22, Jul. 8, Jul. 28)

1921
John Boyd Dunlop, Scottish inventor who was a pioneer of the pneumatic tyre, died in Dublin, Ireland. (☞ Feb. 5)

1944
Charles Glover Barkla, British physicist, Nobel laureate (Physics, 1917), who discovered the characteristic X-ray spectra of chemical elements, died in Edinburgh, Scotland. (☞ Jun. 7)

1986
Edward Adelbert Doisy, American biochemist, Nobel laureate (Physiology or Medicine, 1943), who discovered the chemical nature of vitamin K, died in St. Louis, Missouri, USA. (☞ Nov. 13)

1989
The American space shuttle *Atlantis*, carrying astronauts Donald E. Williams, Michael J. McCulley, Franklin R. Chang-Diaz, Shannon W. Lucid, and Ellen S. Baker landed at the Edwards Air Force Base, California, USA on completion of its 5th flight during which the spacecraft *Galileo* was deployed to explore Jupiter. (☞ Oct. 18)

October 24

1601
Tycho Brahe, Danish astronomer whose work in developing astronomical instruments and in measuring and fixing the positions of stars paved the way for future discoveries, died in Prague, Bohemia (now Czech Republic). (☞ Dec. 14)

1632
Antonie van Leeuwenhoek, Dutch scientist, microscopist and zoologist who was the first to observe bacteria and protozoa, was born in Delft, Netherlands. Coming from a family of tradesmen, Leeuwenhoek received no higher education or university degrees, and knew no languages other than his native Dutch. Yet with skill, diligence, an endless curiosity, and an open mind, he succeeded in making some of the most important

discoveries in the history of biology. Using his home-made simple microscope, he was the first to discover bacteria, free-living and parasitic microscopic protists, sperm cells, blood cells, microscopic nematodes and rotifers, and much more. His researches opened up an entire world of microscopic life to the awareness of scientists. He seems to have been inspired to take up microscopy by the English physicist Robert Hooke's illustrated book *Micrographia*, which depicted Hooke's own observations with the microscope and was very popular. Some time before 1668, Leeuwenhoek learned to grind lenses and made simple microscopes. Leeuwenhoek is known to have made over 500 microscopes. Compared to modern microscopes, it was an extremely simple device, using only one lens mounted in a tiny hole in the brass plate that made up the body of the instrument. The specimen was mounted on the sharp point that stuck up in front of the lens, and its position and focus could be adjusted by turning the two screws. In 1673, Leeuwenhoek began writing letters to the newly formed Royal Society of London, describing what he had seen with his microscopes – his first letter contained some observations on the stings of bees. For the next fifty years he corresponded with the Royal Society; his letters, written in Dutch, were translated into English or Latin and printed in the *Philosophical Transactions of the Royal Society*. In 1680 he was elected a member of the Royal Society. (☞ Aug. 26)

1851

Ariel and Umbriel, moons of Uranus, were discovered by the amateur English astronomer William Lassell. Ariel, with a diameter of about 1,160 kilometres, is named after a Shakespearean character while the 1,170-km diameter Umbriel is named after a character from Alexander Pope's *The Rape of the Lock*. Ariel orbits Uranus at a distance of 191,240 kilometres and Umbriel has an orbital radius of 266,000 kilometres. (☞ Jun. 18)

1889

Sisir Kumar Mitra, Indian radio physicist, Fellow of the Royal Society, was born in Calcutta (now Kolkata). After completing his schooling in Bhagalpur, Bihar, Mitra did his BSc and MSc in physics from the Presidency College, Kolkata. Calcutta University awarded him a DSc in 1919, and soon thereafter, he went to the Sorbonne University in Paris, France, to work under the French physicist and discoverer of the ozone layer, Charles Fabry. He received his second DSc degree there in 1923. Mitra returned to India at the end of 1923 and was appointed the Khaira Professor of Physics in the Calcutta University, which marked the beginning of teaching and research in radio electronics in India. Mitra succeeded in establishing an international level school of teaching and research at the Calcutta University, which later grew up into a separate department known as the Institute of Radio Physics and Electronics. Mitra is renowned for his contributions to the study of the ionosphere – a layer containing electrically charged particles called 'ions' in the upper atmosphere, which reflects radio waves like a mirror. The ionosphere was known to be composed of several layers. In 1930, Mitra was the first to obtain experimental evidence of the region of the ionosphere designated as the 'E-layer,' which is produced by sunlight and, being at a lower

level, disrupts long-distance transmission of radio signals during the day. (☞ Aug. 13)

1932

Pierre-Gilles de Gennes, French physicist, Nobel laureate, was born in Paris, France. Gennes majored from the École Normale in 1955. From 1955 to 1959, he was a research engineer with the French Atomic Energy Commission, working mainly on neutron scattering and magnetism. From 1961 to 1971, he was a professor with the Orsay Liquid Crystals Group of the University of Paris, and in 1971 became a professor at the Collège de France. Gennes investigated how extremely complex forms of matter behave during the transition from order to disorder. On the basis of studies conducted during the 1960s and 1970s, he showed that phase transitions in such apparently widely differing physical systems as magnets, superconductors, liquid crystals and polymer solutions can be described in mathematical terms of surprisingly broad generality. He was able, for example, to describe mathematically how magnetic dipoles, long molecules or molecule chains can under certain conditions form ordered states, and what happens when they pass from an ordered to a disordered state. He showed how electrically or mechanically induced phase changes transform liquid crystals from a transparent to an opaque state, the phenomenon exploited in liquid-crystal displays. His research on polymers contributed to understanding how the long molecular chains in molten polymers move, making it possible for scientists to better determine and control polymer properties. Gennes was awarded the 1991 Nobel Prize for Physics for discovering that methods developed for studying order phenomena in simple systems can be generalised to more complex forms of matter, in particular to liquid crystals and polymers.

2001

American space probe *2001 Mars Odyssey* arrived at Mars as part of NASA's Mars Exploration Program, a long-term effort of robotic exploration of the red planet. It carried on board an instrument called 'Martian radiation environment experiment', or MARIE, which was designed to calculate the radiation exposure humans would experience on the way to, and in orbit around, Mars. *Odyssey* has been mapping the distribution of materials on and near Mars' surface since early 2002, and by early 2004 had completed nearly a full annual cycle on Mars. Besides tracking seasonal changes, such as the advance and retreat of polar dry ice, the orbiter has returned evidence useful for studying longer-term dynamics. *Odyssey* data has shown that frozen water makes up as much as 10 per cent of the top metre of surface material in some regions close to the Martian equator. *Odyssey* also acted as a communications relay for the rovers, *Spirit* and *Opportunity*, which landed on Mars in January 2004, transmitting over 85 per cent of the data from the rovers to Earth.

For the first time, the mission mapped the amount and distribution of chemical elements and minerals that make up the Martian surface. The spacecraft globally mapped many elements, and the maps of hydrogen distribution led scientists to discover vast amounts of water ice in the polar regions buried just beneath the surface. (☞ Apr. 7)

2003
The supersonic airliner *Concorde* carried passengers for the last time. Three flights of *Concorde* landed at the Heathrow airport in London, UK, within five minutes of each other, marking the end of three decades of supersonic travel. They included the last ever passenger flight by *Concorde* across the Atlantic from New York JFK airport in USA, which carried on board an array of 100 stars and celebrities. (☞ Jan. 9, Jan. 21, Mar. 2, Jul. 25, Oct. 1)

October 25

1647
Evangelista Torricelli, Italian physicist and mathematician who invented the barometer and whose work in geometry aided in the eventual development of integral calculus, died in Florence, Italy. (☞ Oct. 15)

1671
Iapetus, one of Saturn's moons, was discovered by the Italian-born French astronomer Giovanni Cassini. Iapetus is the third largest of the 31 moons of Saturn. It is note-worthy as its leading side is dark with a slight reddish colour while its trailing side is bright. The origin of this dark surface is not known but it is theorised that it might be composed of matter that was either swept up from space or oozed out from the moon's interior. In Greek mythology, Iapetus was the son of Uranus, the father of Prometheus and Atlas and an ancestor of the human race. (☞ Jun. 8)

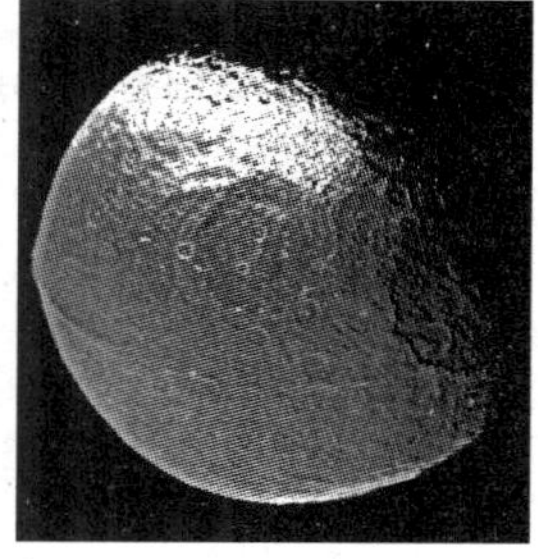

1789
Samuel Heinrich Schwabe, amateur German astronomer who discovered the 10-year sunspot activity cycle, was born in Dessau, Germany. In keeping with family tradition, Schwabe studied pharmacy and took over the family business. But his real love was for astronomy. At the age of 40 he sold his pharmacy business and devoted all his time to the stars. Three years later he started looking for a possible planet within the orbit of Mercury. He observed the Sun regularly with the hope of finding the unseen planet as it passed across the Sun's disc in transit. Schwabe did not discover a new planet. Instead, after 17 years of nearly continuous sunspot observations, he discovered a 10-year periodicity in the number of sunspots visible on the solar disc. A Swiss astronomer J.R. Wolf later confirmed this to be 11.1 years. For 42 years, Schwabe kept a record of sunspots, which he observed virtually every day that the weather allowed. In doing so he accumulated volumes of sunspot drawings. In 1831, Schwabe made the first known detailed drawing of the Great Red Spot on Jupiter. He remained an active astronomer until his death in 1875. (☞ Apr. 11)

1883
Darashaw Nosherwan Wadia, Indian geologist, Fellow of the Royal Society, was born in Surat, Gujarat, India. Wadia had his early education in Surat and Baroda, and obtained his bachelor's degree in 1905 from Baroda College. In 1907, after doing his MSc from the Bombay University, Wadia joined the Prince of Wales College in Jammu, where he set up the department of geology. In 1921 he joined the Geological Survey of India as Scientific Officer and became Assistant Superintendent in 1935. Wadia's work led to the understanding of the geological history of the northwestern Kashmir. He demonstrated that tectonic deformation in the Jammu region is of very recent origin and the topography is very young in nature. In 1935, Wadia, jointly with M. S. Krishnan and P. N. Mukherjee, published the first soil map of India. After his retirement from the GSI in 1938 he went to Ceylon (now Sri Lanka) as a Government Mineralogist. He was appointed Minerals Advisor to the Government of India in 1945. Later in 1949, the Government of India assigned him the responsibility of organising the Atomic Minerals Division for exploring and prospecting raw material for India's atomic energy programme. Wadia was the founder Director of the Indian Bureau of Mines and the Institute of Himalayan Geology, which is now named the Wadia Institute of Himalayan Geology in his honour. He was also intimately associated with the establishment and running of the National Geophysical Research Institute in Hyderabad, Andhra Pradesh, and the National Institute of Oceanography in Goa. (☞ Jun. 15, Jun. 29)

1975
Soviet unmanned spacecraft *Venera-10* landed on Venus. During the 65 minutes of operation, it returned several images of the rocky surface of Venus. The Sun was near zenith during the time *Venera-10* landed and the lighting was about what would be seen on Earth on an overcast summer day. (☞ Jun. 14)

October 26

1877
The first surgical operation under antiseptic conditions was performed by the British surgeon Joseph Lister at King's College, London, England. The operation involved repair of a fractured kneecap. After studying the French bacteriologist Louis Pasteur's germ theory of disease, which states that bacteria cause infections, Lister introduced phenol (carbolic acid) as a form of disinfectant into his surgery; as a result the surgical death rate dropped from 45% to 15%. Lister documented his antiseptic procedures in the British medical journal *The Lancet*, and subsequently published a lengthier treatise, 'On the antiseptic principle in the practice of surgery'. In the course of time, corrosive carbolic acid was replaced by other antiseptic methods. (☞ Mar. 16, Apr. 5)

1957
Gerty Theresa Cori, Czech-born American biochemist, Nobel laureate (Physiology or Medicine, 1947), died in St. Louis, Missouri, USA. (☞ Aug. 15)

1972
Igor Ivan Sikorsky, Russian-born American pioneer in aircraft design who invented the helicopter, died in Easton, Connecticut, USA. (☞ May 25)

1984
The first animal heart was transplanted in a human baby. The baby girl was born prematurely on October 14, suffering from hypoplastic left-heart syndrome, a malformation of the heart, and was brought to the Loma Linda University Medical Center, California, USA. Babies born with this condition usually die a few days after birth. To save the baby, a surgeon named Leonard Bailey replaced the baby's ailing

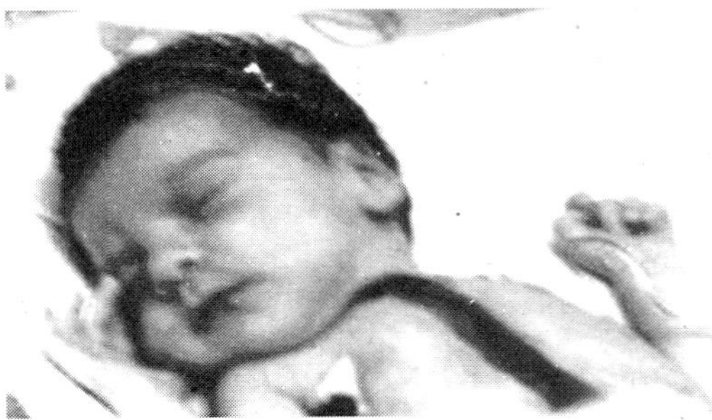

Baby Fae

heart with a healthy baboon's heart. At first the operation seemed successful and the baby, nicknamed Baby Fae, appeared much healthier. But despite the drugs, her immune system began to reject the baboon heart and she died three weeks after the surgery. (☞ Nov. 15)

1989
Charles J. Pedersen, Korean-born American chemist, Nobel laureate (Chemistry, 1987), who synthesised the first molecules with structure-specific reactivity, called crown ethers, died in Salem, New Jersey, USA. (☞ Oct. 3)

October 27

1449
Ulugh Beg, Mongol scientist, mathematician, and the greatest astronomer of his time, died in Samarkand, Timurid Empire (now in Uzbekistan). (☞ Mar. 22)

1728
James Cook, English seaman who was the first of the really scientific navigators, was born in Marton, Yorkshire, England. His first experience at sea came at the age of 18 when he signed on as a deckhand aboard a Whitby collier carrying coal to London. He became an accomplished mathematician in his spare time and was actually offered a command of his own ship but refused and joined the Royal Navy as a seaman. He spent several years surveying the coasts of Labrador and Newfoundland. In 1768 he was placed in command of the *Endeavour*, which was to take a number of scientists to the Pacific Ocean for observance of the Venus transit in June 1769. This would help scientists plot the distance between the Sun and the Earth – the 'astronomical unit'. Cook observed the transit of Venus on this voyage from the island of Tahiti. After visiting Tahiti he discovered New Zealand and then the east coast of Australia and claimed both territories as British possession. The *Endeavour* returned to England in 1771 having added significantly to Britain's potential empire in the Pacific. Cook led two more expeditions to the Pacific and was killed by cannibal natives in Kealakekua Bay, Hawaii in 1779. (☞ Jan. 18, Feb. 14, Jun. 3)

1968
Lise Meitner, Austrian-born Swedish physicist, whose researches led to the discovery of uranium fission and who co-discovered the radioactive element protactinium, died in Cambridge, England. (☞ Nov. 7)

1980
John Hasbrouck Van Vleck, American physicist and mathematician, Nobel laureate (Physics, 1977), who explained the behaviour of electrons in magnetic, non-crystalline solid materials, died in Cambridge, Massachusetts, USA. (☞ Mar. 13)

October 28

1914
Jonas Edward Salk, American physician and medical researcher, who developed the first safe and effective vaccine for poliomyelitis, was born in New York, USA. Salk received an MD from the New York University in 1939 and was appointed assistant professor of epidemiology at the University of Michigan. In 1947 he became head of the virus research laboratory at the University of Pittsburgh and was later research professor of bacteriology, professor of preventive medicine and chairman of the department, and professor of experimental medicine. In the 1940s, Salk's work on an anti-influenza vaccine led him and his colleagues to develop an inactivated vaccine against polio in 1952. He developed the vaccine against polio by cultivating three strains of the virus separately in monkey tissue. The virus was separated from the tissue, stored for a week, killed with formaldehyde, and then tested to make certain that it was dead. A series of three or four injections with the killed virus vaccine was required to confer immunity. After successful wide-scale testing in 1954, the vaccine was distributed nationally. It greatly reduced the incidence of polio. The only disadvantage of the Salk vaccine was the need to be injected subcutaneously. For this reason, the oral polio vaccine, developed by the Polish-born American physician and microbiologist Albert Bruce Sabin became more popular. In 1963 Salk became fellow and director of the Institute for Biological Studies in San Diego, California, later called the Salk Institute. Salk also conducted research into the treatment of acquired immune deficiency syndrome (AIDS). (☞ Jun. 23, Aug. 26)

1914
Richard Laurence Millington Synge, British biochemist, Nobel laureate, was born in Liverpool, England. Synge went to the Winchester College, where he studied mainly classics until 1931, and thereafter natural science. In 1933 he entered the Trinity College, University of Cambridge and studied physics, chemistry, biochemistry, and physiology for the Natural Sciences Tripos. He obtained his PhD degree from Cambridge in 1941 and joined the staff of the Wool Industries Research Association at Leeds where he worked with the British biochemist A. J. P. Martin and developed partition chromatography, a technique that is used to separate mixtures of closely related chemicals such as amino acids for identification and further study. In paper chromatography, a drop of a mixture of substances is dropped onto a strip of filter paper, which is allowed to draw up a suitable solvent like butyl alcohol–water, by capillary action. The spot begins to move, then gradually segregates into several spots. Some spots rapidly follow the solvent, while others lag behind. The result is a resolution of the mixture into its component parts. Thus, one drop of extremely complicated

mixtures can be analysed easily. Synge used paper chromatography to work out the exact structure of the simple protein molecule gramicidin S. Synge and Martin gave the first demonstration of partition chromatography to the Biochemical Society at its meeting at the National Institute for Medical Research, London, in June 1941. From 1942 to 1948, Synge worked almost exclusively with the antibiotic peptides of the gramicidin group. He was made a Fellow of the Royal Society in 1950. Synge also did research at the Lister Institute of Preventive Medicine, London, and at the Rowett Research Institute, near Aberdeen, Scotland. Synge was awarded the 1952 Nobel Prize for Chemistry, jointly with Martin, for their invention of partition chromatography. (☞ Mar. 1, Aug. 18)

1914

The American inventor George Eastman's company announced its original Kodachrome colour photographic process, which made use of two filters, red and green, to make two black-and-white negatives and then combine the dyed positives to produce the final colour photo. The original Kodachrome two-colour process was well suited for portraits, as it accurately reproduced flesh tones. The Kodachrome available today, which Kodak developed in 1935, is a three-colour process. All colour films in use today work on the same principle, but instead of exposing three negatives through three filters, the modern colour film has three layers sensitive to red, green and blue colours. (☞ Jul. 12, Oct. 14)

1969

Two 160-MW units of India's first atomic power station, the Tarapur Atomic Power Project at Tarapur in Maharashtra were commissioned. Both the units are boiling water reactors that use enriched uranium as fuel and water as moderator and coolant. The initial supply of enriched uranium came from USA, but after it was stopped following the Pokhran atomic test of 1974, an indigenously developed mixture of natural uranium and plutonium oxides has been used as fuel for the two reactors.

2004

Discovery of a new human-like species, called *Homo floresiensis*, was reported in the journal *Nature*. Skeletal remains of the new species, which lived 18,000 years ago, show that they were only one metre tall and had a brain one-third the size of that of modern humans. The remains – a near-complete skeleton, thought to be a female, including the skull, jaw and most of the teeth, along with bones and teeth from at least seven other individuals – were found by Australian and Indonesian scientists in a rock shelter called Liang Bua on the Indonesian island of Flores. Bones from a Komodo dragon and an extinct pygmy elephant called *Stegodon* were also found at the same site. *H. floresiensis* is believed to be a long-term, isolated descendant of *H. erectus* that had undergone endemic dwarfing.

2005

Richard Errett Smalley, American chemist and physicist, Nobel laureate (Chemistry, 1996), who discovered fullerenes, died in Houston, USA. (☞ Jun. 6)

October 29

1920
Baruj Benacerraf, Venezuelan-born American immunologist, Nobel laureate, was born in Caracas, Venezuela. Benacerraf completed his primary and secondary education in Paris, France, and obtained his BSc degree from the Columbia University in the School of General Studies in New York, USA, in 1942. He received his MD degree from the Medical College of Virginia in 1945 and after a brief internship at the Queens General Hospital in New York he served in the US Army in 1946–48. He started his research career in 1948 at the Neurological Institute, Columbia University School of Physicians and Surgeons in New York. In 1949 he moved to Paris and joined the Broussais Hospital where he carried out studies on reticuloendothelial function in relation to immunity and developed techniques to study the clearance of particulate matter from the blood by the RES, and formulated the equations that govern this process in mammalian organisms. He was able to demonstrate a very important group of genes within the major histo-compatibility complex, which has several very central functions for regulating various immunological reactions in the body. Through his studies he showed that genetic factors intimately related to the genes that determine an individual's unique constitution of H antigens actually regulate the interaction among the various cells belonging to the immunological system and are thereby important to the strength of an immunological reaction. Benacerraf was awarded the 1980 Nobel Prize for Physiology or Medicine, jointly with American geneticist George Snell and French immunologist Jean Dausset, for their discoveries concerning genetically determined structures on the cell surface that regulate immunological reactions. (☞ Oct. 19, Dec. 19)

1933
Albert Calmette, French bacteriologist and co-developer of the tuberculosis vaccine Bacillus Calmette-Guérin (BCG), died in Paris, France. (☞ Jul. 12)

1958
The first coronary angiogram was obtained by American paediatric cardiologist F. Mason Sones, at the Cleveland Clinic, Cleveland, Ohio, USA. Coronary angiography is a diagnostic X-ray procedure designed to visualise blockages of the small arteries of the heart. Sones discovered the technique by accident. In studies on dog's hearts, dye injected in coronary arteries caused heart fibrillation, and therefore had never been used on humans. When he accidentally injected an angiography dye into the mouth of the coronary arteries for the first time in a living human patient during an attempt to visualise other cardiac structures, there was no defibrillation. Aware of the significance of this observation, Sones went about designing equipment and techniques to routinely X-ray the various coronary arteries in human patients. Since that initial discovery, there have been tremendous refinements in angiographic equipment, technique and safety. Coronary angiography has become the cornerstone for both diagnosis and treatment of coronary artery disease worldwide.

1971
Arne Wilhelm Kaurin Tiselius, Swedish biochemist, Nobel laureate (Chemistry, 1948), who developed the technique of

electrophoresis for separating proteins, died in Uppsala, Sweden. (☞Aug. 10)

1991
The American spacecraft *Galileo* made the closest ever approach to a minor planet Gaspra as it flew by at a distance of 1,900 kilometres. (☞ Oct. 18)

1996
India's first experimental thorium fuelled nuclear reactor, the Kalpakkam Mini Reactor (KAMINI) attained criticality at the Indira Gandhi Atomic Research Centre (IGCAR), Kalpakkam, Tamilnadu. The 30-kW KAMINI reactor was jointly designed and built by the Bhabha Atomic Research Centre and IGCAR and uses Uranium-233 produced from irradiated thorium as fuel. It is used for research in neutron radiography, neutron activation analysis, and the irradiation of samples for experimental and forensic purposes.

October 30

1895
Gerhard Domagk, German bacteriologist and pathologist, Nobel laureate, was born in Lagow, Brandenburg, Germany. Domagk studied medicine at Kiel and, when World War I (1914–1918) broke out, he served in the Army, and was wounded. He resumed his medical studies at Kiel in 1918 and earned a medical degree from the University of Kiel in 1921. After teaching at the universities of Greifswald and Munich, he became director of the I.G. Farben (Bayer) Laboratory for Experimental Pathology and Bacteriology at Wuppertal. There, while testing newly developed dyes for their possible effects against various infections, he noticed the antibacterial action of one of the dyes against streptococcal infection in mice. The dye, 'prontosil red', was then tried clinically against streptococcal infections in humans with great success. The active component of prontosil turned out to be sulphanilamide, which became another important sulpha drug. Ironically, the medical fraternity were reluctant to use the drug, but the drug became a blockbuster once the British Medical Research Council and the American Medical Association confirmed Domagk's findings. Domagk was awarded the 1939 Nobel Prize for Physiology or Medicine for the discovery of the antibacterial effects of prontosil. (☞ Apr. 24)

1895
Dickinson Woodruff Richards, American physiologist, Nobel laureate, was born in Orange, New Jersey, USA. Richards was educated at the Hotchkiss School in Connecticut, and, in 1913, went to the Yale University to study English and Greek. He received his AB degree in 1917. After serving with the US Army during World War I, he entered the Columbia University College of Physicians and Surgeons and received his MA degree in physiology in 1922, and his MD degree in 1923. He spent the years 1923–1927 on the Staff of the Presbyterian Hospital, New York, and then went to work for a year at the National Institute for Medical Research, London, England, on the control of the circulation in the liver. After returning to the Presbyterian Hospital and the College of Physicians and Surgeons, Richards began his researches on pulmonary and circulatory physiology. In 1931 he began to collaborate with French-born American physiologist André

Cournand at the Bellevue Hospital, New York, and this work resulted, in 1940, in the development of a technique for catheterisation of the heart which was used in studies of traumatic shock, the physiology of heart failure. They also developed techniques for the diagnosis of congenital heart diseases. In 1945 Richards was appointed professor of medicine at the Columbia University. Richards was awarded the 1956 Nobel Prize for Physiology or Medicine, jointly with Cournand and German surgeon Werner Forssmann, for their discoveries concerning heart catheterisation and pathological changes in the circulatory system. (☞ Feb. 23, Aug. 20, Sep. 24)

1900

Ragnar Arthur Granit, Finnish-born Swedish physiologist, Nobel laureate, was born in Helsinki, Finland. Granit matriculated from the Helsingfors University in 1919. He studied medicine at the University of Helsinki, graduating in 1924. He obtained his doctorate in medicine in 1927. The topic of his thesis was the theory of colour recognition. In addition to his medical studies, he obtained a bachelor of philosophy degree in 1924 in theoretical and practical philosophy, aesthetics and chemistry. He spent the years 1929–1931 as fellow in medical physics at the Johnson Foundation of the University of Pennsylvania and in 1932 went to the Oxford University in England as a fellow of the Rockefeller Foundation to work with the most notable nerve

physiologist of the time, Charles Scott Sherrington. Granit wanted to understand vision and realised that the underlying fact was that the retina itself functions as a nerve centre which processes visual information and transmits already processed information to the brain's visual centre. He continued his electroretinogram bioelectric research, measuring the electric impulse of a single nerve, on the visual nerve and the retina at the University of Helsinki from 1935 to 1940. Granit also pursued the physiological basis of colour perception and observed that the electric impulses generated in the retina, the electroretinogram, showed that sensitivity to colour is concentrated mainly in three different groups in the area of blue, green and red. This provided the first biological demonstration in support of the Young–Hemholz three-colour theory. In 1945 Granit became the director of the newly founded Neurophysiology Laboratory of the Nobel Institute of Medicine in Stockholm, a position he occupied until his retirement in 1967. Granit was awarded the 1967 Nobel Prize for Physiology or Medicine, jointly with the American biochemist George Wald and physiologist Haldan Hartline, for their discoveries concerning the primary physiological and chemical visual processes in the eye. (☞ Mar. 12, Nov. 18, Dec. 22)

1909

Homi Jehangir Bhabha, Indian nuclear scientist, Fellow of the Royal Society, was born in Bombay (now Mumbai). Bhabha was educated at the Cathedral and John Cannon High School, Elphinstone College and the Royal Institute of Science in Mumbai. He had access to his grandfather's large library, which contained many books on science. At the age of 15, he had already read Einstein's book on relativity. Bhabha went to Cambridge, England, in

1927 to study engineering. He passed the Mechanical Engineering Tripos in first class in 1930 and then went on to study theoretical physics as a research scholar. During the two years of his work at the Cavendish Laboratory in Cambridge, Bhabha won several scholarships and travelled across Europe. In 1937, along with the German physicist Walter Heitler, he presented the cascade theory of electron showers, which came to be known as the 'Bhabha–Heitler cascade theory'. The theory explains the process of electron showers in cosmic rays. Bhabha returned to India in 1939 and joined the Indian Institute of Science in Bangalore as reader in 1940. He was given charge of a special cosmic ray research unit set up for him. In 1944, he wrote a letter to the Dorabji Tata Trust, proposing the setting up of a school of research in the fundamental problems of physics, 'so that when nuclear energy had been successfully applied to power production, India would not have to look abroad'. The Tata Institute of Fundamental Research was inaugurated in Mumbai in 1945. In 1948, the Atomic Energy Commission was set up and Bhabha was appointed its Chairman. When the Atomic Energy Establishment (later re-named Bhabha Atomic Research Centre) was set up at Trombay, near Mumbai, for carrying out research on peaceful application of atomic energy, Bhabha became its director. Bhabha strived to make the country self-reliant in the nuclear field. Reactors like Apsara, uranium and zirconium plants, the Van de Graff and cyclotron equipment at BARC – were all Bhabha's gifts to the nation. (☞ Jan. 12, Jan. 20, Jan. 24)

1928

Daniel Nathans, American microbiologist, Nobel laureate, was born in Wilmington, Delaware, USA. Nathans was educated in public schools of Wilmington and at the University of Delaware. He obtained his MD degree from the Washington University in St. Louis, Missouri, in 1954. After residency at the Columbia-Presbyterian Medical Center in New York, he served as a clinical associate at the National Cancer Institute and as a guest investigator at the Rockefeller University. His first faculty appointment was in 1962, as an assistant professor of microbiology at the Johns Hopkins University School of Medicine, where he remained a faculty member for more than three decades. He became a full professor in 1967, director of the Department of Microbiology in 1972, and director of the Department of Molecular Biology and Genetics in 1981. Nathan's major work was with tumour-producing viruses. He used a restriction enzyme discovered by his colleague and compatriot, Hamilton O. Smith, as a 'biochemical scissors', to analyse DNA of the tumour-producing virus called SV40 and discovered that it cut the DNA in 10 distinct places, creating 11 well-defined fragments. He found ways to use this cutting to help determine where genes began and ended in SV40 DNA. This helped him locate a gene in the virus that gives the order for production of a tumour-making protein. The techniques developed by Nathans in working with animal tumour viruses opened up new avenues to study the organisation and expression of genes of higher animals and to solve basic problems

in developmental biology. The use of restriction enzymes to construct maps of the genome of viruses laid the groundwork for the worldwide effort to map the human genome and also opened new avenues to study the basic problems in developmental biology; and in medicine, to help the prevention and treatment of malformations, hereditary diseases and cancer. Nathan was awarded the 1978 Nobel Prize for Physiology or Medicine, jointly with Smith and Swiss microbiologist Werner Arber, for the discovery of restriction enzymes and their application to problems of molecular genetics. (☞ Jun. 3, Aug. 23, Nov. 16)

1939
Leland H. Hartwell, American scientist, Nobel laureate, was born in Los Angeles, California, USA. Hartwell received his BS degree from the California Institute of Technology, Pasadena, California, in 1961, and a PhD from the Massachusetts Institute of Technology, Cambridge, Masachusetts, in 1964. In 1965 he joined the University of California at Irvine as assistant professor. He moved to the University of Washington in Seattle, Washington in 1968 as associate professor, later to become full professor. In 1996 he joined the Fred Hutchinson Cancer Research Center in Seattle, Washington, becoming president and director in 1997. Hartwell's research mainly concerned cell division and the cell cycle. In the late 1960s he began using baker's yeast, *Saccharomyces cerevisiae*, as a model system, to study how cells control their growth and division. He identified more than 100 genes, termed cell-division-cycle (CDC) genes, involved in cell-cycle control. One of these genes, designated CDC28 by Hartwell, controls the first step in the cell cycle, and was therefore also called 'start'. The 'start' gene was found to have a central role in controlling the first step of each cell cycle. Hartwell also introduced the concept of 'checkpoint', a valuable aid to understanding the cell cycle. Hartwell was awarded the 2001 Nobel Prize for Physiology or Medicine, jointly with the British biologists Paul M. Nurse and Timothy Hunt, for their discoveries of key regulators of the cell cycle. (☞ Jan. 25, Feb. 19)

1975
Gustav Ludwig Hertz, German physicist, Nobel laureate (Physics, 1925), who discovered the laws governing the impact of an electron upon an atom, died in Berlin, Germany. (☞ Jul. 22)

1981
Soviet unmanned spacecraft *Venera-13* was launched towards Venus from the Baikanur cosmodrome in Kazakhstan. It carried a lander equipped with instruments to take chemical and isotopic measurements, monitor the spectrum of scattered sunlight, and record electric discharges during its descent through the thick atmosphere of the planet. (☞ Mar. 1)

October 31

1815
Humphry Davy, the English chemist, patented the miner's safety lamp. In the safety lamp, a copper or iron gauze cylinder, with a mesh size of about 1 mm surrounds the flame inside. Such fine gauze prevents flame passing

through. The wire absorbs or conducts away the heat of the flame contained inside the lamp so that it does not explode gas outside the lamp. If methane gas is present, a pale blue flame appears around the central flame. This warns a miner to leave the area immediately. (☞ Dec. 17)

1835

Adolf von Baeyer, German chemist, Nobel laureate, was born in Berlin, Germany. Even as a child, Baeyer was interested in chemical experiments and at the age of twelve, he found a new double salt of copper. He entered the University of Berlin in 1853 to study physics and mathematics but moved to Heidelberg in 1856 to study chemistry under Robert Bunsen. His studies here on methyl chloride resulted in his first published work, which came out in 1857. During the next year he worked in the private laboratory of August Kekulé in Heidelberg and was associated with his ingenious theory of the structure of benzene. In 1858, in Berlin, he received his doctorate for his work on cacodyl compounds, which had been done in Kekulé's laboratory. In 1860 he became a lecturer in organic chemistry at the Vocational Institute in Berlin. Six years later he joined the Berlin University as lecturer and started his research on indigo, the blue dye that had fascinated him since his youth. His research soon led to the discovery of indole and to the partial synthesis of indigotin. After he joined the Strassburg University as professor in 1871, Baeyer brought to light an important category of dyestuffs – the phthaleins. On the death of the German chemist Justus von Liebig in 1873, Baeyer joined the University of Munich as professor and there, over many years, built up an excellent new chemical laboratory. At Munich, Baeyer carried out his elegant total syntheses of indigo, and derived the famous Baeyer strain theory of the carbon rings that helped explain why carbon rings of five or six atoms are so much more common than carbon rings with other numbers of atoms. He also carried out studies of the constitution of benzene as well as comprehensive investigations into cyclic terpene. Baeyer was awarded the 1905 Nobel Prize for Chemistry in recognition of his services in the advancement of organic chemistry and the chemical industry, through his work on organic dyes and hydro-aromatic compounds. (☞ Aug. 20)

1925

John A. Pople, British mathematician and theoretical chemist, Nobel laureate, was born in Burnham-on-Sea, Somerset, England. Pople graduated with a bachelor's degree in mathematics from the Cambridge University in 1946, and received his doctorate from Cambridge in 1951. In 1958 he left Cambridge to head the basic physics division at the National Physical Laboratory in England, and in 1964 he became a professor of chemistry at the Carnegie Institute of Technology (now part of the Carnegie-Mellon University) in Pittsburgh, Pennsylvania, USA. In 1993 he moved to the Northwestern University in Evanston, Illinois. Pople's interest in the puzzles of physical chemistry, as opposed to abstract mathematics, dated from early in his career. His doctoral thesis, for instance, explored the structure of water. At the Carnegie-Mellon University, using Schrödinger's fundamental laws of quantum mechanics, he developed a computer programme which, when provided with particulars of a molecule or a chemical reaction, outputs a description

of the properties of that molecule or how a chemical reaction may take place. He pioneered the development of more sophisticated computational methods, so-called *ab initio* methods, which used basis sets of either Slater type orbitals or Gaussian orbitals to model the wave function. While in the early days these calculations were extremely expensive to perform, the advent of high-speed microprocessors has made them much more feasible today. He was instrumental in the development of one of the most widely used computational chemistry packages, the 'Gaussian' suite of programmes. Pople was awarded the 1998 Nobel Prize for Chemistry, jointly with Austrian-born American physicist Walter Kohn, for his development of computational methods in quantum chemistry. (☞ Mar. 9, Mar. 15)

1927

Narinder Singh Kapany, Indian-born American physicist who is widely acknowledged as the father of fibre optics, was born in Moga, Punjab, India. Kapany graduated from the Agra University and completed advanced studies in optics at the Imperial College of Science and Technology, London, England. In 1953, after experiments with light propagation in glass fibres, Kapany developed fibres with cladding. The clad fibres greatly improved transmission characteristics and reduced dispersion of light. He coined the term fibre optics for the technology of transmitting light through fine glass strands in devices from endoscopy to high-capacity telephone lines that has changed the medical, communications and business worlds. Kapany received his PhD from the University of London in 1955. In 1960, he founded Optics Technology Inc., in New York, USA, and was Chairman of the Board, President, and Director of Research for twelve years. In 1967, the company went public and in 1973, he founded Kaptron Inc., and was President and CEO until 1990. Kapany holds over 100 patents.

1930

Michael Collins, American astronaut, was born in Rome, Italy. Collins graduated from the US Military Academy at West Point, New York and joined the US Air Force as a test pilot at the Edwards Air Force Base in California. He joined the US space programme in 1963. Collins first went into space in July 1966 on board *Gemini-10* along with astronaut John W. Young and took part in the first docking in space. After docking with an Agena target vehicle, the two astronauts used the Agena's engines to propel them to a record altitude of 764 kilometres. Collins participated in the *Apollo-11* mission in July 1969 with Neil A. Armstrong and Edwin E. Aldrin. During the mission, Armstrong and Aldrin landed on the Moon and Collins remained in the Command Module, circling the Moon at an altitude of 97 to 121 kilometres. In 1971 he joined the administrative staff of the Smithsonian Institution, Washington, DC (☞ Jan. 20, Jul. 21, Aug. 5)

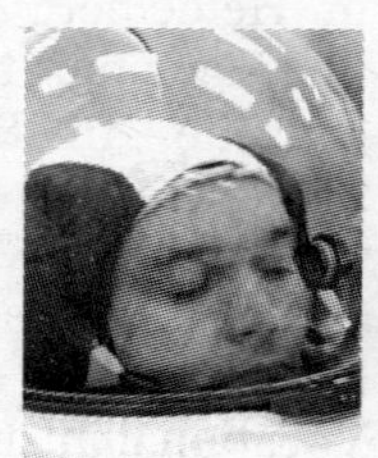

1986

Robert Sanderson Mulliken, American chemist and physicist, Nobel laureate

(Chemistry, 1966), who developed the molecular-orbital theory of chemical bonding, died in Arlington, Virginia, USA. (☞ Jun. 7)

1992

The Vatican admitted erring for over 359 years in formally condemning the Italian natural philosopher, astronomer, and mathematician Galileo Galilei for entertaining scientific truths such as the Earth revolves around the Sun, which the Roman Catholic Church long denounced as anti-scriptural heresy. After 13 years of inquiry, the Pope's commission of historic, scientific and theological scholars brought the Pope a 'not guilty' finding for Galileo. In 1633, at age 69, Galileo was forced by the Roman Inquisition to repent for his heresy – he spent the last eight years of his life under house arrest. (☞ Jan. 8, Feb. 15, Jun. 22)

November 1

1772
Antoine-Laurent Lavoisier, French chemist, demonstrated that when sulphur or phosphorus burns, the gain in weight is due to its combination with air. (☞ Aug. 26)

1815
Crawford Williamson Long, American physician who was the first to use ether as anaesthetic in surgery, was born in Danielsville, Georgia, USA. In 1942, while practising in rural Georgia, he first used ether anaesthetic while he removed a tumour from a patient's neck. Although he continued to use ether for operation, he was apparently unaware of its full significance and did not publish a description of his procedure until 1849. By that time, William Morton, a dental surgeon in Boston, had already demonstrated the use of ether in a similar type of surgery. Despite Morton's claims to the discovery and the publicity that attended his demonstration, Long is accepted as the pioneer of surgical anaesthesia. (☞ Mar. 30, Jun. 16, Oct. 16)

1880
Alfred L. Wegener, German scientist, meteorologist, geophysicist, who formulated the first complete statement of the continental drift hypothesis, was born in Berlin, Germany. Wegener obtained his doctorate in planetary astronomy in 1905 but soon became interested in meteorology. During his lifetime, he participated in several meteorological expeditions to Greenland. Like other scientists before him, Wegener was intrigued about the way the coastlines of eastern South America and western Africa fit together if the two continents were juxtaposed. He speculated that the two continents were once joined; furthermore, he proposed that all of the present-day continents originally formed one landmass, which he called Pangaea (meaning 'all lands' in Greek). Wegener believed that the super-continent began to break into smaller continents around 200 million years ago. In 1915, Wegener published his hypothesis in a book entitled *The Origin of Continents and Oceans* in which he put forth his theory of continental drift. Wegener's theory was supported by scientific evidence, such as the discovery of fossilised tropical plants beneath the Greenland icecap, and the fact that there were glaciated landscapes in the tropics of Africa and South America. Unfortunately, Wegener's book met with considerable hostility. Contemporary scientists held the view that continents and oceans held permanent positions on the globe. But despite the rejection of his theory by his peers, Wegener continued to search tenaciously for further evidence to support his ideas until the end of his life. In September 1930, Wegener made a final expedition to Greenland, to help establish a weather station to study the jet stream. He

froze to death during this expedition in November 1930; his body was recovered the following summer. (☞ May 12)

1884

Greenwich Mean Time (GMT) was adopted universally at a meeting of the International Meridian Conference in Washington, DC, USA. From then, the International Date Line was drawn up and 24 time zones created. Since 1964 GMT has been replaced by the Coordinated Universal Time (UTC), which is widely broadcast by precisely coordinated radio signals and forms the international basis of civil and scientific time.

1919

Hermann Bondi, Austrian-born British mathematician and cosmologist, was born in Vienna, Austria. Bondi obtained his BA in mathematics from the Cambridge University in England in 1940. In 1941 he became a research fellow at the Trinity College, Cambridge and received his MA in 1942. While working on naval radar for the British Admiralty in 1942, Bondi met the British mathematician and astronomer Fred Hoyle. Bondi, Hoyle and the Austrian-born British astronomer Thomas Gold, enjoyed discussing cosmology. After the war, Bondi taught mathematics at Cambridge and King's College while continuing to collaborate with Hoyle and Gold. From 1945 to 1948, he was an assistant lecturer, and from 1943 to 1954 he was a university lecturer. In 1947 he became a British citizen. In 1948, Bondi, in collaboration with Hoyle and Gold, proposed the cosmological theory of a 'Steady State Universe', according to which the universe is ever expanding, without a beginning and without an end. Further, they said that since the universe must be expanding, new matter must be continually created in order to keep the density constant, by the interchange of matter and energy. The theory received a severe setback in 1965 with the discovery of the cosmic microwave background radiation, which gave convincing support to the 'Big Bang' theory of the universe. In 1951 Bondi was a Visiting Professor to the Cornell University, Ithaca, New York, and in 1953 he went to the Harvard University, Cambridge, Massachusetts. From 1967 to 1971 he was director-general of the European Space Research Organization (now European Space Agency), and from 1971 to 1977, the chief scientific adviser at the Ministry of Defence in UK.

1950

Robert B. Laughlin, American physicist, Nobel laureate, was born in Visalia, California, USA. Laughlin did undergraduate and masters-level work at the University of California at Berkeley, and received his PhD from the Massachusetts Institute of Technology in 1979. He worked briefly at the Bell Laboratories before joining the Lawrence Livermore National Laboratory in California as a research physicist in 1982, where he was to work on modelling matter at a temperature of about 10 eV and a density of about $^1/_{10}$ that of ordinary solids. It was here that Laughlin learnt of the work of the German-born American physicist Horst L. Störmer and the Chinese-born American physicist Daniel C. Tsui who had discovered the fractional quantum Hall effect, where groups of electrons act as if they are quarks, or fractions of a single electron. Störmer and Tsui made the discovery in 1982 in an experiment using extremely powerful magnetic fields and low temperatures. By 1983 Laughin worked out

a 'quantum fluid' model to explain the quantum Hall effect. Apart from its superfluidity, which explains the disappearance of Ohmic resistance in the quantum Hall effect, one of the many unusual properties proposed by Laughlin's theory is that the electrons can form a new type of quantum fluid with quasi-particles carrying fractions of an electron's charge. These quasi-particles are not particles in the normal sense but a result of the common dance of electrons in the quantum fluid. Laughlin was the first to demonstrate that the quasi-particles have precisely the correct fractional charge to explain Störmer and Tsui's results. Laughlin was awarded the 1998 Nobel Prize for Physics, jointly with Störmer and Tsui, for their discovery of a new form of quantum fluid with fractionally charged excitations. (☞ Feb. 28, Apr. 6)

1952

The United States tested 'Mike', the world's first hydrogen bomb, on the Enewetak atoll off the Marshall Islands in the Pacific Ocean. It was a bomb unlike any that was used before. The 10.4-megaton bomb was 600 times more powerful than the atomic bombs that destroyed Hiroshima and ended the Japanese front of World War II in August of 1945. The 'mushroom' cloud from the explosion rose to a maximum height of 41,000 metres (the top of the stratosphere) in 5 minutes and eventually spread to an area of 1,600 kilometres. (☞ May 20)

1993

Severo Ochoa, Spanish-American biochemist and molecular biologist, Nobel laureate (Physiology or Medicine, 1959), who co-discovered the mechanisms in the biological synthesis of ribonucleic acid and deoxyribonucleic acid, died in Madrid, Spain. (☞ Sep. 24)

November 2

1833

Mahendra Lal Sircar, Indian doctor and philanthropist, who founded the Indian Association for the Cultivation of Science, was born in Paikpara, near Howrah, West Bengal, India. Sircar topped the MD examination in 1863 but devoted his life wholly to the propagation of the principles of homoeopathy. He was appointed a Fellow of the Calcutta University in 1870 and Sheriff of Calcutta (now Kolkata) in 1887. He was also a member of the Bengal Legislative Council from 1887 to 1893 and was associated with many learned societies of Kolkata. Sircar is better known for his contributions to the development of scientific research in India. He recognised the important role science could play in the regeneration of modern India. Although trained as a doctor of medicine, he visualised an institution for spreading the message of science to the people. It was through his sustained efforts that the Indian Association for the Cultivation of Science was established in Kolkata in 1878, with 'the object of enabling the natives of India to cultivate science in all its departments by original research and to its varied application to the art and comforts of life'. It is the oldest institution in India devoted to the pursuit of research in basic sciences. Sircar was its first Honorary Secretary. (☞ Feb. 23, Jul. 29)

1885
Harlow Shapley, American astronomer who discovered the Sun's position in the galaxy, was born in Nashville, Missouri, USA. Shapley received his master's degree from the Missouri State University in 1911 and his doctorate from the Princeton University, New Jersey, in 1913. He joined the Mount Wilson Observatory near Pasadena, California, as an astronomer in 1914 and started work on calibrating the period vs. luminosity relation for a class of stars called Cepheid variables. In 1912 the American astronomer Henrietta S. Leavitt had discovered a direct relation between the brightness of a Cepheid variable and the periodicity of its brightness change. Shapley used this relation to determine the distances of globular clusters – immense, densely packed groups of stars, some containing as many as 1,000,000 members. Using the 1.5-metre reflecting telescope at Mount Wilson, he made a study of the distribution of globular clusters in the galaxy and found that of the 100 clusters known at the time, one-third lay within the boundary of the constellation Sagittarius. From his measurements, Shapley deduced that the Sun lies at a distance of 50,000 light-years from the centre of the Milky Way galaxy; the number was later corrected to 30,000 light-years. His efforts gave the first realistic estimate for the actual size of the Milky Way galaxy and marked the beginning of modern galactic astronomy. In 1921 Shapely was appointed director of the Harvard College Observatory, and in 1952 he became professor of astronomy at the Harvard University. (☞ Oct. 20)

1929
Richard E. Taylor, Canadian physicist, Nobel laureate, was born in Medicine Hat, Alberta, Canada. Taylor received his bachelor's degree in 1950 and master's degree in physics in 1952 from the University of Alberta in Edmonton, Canada. He received his doctorate from the Stanford University, California, USA, in 1962 and worked for a year as a physicist at the Lawrence Berkeley Laboratory of the University of California. In 1963 he joined the Stanford Linear Accelerator Center (SLAC) to work on the design of the experimental areas for the new accelerator, especially on the requirements for electron scattering apparatus in the larger of two experimental areas. He spent the next decade helping to build equipment and taking part in various electron scattering experiments and became an associate professor at the Stanford University in 1968 and a full professor there in 1970. Between 1967 and 1973 Taylor together with two American physicists from the Massachusetts Institute of Technology – Jerome Isaac Friedman and Henry W. Kendall – carried out a series of experiments using the 3.2-km-long particle accelerator at SLAC to produce a beam of high-energy electrons, which they directed at target protons and neutrons. The resulting data showed protons and neutrons to be composed of still smaller point-like particles – the first empirical evidence for the existence of quarks. This discovery was crucial to the formulation of the currently accepted theoretical description of matter and its interactions, known as the standard model. Taylor was awarded the 1990 Nobel Prize for Physics, jointly with Friedman and Kendall, for their pioneering investigations concerning deep inelastic scattering of electrons on protons and bound neutrons, which have

been of essential importance for the development of the quark model in particle physics. (☞ Mar. 28, Dec. 9)

1931
The DuPont Company, of Wilmington, Delaware, USA, announced the production of the first synthetic rubber. It was known as 'DuPrene', now renamed neoprene. The rubbery white product was accidentally discovered in a jar in which a sample of vinyl acetylene was left with hydrochloric acid for about five weeks. The hydrochloric acid had reacted with the vinyl acetylene, making chloroprene, which then polymerised to become polychloroprene. The new rubber was expensive, but resisted oil and gasoline, which natural rubber did not. It was the first good synthetic rubber.

1932
Melvin Schwartz, American physicist, Nobel laureate, was born in New York, USA. Schwartz studied at the Bronx High School of Science in New York, and received his PhD from the Columbia University, New York in 1958. He taught at the Columbia University from 1958 to 1966 and was then a professor of physics at the Stanford University from 1966 to 1983. During 1960–62 Schwartz, jointly with Leon M. Lederman and Jack Steinberger performed a series of experiments at the Brookhaven National Laboratory on neutrinos. Neutrinos are produced when unstable atomic nuclei or subatomic particles disintegrate. Schwartz and his team wanted to study the 'weak' nuclear force that creates certain kinds of radioactivity. The team used a particle accelerator to generate a stream of high-energy protons, which were then fired at a target made of the metal beryllium. The bombardment produced a stream of different particles, including those called pions (pi mesons) that decayed into muons (mu mesons) and neutrinos as they travelled. The beam of particles was filtered through a steel barrier and the resulting beam of neutrinos was then made to hit other matter. When the scientists studied the reactions produced, they discovered a new kind of neutrino called a muon neutrino, and obtained new information about the structure of particles called leptons. Schwarz was awarded the 1988 Nobel Prize for Physics, jointly with Lederman and Steinberger, for the neutrino beam method and the demonstration of the doublet structure of the leptons through the discovery of the muon neutrino. (☞ May 25, Jul. 15)

1966
Peter Joseph Wilhelm Debye, Dutch physical chemist, Nobel laureate (Chemistry, 1936), who made important contributions in the field of dipole moments, X-rays, and light scattering in gases, died in Ithaca, New York, USA. (☞ Mar. 24)

2000
The first permanent residents in outer space – an American astronaut and two Russian cosmonauts – occupied the *International Space Station* to begin a four-month mission. After their *Soyuz* spacecraft linked up at 11:00 am GMT, William Shepherd, Sergei Krikalev and Yuri Gidzenko entered the station, turned on the lights and life support systems. They were confined to two of the space station's three rooms until space shuttle *Endeavour* arrived in early December with giant solar panels that would provide all the necessary power for the entire station. (☞ Dec. 4)

November 3

1664
The English physicist Robert Hooke showed an advanced copy of his classic book *Micrographia*, to the Royal Society in London. This large volume contained sketches of practically everything Hooke could view with the latest invention of the day, the microscope. It was also history's first treatise on microbiology, and coined the word 'cell' in the biological context. Also included in this volume were many snow crystal drawings, which for the first time revealed the complexity and intricate symmetry of the snow crystal structure. The book was published in 1665 and promptly became a best seller. (☞ Jul. 18)

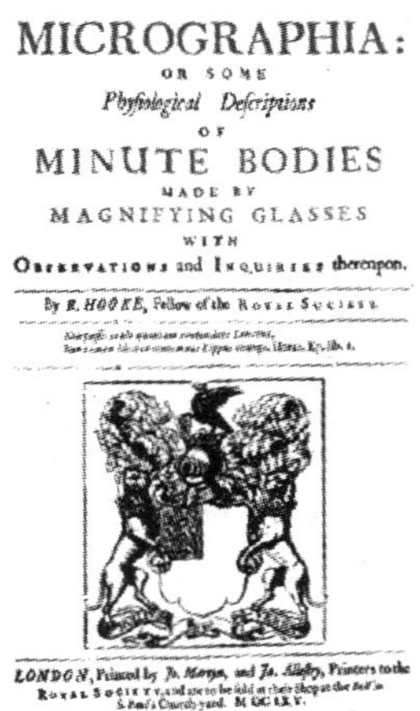
MICROGRAPHIA:
OR SOME
Physiological Descriptions
OF
MINUTE BODIES
MADE BY
MAGNIFYING GLASSES
WITH
OBSERVATIONS and INQUIRIES thereupon.
By R. HOOKE, Fellow of the ROYAL SOCIETY.
LONDON, Printed by Jo. Martyn, and Ja. Allestry, Printers to the ROYAL SOCIETY, and are to be sold at their Shop at the Bell in S. Paul's Church-yard. M DC LX V.

1892
The first automatic telephone exchange, using the electromechanical switching device invented by the American undertaker and inventor Almon Brown Strowger, was opened to the public in LaPorte, Indiana, USA, with about seventy-five subscribers. This early system did not use a dial to enter the desired number. Instead, using three keys, one for each digit of a three-digit number, a subscriber pressed each key the appropriate number of times for each digit. Strowger was apparently motivated to invent an automatic telephone exchange after having difficulties with the local telephone operators. He was convinced that the local manual telephone exchange operators were sending calls to his competitor rather than to his business. (☞ Mar. 10)

1957
The Russian satellite *Sputnik-2* was launched into space carrying a Siberian husky dog named Laika ('barker' in Russian). She was the first living creature launched into space. Because the craft was not designed for recovery, there was no way to return Laika to Earth. She died after a few days in orbit when the batteries of her life support system eventually ran down. Biological data returned for approximately a week showed scientists how Laika was adapting to space – information important to the manned missions already being planned. The satellite remained in orbit for 162 days.

1973
The American space probe *Mariner-10* was launched on a mission to Mercury and Venus from the Kennedy Space Center, Florida, USA. (☞ Feb. 5, Mar. 16, Mar. 29, Sep. 21)

November 4

1847
British surgeon James Young Simpson of Edinburgh, Scotland, became the first person to use chloroform as an anaesthetic to relieve the pain of childbirth because he was dissatisfied with ether – it irritated the eyes, and was highly inflammable. (☞ Mar. 30)

1890
The world's first electric underground railway line was opened between Stock Wall and King William Street in London, England. (☞ Jan. 10)

1862
Richard Jordan Gatling, a trained dentist from North Carolina, USA, received a patent for the machine gun. The Gatling gun consisted of six barrels mounted in a revolving frame and could fire 616 shots in two minutes. The United States Army purchased these guns in 1865 and over the next few years, most major armies in Europe purchased them.

1939
The first air-conditioned automobile was exhibited at the 40th Automobile Show in Chicago, Illinois, USA. Air in the car was cooled, dehumidified, filtered and circulated. Heat was provided for use in winter. The refrigerating coils were located behind the rear seat in an air duct, with heating coils in another compartment of the same duct. The capacity of the unit was equivalent to 1.5 tonnes of ice in 24 hours when the car was driven at 96 km per hour.

1965
The Industrial Toxicology Research Centre (ITRC), a constituent laboratory of the Council of Scientific & Industrial Research (CSIR) was established. Located in Lucknow, Uttar Pradesh, ITRC is dedicated to providing health safeguards to industrial and agricultural workers through its rich knowledge base created painstakingly over the years. The major R&D activities of ITRC include identification of occupational health hazards due to exposure to chemicals in industries, mines, agricultural fields and general environment by undertaking health and environmental surveys; studies for working out the mode of action of toxic chemicals/pollutants; development of simple/rapid diagnostic tests for disorders caused by industrial and environmental chemicals. ITRC also undertakes safety evaluation of chemicals used in industry, agriculture and everyday life, and suggests remedial/preventive measures to safeguard health and environment from pollutants. The institute has made significant contributions in elucidating the mechanisms by which pesticides, heavy metals, monomers, plasticisers and other plastic additives, solvents, food colours and dye intermediates act and produce deleterious effects on human health. It has also developed a method for early diagnosis of manganese poisoning, a peripheral model for monitoring neurological disorders caused by neurotoxic substances, and test kits for the detection of toxic food adulterants. The institute also undertakes the testing of new products reaching the markets including bio-engineered products.

1982
Soviet unmanned spacecraft *Venera-14* was launched towards Venus from the Baikanur Cosmodrome in Kazakhstan. The spacecraft carried instruments to take chemical and isotopic measurements, monitor the spectrum of scattered sunlight, and record electric discharges during

its descent phase through the Venusian atmosphere. (☞ Mar. 5)

November 5

1854

Paul Sabatier, French organic chemist, Nobel laureate, was born in Carcassonne, France. Sabatier graduated from the École Normale Supérieure in Toulouse in 1877 and taught physics for a year in a local school at Nîmes before going to the Collège de France in 1878 as assistant to the French chemist Marcellin Berthelot. He received the degree of Doctor of Science in 1880 and took charge of courses in physics in the Faculty of Sciences at Bordeaux until January 1882, when he accepted a similar post at the University of Toulouse. He had the additional responsibility for courses in chemistry in 1883 and was appointed professor of chemistry in 1884, a post that he retained until his retirement in 1930. Sabatier's earliest researches concerned the thermochemistry of sulphur and metallic sulphates. When he started his investigations into the phenomenon of catalysis, he formulated his chemical theory, which postulated the formation of unstable intermediaries. His subsequent painstaking work and discovery of the use of finely divided metal catalysts for hydrogenation subsequently laid the foundation of the oil hydrogenation and synthetic methanol industries. Sabatier demonstrated the selectivity of catalytic action and also the selectivity of catalysts to poisons. He found that increasing the surface area of catalysts such as copper and nickel by finely dividing them greatly increases their effectiveness. He also made a close study of catalytic hydration and dehydration, examining carefully the feasibility of specific reactions and the general activity of various catalysts. Sabatier was awarded the 1912 Nobel Prize for Chemistry, jointly with the French chemist Victor Grignard, for his method of hydrogenating organic compounds in the presence of finely disintegrated metals whereby the progress of organic chemistry has been greatly advanced in recent years. (☞ May 6, Aug. 14)

1879

James Clerk Maxwell, Scottish physicist and mathematician, whose work united electricity and magnetism into the concept of the electromagnetic field, died in Cambridge, UK. (☞ Jun. 13)

1892

John Burdon Sanderson Haldane, British-born Indian geneticist and biometrician who opened new paths of research in population genetics and evolution, was born in Oxford, Oxfordshire, England. Educated at the Eton College and New College, Oxford, and trained in biochemistry, Haldane was not only broadly knowledgeable in science; he also had a well-developed social conscience. This led him to follow Marxism for a time, but he later became disillusioned and abandoned Marxism in the 1940s. After World War I he taught at the University of Cambridge, England, the University of California, Berkeley in USA, and at the University of London, England. In an essay written in 1929, he first speculated about the chemical origin of life on Earth. His famous book, *The Causes of Evolution*, published in 1932, was the first major work of what came to be known as the 'modern evolutionary synthesis'. He studied

relationships among different disciplines and problems, including the consequence of Mendelian genetics on evolutionary theory, the relationship between enzymology and genetics, and the application of mathematics and statistics to the study of biology. He is also known for an observation from his essay, 'On being the right size', which has since been referred to as the 'Haldane's principle', that sheer size very often defines what bodily equipment an animal must have: 'Insects, being so small, do not have oxygen-carrying bloodstreams. What little oxygen their cells require can be absorbed by simple diffusion of air through their bodies. But being larger means an animal must take on complicated oxygen pumping and distributing systems to reach all the cells'. Haldane became known to a large public as a witty populariser of science and author of several popular books. (☞ Dec. 1)

1906

Fred Lawrence Whipple, American astronomer who proposed the 'dirty snowball' model for comet nuclei, was born in Red Oak, Iowa, USA. After studying at the Occidental College in California, he earned an undergraduate degree at the University of California, Los Angeles, and a doctoral degree in astronomy at the University of California at Berkeley. He helped compute for the first time the orbit of newly discovered Pluto while a graduate student at Berkeley. Whipple worked at the Harvard University, Cambridge, Massachusetts, from 1931 to 1977, and directed the Smithsonian Astrophysical Observatory (SAO) in Cambridge from 1955 to 1973. He was one of the leaders in the 1973 merger of the SAO with the Harvard College Observatory to form the Harvard–Smithsonian Center for Astrophysics. In the 1930s, using a new, two-station method of photography, he determined meteor trajectories and found that nearly all visible meteors were made up of fragile material from comets, and concluded that none can came from beyond the solar system. His pioneering research at Harvard on comets became pivotal to understanding the complexity and makeup of these enigmatic celestial bodies. Whipple originated the 'dirty snowball' model of comet structure in 1950. He theorised that, rather than loose conglomerates of dust and rock, comets were solid chunks of ice with dust and rocky particles mixed in. This model was confirmed 36 years later when the European Space Agency's *Giotto* mission passed by Comet Halley in 1986. In 1982, the former Mt. Hopkins Observatory in Arizona was re-named the Fred L. Whipple Observatory in his honour.

1930

Christiaan Eijkman, Dutch scientist, physician, Nobel laureate (Physiology or Medicine, 1929), whose demonstration that beriberi is caused by poor diet led to the discovery of vitamins, died in Utrecht, Netherlands. (☞ Aug. 11)

1944

Alexis Carrel, French scientist, surgeon, biologist, Nobel laureate (Physiology or Medicine, 1912), who developed techniques for joining and transplantation of blood vessels and organs, died in Paris, France. (☞ Jun. 28)

1948

William D. Phillips, American physicist, Nobel laureate, was born in Wilkes-Barre,

Pennsylvania, USA. Phillips had his schooling at the Camp Hill High School, Camp Hill, Pennsylvania, and did his BS in physics from the Juniata College, Huntington, Pennsylvania, in 1970. He obtained his PhD in physics from the Massachusetts Institute of Technology, Cambridge, Massachusetts, in 1976. In 1978 he joined the staff of the National Bureau of Standards (now the National Institute of Standards and Technology) in Gaithersburg, Maryland. In 1990 he became head of the 'laser cooled and trapped atoms' group of the Atomic Physics Division. Here with his team, he developed methods to use laser light to cool gases to the 10^{-6} Kelvin temperature range near absolute zero, and keep the chilled atoms floating or captured in different kinds of 'atom traps'. Individual atoms can then be studied with very great accuracy and their inner structure can be determined. As more and more atoms are captured in the same volume, a thin gas forms, and its properties can be studied in detail. One result of the development of laser-cooling techniques was the first observation, in 1995, of the Bose–Einstein condensate, a new state of matter originally predicted 70 years earlier by the German-born American physicist Albert Einstein and the Indian physicist Satyendra Nath Bose. In this state atoms are so chilled and so slow that they, in effect, merge and behave as one single quantum entity that is much larger than any individual atom. Phillips was awarded the 1997 Nobel Prize for Physics, jointly with the American physicist Steven Chu and the French physicist Claude Cohen-Tannoudji, for development of methods to cool and trap atoms with laser light. (☞ Feb. 28, Apr. 1)

1975
Edward Lawrie Tatum, American biochemist and molecular geneticist, Nobel laureate (Physiology or Medicine, 1958), who helped demonstrate that genes determine the structure of particular enzymes or otherwise act by regulating specific chemical processes in living things, died in New York, USA. (☞ Dec. 14)

1992
Jan Hendrik Oort, Dutch physicist and astronomer, who proposed that comets originate from a vast cloud of small bodies that orbit the Sun at a distance of about one light-year and after whom the cloud is named, died in Leiden, Netherlands. (☞ Apr. 28)

2003
The American unmanned spacecraft *Voyager-1* set a new milestone of reaching 90 astronomical units (AU) from the Sun after travelling through space for more than 27 years. 90 AU is the equivalent of about 13.5 billion kilometres. It is the only spacecraft to have made measurements in the solar wind from such a great distance from the source of the dynamic solar environment. (☞ Mar. 5, Aug.15, Sep. 5, Nov. 12)

November 6

1822
Claude-Louis Berthollet, French chemist who carried out research into dyes and bleaches and determined the composition of ammonia, died in Arcueil, France. (☞ Dec. 9)

1919
The official results of the British expeditions to observe the total solar eclipse on

May 26, 1919, which confirmed Einstein's relativity theory that predicted the bending of starlight by gravity, was announced at a joint meeting of the Royal Society and the Royal Astronomical Society in London. (☞ Mar. 14)

1928
US Army Lieutenant Colonel Jacob Schick received a patent for the first electric shaver. The early electric shaver was quite cumbersome; it consisted of a small electric motor connected to a flexible shaft arrangement that terminated in a shaving head.

1966
American spacecraft *Lunar Orbiter-2* was launched on a mission to photograph smooth areas of the lunar surface to select safe landing sites for the Apollo missions. The spacecraft was also equipped to collect data on radiation intensity and micro-meteoroid impact on the Moon. The spacecraft photographed the Moon from November 18 to 29. In all five *Lunar Orbiter* missions were sent and all were successful, photographing 99% of the Moon's surface with a resolution of 60 m or better.

November 7

1631
The French astronomer Pierre Gassendi became the first person ever to observe a transit of Mercury across the face of the Sun. Gassendi made the observation from his Paris apartment, using a telescope to project a magnified image of the Sun's disc onto a paper screen in a darkened room. The transit was predicted by the German astronomer Johannes Kepler, who had alerted astronomers to the event and encouraged observations, but died before his predictions were fulfilled. About 13 or 14 transits of Mercury occur every 100 years, usually in May or November. November transits occur at intervals of 7, 13, or 33 years; May transits only occur at intervals of 13 or 33 years. The last two transits were in May 2003 and November 2006; the next two will occur in May 2016 and November 2019. (☞ Dec. 27)

1867
Marie Curie (*nee* Maria Sklodowska), Polish-born French chemist and physicist, Nobel laureate, was born in Warsaw, Poland. After completing her secondary education at the Russian lycée, Maria Sklodowska worked as a teacher. In 1891 she went to Paris to study at the Sorbonne. There she began to study different elements and materials for radioactivity and discovered that the element thorium was also radioactive. She married the French physical chemist Pierre Curie in 1895, and together they started working on pitchblende, an ore of uranium that they found was more radioactive than the uranium it contained. In 1898, Marie discovered the radioactivity to be due to two new elements, which were named polonium (Po), after Poland, and radium (Ra). In 1902, Marie produced radium chloride and later the pure metal. She received her doctorate in 1903. The unit 'curie', defined as the amount of radioactive material undergoing 3.7×10^{10} disintegration per second, is named in her honour. Marie and Pierre Curie were awarded the 1903 Nobel Prize for Physics, jointly with the French physicist Antoine-Henri Becquerel. Pierre and Marie Curie shared

half the prize money for investigations of the radiation phenomena, while Becquerel received the other half for the discovery of spontaneous radioactivity. Marie Curie also received the 1911 Nobel Prize for Chemistry for discovery of polonium and radium and isolation of radium. (☞ Mar. 25, Jul. 4, Jul. 18, Dec. 10)

1878

Lise Meitner, Austrian-born Swedish physicist, Fellow of the Royal Society, was born in Vienna, Austria. Meitner received her doctorate from the University of Vienna in 1906 and went to Berlin, Germany, to work with the German physicist Max Planck and chemist Otto Hahn. In 1913 she joined the Kaiser Wilhelm Institute for Chemistry of which she later became joint director along with Hahn and carried out experiments on the effect of neutron bombardment on uranium. In 1917, Meitner and Hahn discovered the new radioactive element protactinium (at. no. 91). In 1938 she went to Sweden, where she resumed her research, first at the Nobel Institute for Physics in Stockholm and later at the Atomic Energy Association. After Meitner left Germany, Hahn continued his research with Fritz Strassmann and discovered barium in neutron-bombarded uranium. In 1939, after hearing about this, Meitner and her nephew Otto Frisch concluded that the barium resulted from the break-up of the uranium nucleus, a process they named 'nuclear fission'. Meitner visited USA after World War II but returned to Sweden, where she became a citizen in 1949. She was elected to the Royal Society in 1955 and settled in Cambridge, England in 1960. (☞ Oct. 27)

1888

Chandrasekhara Venkata Raman, Indian physicist, Nobel laureate, was born in Trichinopoly (now Tiruchirapalli) in Tamil Nadu, India. Raman did his BA in 1904 and MA in 1907 from the Presidency College, Madras, (now Chennai), obtaining the highest distinctions. His earliest researches were in optics and acoustics – the two fields of investigation to which he has dedicated his entire career – while he was a student. He was deeply interested in musical instruments, especially Indian musical instruments such as the veena, the violin, the mridangam and the tabla, and by 1918, could explain the complex vibrations of the strings of musical instruments. But, in India during the era in which Raman lived, a scientific career did not afford the best opportunities. So, Raman joined the Indian Civil Service in the Finance Department as Assistant Accountant General in Calcutta (now Kolkata) in 1907. Soon he started doing experimental research after office hours in the laboratory of the Indian Association for the Cultivation of Science in Kolkata. In 1917 he was offered the newly endowed Palit Chair of Physics at the Calcutta University, and decided to accept it, giving up his job in the Finance Department. It was during his stay in Kolkata that Raman discovered the 'Raman effect' – the inelastic scattering of photons. In 1933, after 15 years in Kolkata, Raman moved to Bangalore to join the Indian Institute of Science as professor. Raman carried out investigations on the diffraction of light by acoustic waves of ultrasonic and hypersonic frequencies. In 1948 he left IISc to set up his own research institute – the Raman Institute of Research – at Bangalore, of which he became the

Director. He studied the spectroscopic behaviour of crystals, thus approaching the fundamentals problems of crystal dynamics in a new way. He sponsored the establishment of the Indian Academy of Sciences and served as President since its inception. He was elected a Fellow of the Royal Society in 1924, but resigned from the Fellowship in 1968. Raman was awarded the 1930 Nobel Prize for Physics for his work on the scattering of light and for the discovery of the effect named after him. In 1954, Raman was honoured with the 'Bharat Ratna', the highest civilian honour in India. (☞ Feb. 28, Nov. 21)

1903

Konrad Lorenz, Austrian zoologist, Nobel laureate, was born in Vienna, Austria. Lorenz showed an interest in animals at an early age, and kept animals of various species – fish, birds, monkeys, dogs, cats, and rabbits – as pets, many of which he brought home from his boyhood excursions. He also kept detailed records of bird behaviour in the form of diaries. Lorenz graduated from secondary school in 1922 and then spent two semesters studying medicine at the Columbia University, in New York, USA, before returning to Vienna to study. During his medical studies, Lorenz continued to make detailed observations of animal behaviour; a diary about a jackdaw that he kept was published in 1927 in the prestigious *Journal fur Ornithologie*. He received the MD degree from the University of Vienna in 1928 and was awarded a PhD in zoology in 1933. Encouraged by the positive response to his scientific work, Lorenz established colonies of birds, such as the jackdaw and greylag goose, and

published a series of research papers on his observations of them. In 1935 Lorenz observed that at a certain critical stage, soon after hatching, young ducklings and goslings learn to follow real or foster parents. The process, which is called imprinting, involves visual and auditory stimuli from the parent object. His ideas revealed how behavioural patterns may be traced to an evolutionary past. He was also known for his work on the roots of aggression. From 1942 to 1944 Lorenz served as a physician in the German army and was captured as a prisoner of war in the Soviet Union. He was returned to Austria in 1948. He headed the Institute of Comparative Ethology at Altenberg from 1949 to 1951. Lorenz was awarded the 1973 Nobel Prize for Physiology or Medicine, jointly with the Austrian-born German zoologist Karl von Frisch and Dutch-born British zoologist Nicolaas Tinbergen, for their discoveries concerning organisation and elicitation of individual and social behaviour patterns. (☞ Feb. 27, Apr. 15, Nov. 20)

1913

Alfred Russel Wallace, British naturalist, and biogeographer, who was the first westerner to describe some of the most interesting natural habitats in the tropics, died in Old Orchard, Broadstone, England. (☞ Jan. 8)

1929

Eric R. Kandel, Austrian-born American neurobiologist, Nobel laureate, was born in Vienna, Austria. Kandel did his BA in history and literature from the Harvard College in Boston, Massachusetts and obtained an MD from the New York University School of Medicine in 1956. In 1965 he joined the New York University as associate professor in the Department of Physiology and Psychiatry. During 1974–83,

he was Director, Center for Neurobiology and Behavior, Columbia University, New York where in 1984 he became Senior Investigator in the Howard Hughes Medical Institute. Kandel's research mainly concerned the molecular mechanism of learning and memory. By probing the synaptic connections between nerve cells in the humble sea slug, he uncovered some of the basic molecular mechanisms underlying learning and memory in animals ranging from snails to flies to mice and even humans. His groundbreaking studies have demonstrated the fundamental ways that nerve cells alter their response to chemical signals to produce coordinated changes in behaviour. This work is central to understanding not only normal memory but also dementia and other mental illnesses that affect memory. Kandel's research has shown that learning produces changes in behaviour by modifying the strength of connections between nerve cells, rather than by altering the brain's basic circuitry. He went on to determine the biochemical changes that accompany memory formation, showing that short-term memory involves a functional modulation of the synapses while long-term memory requires the activation of genes and the synthesis of proteins to grow new synaptic connections. Kandel was awarded the 2000 Nobel Prize for Physiology or Medicine, jointly with Arvid Carlsson and Paul Greehgard, for their discoveries concerning signal transduction in the nervous system. (☞ Jan. 25, Dec. 11)

1964

Hans von Euler-Chelpin, German-born Swedish biochemist, Nobel laureate (Chemistry, 1929), who worked out the structure of the yeast coenzyme and helped determine the chemical structure of several vitamins which form portions of coenzymes, died in Stockholm, Sweden. (☞ Feb. 15)

1967

American Moon-probe *Surveyor-6* was launched towards the Moon from the Kennedy Space Center, Florida, USA. (☞ Nov. 10)

1996

The American space probe *Mars Global Surveyor* was launched on a mapping mission to Mars. It was designed to observe the planet from a low altitude, nearly polar orbit over the course of one complete Martian year (the equivalent of nearly two Earth years) and also send back scientific data and pictures, including 3-D views of the planet. (☞ Sep. 11)

November 8

1656

Edmond Halley, British astronomer, Fellow of the Royal Society, who was the first to calculate the orbit of a comet later named after him, was born in Haggerston, Shoreditch, near London, England. Halley began his early education at St. Paul's School, London and later studied at Queen's College, Oxford, where he was introduced to the British astronomer John Flamsteed, who was appointed astronomer royal in 1676. He later visited the Royal Greenwich Observatory, where Flamsteed did his work, and was encouraged to study astronomy. In 1676 he undertook a voyage to the southern hemisphere on board a ship of the East India Company to compile an accurate catalogue of stars of the

southern sky. By the time he returned to England in 1678, he had recorded the celestial longitudes and latitudes of 341 stars, and observed a transit of Mercury across the Sun's disc. Halley's star catalogue, published late in 1678, was the first such work to be published containing locations of southern stars. The same year he was granted an MA degree from the Oxford University. In 1684, Halley met the English physicist Isaac Newton in Cambridge for the first time and was instrumental in getting Newton's *Principia* published. Halley is best known for his prediction of the comet named after him. In 1705 he published *A Synopsis of the Astronomy of Comets*, in which he described the orbits of 24 comets that had been observed from 1337 to 1698, of which the orbits of three – seen in 1531, 1607, and 1682 – were so similar that they could only be successive returns of the same comet. He predicted that the same comet would return again in 1758. Halley did not live to see it, but the comet did return as predicted and was named Comet Halley in his honour. (☞ Jan. 14, Dec. 25)

1895

Wilhelm Conrad Röntgen, German physicist, accidentally discovered X-rays. He had set up his experiment using a Crookes tube fitted with an anode and a cathode, separated from each other by a few centimetres in the tube to explore the properties of cathode rays outside the tube He used an induction coil to produce a potential difference of a few thousand volts, and covered the tube with a lightproof cardboard jacket. He laid a screen of fluorescent material on a table a few metres away from the apparatus. When he passed a current through the discharge tube, he noticed a green fluorescence on the screen. He knew the fluorescence could not be produced by the cathode rays since they could not penetrate through the wall of the tube. Visible light could not be the stimulus since the tube was covered with a lightproof jacket. He boldly hypothesised that he must have produced an unknown type of radiation, which he called 'X-rays'. (☞ Jan. 1, Mar. 27)

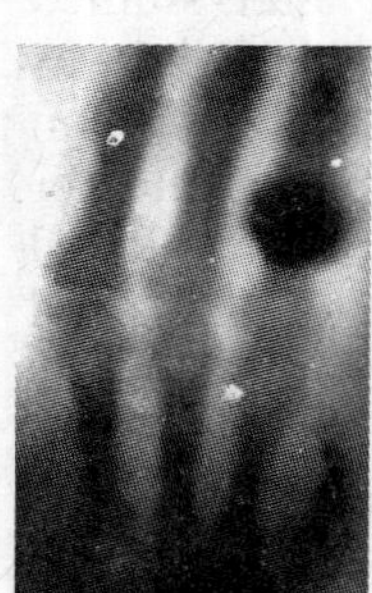

Röntgen's first X-ray

1922

Christiaan Barnard, South African surgeon who performed the first human heart transplant operation, was born in Beaufort West, South Africa. Barnard was educated in medicine at the University of Cape Town, South Africa and did his internship and residency at the Groote Schuur Hospital in Cape Town. He received his master's degree in medicine from the University of Cape Town in 1953. From 1956 to 1958 he attended the University of Minnesota, Minneapolis, Minnesota, USA, to do his doctorate in surgery. After returning to South Africa he became a cardiothoracic surgeon at the Groote Schuur Hospital in 1958, established the hospital's first heart unit and introduced open-heart surgery to South Africa. He also lectured at the University of Cape Town. After the first successful kidney transplant in 1954, Barnard experimented for several years with animal heart transplants. In December 1967 he led a team of 20 surgeons at the Groote Schuur Hospital in replacing the

heart of Louis Washkansky, an incurably ill South African grocer, with a heart taken from a young woman, Denise Darvall, killed in a road accident. Although the transplant itself was successful, Washkansky died 18 days later from double pneumonia, because of the drugs administered to suppress rejection of the new heart, which destroyed his body's immunity. Barnard continued to perform heart transplants (his second patient, operated on in January 1968, survived for 19 months) as well as pioneer new techniques. These included double transplants, joining a healthy heart to the patient's heart to create a 'double pump', designing artificial heart valves and using monkeys' hearts to keep alive desperately ill people. (☞ Sep. 2, Dec. 3)

1923

Jack S. Kilby, American electrical engineer, Nobel laureate, was born in Great Bend, Kansas, USA. After completing high school in Great Bend, Kilby studied electrical engineering at the University of Illinois from where he graduated in 1947, just one year after the invention of the transistor. He then joined an electronics manufacturer company in Milwaukee, Wisconsin, that made parts for radios, televisions and hearing aids. While in Milwaukee, he took evening classes at the University of Wisconsin towards a master's degree in electrical engineering, which he received in 1950. Between 1958 and 1970 he was employed at the Texas Instruments Incorporated in Dallas, Texas. There in 1959, he and the American electrical engineer Robert Norton Noyce independently developed the first monolithic electronic circuit called the integrated circuit, or IC, which later came to be known as the 'chip'. The chip laid the foundation for the entire field of modern microelectronics. But in 1961, Noyce was granted a patent for his invention of a 'semiconductor device-and-lead structure', which made the microchip possible and launched a modern electronics revolution. Kilby pioneered industrial, military, and commercial applications of microchip technology. In 1970 Kilby took leave of absence from the Texas Instruments Company to work as an individual inventor. Much of his later work has been directed toward the development of a novel solar energy system. Kilby was awarded the 2000 Nobel Prize for Physics, jointly with Russian physicist Zhores Ivanovich Alferov and fellow American physicist Herbert Kroemer, for basic work on information and communication technology. Kilby received half the prize money for his part in the invention of the integrated circuit, while Alferov and Kroemer shared the other half for developing semiconductor heterostructures used in high-speed- and opto-electronics. (☞ Mar. 15, Apr. 25, Jun. 20, Aug. 25, Dec. 12)

1968

American unmanned spacecraft *Pioneer-9* was launched from Cape Canaveral, Florida, USA. *Pioneer-9* was the fourth in a series of solar-orbiting satellites designed to obtain measurements of interplanetary phenomena from widely separated points in space on a continuing basis. The spacecraft carried experiments to study the positive ions and electrons in the solar wind, the interplanetary electron density, solar and galactic cosmic rays, the interplanetary magnetic field, cosmic dust, and electric fields. *Pioneer-9*'s accomplishments include determining the structure and flow of the solar wind, the million-km/h stream of ionised gases that spirals out from the Sun. It also measured the twisted magnetic fields threading the solar wind and the high-energy particle streams, which follow the course of the magnetic field out from the Sun. Before the *Pioneer* findings, the solar wind was thought to be a gentle, steady flow, but they were discovered to be variable and sometimes quite rapid.

November 9

1897

Ronald George Wreyford Norrish, British chemist, Nobel laureate, was born in Cambridge, UK. Norrish spent his early years at the local Board school, and obtained a scholarship to the Perse Grammar School in 1910. In 1915 he obtained an entrance scholarship to the Emmanuel College, Cambridge, in natural sciences, but left in 1916 with a commission in the Royal Field Artillery for service in France. He was made prisoner of war in March 1918 and spent the rest of the war in Germany. Norrish returned to the Emmanuel College in 1919 as a student. In 1925, he joined the Chemistry Department of the Cambridge University, first as a demonstrator and in 1937 a professor. From 1949–65 he collaborated with his former pupil George Porter in the development of flash photolysis and kinetic spectroscopy for the investigation of very fast reactions, which had considerable influence on the subsequent development of photochemistry and reaction kinetics. Norrish also corrected Draper's law, propounded in the mid-19th century by the English-born American chemist, John Draper, according to which the amount of photochemical change is proportional to the intensity of the light multiplied by the time for which it acts. Norrish showed that the rate was in fact proportional to the square root of the light intensity. Norrish was awarded the 1967 Nobel Prize for Chemistry, jointly with Porter and the German physicist Manfred Eigen, for their studies of extremely fast chemical reactions, effected by disturbing the equilibrium by means of very short pulses of energy. (☞ May 9, Jun. 7, Dec. 6)

1904

Panchanan Maheshwari, Indian botanist, Fellow of the Royal Society, was born in Jaipur, Rajasthan, India. Maheshwari obtained his BSc degree in 1925 and MSc degree in 1927 from the Allahabad University. In 1931, the Allahabad University awarded him the DSc degree for his thesis on the embryology of the leguminous tree *Albizzia lebbek*. Maheshwari began his teaching career in 1928 at the Ewing Christian College, Allahabad, and subsequently served the universities of Agra, Allahabad, Lucknow, and Dacca, before joining the University of Delhi as head of the department of botany in 1949. Maheshwari's chief interest was plant morphology, particularly the embryology of seed plants on which he carried out intensive research. His book *An Introduction to the Embryology of Angiosperms*, published in 1950, is a standard reference text the world over even today. In 1956 Maheshwari started experimentation in embryology, and soon discovered the vast application of plant tissue and organ culture techniques in studies of embryology. His works on

embryology in relation to taxonomy, and test-tube fertilisation in flowering plants are noteworthy. (☞ May 18)

1934

Carl Edward Sagan, American astronomer, exobiologist and science writer, was born in Brooklyn, New York, USA. Sagan received a bachelor's degree in 1955 and a master's degree in 1956, both in physics, and a doctorate in astronomy and astrophysics in 1960, all from the University of Chicago, Illinois. He taught at the Harvard University, Cambridge, Massachusetts, in the early 1960s before joining the Cornell University, Ithaca, New York, where he became a full professor in 1971. Sagan played a leading role in NASA's *Mariner*, *Viking*, *Voyager* and *Galileo* missions to the planets of the solar system. His research focussed on topics such as the greenhouse effect on Venus; windblown dust as an explanation for the seasonal changes on Mars; organic aerosols on Saturn's moon Titan; the long-term environmental consequences of nuclear war; and the origin of life on Earth. A pioneer in the field of exobiology, he continued to teach graduate and undergraduate students courses in astronomy and space sciences at Cornell. Sagan was co-founder of the Planetary Society, a 100,000-member organisation and the largest space-interest group in the world. It supports major research programmes in the radio search for extraterrestrial intelligence, the investigation of near-Earth minor planets and, with the French and Russian space agencies, the development and testing of balloon and mobile robotic exploration of Mars. Sagan's enthusiasm for space science and the possibility of intelligent life elsewhere in the universe stimulated the public's interest through such works as the television series *Cosmos* and books like the *Cosmic Connection*, and *Contact*. The *Mars Pathfinder* lander, which landed on Mars in 1997, has been named the Carl Sagan Memorial Station in honour of Sagan. (☞ Jul. 4, Dec. 20)

1994

The first atom of the heaviest chemical element with atomic number 110 was detected at the Gesellschaft für Schwerionenforschung (GSI) in Darmstadt, Germany. The isotope of element 110 that was discovered had an atomic mass number of 269 (i.e. 269 times heavier than hydrogen), thereby making it the heaviest man-made atom ever produced. The new element was produced by bombarding a target cf ^{208}Pb with a beam of ^{62}Ni nuclei. Chemically, the element 110 belongs to the same group as nickel, palladium, and platinum, but unlike these lighter atoms, element 110 decays after a small fraction of a thousandth of a second into lighter elements by emitting alpha-particles which are the nuclei of helium atoms. (☞ Aug. 16)

2005

The European Space Agency's *Venus Express* was launched towards Venus by a Soyuz-Fregat rocket from Baikonur, Kazakhstan. The name *Venus Express*

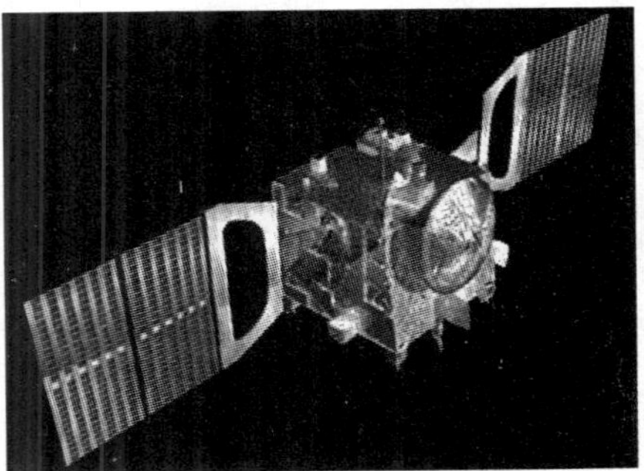

comes from the short time taken to define, prepare and launch the mission. It took less than three years from the approval to the launch of the mission. It went into orbit around Venus in 2006. The spacecraft was designed to carry out the most comprehensive study of the Venusian atmosphere ever. It will study the complex dynamics and chemistry of the Venusian atmosphere and the interactions between the atmosphere and the surface from orbit, which would give clues about the surface's characteristics. It would also study the interactions between the atmosphere and the interplanetary environment such as solar wind to better understand the evolution of the planet. (☞ Apr. 11)

November 10

1885

The world's first motorcycle, designed by the German engineer Gottlieb Daimler, made its debut. The frame and wheels were made of wood. A leather belt transferred power from the engine to large brass gears mounted to the rear wheel. Since there was no suspension (front or rear), the ride was quite bumpy. The top speed for the motorcycle was 12 km/h. This was built as an experimental vehicle to test the new Daimler engine, which was to power Daimler's first motorised carriage the following year.

1918

Ernst Otto Fischer, German theoretical chemist and educator, Nobel laureate, was born in Munich, Germany. Fischer completed four years at elementary school and then went on to grammar school in 1929, from which he graduated in 1937. He served in Poland, France and Russia during World War II and during a period of study leave studied chemistry at the Technical University in Munich from where he received a doctorate in 1952. He lectured there in 1954–57 and became professor of inorganic chemistry and director of the Inorganic Chemistry Institute in 1964. In 1951 Fischer read about a newly developed synthetic compound called ferrocene whose structure was unknown. After studying the substance, he concluded that it contained bonds between metal atoms and a whole molecule, arranged in a way that has not been considered as possible or probable earlier. He found it consisted of an iron atom sandwiched between two plate-like molecules called cyclopentadienyl groups on each side. The British scientist Geoffrey Wilkinson made this same discovery of organometallic sandwich compounds independently of Fischer. Fischer also made a sandwich compound with chromium and benzene molecules, a great surprise for most chemists at that time. Fisher was awarded the 1973 Nobel Prize for Chemistry, jointly with Wilkinson, for their pioneering work, performed independently, on the chemistry of the organometallic, so-called sandwich compounds. (☞ Jul. 14)

1967
American Moon-probe *Surveyor-6* landed in Sinus Medii on the Moon. The spacecraft transmitted 29,952 television pictures and scientific data on the chemical composition of the lunar material. As part of the surface mechanical properties investigation, *Surveyor-6* performed a 'hop' manoeuvre, moving 2.5 metres away from its original landing area. This manoeuvre provided excellent views of the surface disturbances produced by the initial landing and the effects of firing rocket engines close to the lunar surface. (☞ Nov. 7)

1970
Russian Moon-probe *Luna-17* was launched from the Baikanur Cosmodrome in USSR (now Kazakhstan). It carried a remotely controlled lunar rover *Lunokhod-1* that was equipped with four television cameras, and special extendable devices to impact the lunar soil for soil density and mechanical property tests. The lunar rover also carried an X-ray spectrometer, an X-ray telescope, cosmic ray detectors, and a laser device. It was powered by solar cells. (☞ Nov. 17)

1974
The discovery of the 'charmed quark' subatomic particle (also known as J/psi particle) was announced simultaneously by the two American experimental groups – an MIT group led by physicist Samuel Ting at the Brookhaven National Laboratory, New York, and a group led by the physicist Burton Richter at the Stanford Linear Accelerator Centre in Berkeley on the west coast of USA. The new particle, with a mass of 3,095 MeV, had a lifetime about 1,000 times more than that of other particles of comparable mass. Within two years, in 1976, Ting and Richter were awarded the Nobel Prize for Physics for their discovery. (☞ Jan. 27, Mar. 22)

November 11

1493
Paracelsus (full name Theophrastus Phillippus Aureolus Bombastus von Hohenheim), German-Swiss physician and alchemist who condemned medical teaching that was not based on observation and experience, was born in Einsiedeln, Switzerland. As a boy, Paracelsus learned some practical medicine at home through observing his father, who was a physician. It is likely that he learned some folk medicine as well. He also picked up some alchemy from his father who had an interest in the subject. At the age of fourteen he left home to begin a long period of wandering. He visited most of the countries of Central, Northern, and Eastern Europe and may have visited several universities, but there is no proof that he ever took a medical degree. As an adult, however, he picked up practical medical knowledge by working as a surgeon in a number of mercenary armies that ravaged Europe in the seemingly endless wars of the period. He introduced chemical remedies to replace traditional herbal ones, influenced the development of medicine during the Renaissance and gave alchemy a wider perspective. Paracelsus saw illness as having a specific external cause rather than being caused by an imbalance of the humours in the body. He disputed that mental illness was caused by demons, and linked goitre with minerals in drinking water. He gave a clinical description of syphilis in 1530. He published *Der grossen Wundartzney* ('Great Surgery Book') in 1536. (☞ Sep. 24)

1572
The Danish astronomer Tycho Brahe was the first to observe a supernova in the constellation of Cassiopeia. Also called 'B Cassiopeiae', or 'Nova Cassiopeiae 1572', it was one of the few recorded supernovae in the Milky Way galaxy. From his precise measurements, Brahe showed that it was not merely a nearby object, such as a comet, but was in fact at the great distance of the stars. Thus, he demonstrated that the stars were not unchangeable, and in fact that they could change in a remarkable manner. At its brightest, the star was visible to the eye by daylight. (☞ Dec. 14)

1925
The discovery of cosmic rays was announced in Madison, Wisconsin, USA, by the American physicist Robert A. Millikan who coined their name. (☞ Mar. 22)

1952
The first video tape recorder, which for the first time recorded moving pictures through means other than photography, was demonstrated by American electrical engineers John Mullin and Wayne Johnson at the Bing Crosby Enterprises Laboratories in Beverly Hills, California, USA. The first video tape recorder (VTR) captured live images from television cameras by converting the information into electrical impulses and saving the information onto magnetic tape. The machine used 12 recording heads – 10 for the video signal, one for synchronisation, and one for audio. The tape, with tracks laid longitudinally, ran at 100 inch (250 cm) per second and had a recording time of 16 minutes from one reel of tape. Later, an American electronic engineer, Charles Paulson Ginsburg, led a research team at Ampex Corporation in developing the first practical videotape recorder that could run the tape at a much slower rate because the recording heads rotated at high speed, allowing the necessary high-frequency response.

1973
Artturi Ilmari Virtanen, Finnish biochemist, Nobel laureate (Chemistry, 1945), whose investigations led to improving the production and storage of protein-rich green fodder, died in Helsinki, Finland. (☞ Jan. 15)

1982
American space shuttle *Columbia*, carrying astronauts Vance D. Brand, Robert F. Overmyer, Joseph P. Allen, and William B. Lenoir, was launched on its 5th flight from the Kennedy Space Center, Florida, USA. During the five-day mission, the crew deployed two commercial communications satellites, *ANIK C-3* for TELESAT Canada and *SBS-C* for Satellite Business Systems. The astronauts also conducted three Shuttle Student Involvement Program experiments. (☞ Nov. 16)

November 12

1746
Jacques-Alexandre-César Charles, French mathematician, physicist, and inventor, who invented the hydrogen balloon and formulated the gas law known after him, was born in Beaugency, France. Beginning as a clerk in the finance ministry, Charles turned to science and experimented with electricity. He has to his credit several inventions, including a hydrometer and reflecting goniometer. During his lifetime, Charles was better known for inventing the

hydrogen balloon. In 1783, realising that hydrogen was lighter than air, he made the first balloon using hydrogen gas and, in August, ascended to a height of nearly 914 metres. But Charles is now best known for his formulation in 1787 of one of the basic gas laws. Charles's law states that at constant pressure, the volume occupied by a fixed weight of gas is directly proportional to the absolute temperature. He was elected to the French Académie des Sciences in 1785. (☞ Apr. 7)

1842

Lord Rayleigh (John William Strutt), English physical scientist, Nobel laureate, was born in Maldon, Essex, England. At an early age, Rayleigh suffered from poor health and his schooling at both Eton and Harrow was disrupted. He spent four years at the Reverend Warner's boarding school to prepare for university and entered the Trinity College, Cambridge, in 1861 where he graduated in the Mathematical Tripos in 1865 as Senior Wrangler and Smith's Prizeman. In 1866 he obtained a fellowship at Trinity, which he held until 1871. In 1873, after his father's death, he became the third Baron Rayleigh and had to devote part of his time to the management of his estates. But he was determined to devote his life to science.

After reading widely the current scientific literature, he worked out the important problems on which to undertake research. While on a visit to the United States he purchased equipment for undertaking scientific experiments and set it up on the family estate at Terling. Rayleigh's theory of scattering, published in 1871, was the first correct explanation of why the sky is blue. The first volume of his major text *The Theory of Sound*, on the mechanics of a vibrating medium which produces sound, was published in 1877, while the second volume on acoustic wave propagation was published the following year. From 1879 to 1884 Rayleigh was the second Cavendish professor of experimental physics at Cambridge. One of the important pieces of experimental work he carried out during his time as Cavendish professor was a standardisation of the ohm. Rayleigh's greatest single contribution to science is generally considered to have been his discovery and isolation of argon, one of the rare gases of the atmosphere. He made an interesting discovery during his precision measurements of the density of gases conducted during the 1880s. He found that the density of nitrogen obtained from the atmosphere is greater by a small though definite amount than the density of nitrogen obtained from one of its chemical compounds, such as ammonia. Rayleigh decided to explore the possibility that the discrepancy he had discovered resulted from the presence in the atmosphere of a hitherto undetected constituent. In 1895, after a long and arduous experimental program, he finally succeeded in isolating the gas, which was appropriately named argon, from the Greek word meaning 'inactive'. Rayleigh was awarded the 1904 Nobel Prize for Physics for his investigations of the densities of the most important gases and for his discovery of argon in connection with these studies. (☞ Jun. 30, Aug. 13)

1896

Sálim Moizuddin Abdul Ali, better known as Salim Ali, was born in Bombay (now Mumbai). Orphaned at a very young age, Ali was brought up by his maternal uncle, Amiruddin Tyabji, who was a keen *shikari* (hunter) and nature-lover. Under

his guidance, young Ali learnt his first lessons in *shikar* and became aware of the nature around him. When he was ten years old, Ali shot a bird that had a yellow streak below its neck. His uncle, not being able to satisfy his curiosity about the bird, advised him to go to the Bombay Natural History Society (BNHS), Mumbai, where it was identified as the yellow-throated sparrow. Ali became interested in birds by this incident and wanted to pursue a career in ornithology. He did his BSc in zoology from the St. Xavier's College, Mumbai. In 1919, Ali went to Burma to look after the family mining and timber business. In 1930, he started working for the BNHS, but without a salary. He was paid for the expenses that were incurred during his ornithological surveys. In 1941, Ali published *The Book of Indian Birds*, which was highly appreciated and widely read. He also wrote the 10-volume *Handbook of the Birds of India and Pakistan*, which he wrote with the help of zoologist S. Dillon Ripley. Ali received numerous awards including the J. Paul Getty International Award, the Golden Ark of the International Union for Conservation of Nature, the golden medal of the British Ornithology Union, and a Padma Shree and Padma Vibhushan from the Indian Government. (☞ Jul. 27)

1916

Percival Lowell, American astronomer who predicted the existence of the planet Pluto and initiated the search that ended in its discovery, died in Flagstaff, Arizona, USA. (☞ Feb. 18, Mar. 13)

1935

The first modern surgery on the frontal lobes of the brain for treatment of mental disorders was performed by Portugese neurologist Egas Moniz at the Santa Marta Hospital in Lisbon, Portugal. He used a technique that severed neurons and led to the pre-frontal lobotomy techniques of the 1940s. Moniz was later awarded the 1949 Nobel Prize for Physiology or Medicine for the development of pre-frontal lobotomy as a radical therapy for certain psychoses, or mental disorders. Such radical surgery fell out of favour when psychoactive medication became available. (☞ Nov. 29)

1980

The American unmanned spacecraft *Voyager-1* had the closest encounter with Saturn as it flew by at a distance of 124,000 kilometres. During its fly-by, the spacecraft found five new moons and a ring system consisting of thousands of bands, discovered a new ring (the 'G-Ring'), and found 'shepherding' satellites on either side of the F-ring satellites that keep the rings well defined. The spacecraft also photographed Saturn's moons, Titan, Mimas, Enceladus, Tethys, Dione, and Rhea, all of which appeared to be largely composed of water ice. Perhaps the most interesting target was Saturn's largest moon Titan, which *Voyager-1* passed at a range of 4,000 kilometres. The images showed a thick atmosphere that completely hid the surface. The spacecraft found that the moon's atmosphere was composed on 90 per cent nitrogen. (☞ Mar. 5, Aug. 15, Sep. 5, Nov. 5)

1981

American space shuttle *Columbia*, carrying astronauts Joseph H. Engle, and Richard H. Truly, was launched on its second flight from the Kennedy Space Center, Florida, USA. During the three-day mission, the crew carried out Earth observation experiments and tested out a remote manipulator for the first time. (☞ Nov. 14)

November 13

1893

Edward Adelbert Doisy, American biochemist, Nobel laureate, was born in Hume, Illinois, USA. Doisy was educated at the University of Illinois, where he took his BA degree in 1914 and his MS degree in 1916. He obtained a PhD degree from the Harvard University, Cambridge, Massachusetts, in 1920. In 1923 he became professor of biochemistry at the St. Louis University School of Medicine, St. Louis, Missouri. In 1924 he was appointed Director of the Department of Biochemistry. In 1929–1930 he succeeded in isolating oestrone, a feat independently accomplished at about the same time by Adolf Friedrich Johann Butenandt in Germany. Doisy was concerned chiefly with biochemical studies of the sex hormones and vitamins K_1 and K_2. Vitamin K, a substance that encourages blood clotting, had been identified by the Danish biochemist Henrik Dam, and in 1936–39 Doisy isolated two forms of the vitamin, determined their chemical structures, and synthesised the vitamin. In addition to his work on hormones, Doisy also improved the methods used for the isolation and identification of insulin and made important contributions to the knowledge of antibiotics and blood buffer systems, and bile acid metabolism. His writings include *Sex Hormones* (1936) and *Sex and Internal Secretions* (1939), with Edgar Allen and Charles H. Danforth. Doisy was awarded the 1943 Nobel Prize for Physiology or Medicine, jointly with Dam, for his discovery of the chemical nature of vitamin K. (☞ Feb. 21, Oct. 23)

1907

Paul Cornu, French bicycle maker and inventor, made the first flight in a vertical take-off-and-land aircraft at Lisieux, Calvados, France. His helicopter used two 6-metre counter-rotating rotors driven by a 24-hp Antoinette engine and lifted him to about 1.5 metres. It remained aloft for one minute. This helicopter was the first flying machine to have risen from the ground using rotor blades instead of wings. (☞May 13, Sep. 14)

1946

The first artificially induced snowstorm was created in New Hampshire, USA, by the American physicist Vincent J. Schaefer by sprinkling dry ice particles from an aeroplane on clouds containing super-cooled water.

1971

The American unmanned spacecraft *Mariner-9* entered Martian orbit, becoming the first man-made object to orbit another planet. The 506-kg spacecraft circled Mars twice each day for a full year, photographing the surface and analysing the atmosphere with infrared

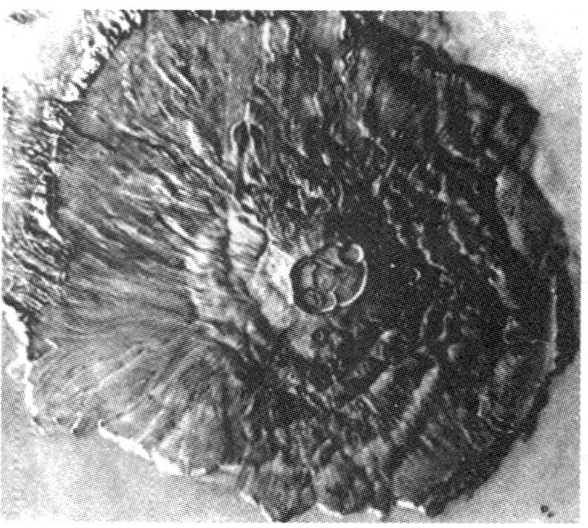

Mariner image of Olympus Mons, a volcano on Mars

and ultraviolet instruments. The spacecraft gathered data on atmospheric composition, density, pressure, and temperature and also surface composition, temperature, and topography of Mars. When *Mariner-9* first arrived, Mars was almost totally obscured by dust storms. Ground controllers sent commands to the spacecraft to wait until the storm had abated, the dust had settled, and the surface was clearly visible before starting its imaging mission of the Martian surface. The storm persisted for a month, but after the dust cleared, *Mariner-9* proceeded to reveal a very different planet than expected – one that boasted gigantic volcanoes and a grand canyon stretching 4,800 kilometres across its surface. More surprisingly, the relics of ancient riverbeds were carved in the landscape of this seemingly dry and dusty planet. During the mission, *Mariner-9* completed photo mapping 100 per cent of the planet's surface. The spacecraft also provided the first close-up pictures of the two small, irregular Martian moons: Phobos and Deimos. (☞ May 30)

1979
American inventor Robert Koffler Jarvik was granted a patent for an artificial heart. Made of Dacron polyester, plastic, and aluminium, the artificial heart called *Jarvik-7* had an internal power system that regulated the pump through a system of compressed air hoses that entered the heart through the chest. The air hoses were connected to the chambers. The heart's power system drove the pumps, which pumped blood through the patients' body. Jarvik and his team first tested the device on cows and other animals, making sure the heart could consistently beat at least 100,000 times a day. (☞ May 11, Nov. 25, Dec. 2)

November 14

1716
Gottfried Wilhelm Leibniz, German philosopher, mathematician, and logician who is probably most well known for having invented the differential and integral calculus independently of Sir Isaac Newton, died in Hanover, Germany. (Jul. 1)

1765
Robert Fulton, American inventor, engineer, and artist, who invented the first practical steamboat, was born in Little Britain, Pennsylvania, USA. As a boy he showed a decided gift for drawing and mechanics. At the age of 17 he went to Philadelphia to take up the study of portrait and landscape painting and mechanical drawing. In 1793, Fulton wrote to the Earl of Stanhope, offering to make known to him his scheme for moving ships by steam and his plan for an inclined plane for the operation of canals. In the same year, he invented a mill for sawing marble and other stone. Fulton also designed a practical submarine or 'plunging' boat, called the '*Nautilus*'. This vessel, 6 feet (about 2 m) in diameter and 20 feet (about 6 m) long, was built of iron and copper, and in general outline resembled the modern submarine. It was provided with a sail for surface navigation, which could be folded back upon the boat when diving. To submerge, water was admitted into the boat, which was removed by pumping when it was desired to come to the surface. Fulton made his first successful steamboat trip from New York to Albany in August 1807. He made the trip in thirty-two hours and

successfully demonstrated to the world the possibilities of steam navigation. (☞ Feb. 9, Feb. 24, Aug. 14)

1863

Leo Hendrik Baekeland, Belgian inventor and industrial chemist who helped found the modern plastics industry through his invention of bakelite, was born in Ghent, Belgium. After completing his doctorate at the University of Ghent, Baekeland taught for several years. In 1889, when he was twenty-six, he travelled to USA on a fellowship (that also allowed him to visit universities in England, Scotland, and Germany) to continue his study of chemistry. While in USA he joined a photographic firm and soon set up his own company to manufacture his invention, Velox, a photographic paper that could be developed under gas light rather than sunlight. Baekeland sold his company and rights to the paper to the Eastman Kodak Company in 1898. In 1905 he began to investigate the reactions of phenol and formaldehyde, and first produced soluble phenol–formaldehyde shellac called Novolak, which never became a market success. Then, by carefully controlling the pressure and temperature applied to an intermediate made from the two reagents, he could produce a polymer that, when mixed with fillers, produced a hard mouldable plastic, which he called bakelite. It was the first thermosetting plastic – a plastic that does not soften when heated. Though relatively ex-pensive, bakelite soon found many uses, especially in the rapidly growing electrical, automobile, and radio industries. It was used for manufacturing the first generation of telephones, amongst many other things. From 1910 to 1939 Baekeland served as president of the General Bakelite Corp. (☞ Feb. 23, Dec. 7)

1891

Birbal Sahni, Indian palaeobotanist, Fellow of the Royal Society, was born at Bahera, Shahpur district, Punjab (now in Pakistan). Sahni received his early education in Lahore and obtained a degree in botany from the Punjab University. Sahni went to England in 1913 to study for the Natural Science Tripos at the Cambridge University. He took the Tripos in 1915. The University of London awarded him the DSc degree in 1919 for his contribution to the study of fossil plants. After his return to India in 1919, and brief stints at the Banaras Hindu University and Punjab University, Sahni was appointed Professor of Botany at the University of Lucknow in 1921. Sahni is best known for his study of fossil plants of India. He devoted much of his time to examine critically the collection of fossil plants in the Geological Survey of India. Besides this, he and his research students discovered many new fossil-rich localities in the country. In 1946, at a meeting of the Royal Society, Sahni reported the discovery of a new type of gymnospermous plant of the Jurassic age, the Pentoxyleae. This was a completely new group of organisms of very great morphological significance. During the decade 1940–1949, Sahni and his research team reported novel types of plant structures and flora of the Intertrappen beds of the Deccan. Sahni initiated a new line of research on microfossils and their use in stratigraphical geology. Sahni founded the Birbal Sahni Institute of Palaeobotany in

Lucknow for which, he bequeathed his private funds, immovable property, his personal library and fossil collection. (☞ Apr. 10)

1891

Frederick Grant Banting, Canadian physician, Nobel laureate, was born in Alliston, Ontario, Canada. Banting was educated at the University of Toronto, served in World War I, and then practised as a surgeon in London, Ontario, where he became interested in diabetes, a chronic disorder of the sugar metabolism due to insufficient production of the insulin hormone. It was known long before Banting's experiments that removal of the pancreas in experimental animals caused a condition similar to diabetes, but all attempts to extract insulin from pancreas had failed because the pancreas' own digestive enzymes destroyed the insulin molecules as soon as the pancreas was ground up. In May 1921 Banting and Charles H. Best, a medical student, began an intensive effort in the laboratories of the Scottish physiologist John James Rickard Macleod, at the University of Toronto, to isolate the hormone. They tied up the pancreatic ducts of some dogs; in others they were removed to create diabetic animals. After a few months, an operation on a dog with tied ducts showed a pancreas about one-third the normal size. It was removed and ground in a solution of sand and saline. The extract was then injected into a diabetic dog near death. The blood sugar decreased so dramatically that the dog lived weeks longer than expected. The success was announced in 1922. In 1923 Banting became head of the new Banting and Best Department of Medical Research created at the University of Toronto. Banting was awarded the 1923 Nobel Prize for Physiology or Medicine, jointly with Macleod, for the discovery of insulin. Best did not share the prize because he did not receive his medical degree till 1925. (☞ Feb. 21, Feb. 27, Apr. 15, Sep. 6)

1905

Robert Whitehead, British engineer who invented the modern torpedo, died at Worth, Kent, England. (☞ Jan. 3)

1969

American manned lunar mission *Apollo-12*, carrying astronauts Charles Conrad, Jr., Richard F. Gordon, and Alan L. Bean, was launched towards Moon by a Saturn-V rocket from the Kennedy Space Center, Florida, USA. (☞ Nov. 19, Nov. 24)

1981

American space shuttle *Columbia*, carrying astronauts Joseph H. Engle, and Richard H. Truly, landed at the Edwards Air Force Base, California, USA, after completing its second flight. (☞ Nov. 12)

1985

The discovery of fullerenes was announced by the American chemists Robert F. Curl, Jr. and Richard E. Smalley, and the English chemist Harold W. Kroto in a paper published in the British journal *Nature*. Fullerenes are soccer-ball-shaped molecules made up of 60 or more carbon atoms linked together in a cage-like structure, which constitute the third elemental form of carbon (after graphite and diamond). Fullerenes have been produced in the form of continuous carbon fibres, which are

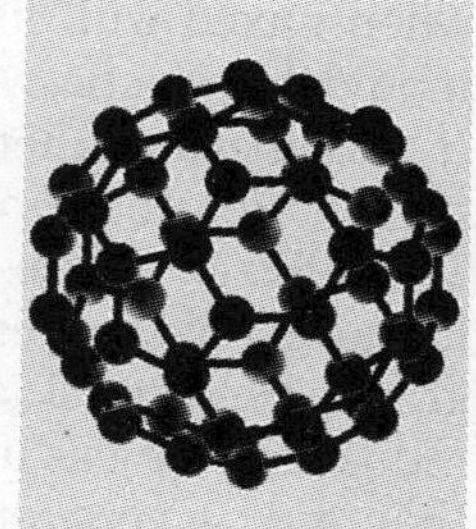

essentially giant single-fullerene molecules. Just a few nanometres in width, but many centimetres in length, these fullerene fibres are expected to be the strongest fibres ever made. (☞ Jun. 6, Aug. 23, Oct. 7)

2003
A new planet-like object was discovered in orbit around the Sun at a distance 90 times greater than that of the Earth from the Sun – about 3 times further away than Pluto. The discovery was made by the team of Michale Brown of Caltech, Chad Trujillo of the Gemini Observatory, and David Rabinowitz of the Yale University with the Samuel Oschin Telescope at the Palomar Observatory in California, USA. Known as 2003 VB12, the object measures about 1,180–2,360 kilometres and circles the Sun every 10,500 years. (☞ Jun. 4)

November 15

1630
Johannes Kepler, German astronomer who discovered three major laws of planetary motion, died in Regensburg, Germany. (☞ Dec. 27)

1738
William Herschel (original name Friedrich Wilhelm Herschel), German-born British astronomer, was born in Hanover, Germany. Right from childhood, Herschel was deeply interested in music. After arrival in England in 1757, he was employed as a military bandmaster, then as a music teacher and organist at the Octagon Chapel in Bath where he started composing and giving concerts. Besides music, Herschel also developed an avid interest in astronomy and this soon became his first love. He rented a small reflecting telescope, but this only whetted his desire to have his own larger telescope. Since he did not have enough money to buy one, he contrived to build it with the help of his brother, Alexander, and his sister Caroline. With a grant of 4,000 pounds sterling from King George III he built his largest telescope, a 120-centimetre reflector. Later, he built a 6-metre-long reflecting telescope with which he made most of his recorded observations. Herschel became one of the most notable observers in the history of astronomy. His systematic survey of the sky was one of the most important of his accomplishments. Herschel's observations and discoveries are numerous. He not only discovered the planet Uranus, but also two of its moons, two of Jupiter's moons and nearly 1,000 double stars. He observed sunspots and confirmed the gaseous nature of the Sun. He was the first to detect the infrared component of sunlight. (☞ Mar. 13, Aug. 25)

1874
Schack August Steenberg Krogh, Danish physiologist, Nobel laureate, was born in Grenå, Denmark. Even as a schoolboy, Krogh was deeply interested in natural sciences and spent a great deal of his time in experimentation. He studied zoology at the University of Copenhagen and after graduation in 1897 began to work in the Laboratory of Medical Physiology. In 1908 he was appointed associate professor in animal physiology at the University of Copenhagen, where he became professor in 1916. Krogh's scientific work embraces a number of different fields. He worked out methods for microscopical analyses of the gas contained in the air bladders of

the Corethra (a kind of gnat) larvae and was able to prove that these organs functioned like the diving tanks of a submarine, their content being regulated until equilibrium with the surrounding water was restored. His interest in problems connected with the gas exchange of the living organism led him to study the gas exchange in the frog. He found that, while skin respiration was relatively constant, great variations occurred with regard to lung respiration. From this he concluded that there must be a physiological mechanism to control the action of the capillaries in response to the needs of the body for oxygen. He found that the capillaries contract or dilate in proportion to the tissue's requirement for blood – that active muscles, for example, have a greater number of open capillaries than the less active ones. Krogh was awarded the 1920 Nobel Prize for Physiology or Medicine for his discovery of the capillary motor regulating mechanism. (☞ Sep. 13)

1904
King Camp Gillette, American inventor, was awarded a US patent for his 'safety razor', which used a cheap disposable blade for shaving. (☞ Jan. 5)

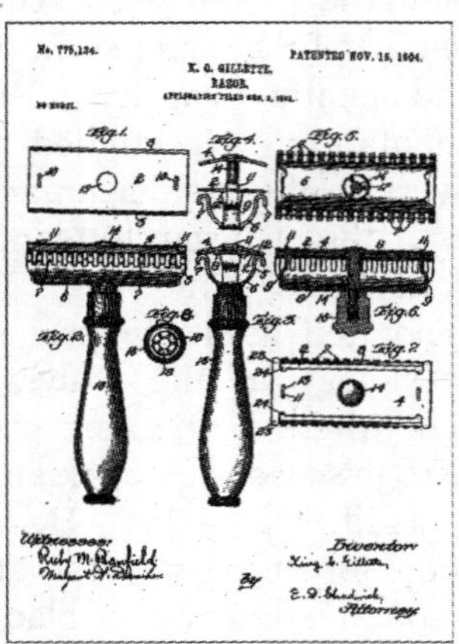

1919
Alfred Werner, French-born Swiss chemist, Nobel laureate (Chemistry, 1913), who developed the coordination theory that permitted a simple classification of inorganic compounds and extended the concept of isomerism, died in Zürich, Switzerland. (☞ Dec. 12)

1959
Charles Thomson Rees Wilson, Scottish physicist, Nobel laureate (Physics, 1927), who invented the cloud chamber – named after him – which makes the paths of electrically charged particles visible by condensation of vapour, died in Carlops, Scotland, UK. (☞ Feb. 14)

1978
Margaret Mead, American anthropologist best-known for her studies of the non-literate peoples of Oceania and author of the bestseller *Coming of Age in Samoa*, died in New York, USA. (☞ Dec. 16)

1984
Baby Fae, the first human with a transplanted animal (baboon) heart, died in Loma Linda, California, USA. (☞ Oct. 26)

November 16

1877
Ardaseer Cursetjee, Indian engineer and ship builder, the first Indian Fellow of the Royal Society, died in Richmond, UK. (☞ Oct. 8)

1904
The first electron tube, a diode thermionic valve, was invented by the English engineer John Ambrose Fleming. The valve consists of an evacuated glass envelope into which a carbon or tungsten filament and a metal plate are sealed. When the filament is heated by applying electric current, it emits electrons which flows to the metal plate when connected to the positive terminal of a battery; but no current flows if the plate is connected to the negative terminal; that is, it conducts current only in one direction. This means that if the valve is connected in a circuit with an oscillating current, it can convert it into a direct current capable of actuating, for example, a telephone receiver.

1945
The discovery of two new elements – americium (at. no. 95) and curium (at. no. 96) – was announced. Curium was discovered in a plutonium isotope, plutonium-239, that had been bombarded by helium ions (alpha particles) in the 60-inch cyclotron at the University of California, Berkeley. It was the third transuranium element to be discovered. It is named in honour of the French chemist couple Pierre and Marie Curie. Americium, named after the Americas, was produced by intense neutron irradiation of pure plutonium by the American nuclear chemists Glenn T. Seaborg, Ralph A. James, Leon O. Morgan, and Albert Ghiorso in a nuclear reactor. It was the fourth transuranium element to be discovered. It is used as a portable source for gamma radiography, and in smoke detectors. (☞ Apr. 19)

1965
Soviet unmanned spacecraft *Venera-3* was launched towards Venus from the Baikanur Cosmodrome in Kazakhstan. (☞ Mar. 1)

1973
'Skylab-4', the last of the *Skylab* missions, carrying a crew of three astronauts – Gerald P. Carr, William R. Pogue and Edward C. Gibson – was launched from the Kennedy Space Center, Florida, USA, on an 84-day mission that remained the longest American space flight for over two decades. They carried out four space walks totalling 22 hours and 13 minutes including one on Christmas Day to observe the comet Kohoutek. The crew returned to Earth after 1,214 orbits. (☞ Feb. 8, May 14, May 25, Jun. 22, Jun. 28, Jul. 11, Sep. 25)

1982
American space shuttle *Columbia*, carrying astronauts Vance D. Brand, Robert F. Overmyer, Joseph P. Allen, and William B. Lenoir, landed at the Edwards Air Force Base, California, USA after completing its second flight. (☞ Nov. 11)

1999
Daniel Nathans, American microbiologist, Nobel laureate (Physiology or Medicine, 1978), who developed the technique of using restriction enzymes, the so-called 'molecular knives', to study viral DNA, died in Baltimore, Maryland, USA. (☞ Oct. 30)

2000
The 220-MW second unit of the Kaiga Atomic Power Station at Kaiga in Karnataka, India, was commissioned. It is a pressurised heavy water reactor that uses natural uranium as fuel and heavy water as moderator and coolant. (☞ Mar. 16)

November 17

1865
William Merriam Burton, American chemist, was born in Cleveland, Ohio, USA. Burton joined the Standard Oil Company refinery overlooking Lake Michigan at Whiting in Indiana in 1890 and became president of the company in 1918. Here he developed the 'thermal cracking' process for reforming longer molecules of less volatile components that more than doubled the potential yield of gasoline from crude petroleum. Conventional process of fractional distillation only separates the more volatile components leaving behind the heavier fractions. Burton's thermal cracking process, patented in 1913, utilised chemical methods that doubled the yield of gasoline by transforming many of the less volatile components into gasoline. Burton's invention paved the way for the automobile era. Although other refining techniques have since been developed, the Burton process is still widely used. (☞ Dec. 29)

1902
Eugene Paul Wigner, Hungarian-born American physicist, Nobel laureate, was born in Budapest, Hungary. Wigner studied chemical engineering and received his PhD from the Institute of Technology in Berlin in 1925. After serving as a lecturer there and at the University of Göttingen, he went to the United States. Apart from two years as professor of physics at the University of Wisconsin, he spent his academic life at the Princeton University, serving as a professor of mathematical physics from 1938 until his retirement in 1971. In the late 1920s, Wigner explored the field of quantum mechanics, then being shaped by the German physicists Werner Heisenberg, Erwin Schrödinger and the English theoretical physicist Paul Dirac. At the University of Göttingen, Wigner formulated his law of the conservation of parity, which implies that it is impossible to distinguish left from right in fundamental physical interactions. This theory became an integral part of quantum mechanics, but in 1956 the Chinese-born American physicists Tsung-Dao Lee and Chen Ning Yang showed that parity is not always conserved in weak interactions of subatomic particles. Wigner also developed the principles involved in applying mathematical group theory to investigate the energy levels of atomic nuclei. In 1936 he worked out the theory of neutron absorption, which later proved useful in building nuclear reactors. Wigner was awarded the 1963 Nobel Prize for Physics, jointly with the German physicist J. Hans D. Jensen and German-born American physicist Maria Goeppert-Mayer. Wigner received half the prize money for his contributions to the theory of the atomic nucleus and the elementary particles, particularly through the discovery and application of fundamental symmetry principles, while Jensen and Goeppert-Mayer shared the other half for their discoveries concerning nuclear shell structure. (☞ Jan. 1, Jun. 25, Jun. 28)

1922
Stanley Cohen, American biochemist, Nobel laureate, was born in Brooklyn, New York, USA. Cohen was educated in the public school system of New York City. His scientific interests throughout his undergraduate days were directed to cell biology and especially the mysteries of embryonic development. After working for a short period as a bacteriologist in a

milk processing plant he joined the Oberlin College, where he received an MA in zoology in 1945. He received his PhD from the University of Michigan in 1948. His PhD thesis concerned the metabolic mechanism by which the end product of nitrogen metabolism in the earthworm is switched from ammonia to urea during starvation. In 1952 he joined the Italian-born American neurologist Rita Levi-Montalcini at the Washington University, St. Louis, Missouri, as a researcher. His training as a biochemist enabled him to help isolate the nerve growth factor (NGF), a natural substance that Levi-Montalcini had found stimulated the growth of nerve cells and fibres. Cohen found another cell growth factor in the chemical extracts which caused the eyes of newborn mice to open and their teeth to erupt several days sooner than normal. Cohen termed this substance epidermal growth factor (EGF), and he went on to purify it and completely analyse its chemistry. The discovery of NGF and EGF opened new fields of widespread importance to basic science and increased understanding of many disease states such as developmental malformations, degenerative changes in senile dementia, delayed wound healing and tumour diseases. Cohen was awarded the 1986 Nobel Prize for Physiology or Medicine, jointly with Levi-Montalcini, for their discoveries of growth factors. (☞ Apr. 22)

1929
Herman Hollerith, American inventor of a tabulating machine that was an important precursor of the modern electronic computer, died in Washington, DC USA. (☞ Feb. 29)

1970
Douglas Engelbart, an American inventor was issued a patent for the computer mouse. It was described as a 'X-Y position indicator for a display system'. (☞ Dec. 9)

1970
Russian Moon-probe *Luna-17*, carrying the remotely controlled lunar rover *Lunokhod-1*, landed in Mare Imbrium on the Moon. During the mission the *Lunokhod-1* travelled almost 10.5 kilometres and transmitted more than 20,000 television pictures and 200 television panoramas of the lunar surface. It also conducted more than 500 lunar soil tests. The *Lunokhod-1* was originally intended to operate through three lunar days but actually operated for eleven lunar days. (☞ Nov. 10)

1990
Robert Hofstadter, American scientist, Nobel laureate (Physics, 1961), who measured the size of the neutron and proton in the nuclei of atoms and who revealed the hitherto unknown structure of these particles, died in Stanford, California, USA. (☞ Feb. 5)

1995
The *Infrared Space Observatory* (*ISO*), an astronomical satellite of the European Space Agency, was launched from Kourou, French Guiana. The *ISO* made observations

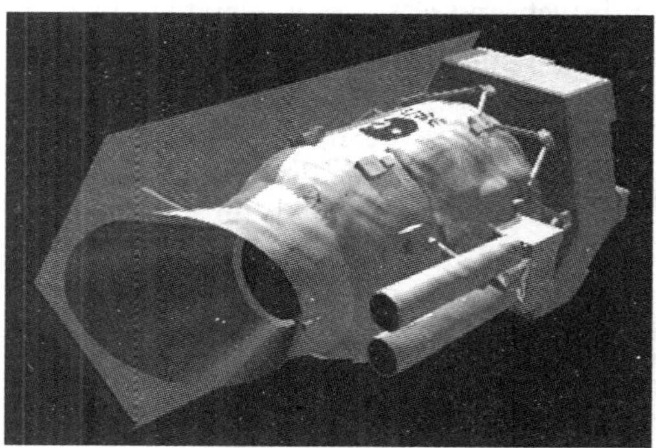

in a very wide range of infrared wavelengths – between 2.5 and 240 microns (much wider than that covered by the Infrared Astronomy Satellite, *IRAS*). It was also thousands of times more sensitive than *IRAS* and viewed infrared sources with much better resolution due to advances in infrared technology. Initially, the *ISO* was supposed to be operational for 20 months, but thanks to meticulous engineering and some good fortune, the satellite's working life was stretched to more than 28 months, until its liquid helium coolant ran out in April 1998. The *ISO* made more than 900 revolutions of the Earth and completed well over 26,000 scientific observations, leading to numerous new discoveries.

2000
Louis Eugène Félix Néel, French physicist, Nobel laureate (Physics, 1970) who made fundamental discoveries concerning anti-ferromagnetism and ferrimagnetism, died in Brive-Corrèze, France. (☞ Nov. 22)

November 18

1787
Louis Jacques Mande Daguerre, French artist, chemist, and photographer, was born in Cormeilles, near Paris, France. Daguerre began work as an apprentice architect, and at the age of sixteen was an assistant stage designer in a Paris theatre, his elaborate stage designs winning him considerable acclaim. He had an astonishing ingenuity in the handling of light and lighting effects, and he supplied the scenic and lighting effects for a number of operas in theatres in Paris. He regularly used a camera obscura (an optical device used in drawing similar to a pin-hole camera) as an aid to painting in perspective, and this led him to seek to make a permanent record of the image. In 1826 he learned of the work of Nicephore Niépce, who had developed a technique for making a permanent photographic image using a kind of bitumen, and signed up a partnership with him. After the death of Niépce in 1833, Daguerre continued to experiment and in 1835 made an important discovery by accident. He had placed an exposed plate in his chemical cupboard, and some days later found, to his surprise, that the latent image had developed. Daguerre eventually concluded that this was due to the presence of mercury vapour from a broken thermometer. Though he now knew how to produce an image, it was not until 1837 that he was able to fix them. This new process he called a 'Daguerreotype.' (☞ Jan. 9, Jul. 10, Aug. 19)

1897
Patrick Maynard Stuart Blackett, English physicist, Nobel laureate, was born in London, UK. Blackett graduated from the Cambridge University 1921 and spent the next 10 years doing research using the Wilson cloud chamber at the Cavendish Laboratory. In 1924, he obtained the first photographs of the transmutation of nitrogen into an oxygen isotope. In 1932, he designed a self-triggered cloud chamber that could make cosmic rays take their own photographs. He became professor at the Birkback College, London, in 1933, and there continued his cosmic ray research work. His varied research interests led to the creation of the first chair of radio astronomy, at the University of Manchester, and to the building of the Jodrell Bank Experimental Station for Radio Astronomy. In 1953 he was appointed

professor and head of the physics department of the Imperial College of Science and Technology in London, where he became senior research fellow in 1965. That year he was named president of the Royal Society. Blackett felt that Britain had an obligation to improve the scientific and technological situations in its former colonies. He played a key role in the establishment of India's Council of Scientific and Industrial Research. Blackett was awarded the 1948 Nobel Prize for Physics for his development of the Wilson cloud chamber method, and his discoveries therewith in the fields of nuclear physics and cosmic radiation. (☞ Jul. 13)

1906

George Wald, American biochemist, Nobel laureate, was born in New York, USA. While studying in Berlin, Germany, as a National Research Council fellow (1932–33), Wald discovered that vitamin A is a vital ingredient of the pigments in the retina and, hence, important in maintaining vision. After further research in Heidelberg, Germany, and at the universities of Zurich in Switzerland, and Chicago in USA, he joined the faculty of the Harvard University, Cambridge, Massachusetts in USA in 1934. By the early 1950s Wald had succeeded in elucidating the chemical reactions involved in the vision process of retina cells called the rods, which are involved in night vision. While researching the biochemistry of vision at Harvard, Wald identified

visual pigments – pigments in the retina that are sensitive to yellow-green, red, and blue light – and their precursors. Wald's work also demonstrated the role of vitamin A in forming the three colour pigments and showed that colour blindness is caused simply by the absence of one of them. Wald became professor emeritus at Harvard in 1977. He was awarded the 1967 Nobel Prize for Physiology or Medicine, jointly with the American physiologist Haldan Keffer Hartline and Finnish-born Swedish physiologist Ragnar Granit, for their discoveries concerning the primary physiological and chemical visual processes in the eye. (☞ Apr. 12, Oct. 30, Dec. 22)

1918

Coluthur Gopalan, Indian physician and nutritionist, Fellow of the Royal Society, was born in Salem, Tamil Nadu, India. Gopalan studied medicine in Madras (now Chennai) and obtained his MD in 1945 from the University of Madras. He did his PhD and DSc from the London University. Gopalan's main contribution has been in the field of nutritional deficiency, especially nutritional deficiency in children. He has also worked on the nutritive value of Indian foods. Gopalan's involvement with the study of nutrition and nutrition-related problems started in the mid-1950s when he joined the Nutrition Research Laboratories in Hyderabad, which under his stewardship blossomed into the present National Institute of Nutrition. He was Director-General of the Indian Council of Medical Research from 1974 to 1979. He founded the Nutrition Foundation of India in New Delhi in 1980 and is at present its President.

1941
Walther Hermann Nernst, German scientist, Nobel laureate (Chemistry, 1920), who formulated the 'third law of thermodynamics', died in Muskau, Germany. (☞ Jun. 25)

1962
Niels Henrik David Bohr, Danish physicist, Nobel laureate (Physics, 1922), who developed the so-called Bohr theory of the atom and was the first to apply quantum theory, which restricts the energy of a system to certain discrete values, to the problem of atomic and molecular structure, died in Copenhagen, Denmark. (☞ Oct. 7)

1989
The Cosmic Background Explorer (*COBE*) satellite of the American space agency NASA was launched. The satellite was developed by NASA's Goddard Space Flight Center to study both infrared and microwave characteristics of the cosmic background radiation (the remains of the extreme heat that was created by the Big Bang). Over the next four years, *COBE* mapped the brightness of the entire sky at several infrared wavelengths and discovered that the cosmic background radiation is not entirely smooth, showing extremely small variations in temperature in patches. Cosmologists believe these variations may have led to the formation of galaxies. *COBE* data showed that 99.97 per cent of the energy of the Universe was released in the first year after the Big Bang, suggesting that only one colossus explosion created the Universe.

November 19

1883
William Siemens (also Carl Wilhelm Siemens), German-born English engineer and inventor, pioneer in the field of undersea cables, died in London, England. (☞ Apr. 4)

1887
James Batcheller Sumner, American biochemist, Nobel laureate, was born in Canton, Massachusetts, USA. Sumner had his schooling at the Eliot Grammar School and Roxbury Latin School. He graduated from the Harvard College, Cambridge, Massachusetts, in 1910, having specialised in chemistry. He obtained his PhD from the Harvard Medical School in 1914 and joined the Cornell Medical School, Ithaca, New York, as assistant professor of biochemistry, a post that he held until 1929, when he was made full professor of biochemistry. In 1926 Sumner isolated and crystallised the enzyme – urease. He was the first to crystallise an enzyme, an achievement that revealed the protein nature of enzymes. In 1937, he went to Uppsala, Sweden, on a Guggenheim Fellowship and worked in the laboratory of Theodore Svedberg for five months, where he crystallised the enzyme catalase and also contributed to the purification of several other enzymes. Meanwhile, the American biochemist John Howard Northrop at the Rockefeller Institute in USA had obtained crystalline pepsin, and subsequently other enzymes. It became clear that Sumner had devised a general crystallisation

method for enzymes. Sumner was awarded the 1946 Nobel Prize for Chemistry, jointly with Northrop and American biochemist Wendell Meredith Stanley. Sumner received half of the prize money for his discovery that enzymes can be crystallised; Northrop and Stanley shared the other half for their preparation of enzymes and virus proteins in a pure form. (☞ Jul. 5, Aug. 12, Aug. 16)

1912

George E. Palade, Romanian-born American cell biologist, Nobel laureate, was born in Iasi, Romania. Palade studied at the School of Medicine of the University of Bucharest from where he obtained his MD in 1940. During World War II, he served in the medical corps of the Romanian Army, and after the war, emigrated to USA and joined the Rockefeller Institute for Medical Research (now Rockefeller University) in New York to work with the Belgian-born American cytologist Albert Claude on cell fractionation procedures. After Claude left in 1949, Palade started working in electron microscopy with the general aim of developing preparation procedures applicable to organised tissue. Palade added important methodological improvements both to differential centrifugation and to electron microscopy. In particular he became instrumental in combining the two techniques, often in combination, in order to obtain biologically basic information. The new layer of biological structure revealed by electron microscopy proved to be unexpectedly rich and surprisingly uniform for practically all eukaryotic cells. Palade performed many studies on the internal organisation of cell structures such as mitochondria, chloroplasts, the Golgi complex, and others. His most important discovery was that microsomes, bodies formerly thought to be fragments of mito-chondria, are actually parts of the endoplasmic reticulum and have a high RNA content. They were subsequently named ribosomes, organelles that carry out the protein synthesis in the cell. Palade became a naturalised citizen of the USA in 1952 and in 1958 he became a professor of cytology at the Rockefeller Institute, which he left in 1972 to direct studies in cell biology at the Yale University Medical School. Palade was awarded the 1974 Nobel Prize for Physiology or Medicine, jointly with Claude and the Belgian cytologist Christian de Duve, for their discoveries concerning the structural and functional organisation of the cell. (☞ Aug. 24, Oct. 2)

1915

Earl W. Sutherland, Jr., American pharmacologist and physiologist, Nobel laureate, was born in Burlingame, Kansas, USA. Sutherland obtained his BS degree from the Washburn College, Topeka, Kansas in 1937 and did his MD from the Washington University School of Medicine, St. Louis in 1942. He continued at the same university in various teaching capacities and in 1953 joined the School of Medicine, Western Reserve University, Cleveland, Ohio, as professor of pharmacology and director. In 1963 he moved to the Vanderbilt University School of Medicine, Nashville, Tennessee, as professor of physiology. Sutherland started his investigations on hormone action in the 1950s. In collaboration with the Czech-born American biochemist Carl Cori he studied the mechanism by which

epinephrine regulates the degradation of glycogen to glucose in the liver. Epinephrine is secreted into the blood stream from the adrenals under conditions of 'stress' and constitutes the signal that adapts the individual to the new situation. In an analogous manner many other hormones act as specific signals for the adaptation of the individual to the varied requirements of his surroundings. Sutherland first discovered that epinephrine acts by activating the enzyme phosphorylase, which leads to the formation of glucose from glycogen. Later he found that this activation took place by means of a hitherto unknown substance, which occurs as an intermediate during the process. The discovery and chemical characterisation of the intermediate was of crucial importance for an understanding of the mechanism of action of epinephrine and of many other hormones. The newly identified substance proved to be a nucleotide, and was named cyclic adenosine phosphate, or cyclic AMP, which has been dubbed as the 'second messenger' in relation to hormonal effects. Sutherland was awarded the 1971 Nobel Prize for Physiology or Medicine for his discoveries concerning the mechanisms of the action of hormones. (☞ Mar. 9)

1940
Chester Carlson, American inventor, was awarded an American patent for 'electron photography', which forms the basis of present-day photocopiers.

1969
Apollo-12 astronauts Charles Conrad and Alan Bean landed in Oceanus Procellarum on the Moon on board the lunar module *Intrepid*, while Richard F. Gordon remained in the command module *Yankee Clipper* in lunar orbit. The two astronauts spent a total of 7 hours 45 minutes outside the lunar module, setting up scientific experiments to measure the Moon's seismicity, solar wind flux and magnetic field. They also took photographs, and collected 34.4 kg of lunar rocks. They examined the nearby *Surveyor-3* spacecraft, which had landed on the Moon in April 1967, and removed pieces for later examination on Earth. (☞ Nov. 14, Nov. 24)

1997
Kalpana Chawla became the first Indian woman to go into space when she along with astronauts Steven W. Lindsey, Kevin R. Kregel, Winston E. Scott, Takao Doi, and Leonid K. Kadenyuk were launched on board the American space shuttle *Columbia* on its 24th flight from the Kennedy Space Center, Florida, USA. During the 16-day mission, astronauts Scott and Doi did a seven-hour, 43-minute space walk and completed a series of activities in preparations for on-orbit assembly of the *International Space Station*. Doi became the first Japanese citizen to walk in space. (☞ Jan. 16, Feb. 1, Jul. 1, Dec. 5)

November 20

1866
French inventor Pierre Lallement of Paris was awarded a patent for his velocipede, the forerunner of today's bicycle. The velocipede did not have a chain-and-sprocket drive, but was propelled by pedals fixed on the front wheel. The patent explains that the 'guiding wheel' is used to move to the left or the right and that greater the velocity, the more able the bicycle is to stay upright. 'After a little practice the rider is

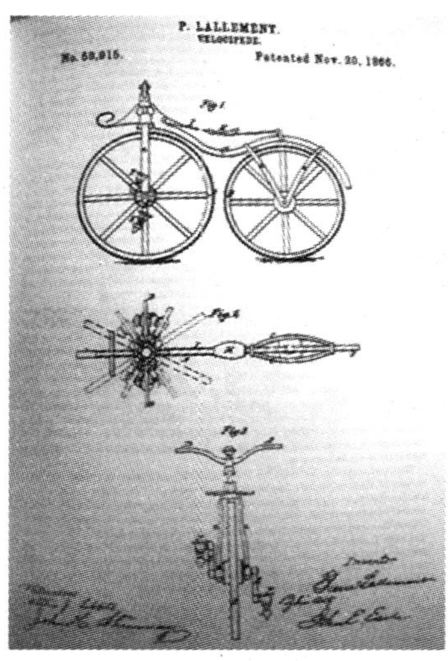

enabled to drive the same at an incredible velocity, with the greatest ease'.

1886

Karl von Frisch, Austrian zoologist, Nobel laureate, was born in Vienna, Austria. Frisch initially studied medicine at the University of Vienna, but later moved to the Munich University in Germany to study zoology. Frisch specialised in bees, discovering how they communicate the location of sources of nectar by movements called dances. In his early work he showed that bees have the ability to recognise different odours and tastes, and discovered their famous 'waggle dance' used by bees to communicate among with others in the hive. This sequence of body movements indicates the direction and distance to food. He found that round dances mean that food is nearby and waggle dances mean that the food is at a distance. The straight component of the waggle dance points the way to the food, and the duration of the dance indicate the distance.

He showed that polarised light is used by the bees as a 'compass' for navigation, even when the sun is not visible. In fish, he demonstrated their acute hearing, and ability to distinguish colours and brightness. Fritsch was awarded the 1973 Nobel Prize for Physiology or Medicine, jointly with fellow Austrian zoologist Konrad Lorenz and the Dutch-born British zoologist Nikolaas Tinbergen, for their discoveries concerning organisation and elicitation of individual and social behaviour patterns. (☞ Apr. 15, Jun. 12, Nov. 7)

1889

Edwin Powell Hubble, American astronomer, who is considered the founder of extragalactic astronomy, was born in Marshfield, Missouri, USA. Hubble attended high school in Chicago, and after graduation from the University of Chicago, won a Rhodes scholarship and earned a law degree at the Oxford University in England. He taught in a high school for a year in Indiana, USA, and then returned to the University of Chicago, receiving his PhD in astronomy in 1917. In 1919 he began to work at the Mt. Wilson Observatory in California,

where he stayed for the rest of his life. At the observatory, he studied nebulae – fuzzy patches of light in the sky. In 1924, he announced the discovery of a Cepheid, a kind of variable star, in the Andromeda Nebula. When he calculated its distance, he found that this Cepheid was nearly a million light-years away – much further away than anyone had thought. So he inferred that the Andromeda Nebula was not a gaseous cloud inside our galaxy, like so many nebulae, but in fact, a galaxy of stars just like the Milky Way outside our galaxy. Hubble went on to find the distances to many other galaxies, eventually pushing the frontiers of the universe to hundreds of millions of light-years. He also devised a classification system for the various galaxies he observed, sorting them by content, distance, shape, and brightness. In 1929 he showed that galaxies are moving away from us with a speed proportional to their distance. By comparing the distances to the speeds with which the galaxies were racing away, he deduced that the farther away the galaxy, the faster it moved. This relation, known as Hubble's Law, was observational proof of an expanding universe. (☞ Jun. 18, Sep. 28, Dec. 30)

1945
Francis William Aston, British physicist, Nobel laureate (Chemistry, 1922), who invented the mass spectrograph and discovered more than two hundred naturally occurring isotopes, died in Cambridge, UK. (☞ Sep. 1)

1967
India's first indigenously designed and built sounding rocket, *Rohini-75*, was launched from the Thumba Equatorial Rocket Launching Station (TERLS) near Thiruvananthapuram in Kerala. With a diameter of 75 millimetres and weighing just 10 kilograms, the solid-fuelled rocket attained an altitude of 4.2 kilometres.

1974
Rustam Jal Vakil, Indian cardiologist, who popularised the use of reserpine, extracted from the roots of the plant *Rauwolfia serpentina*, died in Mumbai. (☞ Jul. 17)

1981
Bhaskara-II, the second Indian experimental Earth observation satellite, was launched by a Soviet Cosmos-C rocket from Kapustin Yar in Russia. Named

after Bhaskara-II, the famous Indian mathematician of 12th century AD, the 436-kg satellite was placed in a low-Earth orbit ranging in height between 541 and 557 kilometres, and went round the Earth once every 92 minutes. The satellite carried two TV cameras, one working in visible and the other in near-infrared band, and a 3-frequency passive microwave radiometer for remote sensing.

November 21

1783
The first flight of the Montgolfier hot air balloon carrying human passengers, Frenchmen Pilatre de Rozier of the Royal Academy and his friend François Laurent, Marquis d'Arlandes, was made at Bois de Boulogne, on the outskirts of Paris. Among the spectators present were Benjamin Franklin, American scientist, inventor, printer, publisher, and the then US Minister to France. The balloon rose to a height of more than 150 metres and carried the passengers beyond

the old ramparts of Paris, covering a distance of 9 kilometres in about 25 minutes before making a gentle landing. (☞ Jan. 6, Jun. 4, Aug. 26, Sep. 19)

1953
The 40-year-long hoax of the Piltdown Man ended when the British Museum revealed that it was a 'perfectly executed and carefully prepared fraud'. The Piltdown forgery was conceived, planned and executed sometime between 1907 and 1911. In 1912 Charles Dawson, an amateur archaeologist, geologist, and a collector of fossils for the British museum, discovered the first of two skulls found in the Piltdown quarry in Sussex, England, which were claimed to be skulls of an apparently primitive hominid, an ancestor of man. It was named Piltdown Man or *Eoanthropus dawsoni* to use its scientific name, and was a sensation. It was supposed to be the expected 'missing link', a combination of human and ape with the noble brow of *Homo sapiens* and a primitive jaw. As was discovered later, the fake hominid skull was constructed from the remains of a recent human cranium, half the lower jaw of an orang-utan from which telltale parts had been removed and whose teeth had been filed to resemble worn human teeth and a doctored canine tooth, probably from the same lower jaw. In all, thirty-seven pieces of bone and stone were involved – each carefully selected, altered and stained.

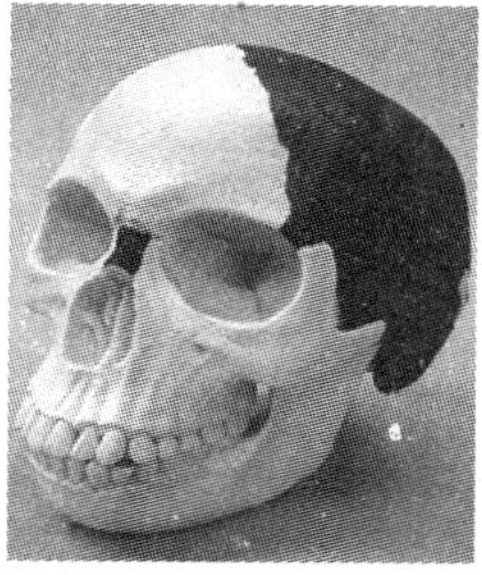

1963
India's first rocket launching station was inaugurated at Thumba near Thiruvananthapuram in Kerala with the launching of an American 2-stage Nike-Apache rocket. The facility, named Thumba Equatorial Rocket Launching Station (TERLS), was dedicated to the United Nations in 1968 and is now an international rocket launching facility for the study of the upper atmosphere near the equator.

1970
Chandrasekhara Venkata Raman, Indian physicist, Nobel laureate (Physics, 1930), who discovered the Raman Effect, died in Bangalore, Karnataka, India. (☞ Nov. 7)

1993
Bruno Rossi, Italian pioneer in the study of cosmic radiation, whose experimental investigations of cosmic rays and their interactions with matter in the 1930s laid the foundation for high-energy particle physics, died in Cambridge, Massachusetts, USA. (☞ Apr. 13)

1996
Abdus Salam, Pakistani nuclear physicist, Nobel laureate (Physics, 1979), who independently formulated a theory explaining the underlying unity of the weak nuclear force and the electromagnetic force, died in Oxford, England. (☞ Jan. 29)

November 22

1904
Louis Eugène Félix Néel, French physicist, Nobel laureate, was born in Lyon, France. Néel studied at the Ecole Normal Supérieure in Paris, where he was appointed lecturer in 1928. In 1932 he obtained his DSc at the University of Strasbourg, where he was appointed professor at the Faculty of Science in 1937. He was professor in the University

of Grenoble since 1945. From 1954 until 1970 he was director of the Institut Polytechnique de Grenoble and of the Ecole Française de Papeterie. In 1970 he was appointed President of the Institut National Polytechnique in Grenoble. Néel began his first research work on magnetism between 1928 and 1939 in Strasbourg. Called up for war service in 1939, he worked on ways to defend the ships of the French fleet against German magnetic mines and invented an effective new method of protection (neutralisation). Néel's research mostly concerned magnetism. Three states of magnetism had long been recognised, di-, para- and ferromagnetism. In 1932 Néel added a fourth – anti-ferromagnetism. He found that for certain crystals, adjacent elementary magnets could align themselves anti-parallel and not parallel as in ferromagnetic materials. Anti-ferromagnetism is an ordered state with important properties. In 1948, Néel made another fundamental discovery with his explanation of the strong magnetism found in the ferrite materials, of which magnetite is one; Néel called such material 'ferrimagnetic.' Néel's work could also explain the behaviour of the new synthetic magnetic materials. These developments have been of considerable technical importance, e.g. in computer memories and in high-frequency techniques. Néel also gave an explanation of the weak magnetism of certain rocks, making possible the study of the past history of the Earth's magnetic field. Néel was awarded the 1970 Nobel Prize for Physics, jointly with the Swedish astrophysicist Hannes Alfvén, for fundamental work and discoveries concerning anti-ferromagnetism and ferrimagnetism, which have led to important applications in solid-state physics. (☞ May 30, Nov. 17)

1907

Asaph Hall, American astronomer, who discovered and named the two moons of Mars and calculated their orbits, died in Annapolis, Maryland, USA. (☞ Oct. 15)

1917

Andrew Fielding Huxley, English physiologist, Nobel laureate, was born in Hampstead, London, England. Huxley was educated at the University College School and the Westminster School. He received his MA from the Trinity College, Cambridge, where later, from 1941 to 1960, he was a fellow and then director of studies, a demonstrator, an assistant director of research, and finally a reader in experimental biophysics in the Department of Physiology. In 1960 he became head of the Department of Physiology at the University College London. Huxley's researches mainly concerned mechanism of nerve conduction. From 1946 to 1951, he worked mostly in collaboration with the English physiologist and biophysicist Alan Hodgkin on nerve conduction, forming their theory on action potentials in nerves – the electrical impulses that enable the activity of an organism to be coordinated by a central nervous system. The experimental measurements on which they based their action potential theory represent one of the earliest applications of a technique of electrophysiology known as the 'voltage clamp'. Working on the nerve cells of the giant Atlantic squid (*Loligo pealei*) implanted with fine electrodes, Huxley and Hodgkin recorded ionic currents, as they would not have been able to do in almost

any other neuron, such cells being too small to study by the techniques of known during their time. On the basis of their findings, they hypothesised the existence of 'ion channels,' which were confirmed decades later. Apart from the researches on nerve conduction, Huxley made contributions of fundamental importance to the knowledge of muscle contraction and developed an interference microscope for studying the striation pattern in isolated muscle fibres. Huxley was awarded the 1963 Nobel Prize for Physiology or Medicine, jointly with Hodgkin and the Australian physiologist John Carew Eccles, for their discoveries concerning the ionic mechanisms involved in excitation and inhibition in the peripheral and central portions of the nerve cell membrane. (☞ Jan. 27, Feb. 5, Dec. 20)

1944
Arthur Stanley Eddington, British astrophysicist, who first explained the internal structure of stars, died in Cambridge, UK. (☞ Dec. 28)

1966
Stephanie Louise Kwolek of Wilmington, Delaware and Paul Winthrop Morgan of West Chester, Pennsylvania, American chemists, working with the DuPont Company were awarded a patent for the high-strength polymer 'Kevlar'. The patent related to a 'novel class of crystalline, linear condensation polyamides and to a manufacturing process thereof'. Kevlar (poly-paraphenylene terephthalamide) is the trademark of the DuPont Company for a synthetic material constructed of para-aramid fibres that the company claims is five times stronger than the same weight of steel, while being lightweight, flexible and comfortable. It is also very heat resistant. Originally intended to replace the steel belts in tires, it is extensively used for making bullet-proof vests.

1981
Hans Adolf Krebs, German-born British biochemist, Nobel laureate (Physiology or Medicine, 1923), who discovered the Krebs cycle, died in Oxford, England. (☞ Aug. 25)

2004

Swift, the first space mission dedicated to gamma ray bursts was launched from the Kennedy Space Center in Florida, USA. Gamma ray bursts (GRB) are the most energetic events in the universe, but they continue to baffle astrophysicists more than 30 years after they were first discovered. The *Swift* mission, which was slated to run for at least two years, is expected to observe at least 100 events per year. To date, Swift has detected over 200 GRBs.

November 23

1826
Johann E. Bode, German astronomer, who confirmed the empirical law that gave a mathematical expression to relate mean distances between the Sun and its planets, died in Berlin, Germany. (☞ Jan. 19)

1837
Johannes Diederik van der Waals, Dutch physicist, Nobel laureate, was born in Leiden, Netherlands. After finishing elementary education at his birthplace, van der Waals became a schoolteacher. He continued studying at the Leiden University in his spare time during 1862–65. In this way he also obtained teaching certificates in mathematics and physics.

In 1864 he was appointed teacher at a secondary school at Deventer; in 1866 he moved to The Hague, first as a teacher and later as director of one of the secondary schools in that town. In 1873 he obtained his doctor's degree for a thesis entitled *Over de Continuïteit van den Gas – en Vloeistoftoestand* ('On the continuity of the gas and liquid state'). In this thesis he put forward an 'Equation of State' (also known as van der Waals equation) embracing both the gaseous and the liquid state. He demonstrated that these two states of aggregation not only merge into each other in a continuous manner, but that they are in fact of the same nature. In 1880, he enunciated the 'Law of Corresponding States', and showed that if pressure is expressed as a simple function of the critical pressure, volume as one of the critical volume, and temperature as one of the critical temperature, a general form of the equation of state is obtained which is applicable to all substances. It was this law which served as a guide during experiments which ultimately led to the liquefaction of hydrogen in 1898 and of helium in 1908. The weak electrostatic attractive forces between molecules and between atoms are called van der Waals forces in his honour. van der Waals was awarded the 1910 Nobel Prize for Physics for his work on the equation of state for gases and liquids. (☞ Mar. 9)

1887

Henry Gwyn Jeffreys Moseley, English physicist who firmly established the relationship between atomic number and the charge of the atomic nucleus, was born in Weymouth, Dorset, England. Mosley graduated from the Trinity College, Oxford, in 1910. He began his research under physicist Ernest Rutherford while serving as lecturer at the University of Manchester and soon devoted himself entirely to research. In 1913, while experimenting with X-ray spectroscopy, Moseley found that the X-ray spectra of the elements were similar but with a deviation that changed regularly through the series. A graph of the square root of the frequency of each radiation against the number representing the element's position in the periodic table gave a straight line. He called this number the 'atomic number' of the element. The equation which describes the relationship between atomic number and wavelength of a spectral line is known as 'Moseley's law'. When the elements are arranged by atomic number instead of atomic mass, problems appearing in the Mendeleyev version based on atomic masses are resolved. The numbering system also enabled Moseley to predict correctly the existence of three missing elements –those with numbers 43, 61, and 75 – in the periodic table, all of which were subsequently identified. (☞ Aug. 10)

1927

Erik Rotheim, Norwegian engineer, patented the first aerosol can and valve that could hold and dispense products and propellant systems. This was the forerunner of the modern aerosol spray can and valve. (☞ Aug. 22)

1937

Jagadis Chandra Bose, Indian plant physiologist and physicist, Fellow of the Royal Society, who was the first to transmit radio signals and measure plant growth, died in Giridih, Bihar, India. (☞ Nov. 30)

1948
Frank Gerard Back, American inventor, was granted a patent for a 'varifocal lens for cameras', an early version of today's zoom lenses used in television and movie cameras for taking close-up shots of distant scenes. Previously, moving into a close-up shot required moving the entire camera toward the actor or object, or away for a long shot. Now the same zoom effects can be produced using the zoom lens without moving the camera.

November 24

1639
Jeremiah Horrocks, English astronomer and clergyman, predicted and observed a transit of Venus, the first one ever observed, from the tiny village of Much Hoole, Lancashire, England. From the observation, he corrected the solar parallax, indicating a much greater distance of the Sun than anyone before him had admitted. (☞ Apr. 8, Jun. 3, Jun. 8)

1859
The Origin of Species by Means of Natural Selection by the British naturalist Charles Darwin was published in England. In this groundbreaking book Darwin argued that species are the result of a gradual biological evolution in which nature encourages, through natural selection, the propagation of those species best suited to their environments. This book is unquestionably one of the most influential in the history of science. (☞ Feb. 12, Apr. 19)

1874
Joseph F. Glidden of De Kalb, Illinois, American inventor was granted a patent for an improved design of barbed wire for fencing. It was primarily meant for keeping

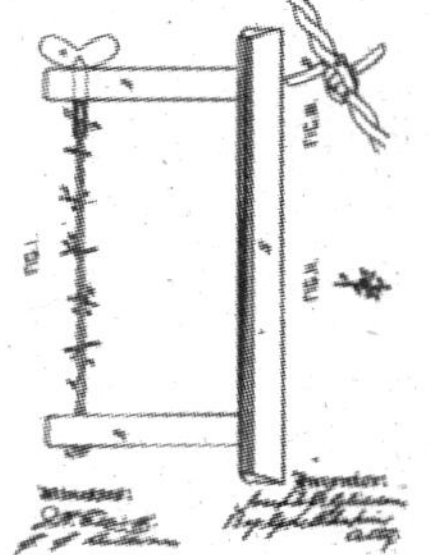

Detail from Glidden's patent

livestock out of arable farmlands. Glidden fashioned barbs on an improvised coffee bean grinder, placed them at intervals along a smooth wire, and twisted another wire around the first to hold the barbs in a fixed position. The advent of Glidden's successful invention set off a creative frenzy that eventually produced over 570 barbed wire patents. It also set the stage for a three-year legal battle over the rights to these patents. When the legal battles were over, Glidden was declared the winner and the inventor of barbed wire. He established the Barb Fence Company at De Kalb in Illinois and went on to become one of the richest men in the United States. Barbed wire was used for the first time in war in 1898 in Cuba.

1925
Simon van der Meer, Dutch engineer and physicist, Nobel laureate, was born in The Hague, Netherlands. Van der Meer passed out from the Gymnasium in The Hague in 1943 but could not enter college immediately because of the German occupation of Netherlands. From 1945 onwards, he studied at the University of Technology, Delft, specialising in measurement and regulation technology. After obtaining his engineering degree in 1952, he worked in the Philips Research Laboratory, Eindhoven, mainly on high-voltage equipment and electronics for electron microscopes. In 1956 he moved to Geneva, Switzerland to join the recently

founded European Organization for Nuclear Research (CERN), where he has been working ever since on many different projects. At CERN, van der Meer improved the design of particle accelerators used to produce collisions between beams of subatomic particles. He invented a device that would monitor and adjust the particle beam with correcting magnetic fields by a system of 'kickers' placed around the accelerator ring. These improvements in accelerator design ultimately led to the discovery of the short-lived sub-atomic particles W and Z, carriers of weak interaction. The discovery of the W and Z particles was crucial to the unified electroweak theory postulated in the 1970s by the American physicists Steven Weinberg and Sheldon Glashow, and the Pakistani physicist Abdus Salam. Van der Meer was awarded the 1984 Nobel Prize for Physics, jointly with the Italian physicist Carlo Rubbia, for their decisive contributions to the large project which led to the discovery of the field particles W and Z, communicators of weak interaction. (☞ Mar. 31)

1969
The command module of *Apollo-12* carrying astronauts Charles Conrad, Jr., Richard F. Gordon, and Alan L. Bean splashed down safely in the Pacific Ocean. (☞ Nov. 14, Nov. 19)

November 25

1844
Karl Benz, German automobile engineer, was born in Karlsruhe, Germany. Benz attended high school before going on to study at a technical college in Karlsruhe. He then underwent a two-year traineeship at a mechanical engineering company in the same city. After completing his training he joined a scales manufacturing factory in Mannheim as a draftsman and designer. When he lost his job in 1868 he moved to a bridge construction company. In 1871, the technically minded Benz founded his first company in Mannheim and started work on two-stroke engines. In 1883 he founded his company Benz & Co. and started work on a car engine. In 1886 he was awarded the first patent for a petrol-driven vehicle – a three-wheeler. He built his first four-wheeled car in 1891. At the turn of the century, Benz & Co. had grown into the world's leading automotive manufacturer. In 1903 Benz announced his retirement from active work within the company, but continued in an advisory role till his death. (☞ Jan. 29, Apr. 4)

1867
Alfred Nobel, the Swedish chemist and inventor, patented dynamite. Nobel had produced dynamite as a safe and manageable form of nitroglycerine. (☞ Jul. 14, Oct. 21, Dec. 10)

1926
Tsung-Dao Lee, Chinese-born American physicist, Nobel laureate, was born in Shanghai, China. Lee was educated at the Kiangsi Middle School in Kanchow, Kiangsi, from which he graduated in 1943. He did his matriculation at the National Chekiang University in Kweichow province. In 1946, he went to the University of Chicago, USA, on a Chinese Government Scholarship to do his PhD, which he completed in 1950

and then served as research associate at the Yerkes Astronomical Observatory, Lake Geneva, Wisconsin. From 1950 to 1951, Lee was a research associate and lecturer at the University of California in Berkeley, and then accepted a fellowship at the Institute of Advanced Study at Princeton, New Jersey where he worked with the Chinese-born American physicist Chen Ning Yang on parity conservation. Lee was appointed assistant professor of physics at the Columbia University in 1953. In 1956 Lee and Yang concluded that the theta-meson and tau-meson, previously thought to be different because they decayed by modes of differing parity, were in fact the same particle (now called the K-meson). Because the law of parity conservation prohibits a single particle from having decay modes exhibiting opposite parity, the only possible conclusion was that for weak interactions, at least, parity is not conserved. They suggested experiments to test their hypothesis, and in 1956–57 the Chinese-born American physicist Chien-Shiung Wu, working at the Columbia University, experimentally confirmed their theoretical conclusions. Lee was awarded the 1957 Nobel Prize for Physics, jointly with Yang, for their penetrating investigation of the so-called parity laws, which has led to important discoveries regarding the elementary particles. (☞May 29, Sep. 22)

1961

The world's first nuclear-powered aircraft carrier, *USS Enterprise* was commissioned with the US Navy. (☞ Sep. 24)

1975

Robert S. Ledley, American inventor, was issued the first US patent for a whole-body X-ray scanner. Then known as the automatic computerised transverse axial (ACTA) diagnostic X-ray scanner, the machine could make a three-dimensional analysis of all organs and parts of the body in a series of cross-section images using thin X-ray beams and high power computer processing of the collected data. The ACTA scanner revolutionised medical diagnosis and set the fundamental design for modern CT scanners. It made the diagnosis of tumours, infection or bleeding even deep within large organs possible.

1984

William Schroeder, retired American Government quality-control inspector, was implanted with a *Jarvik-7* artificial heart at Humana Audubon Hospital in Louisville, Kentucky, USA. He was the second recipient of the *Jarvik-7* heart and lived a record 620 days. (☞ Aug. 18, Nov. 13)

November 26

1801

Discovery of the element columbium (Cb, at. no. 41) was announced in the Royal Society by the English chemist Charles Hatchett. He had found it while analysing a piece of a black rock called columbite from the collection of the British Museum in London. Forty years later a German chemist, Heinrich Rose, re-discovered the element and named it niobium (Nb), its present name.

1876

Willis Haviland Carrier, American engineer and inventor of modern air conditioning, was born in Angola, New York, USA. Carrier loved mathematics, and studied it whenever he was not inventing his own devices. In 1895 he received a scholarship to the Cornell University and graduated in 1901 with a degree in mechanical engineering. After leaving college, he joined the Buffalo Forge Company, a company which manufactured heaters, blowers and air exhaust systems. He worked in their heating engineering department designing heating systems to dry lumber and coffee. Carrier soon developed a better way to measure the capacity of heating systems and was named director of the company's experimental engineering department. At the age of 25, he devised his first important invention, a system to control heat and humidity for a printing firm in Brooklyn, which had been unable to print reliable colours at times because of the effects of heat and humidity on paper and ink. By controlling humidity as well as temperature, Carrier invented air conditioning as we know it today. In 1906 he received a patent for his method. He went on to work on other cooling and humidity-control inventions, and was eventually made the head of the Buffalo Forge subsidiary named Carrier Air Conditioning Company in his honour. His development of the first safe, low pressure centrifugal refrigeration machine using non-toxic, non-flammable refrigerant marked the beginning of the era of comfort cooling. (☞ Jan. 2, Oct. 9)

1885

Debendra Mohan Bose, Indian physicist, Foundation Fellow, Indian National Science Academy, was born in Calcutta (now Kolkata). Bose did his BSc in 1905 and MA in 1906 from the Calcutta University, and his PhD from the Berlin University, Germany, in 1919. He was Rash Behari Ghosh Professor of Physics from 1914 to 1935, and Palit Professor of Physics from 1935 to 1938 at the Calcutta University. He succeeded Jagadis Chandra Bose as Director of the Bose Institute, Kolkata, after the latter's death in 1938. Bose's major contribution was in the use of cloud chamber and photographic emulsion technique for the study of cosmic rays. He also carried out plant physiological investigations. He was the editor-in-chief of the *Indian Journal of History of Science*. (☞ Jun. 2)

1894

Norbert Wiener, American mathematician, who established the science of cybernetics, was born in Columbia, Missouri, USA. Wiener was a precocious child and finished his schooling at an early age. At the age of eleven he entered the Tufts College in Medford, Massachusetts. In 1909 Wiener graduated from Tufts at the age of fourteen and entered the Harvard University, Cambridge, Massachusetts, to begin graduate studies in zoology. He completed his PhD at Harvard in 1913 with a dissertation on mathematical logic. From Harvard, Wiener went to Cambridge, England, to study under the English logician and philosopher Bertrand Russell. He also attended courses by the English mathematician G. H. Hardy. In 1914 he went to Göttingen, Germany to study differential equations under the German mathematician David Hilbert. During the 1920s Wiener did highly innovative and fundamental work on

what are now called stochastic processes and, in particular, on the theory of Brownian motion. Wiener had an extraordinarily wide range of interests and contributed to many areas including communication theory, cybernetics (a term he coined), and quantum theory. During World War II he worked on gunfire control. In 1948 his book *Cybernetics; or, Control and Communication in the Animal and the Machine* was published. Wiener's other works include *Nonlinear Problems in Random Theory,* and *God and Golem, Inc.: A Comment on Certain Points Where Cybernetics Impinges on Religion.* (☞ Mar. 18)

1896

Karl Ziegler, German chemist, Nobel laureate, was born in Helsa, near Kassel, Germany. Ziegler graduated from the University of Marburg in 1920, and qualified as a lecturer in 1923. After working for a short period at the University of Frankfurt am Main, he spent 10 years as a lecturer at the Heidelberg University. In 1936 he became professor and director of the Chemisches Institut at the University of Halle. In 1943 he became director of the Kaiser Wilhelm (later Max Planck) Institute for Coal Research at Mülheim an der Ruhr. Ziegler's research works include syntheses and reactions in the chemistry of organoaluminium compounds, and discovery of organometallic mixed catalysts for the polymerisation of olefins. His researches with lithium in organic chemistry produced compounds more reactive than the Grignard reagents on which they were modelled. His work on cyclic carbon compounds found use in synthesising the aroma of musk for perfumes. The catalyst derived from aluminium alkyl and titanium tetrachloride proved especially useful. It formed the basis of nearly all later developments in the production of long-chain polymers of hydrocarbons from such olefins as ethylene and butadiene; the resulting products came into widespread use as plastics, fibres, rubbers, and films. The catalyst developed by Ziegler and the Italian chemist Giulio Natta for the synthesis of polyolefins is known as the Ziegler–Natta catalyst. Ziegler was awarded the 1963 Nobel Prize for Chemistry, jointly with Natta, for their discoveries in the field of the chemistry and technology of high polymers. (☞ Feb. 26, Aug. 12)

1950

The National Metallurgical Laboratory (NML), a constituent national laboratory of the Council of Scientific and Industrial Research (CSIR), was inaugurated. Located in Jamshedpur, Jharkhand, NML carries out R&D in the areas of minerals, metals, and materials including metallurgical wastes and pollutants. The laboratory has been rendering its expertise to industries in the areas of materials processing and engineering; ferrous and non-ferrous metal extraction, metal casting and forming; materials characterisation and evaluation; refractories and advanced ceramics; corrosion protection of metals/materials; process modelling and simulation; pollution mitigation; and project and process engineering. NML has developed more than 150 processes relating to gainful utilisation of minerals and waste materials.

1966

The world's first tidal power station was opened at Rance estuary, in Brittany, France. The plant uses reversible turbines to be able to operate regardless of the direction of water flow. La Rance is the most powerful tidal power plant in the world and each year it generates 500 million kWh of electricity.

1987

The Centre for Cellular and Molecular Biology (CCMB), a constituent national laboratory of the Council of Scientific and Industrial Research (CSIR), was dedicated to the nation. Located in Hyderabad, Andhra Pradesh, CCMB carries out research in three major categories – high quality basic research in the frontier areas of modern biology; research relevant to societal needs; and application-oriented research towards commercialisation. These include the areas of biomedicine and diagnostics; evolution and development; gene regulation in prokaryotes and eukaryotes; host–parasite interactions; membrane biology; protein structure; bioinformatics; functional genomics; and theoretical biology. CCMB has also taken lead in the dissemination of modern biological information through popularisation of science, science education in schools, and has been a meeting point for art and science. In recognition of its contribution to modern biology, CCMB has been chosen as a Centre of Excellence by UNESCO Global Network for Molecular and Cell Biology and has been designated as a South Centre for Excellence for Research and Training by the Third World Academy of Sciences (TWAS), Italy.

November 27

1701

Anders Celsius, Swedish astronomer, physicist and mathematician, who developed the centigrade temperature scale, was born in Uppsala, Sweden. In 1730 he succeeded his father as professor of astronomy at the Uppsala University. He built Sweden's first observatory in Uppsala in 1741. He described his 'fixed scale' thermometer in a paper read before the Swedish Academy of Sciences in 1742. His centigrade scale for measuring temperature defines zero degrees as the temperature at which water freezes, and 100 degrees as the temperature at which water boils. This scale, an inverted form of Celsius' original design, was adopted as the standard and is still used in almost all scientific work. (☞ Apr. 24)

1857

Charles Scott Sherrington, English physiologist, Nobel laureate, was born in London, England. Sherrington began medical studies at the St. Thomas's Hospital in 1876. In 1878 he passed the primary examination of the Royal College of Surgeons, and a year later the primary examination for the Fellowship of that College. He received his BA at Gonville and Caius College, Cambridge, graduating in 1883. In 1884 he obtained his MRCS, and in 1885 a first class in the Natural Sciences Tripos at Cambridge with distinction. In 1887 Sherrington joined the faculty of the St Thomas' Medical School. He joined the Oxford University as professor of physiology in 1913. Sherrington retired from Oxford in 1935, but continued to lecture and write. He served as president of the Royal Society from 1920 to 1925. Sherrington's researches mainly concerned the functioning of nerves. He used reflexes in the spinal cord as a way of

investigating the general properties of neurons and the nervous system. Working with cats, dogs, monkeys, and apes that had been deprived of their cerebral hemispheres, he found that reflexes must be regarded as integrated activities of the total organism, not as the result of the activities of isolated 'reflex arcs', a notion that was accepted at that time. Sherrington also coined the terms 'neuron' and 'synapse' to denote the nerve cell and the point at which the nervous impulse is transmitted from one nerve cell to another, respectively. Sherrington was awarded the 1932 Nobel Prize for Physiology or Medicine, jointly with the English electrophysiologist Edgar Douglas Adrian, for their discoveries regarding the functions of neurons. (☞ Mar. 4, Nov. 30)

1903

Lars Onsager, Norwegian-born American chemist, Nobel laureate, was born in Kristiania (now Oslo), Norway. Onsager finished his schooling at the Frogner School in Oslo in 1920. He went on to study chemical engineering at the Norwegian Institute of Technology in Trondheim, from where he graduated in 1925. His early work in statistical mechanics attracted the attention of the Dutch chemist Peter Debye, under whose direction Onsager studied at the Federal Institute of Technology, Zürich, Switzerland. He then went to USA and taught at the Johns Hopkins University, Baltimore, and Brown University, Providence, Rhode Island. While at the Brown University, Onsager made significant contributions to statistical mechanics and thermodynamics. His work was mainly concerned with the effects on diffusion of temperature gradients. He produced the Onsager reciprocal relations, a set of equations published in 1929 and, in an expanded form, in 1931, in statistical mechanics. In 1925 he modified the Debye–Hückel theory of electrolytic dissociation, which describes the motions of ions in solution, to take into account Brownian movement. Onsager was also able to formulate a general mathematical expression about the behaviour of non-reversible chemical processes that has been described as the 'fourth law of thermodynamics'. He received his PhD from the Yale University, New Haven, Connecticut, in 1935. He joined the faculty of Yale in 1933 and became professor of theoretical chemistry there in 1945. Onsager was awarded the 1968 Nobel Prize for Chemistry for the discovery of the reciprocal relations bearing his name, which are fundamental for the thermodynamics of irreversible processes. (☞ Oct. 5)

1971

Soviet unmanned spacecraft *Mars-2* went into orbit around Mars and released a descent module carrying a tiny rover. Although the lander crashed on the surface after its braking rockets failed and returned no data, it became the first human-made object to reach the surface of Mars. The *Mars-2* orbiter successfully completed its mission and returned useful data to Earth. (☞ May 19)

November 28

1893

Alexander Cunningham, British army officer and archaeologist who excavated many sites in India, including Sarnath and Sanchi, and served as the first director of the Indian Archaeological Survey, died in London, England. (☞ Jan. 23)

1950
Russell Alan Hulse, American physicist, Nobel laureate, was born in New York, USA. Hulse showed an interest in science at a very early age, playing with chemistry sets, mechanical engineering construction sets, biology dissection kits, photography, telescopes and electronics over the years. After attending the Bronx High School of Science in New York, Hulse received his bachelor's degree in physics from the Cooper Union College in 1970 and then started graduate school at the University of Massachusetts in Amherst. In 1974, using the large radio telescope at Arecibo, Puerto Rico, Hulse and his research guide, astrophysicist Joseph H. Taylor, discovered dozens of pulsars – rapidly spinning neutron stars that emit rapid, regular bursts of radio waves. Irregularities in the radio emissions of the pulsar PSR 1913+16 led them to deduce that it was a binary pulsar, with a companion neutron star with which it was locked in a tight orbit. The discovery provided the first means of detecting gravity waves, predicted by Albert Einstein in his theory of relativity. Hulse and Taylor found that the stars were rotating ever faster around each other in an increasingly tight orbit, presumably because they were losing energy in the form of gravity waves. After completing his PhD in 1975, Hulse had a post-doctoral appointment at the National Radio Astronomy Observatory in Charlottesville, Virginia from 1975 to 1977 and then joined the Plasma Physics Laboratory at the Princeton University

where he conducted research associated with the Tokamak Fusion Test Reactor, an experimental nuclear-fusion facility. Hulse was awarded the 1993 Nobel Prize for Physics, jointly with Taylor, for the discovery of a new type of pulsar, a discovery that has opened up new possibilities for the study of gravitation. (☞ Mar. 24)

1954
Enrico Fermi, Italian-born American physicist who was one of the chief architects of the nuclear age, died in Chicago, Illinois, USA. (☞ Sep. 29)

1964
The American unmanned spacecraft *Mariner-4* was launched from the Kennedy Space Center, Florida, USA, towards Mars.

It was the first spacecraft to send close-up photographs of Mars. (☞ Jul. 14)

1967
The first pulsating radio source, or 'pulsar', was detected by an Irish graduate assistant, Jocelyn Bell, then working under the direction of the British astrophysicist Antony Hewish at the Mullard Radio Astronomy Observatory, Cambridge, England, using a special radio telescope and a large array of 2,048 aerials covering an area of 4.4 acres. These fascinating objects turned out to be fast spinning neutron stars that emitted radio waves like a beacon. The rotational period of pulsars ranges from 1.57 milliseconds to 5.1 seconds. (☞ Feb. 24)

1983
American space shuttle *Columbia*, carrying for the first time a six-person crew of astronauts – John W. Young, Brewster

H. Shaw, Owen K. Garriott, Robert A. Parker, Byron K. Lichtenberg, and Ulf Merbold, was launched on its sixth flight from the Kennedy Space Center, Florida, USA. *Columbia* carried the first Spacelab mission, *Spacelab-1*, jointly sponsored by the European Space Agency and the American space agency NASA to conduct investigations which demonstrate capability for advanced research in space. It is an orbital laboratory and observation platform composed of cylindrical pressurised modules and U-shaped unpressurised pallets, which remained in the orbiter's cargo bay during flight. Altogether 73 separate investigations were carried out in astronomy and physics, atmospheric physics, Earth observations, life sciences, materials sciences, space plasma physics and technology. (☞ Feb. 1, Apr. 12, Dec. 8)

November 29

1803

Christian Doppler, Austrian physicist who first described how the observed frequency of sound waves is affected by the relative motion of the source and the detector, was born in Salzburg, Austria. Doppler attended primary school in Salzburg and then secondary school in Linz. Doppler joined the Vienna Polytechnic Institute in 1822 to study mathematics. He excelled in his studies and graduated in 1825. In 1842, Doppler developed an equation about frequency and relative movement of sound source and observer. To test his theories, he conducted a unique experiment.

For two days, Doppler had a train pull repeatedly, at different speeds, a freight car with trumpeters playing on top of it. He then had a musician capable of understanding the differences in sounds record the pitch of the notes played as the train moved closer or further away. The observed effect was just as predicted by Doppler's theory. The pitch rose as the train moved nearer and fell as it moved away, the effect known as Doppler Effect. The Doppler Effect, as applied to light, was used to confirm the expansion of the universe and is a valuable tool of astronomy. (☞ Mar. 17)

1874

António Egas Moniz, Portugese neurologist, Nobel laureate, was born in Avanca, Portugal. Moniz received his early education from his uncle before joining the faculty of medicine at the Coimbra University. He received further education at Bordeaux and Paris in France and became professor at the Coimbra University in 1902. In 1911 he moved to the University of Lisbon as its first professor of neurology and remained there until his death. At Lisbon, Moniz introduced and developed cerebral angiography (arteriography), a method of making visible the blood vessels of the brain by injecting into the carotid artery substances that are opaque to X-rays. Moniz's most notable contribution was his development of a surgical procedure known as 'pre-frontal lobotomy', which involves the severing of the nerve fibres between the frontal lobes and the thalamus in the brain to treat certain psychological disorders. He had observed that certain psychoses, particularly schizophrenia and severe paranoia, involve recurrent

thought patterns that dominate normal psychological processes. The frontal lobes were known to be closely associated with psychological responses. So Moniz reasoned that severing the nerve fibres between the frontal lobes and the thalamus might force a transformation of existing thought patterns into more normal ones. Although lobotomy was initially found to be quite effective in treating persons suffering from apparently incurable psychoses, the procedure is now known to have serious side-effects, and has been largely replaced by the use of tranquilising drugs. Moniz was awarded the 1949 Nobel Prize for Physiology or Medicine, jointly with the Swiss physiologist Walter Rudolf Hess, for his discovery of the therapeutic value of leucotomy in certain psychoses. (☞ Mar. 17, Nov. 12, Dec. 13)

1936
Yuan Tseh Lee, Taiwan-born American chemist, Nobel laureate, was born in Hsinchu, Taiwan. Lee had his elementary and secondary education in Hsinchu and obtained his BS degree from the National Taiwan University, Taipei in 1959. He received his master's degree from National Tsinghua University and later carried out the X-ray structure determination of tricyclopentadienyl samarium at the same university. He received his PhD in 1965 from the University of California at Berkeley, California, USA. He did post-doctoral work with the American chemist Dudley R. Herschbach at the Harvard University, Cambridge, Massachusetts, where he invented what is known as the 'crossed molecular beam technique', a technique in which beams of molecules are brought together at supersonic speed under carefully controlled conditions. This procedure enabled a detailed, molecule-by-molecule examination of a chemical reaction event. From 1968, he taught at the University of Chicago, Illinois. In 1974 he returned to Berkeley as professor of chemistry and principal investigator at the Lawrence Berkeley Laboratory of the University of California. Lee was awarded the 1986 Nobel Prize for Chemistry, jointly with Herschbach and the German chemist John C. Polanyi, for their contributions concerning the dynamics of chemical elementary processes. (☞ Jan. 23, Jun. 18)

November 30

1694
Marcello Malpighi, Italian physician and biologist who founded the science of microscopic anatomy, died in Rome, Italy. (☞ Mar. 10)

1850
Germain Henri Hess, Swiss-born Russian chemist whose studies of heat in chemical reactions formed the foundation of thermochemistry, died in St. Petersburg, Russia. (☞ Aug. 7)

1858
Jagadis Chandra Bose, Indian plant physiologist and physicist, Fellow of the Royal Society, was born in Myemsingh, Bengal, India (now in Bangladesh). Till the age of 11, Bose studied in a local school in his native language, Bengali. In 1869, he was sent to Calcutta (now Kolkata) to learn English and was educated at the St.Xavier's School and College, passing his BA in physical sciences in 1879. In 1880 Bose went to London, where he joined a medical

school, but had to give up the study of medicine due to ill health. He then joined the Christ's College, Cambridge. Bose passed the Tripos examination with distinction. In 1884, he was awarded a BA degree from Cambridge and next year a BSc degree from the London University. After returning to India in 1885, Bose became professor of physical science at the Presidency College, Calcutta. When Bose first joined the Presidency College, there was no laboratory worth the name. However, he went ahead with his research in a makeshift laboratory. The experiments performed in the makeshift laboratory finally resulted in the invention of a device for producing electromagnetic waves. In November 1894, Bose gave the first public demonstration of wireless transmission using electromagnetic waves to ring a bell and to explode a small charge of gunpowder from a distance. His experiments on the quasi-optical properties of very short radio waves in 1895 led him to make improvements on the coherer, an early form of radio detector, which enabled the Italian inventor Guglielmo Marconi to transmit radio signals across the Atlantic in December 1901. In 1896, at a meeting of the British Association of Advancement of Science, in Liverpool, England, Bose demonstrated his apparatus for the study of the properties of electric waves. In January 1897, Bose delivered at the Royal Institution, London, the famed 'Evening Discourse' on 'the electromagnetic radiation and the polarisation of the electric ray'. The University of London conferred on him the DSc degree for his work on electric waves. During 1897–1900, Bose turned his interest to comparative physiology, plant physiology in particular. He performed delicate experiments on plant life and invented recorders of high magnification and great precision. His invention of highly sensitive instruments for the detection of minute responses by living organisms to external stimuli enabled him to anticipate the parallelism between animal and plant tissues noted by later biophysicists. Bose left the Presidency College in 1917 to found the Bose Research Institute, Calcutta, and remained its director till his death. (☞ Jan. 29, Nov. 23)

1869

Nils Gustaf Dalén, Swedish engineer, Nobel laureate, was born in Stenstorp, Sweden. Dalén gradua-ted as an engineer from the Chalmers Institute at Gothenburg in 1896 and spent a year in Switzerland. On his return to Sweden, he carried out research at Gothenburg and worked as a consulting engineer. In 1909, he became managing director of the company Svenska Aktiebolaget Gasaccumulator (AGA). His subsequent invention of the sun-valve, which causes a beacon to light automatically at dusk and extinguish itself at dawn, enabled lighthouses to function perfectly and unattended for periods of up to a year. His invention of a cylinder filled with a porous mass of asbestos and diatomaceous earth for storage of acetylene considerably reduced the hazards in handling this material and its use in welding became safe. He also invented a mixer for providing a constant and correct balance of gas and air for use in the incandescent mantle and a device for removing broken mantles and replacing them by new ones. Dalén was awarded the 1912 Nobel Prize for Physics for his invention of automatic regulators for use in

conjunction with gas accumulators for illuminating lighthouses and buoys. (☞ Dec. 9)

1889

Edgar Douglas Adrian, English electro-physiologist, Nobel laureate, was born in London, UK. Adrian was educated at the Westminster School, London, and in 1908 went to the Trinity College, Cambridge, to study physiology and the other subjects of the Natural Sciences Tripos. He took his BA degree in 1911 and obtained his medical degree from the Trinity College in 1915. After working for a time on clinical neurology at the St. Bartholomew's Hospital, London, he returned to Cambridge in 1919, to lecture on the nervous system. He was elected Fellow of the Royal Society in 1923. In 1925 he began investigating sense organs by electrical methods. He developed the capillary electrometer to measure minute signals from nerve fibres and discovered the nature of coding of the motor and sensory impulses in afferent and efferent fibres. With this apparatus he was able to record the electrical discharges in single nerve fibres, which were produced by tension on the muscle, pressure on it, touch, the movement of a hair and pricking with a needle. By 1928 he was able to publish his conclusion that a stimulus of constant intensity applied to the skin immediately excites the end organ, but that this excitation progressively decreases for as long as the stimulation continues. In the early 1930s, Adrian studied the electrical activity of the brain; his work on the variations and abnormalities of the changes opened new fields of investigation in epilepsy and in the location of cerebral lesions. Adrian was awarded the 1932 Nobel Prize for Physiology or Medicine, jointly with the English physiologist Charles Scott Sherrington, for their discoveries regarding the functions of neurons. (☞ Aug. 4, Nov. 27)

1915

Henry Taube, Canadian-born American chemist, Nobel laureate, was born in Neudorf, Saskatchewan, Canada. Taube received his BS degree in 1935 and his MS degree in 1937 from the University of Saskatchewan. He received his PhD at the University of California, Berkeley, USA, in 1940. He taught at the Cornell University in Ithaca, New York, and at the University of Chicago, before joining the faculty of the Stanford University, California, in 1961. He became a US citizen in 1942. Taube's research mainly concerned the behaviour of metal ions in water solution. In the late 1940s, he carried out experiments with isotopes to show that in water solution, the ions of metals form chemical bonds with several molecules of water forming hydrates or coordination compounds, the stability and geometric arrangement of which varied widely depending on the metal and oxidation state of the ion. He also devised an interpretation of their properties in terms of their electronic configurations and helped develop techniques for studying such substances. Taube's findings have been applied in selecting metallic compounds for use as catalysts, pigments, and superconductors and in understanding the function of metal ions as constituents of certain enzymes. Taube was awarded the 1983 Nobel Prize for Chemistry for his work on the mechanisms of electron transfer reactions, especially in metal complexes.

1917
The Bose Institute was inaugurated in Calcutta (now Kolkata), West Bengal. The institute was founded by the Indian physicist Jagadis Chandra Bose, to provide a good laboratory with proper infrastructure and equipment for Indian scientists to conduct research freely and undisturbed. The institute's main objective is the advancement of science and diffusion of knowledge. At the time of its inception, care was taken to ensure that its scope was not confined to any one area of science but covered all its cognate branches. Consequently, research undertaken was made increasingly more broad-based, with emphasis on finding solutions to problems in agriculture, industry and medicine. Today the institute has departments of physics, chemistry, botany, microbiology, biochemistry, biophysics, animal physiology, environmental science, immunotechnology, and plant molecular and cellular genetics, and animal physiology. The institute also houses a museum which attracts people from within the country and abroad. Most of the instruments invented by Bose are on display here in their original form. (☞ Nov. 30)

1926
Andrew V. Schally, Polish-born American endocrinologist, Nobel laureate, was born in Wilno, Poland. Schally received his high school diploma in Scotland in 1946 and afterwards studied chemistry in London. He worked for three years at the National Institute for Medical Research in London before travelling to Montreal, Quebec, Canada, to enter the McGill University, from where he graduated in 1955. He received a PhD in biochemistry from McGill in 1957 and then moved to the Baylor University in Houston, Texas, USA, to work on corticotropin-releasing factors. In 1962 he joined the Veterans Administration Medical Center and the Department of Medicine at the Tulane University School of Medicine in New Orleans, Louisiana, where he became professor in 1967. Schally's researches mainly concerned the pituitary hormones. Towards the end of the 1950s, Schally and his colleagues extracted from the hypothalamus of sheep and pigs some compounds which, when administered to pituitary tissue, brought about a release of its hormones. They isolated, analysed, or synthesised several hormones including ACTH (adrenocorticotrophic hormone), TSH (thyroid stimulating hormone), and LH and FSH (the gonadotrophic hormones), which were termed 'releasing factors' (RF) or 'releasing hormones' (RH). The hormone inducing the release of TSH was thus called TSH-RF or TRF. Schally's chief accomplishments were the synthesis of TRH (thyrotropin-releasing hormone), the isolation and synthesis of LH-RH (luteinising hormone-releasing hormone), and studies of the action of somatostatin. Schally was awarded the 1977 Nobel Prize for Physiology or Medicine, jointly with the French-born American physiologist Roger Guillemin and the American medical physicist Rosalyn S. Yalow. Schally and Guillemin shared half of the prize money for their discoveries concerning the peptide hormone production of the brain, while Yalow received the other half for the development of radioimmunoassays of peptide hormones. (☞ Jan. 11, Jul. 19)

December 1

1743
Martin Heinrich Klaproth, German chemist, was born at Wernigerode, Germany. Klaproth studied pharmacy in Quedlinburg, Hamburg, Berlin, and Danzig. In 1780 he started his own establishment in Berlin. In 1787 he was appointed lecturer in chemistry to the Royal Artillery, and in 1810 was appointed professor of chemistry at the newly founded University of Berlin. Klaproth did much to improve and systematise the processes of analytical chemistry and mineralogy. He was the first to discover uranium, zirconium and cerium, and to characterise them as distinct elements, though he did not obtain any of them in the pure metallic state. He re-discovered titanium in 1795, about four years after its initial discovery and named it. He also elucidated the composition of numerous substances which were till then imperfectly known, including compounds of the then newly recognised elements tellurium, strontium, cerium and chromium. (☞ Jan. 1)

1866
George Everest, British surveyor and geographer, Fellow of the Royal Society, after whom Mount Everest, the highest mountain peak in the world is named, died in London, England. (☞ Jul. 4)

1898
Valdemar Poulsen, Danish telephone engineer, was issued a patent for the first working magnetic recorder called the telegraphone with wire wrapped around a drum and a recording/playback head that moved by a screw thread on top. Later, Poulsen and his partner discovered that the application of a direct current to the recording head, called dc bias, improved the sound quality on a steel tape version of the telegraphone. This modified edition was exhibited at the Paris Fair in 1900.

1913
Henry Ford, the American industrialist, introduced the continuous moving assembly line in his factory at Detroit, Michigan. It was capable of delivering a car every 2 min. 38 sec. By the summer of 1913, three subassemblies (magnetos, motors, and transmissions) were assembled on moving lines using conveyor belts. The method was so successful that the Ford Motor Co. became the world's largest car manufacturer. Businessmen the world over flocked to Detroit to glean the secrets of the moving assembly line. (☞ Jul. 30, Oct. 1)

1925
Martin Rodbell, American biochemist, Nobel laureate, was born in Baltimore, Maryland, USA. Rodbell matriculated from the Johns Hopkins University at Baltimore in 1943 but was soon drafted into the US Navy. He served with the Navy in the South Pacific, the Philippines, Korea, and China before returning to the Johns Hopkins University and earning his BA degree in biology in 1949. He received

his PhD degree from the University of Washington at Seattle in 1954. Rodbell began his career as a biochemist at the National Institutes of Health in Bethesda, Maryland in 1956, and moved to the National Institute of Arthritis and Metabolic Diseases (NIAMD) in 1961 where his research on hormone-sensitive fat cells led to his landmark theory of transducers as mediators of hormone action. Prior to Rodbell's award-winning research, scientists believed that only two substances – a hormone receptor and an interior cell enzyme – were responsible for cellular communication. In 1970, Rodbell discovered that signal transmission requires a small intracellular molecule called guanosine triphosphate (GTP) and that the breakdown of GTP to guanosine diphosphate (GDP) was an integral regulatory step in the control of hormone action. GTP was subsequently found to bind to G-proteins, a family of proteins that serve as intermediaries between incoming signalling molecules, such as hormones and drugs, and the cellular proteins that respond to the signals. Rodbell was awarded the 1994 Nobel Prize for Physiology or Medicine, jointly with the American pharmacologist Alfred G. Gilman, for their discovery of G-proteins and the role of these proteins in signal transduction in cells. (☞ Jul.1, Dec. 7)

1964
John Burdon Sanderson Haldane, British-born Indian geneticist and biometrician who opened new paths of research in population genetics and evolution, died in Bhubaneshwar, Orissa, India. (☞ Nov. 5)

1990
British and French workers digging the Channel Tunnel between the United Kingdom and France finally met in the service tunnel after knocking out a passage large enough to walk through and shake hands. The 38-km undersea rail tunnel beneath the English Channel linked the two countries for the first time in 8,000 years. The tunnelling began three years earlier, on the same day in 1987.

December 2

1885
George Richards Minot, American physician, Nobel laureate, was born in Boston, Massachusetts, USA. Minot studied medicine at the Harvard University, Cambridge, Massachusetts, and received his MD in 1912 and an honorary degree of ScD in 1928. He did his hospital training at the Massachusetts General Hospital and then worked at the Johns Hopkins Hospital and Medical School at Baltimore, Maryland. He was appointed assistant in medicine at the Harvard Medical School and the Massachusetts General Hospital. In 1928 he was elected professor of medicine at the Harvard University, director of the Thorndike Memorial Laboratory and visiting physician to the Boston City Hospital. When he was a medical student, Minot became interested in the disorders of the blood. He studied the coagulation of blood, blood transfusion, blood platelets and reticulocytes as well as certain blood disorders. He described an atypical familial haemorrhagic condition associated with prolonged anaemia. He also studied the condition of the blood in certain cases of industrial poisoning. He investigated the anaemia occurring among

New Jersey ammunition workers. From studies of their blood, he found that the trinitrotoluene (TNT) used to fill shells, destroyed the worker's red cells, often producing anaemia. Minot was the first to suggest, in 1914, a raw-liver diet to regenerate blood haemoglobin in the treatment of pernicious anaemia, which was previously an invariably fatal disease. Later, he helped develop liver extract for oral use (now replaced by vitamin B_{12} injections). Minot was awarded the 1934 Nobel Prize for Physiology or Medicine, jointly with the American pathologist George Hoyt Whipple and physician William Murphy, for their discoveries concerning liver therapy in cases of anaemia. (☞ Feb. 6, Feb. 25, Aug. 28)

1942

The first self-sustaining nuclear chain reaction was demonstrated at the University of Chicago where the Italian physicist Enrico Fermi had set up an 'atomic pile'. The atomic pile consisted of uranium pellets as the neutron producing element, separated by graphite blocks to slow the neutrons. Cadmium rods were used to absorb the neutrons thus serving as a control element of the reaction. The equipment was set up in a squash court at the University. As a few of the cadmium rods were pulled out from the pile, the reaction began. It was soon proved to be self-sustaining. The team could also increase or decrease the energy output by adjusting the rods. (☞ Sep. 29, Nov. 28)

Chicago atomic pile

1971

The Russian spacecraft *Mars-3* arrived at Mars and released a descent module that entered the Martian atmosphere. Through aerodynamic braking, parachutes, and retro-rockets, the lander achieved a soft landing, becoming the first spacecraft to land successfully on Mars. The lander failed after relaying 20 seconds of video data to the orbiter. The *Mars-3* orbiter returned data until August 1972, sending measurements of surface temperature and atmospheric composition. (☞ May 28)

1974

American unmanned spacecraft *Pioneer-11* flew by Jupiter at a distance of 46,400 kilometres and sent back dramatic images of the Great Red Spot. It also made the first observation of Jupiter's immense polar regions, and determined the mass of its moon, Callisto. (☞ Apr. 5, Sep. 1)

1982

Barney Clark, a 61-year-old retired dentist became the world's first recipient of a permanent artificial heart, the *Jarvik-7*. William DeVries performed the surgery in cooperation with the inventor of the artificial heart, Robert Jarvik, at the University of Utah Medical Center. Clark lived for 112 days on the artificial heart. Of the next four implants, the longest survivor was William Schroeder, who lived for 620 days. (☞ Mar. 23, May 11, Nov. 13, Nov. 25)

1987

Luis Federico Leloir, Argentinean biochemist, Nobel laureate (Chemistry, 1970), who discovered sugar nucleosides and their role in the biosynthesis of carbohydrates, died in Buenos Aires, Argentina. (☞ Sep. 6)

1989

The *Solar Maximum Mission* satellite, or *Solar Max*, launched by the American space agency, NASA to provide coordinated observations of solar activity, in particular solar flares, during a period of maximum solar activity, burned up in the Earth's atmosphere. (☞ Feb. 14)

1993

American space shuttle *Endeavour*, carrying astronauts Richard O. Covey, Kenneth D. Bowersox, F. Story Musgrave, Kathryn C. Thornton, Claude Nicollier, Jeffrey A. Hoffman, and Thomas D. Akers, was launched on its 5th flight from the Kennedy Space Center, Florida, USA. The main objective of the mission was to service the Hubble Space Telescope. (☞ Dec. 8, Dec. 13)

1995

The *SOHO* (*Solar & Heliospheric Observatory*) was launched from Cape Canaveral, Florida, USA. The project is a collaborative effort between the European Space Agency (ESA) and the US National Aeronautics and Space Administration (NASA) as part of the Solar Terrestrial Science Program (STSP). The *SOHO* spacecraft was built in Europe by an industrial team, and instruments were provided by European and American scientists. NASA was responsible for the launch and mission operations. *SOHO* has been placed in orbit between the Earth and the Sun, at a point about 150,703,456 kilometres from the Sun

and about 1,528,483 kilometres from the Earth (three times farther than the Moon). This orbit is around a mathematical point between the Earth and the Sun known as the Lagrange point or the L1 point – a point where the gravitational pull is equal from both the Sun and the Earth. From this vantage point, *SOHO* was designed to give solar physicists their first long-term, uninterrupted view of the Sun, which would include study of the internal structure of the Sun, its extensive outer atmosphere and the origin of the solar wind – the stream of highly ionised gas that blows continuously outward through the Solar System. *SOHO* data will enable scientists to solve some of the most perplexing riddles about the Sun, including the heating of the solar corona, the acceleration of the solar wind, and the physical conditions of the solar interior.

December 3

1886

Karl Manne Georg Siegbahn, Swedish physicist, Nobel laureate, was born in Örebro, Sweden. Siegbahn was educated at the University of Lund and obtained his doctorate there in 1911. From 1907 to

1911 he served as assistant to the Swedish physicist J. R. Rydberg in the Physics Institute of the University; afterwards he was appointed lecturer and in 1915 he was made deputy professor of physics. When Rydberg died in 1920, Siegbahn was appointed professor. In 1923 he became professor of physics at the University of Uppsala. In 1937 he was appointed as Research Professor of Experimental Physics at the Royal Swedish Academy of Sciences. When the physics department of the Nobel Institute of the Academy came into being that same year, Siegbahn was made its first director. Siegbahn's early work was concerned with problems of electricity and magnetism. From 1912 to 1937, his research work was mainly devoted to X-ray spectroscopy. He developed new methods, and designed instruments for this purpose. His improvements and new constructions of air pumps and X-ray tubes enabled a considerable increase of the radiation intensity. The numerous spectrographs and crystal or linear gratings which he constructed, resulted in a highly increased accuracy of his measurements and led to the discovery of a large number of new series within the characteristic X-radiations of elements. The new precision technique thus developed by Siegbahn led to a practically complete knowledge of the energy and radiation conditions in the electron shells of the atoms. Siegbahn was awarded the 1924 Nobel Prize for Physics for his discoveries and research in the field of X-ray spectroscopy. (☞ Sep. 26)

1900

Richard Kuhn, German biochemist, Nobel laureate, was born in Vienna, Austria. Kuhn studied chemistry at the Vienna University and later at the University of Munich, Germany, from where he received his PhD in 1922 on a thesis on the specificity of enzymes. From 1926 to 1929, he held the post of professor for general and analytical chemistry at the Eidgenössische Technische Hochschule in Zurich, Switzerland. He then became principal of the Institute for Chemistry at the newly founded Kaiser-Wilhelm-Institut (later Max Planck Institut) for Medical Research in Heidelberg, Germany. Kuhn's research interests encompassed a wide range – theoretical problems of organic chemistry including stereochemistry of aliphatic and aromatic compounds; syntheses of polyenes and cumulenes; constitution and colour; the acidity of hydrocarbons, as well as extensive fields in biochemistry including carotenoids, flavins, vitamins, and enzymes. For a period of twenty years Kuhn investigated compounds containing double bonds which proved to be of great interest in connection with the study of the chemical nature of the carotenoids – the fat-soluble yellow colouring agents widely distributed in nature. He discovered eight new types of carotenoids and was able to analyse their constitution. He also carried out important work on vitamin B_2 and the antidermatitis vitamin B_6. Kuhn was awarded the 1938 Nobel Prize for Chemistry for his work on carotenoids and vitamins. (☞ Aug. 1)

1903

John von Neumann, Hungarian-born American mathematician who made important contributions in quantum physics, logic, meteorology, and computer science, was born in Budapest, Hungary. Von Neumann was a child prodigy. He could divide eight-digit numbers in his head even at the age of six. He received his early education in Budapest, where he published his first paper at the age of 18. He studied chemistry at the University of Budapest, and then moved to Berlin, Germany, and Zurich, Switzerland, from where he received his diploma in chemical engineering in 1925. He obtained his PhD in mathematics from the University of Budapest in 1926 with a thesis on set theory. He was invited to visit the Princeton University, New Jersey, USA, in 1930, and when the Institute for Advanced Studies was founded there in 1933, he was appointed as one of the first six professors of mathematics, a position that he retained for the remainder of his life. During and after World War II, von Neumann served as a consultant to the armed forces. His valuable contributions included a proposal of the implosion method for bringing nuclear fuel to the point of explosion and his participation in the development of the hydrogen bomb. He invented game theory, the branch of mathematics that analyses strategy and is now widely employed for military and economic purposes. He also set quantum theory upon a rigorous mathematical basis. In computer theory, von Neumann did much of the pioneering work in logical design, in the problem of obtaining reliable answers from a machine with unreliable components, the function of 'memory', and machine imitation of 'randomness'. In the 1950s, von Neumann was employed as a consultant to IBM to review proposed and ongoing advanced technology projects. (☞ Feb. 8)

1907

Reinhold Burger, a German glass blower, was issued a patent for the vacuum flask. He had applied for a patent for a 'double walled vessel with a space for a vacuum between the walls', in 1903. Burger's company, Thermos GmbH, made the first vacuum flasks for commercial use in 1904. 'Thermos', their trade name for their flasks, remains a registered trademark. (☞ Sep. 20)

1933

Paul Crutzen, Dutch chemist, Nobel laureate, was born in Amsterdam, Netherlands. After attending the middle technical school in Amsterdam, Crutzen joined the Bridge Construction Bureau of the City of Amsterdam in 1954. In 1959 he moved to Stockholm, Sweden and joined the department of meteorology of the Stockholm University where, until about 1966, he was mainly involved in various meteorological projects, especially helping to build and run some of the first numerical weather prediction models. He also programmed a model of a tropical cyclone. Around 1965 Crutzen was given the task of helping the American chemist F. Sherwood Rowland develop a numerical model of the ozone distribution in the stratosphere, mesosphere and lower thermosphere. In 1970 he showed that the nitrogen oxides NO and NO_2 react catalytically (without themselves being consumed) with ozone, thus accelerating the rate of reduction of the ozone content. The nitrogen oxides are formed in the atmosphere through the decay of the chemically stable nitrous oxide N_2O, which originates from soil bacteria. The connection demonstrated by Crutzen between micro-organisms in the soil and the thickness of the ozone layer was one of the motives for the rapid development of research on global biogeochemical

cycles. For his research, he received a PhD in meteorology from the Stockholm University in 1973. Crutzen was awarded the 1995 Nobel Prize for Chemistry, jointly with the Mexican-born American chemist Mario Molina and Rowland, for their work in atmospheric chemistry, particularly concerning the formation and decomposition of ozone. (☞ Mar. 19, Jun. 28)

1967
The first human heart transplant was performed on a South African businessman, 54-yr-old Louis Washkansky by a team of doctors led by Christiaan Barnard at the Groote Schuur Hospital, Cape Town, South Africa. Washkansky's diseased heart was replaced with the healthy heart of a 25-year old woman who had died in a car crash. Washkansky lived for only 18 days with the transplanted heart before succumbing to double pneumonia, contracted as a result of the destruction of his body's immunity mechanism by drugs administered to suppress rejection of the new heart. However, the next transplant patient, a retired South African dentist named Philip Blaiberg, lived for nearly two years. Since then, more than 100,000 human heart transplants have been performed. (☞ Nov. 8)

1973
The American space probe *Pioneer-10* flew by Jupiter at a distance of 130,000 kilometres and sent back pictures and scientific data. The probe revealed the severe effects of Jupiter's radiation belt on the spacecraft. In addition, the probe discovered that Jupiter has an enormous magnetosphere. After its encounter with Jupiter, *Pioneer-10* explored the outer regions of the solar system, studying energetic particles from the Sun, and cosmic rays entering our portion of the Milky Way. The spacecraft continued to make valuable scientific investigations in the outer regions of the solar system until its science mission ended in March 1997. The spacecraft was tracked till its power source became too feeble and the last contact was made in January 2003. (☞ Jan. 22, Mar. 2, Jun. 13)

1984
The world's worst industrial disaster happened in Bhopal, Madhya Pradesh, India, as a result of leakage of poisonous gas. Shortly after midnight, over 40 tonnes of highly poisonous methyl isocyanate gas leaked out of the Union Carbide pesticide factory in the city, engulfing an area of 40 square kilometres. Of the 800,000 people living in Bhopal at the time, 2,000 died immediately, and as many as 300,000 were injured. In addition, about 7,000 animals were injured, of which about one thousand were killed. Over 400,000 people suffered from acute breathlessness, pain in the eyes and vomiting. Poison clouds hung close to the ground for more than four hours. In 1989, after years of litigation, the Union Carbide agreed to pay the Indian government $470 million in damages.

December 4

1798
Luigi Galvani, Italian physician and physicist, who studied the structure of organs and the physiology of tissues and is best known for his investigation of the nature and effects of what he conceived to be electricity in animal tissue, died in Bologna, Italy. (☞ Sep. 9)

1893
John Tyndal, British physicist who demonstrated why the sky is blue, died in Hindhead, Surrey, England. (☞ Aug. 2)

1898

Kariamanikkam Srinivasa Krishnan, Indian physicist, Fellow of the Royal Society, was born in the village of Watrap in the Tirunelveli District of Tamil Nadu, India. After schooling in his village and neighbouring Srivilliputtur, Krishnan studied in the American College, Madurai, and the Christian College, Madras (now Chennai). After graduating in physics from the Christian College, he was, for some time, a demonstrator in chemistry at the same college. In 1920, Krishnan went to Kolkata to do his MSc in physics under Chandrasekhara Venkata Raman. He stayed in Kolkata to work as a research scholar with Raman in the Indian Association for Cultivation of Science (IACS). There he did the crucial experiments that led to the discovery of the Raman effect. He collaborated with Raman in making a detailed examination of the changes of frequency and polarisation of light associated with molecular scattering. It was in this period that the now famous Raman effect was announced by Raman and Krishnan in a paper, titled 'A new type of secondary radiation' in the British journal *Nature*. Krishnan served as a Reader in Physics at the Dacca University from 1929 to 1933, where he developed precise and delicate experimental techniques for the measurement of magnetic anisotropies in dia- and paramagnetic crystals. His work in this period laid the foundations of modern crystal magnetism and magneto-chemistry.

In 1933, Krishnan went back to IACS, Kolkata, as the Mahendra Lal Sircar Professor of Physics. In 1942, he moved to the University of Allahabad as Professor and Head of the Department of Physics. After India's Independence, Krishnan joined the National Physical Laboratory in New Delhi as Director. (☞ Jun. 14)

1908

Alfred Day Hershey, American biologist, Nobel laureate, was born in Lansing, Michigan, USA. Hershey studied at the Michigan State College in Lansing, where he obtained his BS degree in 1930, and PhD in 1934. In 1967 he got an honorary DSc at the University of Chicago, Chicago, Illinois. From 1934 till 1950 he was engaged in teaching and research, at the department of bacteriology, Washington University School of Medicine, St. Louis, Missouri. He joined the department of genetics at the Carnegie Institution of Washington in Cold Spring Harbor on Long Island, New York, in 1950 after giving up his position as professor at the Washington University.

Earlier, while Hershey was working on a little-studied organism called bacteriophage – a group of viruses that infect bacteria rather than ordinary cells – at the Washington University in St. Louis, he received a letter from the German biologist Max Delbrück, who was also studying bacteriophages with the Italian-born American biologist Salvador Luria at the Vanderbilt University in Nashville, Tennessee. At Cold Spring Harbor, in 1952, Hershey, with his assistant, Martha Chase, showed that the phage DNA and not its protein component contain the phage genes. After a tadpole-shaped phage particle attaches to a bacterium, its DNA enters through a tiny hole while its protein coat remains outside. Key to Hershey's success was showing that viral infection is not affected by violent agitation in a kitchen blender, which removes the empty viral protein shells from the bacterial surface but not the DNA which has already entered the bacterium. This proved that the DNA was sufficient to transfer to the bacteria all the genetic information needed to produce more phage. Hershey was awarded the 1969 Nobel Prize for Physiology or Medicine, jointly with Delbrück and Luria, for their discoveries concerning the replication mechanism and the genetic structure of viruses. (☞ May 22, Aug. 13, Sep. 4)

1935
Charles Richet, French physiologist, Nobel laureate (Physiology or Medicine, 1913), who discovered and coined the term anaphylaxis, died in Paris, France. (☞ Aug. 26)

1945
Thomas Hunt Morgan, American zoologist and geneticist, Nobel laureate (Physiology or Medicine, 1933), famous for his experimental research with the fruit fly by which he established the chromosome theory of heredity, died in Pasadena, California, USA. (☞ Sep. 25)

1978
American space probe *Pioneer Venus-1* entered an elliptical orbit around Venus and began a lengthy reconnaissance of the planet. The spacecraft returned global maps of the Venusian clouds, atmosphere, and ionosphere, measurements of the interaction between the atmosphere and the solar wind, and radar maps of 93% of the planet's surface. In 1991 the Radar Mapper was re-activated to investigate previously inaccessible southern portions of the planet. In May 1992 *Pioneer Venus-1* began the final phase of its mission, during which it mapped the surface from as close as 150–250 kilometres until the fuel ran out and atmospheric entry destroyed the spacecraft in October 1992. (☞ May 20)

1996
American space probe *Mars Pathfinder* was launched towards Mars to demonstrate a new way to land a spacecraft on the planet's surface using a rover and the operation of the independent robotic rover. (☞ Jul. 4, Sep. 27)

1998
The space shuttle *Endeavour*, carrying a crew of six comprising Robert D. Cabana, Frederick W. Sturckow, J. Currie, Jerry L. Ross, James H. Newman, and Sergei K. Krikalev, was launched on its 13th flight. It was the first shuttle mission to begin assembling the *International Space Station*. The primary objective of the seven-day mission was to integrate the US-built 'Node-1' station element to the 'functional energy block' (FGB), which was already in orbit, and two space walks to connect power and data transmission cables between the Node and the FGB. Node-1 was the first space station hardware delivered by the space shuttle. (☞ May 7, Dec.15)

December 5

1896
Carl Ferdinand Cori, American biochemist, Nobel laureate, was born in Prague, Czech Republic. Cori studied at the gymnasium in Trieste, Italy, and obtained his doctorate in medicine in 1920 from the German University of Prague, where he studied with his future wife, Gerty. He spent a year at the University of Vienna and a year as assistant in pharmacology at the University of Graz in Austria. In 1922, he emigrated to USA and joined the State Institute for the Study of Malignant Diseases in Buffalo, New York, as a biochemist. In 1931, he was appointed professor of pharmacology at the Washington University Medical School

in St. Louis, Missouri, where he later became professor of biochemistry. His wife also joined him at St. Louis and they collaborated in most of their research. The Coris first studied the fate of sugar in the animal body and the effects of insulin and epinephrine. In 1936, they isolated glucose-1-phosphate, now known as the 'Cori ester', and traced its presence to the activity of the phosphorylase, which catalyses the breakdown and synthesis of polysaccharides. This discovery made possible the enzymatic synthesis of glycogen and starch *in vitro*. Six years later they isolated and purified the enzyme responsible for catalysing the glycogen—Cori ester reaction, and with it they achieved the test-tube synthesis of glycogen in 1943. Carl Cori was awarded the 1947 Nobel Prize for Physiology or Medicine, jointly with his wife Gerty and the Argentinean physiologist Bernardo Houssay, for their discovery of the course of the catalytic conversion of glycogen. (☞ Apr. 10, Aug. 15, Oct. 20)

1901

Werner Karl Heisenberg, German physicist, Nobel laureate, was born in Wurzburg, Germany. Heisenberg studied theoretical physics at the University of Munich under Arnold Sommerfeld and obtained his doctor's degree in 1923. In the 1920s Heisenberg proposed a radical re-interpretation of the basic concepts of mechanics as applied to atomic particles. He gave up the existing model of discrete particles moving in prescribed paths for an approach in which such phenomena merely represented observable, or measurable, quantities called the 'quanta' that could be expressed by arrays of numbers obeying the rules of matrix algebra. Heisenberg's work on the quantum theory profoundly influenced the development of atomic and nuclear physics. In 1927 he published his 'indeterminacy', or 'uncertainty', principle, upon which he built his philosophy and for which he is best known. According to the Heisenberg's uncertainty principle, the position and the velocity of an object cannot both be measured exactly, at the same time, even in theory. It implied that the very concepts of exact position and exact velocity together, have no meaning in nature. Heisenberg also made important contributions to the theories of the hydrodynamics of turbulence, the atomic nucleus, ferromagnetism, cosmic rays, and elementary particles. He planned the first post-World War II German nuclear reactor at Karlsruhe, then in West Germany. Heisenberg was awarded the 1932 Nobel Prize for Physics, for the creation of quantum mechanics. (☞ Feb. 1)

1903

Cecil Powell, English educator, physicist, Nobel laureate, was born in Tonbridge, Kent, England. Powell attended a local elementary school and graduated in the Natural Science Tripos from the Sidney Sussex College, Cambridge in 1925. As a postgraduate student, he worked in the Cavendish Laboratory at Cambridge under the Scottish physicist C.T.R. Wilson and British nuclear physicist Ernest Rutherford. He received his PhD in 1927. He then moved to the University of Bristol as research assistant. He became professor of physics at Bristol in 1948 and director of the Henry Herbert Wills Physical Laboratory at the same university in 1964.

Between 1939 and 1945, at Bristol, he devoted years of patient work to the development of accurate techniques for measuring the mobility of positive ions and to establishing the nature of the ions in most common gases. He also developed techniques for using sensitive photographic emulsions to record the paths of cosmic rays. In plates exposed at the top of high mountains or sent up in high-altitude balloons, cosmic ray interactions were recorded, and, in 1947, the data revealed the existence of the pi-meson (π^+) as well as the process whereby it decays into two other particles, an anti-muon (mu-meson) and a neutrino. Powell was awarded the 1950 Nobel Prize for Physics for his development of the photographic method of studying nuclear processes and his discoveries regarding mesons made with this method. (☞ Aug. 9)

1932
Sheldon Lee Glashow, American theoretical physicist, Nobel laureate, was born in New York, USA. Glashow was educated at the Bronx High School of Science in New York and graduated from the Cornell University, Ithaca, in 1954. He received his PhD in physics from the Harvard University, Cambridge, Massachusetts, in 1959. He joined the faculty of the University of California at Berkeley in 1961 and worked at the Niels Bohr Institute in Copenhagen, Denmark before becoming professor at Harvard in 1967. Glashow helped develop important theories of electromagnetic and nuclear particle interaction that affected subsequent research on quarks and leptons. In 1964 he proposed the existence of a fourth, 'charmed' quark, and later argued that quarks must be 'coloured'. Insights gained from these theoretical studies enabled Glashow to consider ways in which the weak nuclear force and the electromagnetic force – two of the fundamental forces of nature – could be unified as a single force now called the electroweak force. Glashow was awarded the 1979 Nobel Prize for Physics, jointly with the American nuclear physicist Steven Weinberg and the Pakistani nuclear physicist Abdus Salam, for their contributions to the theory of the unified weak and electromagnetic interaction between elementary particles, including, *inter alia*, the prediction of the weak neutral current. (☞ Jan. 29, May 3)

1965
Joseph Erlanger, American physiologist, Nobel laureate (Physiology or Medicine, 1944), who discovered that fibres within the same nerve cord possess different functions, died in St. Louis, Missouri, USA. (☞ Jan. 5)

1986
Nil Ratan Dhar, Indian chemist and soil scientist and founder of the Sheila Dhar Institute of Soil Science, died in Allahabad, India. (☞ Jan. 2)

1997
American space shuttle *Columbia*, carrying astronauts Indian-born Kalpana Chawla, Steven W. Lindsey, Kevin R. Kregel, Winston E. Scott, Takao Doi, and Leonid K. Kadenyuk, landed at the Kennedy Space Center, Florida, USA after completing its 24th flight. (☞ Jan. 16, Feb. 1, Nov. 19)

December 6

1631
The transit of Venus occurred as predicted by the German astronomer Johannes Kepler. He was the first to correctly predict that an ascending node transit of Venus would occur in December 1631. But no one observed it – due to the fact that it occurred after sunset for most of Europe. Kepler himself died in 1630. He not only predicted this particular transit but also worked out that transits of Venus involve a cyclical period of approximately 120 years. When such a transit is observed, Venus appears as a small black circle moving across the face of the Sun. None occurred during the 20th century, but during the 21st century one was seen on June 8, 2004 and another will occur on June 6, 2012. (☞ Jun. 8, Nov. 24, Dec. 27)

1742
Nicolas Leblanc, French surgeon and chemist, was born in Ivoy-le-Pre, France. Leblanc studied medicine and after graduating with a master's degree in surgery, opened a medical practice. In 1780 he became a private physician to the Duke of Orléans. Five years earlier the French Academy of Sciences had offered a prize for a process to convert common salt (sodium chloride) to soda ash (sodium carbonate), which at that time was extracted by crude methods from wood or seaweed ashes. Soda ash was an essential ingredient for making paper, glass, soap, and porcelain; if these industries were to expand, a cheaper process was needed. In 1790 Leblanc developed a simple process for making soda ash from common salt. This process, which came to be known as the 'Leblanc process', became one of the most important industrial–chemical processes of the 19th century. In the Leblanc process, salt is first converted into sodium sulphate by the action of sulphuric acid, which is then roasted with chalk or limestone (calcium carbonate) and coal to produce a mixture of sodium carbonate and sulphide. The carbonate is leached out with water and the solution crystallised. The Leblanc process was simple, cheap, and direct, but because the French Revolution had begun by the time Leblanc completed his experiments in 1790, he never received his prize. The French National Assembly awarded him a 15-year patent in September 1791 but confiscated his patent and factory three years later with only token compensation. (☞ Jan. 16)

1778
Joseph Louis Gay-Lussac, French chemist best known for his work on gases, was born in Saint-Léonard-de-Noblat, France. Gay-Lussac studied at the École Polytechnique in Paris where in 1809 he became professor of chemistry. He was a professor of physics at the Sorbonne, Paris from 1808 to 1832. In 1802, Gay-Lussac formulated the law that a gas expands linearly with a fixed pressure and rising temperature, but the discovery later became known as Charles's Law, after the French physicist Jacques-Alexandre-César Charles, who had independently formulated the law in 1787. In 1804, working at the Institute of France, Gay-Lussac performed two balloon ascents with the purpose of studying the chemistry and physics of the outer layers of the atmosphere, and during 1805–06 accompanied the German naturalist and explorer Alexander von Humboldt on an expedition to measure terrestrial magnetism. With Humboldt he accurately determined the proportions of hydrogen and oxygen in water, showing the volume ratio to be 2:1. In 1808 he formulated his 'law of combining volumes' that now bears his name, which states that gases combine

in simple proportions by volume and that the volumes of the products are related to the original volumes. He developed techniques of quantitative chemical analysis, confirmed that iodine was an element, discovered cyanogen, improved the process for manufacturing sulphuric acid, and prepared potassium and boron. (☞ May 9)

1863

Charles Martin Hall, American chemist and inventor who discovered the electrolytic method of producing aluminium, was born in Thompson, Ohio, USA. Hall graduated from the Oberlin College in 1885, but continued to use the college laboratory. Eight months later he discovered a method for extracting pure aluminium from its ore. He dissolved the alumina ore in a bath of molten cryolite, a mineral containing fluorine, sodium and aluminium, and passed electric current through the solution using carbon electrodes and home-made batteries, producing his first small globules of aluminium. In 1889 he patented the process, which was the first to become an inexpensive, commercial process for extracting aluminium. Up to that time, pure aluminium was considered a precious metal, used only in fine jewellery. In 1888, together with financier Alfred E. Hunt, Hall founded the Pittsburgh Reduction Company now know as the Aluminium Company of America (ALCOA). By 1914, Hall's process had brought the cost of aluminium, once a precious metal costing around $1,200 per kilogram in 1852, down

to 36 cents a kilogram. Most importantly, the use of aluminium allowed automobiles and airplanes to be lighter, more fuel-efficient, more rust-resistant, and less expensive. (☞ Feb. 23, Apr. 2, Dec. 27)

1877

Thomas Alva Edison, American inventor, demonstrated the first sound recording at West Orange, New Jersey, by reciting 'Mary had a little lamb' into a recording device using a needle to scratch a sound track onto a cylinder wrapped in tinfoil. (☞ Jan. 27, Feb. 11, Feb. 19, Apr. 15, Aug. 8, Aug. 9, Aug. 27, Oct. 21)

1882

The first transit of Venus to acquire international interest occurred. The 19th century's second transit, it stimulated considerable efforts by astronomers from countries worldwide. However, some astronomers did not take part in the international collaboration, most notably Russia and Austria, because they were not convinced that the transit was the best way to determine the solar parallax. Astronomers from the United States went on eight expeditions. British observers went to Southern Africa, Australia, Canada and the Caribbean. French observers mounted ten expeditions ranging from Fort Marion in Florida to Cape Horn and from Puebla in Mexico to Fort Tartenson in Martinique. (☞ Jun. 3, Jun. 8)

1920

George Porter, English chemist, Nobel laureate, was born in Stainforth, Yorkshire, England. Porter was educated at the Thorne Grammar School, and graduated from the Leeds University in 1941. During his final honours year he took a special course in radio physics that led to his service, during 1941–1945, as a Royal Naval Volunteer Reserve Radar Officer.

His wartime training in electronics and pulse techniques was to prove useful later in suggesting new approaches to chemical problems. In 1945 he went to the University of Cambridge to undertake postgraduate research under the English chemist Ronald G.W. Norrish in the field of chemical kinetics and photochemistry and obtained his doctorate in 1949. With Norrish, he developed the technique of flash photolysis. In this technique, a gas or liquid in equilibrium is illuminated with an ultra-short burst of light that causes photochemical reactions in the substance. The extremely short-lived intermediate products of these reactions – mainly free radicals – are illuminated by a second burst of light that enables an absorption spectrum to be taken of the reaction products before the gas has returned to a state of equilibrium. Porter's collaboration with Norrish continued until 1954 when Porter left Cambridge. His subsequent work was mainly concerned with showing how the flash photolysis method could be extended and applied to a great variety of problems in physics, chemistry and biology. After leaving Cambridge, Porter spent a year as assistant director of the British Rayon Research Association in Manchester where he applied the new methods to practical problems of dye fading and the phototendering of fabrics. He then moved to the University of Sheffield as professor of physical chemistry. Between 1963 and 1966 he was also professor of chemistry at the Royal Institution. Porter was awarded the 1967 Nobel Prize for Chemistry, jointly with Norrish and the German chemist Manfred Eigen, for their studies of extremely fast chemical reactions, effected by disturbing the equilibrium with very short pulses of energy. (☞ May 9, Aug. 31, Nov. 9)

1981

The first Indian scientific expedition to Antarctica set sail from Mormugao harbour in Goa. The expedition team of 21 scientists, technicians, and naval personnel selected from seven different institutions was led by S.Z. Qasim, Secretary, Department of Ocean Development, Government of India. The expedition code-named 'Operation Gangotri', was undertaken in an icebreaker *MV Polar Circle* chartered from A/S G.C. Reiber Bergen, Norway. The expedition team returned home after 77 days. Since this first expedition, India has been sending multi-disciplinary scientific expeditions to Antarctica on an annual basis. The first Indian research station 'Dakshin Gangotri' was commissioned in 1983 and replaced in 1988 by the indigenously designed second Indian permanent station, 'Maitri'. (☞ Jan. 9)

1993

Wolfgang Paul, German physicist, Nobel laureate (Physics, 1989), who developed the 'Paul trap', an electromagnetic device that captures ions and holds them long enough for study and precise measurement of their properties, died in Bonn, Germany. (☞ Aug. 10)

December 7

1810

Theodor Schwann, German physiologist, who founded modern histology by defining the cell as the basic unit of animal structure, was born in Neuss, Prussia. Schwann studied medicine at the University of Berlin. In 1836, while at the

University of Berlin, he discovered pepsin, an enzyme responsible for digestion, in extracts from the stomach lining. Pepsin was the first enzyme prepared from animal tissue. Between 1838 and 1848 Schwann taught at the Universities of Louvain, and Liège, in Belgium. In 1839, together with the German physiologist Matthias Schleiden, he developed the 'cell theory', which identified cells as the fundamental particles of plants and animals. Schwann and Schleiden recognised that some organisms are unicellular while others are multicellular. They also recognised membranes, nuclei and cell bodies to be common cell features and described them after comparing various animal and plant tissues. Schwann was also the first to identify eggs as single cells, coined the term 'metabolism' to describe the chemical process in living things and discovered 'Schwann cells', which formed the sheath around nerve processes that enabled nerves impulses to move very rapidly through the body. Schwann's animal cell theory stimulated a great deal of research. He is now recognised as the founder of modern histology. (☞ Jan. 11)

1872

The British research ship *HMS Challenger* sailed from Portsmouth, England on the world's first scientific voyage around the world. Physicists, chemists, and biologists collaborated with expert navigators to map the sea. The *Challenger* was a corvette class ship, a military vessel that travelled under sail but had auxiliary steam power. The ship was fitted with a natural history laboratory where specimens could be examined, identified, dissected and drawn; a chemistry laboratory; and scientific equipment. During its 5-year voyage, the *Challenger* zigzagged around the globe to visit every continent, sounded the ocean bottom to a depth of 8,200 m, found many new species, and provided collections for scores of biologists. The *Report on the Scientific Results of the Voyage of H.M.S Challenger* was issued in 50 volumes between 1880 and 1895; many of the data gathered at that time are still used today.

1889

John Boyd Dunlop, a Scottish inventor, was issued a patent for pneumatic tyre. Dunlop experimented with his son's tricycle and devised the first pneumatic tyre in 1887 by fitting rubber air tubes around a wooden ring and inflating them with a football pump for a cushioning effect. Dunlop continued development, until he patented the idea. (☞ Feb. 5)

1905

Gerard Peter Kuiper, Dutch-born American astronomer, was born in Harenkarspel, Netherlands. Kuiper graduated from the University of Leiden in 1927 and received his PhD from the same university in 1933. The same year he emigrated to USA. He joined the staff of the Yerkes Observatory (affiliated to the University of Chicago) and was its director from 1947 to 1949, and again from 1957 to 1960. He founded the Lunar and Planetary Laboratory at the University of Arizona in 1960 and served as its director until his death. Kuiper is considered to be the father of modern planetary science for his wide-ranging studies of the solar system. In 1948, he correctly predicted that carbon dioxide was one of the chief constituents of the Martian atmosphere. He discovered the fifth moon of Uranus, which he called Miranda, also in 1948; and in 1949 he discovered the second moon of Neptune, Nereid. Kuiper's spectroscopic studies of Uranus and Neptune led to the discovery of features subsequently named 'Kuiper bands', which indicate the presence of methane. His other discoveries include the atmosphere of Saturn's moon Titan, and measurement of the diameter of Pluto. In 1951 he proposed the existence of a disc-shaped region of minor planets outside the orbit of Neptune – now called the Kuiper belt – as a source for short-period comets that complete orbits around the Sun in less than 200 years. During the 1960s, Kuiper served as chief scientist for the *Ranger* lunar-probe program, choosing crash-landing sites on the Moon, which helped the National Aeronautics and Space Administration (NASA) to identify sites for the later *Surveyor* and *Apollo* programs. A pioneer in the development of infrared astronomy, he was honoured posthumously when, in 1975, NASA named its airborne infrared telescope the *Kuiper Airborne Observatory*. (☞ Dec. 23)

1905

The first successful eye (cornea) transplant was done on a human patient. Czech ophthalmic surgeon Eduard Zirm carried out the transplant in Olomouc, a town in the east of the Czech Republic. The patient was 43-year-old Alois Gloger, a labourer who had been blinded in an accident while working with lime. He was given no antibiotics, no drugs to stop him rejecting the tissue, and had to endure his eyelids being sewn shut for 10 days before he knew if the procedure had worked. The corneas were taken from the eyes of an 11-year-old boy who had been blinded by deep injuries to his eyes. The operation was a success, and Gloger could see again. He retained his eyesight for the rest of his life and was back working on his farm within three months.

1909

Leo Baekeland, American chemist, was issued the first American patent for a thermosetting artificial plastic, which he called bakelite. The patent was for 'an improvement in methods of making insoluble condensation products of phenol-formaldehyde'. Bakelite was cheaper and more versatile than other known plastics and could withstand heat. It gave birth to the modern plastics industry. (☞ Nov. 14)

1926

The first American patent for a household gas absorption refrigerator, operating on heat, was issued to the Electrolux Servel Corporation of Newburgh, New York. These refrigerators have no moving parts and use kerosene or propane gas as their primary source of energy with liquid ammonia as the coolant placed in a hermetically sealed evaporator. Water, ammonia and hydrogen gas is used to create a continuous liquid–gas–liquid cycle for the ammonia. Liquid ammonia is allowed to boil, removing the heat away from the evaporator and producing the cooling effect. It is then condensed using water. The heat from the burning kerosene or propane is used to vapourise the ammonia in order to separate it from water and make it suitable for re-cycling. Swedish engineering students Carl G. Munters and Baltzar von Platen invented the heat-driven absorption refrigerator.

1972
Apollo-17, the sixth and last American manned Moon mission, was launched by a Saturn-V rocket from the Kennedy Space Center, Florida, USA. It carried astronauts Eugene A. Cernan, Ronald E. Evans, and Harrison H. Schmitt. Four days after reaching the Moon, the lunar module *Challenger*, carrying Evans and Schmitt, landed on the lunar surface while Cernan in *Apollo-17* remained in orbit. During their three-day stay on the Moon, the astronauts set up scientific experiments, took photographs, and collected lunar samples. (☞ Dec. 11, Dec. 19)

1995
INSAT-2C, the third of India's second generation INSAT series of satellites, was launched by the European Space Agency's Ariane rocket from Kourou in French Guiana. Designed and built by the Indian Space Research Organisation, *INSAT-2C* was exclusively used for telecommunication and carried three S-band transponders for mobile communications, 18 C-band transponders for TV broadcasting, and three Ku-band transponders for business communications between the four Metro cities of India. The satellite was placed in the geostationary orbit at 93.5°E longitude.

1995
The American *Galileo* spacecraft arrived at Jupiter. It performed several important tasks to place itself in the correct orbit around Jupiter. Later it released an atmospheric probe that descended into Jupiter's atmosphere and sent back data. (☞ Jul. 13, Oct. 18)

1998
Martin Rodbell, American biochemist, Nobel laureate (Physiology or Medicine, 1994), who discovered G-proteins and the role of these proteins in signal transduction in cells, died in Chapel Hill, North Carolina, USA. (☞ Dec. 1)

December 8

1931
Lloyd Espenschied of Kew Gardens, New York and Herman A. Affel of Ridgewood, New Jersey were issued an American patent for coaxial cable. In the early 1920s, engineers of the American Telegraph and Telephone Co. of New York City recognised that the open wire and cable in use at the time would be unable to carry the high frequencies needed for the broadband systems of the future. So Espenschied and Affel developed a new kind of wire system that could transmit a continuous range of high frequencies over long distances. Described in the patent as a 'concentric conducting system', this new transmission system was based on the use of a coaxial conductor: two concentric cylinders of conducting material separated mainly by air. This structure reduced frequency losses and provided freedom from outside interference.

1947
Thomas Robert Cech, American biochemist and molecular biologist, Nobel laureate, was born in Chicago, Illinois, USA. After schooling in Iowa City, Cech did his graduation from the Grinnell College in Grinnell, Iowa, and obtained his PhD in chemistry from the University of California at Berkeley in 1975. He served as a National Cancer Institute fellow at the Massachusetts Institute of Technology, Cambridge, Massachusetts, from 1975 to 1977 and then joined the department of chemistry at the University of Colorado at Boulder

as professor in 1978, where he discovered that RNA could have an independent catalytic function and act as a 'ribozyme', aiding a chemical reaction without being consumed or changed. It was previously believed that enzymatic activity was only possible in protein molecules and that RNA was merely a genetic messenger – an intermediate in the synthesis of proteins from DNA. Cech's discovery had major implications for genetic engineering as well as for understanding how life arose. Cech was awarded the 1989 Nobel Prize for Chemistry, jointly with the Canadian-born American molecular biologist Sidney Altman, for their discovery of catalytic properties of RNA. (☞ May 7)

1983
American space shuttle *Columbia*, carrying a six-person crew of astronauts John W. Young, Brewster H. Shaw, Owen K. Garriott, Robert A. Parker, Byron K. Lichtenberg, and Ulf Merbold, landed at the Edwards Air Force Base, California, USA after completing its 6th flight. The shuttle carried *Spacelab-1*'s orbital laboratory during the mission. (☞ Nov. 28)

1993
The crew of the American space shuttle *Endeavour* completed the mission of servicing the Hubble Space Telescope. During a record five back-to-back space walks, totalling 35 hours and 28 minutes, the astronauts installed new instruments and corrected the optics of the flaw in Hubble's primary mirror. They also installed new solar arrays. (☞ Dec. 2, Dec. 13)

1994
The creation of element 111 was announced by a team of German scientists led by Peter Armbruster at the Gesellschaft für Schwerionenforschung (GSI) facility at Darmstadt, Germany. The element was made by accelerating nickel atoms (at. no. 28) to high speed and bombarding them into bismuth (at. no. 83). When an atom of each fused to make the new nucleus (at. no. 111), it lasted for about four-thousandths of a second before decaying into smaller nuclei. The new element had 111 protons and 161 neutrons in its nucleus, giving it a mass number of 272. It was named unununium, symbol Uuu, according to an internationally adopted system for naming new elements.

December 9

1742
Carl Wilhelm Scheele, Swedish chemist who discovered oxygen, was born in Stralsund, Pomerania (now Pomorze, Poland). Scheele received very little formal education and no training whatsoever in science. At age 14, he became an apprentice apothecary of a firm in Gothenburg. The supply of chemicals present in the pharmacy afforded him with a starting point for many investigations and discoveries. He worked as a pharmacist in Stockholm, Uppsala, and later in Köping. In 1770, Scheele moved to Uppsala where he discovered oxygen in 1773. He produced oxygen by several methods – by reacting nitric acid with potassium nitrate, and by heating mercuric oxide and manganese dioxide. However, due to his delay in publishing the discovery, which he did only in 1777, he lost priority to the British

chemist Joseph Priestley, who independently discovered oxygen in 1774. Scheele also discovered other chemical elements such as barium, chlorine, manganese, molybdenum, and tungsten, as well as several chemical compounds, including citric acid, glycerol, hydrogen cyanide, hydrogen fluoride, and hydrogen sulphide. In addition, he discovered a process similar to pasteurisation. (☞ May 21)

1748

Claude-Louis Berthollet, French chemist, was born in Talloires, near Annecy, France. Berthollet studied medicine at Turin, Italy, receiving his degree in 1768. He studied chemistry in Paris from 1772, and conducted chemical experiments in a private laboratory in the Palais Royal, where he met Antoine Lavoisier. He received his doctorate in medicine from the University of Paris in 1778. In 1784 he became director of the national Gobelins tapestry factories and in 1794 was appointed professor of chemistry at the École Normale in Paris. Berthollet was the first to note that the completeness of chemical reactions depends in part upon the masses of the reacting substances; he thus came close to formulating the law of mass action, though he incorrectly concluded that elements unite in all proportions. His investigations of ammonia and halogen salts were exemplary. He introduced chlorine as a bleaching agent and was the first to show that ammonia was a compound of hydrogen and nitrogen. He also prepared potassium chlorate and attempted to use it in the manufacture of gunpowder as a substitute for saltpetre, which had to be imported from Chile. Berthollet collaborated with Lavoisier in his researches and in reforming chemical nomenclature. He supported him in his theory of combustion. In 1803 he published his *Essai de statique chimique*, in which he presented his speculations on chemical affinity and his discovery of the reversibility of reactions. (☞ Nov. 6)

1868

Fritz Haber, German physical chemist, Nobel laureate, was born in Breslau, Silesia, Prussia (now Wroclaw, Poland). Haber studied at the St. Elizabeth School at Breslau. From 1886 until 1891, he studied chemistry at the Universities of Heidelberg and Berlin, and at the Technical School at Charlottenberg. He took up organic chemical research at the University of Jena, but left in 1894 to take up an assistantship at the Technische Hochschule of Karlsruhe, where he taught physical chemistry and also did research. In 1906 he was appointed professor of physical chemistry and electrochemistry and director of the Institute established at Karlsruhe to study these subjects. In 1911 he was appointed director of the Institute for Physical and Electrochemistry at Berlin-Dahlem, where he remained until 1933 when he went to Cambridge, England to work in the Cavendish Laboratory. Haber developed a method for synthesising ammonia from nitrogen and hydrogen, and by 1909 he had established conditions for the large-scale synthesis of ammonia. The process was handed over to Carl Bosch of Badische Anilin- und Soda-Fabrik for industrial development, leading to the Haber–Bosch

process, which led to commercial large-scale production of nitrogen fertilisers. The Haber–Bosch high-pressure process followed in the 1920s. Haber was awarded the 1918 Nobel Prize for Chemistry for the synthesis of ammonia from its elements. (☞ Jan. 29)

1917

Leo James Rainwater, American physicist, Nobel laureate, was born in Council, Idaho, USA. Rainwater graduated from the California Institute of Technology, Pasadena, as physics major in 1939 and began graduate study in physics as a teaching assistant at the Columbia University in New York, from where he received his doctorate in 1946. He was a professor of physics at the Columbia University from 1952 and Pupin professor of physics from 1982. Rainwater worked on the Manhattan Project to develop the atomic bomb during World War II. In 1949 he began formulating a theory that not all atomic nuclei are spherical, as was then generally believed. The theory was tested experimentally and confirmed by Danish physicists Aage N. Bohr and Benjamin Roy Mottelson. Rainwater also conducted valuable research on X-rays and worked on projects of the US Atomic Energy Commission and US Navy. Rainwater was awarded the 1975 Nobel Prize for Physics, jointly with Bohr and Mottelson, for the discovery of the connection between collective motion and particle motion in atomic nuclei and the development of the theory of the structure of the atomic nucleus based on this connection. (☞ May 31, Jun. 19, Jul. 9)

1919

William Nunn Lipscomb, Jr., American physical chemist, Nobel laureate, was born in Cleveland, Ohio, USA. Lipscomb received his BSc degree from the University of Kentucky in 1941 and his PhD degree from the California Institute of Technology, Pasadena, California, in 1946. He worked as a physical chemist in the Office of Science Research and Development from 1942 to 1946 and then joined the University of Minnesota as assistant professor. Since 1959, he has been a professor of chemistry at the Harvard University, Cambridge, Massachusetts. He deduced the molecular structure of hydrides of boron, commonly known as boranes, using X-ray crystallography in the 1950s and developed theories to explain their bonds. The stability of boranes could not be explained by traditional concepts of electron bonding, in which each pair of atoms is linked by a pair of electrons, because boranes lacked sufficient electrons. Lipscomb showed how a pair of electrons could be shared by three atoms. Boranes became important in chemical research in the 1940s and 1950s because of the need to find volatile uranium compounds (borohydrides) for isotope separation, as well as the need to develop high-energy fuels for rockets and jet aircraft. To map the molecular structures of boranes, Lipscomb also developed X-ray techniques that later found application in many other areas of chemical research. Later he applied the same methods to related problems, including the structure of carboranes – cage compounds composed of two carbons and 10 borons. His later research focussed on the relationship of three-dimensional structure and mechanisms of enzymes and

other proteins. Lipscomb was awarded the 1976 Nobel Prize for Chemistry for his studies on the structure of boranes illuminating problems of chemical bonding.

1926

Henry Way Kendall, American nuclear physicist, Nobel laureate, was born in Boston, Massachusetts, USA. Kendall received his BS degree in mathematics from the Amherst College, Massachusetts, in 1950 and obtained a PhD in nuclear and atomic physics from the Massachusetts Institute of Technology, Cambridge, Massachusetts, in 1954. He was a National Science Foundation Postdoctoral Fellow at the Brookhaven National Laboratory and MIT from 1954 to 1956. He joined the faculty at the Stanford University, California, in 1956, serving for the next five years as a lecturer and assistant professor in the University's physics department before returning to the MIT physics department in 1961. He served as a professor of physics at MIT from 1967 to 1991, when he was named the Julius A. Stratton Professor of Physics. During 1967–73 Kendall collaborated with the American physicist Jerome Isaac Friedman and Canadian physicist Richard E. Taylor to study the internal structure of the proton using the 3-km linear accelerator at the Stanford Linear Accelerator Center, California, in which electrons were accelerated to an energy of 20,000 million electron volts and directed against a target of liquid hydrogen. In course of the experiments they found experimental evidence for the existence of the subatomic particles known as quarks. In 1969 Kendall helped found the Union of Concerned Scientists. In 1997, in connection with the Kyoto Climate Summit, he helped produce a statement signed by 2,000 scientists calling for action on global warming. Kendall was awarded the 1990 Nobel Prize for Physics, jointly with Friedman and Taylor, for their pioneering investigations concerning deep inelastic scattering of electrons on protons and bound neutrons, which have been of essential importance for the development of the quark model in particle physics. (☞ Feb. 15, Mar. 28, Nov. 2)

1937

Nils Dalén, Swedish engineer, Nobel laureate (Physics, 1912), who invented automatic regulators for lighting coastal beacons and light buoys, died in Stockholm, Sweden. (☞ Nov. 30)

1961

Eric A. Cornell, American physicist, Nobel laureate, was born in Palo Alto, California, USA. Cornell had his schooling in Cambridge, Massachusetts and San Fransisco, California. He did his BS in physics from the Stanford University in California in 1985 and obtained his PhD in physics from the Massachusetts Instituite of Technology, Cambridge, Massachusetts in 1990. During 1990–1992 Cornell did post-doctoral work at the Joint Institute for Laboratory Astrophysics, Boulder, Colorado and then joined the physics department of the University of Colorado as assistant professor, becoming professor in 1995. He has been a senior scientist at the National Institute of Standards and Technology (NIST), Boulder, since 1992. In 1995, Cornell together with fellow American physicist Carl Wieman

succeeded in achieving a new state of matter known as the Bose–Einstein condensate, first predicted by the Indian physicist Satyendra Nath Bose and the German-born American physicist Albert Einstein in 1924. Cornell and Wieman produced a pure condensate of about 2,000 rubidium atoms at a temperature of 20 degrees nanokelvin, i.e., 20 billionths of a degrees above absolute zero, using the technique of laser cooling. Cornell was awarded the 2001 Nobel Prize for Physics, jointly with Wiemann and German physicist Wolfgang Ketterle, for the achievement of Bose–Einstein condensation in dilute gases of alkali atoms, and for early fundamental studies of the properties of the condensates. (☞ Mar. 26, Oct. 21)

1968
The first demonstration of the use of a computer mouse was given at the American Federation of Information Processing Societies' Joint Computer Conference at the Stanford University, California, USA. The mouse was invented by Douglas Engelbart, principal investigator at the Augmentation Research Center (ARC) of the Stanford Research Institute. The ARC mouse could be used as a graphical user interface and to display editing and integrated text and graphics. It had three buttons which could be used for typing, so the user's hand never had to leave it. (☞ Nov. 17)

December 10

1603
William Gilbert, English scientist, who pioneered the study of electricity and magnetism and spent years investigating magnetic and electrical attractions, died in London, England. (☞ May 24)

1831
Thomas Johann Seebeck, German physicist who discovered that an electric current flows between different conductive materials kept at different temperatures, died in Berlin, Prussia (now Germany). (☞ Apr. 9)

1896
Alfred Bernhard Nobel, Swedish chemist and inventor of dynamite and other, more powerful explosives, died in San Remo, Italy. (☞ Oct. 21)

1901
The first Nobel Prizes were awarded. The king of Sweden distributed the first Nobel Prizes, in Stockholm, in accordance with the will of inventor Alfred Nobel. There were six recipients – one each for chemistry, physics, physiology or medicine, and literature, and two recipients for the prize for peace. (☞ Oct. 21)

1903
Marie Curie, Polish-born French chemist, became the first woman to receive the Nobel Prize (Physics). (☞ Nov. 7)

1934
Howard Martin Temin, American virologist, Nobel laureate, was born in Philadelphia, Pennsylvania, USA. Temin had his elementary and high school education in the public schools of Philadelphia and studied biology at the Swarthmore College, Pennsylvania, and the California Institute of Technology in Pasadena, California, from where he received his PhD in 1959. From 1960 he worked at the University of Wisconsin in Madison, becoming professor 1969. Temin began his cancer research while still a student, working with his professor Renato Dulbecco and fellow student David Baltimore. In 1970 they experimentally verified Temin's

hypothesis that cancer cells affect genetic material. According to this hypothesis, the protein coat of certain viruses contains an enzyme that facilitates the copying of viral genes into the deoxyribonucleic acid (DNA) of the host cell. In 1970 Temin and Baltimore both independently isolated the enzyme, now called 'reverse transcriptase'. The viruses that contain the enzyme are known as retroviruses. Temin also investigated how genetic information in the provirus transforms a normal animal cell into a tumour cell. Temin was awarded the 1975 Nobel Prize for Physiology or Medicine, jointly with Dulbecco and Baltimore, for their discoveries concerning the interaction between tumour viruses and the genetic material of the cell. (☞ Feb. 9, Feb. 22, Mar. 7)

1993
The crew of the space shuttle *Endeavour* deployed the repaired Hubble Space Telescope into Earth orbit. During the mission *Endeavour* astronauts installed a module called 'corrective optics space telescope axial replacement' (COSTAR) to correct the optical aberration in Hubble's main mirror. They also fitted several new gyroscopes, new solar panels, and an upgraded wide-field planetary camera. (☞ Dec.2, Dec. 8, Dec. 13)

December 11

1796
Johann D. Titius, Prussian astronomer and physicist, who was the first to propose a law relating distances of the planets from the Sun, died in Wittenberg, Saxony (now in Germany). (☞ Jan. 2)

1843
Robert Koch, German physician, Nobel laureate, was born in Clausthal, Germany. Koch studied medicine at the University of Göttingen and obtained his MD degree in 1866. He passed the District Medical Officer's Examination in 1869 and, after volunteering for service in the Franco-Prussian War, was appointed District Medical Officer for Wollstein in 1872. It was here that he carried out his epoch-making researches. Anthrax was, at that time, prevalent among the farm animals in the Wollstein district and Koch embarked on a study of the disease. His studies confirmed the work of others who had shown that the disease could be transmitted by means of the blood of animals suffering from anthrax. In 1880 Koch was appointed a member of the Imperial Health Bureau in Berlin where he discovered the tubercle bacillus that causes tuberculosis and also devised a method of growing it in pure culture. In 1882 he published his classical work on this bacillus. In 1883 he was sent to Egypt as leader of the German Cholera Commission to investigate an outbreak of cholera in that country. Here he discovered the vibrio that causes cholera and brought back pure cultures of it to Germany. He also studied cholera in India. In 1885 Koch was appointed professor of hygiene in the University of Berlin and director of the newly established Institute of Hygiene there. In 1891 he became an honorary professor of the Medical Faculty of Berlin and director of the new Institute for Infectious Diseases. He studied the 1897 bubonic plague outbreak in Bombay, and malaria and sleeping sickness in Africa. He also investigated tropical dysentery, and the Egyptian eye disease (trachoma),

and typhus in tropical Africa, and carried out work on destructive tropical cattle diseases, such as rinderpest, coast fever in cattle and the trypanosome disease carried by the tsetse fly. Koch was awarded the 1905 Nobel Prize for Physiology or Medicine for his investigations and discoveries in relation to tuberculosis. (☞ Mar. 24, May 27)

1882
Max Born, German physicist, Nobel laureate, was born in Breslau, Germany. Born attended the König Wilhelm's Gymnasium in Breslau and continued his studies at the Universities of Breslau, Heidelberg, Zurich in Switzerland, and Göttingen. He graduated from the University of Göttingen in 1907 and taught there for some time. After a brief stint as professor at the University of Frankfurt-on-Main, Born returned to Göttingen as professor in 1921 and remained there for twelve years, interrupted only by a trip to America in 1925. During these years his most important works on the quantum theory were created. Among his collaborators at this time were many physicists, later to become well known, such as Wolfgang Pauli, Werner Heisenberg, Pascual Jordan, Enrico Fermi, and Paul Dirac. During the years 1925 and 1926 he published, with Heisenberg and Jordan, investigations on the principles of quantum mechanics. When Erwin Schrödinger put forward his quantum mechanical wave equation, Born's studies of the wave function led to the replacement of the

original quantum theory, which regarded electrons as particles, with a mathematical description. During the winter of 1935–1936, Born spent six months at the Indian Institute of Science, Bangalore, India, where he worked with the Indian physicist C.V. Raman and his pupils. In 1936 he was appointed Tait professor of natural philosophy at the University Edinburgh, Scotland, where he worked until his retirement in 1953. Born was awarded the 1954 Nobel Prize for Physics, jointly with the German physicist Walther Bothe, for his fundamental research in quantum mechanics, especially for his statistical interpretation of the wave function. (☞ Jan. 5, Jan. 8)

1910
Neon lighting, developed by French physicist Georges Claude, made its public debut at the Paris Motor Show. Claude was the first to apply an electrical discharge to a sealed tube of neon gas to create a lamp, around 1902. The success came after decades of experiments to create a practical alternative to incandescent lighting. In 1923, Claude and his French company Claude Neon, introduced neon gas signs to the United States, by selling two to a Packard car dealership in Los Angeles. Neon lighting soon became a popular fixture in outdoor advertising. (☞ Sep. 24)

1925
Paul Greengard, American neurologist, Nobel laureate, was born in New York, USA. Greengard attended public schools in Brooklyn and Queens. During World War II, he spent three years in the Navy as an electronics technician and was assigned to a team at the Massachusetts Institute of Technology, Cambridge, Massachusetts, that was involved in developing an early-warning system to intercept Japanese

kamikaze planes. After the war, he graduated from the Hamilton College, Clinton, New York, in 1948. Thereafter he joined the newly set-up department of biophysics at the Johns Hopkins University in Baltimore, Maryland, from where he obtained his PhD in 1953. Between 1953 and 1959, Greengard did post-doctoral studies in biochemistry at the University of London, Cambridge University, and National Institute for Medical Research, England, and the National Institutes of Health, Bethesda, Maryland, USA. During this period he continued to explore the relationships between biochemistry and electrophysiology in the nervous system. He traced the biochemical reactions that occur in nerve cells in response to neurotransmitters such as dopamine. Abused stimulants, such as methamphetamine, are known to alter nerve cells' exposure or reaction to neurotransmitters, which produces feelings of pleasure and leads to addiction. Greengard's continuing research on how cocaine and amphetamine change neurotransmitter function could lead to development of medications to prevent or treat the addictive effect. In 1959 Greengard became director of the department of biochemistry at the Geigy Research Laboratories, Ardsley, New York. Since 1983, he has been professor and head, Laboratory of Molecular and Cellular Neuroscience, the Rockefeller University, New York. Greengard was awarded the 2000 Nobel Prize for Physiology or Medicine, jointly with the Swedish pharmacologist Arvid Carlsson and Austrian-born American neuroscientist Eric R. Kandel, for their discoveries concerning signal transduction in the nervous system. (☞ Jan. 25, Nov. 7)

1972

Apollo-17 astronauts Gene Cernan and Harrison Schmitt landed on the Moon for a three-day exploration, which would be the last Apollo mission to the Moon. Their lunar module *Challenger* landed in the Taurus-Littrow Valley on the eastern edge of Mare Serenitatis. This site was picked for *Apollo-17* as a location where rocks both older and younger than those previously returned from other Apollo missions and from the Luna missions might be found. *Apollo-17* carried a lunar roving vehicle that was used by the astronauts to extend their area of study. The *Apollo-17* crew collected 741 individual rock and soil samples, including a deep drill core that included material from 3 metres below the lunar surface, with a total mass of 111 kilograms. (☞ Dec. 7, Dec. 19)

1978

Vincent du Vigneaud, American biochemist, Nobel laureate (Chemistry, 1955), who for the first time synthesised the polypeptide hormone vasopressin, died in White Plains, New York, USA. (☞ May 18)

December 12

1866

Alfred Werner, French-born Swiss chemist, Nobel laureate, was born in Mulhouse, France. Werner showed an interest in chemistry while he was at school and did his first independent chemical research when he was only 18. He obtained a diploma in technical chemistry from the Federal Technical High School at Zurich in 1889. In 1890 he obtained his PhD degree from the University of Zurich in Switzerland, with a thesis on the spatial arrangements of the atoms in molecules

containing nitrogen. In this work he formulated the idea that, in the numerous compounds of trivalent nitrogen, the three valence bonds of the nitrogen atom are directed towards the three corners of a tetrahedron, the fourth corner being occupied by the nitrogen atom. In 1892 Werner was appointed as a lecturer in the Technical High School, Zurich, and in 1893 he was appointed associate professor in the University of Zurich. In 1893 he stated, in a paper on mineral compounds, his theory of variable valence, according to which inorganic molecular compounds contain single atoms which act as central nuclei around which are arranged a definite number of other atoms, radicals or other molecules in a simple, spatial, geometric pattern. The figure which expresses the number of atoms thus grouped round a central nucleus – called 'ligands' – was called by Werner the 'coordination number', the most important of these coordination numbers being 3, 4, 6 and 8. In 1895 he became professor of chemistry in the University of Zurich, giving the lectures on organic chemistry until, in 1902, he took over the lectures on inorganic chemistry as well. Werner was awarded the 1913 Nobel Prize for Chemistry in recognition of his work on the linkage of atoms in molecules by which he has thrown new light on earlier investigations and opened up new fields of research especially in inorganic chemistry. (☞ Nov. 15)

1901

Guglielmo Marconi, Italian physicist and inventor, sent the first transatlantic radio signal – the alphabet 'S' – from Poldhu in Cornwall, England, which was received in St John's, Newfoundland, Canada, over a distance of 3,419 kilometres. (☞ Apr. 25)

1927

Robert Norton Noyce, American engineer and co-inventor of the integrated circuit, was born in Burlington, Iowa, USA. Noyce received a BA degree from the Grinnell College, Iowa in 1949 and a PhD from the Massachusetts Institute of Technology, Cambridge, Massachusetts, in 1953. He then conducted research at the Philco Corporation until 1956, when he joined the Shockley Semiconductor Laboratory in Palo Alto, California, to work on transistor technology. In 1957 Noyce co-founded the Fairchild Semiconductor Corporation in Mountain View, California, of which he was research director until 1959 when he became vice president and general manager. As research director of Fairchild Semiconductor, he was responsible for initial development of transistors for mass production. In 1959, Noyce and the American inventor Jack Kilby, who was working at Texas Instruments, independently invented the integrated circuit computer chip. But in 1961, Noyce was granted a patent for his invention of a 'semiconductor device-and-lead structure', which made the microchip possible and launched a modern electronics revolution with the development of his integrated circuit. In July 1968 he co-founded the Intel Corporation with Gordon E. Moore, who had also been a co-founder of Fairchild Semiconductor and a member of the Shockley Laboratory staff. Noyce served as president of Intel until 1975 and chairman of the board from 1975 to 1979. (☞ Apr. 25, Jun. 3, Nov. 8)

1955
British inventor Christopher Cockerell obtained a patent for the hovercraft, which he described as a vehicle that is 'neither an airplane, nor a boat, nor a wheeled land craft'. (☞ May 30, Jun. 4)

December 13

1835
Samanta Chandra Sekhar (also known as Pathani Samanta), Indian astronomer, was born in Khandpara, Orissa, India. From his childhood, Samanta showed extraordinary acuity. He is said to have spotted the planet Venus in the sky during the day at the age of four. Samanta received his primary education in Sanskrit and studied Sanskrit grammar, Smritis, Puranas, Darshan and the original texts of many Kavyas. When he was ten years old, one of his uncles interested him in the observation of prominent stars in the sky. Soon Samanta was studying ancient astronomical works like *Surya Siddhanta* and *Siddhanta Shiromani* from the family library. At the age of fifteen, when he began to check the predictions of the *Siddhantas* with his own astronomical observations, he was surprised to find that the predictions of the ancient classics did not agree with his observations. The stars and planets neither appeared at the right place in the sky, nor at right time, as per the calculations of the *Siddhantas*. Young Samanta resolved to rectify all the errors accumulated in Indian astronomy over thousands of years. He devised his own instruments for the measurement of time, and the positions of the celestial bodies. After making observations for eight years, Samanta wrote down his findings in the form of Sanskrit shlokas in a treatise called *Siddhanta Darpana*, which was completed in 1869. Samanta played a pivotal role in the revival of traditional Indian astronomy. (☞ Jun. 11)

1923
Philip Warren Anderson, American physicist, Nobel laureate, was born in Indianapolis, Indiana, USA. Anderson received his BS and PhD degrees from the Harvard University, Cambridge, Massachusetts, in 1943 and 1949, respectively. After completing his studies, he began a long career with the Bell Laboratories in Murray Hill, New Jersey. From 1967 to 1975 he was professor of theoretical physics at the Cavendish Laboratory at the University of Cambridge and from 1975 he taught at the Princeton University in New Jersey, USA. Following his retirement from the Bell Labs in 1984, he began working full time at Princeton. In 1985, Anderson became involved with the newly established Santa Fe Institute in Santa Fe, New Mexico. During 1993–94 he was professor of physics at the Oxford University in England. Anderson's main contri-butions have been in the field of condensed matter physics. He conceived a model – known as the Anderson model – to describe what happens when an impurity atom is present in a metal. He also investigated magnetism and high-temperature superconductivity, and his work is of fundamental importance for modern solid-state electronics, making possible the development of inexpensive electronic switching and memory devices

in computers. Anderson was awarded the 1977 Nobel Prize for Physics, jointly with the American physicist and mathematician John H. Van Vleck and the English physicist Nevill F. Mott, for their fundamental theoretical investigations of the electronic structure of magnetic and disordered systems. (☞ Mar. 13, Sep. 30)

1930
Fritz Pregl, Austrian chemist, Nobel laureate (Chemistry, 1923), who developed the technique of microanalysis, died in Graz, Austria. (☞ Sep. 3)

1935
Victor Grignard, French chemist, Nobel laureate (Chemistry, 1912), who developed the use of alkyl magnesium halide reagents for the synthesis of a large number of different classes of compounds, particularly secondary and tertiary alcohols, hydrocarbons, and carboxylic acids, died in Lyon, France. (☞ May 6)

1955
António Egas Moniz, Portugese neurologist, Nobel laureate (Physiology or Medicine, 1949), who discovered the therapeutic value of leucotomy in certain psychoses, died in Lisbon, Portugal. (☞ Nov. 29)

1977
A patent for 'a multipoint data communication system with collision detection', commonly known as the Ethernet, was awarded to the Xerox Corporation. The idea was mooted by American technology pioneer Robert Metcalfe who wrote a memo describing a way to transmit data from the early generation of personal computers to a new device, the laser printer. He called his multipoint data communications system Ethernet. Metcalfe's idea was further developed by his team at the Xerox Palo Alto Research Center in California, USA. Reliable and easy to install and set up, Ethernet enabled companies to rapidly interconnect computers.

1993
American space shuttle *Endeavour*, carrying astronauts Richard O. Covey, Kenneth D. Bowersox, F. Story Musgrave, Kathryn C. Thornton, Claude Nicollier, Jeffrey A. Hoffman, and Thomas D. Akers, landed at the Kennedy Space Center, Florida, USA, after repairing the Hubble Space Telescope in orbit. (☞ Dec. 2, Dec. 8, Dec. 10)

December 14

1546
Tycho Brahe, Danish astronomer whose work in developing astronomical instruments and in measuring and fixing the positions of stars paved the way for future discoveries, was born in Knudstrup, Denmark. In 1559 Tycho began his studies at the University of Copenhagen, where he studied law and a variety of other subjects but became interested in astronomy. It was, however, the solar eclipse of 21 August 1560, particularly the fact that it had been predicted, that so impressed him that he began to make his own studies of astronomy helped by some of the professors. In 1562 he joined the University of Leipzig in Germany where he began making observations and kept a record of these observations. He then travelled through Germany, studying further at the universities of Wittenberg, Rostock, and Basel. During this period he bought several astronomical instruments. In a duel with another student, in Wittenberg in 1566, Tycho lost part of his

nose. For the rest of his life he wore a metal insert over the missing part. He returned to Denmark in 1570. In 1572 he observed the new star (a nova) in the constellation of Cassiopeia and published a brief tract about it the following year. In 1574 he gave a course of lectures on astronomy at the University of Copenhagen. Convinced that the improvement of astronomy hinged on accurate observations, he went on to set up a purpose-built observatory, on the island of Hven, with financial help from the king of Denmark. The observatory, called Uraniborg, was equipped with exceptionally large and accurate instruments. At Uraniborg Tycho made twenty years' worth of astronomical observations, which later helped the German astronomer Johannes Kepler in formulating his laws of planetary motion. In 1599 Tycho moved to Prague, Bohemia (now Czech Republic) with Kepler as his assistant. (☞ Oct. 24, Nov. 11)

1873

Louis Agassiz, Swiss-born American naturalist, geologist, and teacher who made revolutionary contributions to the study of natural science with landmark work on glacier activity and extinct fishes, died in Cambridge, Massachusetts, USA. (☞ May 28)

1909

Edward Lawrie Tatum, American biochemist and molecular geneticist, Nobel laureate, was born in Boulder, Colorado, USA. Tatum was educated at the University of Chicago from where he obtained his BA degree in chemistry in 1931 and MSc degree in microbiology in 1932. He obtained his PhD in biochemistry from the University of Wisconsin in 1934. In 1937 he joined the Department of Biological Sciences at the Stanford University, California, as a research associate and later became assistant professor. From 1945 until 1948 he taught at the Yale University, New Haven, Connecticut, and returned to Stanford in 1948. He joined the staff of the Rockefeller Institute for Medical Research (now Rockefeller University), New York, in 1957. Tatum's research was primarily on the biochemistry, nutrition and genetics of micro-organisms and of the fruit fly, *Drosophila melanogaster*. While at Stanford he collaborated with the American geneticist George Wells Beadle in the study of the genetics of eye colour in *Drosophila*. Later they studied the pink bread mould, *Neurospora crassa*. They used X-rays to induce mutations in the mould and showed that when a genetic mutation can be shown to affect a specific chemical reaction, the enzyme catalysing that reaction is altered or missing. Thus, they showed that each gene in some way determines the structure of a specific enzyme, which came to be known as the 'one-gene–one-enzyme' hypothesis.

Tatum was awarded the 1958 Nobel Prize for Physiology or Medicine, jointly with Beadle and Joshua Lederberg. Beadle and Tatum shared half the prize money for their discovery that genes act by regulating definite chemical events; the other half was received by Lederberg for his discoveries concerning genetic recombination and the organisation of the genetic material of bacteria. (☞ May 23, Oct. 22, Nov. 5)

1922

Nicolay Gennadiyevich Basov, Soviet physicist, Nobel laureate, was born in Usman, near Voronezh, Russia. After

finishing secondary school in 1941 in Voronezh, Basov served in the Soviet Army and took part in World War II. In 1945, after the war, he entered the Moscow Institute of Physical Engineers where he studied theoretical and experimental physics, graduating in 1950. He then began working at the P.N. Lebedev Physical Institute of the Academy of Sciences of the Soviet Union (now the Russian Academy of Sciences). In 1955, while working as a research student with Aleksandr Prokhorov at the Soviet Academy of Sciences, he devised a microwave amplifier based on ammonia molecules. The maser principle is based on the phenomenon of stimulated emission, in which atoms emit radiation when their electrons make the transition from a higher to a lower energy level. Later, Basov and Prokhorov worked to develop a practical device based on this principle that would produce cohe-rent microwave radiation. Beginning from 1961 Basov carried out theoretical and experimental research in the field of powerful lasers. He has also made significant contributions in the field of chemical lasers. Basov was awarded the 1964 Nobel Prize, jointly with Prokhorov and the American physicist Charles Townes, for fundamental work in the field of quantum electronics, which has led to the construction of oscillators and amplifiers based on the maser–laser principle. Basov and Prokhorov shared half of the prize money while Townes received the other half. (☞ Jul. 11, Jul. 28)

1962

The American space probe *Mariner-2* became the first spacecraft to fly-by Venus at its closest distance of 34,773 kilometres and sent back photographs and scientific data. Scientific discoveries made by *Mariner-2* included a slow retrograde rotation rate for Venus, hot surface temperatures and high surface pressures, a predominantly carbon dioxide atmosphere, continuous cloud cover with a top altitude of about 60 kilometres, and no detectable magnetic field. (☞ Aug. 27)

1967

The first synthesis of biologically active DNA in a test tube was announced at a press conference by the American biochemist Arthur Kornberg. He chose to replicate the relatively simple DNA chain of the Phi X174 virus which infects bacteria. It has a single strand of DNA only about 5,500 nucleotide building blocks long, and with about 11 genes. Having isolated the Phi X174 DNA, they used the DNA from *E. coli*, a common bacterium found in the human intestine to copy the viral DNA. The viral DNA thus copied was found to be able to infect bacteria showing that it was active DNA. (☞ Mar. 3)

1986

The experimental aircraft *Voyager*, piloted by American aviators Dick Rutan and Jeana Yeager, took off from the Edwards Air Force Base in California, USA, on a record-setting mission to fly around the world without stopping and without refuelling. *Voyager*'s main structure/fuselage was built of space-age composite material consisting mainly of graphite, Kevlar, and fibreglass. The structural weight of *Voyager* was only about 426 kilograms, but when its 17 fuel tanks were full, its takeoff weight exceeded 4,400 kilograms

– more than 10 times its structural weight. (☞ Dec. 23)

1989
Andrey Dmitriyevich Sakharov, Soviet nuclear physicist, an outspoken advocate of human rights and disarmament, died in Moscow, Russia. (☞ May 21)

December 15

1654
The world's first meteorological office was established by Duke Ferdinand II of Tuscany in Tuscany, Italy, with Luigi Antinori in charge. Daily reports and temperature readings were provided by observers in Parma, Milan, Bologna, and Florence. The oldest continuously reporting meteorological station in operation today started in 1763 at the Benedictine Monastery of Kremsmünster in Austria.

1852
Antoine-Henri Becquerel, French physicist, Nobel laureate, was born in Paris, France. Becquerel received his formal scientific education at the École Polytechnique and engineering training at the École des Ponts et Chaussées in Paris. In 1892 he was appointed professor of applied physics in the department of natural history at the Paris Museum. He became a professor at the École Polytechnique in 1895. Becquerel's earliest work was concerned with the plane polarisation of light, with the phenomenon of phosphorescence and with the absorption of light by crystals. He also worked on the subject of terrestrial magnetism. After the discovery of X-rays by the German physicist Wilhelm Röntgen in 1895, Becquerel decided to investigate whether there was any connection between X-rays and naturally occurring phosphorescence. He had inherited from his father a supply of uranium salts, which phosphoresced on exposure to light. When the salts were placed near a photographic plate covered with opaque paper, the plate became fogged. The phenomenon was found to be common to all the uranium salts studied. Later, Becquerel showed that the rays emitted by uranium caused gases to ionise and that they differed from X-rays in that they consisted of charged particles that were deflected by a magnetic field. He had discovered a new phenomenon of spontaneous emission of ionising radiation that was different from X-rays. Becquerel was awarded the 1903 Nobel Prize for Physics, jointly with the French scientist couple Pierre and Marie Curie. Becquerel received half the prize money in recognition of the extraordinary services he has rendered by his discovery of spontaneous radioactivity, while the Curie couple shared the other half in recognition of the extraordinary services they have rendered by their joint researches on the radiation phenomena discovered by Henri Becquerel. (☞ May 15, Aug. 25, Nov. 7)

1860
Niels Ryberg Finsen, Danish physician, was born in Tórshavn, Faroe Islands, Denmark. After schooling at Reykjavik, Finsen went to Copenhagen in 1882 to study medicine, obtaining his MD in 1890. He then taught anatomy at the University of Copenhagen's department of surgery, which he left in 1893 in order to devote himself full-time to studying 'phototherapy', or the therapeutic effects of light. He began to study the use

of filtered sunlight in the treatment of skin lesions caused by smallpox, a viral disease. Red light proved successful in promoting the healing of smallpox lesions. After publishing key papers on phototherapy in 1893 and 1894, Finsen began research into the treatment of *lupus vulgaris*, a disfiguring skin disease caused by bacteria. Using artificial light, Finsen exposed diseased tissue to high concentrations of ultraviolet light, which he called 'chemical rays'. It proved highly effective in treating *lupus vulgaris*. In 1896 Finsen established his Medical Light Institute in Copenhagen, where hundreds of *lupus vulgaris* patients were successfully treated over the next few years. The use of ultraviolet light remained the central treatment for *lupus vulgaris* for decades. Although other forms of radiation and drug therapy have largely superseded phototherapy, Finsen's work did much to encourage the radiation therapy then being developed and led to the use of ultraviolet sterilisation techniques in bacteriological research. Finsen was awarded the 1903 Nobel Prize for Physiology or Medicine in recognition of his contribution to the treatment of diseases, especially *lupus vulgaris*, with concentrated light radiation, whereby he has opened a new avenue for medical science. (☞ Sep. 24)

1916

Maurice Hugh Frederick Wilkins, New Zealand-born British biophysicist, Nobel laureate, was born in Pongaroa, New Zealand. Wilkins was educated at the King Edward's School, Birmingham. He studied physics at the St. John's College, Cambridge, taking his bachelor's degree in 1938, and obtained his PhD from the Birmingham University in 1940. His thesis contained his original formulation of the electron-trap theory of phosphorescence and thermoluminescence. During World War II he participated in the Manhattan Project at the University of California, Berkeley, working on mass spectrographic separation of uranium isotopes for use in the atomic bomb. His interest shifted to biophysics after he joined the St. Andrews' University, Scotland, where a course on biophysical studies was being organised. When the biophysics project moved in 1946 to the King's College, London, he became a member of the staff of the newly formed Medical Research Council Biophysics Research Unit. He became assistant director of the Medical Research Council Unit in 1950 and its deputy director in 1955. At the King's College, Wilkins carried out a series of investigations that led ultimately to his X-ray diffraction studies of DNA, which proved crucial to the determination of the molecular structure of DNA by the American geneticist James Watson and British biophysicist Francis Crick. He later applied X-ray diffraction techniques to the study of ribonucleic acid. Wilkins was awarded the 1962 Nobel Prize for Physiology or Medi-cine, jointly with Watson and Crick, for their discoveries concerning the molecular structure of nucleic acids and its significance for information transfer in living material. (☞ Apr. 6, Apr. 25, Jun. 8)

1923

Freeman John Dyson, English-born American physicist and mathematician, Fellow of the Royal Society, was born in Crowthorne, Berkshire, England. Dyson

studied at the Winchester College, and obtained his BA degree in mathematics from the University of Cambridge in 1945. During World War II he worked as an analyst for the British Bomber Command. Dyson returned to England in 1949 to become a research fellow at the University of Birmingham, but he was appointed professor of physics at the Cornell University, Ithaca, New York, USA, in 1951 and two years later he became professor of physics at the Institute for Advanced Study, Princeton, New Jersey, where he remained till his retirement in 1954; since then he has been Professor Emeritus. He became an American citizen in 1957. Dyson is best known for his speculative work on extraterrestrial civilisations. A long-time advocate of exploration and colonisation of the solar system and beyond, Dyson studied ways of searching for evidence of intelligent extraterrestrial life. He proposed that a highly advanced technological civilisation would ultimately completely surround its host star with a huge shell to capture 100% of the useful radiant energy. This 'Dyson shell', viewed from other galaxies would have a highly distinctive, unnatural light. He suggested that a search for such telltale coloured stars, might reveal advanced, intelligent life. Dyson was a prolific author. He wrote several books, including *Disturbing the Universe, an Autobiography*; *Weapons and Hope*; *Origins of Life*; *Imagined Worlds*; *From Eros to Gaia;* and *Infinite in All Directions*.

1958
Wolfgang Pauli, Austrian-born American physicist, Nobel laureate (Physics, 1945), who discovered the exclusion principle named after him, died in Zurich, Switzerland. (☞ Apr. 25)

1970
Soviet unmanned spacecraft *Venera-7* entered the atmosphere of Venus and released a landing capsule, which after aerodynamic braking deployed a parachute system for landing. It sent back data for 23 minutes after landing before it was destroyed by atmospheric pressure on Venus. *Venera-7* was the first man-made object to return data after landing on another planet. (☞ Aug. 17)

1984
Soviet spacecraft *Vega-1* was launched on a mission to Venus and Comet Halley. (☞ Mar. 6, Jun. 11)

1998
The space shuttle *Endeavour*, carrying a crew of six comprising Robert D. Cabana, Frederick W. Sturckow, J. Currie, Jerry L. Ross, James H. Newman, and Sergei K. Krikalev landed at the Kennedy Space Center, Florida, USA, after beginning the assembly of the *International Space Station* in orbit. (☞ May 7, Dec. 4)

December 16

1901
Margaret Mead, American anthropologist best known for her studies of the non-literate peoples of Oceania, was born in Philadelphia, Pennsylvania, USA. Mead graduated from the Barnard College in 1923 and entered the graduate school of the Columbia University, New York, from where she received an MA in 1924 and a PhD in 1929. A keen observer, Mead went on her first field trip to the South Pacific in 1925, during which she gathered material for her bestseller *Coming of Age in Samoa* that was published in 1928. The book dealt

with various aspects of comparative child psychology, oceanic ethnology, cooperation and competition among primitive peoples, and cross-cultural communications. Her other books include *Male and Female: A Study of the Sexes in a Changing World, Anthropology: A Human Science,* and *Culture and Commitment*. She continued her research throughout her life in such locations as New Guinea, Samoa, Bali, and many other places.

Mead served the American Museum of Natural History in New York in various capacities – as assistant curator from 1926 to 1942; associate curator from 1942 to 1964; curator of ethnology from 1964 to 1969, and curator emeritus from 1969 to 1978. In 1973 she was elected president of the American Association for the Advancement of Science. (☞ Nov. 15)

1917

Arthur Charles Clarke, British author and inventor, was born in Minehead, Somerset, England. Clarke attended secondary school, but was unable to afford university and consequently acquired a job as an auditor in the pensions section of the Board of Education.

In 1936 he moved to London, where he joined the British Interplanetary Society. As an RAF officer during World War II, he was in charge of the first radar talk-down equipment, the Ground Controlled Approach, during its experimental trials. In the early 1940s, while in the RAF, Clarke began selling his science-fiction stories to magazines. He worked briefly as assistant editor of *Science Abstracts* before devoting himself to writing full-time from 1951. His most important contribution was the idea that geostationary satellites could be ideal telecommunications relays. He proposed this concept in a scientific paper titled 'Can rocket stations give worldwide radio coverage?' published in *Wireless World* in October 1945. The geostationary orbit is now known as the 'Clarke orbit' in his honour. He obtained first class honours in physics and mathematics at the King's College, London in 1948. In 1964, Clarke started to work with Stanley Kubrick in a science fiction movie script. After four years, he shared an Oscar Academy Award nomination with him for the film version of *2001: A Space Odyssey*. Clarke has lived in Colombo, Sri Lanka since 1956 and has been doing underwater exploration along that coast and the Great Barrier Reef. There is an asteroid named in his honour, '4923 Clarke', as well as a species of Ceratopsian dinosaur, *Serendipaceratops arthurcclarkei*, discovered in Inverloch in Australia. Apart from his fiction, Clarke has written two autobiographies. *Ascent to Orbit* is what he calls his scientific autobiography and *Astounding Days* his science-fictional autobiography.

1973

The first 200-MW unit of the Rajasthan Atomic Power Project at Rawatbhata, near Kota, Rajasthan, was commissioned. It is a pressurised heavy water reactor that uses natural uranium as fuel and heavy water as moderator and coolant. (☞ Apr. 1, Jun. 1, Dec. 23)

December 17

1778
Humphry Davy, English chemist who discovered several chemical elements and compounds, and invented the miner's safety lamp, was born in Penzance, Cornwall, England. Davy received his education in Penzance and in Truro. In 1794 he was apprenticed to a surgeon and apothecary, and hoped eventually to qualify in medicine. In 1798 he became superintendent of the Medical Pneumatic Institution of Bristol. This organisation was devoted to the study of the medical value of various gases. Davy studied the oxides of nitrogen and discovered the physiological effects of nitrous oxide, which became known as laughing gas. He also investigated gases like ammonia and water gas and published the results in *Researches, Chemical and Philosophical*, in 1800. While studying the chemical effects of electricity, Davy found that when he passed electrical current through some substances, it led to chemical decomposition. He concluded that the production of electricity in simple electrolytic cells also resulted from chemical action. In 1807 Davy discovered the alkali metals, potassium and sodium, isolating them with electric current for the first time; he discovered the alkaline earth metals calcium, strontium, barium, and magnesium in 1808. He invented the miner's safety lamp around 1815 after studying the conditions under which mixtures of firedamp and air explode in coalmines. In 1820 he became president of the Royal Society, a position he held until 1827. (☞ Apr. 9, May 29, Oct. 6, Oct.13, Oct. 31)

1903
The first powered flight of an aeroplane took place. The American inventors and aviation pioneers, Wilbur and Orville Wright, flew their first heavier-than-air machine near Kitty Hawk in North Carolina, USA. During the first flight, with Orville at the controls, the plane flew 37 metres in 12 seconds. The brothers made three more flights that day. The longest, by Wilbur, was 260 metres in 59 seconds. (☞ Apr. 16, Aug. 19)

1907
Lord Kelvin, Scottish physicist, mathematician and engineer, Fellow of the Royal Society, after whom the absolute scale of temperature is named, died in Netherhall, near Largs, Ayrshire, Scotland. (☞ Jun. 26)

1908
Willard Frank Libby, American chemist, Nobel laureate, was born in Grand Valley, Colorado, USA. Libby attended grammar and high schools near Sebastopol, California, moving to the University of California at Berkeley in 1927, where he received his BSc in 1931 and PhD in 1933 after which he was appointed instructor in the department of chemistry at the same University, and became associate professor of chemistry later. While working for the Manhattan Project at the Columbia University in

New York during World War II, Libby helped develop a method for separating uranium isotopes, an essential step in the creation of the atomic bomb. In 1946 he showed that tritium, the heaviest isotope of hydrogen, was produced by cosmic radiation. Libby specialised in radiochemistry, particularly hot atom chemistry, tracer techniques, and isotope tracer work. He became well known for his work on natural carbon-14 (radiocarbon) and its use in dating archaeological artefacts, and natural tritium, and its use in hydrology and geophysics. The radiocarbon dating method can be used to date material derived from carbon-containing artefacts as old as 50,000 years. In 1945, at the end of World War II, , Libby was appointed professor of chemistry in the department of chemistry and Institute for Nuclear Studies (now the Enrico Fermi Institute for Nuclear Studies) of the Chicago University, where he remained till 1954, when he was appointed a member of the US Atomic Energy Commission. In 1959, Libby became professor of chemistry in the University of California at Los Angeles, and in 1962 he was appointed director of the Institute of Geophysics and Planetary Physics at the same university. Libby was awarded the 1960 Nobel Prize for Chemistry for his method of using carbon-14 for age determination in archaeology, geology, geophysics, and other branches of science. (☞ Sep. 8)

1931

The Indian Statistical Institute (ISI) was founded in Calcutta (now Kolkata), West Bengal. The first important work in statistics in the modern sense in India was undertaken by the Indian historian of science B.N. Seal, who sought the help of the Indian statistician Prasanta Chandra Mahalanobis in the statistical analysis of the examination results of the Calcutta University. Mahalanobis founded the ISI as a learned society housed in the Statistical Laboratory of the Presidency College in the city. Until the mid-1930s, all or nearly all the statistical work done in India was done single-handedly by Mahalanobis. The early statistical studies included analyses of data on stature of Anglo-Indians, meteorological data, rainfall data, data on soil conditions, etc. Some of the findings of these early studies were of great impact in the control of floods, development of agriculture, etc., and led to the recognition of statistics as a key discipline. Theoretical research in statistics began to flourish in the Institute. Research on large-scale sample surveys won Mahalanobis a Fellowship of the Royal Society. The first exploratory sample survey in India was undertaken at the institute as early as in 1937 for improving estimates of jute crop. The pioneering work on crop yield estimation done by the institute attracted the attention of the then Prime Minister Jawaharlal Nehru, who desired that comprehensive information relating to social, economic and demographic characteristics be collected by conducting sample surveys on a countrywide basis. The institute was asked by the Government to take charge of the design of these surveys and of processing of data thereof. The National Sample Survey (NSS) was established in 1950. This led to the regular conduct of the most comprehensive sample surveys in the world at that time, on a regular basis. At present, research in statistics and related disciplines is the primary activity of the Institute. It has two centres located in Delhi and Bangalore, Karnataka. Teaching activities are undertaken at Kolkata, Delhi and Bangalore. (☞ Jun. 29)

1947
Johannes Nicolaus Brønsted, Danish physical chemist, Fellow of the Royal Society, known for the widely applicable acid–base concept that an acid gives up a hydrogen ion, while a base combines with one, died in Copenhagen, Denmark. (☞ Feb. 22)

1964
Victor Francis Hess, Austrian-born physicist, Nobel laureate (Physics, 1936), who discovered cosmic rays, died in Mount Vernon, New York, USA. (☞ Jun. 24)

December 18

1829
Jean-Baptiste Lamarck, French biologist who is best known for his idea that acquired traits are inheritable, died in Paris, France. (☞ Aug. 1)

1856
Joseph John Thomson, English physicist, Nobel laureate, was born in Cheetham Hill, near Manchester, England. Thomson was educated at Owens College, Manchester, and in 1876 entered the Trinity College, Cambridge as a minor scholar, where he remained for the rest of his life, becoming lecturer in 1883 and master in 1918. He was Cavendish Professor of Experimental Physics at the Cambridge University from 1884 to 1918. After taking his BA degree in mathematics in 1880, the opportunity of doing experimental research drew him to the Cavendish Laboratory. He also began to develop the theory of electromagnetism. In 1896, Thomson visited USA to give a course of four lectures at the Princeton University, New Jersey, which summarised his current researches. These lectures were subsequently published as *Discharge of Electricity through Gases* in 1897. On his return from USA, he started experimenting with currents of electricity inside empty glass tubes, investigating a long-standing puzzle known as the 'cathode rays'. His experiments prompted him to make a bold proposal that these mysterious rays were streams of particles much smaller than atoms. He called these particles 'corpuscles', and suggested that they might make up all of the matter in atoms. Thomson's study of cathode rays culminated in the discovery of the electron, which was announced during the course of his Friday evening lecture at the Royal Institution, London in April 1897. Thomson was awarded the 1906 Nobel Prize for Physics in recognition of the great merits of his theoretical and experimental investigations on the conduction of electricity by gases. (☞ Apr. 30, Aug. 30)

1936
Andrija Mohorovicic, Croatian meteorologist and geophysicist who discovered the boundary between the Earth's crust and mantle, died in Zagreb, Yugoslavia. (☞ Jan. 23)

1939
Harold E. Varmus, American virologist, Nobel laureate, was born in Oceanside, New York, USA. Varmus graduated from the Amherst College, Massachusetts in 1961, and obtained his master's degree from the Harvard University, Cambridge, Massachusetts, in 1962. After doing his MD from the Columbia University, New York in 1966, Varmus worked as a medical house officer at the Columbia-Presbyterian Hospital from 1966 to

1968, and then joined the National Cancer Institute, Bethesda, Maryland, as a clinical associate. There he studied bacterial gene regulation by cyclic AMP. In 1970 he joined the University of California, San Francisco (UCSF) as post-doctoral fellow and was appointed lecturer shortly thereafter. In 1972 he became a regular member of the faculty in the department of microbiology and immunology, ascending to the rank of professor by 1979. Most of the research of Vermus at UCSF concerned the behaviour of retroviruses: various aspects of their unusual life cycle, the nature and origin of their transforming genes, and their potential to cause genetic change. Much of this work was performed in collaboration with fellow American virologist J. Michael Bishop. Their partnership and subsequent research into cancer genes would change the course of cancer research. Varmus and Bishop found that, under certain circumstances, normal genes in healthy cells of the body can cause cancer; these genes, termed 'oncogenes', ordinarily control cellular growth and division, but, if they are picked up by infecting viruses or affected by chemical carcinogens, they can cause cancer. This finding was contrary to the earlier held theory that cancer is caused by viral genes, distinct from a cell's normal genetic material, that lie dormant in body cells until activated by carcinogens. Varmus was awarded the 1989 Nobel Prize for Physiology or Medicine, jointly with Bishop, for their discovery of the cellular origin of retroviral oncogenes. (☞ Feb. 22)

December 19

1852

Albert Abraham Michelson, German-born American physicist, Nobel laureate, was born in Strelno, Prussia (now Strzelno, Poland). Michelson matriculated from the High School in San Francisco, USA, in 1869. He graduated from the US Naval Academy at Annapolis, Maryland in 1873 and became an instructor in physics and chemistry at the Academy in 1875. He travelled to Europe in 1880 and spent two years in Berlin and Heidelberg in Germany, and Paris in France. He resigned from the US Navy in 1881. While in Europe, Michelson began constructing an interferometer, a device designed to split a beam of light in two, send the parts along perpendicular paths, then bring them back together. He wanted to use the interferometer to measure the Earth's velocity against the 'aether' that was then thought to permeate the entire universe. If the Earth were travelling through the aether, then the speed of the light travelling in the same direction would be expected to be equal to the velocity of light plus the velocity of the Earth, whereas the speed of light travelling at right angles to the Earth's path would be expected to travel only at the velocity of light. After returning to USA in 1883, Michelson joined the Case School of Applied Science in Cleveland as a professor of physics and concentrated his efforts on improving the accuracy of his interferometer experiment. Here, in cooperation with Edward Williams Morley, and using the interferometer, he showed that light travels at a constant speed in all inertial systems of reference, thus disproving the existence of aether. This

Michelson–Morley experiment eventually led the German physicist Albert Einstein to his theory of relativity. The interferometer also enabled distances to be measured with greater accuracy by means of the length of light-waves. Michelson was awarded the 1907 Nobel Prize for Physics for his optical precision instruments and the spectroscopic and metrological investigations carried out with their aid. (☞ Jan. 29, May 9)

1871
Albert L. Jones of New York, USA, was issued the first US patent for corrugated paper as an 'improvement in paper for packing'. He used it as a packing material, for shipping kerosene-lamp chimneys and other glass, in place of messy sawdust and straw. It was not until later that a facing sheet was applied to one or two sides to make single-face or double-face corrugated cardboard. Boxes made from corrugated paper came into use in about 1890.

1903
George Davis Snell, American geneticist, Nobel laureate, was born in Bradford, Massachusetts, USA. Snell obtained his bachelor's degree in biology from the Dartmouth College in 1926 and received a PhD in genetics from the Harvard University, Cambridge, Massachusetts, in 1930. During 1933–34, with a National Research Council Fellowship, he worked under the American geneticist Hermann J. Muller at the University of Texas, and then moved to the Washington University in St Louis as an assistant professor. In 1935, Snell joined the Jackson Laboratory in Bar Harbor, Maine, where he carried out his research in mouse genetics, focussing on the genetics of transplantation. His collaboration with British geneticist, Peter Gorer led to the identification of a group of genes in the mouse called the H-2 gene complex, where H stands for histo-compatibility – the suitability of cells from one tissue to withstand the presence of cells from another tissue. The histo-compatibility genes encode cell surface proteins that allow the body to distinguish its own cells from those that are foreign and play a major role in rejection of transplanted tissue. Snell's studies of histo-compatibility in mice resulted in the identification of the H-2 gene complex and subsequently to the recognition of the major histo-compatibility complex (MHC), an assortment of antigens (substances that cause the production of antibodies) that is common to the genetic makeup of all vertebrates. His work also increased the understanding of autoimmune diseases. Snell was awarded the 1980 Nobel Prize for Physiology or Medicine, jointly with the French immunologist Jean Dausset and Venezuelan-born American immunologist Baruj Benacerraf, for their discoveries concerning genetically determined structures on the cell surface that regulate immunological reactions. (☞ Oct. 19, Oct. 29)

1910
Rayon, also known as artificial silk, was first commercially produced, in Marcus Hook, Philadelphia, USA, by the American Viscose Company. It was not until 1924 that the term 'rayon' was adopted. Rayon is a re-generated fibre – cellulose is converted to a liquid compound and then

back to cellulose in the form of fibre. The cellulose, obtained from soft woods or from the short fibres adhering to cottonseeds, is chemically treated to form a solution that is forced through tiny holes in a nozzle. The emerging filament is hardened by drying in air or by chemical means. The filament is sometimes cut into shorter pieces having uniform length, called staple, and twisted together to make yarn. (☞ Sep. 30)

1915
Alois Alzheimer, German psychiatrist, who recognised the progressive, degenerative disorder of the brain named after him, died in Breslau (now Wroclaw, Poland). (☞ Jun. 14)

1946
Paul Langévin, French physicist and inventor of the sonar, died in Paris, France. (☞ Jan. 23)

1953
Robert Andrews Millikan, American physicist, Nobel laureate (Physics, 1923), who measured the charge of single electrons with his famous oil-drop experiment and coined the term 'cosmic rays' during his study of the radiation from outer space, died in San Marino, California, USA. (☞ Mar. 22)

1958
The first radio broadcast from outer space was transmitted. The American president Dwight D. Eisenhower's voice issued a Christmas greeting from a pre-recorded tape on a recorder aboard the world's first communications satellite, *Project SCORE* (*S*ignal *C*ommunication by *O*rbiting *R*elay *E*quipment), which had been launched two days earlier. The battery-operated 132 MHz all-vacuum-tube transmitter had an output of eight watts.

1972
The command module of the American spacecraft *Apollo-17*, carrying astronauts Eugene A. Cernan, Ronald E. Evans, and Harrison H. Schmitt, safely splashed down in the Pacific Ocean. (☞ Dec.7, Dec. 11)

1997
Masaru Ibuka, Japanese industrialist who patented the transistor and launched the modern electronics industry, died in Tokyo. (☞ Apr. 11)

December 20

1805
Thomas Graham, Scottish chemist often referred to as 'the father of colloid chemistry', was born in Glasgow, Scotland. Graham graduated from the Glasgow University in 1826, did his postgraduate study in Edinburgh. In 1828, he returned to Glasgow to work as an industrial consultant, and was appointed lecturer in chemistry and mechanics at the Glasgow Mechanics Institution the following year. In 1830, he became a professor of chemistry at the Anderson's College (later the Royal College of Science and Technology) in Glasgow. In 1933 Graham proposed the law named after him, which stated that the rate of diffusion of a gas is inversely proportional to the square root of its molecular weight. Later he extended this work to the diffusion of one liquid into another and these investigations eventually led to his division of solutes into two classes – crystalloids, such as salt and sugar; and colloids, such as gum arabic and the finely divided gold suspensions. This marked the beginning of colloid chemistry. Colloids

are solutions in which the dispersed particles are between 10^{-7} and 10^{-4} cm in diameter and cannot be separated by filtration or gravity alone. He developed dialysis as a means of separating colloidal solutions from electrolytes – the technique used in hospitals to treat patients with malfunctioning kidney. Graham studied the three forms of phosphoric acid, and from this work the concept of polybasic acids developed. He was also the first to observe that palladium metal is able to absorb large amounts of hydrogen gas, especially at low temperatures. (☞ Sep. 11)

1890

Jaroslav Heyrovsky, Czech chemist, Nobel laureate, was born in Prague, Bohemia (now Czech Republic). Heyrovsky joined the Czech University, Prague in 1909. From 1910 he continued his studies at the University College, London, taking his BSc degree in 1913. He obtained his PhD degree from the Czech University, Prague in 1918 and DSc from the University of London in 1921. Heyrovsky started his university career as assistant in the Institute of Analytical Chemistry of the Charles University, Prague. He was promoted to associate professor in 1922 and in 1926 he became the first professor of physical chemistry at this university. Heyrovsky invented the polarographic method of analysis in 1922 and concentrated his whole further scientific activity on the development of this new branch of electrochemistry.

Polarography is an instrumental method of chemical analysis used for qualitative and quantitative determinations of reducible or oxidisable substances. Heyrovsky's instrument measures the current that flows when a predetermined potential is applied to two electrodes immersed in the solution to be analysed. The cathode used is a small drop of mercury (constantly forming and dropping to keep the surface clean). The voltage is increased slowly and the current plotted against voltage. The current increases in steps, each corresponding to a particular type of positive ion in the solution. The height of the steps indicates the concentration of the ion. Polarography has become one of the most versatile of all analytical techniques. Heyrovsky was awarded the 1959 Nobel Prize for Chemistry for his discovery and development of the polarographic methods of analysis. (☞ Mar. 27)

1901

Robert Jemison Van de Graaff, American physicist and inventor of the high-voltage electrostatic generator and particle accelerator named after him, was born in Tuscaloosa, Alabama, USA. Van de Graaff studied in public schools in Tuscaloosa and then attended the University of Alabama where he received a BS degree in mechanical engineering in 1922 and an MS degree in the same subject in 1923. He studied at the Sorbonne in Paris from 1924 to 1925 and in 1925 went to the Oxford University in England as a Rhodes Scholar. He received a BS in physics in 1926 and a PhD in physics in 1928 at Oxford. While at Oxford, he became aware that subatomic particles could someday be accelerated to speeds sufficient to disintegrate nuclei. By disintegrating atomic nuclei much could be learned about the nature of individual atoms. It is from these ideas that the idea of a particle accelerator occurred to him. Van de Graaff returned to the United States in 1929 to join the Palmer Physics Laboratory at the Princeton University, New Jersey, as a national research fellow. The same year he constructed the first working model of his electrostatic accelerator, which

developed 80,000 volts. He built an improved version in 1931 that could produce over 1,000,000 volts. By 1933 he built a machine that could produce a whopping 7,000,000 volts. In 1935 he received a patent for his invention. Van de Graaff's electrostatic accelerator has found widespread use not only in atomic research but also in medicine and industry. Van de Graaff became a research associate in 1931 and an associate professor in 1934 at the Massachusetts Institute of Technology, Cambridge, Massachusetts. In 1960 he left MIT to devote his whole time to the High Voltage Engineering Corporation, which he had co-founded to manufacture his accelerator. (☞ Jan. 16)

1996
Carl Edward Sagan, American astronomer and exobiologist and science writer, who stimulated popular interest in these subjects through his enthusiastic writings, lectures, and the television series *Cosmos*, died in Seattle, Washington, USA. (☞ Nov. 9)

1998
Alan Lloyd Hodgkin, British physiologist, Nobel laureate (Physiology or Medicine, 1963), who discovered the chemical processes involved in nerve conduction, more specifically, discoveries concerning the ionic mechanisms involved in excitation and inhibition in the peripheral and central portions of the nerve cell membrane, died in Cambridge, England. (☞ Feb. 5)

2002
Grote Reber, American astronomer and radio engineer who built the first radio telescope and was largely responsible for the early development of radio astronomy, died in Tasmania, Australia. (☞ Dec. 22)

December 21

1773
Robert Brown, Scottish botanist best known for his description of the natural continuous motion of minute particles in solution, commonly known as 'Brownian movement', was born in Montrose, Scotland. He was educated at the Marischal College, Aberdeen, and studied medicine at the Edinburgh University but did not obtain his degree. In 1800, Brown joined as a naturalist on board the British Navy ship *Investigator*, which was to embark on a surveying voyage along the northern and southern coasts of Australia. During the next one and a half years, while the ship circumnavigated Australia, Brown made extensive plant collections. During the first three weeks, he collected more than 500 species of plants. Returning to England in October 1805 with vast collections of drawings and notes and many zoological specimens, Brown devoted his time to classifying the approximately 3,900 species he had gathered, almost all of which were new to science. He was the first to recognise the fundamental distinction between the conifers and their allies (gymnosperms) and the flowering plants (angiosperms). He also recognised and coined the name for the cell 'nucleus'. Brown is best known for his description of the natural

continuous motion of fine particles that is named after him. His discovery of the Brownian movement was published in a pamphlet in 1828. In it he recorded that, after having noticed moving particles suspended in the fluid within living pollen grains of *Clarkia pulchella*, he examined both living and dead pollen grains of many other plants and observed a similar motion in the particles of all fresh pollen. Brown's experiments with organic and inorganic substances reduced to a fine powder and suspended in water, revealed such motion to be a general property of matter in that state. He was the President of the Linnaean Society from 1849 to 1853. (☞ Jun. 10)

1824

James Parkinson, English physician and amateur palaeontologist, who was the first to describe the neuromuscular disease now known by his name, died in London, UK. (☞ Apr. 11)

1890

Hermann Joseph Müller, American geneticist, Nobel laureate, was born in New York, USA. Müller was brought up in Harlem, attending the Morris High School in the Bronx. There he and his classmates founded what was perhaps the first high-school science club. Müller attended the Columbia University from 1907 to 1909 earning his BA in 1910 and his PhD in 1916. Müller began his career with the American zoologist and geneticist Thomas Hunt Morgan, who had just introduced the fruit fly *Drosophila* as a tool in experimental genetics. While studying mutations in fruit flies, Müller grew impatient with the mutation rate in Drosophila. He found that the mutation rate could be increased by heat. Next, he experimented with X-rays as a means of inducing mutations, and by 1926 he had proved the method successful. In doing so, he was the first to demonstrate radiation-induced genetic alterations! Moreover, he did so in a quantitative manner that determined the mutation frequency. His discovery of artificially induced mutations in genes had far-reaching consequences. Concerned about the deleterious effect of X-rays, Müller campaigned against the needless use of X-rays in diagnosis and treatment, and pressed for safety regulations to ensure that people who were regularly exposed to X-rays were adequately protected. Müller was awarded the 1946 Nobel Prize for Physiology or Medicine for the discovery of the production of mutations by means of X-ray irradiation. (☞ Apr. 5)

1968

American manned lunar mission *Apollo-8*, carrying astronauts Frank Borman, James A. Lovell, Jr., and William Anders, was launched towards the Moon by the Saturn-V rocket from the Kennedy Space Center, Florida, USA. The spacecraft was to orbit the Moon and return to Earth without landing. (☞ Dec. 23, Dec. 27)

1978

Soviet unmanned spacecraft *Venera-12* made a soft landing on Venus after parachuting through the planet's dense atmosphere and sent back data for 110 minutes. Although the lander carried two colour cameras, no picture was returned. After releasing the lander, the main spacecraft flew past the planet and investigated the nature of interplanetary space. (☞ Sep. 14)

1984
Soviet spacecraft *Vega-2* was launched on a mission to Venus and Comet Halley. (☞ Mar. 9, Jun. 15)

1988
Nikolaas Tinbergen, Dutch-born British zoologist and ethologist, Nobel laureate (Physiology or Medicine, 1973), one of the founders of ethology, the branch of biology that studies animal behaviour, died in Oxford, UK. (☞ Apr. 15)

1988
The world's largest aircraft *AN-225* was flown for the first time. With a wingspan of 88.4 metres and a length of 84 metres, the six-engine plane can carry ultra heavy and/or oversize freight weighing up to 250 tonnes and has a range of 13,000 km. The *AN-225* was primarily designed to carry the Soviet *Buran* shuttle on its back for aerodynamic tests before actual flight. The *Buran* was to be the Soviet counterpart of America's space shuttle. But the collapse of the Soviet Union in 1991 led to the scrapping of its space shuttle programme and the *AN-225* was mothballed. However, by the late 1990s it became apparent that there was a need for a transport aircraft larger than the *AN-124*, the largest aircraft in service at that time. In response to this need, the original *AN-225* was re-fitted with an engine, modified for heavy cargo transportation, and placed back in service under the management of Antonov Airlines. It still remains the world's largest aircraft in service. It is used for transporting objects once thought impossible to move by air, such as locomotives, and 150-tonne generators, and has become a valuable asset to international relief organisations for its ability to quickly transport huge quantities of emergency supplies during disaster relief operations.

December 22

1818
The first successful human-to-human blood transfusion was performed by James Blundell, a British obstetrician, at Guy's Hospital in London, for the treatment of postpartum haemorrhage. Blundell became interested in blood transfusions after having witnessed women going into shock and dying from haemorrhage within a few days of giving birth. Blundell extracted a small amount of blood (about 350 ml) from a man's arm and, using a syringe, he successfully transfused the blood to the wife, and it worked. Between 1825 and 1830, he performed ten documented transfusions, five of which proved beneficial to his patients, and published these results in the medical journal *The Lancet*, advocating blood transfusions only in cases where haemorrhaging endangered life. He also devised various instruments for performing blood transfusions.

1872
Camille Guérin, French veterinarian and co-developer, with the French bacteriologist Albert Calmette, of 'Bacillus Calmette-Guérin', or BCG vaccine that was widely used in combating tuberculosis, was born in Poitiers, France. After graduation, he entered the prestigious Ecole Vétérinaire de Maisons-Alfort in Paris. He joined Calmette at the Pasteur Institute in Lille

in 1897, and from then on he devoted his efforts to vaccine research. The two scientists worked with a strain of bacteria that cause bovine tuberculosis to reduce their virulence. After 13 years they produced a strain in 1921 that they believed was harmless to humans but retained its power to stimulate antibody formation. In 1922 they first used it to vaccinate newborn infants at the Charité Hospital in Paris for immunisation against tuberculosis. In 1948 Guérin became chairman of the first international congress on BCG. He was elected president of the Académie Vétérinaire of France in 1949, and president of the Académie de Médecine in 1951. (☞ Jun. 9, Jul. 12, Oct. 29)

1877

The liquefaction of oxygen was announced by the Swiss chemist and physicist Raoul Pierre Pictet. He sent a telegram from Geneva to the French Academy, which stated, 'oxygen liquefied today under 320-atm and 140 degrees of cold by combined use of sulfurous and carbonic acid'. Similar success was reported by French physicist Louis Cailletet two days later. Both used what is known as the Joule–Thomson effect, which involves the expansion of a compressed gas at a temperature below its inversion temperature (the temperature below which the Joule–Thomson effect produces cooling), leading to cooling of the gas. This method is generally not able to bring about liquefaction in one step, but Pictet and Caillietet succeeded liquefying oxygen by repeating the process several times, cascading the effects. Pictet could produce only a few drops of liquid oxygen, but the success marked the end of the idea of permanent gases and established the possibility of liquefying any gas by moderate compression at temperatures below the critical temperature.

1887

Srinivasa Ramanujan, Indian mathematician, Fellow of the Royal Society, was born in Erode, Tamil Nadu. Ramanujan's mathematical genius was evident from early childhood and at the age of 13, he had already plunged into serious arithmetic and geometry. The turning point in his life came when he chanced upon the book *Synopsis of Elementary Results in Pure and Applied Mathematics*, by George Shoobridge Carr. The book contained theorems, formulae and short mathematical proofs. It was this book that triggered the mathematical genius in him. Ramanujan was so fascinated by mathematics that he started to solve problems on loose sheets of paper and jot down the results in his notebooks. But that led him nowhere. He could not pass exams and nobody would give him a job. After much effort, Ramanujan got a job as a clerk at the Madras Port Trust. The turning point in Ramanujan's life came in 1913, when he sent a letter to the renowned British mathematician G. H. Hardy. He set out 120 theorems and formulae and also gave a key formula in the hypergeometric series, which came to be known after him. After going through the theorems and formulae, Hardy immediately realised Ramanujan's genius and invited him to visit Cambridge, England. Ramanujan arrived in Cambridge in mid-April of 1914. During his five years stay in Cambridge, he published 21 papers, five of which were in collaboration with G.H. Hardy. His achievements at Cambridge included the Hardy–Ramanujan

circle method in number theory and the Roger–Ramanujan identities in partition of integers. He also worked on composite numbers, algebra of inequalities, probability theory, and continued fractions. In 1916 Ramanujan was awarded the BA degree by research of the Cambridge University. He was elected a Fellow of the Royal Society of London in 1918. In October the same year he was elected to a Trinity College Fellowship – the first Indian to be elected Fellow of Trinity College. (☞ Apr. 26)

1903

Haldan Keffer Hartline, American physiologist, Nobel laureate, was born in Bloomsburg, Pennsylvania, USA. Hartline began the study of the electrical responses of the retina of the eye as a National Research Council Fellow at the Johns Hopkins University, Baltimore, receiving his MD in 1927. After attending the universities of Leipzig and Munich in Germany as an Eldridge Johnson travelling research scholar, he became professor of biophysics and chairman of the department at Johns Hopkins in 1949. He joined the Rockefeller University, New York City, in 1953 as professor of neurophysiology. Using minute electrodes in his experiments with the horseshoe crab and frog, he obtained the first record of the electrical impulses sent by a single optic nerve fibre when the receptors connected to it are stimulated by light. He found that the receptor cells in the eye are interconnected in such a way that when one is stimulated, others nearby are depressed, thus enhancing the contrast in light patterns and sharpening the perception of shapes. Hartline thus built up a detailed understanding of the workings of individual photoreceptors and nerve fibres in the retina, and showed how simple retinal mechanisms constitute vital steps in the integration of visual information. Hartline was awarded the 1967 Nobel Prize for Physiology or Medicine, jointly with American biochemist George Wald and Finnish-born Swedish physiologist Ragnar Granit for their discoveries concerning the primary physiological and chemical visual processes in the eye. (☞ Mar. 17, Oct. 30, Nov. 18)

1911

Grote Reber, American astronomer and radio engineer who built the first radio telescope and was largely responsible for the early development of radio astronomy, was born in Wheaton, Illinois, USA. Reber graduated from the Illinois Institute of Technology in 1933 with a degree in radio engineering. He was a ham radio operator, and worked for various radio manufacturers in Chicago from 1933 to 1947. When he learned of Karl Jansky's discovery of radio waves from space in 1933, he decided that this was the field he wanted to work in. Unable to get a job, he decided to build his own radio telescope in his back yard. His design was considerably more advanced than Jansky's, consisting of a parabolic sheet metal mirror 9 metres in diameter, focussing to a radio receiver 8 metres above the mirror. The entire assembly was mounted on a tilting stand, which allowed it to be pointed in various directions, although not turned. The telescope was completed in 1937. In 1941, Reber produced a radio map of the sky, based on a series of systematic observations. His radio astronomy work continued

over the next several years. In 1951 he built a new radio telescope in Hawaii and concentrated on mapping celestial sources of low-frequency long-wave signals, 5.5 to 14 metres in wavelength. In 1954 he joined the Commonwealth Scientific and Industrial Research Organization in Tasmania, Australia. Reber was the first to express received radio signals in terms of flux density and brightness, first to find evidence that galactic radiation is non-thermal, and first to produce radio maps of the sky. (☞ Dec. 20)

1938
A coelacanth, a primitive fish thought extinct, was discovered off the coast of South Africa. The strange blue-coloured fish was caught at the mouth of the Chalumna River on the east coast of South Africa. The 1.5-m long fish was caught at a depth of 70 metres in a shark gill net by the captain of a fishing trawler, Hendrik Goosen and his crew, who had no idea of the significance of their find. They thought the fish was bizarre enough to alert the local museum in the small town of East London. The director of the East London Museum at the time, Marjorie Courtney-Latimer, alerted the prominent South African ichthyologist J.L.B. Smith to this amazing discovery. This 'living fossil' comes from a lineage of fishes that was thought to have been extinct since the time of the dinosaurs. It was hailed as the zoological discovery of the century and equated to finding a living dinosaur! The coelacanth was eventually named (scientific name: *Latimeria chalumnae*) in honour of Courtney-Latimer. This coelacanth specimen led to the discovery of the first documented population, off the Comoros Islands in Indian Ocean, between Africa and Madagascar.

2004
Autar Singh Paintal, Indian physiologist, Fellow of the Royal Society, who discovered the 'J-receptors', died in New Delhi, India. (☞ Sep. 24)

December 23

1672
Saturn's moon Rhea, the fifth major moon of Saturn, was discovered by the Italian astronomer Giovanni Cassini. Measuring 1,500 kilometres in diameter, Rhea is one of the most heavily cratered moons in the solar system. Its surface appears to be saturated with craters, the largest being 300 kilometres in diameter. Cassini also discovered three more of Saturn's major moons – Iapetus, Tethys, and Dione. (☞ Jun. 8)

1911
Niels K. Jerne, Danish immunologist, Nobel laureate, was born in London, England. Jerne studied physics at the University of Leiden, Netherlands and then switched to medicine at the University of Copenhagen, Denmark. He received his medical degree from the University of Copenhagen in 1951, and in 1956 was appointed chief medical officer of the World Health Organization at Geneva, Switzerland, a position he held until 1962. During the 1960s he taught at the Universities of Geneva, Switzerland, and Pittsburgh in Pennsylvania, USA; was professor of experimental therapy at the

Johann Wolfgang Goethe University in Frankfurt am Main, Germany, and director of the Paul Ehrlich Institute, also in Frankfurt. Jerne's main research contributions have been in explaining various aspects of how the immune system defends the body against disease. He proposed in 1955 that from early in its life, the body has a full complement of antibodies, one of which can combine with and eliminate the antigen. Jerne's second theory, put forth in 1971, postulated that the body learns to distinguish between its own components and those that are foreign. According to Jerne's third theory, called the 'network theory,' the immune system is a complex, self-regulating network that can turn itself on or off when necessary. Jerne was awarded the 1984 Nobel Prize for Physiology or Medicine, jointly with Argentinean-born British immunologist César Milstein and German immunologist Georges Köhler, for theories concerning the specificity in develop-ment and control of the immune system and the discovery of the principle for production of mono-clonal antibodies. (☞ Apr. 17, Oct. 7, Oct. 8)

1947

American inventors William B. Shockley, John Bardeen and Walter Houser Brattain announced the invention of the 'transistor' – a tiny bit of germanium, plastic and gold that could act as an electronic amplifier – at a group meeting at the Bell Telephone Laboratories in Murray Hill, New Jersey, USA. (☞ Feb. 10, Feb. 13, May 23)

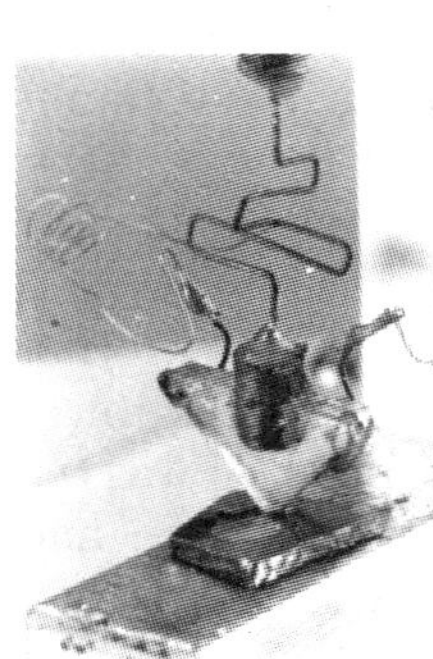

The first transistor

1954

The first successful kidney transplant was carried out between identical twins by surgeons at the Peter Bent Hospital, Boston, Massachusetts, USA. Ronald donated a kidney to his 23-year-old identical twin, Richard Herrick. Because they were identical twins, Richard's body, did not reject the organ as a foreign matter. The operation was led by Dr. John P. Merrill. Richard lived until 1962.

1968

American astronauts on board *Apollo-8* became the first men to orbit the Moon. The three-man crew comprised Frank Borman, James A. Lovell, Jr. and William Anders. This was the first manned flight to and from the Moon, and although there was no landing, the mission served to validate many of the technical procedures necessary for actual lunar landing. During ten lunar orbits, the astronauts took star sightings to pinpoint landmarks, surveyed landing sites, took both still and motion pictures and made two television transmissions to Earth. It was also the world's first manned space flight to escape the influence of Earth's gravity. (☞ Dec. 21, Dec. 27)

1973

Gerard Peter Kuiper, Dutch-born American astronomer who is considered to be the father of modern planetary science for his wide ranging studies of the solar system, died in Mexico City, Mexico. (☞ Dec. 7)

1986

The experimental aircraft *Voyager*, piloted by American aviators Dick Rutan and Jeana Yeager, completed the first non-stop, round-the-world flight without refuelling, landing safely at the Edwards Air Force

Base in California, USA. For nine days, they piloted the aircraft from a cramped 2.3-metre long, 1-metre wide and under 1-metre tall cockpit. The *Voyager* covered 40,200 kilometres in 216 hours, at an average speed of 186.4 km/h. The aircraft's design and lightweight structural materials allowed it to carry an unprecedented amount of fuel. (☞ Dec. 14)

2000
The 220-MW fourth unit of the Rajasthan Atomic Power Project at Rawatbhata, near Kota, Rajasthan, was commissioned. It is a pressurised heavy water reactor that uses natural uranium as fuel and heavy water as moderator and coolant. (☞ Apr. 1, Jun. 1, Dec. 16)

December 24

1818
James Prescott Joule, English physicist who established that the various forms of energy are basically the same and can be changed, one into another, was born in Salford, Lancashire, England. Joule was educated at home, and sent to Cambridge at the age of sixteen to study with the eminent British chemist John Dalton. He soon began to conduct independent research at a laboratory built in the cellar of his father's home. By the 1840s, scientists had realised that heat, electricity, magnetism, chemical change and the energy of motion were all inter-convertible. Joule was extremely involved with this work, and between 1837 and 1847, he established the principle of conservation of energy, and the equivalence of heat and other forms of energy, including the relationship between electric current, resistance, and the amount of heat produced. In 1840, he stated a law, now called Joule's law, that heat produced in an electrical conductor by an electric current is proportional to the product of the resistance of the wire and the square of the current. In 1849 he devised the kinetic theory of gases, and a year later announced the mechanical equivalent of heat with his famous 'paddle-wheel experiment'. Joule's experiments showed that heat is produced by motion, contradicting the caloric theory. Later, with William Thomson (Lord Kelvin), he discovered the Joule–Thomson effect – the cooling effect that occurs when a gas is rapidly expanded – an effect that is used in refrigeration systems. The SI unit of energy or work, the joule (symbol J), is named after him. It is defined as the work done when a force of 1 newton moves a distance of 1 metre in the direction of the force. (☞ Oct. 11)

1868
Linus Yale, American inventor and designer of the compact cylinder pin-tumbler lock, died in New York, USA. (☞ Apr. 4)

1936
The first nuclear medicine – a radioactive isotope of phosphorus-32 as phosphate – was administered to a 28-year-old woman with chronic leukaemia, at the Donner Laboratory in Berkeley, California, USA. The radioactive isotope was artificially produced in a 37-in cyclotron and administered by John Lawrence, brother of the American physicist and inventor of the cyclotron, Ernest Lawrence. It was the first time that a radioactive isotope had been used in the treatment of a human disease. John Lawrence became known as the father of nuclear medicine.

2004

European Space Agency's *Huygens* probe was released from NASA's *Cassini* spacecraft on its mission to Saturn's largest moon Titan. The instrument-laden space probe was to enter the thick atmosphere of Titan and land on its surface. *Huygens* was designed to make the first on-site study of the unique environment of Titan, whose chemistry is thought to be very similar to that of early Earth, before life formed. (☞ Jan. 14, Apr. 26, Jun. 24, Jul. 1, Aug. 18, Oct. 15, Dec. 30)

December 25

1642

Isaac Newton, English physicist and mathematician, who formulated the law of universal gravitation and discovered the composition of white light, laying the foundation for modern physical optics, was born in Woolsthorpe, Lincolnshire, England. Newton was educated at the Trinity College, Cambridge, and lived there from 1661 till 1696, during which time he produced the bulk of his work in mathematics and physics; in 1696 he was appointed to a Government office, and moved to London, where he resided till his death. He obtained his BA degree in 1665. It was about the same time that he discovered the binomial theorem. For several months during 1665–66, Newton had to stay at home in the countryside because of an epidemic of plague in the city. This period was crowded with brilliant discoveries. He worked out the fundamental principles of his theory of gravitation, namely, that every particle of matter attracts every other particle, and that the attraction varied as the product of their masses and inversely as the square of the distance between them. He formulated his three laws of motion during this period. It was also during this time that he devised some instruments for grinding lenses to particular forms other than spherical. His discovery of the composition of white light integrated the phenomena of colours into the science of light and laid the foundation for modern physical optics. Unable to find a means to remove the chromatic aberration in refracting telescopes, Newton designed a reflecting telescope in 1668, for which he used a parabolic metal (speculum) mirror as objective and a diagonal mirror to take the image out for viewing. The design he used is still known by his name. In 1669, Newton was appointed Lucasian professor of mathematics at the Cambridge University. In mathematics, he was the original discoverer of the infinitesimal calculus. Newton's *Philosophiae Naturalis Principia Mathematica* (usually referred to simply as *Principia*), published in 1687, was one of the most important single works in the history of modern science. Newton became a member of the Royal Society in 1672 and was its president from 1703 till his death. (☞ Mar. 20)

1758

The predicted return of Comet Halley was first sighted by German farmer and amateur astronomer, Johann Georg Palitzsch, as a faint object in the constellation of Pisces, proving for the first time a scientific prediction. The British astronomer Edmond Halley had predicted in 1705 that the comet would return to the Earth's vicinity every 75.5 years but did not live to see its return. (☞ Feb. 9, Nov. 8)

1876
Adolf Windaus, German organic chemist, Nobel laureate, was born in Berlin, Germany. Windaus attended the renowned French Grammar School in Berlin, and took up medicine in 1895. In 1899 he obtained his PhD degree on a thesis dealing with the cardiac poisons of the *Digitalis* plant. Windaus started his research on cholesterol and related sterols at the University of Freiburg in 1901. He was professor of chemistry and director of the chemistry laboratories at the University of Göttingen from 1914 to 1944. In 1919 he succeeded in transforming cholesterol into cholanic acid, which had previously been isolated from bile acids. Windaus thus demonstrated that bile acids are closely related to sterols. His research represented the basis for later discoveries in the bile acid metabolism. But his most significant work was on vitamin D. It was already known that mere exposure of certain foods to sunlight could make them active in preventing rickets, as could the vitamin D in cod liver oil. Clearly something in the food is converted photochemically into vitamin D but nobody knew what. In 1926 Windaus succeeded in showing that the precursor of vitamin D, the provitamin, is present as an impurity of cholesterol, ergosterol, which is converted into vitamin D by the action of sunlight. Windaus was awarded the 1928 Nobel Prize for Chemistry, for the services rendered through his research into the constitution of sterols and their connection with vitamins. (☞ Jun. 9)

1904
Gerhard Herzberg, German-born Canadian physicist, Nobel laureate, was born in Hamburg, Germany. Herzberg received his early training in Hamburg and subsequently studied physics at the Darmstadt Institute of Technology where in 1928 he obtained his doctorate degree. From 1928 to 1930, he carried out post-doctoral work at the University of Göttingen in Germany and the University of Bristol in England. In 1935 Herzberg left Germany and took up a guest professorship at the University of Saskatchewan in Canada. A few months later he was appointed research professor of physics, a position he held until 1945. From 1945 to 1948 he was professor of spectroscopy at the Yerkes Observatory of the University of Chicago in USA. He returned to Canada in 1948 and was made Principal Research Officer and shortly afterwards Director of the Division of Physics at the National Research Council. He was elected a Fellow of the Royal Society of London in 1951. Herzberg's main contributions have been in the field of atomic and molecular spectroscopy. He and his associates determined the structures of a large number of diatomic and polyatomic molecules, including the structures of many free radicals, e.g., those of free methyl and methylene, which are difficult to determine in any other way. Free radicals are groups of atoms that contain odd numbers of electrons. Herzberg also applied these spectroscopic studies to the identification of certain molecules in planetary atmospheres, in comets, and in interstellar space. Herzberg was awarded the 1971 Nobel Prize for Chemistry for his contributions to the knowledge of electronic structure and geometry of molecules, particularly free radicals. (☞ Mar. 3)

1906
Ernst August Friedrich Ruska, German electrical engineer, Nobel laureate, was born in Heidelberg, Germany. Ruska studied electronics at the Technical University in Munich during 1925–27 and later in Berlin. Whilst still a student at the Technical College in Berlin he began working with high voltage and vacuum technology at the Institute of High Voltage where, together with other doctoral students, he worked on the development of a high-performance cathode ray oscilloscope. During 1928–29, he worked on the development of an electromagnetic lens for focussing electron beams, just as a glass lens focuses a beam of light. In 1928 Ruska attempted to focus an electron beam with an electromagnetic lens. He went on to add a second lens and thus produced the first electron microscope; it had a magnifying power of about 17×. Improvements, however, came quickly and by 1933 the magnifying power had been increased to 7,000×. In an electron microscope, electrons are passed through a very thin slice of the object under study and are then deflected onto photographic film or onto a fluorescent screen, producing a greatly magnified image. Since electrons have wavelengths about 100,000 times shorter than those of light, an electron microscope could theoretically produce magnifications up to 100,000 times those of the most powerful optical microscopes. In 1937 Ruska joined Siemens–Reiniger–Werke AG as a research engineer and began to work on the production of commercial models. The first such model appeared in the market in 1939. It had a resolution of about 250–500 angstroms. Ruska served as director of the Institute for Electron Microscopy of the Fritz Haber Institute of the Max Planck Society in Berlin-Dahlem from 1957 to 1972. He also gave regular lectures on the basic principles of electron optics and electron microscopy at both the Free University and the Technical University of Berlin. Ruska was awarded the 1986 Nobel Prize for Physics, jointly with Heinrich Rohrer and Gerd Binnig. Ruska received half the prize money for his fundamental work in electron optics, and for the design of the first electron microscope, while Rohrer and Binnig shared the other half for their design of the scanning tunnelling microscope. (☞ Jun. 6, Jul. 20)

1961
Otto Loewi, German-born American physician and pharmacologist, Nobel laureate (Physiology or Medicine, 1936), who discovered the role of acetylcholine in transmission of nerve signals, died in New York, USA. (☞ Jun. 3)

1978
Soviet unmanned spacecraft *Venera-11* made a soft landing on Venus after parachuting through the planet's dense atmosphere and sent back data for 95 minutes. Although the lander carried two colour cameras, no pictures were returned. After releasing the lander, the main spacecraft flew past the planet and investigated the nature of interplanetary space. (☞ Sep. 9)

December 26

1791
Charles Babbage, British inventor of the adding machine, was born in Teignmouth, Devonshire, England. Babbage graduated from the Cambridge University in 1814 and received his MA degree from the same

university in 1817. He started work on the 'difference engine' with funding from the British Government in 1823 In 1828 he was appointed to the Lucasian Chair of Mathematics at Cambridge, a position he held for 12 years. In 1831 he founded the British Association for the Advancement of Science. The idea of a calculating machine had come to Babbage as early as 1812 when he found a high error rate in the manual calculation of mathematical tables. He sought to find a method by which they could be calculated by machine – this would eliminate the mistakes caused by fatigue and boredom, conditions most likely to cause errors in humans. In 1823 he received a gold medal from the Astronomical Society for his development of the 'difference engine', which was a special purpose device intended for the production of tables, and in 1827 he published a table of logarithms from 1 to 108000. In 1833 he started work on his 'analytical engine' that would use input devices based on punched cards, an arithmetic processor that calculated numbers, a control unit that determined that the correct task was carried out, an output mechanism, and a memory where numbers could be stored whilst waiting their turn to be processed. By 1834 he had completed the first drawings of the analytical engine, the forerunner of the modern electronic computer. Although Babbage only produced prototypes of portions of the difference engine and never built an operational machine, his design concepts have been proved correct and such a computer was later built following Babbage's own design criteria. Babbage's other inventions included the cowcatcher, dynamometer, standard railroad gauge, occulting lights for lighthouses, Greenwich time signals, heliograph, and opthalmoscope. He also had an interest in cyphers and lock picking. (☞ Oct. 18)

1865
American inventor James H. Mason of Franklin, Massachusetts, was issued the first US patent for a coffee percolator, described as 'an apparatus for preparing the extract or infusion of coffee'. Mason's design used fluid joints 'to prevent the escape of aroma during or subsequent to the process of preparation'.

1898
The radioactive element radium (Ra, at. no. 99) was discovered by the Polish-born French scientist Marie Curie while experimenting with pitchblende, a common uranium ore. She had observed that this ore was more radioactive than refined uranium, indicating that there must be another element, even more radioactive than uranium, mixed in with this ore. Curie spent almost three years, working with nearly three tonnes of the ore, and finally isolated 0.1 gram of the new element. This was enough for spectroscopic examination, and to determine the exact atomic weight of radium. (☞ Nov. 7)

1973
Comet Kohoutek attained its perihelion. Kohoutek is a long period comet, with an orbital period of approximately 75,000 years. It was hyped by the media as the 'comet of the century' because astronomers believed that it was the comet's first visit to the inner solar system – a comet so bright that it can be noticed by a casual observer. But the comet did not live up to expectations and put up a disappointing display. (☞ Mar. 8)

1985
Dian Fossey, American zoologist who became the world's leading authority on the mountain gorillas of central Africa, died in Karisoke, Rwanda. It was suspected that poachers, whose devastating attacks on the gorillas she had tried to stop, were responsible for her murder, although this was never proved. (☞ Jan. 16)

December 27

1571
Johannes Kepler, German astronomer who discovered three major laws of planetary motion, was born in Weil der Stadt, Württemberg, Germany. Kepler attended the University of Tübingen where he was introduced to the ideas of Copernicus. In 1596, while working as a mathematics teacher in Graz, he published *Mysterium Cosmographicum* ('Cosmographic Mystery'), in which he defended the Copernican theory and described his ideas on the structure of the planetary system. In 1600 Kepler moved to Prague (now Czech Republic) to work with the renowned Danish astronomer, Tycho Brahe. He inherited Tycho's post as Imperial Mathematician when Tycho died in 1601. Using the precise data that Tycho had collected, Kepler discovered that the orbit of Mars was an ellipse. In 1609 he published *Astronomia Nova* ('New Astronomy'), delineating his discoveries, which are now called Kepler's first two laws of planetary motion: (1) Planets move in elliptical orbits with the Sun as one of the foci; (2) A planet sweeps out equal areas in equal times. In 1619 he published *Harmonice Mundi* ('Harmony of the World'), in which he derived the heliocentric distances of the planets and their periods. In this work he stated his third law, relating the periods of the planets to their mean orbital radii. Kepler published the seven-volume *Epitome Astronomiae* in 1621. This was his most influential work and discussed all of heliocentric astronomy in a systematic way. He then went on to complete the *Rudolphine Tables* that Tycho had started long ago. These included calculations using logarithms, which he developed, and provided perpetual tables for calculating planetary positions for any past or future date. Kepler used the tables to predict a pair of transits by Mercury and Venus of the Sun, although he did not live to witness the events. (☞ Mar. 8, Nov. 15, Dec. 6)

1822
Louis Pasteur, French chemist and microbiologist whose contributions were among the most varied and valuable in the history of science and industry, was born in Dole, France. Pasteur earned his BA degree in 1840 and BSc degree in 1842 from the Royal College in Besançon. He obtained his MSc in 1845 and PhD in 1847 from the École Normale Supérieure, the famous teachers' college in Paris. In 1847 at the age of 26, Pasteur did his first work on molecular asymmetry while working with tartaric acid crystals, bringing together the principles of crystallography, chemistry and optics. He formulated a fundamental law: asymmetry differentiates the organic world from the mineral world. In other words, asymmetric molecules are always the product of life forces. His work became the basis of a new science – stereochemistry. He presented his discovery before the Paris Academy of Sciences in 1848. In

1857 Pasteur published his first paper on the formation of lactic acid and its function in souring milk. With further research, he developed the technique of 'pasteurisation', which would go on to revolutionise not only the dairy industry, but also the food processing industry in general. Later, most of Pasteur's efforts went towards convincing other scientists that germs do not originate spontaneously in substances, but enter from the outside. He was able to establish the basic rules of sterilisation or asepsis, and his methods revolutionised surgery and obstetrics. His work in understanding the functioning of bacteria led him to find a cure for a mysterious disease that was attacking silk worms. He discovered the concept of immunisation for curing anthrax. He later used the same immunisation process on humans to prevent rabies that he used on cattle to prevent anthrax and was tremendously successful. His 'germ theory of disease" is one of the most important in medical history, and became the foundation for the science of microbiology, and a cornerstone of modern medicine. (☞ May 22, Sep. 28)

1831

Charles Darwin set sail from Plymouth harbour in England on his voyage of scientific discovery aboard the *HMS Beagle*, a British Navy ship, as a naturalist. The captain of the ship was Robert FitzRoy. The *Beagle*'s mission was to sail to the southern coast of South America in order to complete a government survey. Originally planned to be at sea for two years, the voyage lasted five years, making stops in Brazil, the Galapagos Islands, and New Zealand, during which Darwin made extensive study of the flora and fauna of the places visited. From the observations he made and the specimens he collected on that voyage, Darwin developed his theory of biological evolution through natural selection, which he published in 1859 under the title *On the Origin of Species by Means of Natural Selection*, laying the foundation of modern evolutionary theory. (☞ Feb. 12, Apr. 19)

1914

Charles Martin Hall, American chemist and inventor who discovered the electrolytic method of producing aluminium, thus bringing the metal into wide commercial use, died in Daytona Beach, Florida, USA. (☞ Apr. 2, Dec. 6)

1915

William Howell Masters, American gynaecologist who is known for his studies of human sexuality, was born in Cleveland, Ohio, USA. Masters graduated from the Hamilton College of Clinton, New York, in 1938 and obtained his medical degree from the University of Rochester Medical School. A student of gynaecological and obstetrics, Masters worked at the St. Louis Maternity Hospital and Barnes Hospital, both of St. Louis, Missouri. He also trained in pathology at the St. Louis's Washington University Medical School where he was appointed on its teaching staff in 1947. He began research on human sexual behaviour and dysfunction, which was funded initially by grants from the United States Institutes of Health. Masters did his long series of studies together with Virginia E. Johnson, a former student of psychology. Their investigation of the physical aspects of sexuality produced some of the first reliable data in the field.

One of the most enduring and important aspects of their work has been the four stage model of sexual response, which they described as the human sexual response cycle. The Masters & Johnson Study, as it became known, used clinical and physical data gathering analysis as opposed to purely psychological data extrapolation and analysis. Through emerging medical technology, Masters employed such physiological data gathering tools as surgical cameras, electrocardiographs, electroencephalograms and serology to monitor 276 married couples acting as volunteers in his study. The first 10 years of the Masters & Johnson Study culminated in the 1966 publication of their findings, the report *Human Sexual Response*, which was published for the medical community but became a general bestseller. Their other books include *Human Sexual Response, Human Sexual Inadequacy, The Pleasure Bond*, and *Crisis: Heterosexual Behavior in the Age of AIDS*. (☞ Feb. 16)

1968

American astronauts Frank Borman, James A. Lovell, Jr. and William Anders returned to Earth after orbiting the Moon on board *Apollo-8*. (☞ Dec. 21, Dec. 23)

December 28

1869

William Finley Semple, a dentist from Mount Vernon, Ohio, USA was issued the first patent for chewing gum. His chewing gum was made of 'the combination of rubber with other articles adapted to the formation of an acceptable chewing gum'. He believed his gum would exercise the jaws and clean the teeth at the same time.

1882

Arthur Stanley Eddington, British astrophysicist, was born in Kendal, UK. Eddington graduated from the Cambridge University in 1905 and was appointed chief assistant to the Astronomer Royal at the Royal Greenwich Observatory, where he excelled in practical astronomy. In 1913 he was appointed Plumian Professor of Astronomy at Cambridge, and a year later he was made director of the Royal Greenwich Observatory. Eddington was the first to model the interior of a star under radiative equilibrium, pointing out that the condition for stellar equilibrium involved three forces: gravity, gas pressure, and radiation pressure. He demonstrated that energy could be transported by radiation as well as by convection, and that the centres of stars must be at very high temperatures – in the range of millions of degrees. An extremely important result that emerged from his research was his theoretical formulation of the mass–luminosity relation, verified later by stellar data. He found that a star's luminosity is finite for a supplied mass. Part of the equation for finding a star's maximum luminosity is now called the 'Eddington Limit'. In 1919, Eddington organised a solar-eclipse expedition to Brazil to photograph the stars in the neighbourhood of the eclipsed Sun. This was to look for any apparent shift in the stars' positions in the neighbourhood of the eclipsed Sun, as predicted by Einstein's general theory of relati-vity. Although difficult to measure, the observed shift in position was in rough agreement with Einstein's predicted value. It was the first observational test of the relativity theory. (☞ Nov. 22)

1895

Louis Lumière, French inventor, who invented the Cinématographe, organised the first public screening of his film at the Grand Café on the boulevard des Capucines in Paris. It brought wide public acclaim and marked the beginning of cinema history. (☞ Oct. 5)

1944

Kary Banks Mullis, American biochemist, Nobel laureate, was born in Lenoir, North Carolina, USA. Mullis received a bachelor's degree in chemistry from the Georgia Institute of Technology, Atlanta, Georgia, and a PhD in biochemistry from the University of California, Berkeley in 1973. After graduation and several post-doctoral tours, he joined the Cetus Corporation of Emeryville, California as a scientist in 1979. In 1983, while working at Cetus, Mullis conceived and developed the idea of 'polymerase chain reaction', better known as PCR. Prior to his work, it was extremely difficult and time-consuming to obtain adequate fragments from a large mass of DNA. They also could not be produced artificially in the laboratory. To multiply DNA, Mullis first heated his DNA sample to separate the two complementary strands of the double helix that make up the DNA. As the sample cooled, he added two short DNA sequences, which bonded to the complementary sites on the separated strands. This marked the sequence of nucleotides that needed to be multiplied. He then added free nucleotides and a polymerase enzyme, which joined the nucleotides in the same order as the targeted segment and produced a copy of the desired DNA sequence. By continually repeating the process on the copy (doubling it with each procedure), Mullis was able to produce billions of copies of the original DNA segment in just a few hours. PCR thus enables scientists to create sequences of genetic material in quantities that are large enough to study. The technique has played a major role in the International Human Genome Project, which has major health and anti-aging implications. It has also become invaluable in disease diagnosis, forensic-science analysis in convicting the guilty and freeing the falsely accused, and the study of DNA from ancient or fossil tissues. In 1986 Mullis became director of molecular biology at Xytronyx Inc., a plastics manufacturer in San Diego, California. Since 1988 he has worked as an independent consultant for various firms. Mullis was awarded the 1993 Nobel Prize for Chemistry, jointly with the British-born Canadian biochemist Michael Smith, for his invention of the polymerase chain reaction (PCR) method. (☞ Apr. 26)

1995

India's third operational remote sensing satellite *IRS-1C* was launched by Molnya launch vehicle of the Russian space agency Glavkosmos from Baikanur Cosmodrome in Khazakistan. It was placed in a polar, Sun-synchronous orbit at an altitude of 817 kilometres. *IRS-1C* carried a panchromatic camera, a linear imaging self scanner (LISS-III), and a wide field sensor (WiFS) camera. It provided enhanced coverage through a programmable on-board tape recorder.

December 29

1766

Charles Macintosh, Scottish chemist and manufacturer, Fellow of the Royal Society, was born in Glasgow, Scotland. Macintosh took an early interest in science. In 1786 he started work in a chemical factory and in 1797 he opened the first alum works in Scotland. He invented an easily-transportable bleaching powder and a conversion process which used carbon gases to convert iron to steel which was much faster than the method used until that time. But Macintosh is better known as the inventor of waterproof clothing, known after his name. In 1818, while trying to find uses for waste products generated by gasworks, he used naphtha, a by-product of the distillation of coal tar, as a solvent for rubber. When he joined two sheets of cloth together with this solution and allowed them to dry, he discovered that the new material could not be penetrated by water – the first rainproof cloth! Soon after, he began the manufacture of raincoats and other garments. Despite early problems with the strong smell of rubber, he took out a patent on the process in 1823. Although other forms of waterproof clothing have been developed since, his name is often used to describe them. (☞ Jul. 25)

1800

Charles Goodyear, American inventor of the vulcanisation process that made possible the commercial use of rubber, was born in New Haven, Connecticut, USA. Goodyear was neither a chemist nor scientist. He recognised rubber's valuable properties – elasticity, plasticity, strength, durability, electrical non-conductance, and resistance to water. But the use of raw rubber was limited since it froze hard in winter, become sticky in summer and was attacked by various solvents. In 1839, after years of persistent experimentation, Good-year accidentally mixed and baked sulphur and rubber together, creating a tough, cured compound that could withstand heat and stress. The process, which came to be known as vulcanisation, transformed rubber from a smelly, virtually useless substance into a stable, versatile commercial product with literally hundreds of uses. Goodyear also suggested several applications of his new product in 1850. Rubber paint, car springs, ferryboat bumpers, wheelbarrow tires, inflatable life rafts, and 'frogmen' suits are other recent innovations he described a century and a half ago. Unfortunately, he was a poor businessman, unable to profit from his invention or effectively patent it abroad. (☞ Jun. 15, Jul. 1)

1952

The first transistorised hearing aid went on sale. The model '1010', manufactured by the Sonotone Corporation of USA, weighed about 100 g and measured 7.6 cm × 3.8 cm × 1.5 cm. It was a hybrid device consisting of two sub-miniature pre-amplifier vacuum tubes and a single transistor as the final audio amplifier and cost $299. Within months, transistor production techniques were improved largely eliminating no-ise, so the hybrid model was shortly discontinued.

1954

William Merriam Burton, American ch-emist who developed a thermal cracking process for increasing the proportion of gasoline obtainable from crude petroleum, died in Miami, Florida, USA. (☞ Nov. 17)

1989

Hermann Oberth, German scientist who is considered to be one of the founders of modern astronautics, died in Feucht, near Nuremberg, Germany. (☞ Jun. 25)

1993
The UN Convention on Biological Diversity (CBD) came into force. The objectives of this convention include the conservation of biological diversity, the sustainable use of its components and the fair and equitable sharing of the benefits arising out of the utilisation of genetic resources, including appropriate access to genetic resources and appropriate transfer of relevant technologies, taking into account all rights over those resources and to technologies, and by appropriate funding. The convention provides that each signatory shall, in accordance with its particular conditions and capabilities develop national strategies, plans or programmes for the conservation and sustainable use of biological diversity or adapt for this purpose existing strategies, plans or programmes, and also integrate, as far as possible and as appropriate, the conservation and sustainable use of biological diversity into relevant sectoral or cross-sectoral plans, programmes and policies. The convention was negotiated under the auspices of the United Nations Environment Programme (UNEP). The CBD Secretariat is located in Montreal, Canada.

December 30

1691
Robert Boyle, Anglo-Irish chemist and natural philosopher, noted for his pioneering experiments on the properties of gases and his espousal of a corpuscular view of matter that was a forerunner of the modern theory of chemical elements, died in London, England. (☞ Jan. 25)

1850
John Milne, English seismologist, was born in Liverpool, England. Milne was educated at the King's College, and the Royal School of Mines in London. He worked as a mining engineer in Newfoundland, and Labrador in Canada and served as a geologist on a mining expedition to northwest Arabia. In 1875 Milne joined the Imperial College of Engi-neering, Tokyo, Japan, as professor of geology and mining. He invented the seismograph in 1880 and travelled widely in Japan to set up 968 seismological stations for a survey of Japan's widespread earthquakes. After his return to England he served as secretary of the Seismological Committee of the British Association and organised a worldwide network of observation stations. Many of his findings were published in his books *Earthquakes* (1883), and *Seismology* (1898). (☞ Jul. 30)

1913
The American engineer, William David Coolidge received a patent for a method for making ductile tungsten for the purpose of making filaments for electric lamps. When Coolidge joined the General Electric Research Laboratory in Schenectady, New York, in 1905, he was given the task of replacing the fragile carbon filaments in electric light bulbs with tungsten filaments, although tungsten was difficult to work with. After three years of painstaking research on this intractable metal, Coolidge developed a process by means of which tungsten was made sufficiency ductile at room temperatures to permit drawing through diamond dies. Close control of working temperatures, of tungsten powder grain size, and of trace metal additions, particularly thorium, contributed to the final successful result. (☞ Jul. 4, Oct. 23)

1924
American astronomer Edwin Powell Hubble announced the existence of another galaxy outside the Milky Way. Until then, scientists were not certain whether certain fuzzy clouds of light called 'nebulae' that had been seen with telescopes were small clusters of clouds or galaxies within the Milky Way. While observing at Mount Wilson Observatory, Hubble measured the distance to the Andromeda nebula and showed it to be a hundred thousand times as far away as the nearest stars, which proved it was a separate galaxy, as large as our own Milky Way, but very far away. Thousands more galaxies have been found since Hubble's discovery; some of them have a spiral form like the Milky Way; others are spheroidal, without the spiral arms, or are of irregular shape. (☞ Nov. 20)

1975
India's first oceanographic research vessel *Gaveshani* was commissioned. The 68.5-m long vessel was well equipped for all-weather oceanographic research, which enabled Indian oceanographers to successfully carry out extensive surveys for offshore oil fields, mineral deposits, monitoring oil pollution and developmental activities. It was *Gaveshani*, which put India on the world map for deep-sea mining with a successful haul of nodules. *Gaveshani* visited Malaysia, Sri Lanka, Mauritius and Seychelles and made 246 cruises before being decommissioned in 1994.

2000
The *Cassini–Huygens* spacecraft swung by Jupiter at a distance of 10 million kilometres, en route to Saturn. (☞ Jan. 14, Apr. 26, Jun. 24, Jul. 1, Aug. 18, Oct. 15, Dec. 24)

December 31

1514
Andreas Vesalius, Belgian anatomist and physician, was born in Brussels, Belgium. Vesalius studied at the University of Louvain in Belgium. From 1533 to 1536 he studied medicine at the University of Paris in France, where he developed an interest in anatomy. Vesalius obtained his medical degree in 1537 from the University of Padua in Italy and a got a job as a lecturer in surgery. In 1543 he published his *De Humani Corporis Fabrica* ('On the Workings of the Human Body'), the first accurate book on human anatomy, notable for its excellent descriptions and illustrations of the skeletal, muscular, and nervous systems. The book was based largely on human dissection, and transformed anatomy into a subject that relied on observations taken directly from human dissections. In the preface to the book, Vesalius stated that during his studies at Paris he had himself dissected a corpse in the presence of undergraduates. His work marked a real departure from the work of the 2nd-century anatomist Galen, and provided detailed and accessible information against which future anatomists could compare their observations. After publication of *Fabrica*,

Vesalius decided to leave anatomical research to take up medical practice. Maintaining the long tradition in his family of imperial service, he received an appointment as physician to the imperial court of Emperor Charles V. He remained there until the Emperor's abdication thirteen years later, during which time he revised *Fabrica* and earned himself a formidable reputation as a physician. (☞ Oct. 15)

1719
John Flamsteed, English astronomer who established the Greenwich Observatory and was appointed the first Astronomer Royal of England, died in Greenwich, London, England. (☞ Aug. 19)

1937
Avram Hershko, Israeli biologist, Nobel laureate, was born in Karcag, Jász-Nagykun-Szolnok County, Hungary. Hershko received his MD in 1965 and his PhD in 1969 from the Hadassah Medical School of Hebrew University in Jerusalem. He is currently a Distinguished Professor of Biochemistry at the Rappaport Family Institute for Research in Medical Sciences at the Technion (Israel Institutite of Technology), in Haifa. Starting in the late 1970s, during a series of sabbatical leaves, Hershko together with the Israeli biochemist Aaron J. Ciechanover and American biochemist Irwin A Rose did a series of epoch-making biochemical studies in Rose's laboratory at the Fox Chase Cancer Center in Philadelphia, USA, which ultimately led to the discovery of a molecule called ubiquitin that plays a key role in protein breakdown within cells. Their work showed that the breakdown is not indiscriminate but takes place through a process that is controlled in detail so that the proteins to be broken down at any given moment are given a molecular label by the ubiquitin molecule. The labelled proteins are then fed into proteasomes, the cells' so-called 'waste disposers', where they are chopped into small pieces and destroyed. The ubiquitin–proteasome pathway has a critical role in maintaining the homeostasis of cells and is believed to be involved in the development and progression of diseases such as cancer, muscular and neurological diseases, immune and inflammatory responses, and cystic fibrosis. Hershko was awarded the 2004 Nobel Prize for Chemistry, jointly with Ciechanover and Rose, for the discovery of ubiquitin-mediated protein degradation. (☞ Jul. 16, Oct. 1)

1968
The first flight of a supersonic passenger plane, the Russian TU-144, took place at an airfield near Moscow. The delta-wing aircraft was designed by Andrei Nikolaevich Tupolev, a leading designer at the Moscow-based Central Aero-Hydrodynamic Institute. However, the TU-144 never went into commercial service, as it was never manufactured on a commercial scale following a crash at the Paris air show in 1973. (☞ Jan. 9)

1971
Vikram Ambalal Sarabhai, Indian physicist, main architect of India's space programme, died in Kovalam, near Thiruvananthapuram, Kerala, India. (☞ Aug. 12)

1974
The charge-coupled device (CCD) was patented by American inventors Willard S. Boyle and George E. Smith of the Bell Labs, New Jersey, USA. Boyle and Smith invented the CCD while attempting to create a new kind of semiconductor memory for

computers. They are said to have sketched out the CCD's basic structure, defined its principles of operation, and outlined its applications including imaging as well as memory in the space of an hour in October 1969. A CCD is a solid-state device that works much like a photocell, translating brightness values into electrical signals. By 1970, the Bell Labs researchers had built the CCD into the world's first solid-state video camera. CCDs are used today in facsimile machines, copying machines, image scanners, digital still cameras, and bar code readers.

2002

The world's first magnetic levitation or 'Maglev' train started its commercial run between Shanghai city centre and Pudong airport in China, a distance of 30 kilometres. Held up by powerful magnets, a Maglev train literally floats above a specially designed track and can travel at speeds of over 430 km/h.

Bibliography

Hellemans, Alexander and Bryan Bunch, *The Timetables of Science*. London: Sidgwick & Jackson, 1989.

Porter, Roy and Marilyn Ogilvie, eds., *The Hutchinson Dictionary of Scientific Biography* (2 vols.). Oxford: Helicon Publishing Ltd., 2000.

Bruno, Leonard C. and Donna Olendorf, *Science and Technology Firsts*. Michigan: Thomson Gale, 1997.

Kurian, George Thomas. *The Nobel Scientists: A Biographical Encyclopaedia*. New York: Prometheus Books 2002.

Britannica Yearbook of Science and the Future (1979–2000). Encyclopaedia Britannica Inc.

van Dulken, Stephen, *Inventing the 19th Century*. New York: New York University Press, 2001.

van Dulken, Stephen, *Inventing the 20th Century*. New York: New York University Press, 2002.

Oxford Illustrated Encyclopedia of Invention and Technology. New York: Oxford University Press, 1992.

Harrison, Ian, *The Book of Inventions*. London: Cassell Illustrated, 2004.

Inventors and Discoverers. Washington, D.C.: National Geographic Society, 1988.

Encyclopaedia Britannica 2005. Ultimate Reference Suit CD, Encyclopaedia Britannica Inc., 2005.

The 20th Century Year by Year. London: Marshall Publishing Ltd., 2000.

Web site of the Nobel Foundation. http://www.nobelprize.org.

Web site of NASA. http://www.nasa.gov.

Web site of Today in Science History. http://www.todayinsci.com.

Web site of the European Space Agency. http://www.esa.int.

INDEX

D

G

H

I

N

O

P

S

T

U

V

W

X

Y

Z

People Index

C

D

E

F

G

H

I

J

K

L

N

O

P

Q

R